Differential Equations
A MODELING APPROACH

Frank R. Giordano

U.S. MILITARY ACADEMY

Maurice D. Weir

U.S. NAVAL POSTGRADUATE SCHOOL

ADDISON-WESLEY PUBLISHING COMPANY

Reading, Massachusetts · Menlo Park, California · New York
Don Mills, Ontario · Wokingham, England · Amsterdam
Bonn · Sydney · Singapore · Tokyo · Madrid · San Juan

Sponsoring Editor: Jerome Grant
Production Supervisor: Jack Casteel
Text Design: Deborah Schneck
Copy Editor: Susan Middleton
Illustrator: Scot Graphics Center
Technical Art Consultant: Joseph Vetere
Manufacturing Supervisor: Roy Logan
Cover Design: Marshall Henrichs

Photo Credit: Tom Anderson, © 1988 Tacoma (WA) p. iii

Reprinted with corrections, June 1991

ISBN 0-201-17208-9

2 3 4 5 6 7 8 9 10 DO 9594939291

IN MEMORY OF
MARDIE WEIR KALBFLEISCH
(1943–1989)

In affirmation of
her love of life.

Preface

The study of differential equations is a key course for applied mathematics, science and engineering students at most colleges and universities. It is important to motivate the ideas and results of differential equations by real-world applications that an undergraduate student is likely to encounter in physics, mechanics, chemistry, electrical engineering and other courses in the natural sciences. Rapidly changing technologies require that students learn to solve problems often modeled by a differential equation. This process includes recognizing that a situation or behavior of interest can be modeled by a differential equation, formulating a model representing the behavior, and solving or approximating a solution to the model. It is important also to test the validity of the results before they are used in a significant way.

Prerequisites: We emphasize that this text is written *for the student* and only basic calculus is required as a prerequisite. Since relatively few students actually master all the calculus required for a full understanding of differential equations and their applications, we review briefly key calculus concepts as they naturally arise. Brief reviews of integration by parts (including the tableau method) and partial fraction decomposition (including the Heaviside method) are given in Appendix A.

Linear algebra is *not* a prerequisite, nor do we attempt to summarize or present the elements of linear algebra in the text. Concepts such as linear independence, fundamental sets of solutions, and the Wronskian are defined as they relate to differential equations, with sufficient examples to enhance student understanding. Only the knowledge of evaluating determinants and solving (small) systems of linear equations from high school linear algebra is assumed. Although matrix notation is used in Chapter 7 on linear systems of differential equations, only Chapter 8 actually requires matrix operations. A review of matrix arithmetic, including matrix inverses, determinants, and Cramer's rule, is given in Appendix D. It is assumed that the student has the basic knowledge of vectors in the plane and space, studied in basic calculus. Synthetic division is reviewed in Appendix C.

The presentation is fully exposed and entirely self-contained. Students can read and study the fundamentals outside of class, thereby freeing the instructor to interpret and stress in class those ideas and applications of primary interest, need, and preference.

Mathematical Modeling: We provide a brief introduction to the process of mathematical modeling as it applies to differential equations to give the student a context for the applications presented in the text. Thus the student will understand

- how the derivative is used to model change, and
- how models are made more precise (and more difficult to solve) through the model refinement process.

The model refinement process structures the book. As the text unfolds, models are refined to make them more realistic, which requires in turn the introduction of more powerful mathematical techniques to analyze and solve them. The student learns not only how to solve differential equations, but also how to formulate and analyze them.

Geometrical Intuition: The study of differential equations is a natural extension of basic calculus, and some review of key calculus concepts is usually required. We build upon, and seek to improve, the student's geometric intuition gained from calculus. Graphical analysis is presented in the first chapter to provide insight into the nature of solutions to important differential equations. The student also sees that many problems can be analyzed graphically before an analytical solution is considered or attempted. Graphical ideas are used as well for systems of equations, Laplace transform methods, and partial differential equations.

Numerical Approximations: Numerical solution techniques are integrated throughout the text, although (with the exception of the basic Euler's method) they are marked as optional sections. Beginning in Chapter 2, numerical methods are introduced for first-order equations from a geometric point of view. These methods broaden the student's understanding of how differential equations are often solved in practice. Early on, the student with access to a computer and readily obtainable software can find solutions to models of considerable complexity, even nonlinear ones. In subsequent chapters, numerical methods are introduced for second-order, systems of, and partial differential equations.

Algorithmic Format: After an analytical or numerical approximation solution procedure is motivated and developed, a step-by-step summary in algorithmic format is given to aid the student in learning the method. Examples follow the summary to illustrate its use. Throughout we emphasize solution *processes,* not rote use of a formula or tabled information to produce analytical results.

Applications: Each major type of equation is motivated by a wide range of real-world applications before solution techniques are considered. The student sees that the equation comes from modeling a real behavior he or she relates to (rather than an abstraction), and is challenged to find a solution that helps explain or make predictions about that behavior.

Students are often intimidated by applications, especially nontrivial ones, because they believe they have insufficient background to understand them. We assume only the most elementary ideas from physics, not going beyond those encountered in basic engineering calculus. Furthermore, we develop the needed physical concepts and their interrelationships as they pertain to a model under development, assuming the student may never have had a previous detailed exposure to those concepts. Thus the physical nature of the problem can be better understood along with its mathematical underpinnings.

Many more applications are presented than can be covered in a single course, so instructors and students can freely choose among them as their interests dictate and time permits. We also present a number of applications in considerable detail—clearly marked as optional sections—to allow for a greater in-depth study of the particular behavior. An instructor may wish to present one or two of these extended applications, or assign some of them for student reading projects and reports.

Exercises: Many exercises are given at the end of each section so students can practice solution techniques and work with the mathematical concepts discussed. The unassigned exercises can be used as test questions. Exercises that extend theoretical results are usually near the end. Each chapter ends with a set of review exercises, some of which combine elements from several of the chapter sections. Answers to all odd-numbered exercises are given in the back of the text.

Supplements: A *Student Solutions and Study Guide Manual* is available with complete written solutions to all the odd-numbered exercises, answers to the even-numbered exercises, and study hints and questions coordinated with the various sections of the text. An *Instructor's Manual* is also available for professors. A computer supplement *DERIVE Laboratory Manual for Differential Equations* by David C. Arney is also available. It presents additional aspects of computing solutions to the differential equations studied in this text.

Possible Courses: Several different courses can be designed around this text. The accompanying chart showing the logical dependency of the various chapters should aid with the course design. Each section of the text is written to correspond to one or two hour-long lectures; optional materials are clearly marked. There is ample material for a semester-long course, or even a two-quarter sequence, covering an introduction to differential equations and their applications.

Acknowledgments: We would like to express our appreciation to a number of individuals who have played a role in the development of this book. We are particularly grateful to the following:

Chris Arney for an insightful review, his contributions to the Study Guide portion of the Student Solutions Manual, and his authorship of the computer supplement *DERIVE Laboratory Manual for Differential Equations.*

John Gallo for contributing to the exercise sets and their solutions, and for editing the Chapter Review Exercises.

John Robertson for his excellent ideas on the supplements, application problems, and graphics for partial differential equations.

Jim Hayes for his assistance with the graphs on partial differential equations.

David Cameron and John Edwards for their encouragement and insight in motivating differential equations for the student, and their ideas for numerical solution techniques.

Fletcher Lamkin for his careful and insightful review and suggestions for applications.

Judi and Gale, our wives, and our families for their loving support of this project.

Frank R. Giordano
Maurice D. Weir

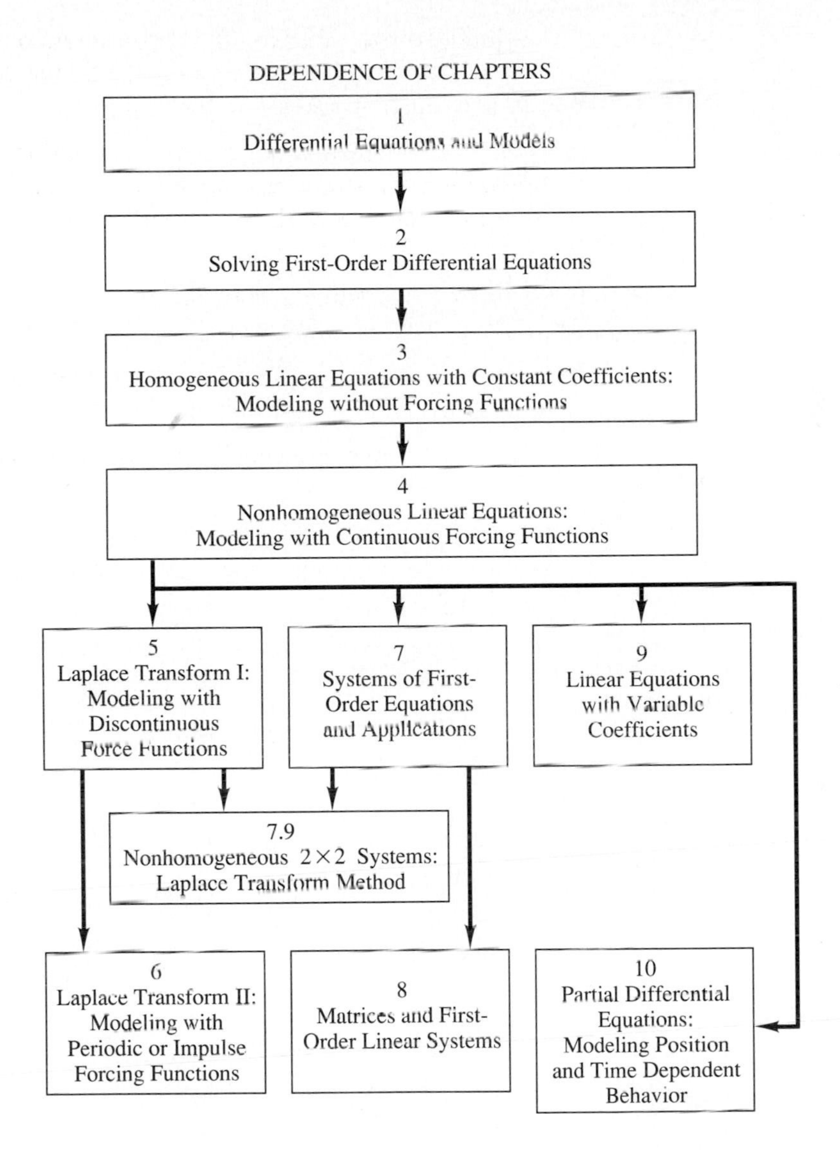
DEPENDENCE OF CHAPTERS
1
Differential Equations and Models
2
Solving First-Order Differential Equations
3
Homogeneous Linear Equations with Constant Coefficients: Modeling without Forcing Functions
4
Nonhomogeneous Linear Equations: Modeling with Continuous Forcing Functions
5
Laplace Transform I: Modeling with Discontinuous Force Functions
7
Systems of First-Order Equations and Applications
9
Linear Equations with Variable Coefficients
7.9
Nonhomogeneous 2×2 Systems: Laplace Transform Method
6
Laplace Transform II: Modeling with Periodic or Impulse Forcing Functions
8
Matrices and First-Order Linear Systems
10
Partial Differential Equations: Modeling Position and Time Dependent Behavior

Contents

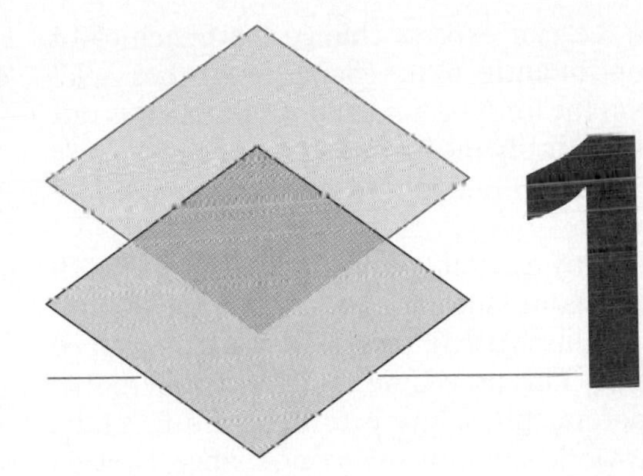

1

Differential Equations and Models

Historically, differential equations arose from humanity's interest in and curiosity about the nature of the world in which we live. This interest includes both physical processes occurring on the earth as well as the movements of heavenly bodies. We needed to know when and how to plant our crops, how to construct bridges and cathedrals so that they would not collapse, and how to interpret the motions of the stars and planets so that we could navigate the seas and oceans in order to carry on commerce and trade. This book mainly concerns the topics and problems you will study in your undergraduate science, mathematics, and engineering courses.

Most of the scientific knowledge we have is empirical; that is, we are guided by our experiences and observations. Today the ways in which we can see and observe may be very sophisticated, using powerful instruments often aided by computers. But if we look at times when such sophisticated instruments were not available, it might be easier to grasp what we were observing. A general answer is simple. We were observing change in everything we saw: in the positions of the stars and planets, in the climate, in living things, and in geological formations. Some of these changes seemed permanent, others appeared to be cyclic like the seasons, and some even seemed to lessen over time.

Calculus is the study of how we can express change mathematically: taking the ratio of the change in one quantity to the change in another yields an average rate of change, which in the limit becomes an instantaneous rate of change or *derivative.* So mathematically the investigation of change produces equations and expressions involving derivatives, *differential equations* which you will study in this book.

One way to investigate change is by differentiating a known function. If the function contains arbitrary constants (such as appear when the antidifferentiation process in integral calculus is used), then differentiation can be used to "eliminate" those constants. This procedure produces a differential equation. However, a more important procedure is to build a differential equation from investigating a real-world situation involving change. How do we translate what we are observing into differential equations? This translation is part of the process of *mathematical modeling.* It is a creative process having some general and widely accepted principles that you can learn.

In this text many differential equations are constructed from real-world problems already familiar to you or that you will encounter as a science or engineering student. Then we present methods for solving the differential equations we create. Both the qualitative and quantitative aspects of a solution may reveal the underlying behavior being studied. Through this construction and solution process you can become a better problem solver. Also discussed in this text are some of the methods and tools currently used by science and engineering practitioners in industry, government, and research laboratories. We assume that you have already had two or three semesters of calculus, including functions of two and three variables. Because you may not have mastered all of this material, we provide some review of the essential ideas as they arise.

Chapter 1 begins by relating the concept of a differential equation to the calculus you have studied. First we preview many of the ideas and terms used throughout the text. Then a brief introduction to the process of mathematical modeling is presented and applied to formulate differential equations from real-world situations.

1.1 SOLUTIONS AND INITIAL VALUE PROBLEMS

In studying the calculus you learned how to find the derivative function $dy/dx = y' = f'(x)$ for the function $y = f(x)$. For example, if

$$y = \sin 2x + 3e^{-x},$$

then

$$\frac{dy}{dx} = 2 \cos 2x - 3e^{-x}. \tag{1}$$

Or, given an equation of the form $g(x, y) = \text{constant}$, you differentiated implicitly to find dy/dx. For instance, differentiating the function implicitly defined by the equation

$$x^2 + y^2 = 4$$

results in

$$2x + 2y\frac{dy}{dx} = 0$$

or

$$\frac{dy}{dx} = \frac{-x}{y}. \tag{2}$$

Equations (1) and (2) are examples of differential equations.

DEFINITION 1.1

A **differential equation** is an equation relating an unknown function and any of its derivatives or differentials.

If only one independent variable is assumed, the equation is called an **ordinary differential equation.** Equations (1) and (2) are examples of ordinary differential equations, as are the following:

$$\frac{dy}{dx} + 2xy = e^x,$$

$$y\,dy - xe^y\,dx = 0,$$

$$\frac{d^2y}{dx^2} - \frac{dy}{dx} - 2y = \cos x,$$

$$\left(\frac{dy}{dx}\right)^2 - x^2e^y = 1.$$

If two or more independent variables appear, the equation is called a **partial differential equation.** For example,

$$\frac{\partial v}{\partial x} + \frac{\partial v}{\partial t} + 2v = 0,$$

$$\frac{\partial^2 u}{\partial x^2} = \frac{\partial u}{\partial t},$$

$$\frac{\partial^2 u}{\partial x^2} + \frac{\partial^2 u}{\partial t^2} = 0$$

are partial differential equations. For most of this text we will be concerned with formulating and investigating ordinary differential equations. In Chapter 10 we study certain partial differential equations.

Notation and Order

The most general form of an ordinary differential equation is

$$f(x, y, y', \ldots, y^{(n)}) = 0. \tag{3}$$

Expression (3) simply says that there is a relationship between the independent variable x and the dependent variable y and its various derivatives, in the form of an equation set identically equal to zero.

The **order** of a differential equation is the order n of the highest derivative appearing in the equation. Thus

$$\frac{d^2y}{dx^2} + y = 0$$

is a **second-order** equation, whereas

$$\frac{dy}{dx} - xy = \sin x$$

is an example of a **first-order** equation.

Solutions

One of the chief concerns about a differential equation is how to solve it. What is meant by a solution to an ordinary differential equation? Basically, a *solution* is a function $y(x)$ that satisfies the differential equation. As the course develops, you will see different ways to represent or approximate a solution.

DEFINITION 1.2

A function $y(x)$ defined over an interval is said to be a **solution** to a differential equation if, for any allowable value of the independent variable x, an identity results when the corresponding values for $y(x)$ and its derivatives are substituted into the equation.

EXAMPLE 1 For any constant k the function $y = ke^{x/2}$ is a solution to the equation

$$\frac{dy}{dx} = \frac{1}{2}y$$

over the interval $-\infty < x < \infty$. Differentiating y we obtain

$$\frac{d}{dx}(ke^{x/2}) = \frac{k}{2}e^{x/2}.$$

Then substitution into the differential equation gives

$$\frac{k}{2}e^{x/2} = \frac{1}{2}(ke^{x/2}),$$

which is true for every real number x.

A solution to a differential equation must be continuous since a derivative appears in the equation. This requirement follows from a result you studied in calculus:

> **THEOREM 1.1**
>
> If a function f is differentiable over an interval, then it is continuous there.

A solution to a differential equation may be represented in different forms, often depending on the method used to obtain it. The forms may be analytical, graphical, or numerical in nature. Each of these forms is discussed in turn.

Analytical Solutions An **analytical representation** of a solution may take one of two forms:

1. In the **explicit form** $y = f(x)$, the dependent variable is completely isolated and appears only to the first power on one side of the equation. The other side of the equation is an expression involving only the independent variable x and constants.

2. The **implicit form** is an equation $h(x, y) = 0$ involving both the dependent and independent variables but no derivatives. In this form the dependent variable y is not expressly given as a function of the independent variable x. We assume that the implicit form is satisfied by at least one function that also satisfies the differential equation.

The solution in Example 1 is expressed in explicit form. To check that an explicit form is a solution you simply differentiate the function (as often as the order of the equation dictates) and then substitute the results into the differential equation to see that it is satisfied.

Following is an example in which the solution is given implicitly.

EXAMPLE 2 The equation $x^2 + y^2 = C$, for any constant $C > 0$, represents an implicit solution to the differential equation

$$\frac{dy}{dx} = -\frac{x}{y}.$$

To establish this result, differentiate both sides of the original equation implicitly with respect to x:

$$\frac{d}{dx}(x^2 + y^2) = \frac{d}{dx}(C),$$

$$2x + 2y\frac{dy}{dx} = 0.$$

Then solving for the derivative gives the differential equation

$$\frac{dy}{dx} = -\frac{2x}{2y} = -\frac{x}{y}.$$

In Example 2 notice that the dependent variable y is not expressly given as a function of the independent variable x in the equation $x^2 + y^2 = C$. In general, you may or may not be able to solve the equation $h(x, y) = 0$ algebraically (or analytically) for y in terms of x. In Example 2, however, it is possible to solve for y. Each of the functions

$$y = \sqrt{C - x^2} \qquad \text{and} \qquad y = -\sqrt{C - x^2}$$

provides a continuous solution over the interval $x^2 < C$ or $-\sqrt{C} < x < \sqrt{C}$. (Of course C must be nonnegative or no real-valued solution exists.)

Although it is possible to solve the equation $x^2 + y^2 = C$ from Example 2 for y explicitly in terms of x, such is not always the case. For instance, by implicitly differentiating the equation

$$xy + e^x \sin y = 1, \tag{4}$$

you will see that it satisfies the first-order equation

$$\frac{dy}{dx} = -\frac{y + e^x \sin y}{x + e^x \cos y}. \tag{5}$$

However, there is no algebraic method for expressing y in Eq. (4) explicitly in terms of x. Thus, we refer to Eq. (4) as an *implicit solution* to the differential Eq. (5) and leave the solution in its implicit form. To verify that an implicit solution satisfies the differential equation, we differentiate the solution equation implicitly to produce the differential equation.

A solution to a differential equation is sometimes approximated by a continuous function represented symbolically. We will demonstrate one approximation process, known as Picard's method, in Chapter 2.

Graphical Solutions What do we mean by a **graphical solution** to a differential equation? Geometrically, a solution of a first-order differential equation is a **curve** whose slope at any point is the value of the derivative there as given by the differential equation (see Figs. 1.1 and 1.2 in Examples 3 and 4). Graphical solutions may be *quantitative* in nature; that is, the graph may be sufficiently precise so that the values of the solution function can be read directly from the graph. Or the solution may be *qualitative,* where the graph is imprecise as far as numerical values are concerned yet still revealing of the general shape and features of the solution curve. Moreover, graphical solutions can be produced in different ways. One way is to plot from a table of numerical values for the solution function. Another is by evaluating an analytic expression for the solution at specified values of the independent variable. A third uses a direction or tangent field of the differential equation. These qualitative and quantitative methods are explored as the text develops.

Graphical representations of solutions are easy to interpret and they usually reveal considerable information concerning the nature of the solution. For instance, the graph may quickly reveal whether the solution has local extrema and if so where these occur, whether the solution is ever zero, where it is increasing and decreasing, and whether it is concave up or down. When the differential equation is higher than first-order, geometrically a solution is still a curve, but since higher-order derivatives are present, the slope at each point is not so readily obtained from the differential equation. A discussion of graphical solutions is given in Section 1.4.

EXAMPLE 3 The family of concentric circles in Fig. 1.1 is a graphical representation of the family of solutions to the differential equation

$$\frac{dy}{dx} = -\frac{x}{y}$$

discussed in Example 2.

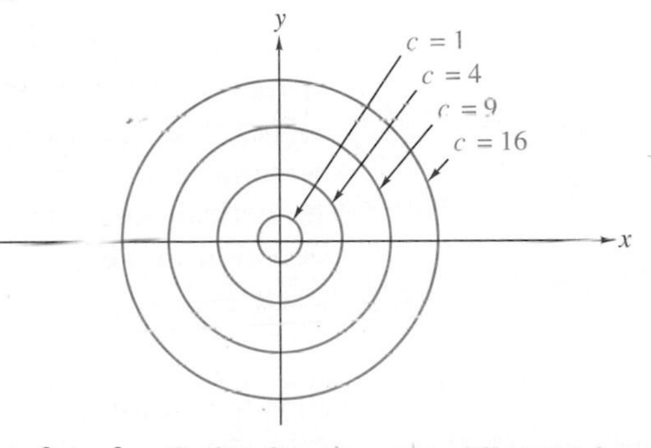

FIGURE 1.1
The family of circles $x^2 + y^2 = C$, $C > 0$, solves the differential equation $y' = -x/y$.

EXAMPLE 4 In Fig. 1.2 we present several graphical solutions to the differential equation

$$\frac{dy}{dx} = r(M - y)y$$

where r and M are positive constants. Notice the qualitative nature of the solution curves. The figure reveals their general shape and features, but you could not use it to extract actual values for a particular solution curve. For example, if $0 < y < M$, the curves have positive slope and hence are increasing. If $y > M$ the curves have negative slope and are decreasing. Further note the suggestion of a point of inflection occurring when $y = M/2$. In Section 1.4 we will show how these solution curves are obtained without actually solving analytically the differential equation.

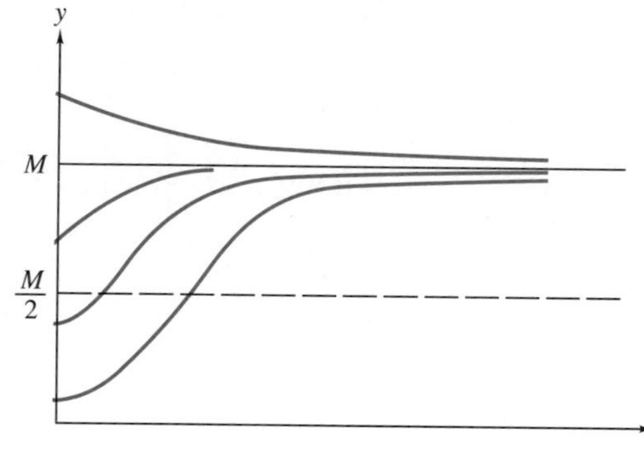

FIGURE 1.2
Solution curves for $y' = r(M - y)y$, $r > 0$, $M > 0$.

Numerical Approximations A solution to a differential equation may also be approximated **numerically.** In this case the form of the solution is a *table of values* of the dependent variable y for preselected values of the independent variable x. Numerical solutions are necessarily approximations to the true value of the solution and, as you might expect with any approximation method, we must be concerned with the accuracy of the numerical values. We will investigate numerical methods for first-order ordinary differential equations in Chapter 2. Later in the text we will also consider numerical solutions for second-order equations, for partial differential equations, and systems of differential equations.

EXAMPLE 5 The following table of values is a numerical solution to the differential equation

$$y' - x = 1 \tag{6}$$

subject to the requirement that $y = 0$ when $x = 0$. The table is valid over the interval $0 \leq x \leq 1$. Figure 1.3 shows a scatterplot of the solution values. The table was produced using an approximation technique presented in Chapter 2.

x	0.0	0.1	0.2	0.3	0.4	0.5	0.6	0.7	0.8	0.9	1.0
y	0.0	0.105	0.22	0.345	0.48	0.625	0.78	0.945	1.12	1.305	1.5

If the differential equation (6) is rewritten in the form

$$\frac{dy}{dx} = x + 1,$$

integration produces the analytical solution

$$y = \frac{x^2}{2} + x + C.$$

Then the requirement that $y = 0$ when $x = 0$ gives

$$0 = \frac{0^2}{2} + 0 + C$$

or

$$C = 0.$$

Therefore

$$y = \frac{x^2}{2} + x \tag{7}$$

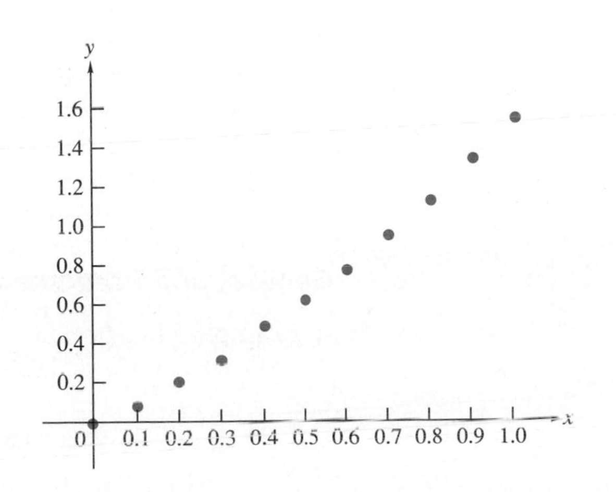

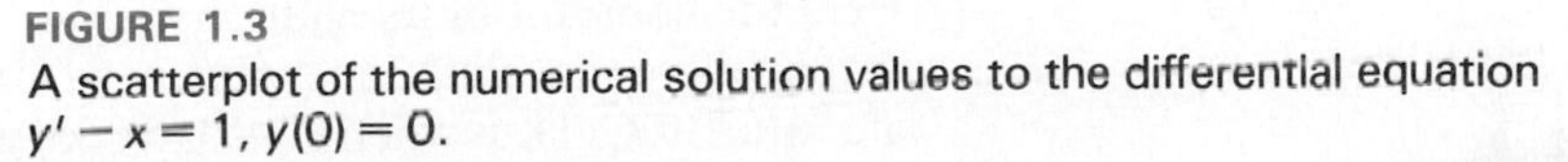
FIGURE 1.3
A scatterplot of the numerical solution values to the differential equation $y' - x = 1$, $y(0) = 0$.

is an explicit solution to Eq. (6) satisfying $y(0) = 0$. In this example it was easy to find an analytic solution, but this situation is rarely the case. Most differential equations arising from real-world problems cannot be solved analytically or by direct integration. Instead, numerical solutions are found with the aid of a computer. You can check that the numerical solution of Eq. (6), exhibited in our table of values above and found by a numerical method you will study in Chapter 2, agrees almost exactly with the analytic solution for each value of x in the table. A graph of the analytic solution superimposed on the plot of the numerical solution values is presented in Fig. 1.4. Note the quantitative nature of the graph. We can read approximate values for the dependent variable directly from the graph itself. For example, when $x = 0.55$, y is approximately 0.7.

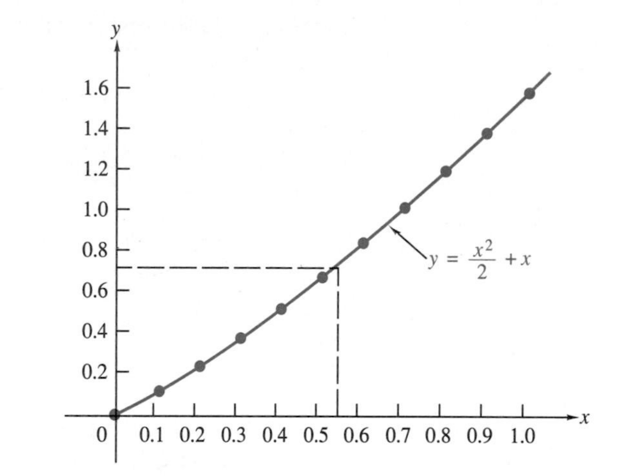

FIGURE 1.4

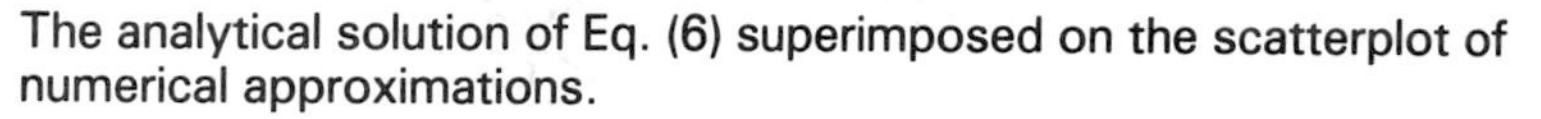

The analytical solution of Eq. (6) superimposed on the scatterplot of numerical approximations.

General and Particular Solutions

In Example 5 integration of the differential equation $y' = x + 1$ resulted in the following analytical solution

$$y = \frac{x^2}{2} + x + C. \tag{8}$$

Here the constant of integration C is an *arbitrary constant.* Every solution to the differential equation is obtained from Eq. (8) by assigning an appropriate value to C. Thus when $C = 0$, we get particular solution (7) whose graph passes through the origin. Solution (8) is called the *general solution* to $y' = x + 1$. Likewise, the equation $x^2 + y^2 = C$ in Example 2 is the general solution to $y' = -x/y$.

DEFINITION 1.3

The **general solution** to an nth-order ordinary differential equation is a solution (expressed explicitly or implicitly) that contains all possible solutions over an interval I. The general solution contains n arbitrary constants.

The question of determining whether a solution to a differential equation is the general solution is nontrivial and requires a deep investigation into the theory of ordinary differential equations. Later on we will say more about general solutions to the special kinds of differential equations you will study in this text.

DEFINITION 1.4

If a solution to an nth-order ordinary differential equation is free of arbitrary constants, then it is called a **particular solution** to the differential equation.

For the first-order equation in Example 5, we found the particular solution (7) from the general solution by requiring that $y = 0$ when $x = 0$. Thus, specifying a particular solution to a first-order equation is equivalent to prescribing a point (x_0, y_0) through which the solution curve must pass. That is, we seek a solution $y = y(x)$ satisfying

$$y(x_0) = y_0. \tag{9}$$

Equation (9) is called an **initial condition** of the first-order differential equation. A first-order equation together with an initial condition is called a **first-order initial value problem.** Therefore

$$y' - x = 1, \quad y(0) = 0$$

is an example of an initial value problem.

Solution Issues

There are some deep questions regarding solutions to differential equations. For example, does a given differential equation have a solution over an interval and satisfying an initial condition $y(x_0) = y_0$? For example,.

$$(y')^2 + x^2 = 0 \tag{10}$$

has no solution over any interval because any function satisfying Eq. (10) would have a slope whose square is a negative number at every value of x. This requirement is impossible in the real number system. So there is a concern over whether solutions exist to a given equation. Moreover, if a solution $y = y(x)$ satisfying the initial condition $y(x_0) = y_0$ does exist, is it the *only* solution?

The questions of *existence* and *uniqueness* of solutions lead quickly into the *theory of ordinary differential equations*. A full discussion is more appropriate for an advanced course, but we will discuss these questions briefly in Chapter 2 after you have acquired some experience solving first-order equations. In Section 1.2 we take a brief excursion into mathematical modeling so you will have some perspective on how real-world behaviors are formulated and studied in mathematical terms.

EXERCISES 1.1

1. Define in your own words the following terms:
differential equation
solution
order of a differential equation
initial value problem

2. In your own words, discuss the following concepts:
ordinary versus partial differential equations
general versus particular solutions

3. Discuss the possible forms of a solution to an ordinary differential equation: analytical (explicit versus implicit), graphical, and numerical.

In problems 4–11, classify the given differential equation as ordinary or partial and determine its order.

4. $(x^2 + 1)y' = x^2 + 2x - 4xy - 1$

5. $xy' - y = \sqrt{xy}$

6. $y'' - 5y' + 6y = e^{3x} - x^2$

7. $(x^2 - x)y'' + xy' + 7y = 0$

8. $y''' - 3y'' + 4y = xe^{2x} - \cos x$

9. $4u_{xx} = u_t$

10. $u_{xx} + u_{yy} + u_{zz} = 0$

11. $(xe^{xy} \cos 2x - 3)y' = ye^{xy} \cos 2x - 2e^{xy} \sin 2x + 2x$

In problems 12–26 verify that the given function is a solution to the differential equation. If an initial condition is given, verify that it is satisfied also.

12. $y' = 2y, \quad y = Ce^{2x}$

13. $y' = y - e^{2x}, \quad y = Ce^x - e^{2x}$

14. $y'' = -y, \quad y = A \cos x + B \sin x$

15. $y'' + 2y' + y = 0, \quad y = (A + Bx)e^{-x}$

16. $y' = -\dfrac{y}{x}, \quad xy = 1, \quad y(1) = 1$

17. $y' = \dfrac{x}{4y}, \quad x^2 - 4y^2 = 16, \quad y(4) = 0$

18. $x^2y' + xy = 2, \quad y = \dfrac{2 \ln x + C}{x}, x > 0$

19. $y' = xy^2, \quad y(x^2 + C) + 2 = 0$

20. $2xyy' = y + 1, \quad e^{2y} = Cx(y + 1)^2$

21. $y' = y - xy^3e^{-2x}, \quad e^{2x} = y^2(x^2 + 1). \quad y(0) = 1$

22. $(x+1)y'' + xy' - y = 0, \quad y = Ax + Be^{-x}$

23. $y'' + y = \sec^3 x, \quad y = \frac{1}{2}\sec x$

24. $y'' + y' - 2y = xe^x, \quad y = \dfrac{x^2e^x}{6} - \dfrac{xe^x}{9}$

25. $y'' - 3y' + 2y = \sin e^{-x}, \quad y = Ae^x + (B - \sin e^{-x})e^{2x}$

26. $x^2y'' + 2xy' - 12y = 0, \quad y = Ax^3 + Bx^{-4}$

1.2

AN OVERVIEW OF MATHEMATICAL MODELING

This section presents the process of formulating real-world behaviors in mathematical terms. We do not intend that you master the modeling process. Rather, we are providing sufficient background so you can appreciate how differential equations are actually formulated and refined to address real-world behaviors. In more advanced courses you will learn to develop in detail many of the models presented in this text. Simply read through this section to get an idea of the modeling process. You may want to return to reread this section from time to time as the text unfolds.

In carrying out a mathematical modeling study, we are normally constructing a description of some real-world behavior or phenomenon in mathematical terms, thus creating a second "world" in which to view the situation, as depicted in Fig. 1.5. We may wish to use information obtained in this way to explain the behavior, to predict what will happen in the future, or to analyze the effects various situations have on that behavior.

Models and Real-world Systems

A **system** is an assemblage of objects joined by some regular interaction or interdependence. The solar system, the U.S. economy, a fish population living in a lake, a satellite orbiting the earth—all are examples of a system.

Real-World Systems	Mathematical World
	Models
Observed behavior or phenomenon	Mathematical operations and rules
	Mathematical conclusions

FIGURE 1.5
Real and mathematical worlds.

Often the modeler is interested in *understanding*

- how a particular system works
- what causes change in the system
- the sensitivity of the system to certain changes.

The modeler may also be interested in *predicting*

- what changes will occur in the system
- when change will occur.

A basic technique used in constructing a mathematical model of some physical system is a combined mathematical-physical analysis. In this approach we start with some known physical principles or reasonable assumptions about the system. Then we reason logically to obtain conclusions. This approach is characteristic of the mathematical modeling that leads to differential equations.

There are various kinds of models, but we are interested in using the modeling process to construct differential equations as models. These models take the form of an equation or system of simultaneous equations involving the derivatives of a function of one or more variables. One of our tasks is building a library of models given as differential equations and recognizing various real-world situations to which they apply. Another task is formulating and analyzing new models. Still another task is learning to solve an equation or system in order to find more revealing or useful expressions relating the variables. We often study the models graphically to gain insight into the behavior under investigation. Through these activities we may develop a strong sense of the mathematical aspects of the problem, its physical underpinnings, and the powerful interplay between them.

Construction of Models

We now focus attention on the construction of mathematical models. As an illustrative example, suppose an object is dropped from a great height, say from a hovering helicopter. There are a number of questions we might ask. For instance, does a heavier object fall faster than a lighter one? Does the object fall at a constant speed? If its speed changes during the course of the fall, how fast is the object moving when it strikes the earth? Does it reach a constant maximum speed? Does the object fall halfway to the earth in half the time of the total fall? How long does it take the object to reach the earth? The first steps in constructing a model involve being very specific about what we are investigating. The later steps involve solving the model and testing it out against real-world observations. Following is a summary of the steps for constructing a mathematical model.

CONSTRUCTING MATHEMATICAL MODELS

Step 1. Identify the problem.
Step 2. Make assumptions.

- **a)** Identify and classify the variables.
- **b)** Determine interrelationships among the variables and submodels.

Step 3. Solve or interpret the model.
Step 4. Verify the model.

- **a)** Does it address the problem?
- **b)** Does it make common sense?
- **c)** Does it hold up when tested with real-world data?

Let us discuss each of these steps and apply them to the falling-body example.

Step 1: Identify the Problem

What is it you would like to do or find out? The modeler must be sufficiently precise (ultimately) in verbally formulating the problem in order to be successful in translating it into mathematics. This translation is accomplished through the Steps 2–4.

In our example of the falling body, we may choose to define the problem as follows: *Given an object of specified mass or weight being dropped from a known height, predict the height and speed of the object at any future time.*
Notice that if we could accurately predict the height and speed of the falling object, then it appears we could answer most of the questions we posed concerning its fall. However, we still need to be more precise about the situation. Are we assuming that the object is simply being released from rest, for example, or is it being thrown from a helicopter? Are there forces that tend to slow the object down as it falls through the atmosphere? If so, can we formulate their effects? And so forth. We need to decide on the scope of our investigation.

Step 2: Make Assumptions

Generally you cannot hope to capture in a usable mathematical model all the factors you have identified as influencing the behavior under investigation. The task is simplified by reducing the number of factors under consideration. Then relationships among the remaining variables must be determined. Again, the complexity of the problem can be reduced by assuming relatively simple relationships. Thus the assumptions fall into two main categories:

a) Classification of the variables. What things influence the behavior you identified in Step 1? List these things as variables. Then classify each variable as either dependent or independent. Those variables the model seeks to explain are the **dependent variables,** and there may be several of these; the remaining variables are the **independent variables.** You may also choose to neglect a given variable altogether.

b) Determination of the interrelationships among the variables selected for study. Before you can hypothesize a relationship among the variables, you generally must make some additional simplifications. The problem may be sufficiently complex that a relationship among all the variables cannot be seen initially. In such cases it may be possible to study **submodels.** That is, one or more of the independent variables are studied separately. Eventually the submodels are connected together under the assumptions of the main model.

For example, in the falling-body problem height and speed of the object can be identified as dependent variables with time as the independent variable. Since we know from calculus that speed is the derivative of position with respect to time, we would probably elect simply to model height. Thus, ultimately we want *to find position as a function of time.*

To simplify the problem we assume that the body is moving vertically only. What factors affect the object's position? In order to change its position some kind of force must act on the object. **Propulsion forces** tend to move the object in some direction (downward for a falling body) and **resistive forces** tend to retard that motion. The resistive forces are caused by **drag** on the object as it falls through the atmosphere and by **buoyancy,** which supports the object and tends to hold it in place. Symbolically we can state the relationship between an object's position and the forces affecting it in function notation as

$$\text{position} = f(\text{propulsion forces, resistive forces})$$

with the submodel

$$\text{resistive forces} = f(\text{drag, buoyancy}).$$

The notation $f(\ldots)$ means "function of" the variables specified within the parentheses. To simplify the notation we always use the same symbol f to connote the generic function idea, although each model or submodel usually has a different functional relationship among its indicated variables.

To simplify the model we may choose to neglect some of the independent variables for several reasons. First, the effect of the variable may be relatively small compared to other factors involved in the behavior. For example, in the falling-body problem we might initially assume the object is a steel ball bearing falling from rest and choose to neglect all resistive forces. Or we may choose to neglect a variable because we want to investigate the problem without the influence of that variable. For instance, we might want to study the falling-body problem by neglecting buoyancy but considering the effects of drag within the earth's atmosphere. We can simplify the submodel for the resistive force by making a reasonable assumption that drag equals some constant times the speed of the object. If subsequent tests of the model prove it to be unsatisfactory, we can later refine the model by changing the assumption concerning the drag effect or by including the effects of buoyancy if it turns out to be a significant factor.

Next we want to construct submodels for the propulsion, drag, and buoyancy forces. The propulsion force acting on an object falling from rest is due to gravity. This gravitational attraction in turn depends on the mass of the object and its height (distance) above the surface of the earth. (The problem of gravitational attraction is studied in detail in Section 1.3.) Thus

$$\begin{aligned}\text{propulsion force} &= F_p\\ &= \text{gravitational attraction}\\ &= f(\text{mass, height}).\end{aligned}$$

Next we consider the resistive force, which is the sum of the drag and buoyancy forces:

$$\begin{aligned}\text{drag} &= F_d\\ &= f(\text{speed, air density, cross-sectional area of object,}\\ &\qquad \text{aerodynamic shape of object}),\\ \text{buoyancy} &= F_b\\ &= f(\text{air density, density of object}).\end{aligned}$$

Recall that a free-body diagram displays all the directed forces acting on the body within a coordinate system indicating the positive direction or directions. From the free-body diagram in Fig. 1.6 it can be seen that the total force F acting on the object is given by

$$F = F_p - F_d - F_b.$$

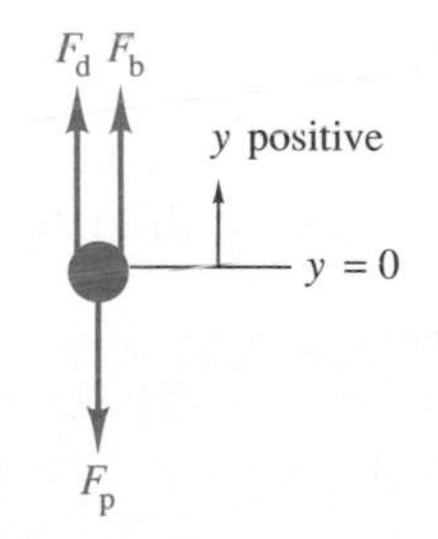

FIGURE 1.6
Free-body diagram for a falling body.

This total force equation connects the various submodels together.

From Newton's second law, $F = ma$ (itself a mathematical model), and neglecting all resistive forces, we find

$$F = F_p - F_d - F_b$$

leads to

$$ma = -mg - 0 - 0$$

or

$$my'' = -mg, \tag{1}$$

subject to the conditions

$$y(0) = h, \quad y'(0) = 0. \tag{2}$$

Here m is the mass of the object, g is the acceleration due to gravity, y is the height of the object measured from the surface of the earth at any time t, and h is the initial height of the object above the surface of the earth. The acceleration a of the object is the second derivative of y with respect to time, and it is this idea of change that led to differential equation (1). Since the propulsion force F_p on the object is acting in the downward direction, it is assigned a negative value (because positive distance is being measured upward from the surface of the earth).

In summary, there are basically two types of simplifying assumptions. We may choose *to neglect some of the independent variables,* such as neglecting drag and buoyancy in our falling-body example. Or we may choose *to simplify relationships* among the variables, such as assuming the propulsion force F_p is a constant multiple of mass (although gravity actually varies with distance from the earth).

Step 3: **Solve or interpret the model**

Now put together all the submodels to see what the model is telling you. For the models we will construct and study in this course, you will usually have to solve a differential equation. Often you will find that you are not quite ready to complete this step, or you may end up with a model so unwieldy you cannot solve or interpret it. In such situations you might return to Step 2 and make additional *simplifying assumptions.* Sometimes you will even want to return to Step 1 to redefine the problem. The ability to simplify or refine a model is an important part of mathematical modeling and will be discussed separately.

In the case of our falling-body model (1), the solution of the differential equation simply requires two integrations as you learned from the calculus. Thus

$$y = -\frac{gt^2}{2} + c_1 t + c_2. \tag{3}$$

We can then use the conditions (2) to evaluate the constants of integration:

$$h = y(0) = 0 + c_1 \cdot 0 + c_2$$

so that

$$c_2 = h.$$

Also, differentiating Eq. (3), we get

$$y' = -gt + c_1. \tag{4}$$

Thus

$$0 = y'(0) = -g \cdot 0 + c_1$$

or

$$c_1 = 0.$$

Substituting the values for these constants into Eqs. (3) and (4) gives

$$y = -\frac{gt^2}{2} + h \tag{5}$$

and

$$y' = -gt. \tag{6}$$

Equations (5) and (6) give the height and velocity of the falling body, respectively, at any time t. Notice that the mass m canceled from both sides of equation (1), so the mass does not affect the speed of the object or the distance it falls.

Step 4: Verify the model

Before using a model to reach real-world conclusions, it must be tested out. There are several questions you need to ask before designing these tests and collecting data—a process that can be expensive and time-consuming. First, does the model answer the problem you identified in Step 1, or did you stray from the key issue as you constructed the model? Second, does the model make common sense? Third, can you gather the data necessary to test and operate the model, and does the model hold up when tested?

In designing a test for your model using actual data obtained from empirical observations, be careful to include observations made over the *same range of values* for the various independent variables that you expect to encounter when actually using the model. The assumptions you made in Step 2 may be reasonable over a restricted range of the independent variables yet very poor outside of those values.

In our falling-body example, the solution equations (5) and (6) do predict the position and speed of the falling object as a function of time, which is what we set out to determine. We notice from Eq. (5) that as time advances the height y of the object decreases (which is certainly what we would expect). Moreover, we can use these equations to answer the questions we asked about the falling body. For instance, we can see from Eq. (6) that the speed increases with time and never reaches a constant maximum value. Don't forget, however, that our model neglected all resistive forces. Inclusion of some of those forces might very well (and indeed does) alter this conclusion.

How might we test our model for the falling body? We could conduct some experiments by dropping a steel ball bearing initially at rest from some given height and measuring its position (height) every half second. Then we could compare these experimental values against the values predicted by Eq. (5). Would solution (5) be an accurate predictor if we dropped a sheet of paper from a given height? Recall that our model neglected drag effects. Consider the situation of dropping an object, like a pearl, into a container of heavy oil. Would the buoyancy effect now be significant in addition to drag? Would you expect Eq. (5) to predict accurately the position of the pearl as it falls to the bottom of the container? In Section 1.3 we construct submodels to take such factors into account for the falling-body problem.

Graphs can be very useful in assessing a model. For experiments involving dropping a steel ball bearing from a given height, Eq. (5) reveals that a plot of position versus time should yield a parabolic curve opening downward (see Fig. 1.7a). Or we could plot the quantity $h - y$ versus t^2. Again from Eq. (5) we see that the graph should approximate a straight line with a slope of approximately $g/2$ that passes through the origin, as illustrated in Fig. 1.7(b). If our experimental data do not lie reasonably close to that line (that is, within tolerable limits of measurement error), we would conclude from this test that the model does not reflect accurately the real-world behavior.

For example, assume an object was dropped from rest at a height of 200 feet and the following data were collected:

t (time, sec)	0	1	1.5	2	2.5	3	3.5	4
y (height, ft)	200	186	172	153	131	107	82	55

Figure 1.8(a) shows a scatterplot of the data superimposed on the model $y = 200 - 16t^2$ (assuming $g = 32$ ft/sec²). Figure 1.8(b) shows a scatterplot

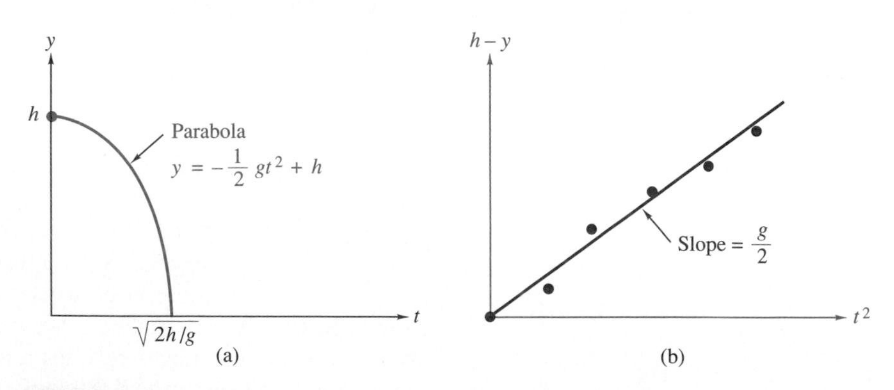

FIGURE 1.7
(a) From Eq. (5) a plot of y versus t yields a parabola opening downward.
(b) A plot of $h - y$ versus t^2 for experimental observations should yield approximately a straight line passing through the origin.

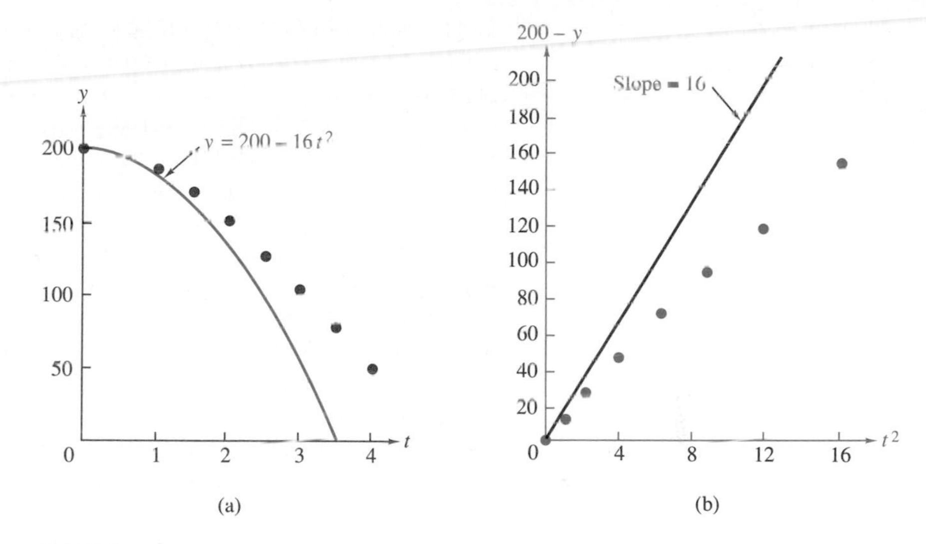

FIGURE 1.8
(a) The predicted values along the parabolic curve $y = 200 - 16t^2$ give y values that are consistently lower than the measurements. Graph (b) shows that the actual $200 - y$ values fail to lie along the predicted straight line. There is some force resisting the downward motion that the present model fails to take into account.

of $200 - y$ versus t^2 for the actual data superimposed on the predicted straight line. Both graphs reveal that the model predicts values for y that are too low; that is, the object is not falling as fast as the model predicts. From this we conclude there is something wrong with the assumptions. Perhaps the object is experiencing some force that tends to retard the downward motion, such as atmospheric drag. (In our tabled hypothetical data there is in fact a resisting force equal to a constant times the velocity of the object.) In Section 1.3 you will see how such factors might be incorporated into the falling-body model.

Be very careful about the conclusions you draw from any tests. Just as you cannot prove a theorem simply by demonstrating many cases in which the theorem does hold, likewise you *cannot extrapolate broad generalizations from the particular evidence you gather about your model.* A model does not become a law just because it is verified repeatedly in some specific instances. Rather, you *corroborate the reasonableness of your model through the data you collect.*

Iterative Nature of Model Construction

Model construction is an iterative process. You begin by examining some system and identifying the particular behavior you wish to predict or explain.

Next you identify the variables and simplifying assumptions, and then you generate a model. You will generally start with a rather simple model, progress through the modeling process, and then refine the model as the results of your validation procedures dictate. If you cannot come up with a model or solve the one you have, you must *simplify* it (see Fig. 1.9). This is done by treating some variables as constants, by neglecting or aggregating some variables, by assuming simple relationships (such as linearity) in any submodel, or by further restricting the problem under investigation. On the other hand, if your results are not precise enough, you must *refine* the model (see Fig. 1.9). Refinement is generally achieved in the opposite way to simplification: you introduce additional variables, assume more sophisticated relationships among the variables, or expand the scope of the problem. By simplification and refinement, you determine the generality, realism, and precision of your model. This process cannot be overemphasized and constitutes the art of modeling. These ideas are summarized in the following table.

THE ART OF MATHEMATICAL MODELING: SIMPLIFYING OR REFINING THE MODEL AS REQUIRED

Model Simplification	Model Refinement
1. Restrict problem identification.	1. Expand the problem.
2. Neglect variables.	2. Consider additional variables.
3. Conglomerate effects of several variables.	3. Consider each variable in detail.
4. Set some variables to be constant.	4. Allow variation in the variables.
5. Assume simple (linear) relationships.	5. Consider nonlinear relationships.
6. Incorporate more assumptions.	6. Reduce the number of assumptions.

The purpose of this section has been to provide a background for understanding the differential equations presented and refined as the text progresses. It is not expected that you be proficient in modeling at this stage in your academic studies since your modeling abilities will improve in later courses; however, you may want to think about a few of the optional exercises before continuing with your study of differential equations.

In the next two sections we apply the modeling principles presented here to real-world systems modeled by an appropriate differential equation. These models, together with suitable refinements, form the basis for many of the illustrative examples used throughout the text.

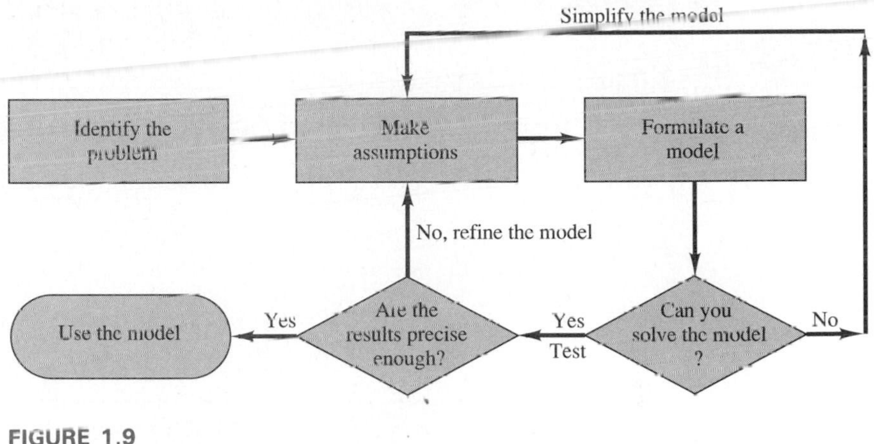

FIGURE 1.9
The iterative nature of model construction.

EXERCISES 1.2 (Optional)

In the following problems the scenarios are stated somewhat vaguely. For these problems consider the first two steps in the modeling process. Identify a problem you would like to study. What variables affect the behavior you have identified in the problem identification? Which variables are the most important? Can you identify any submodels you may wish to consider? What variables might you initially consider to be constant? Can you formulate some simple relationships among the variables? We will consider some of these problems in detail later on in the text.

1. How fast can a skier ski down a mountain slope?

2. Suppose the manager of a skyscraper asks you (the consultant), "How long does it take for the elevator to get to the top floor?"

3. Consider the rule of thumb often given in driver education classes: Allow one car length for every 10 mph of speed under normal driving conditions. How good is this rule?

4. A newspaper article states that we should observe the highway speed limit of 55 mph because for every 5 mph over 50 mph there is a loss of one mile per gallon. Is this a valid argument?

5. How do biological populations grow?

6. What is the required dosage level and the interval of time between doses for a particular drug to be safe yet effective? Consider a submodel for absorption of the drug into the bloodstream and a submodel for the dissipation of the drug.

7. A famous round table fixed on the wall in the great hall of Winchester Castle is thought to be the actual round table of King Arthur, but there is speculation over its authenticity. King Arthur lived in the fifth century A.D. How would you test the age of the table?

8. How long will it be before the world runs out of oil?

9. The orbit of a space satellite decays and the satellite falls back into the earth's atmosphere. Will the satellite burn up before it reaches the ground?

10. A single-stage rocket is fired vertically from the surface of the earth. Will the rocket escape the earth's gravitational field?

11. How far can a golfer hit a golf ball?

12. A chemical has spilled into a lake causing unsafe levels of pollution. Fresh water runs into the lake from a nearby stream, and water runs out of the lake at the same rate. How long will it be before the lake is cleaned to 10% of the initial pollution level?

MODELING WITH FIRST-ORDER DIFFERENTIAL EQUATIONS

Suppose $y = f(x)$ is a differentiable function at $x = x_0$. That is, the limit

$$f'(x_0) = \lim_{x \to x_0} \frac{f(x) - f(x_0)}{(x - x_0)}$$

exists. In elementary calculus you investigated the idea of the derivative from two points of view:

> 1. The derivative $f'(x_0)$ is the **instantaneous rate of change** in y relative to x at the point $x = x_0$.
> 2. The derivative $f'(x_0)$ is the **slope of the tangent line** to the graph of $y = f(x)$ at the point $(x_0, f(x_0))$.

The interpretation of the derivative as an instantaneous rate of change is a fundamental concept used in many modeling applications. In this interpretation, the ratio

$$\frac{f(x) - f(x_0)}{x - x_0} = \frac{\Delta y}{\Delta x}$$

is the **average rate of change** in the function. Passage to the limit as $\Delta x \to 0$ (or equivalently $x \to x_0$) then gives the instantaneous rate of change $f'(x_0)$. For instance, if $y = f(x)$ denotes the position of a particle along a straight line (the y-axis) and x is a measure of time, then $\Delta y / \Delta x$ is the *average velocity* of the particle during the Δx time period. The derivative value $f'(x_0)$ gives the *exact velocity* at the instant when the particle is located at the point $f(x_0)$ along the y-axis.

The geometric interpretation of the derivative as the slope of the line tangent to the curve is useful for developing intuition and gaining insight into a problem as well as for constructing numerical solutions.

We are going to use the derivative to construct several models for situations involving instantaneous rates of change or to approximate average rates of change. The behaviors to be modeled are basic to studies in science and engineering: radioactive decay, heating and cooling, motion in a gravitational field, and population growth with unlimited and then limited resources. Later in the text, we will develop models for oscillatory motion (like the motion of a pendulum), mixing of fluids, electricity and magnetism, and systems like coupled spring systems. All of these models are derived by assuming that some observed change is proportional to a hypothesized variable or function. The reasonableness of the proposed proportionality can then be tested graphically, as we did with the falling-body example in Fig. 1.8 of Section 1.2. The idea of proportionality is very useful in constructing models. It is well worth the effort to review this elementary concept before constructing our differential equation models.

Proportionality

In developing submodels to approximate a behavior we shall often find very revealing the concept of proportionality.

DEFINITION 1.5

> Two quantities P and Q are said to be **proportional** (to one another) if one quantity is a constant multiple of the other, that is, if
>
> $$P = kQ$$
>
> for some constant k. We write $P \propto Q$ in that situation, and say that P "is proportional to" Q.

Thus

$$y \propto x \quad \text{if and only if} \quad y = kx \quad \text{for some constant } k. \tag{1}$$

Other examples of proportionality relationships include the following:

$$y \propto x^2 \quad \text{if and only if} \quad y = k_1 x^2 \quad \text{for } k_1 \text{ a constant,} \tag{2}$$

$$y \propto \ln x \quad \text{if and only if} \quad y = k_2 \ln x \quad \text{for } k_2 \text{ a constant,} \tag{3}$$

$$y \propto e^x \quad \text{if and only if} \quad y = k_3 e^x \quad \text{for } k_3 \text{ a constant.} \tag{4}$$

Geometric Interpretation of Proportionality

Now let us explore a geometric interpretation of proportionality. In model (1), $y = kx$ yields $k = y/x$. Thus k may be interpreted as the tangent of the angle θ depicted in Fig. 1.10, and the relation $y \propto x$ defines a set of points along a line in the plane with angle of inclination θ.

Comparing the general form of a proportionality relationship $y = kx$ with the equation $y = mx + b$ for a straight line, you can see that the graph of a proportionality relationship is a line (possibly extended) that *passes through the origin.* If you plot the proportionality variables for models (2), (3), and (4), you obtain the straight-line graphs presented in Fig. 1.11.

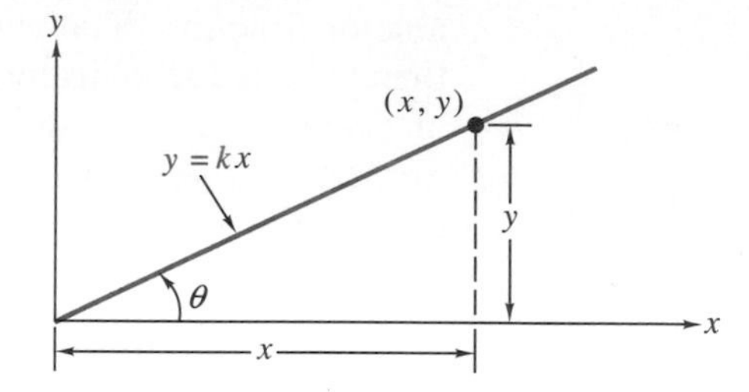

FIGURE 1.10
Geometrical interpretation of $y \propto x$.

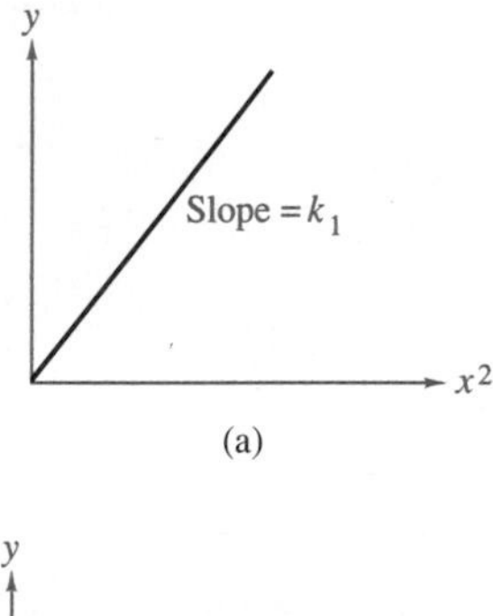

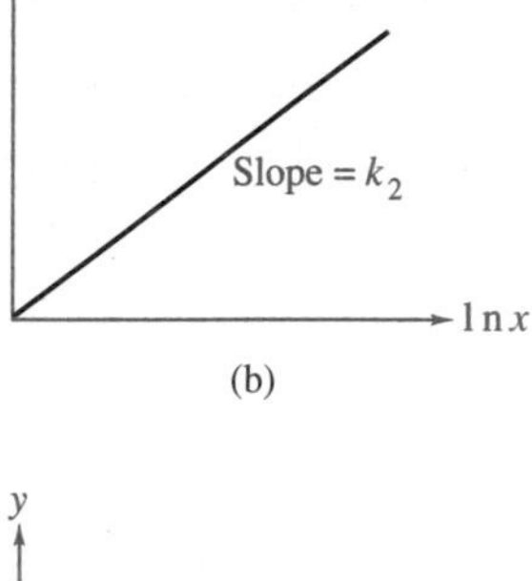

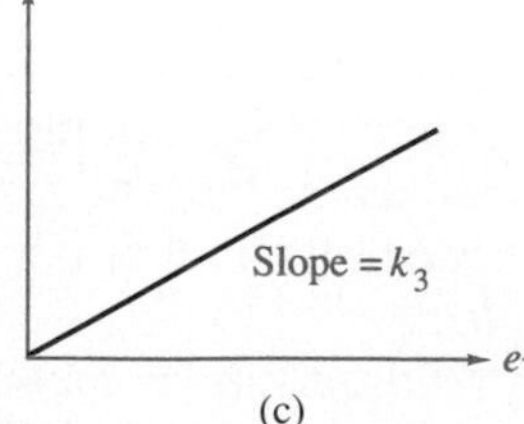

FIGURE 1.11
Geometrical interpretations of models (2), (3), and (4).

The graphing principle used in the plots of Fig. 1.11 may be new to you. In calculus you were accustomed to plotting a dependent variable y versus the independent variable x, not versus a *function* of x. However, the procedure really is no different. For example, to plot y versus x^2 from a table of values relating x and y, first calculate the value of x^2 for each value of x. Next draw a coordinate system in which the horizontal axis represents x^2 values (not x values) and the vertical axis represents y values (see Fig. 1.12). Finally, plot the ordered pair (x^2, y) in this rectangular coordinate system for each x and y related in your original table. For instance, the original table

x	1	2	3	4	5
y	1	5	8	16	21

produces the table

x^2	1	4	9	16	25
y	1	5	8	16	21

The pairs (x^2, y) from this last table are plotted in Fig. 1.12. A similar procedure would be followed to plot y versus $\ln x$, y versus e^x, and y versus $f(x)$ for any function of x.

Testing a Model for Proportionality

The geometric interpretation of a proportionality provides a way to test the validity of a hypothesized proportionality in a model or submodel and to estimate the constant of proportionality. If the model asserts that $P \propto Q$, then a plot of P versus Q for experimentally collected data should lie along a straight line passing through the origin (to within tolerable limits of experi-

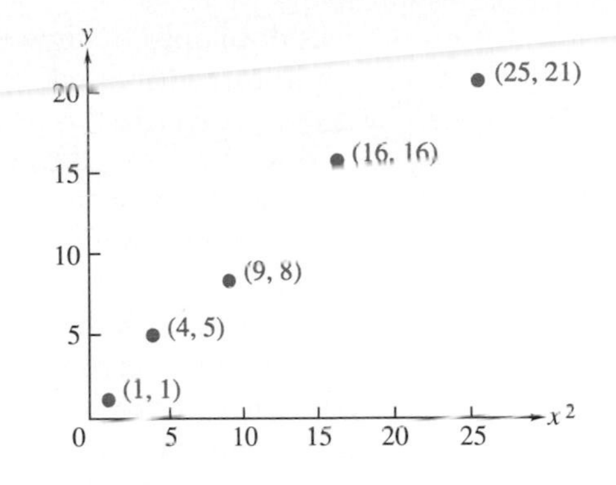

FIGURE 1.12
A plot of some points (x^2, y).

mental error). In the falling-body model, for instance, for which we neglected all resistive forces, we found the solution

$$y = -\frac{gt^2}{2} + h.$$

This model predicts that $h - y$ is proportional to t^2, with $g/2$ equal to the constant of proportionality. Thus a plot of $h - y$ versus t^2 for experimental data should give approximately a straight line through the origin. For the (hypothesized) experimental data presented in Section 1.2, the plot of $h - y$ versus t^2 in Fig. 1.8(b) did not lie approximately along a straight line through the origin. Thus we concluded that the model itself did not adequately describe the vertical motion measured by the data.

In the remainder of this section we model four different types of behaviors using differential equations: radioactive decay, heating and cooling, motion in a gravitational field, and population growth. You may elect to study some or all of these applications, as prescribed by your course requirements and/or your own interests.

APPLICATION 1 Modeling Radioactive Decay

The British physicist Lord Ernest Rutherford (1871 – 1937) established that an atom consists of a nucleus surrounded by electrons. The nucleus itself consists of protons and neutrons, and two nuclei with identical numbers of protons and identical numbers of neutrons belong to the same nuclear species or **nuclide**.

The accidental discovery of radioactivity by Henri Becquerel in 1896 followed quite closely that of x-rays. Becquerel had left a substance containing uranium on a photographic plate wrapped in black paper. After the plate was developed, it showed images of crystals of uranium compounds. The uranium had emitted penetrating rays, and Becquerel coined the word *radioactive* to describe this property. Later research by Becquerel, Madame Curie, and Rutherford led to the discovery that there are three different kinds of radioactive emissions that occur: alpha, beta, and gamma radiation. (Alpha and beta emissions consist of subatomic particles, but gamma radiation takes the form of electromagnetic waves.) Elements that emit such radiations are said to be radioactive elements or **radionuclides.*** Ordinarily it is of interest or concern to know how long it will take for a certain radioactive substance to lose some fraction of its total mass through the emission process, or to know how much of the mass will remain after a specified time period has elapsed. Let us construct a model to provide answers to these questions.

Assumptions Suppose you have a sample containing a very large number of radionuclides. Assume that they are all of the same kind (so that you have a "pure" mass of the radioactive element) and that they all decay by the same process (alpha, beta, or gamma emissions). The decaying process is such that once a particular nucleus decays, it cannot repeat the process again. Moreover, as long as the nucleus has not decayed, the probability or likelihood that it would decay in the next second can be expressed as a constant between 0 and 1. This means that if a nucleus happens to survive for an hour, the probability that it will decay in the very next second is still the same constant as the probability of its decaying in the first second at the beginning of the hour. Thus a nucleus does not "age" like a biological system.

Suppose that at time t there are N undecayed nuclei present in the sample. Then the number of nuclei that decay during the time interval from t to $t + \Delta t$ must be proportional to the product of N and Δt. That is,

$$\Delta N = -kN\,\Delta t$$

where $k > 0$ is the proportionality coefficient, called the **decay constant.** Let us assume that the number of undecayed nuclei is a continuous function of time. (This assumption is not unreasonable if there are many billions of nuclei in our sample.) Then dividing both sides of the last equation by Δt and passing to the limit as $\Delta t \to 0$ results in the differential equation

$$\frac{dN}{dt} = -kN. \tag{5}$$

If N is a single nucleus, then from Eq. (5) we could interpret k as the probability for decay per unit of time. For a given radioactive element, the number k is determined from experimental data. More generally, Eq. (5) hypothesizes that the instantaneous rate of change in the quantity or mass of the radioac-

* Chapters 9–11 of *Introduction to Nuclear Physics,* by H. A. Enge (Reading, Mass.: Addison-Wesley, 1966), provide an excellent discussion of the three kinds of radiation and theories on their emission.

tive substance is directly proportional to the mass that is present. The decay constant k together with the type of decay (alpha, beta, or gamma emissions) and the energy of decay characterize a given radioactive nucleus.

Solving the Model Differential equation (5) is easily solved by simple integration from calculus. Collecting like variables in (5) gives

$$\frac{dN}{N} = -k\,dt.$$

Integration then yields

$$\ln|N| = -kt + C$$

where C is an arbitrary constant of integration. Since $N > 0$ represents mass, $|N| = N$ and the general solution becomes

$$\ln N = -kt + C.$$

Exponentiating each side of this equation and equating the results then gives

$$N = e^{-kt+C} = e^{C} \cdot e^{-kt}$$

or

$$N = C_1 e^{-kt} \qquad \text{where} \qquad C_1 = e^{C}.$$

If we assume that at time $t = 0$, $N(0) = N_0$ is the original mass of the radioactive substance, then

$$N_0 = C_1 e^0,$$

and substitution for C_1 gives the particular solution

$$N = N_0 e^{-kt}. \tag{6}$$

The particular solution (6) is a decaying exponential function whose graph is depicted in Fig. 1.13.

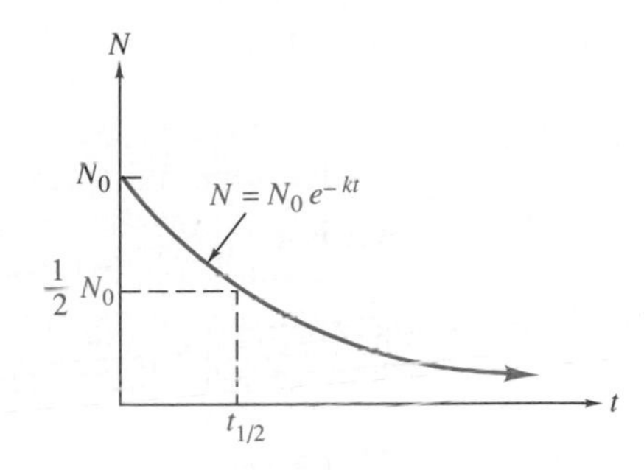

FIGURE 1.13

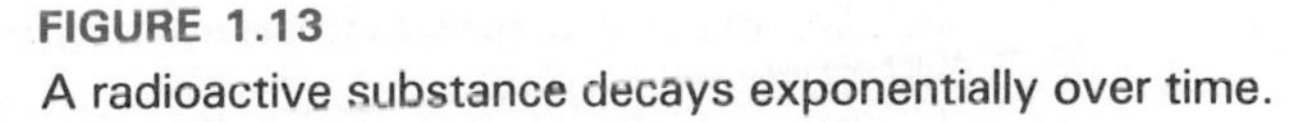
A radioactive substance decays exponentially over time.

Half-life The **half-life** of a radioactive nucleus is defined as the time $t_{1/2}$ during which the number of nuclei reduces to one-half the original value N_0 (see Fig. 1.13). From Eq. (6) we find

$$\frac{N_0}{2} = N_0 e^{-kt_{1/2}},$$

resulting in the half-life value

$$t_{1/2} = \frac{\ln 2}{k} \approx \frac{0.693}{k}. \tag{7}$$

For instance, the half-life of radium is 1620 years, for uranium-238 it is 4.5×10^9 years, and for the isotope plutonium-239 it is 24,180 years. The longer the half-life of a substance the more stable it is said to be. Thus, uranium-238 is highly stable.

APPLICATION 2 Modeling Newton's Law of Cooling

Suppose a space capsule splashes down in the ocean. Its surface would be very hot from reentering and traveling through the atmosphere. Given the temperature at the time of splashdown we would like to predict when the surface reaches a certain temperature. That is, *we desire a function that predicts the surface temperature as a function of time.* (There are many other questions even more important for the space capsule situation. For instance, how hot does the surface actually get in traveling through the atmosphere? How can a shield be designed to dissipate the heat energy to an acceptable level? It is beyond the scope of this text to construct models to answer these more difficult questions. The following discussion is intended only as a first introduction to heating and cooling of an object.)

Assumptions Consider what happens when we immerse a heated object into a cooler fluid. We expect the fluid to heat up and the object to cool down until an equilibrium temperature is reached. The final temperature is going to depend on many factors. Temperature is a measure of heat energy per unit volume. Thus we need to know the initial temperatures, the relative masses involved, and the amount of heat energy required to raise a unit mass of each material by one degree. We assume that a few units of heat energy dissipated into the ocean will have a negligible effect on the temperature of the ocean if there is good local circulation. Since heat energy is dissipated through the surface area of the object, the shape and conductivity of the object, as well as the conductivity of the fluid, are also important. Finally, the temperature difference between the object and the medium appears to be significant.

Certainly it is clear from these assumptions that heat transfer is a complex behavior, which will be described more thoroughly in Chapter 10. For now we desire only a simple first model, so we treat the space capsule as being of uniform temperature throughout. That is, as it cools down, the temperature is (instantaneously) the same everywhere throughout the capsule, with no variation in any direction.

Let us guess at the solution to our problem. Assume the temperature of the mass m, T_m, is initially α. That is $T_m(0) = \alpha$ at time $t = 0$. Further, assume that the temperature of the ocean at any time t is $T_o(t)$ and that initially $T_o(0) = \beta$, where β is much smaller than α. If the ocean circulation is reasonably good, the energy dissipated by the (relatively) small capsule has a negligible effect on the ocean's temperature. We thus predict the final temperature of the capsule to be β. Now we examine the *rate* at which $T_m(t)$ approaches β. It seems logical that the rate of decrease of $T_m(t)$ would diminish as $T_m(t)$ approaches the temperature β. Convince yourself that the graph of Fig. 1.14 qualitatively possesses the characteristics just described.

Constructing a Model Now let us construct a mathematical model that captures the above argument. For a simple model we choose not to take into account the effects of shape and conductivity of the capsule. Instead we represent those effects by a constant of proportionality k. Assume that the average rate of temperature change is proportional to the difference between the temperature of the mass, T_m, and the temperature of the ocean, T_o. (This proportionality has been validated experimentally under our assumptions.) That is, over the time interval Δt,

$$T_m(t + \Delta t) - T_m(t) = -k[T_m(t) - T_o(t)]\,\Delta t, \quad k > 0$$

or

$$\frac{T_m(t + \Delta t) - T_m(t)}{\Delta t} = -k[T_m(t) - T_o(t)]$$

where the negative sign for $-k$ indicates a decrease in temperature. We assume the temperature is changing continuously. This assumption allows

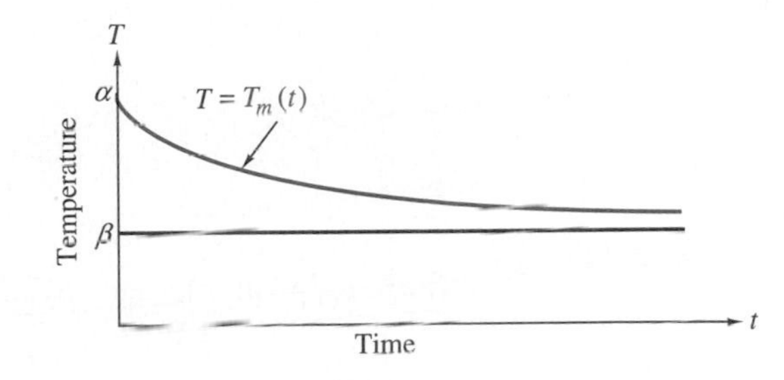

FIGURE 1.14
The temperature of the object approaches the temperature of the ocean asymptotically.

us to take the limit as Δt approaches zero, yielding

$$\lim_{\Delta t \to 0} \left[\frac{T_m(t + \Delta t) - T_m(t)}{\Delta t} \right] = \lim_{\Delta t \to 0} (-k[T_m(t) - T_o(t)])$$

or

$$\frac{dT_m}{dt} = -k[T_m(t) - T_o(t)].$$

We make the further simplification that the temperature of the ocean remains constant, yielding the following model:

$$\frac{dT_m}{dt} = -k(T_m - \beta), \quad k > 0 \tag{8}$$
$$T_m(0) = \alpha.$$

Equation (8) hypothesizes that the **instantaneous rate of change of temperature** is directly proportional to the temperature difference between the object and its surrounding medium. This model is called **Newton's law of cooling.** Just like the equation for radioactive decay, the solution to Eq. (8) is a **decaying exponential function** whose graph qualitatively approximates that depicted in Fig. 1.14. The solution details are presented in Section 2.1, Example 9.

Model Refinement The ocean in our example is a **heat sink,** a reservoir that can absorb a relatively large amount of heat energy without appreciably raising its own temperature. Similarly, a reservoir that can supply a seemingly unlimited amount of energy without significantly lowering its temperature is called a **heat source.** We will use both sinks and sources for simplifying assumptions when we consider heat transfer in Chapter 10. In this example we have ignored the physical dimensions of the space capsule and considered it as having no variation in temperature in any direction. In refining our model in Chapter 10, we will consider temperature to vary not only with time (as here), but also with position within the object (which then has some dimension). We may choose to represent the physical dimensions of the object with one variable (as in a thin wire), two variables (as in a thin flat plate), or three variables (as in a solid object where the variation of the temperature in all three dimensions is important). Of course, the more variables we consider, the more difficult it will be to solve the resulting model.

As an example, consider a long thin filament where x measures the position along the filament as illustrated in Fig. 1.15. If $u(x, t)$ represents the temperature u at any position x at any time t, we develop the following model in Chapter 10:

$$\frac{\partial u}{\partial t} = k \frac{\partial^2 u}{\partial x^2}$$

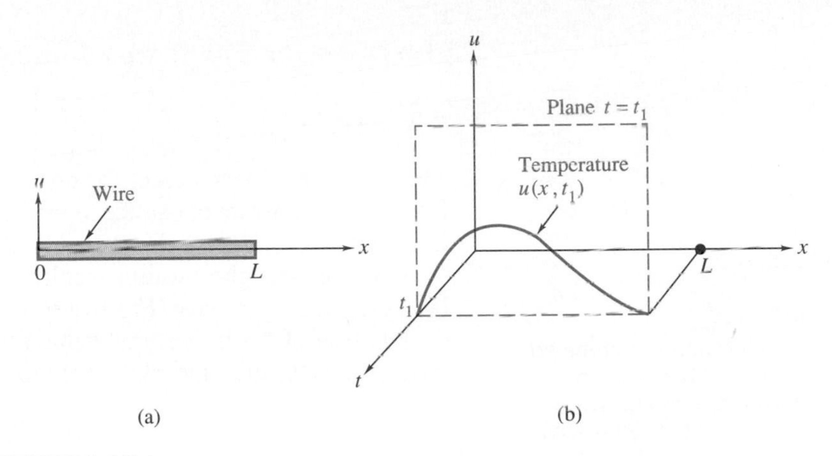

FIGURE 1.15
The temperature $u(x, t)$ in a long thin wire of length L varies with respect to both position x and time t.

Note that since u is a function of two variables (x and t) the derivatives representing the rates of change are partial derivatives. The resulting equation is called a **partial differential equation.** Partial differential equations are powerful models for describing myriad phenomena in the physical universe such as heat conduction, fluid motion, electrostatic and gravitational fields, elasticity, and quantum physics. We will study solution methods for partial differential equations in Chapter 10.

APPLICATION 3 Modeling Motion in a Gravitational Field

Galileo (1564–1642) and Newton (1642–1727) both observed that the rate of change in momentum encountered by a moving object is proportional to the net force applied to it. A general mathematical formulation for this concept is

$$F = \frac{d}{dt}(mv) = v\frac{dm}{dt} + m\frac{dv}{dt} \tag{9}$$

where F is the net external force acting on the object, m is the mass of the object, v is its velocity, and mv is its linear momentum. A rocket burning its fuel is an example of an object that changes its mass while in motion. However, in many situations mass can be considered as a constant. If the mass is assumed constant, Eq. (9) gives the more familiar expression of Newton's second law:

$$F = m\frac{dv}{dt} = ma \tag{10}$$

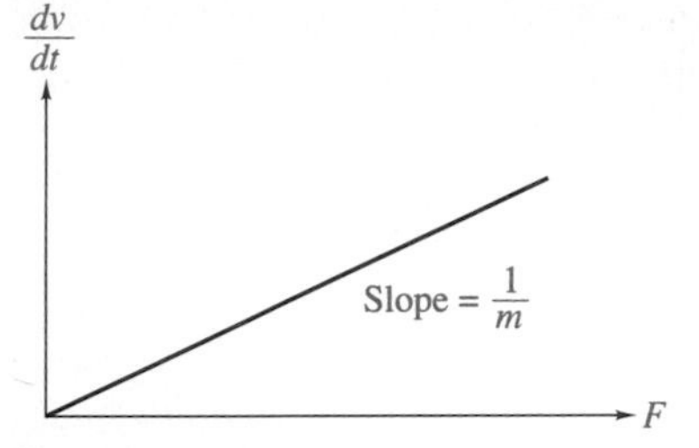

FIGURE 1.16
The acceleration experienced by a moving object having constant mass m is proportional to the net external force F.

or

$$a = \frac{1}{m} F \tag{11}$$

where a is the object's acceleration. Thus the motion modeled by Eq. (11) suggests that the rate of change in velocity is proportional to the net external force applied, as depicted in Fig. 1.16. How would you test this model?

Now consider the motion resulting when the external force is gravity. How does gravity behave? From empirical observations (those made without explanation of the underlying behavior) Newton formulated the following relationship for gravitational attraction:

$$F \propto \frac{m_1 m_2}{r^2}$$

where F is the mutual force of attraction between two bodies of masses m_1 and m_2 whose centers are separated by a distance r. (For a first approximation we assume the mass of an object to be concentrated at its center.) If we name the constant of proportionality G, we obtain **Newton's law of gravitational attraction:**

$$F = \frac{G m_1 m_2}{r^2}. \tag{12}$$

Equation (12) is interpreted in Fig. 1.17 for two objects of constant mass. Now we shall use both of Newton's laws, (11) and (12) to model the motion of an object under the earth's gravitational attraction.

Case 1: A Falling Body in Motion Under the Influence of Gravity Without Resistance

Suppose an object is dropped from a hovering helicopter. Equation (12) suggests that the earth is attracted to the object and vice versa. Model (11)

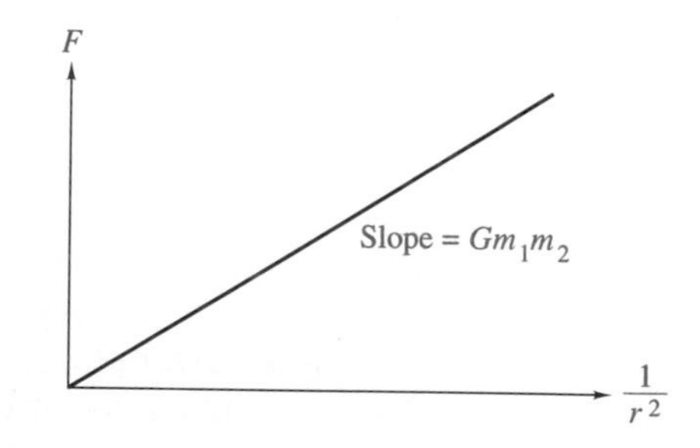

FIGURE 1.17
The mutual force of gravitational attraction between two bodies of constant mass is proportional to the reciprocal of the square of the distance between them.

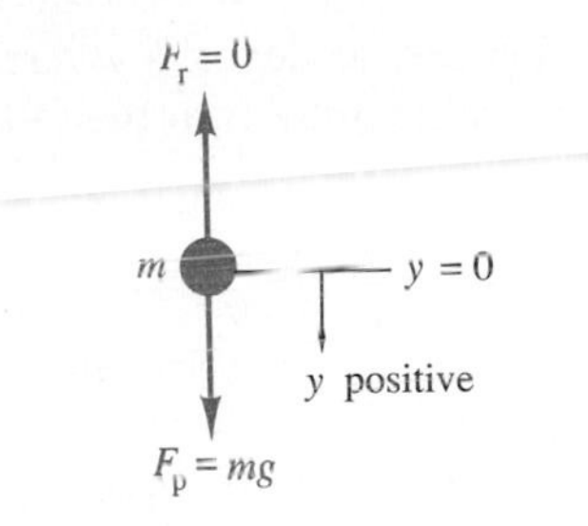

FIGURE 1.18
An object under the influence of earth's gravity.

proposes that the resulting acceleration is proportional to the reciprocal of the mass. The effect of a small object on the earth's motion is imperceptible, so we are going to ignore that effect.

We assume that the object is initially at rest, so that $y'(0) = 0$, and that the positive y direction is measured downward from the position of the helicopter, so $y(0) = 0$. A positive value for dy/dt indicates the object is moving in the positive y (downward) direction. A negative value for a force F indicates it acts in the negative y direction, and so forth. Model (11) requires a consideration of *all* the forces acting on the falling body, including forces of propulsion F_p and resistance F_r, as indicated by the free-body diagram in Fig. 1.18. For the time being we assume there are no resistive forces (as found in a vacuum), and that the only force of propulsion is due to gravitational attraction. Thus from model (12) the total force acting on the falling body is given by

$$F = \frac{Gmm_E}{r^2} \tag{13}$$

where m is the mass of the object, m_E is the mass of the earth, and r is the distance of the object from the center of the earth. For many problems we can assume $1/r^2$ to be constant because the distance of a body dropped from a helicopter to the surface of the earth is very small compared to the radius of the earth. Letting $g = Gm_E/r^2$ results in the following familiar equation for the propulsion force:

$$F_p = mg \tag{14}$$

where $g = 9.8 \text{ m/sec}^2$ or 32.2 ft/sec^2. Thus the net external force F acting on the object in the coordinate system drawn in Fig. 1.18 is given by

$$F = F_p - F_r,$$
$$ma = mg - 0,$$
$$mv' = mg,$$

or

$$v' = g. \tag{15}$$

In the next section we graphically analyze model (15). Of course, it is easy to solve Eq. (15) by integration to obtain

$$v = gt + C.$$

If $v = 0$ when $t = 0$, then the arbitrary constant evaluates to $C = 0$.

In Case 1 we considered the falling-body problem in the absence of any resistive forces. However, that is not a very realistic situation—after all, how often do we encounter an object falling in a vacuum? Normally various types of resistive forces need to be considered if the model is to be reasonably accurate. We now employ the model refinement process to develop several simple submodels for resistive forces that we will incorporate into the falling-body problem.

Viscous and Collision Forces in Fluids

The resistance to motion offered by a fluid to an object moving through it depends on many variables: the velocity of the object, its total surface area, cross-sectional surface area, the relative densities of the fluid and the object, the object's shape and surface smoothness, and fluid compressibility. There are also other factors that you will study in fluid mechanics.

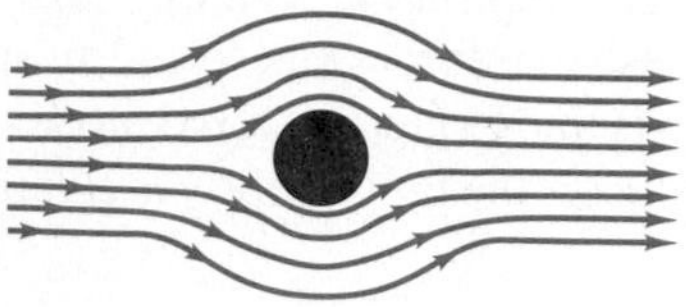

At low velocities the dominant effect providing resistance to motion is the *sliding* of the fluid's molecules over the object's surface. We assume the number of molecules contacting the surface per unit of time is proportional to the product of the object's velocity and its surface area (see Fig. 1.19). Thus a simple submodel for the **frictional force** is the proportionality relation

FIGURE 1.19
The number of molecules contacting the surface is proportional to the object's surface area times the velocity.

$$F_f = k_1 v \tag{16}$$

where the positive constant of proportionality k_1 depends on many factors of both the fluid and object (such as viscosity of the fluid and surface area of the object). Here it is assumed these factors are constant.

At higher velocities other effects become important. Intuitively we understand that as the velocity increases, more collisions of the object with the fluid's molecules take place. Eventually the effect of the energy transfer during collisions comes to dominate the "sliding" or viscous effect predominant at low velocities. One simple submodel observed experimentally is that the resistive drag force F_d due to these collisions is proportional to the square of the velocity:

$$F_d = k_2 v^2 \tag{17}$$

where the positive constant k_2 again depends on many factors (including fluid density, cross-sectional area, and other factors you will study in physics and mechanics courses). In this v^2 submodel the fluid is assumed to be **incompressible;** that is, a complete transfer of energy occurs during the collision.

Finally, as velocities approach the speed of sound the **compressibility** of the fluid must be considered. That is, some of the energy is being absorbed by the fluid as it undergoes deformations. We will not address these near-sonic velocities here but instead make two important observations concerning resistive forces. First, in most practical examples both sliding and collision effects are present, yielding a quadratic submodel for resistance obtained by summing Eqs. (16) and (17). Normally for simplicity, one of the two effects is assumed to be dominant. Second, if the v^2 term is present, the resulting model no longer satisfies the definition of a *linear differential equation.* (We discuss the concept of linearity in Chapter 2.) In fact, the nonlinear equation modeling the motion is generally difficult or impossible to solve analytically. However, graphical and numerical solution techniques may be used, and some of these methods will be investigated later on in the text.

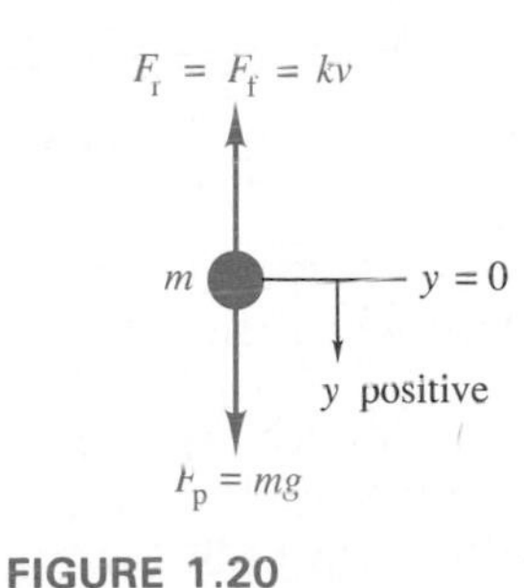

FIGURE 1.20
At low velocities, the principal resistance is assumed to be viscous friction, which is proportional to the velocity.

Case 2: A Falling Body in Motion Under the Influence of Gravity with Only Viscous Friction

Let us now develop a model for an object moving through a dense fluid at low velocities, where the force of viscous friction is the dominant resistance and gravity is the propulsion force. All other forces are to be neglected. For example, consider a steel ball bearing sinking in oil. The situation is suggested in Fig. 1.20.

Using Newton's second law we have,

$$F = F_p - F_r,$$
$$ma = mg - kv,$$
$$mv' = mg - kv,$$

or

$$mv' + kv = mg. \tag{18}$$

We analyze Eq. (18) graphically toward the end of Section 1.4.

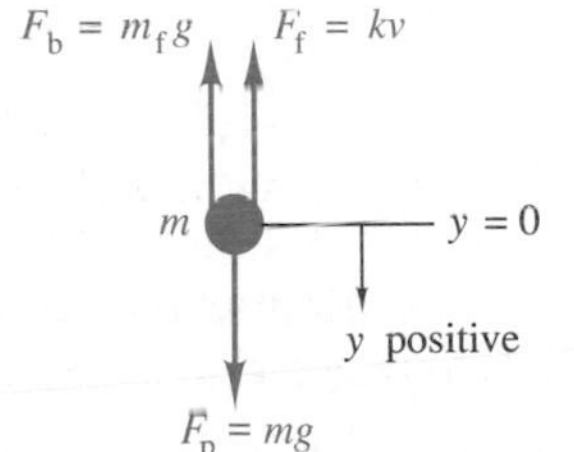

FIGURE 1.21
The buoyant force F_b and fluid friction F_f both resist motion.

Case 3: A Falling Body in Motion Under the Influence of Gravity with Viscous Friction and a Resistive Buoyant Force

In this refinement suppose there is also a resistive buoyant force exerted by the fluid. According to *Archimedes' principle,* this force is equal to the weight of the fluid displaced by the body. If m_f denotes the mass of the *displaced fluid,* then the buoyant force F_b is given by (see Fig. 1.21):

$$F_b = m_f g. \tag{19}$$

Again using Newton's second law we have

$$F = F_p - F_r,$$
$$F = F_p - (F_f + F_b),$$
$$ma = mg - kv - m_f g,$$
$$mv' = mg - kv - m_f g$$

or

$$mv' + kv = (m - m_f)g. \tag{20}$$

A graphical analysis of model (20) is presented in the next section.

When is it reasonable to neglect the buoyant force of the fluid for a nonfloating object? The weight of the displaced fluid is the volume V of the object times the weight density of the fluid δ_f. The weight mg of the object equals its volume V times its weight density δ. Thus the difference between the propulsion and buoyant forces is

$$F_p - F_b = V\delta - V\delta_f.$$

Hence if δ_f is very small relative to δ, it is reasonable to neglect the buoyant force.

In Section 2.8 (which is optional) we will consider a refinement in which the moving body has variable mass, as is the situation when launching a rocket into outer space.

APPLICATION 4 Modeling Population Growth*

Suppose that $P = P(t)$ represents the number of individuals in a particular population at time t. Then for ecological, economic, and other practical reasons it may be desirable to predict future values of P. Knowing the function relating P to t would provide this capability. However, if we start with an assumption relating the *rate of change of P* with respect to t to the dependent variable P, then we do not know immediately the functional relationship between P and t. Instead we must determine the function $P(t)$ subject to the condition that $P(0) = P_0$ is the starting population. We now carefully examine this population problem.

The growth of a population depends on many factors: birth rate, death rate, availability of resources, competition with other species, predators, immigration, emigration, and many other factors. The submodel for the birth rate itself depends on many factors including gestation period, gender and age distributions, medicine, and so forth. It would be difficult to keep the mathematical model solvable and still consider many factors. In this example we consider a simple first model and one refinement. Initially we will assume that the population has unlimited environmental resources, and we will model only the birth and death rates. Then we will consider limitations on the resources.

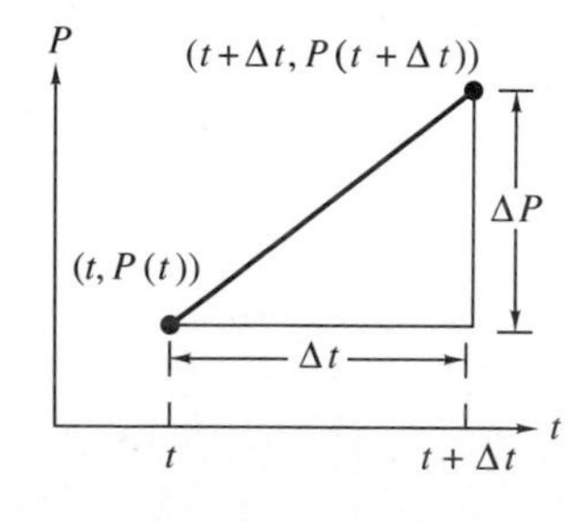

FIGURE 1.22
Geometric interpretation of ΔP, the change in P as t varies from t to $t + \Delta t$.

Assumptions: Growth with Unlimited Resources Let ΔP denote the change in a population size during a given time interval $[t, t + \Delta t]$:

$$\Delta P = P(t + \Delta t) - P(t). \tag{21}$$

A geometric interpretation of ΔP is shown in Fig. 1.22 for a population that is growing in size during the given time interval. The graph shows a population curve that varies continuously with time. Although continuous functions may not describe actual population size in the strictest sense, they do

* The material in this example is freely adapted, with permission, from the authors' text, *A First Course in Mathematical Modeling* (Pacific Grove, Calif.: Brooks/Cole Publishing Company, 1985), pp. 305–306.

provide sufficiently accurate information for many large populations that vary nearly continuously.

For now let us neglect factors such as immigration, emigration, age and gender, and merely assume that for sufficiently small values of Δt, a certain percentage of the population is reproduced and another percentage dies during the interval $[t, t+\Delta t]$.

Constructing the Model If we let $k > 0$ be a constant that represents the birth rate per unit time minus the death rate per unit time, we have the following model:

$$P(t+\Delta t) = P(t) + kP(t)\,\Delta t. \tag{22}$$

Note that since k is a percentage *per unit time,* we have multiplied k by both P and Δt. Applying the definition (21) and dividing ΔP by Δt in Eq. (22), we obtain the **average time rate of change** of P over any interval $[t, t+\Delta t]$:

$$\frac{\Delta P}{\Delta t} = kP. \tag{23}$$

Equation (23) states that the average change $\Delta P/\Delta t$ is proportional to the amount present P. At this point we need to ask whether it makes sense in the physical situation to allow Δt to become arbitrarily small. Even though many populations vary in discrete time periods (like spawning seasons, for example), a reasonable approximation to Eq. (23) in some instances can be obtained by using the continuous model:

$$\frac{dP}{dt} = \lim_{\Delta t \to 0} \frac{\Delta P}{\Delta t} = kP \tag{24}$$

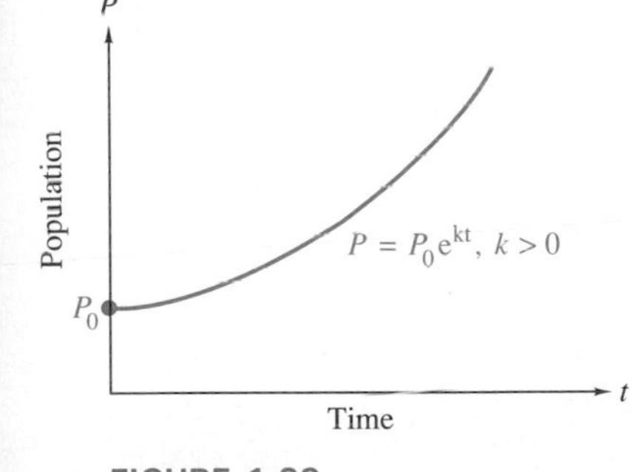

FIGURE 1.23
Exponential model for population growth.

where dP/dt represents the **instantaneous rate of change** of P with respect to t. Differential equations (24) and (5) are identical in form except for the sign in front of the constant k. Since for any constant C

$$\frac{d}{dt}(Ce^{kt}) = Cke^{kt}$$

you can see that any function of the type $P = Ce^{kt}$ solves Eq. (24). The initial condition $P(0) = P_0$ implies that $C = P_0$, to yield the particular solution

$$P(t) = P_0 e^{kt}. \tag{25}$$

The graph of $P(t)$ is presented in Fig. 1.23. If k happened to be negative for a particular species, its population would decline.

Discussion Equations (23) and (24) represent two distinct but closely related models of the population growth problem. Equation (23) is called a **difference equation** since it is formulated in terms of the differences Δt and ΔP; Eq. (24) is a differential equation since it is formulated in terms of the differentials dt and dP. Difference equations are important in modeling many real-world behaviors, and methods do exist for solving some difference equations just as methods exist for solving certain differential equations. Difference equations model *discrete behavior* as opposed to the continuous behavior modeled by differential equations. Each approximates the other, and sometimes one is more amenable to solution than, and hence is used in place of, the other. You will see later that numerical solution methods actually convert differential equations into difference equations that are easily solved with the aid of a computer. The concern in this text is with differential equations, and time and space do not permit a presentation of difference equations as well. Nevertheless, they are important in modeling, so we encourage you to investigate them after you complete this introductory study of differential equations.*

In many cases we will model rates of change that are in fact instantaneous. For example, the flow of heat from a warm to a cold body occurs continuously (for all practical purposes), so there is an instantaneous rate of change of temperature. In other cases, such as population modeling, we will approximate an average rate of change using an instantaneous rate of change. Approximating an average rate of change with a derivative allows us to use various tools of calculus to find the relationship among the variables in question.

Model Refinement: Growth with Limited Available Resources So far we have developed a model considering only the birth and death rate to obtain

$$P(t) = P_0 e^{kt}$$

where $P(t)$ is the population at time t, P_0 is the population at time $t = 0$, and k is the difference between the natural birth rates and death rates. This model is known as the **Malthusian model of population,** named after Thomas Malthus (1766–1834), who first studied the effects of population growth. As simple as the model is, it works well for populations with no limitations on environmental resources. Eventually, however, resources become scarce. We now consider a refinement that models the availability of resources.

Assumptions Let us suppose the natural environment is capable of sustaining some maximum population M. As the population approaches M, resources become scarce and we expect the growth rate k to decrease. Thus k is no longer constant as in our previous model but is a decreasing *function* of the population. One simple submodel we might choose for k to reflect this decreasing behavior is the proportionality

$$k = r(M - P), \quad r > 0$$

where r is a constant. (Other choices for submodels for k are presented in the exercises at the end of Section 1.4.) Note that k decreases as P increases to

* A classic text is *Differential and Difference Equations* by Louis Brand (New York: John Wiley & Sons, 1966), chaps. 8 and 9.

approach M and is negative if P is greater than M. (A real-life example of this occurs in "overstocking" a reserve with animals.) Substituting the above relation into Eq. (24) gives us

$$\frac{dP}{dt} = r(M - P)P = rMP - rP^2, \tag{26}$$

subject to $P(0) = P_0$.

Model (26) was first proposed by the mathematical biologist Pierre-François Verhulst (1804–1849) and is referred to as **logistic growth.** The model is a *nonlinear differential equation* due to the presence of the P^2 term. In Chapter 2 we will derive the following analytical solution, known as the **logistic curve:**

$$P(t) = \frac{P_0 M e^{rMt}}{M - P_0 + P_0 e^{rMt}} \tag{27}$$

From algebraic manipulation (or *l'Hôpital's rule*) it is easy to show that as t becomes large, P approaches the constant M, known as the **carrying capacity** of the environment. In Section 1.4 we will discuss the graphical solution to this limited growth model. Figure 1.24 illustrates the logistic curve superimposed on some data points for the growth of yeast in a culture.

It is thought that many species (such as the whale) will become extinct if their population falls below a minimum survival level. In Exercises 1.4 you will be asked to consider further refinements that incorporate survival levels.

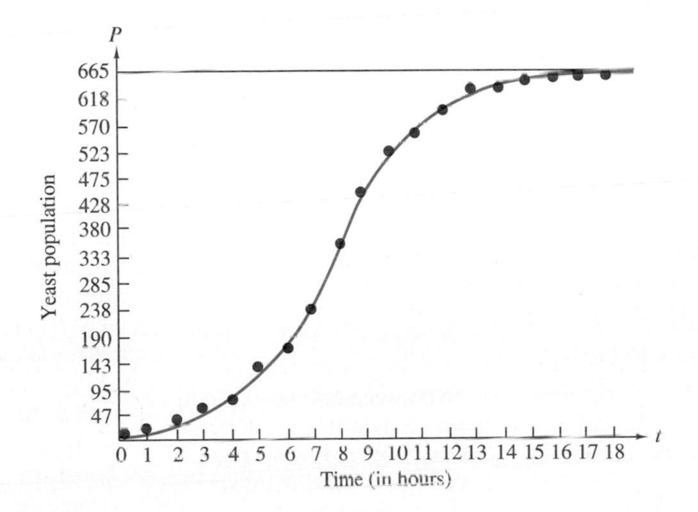

FIGURE 1.24
Logistic curve showing the growth of yeast in a culture. Small circles indicate the observed values. Data are from R. Pearl, "The Growth of Population," *Quart. Rev. Biol.* 2 (1927): 532–548.

Solutions and Exponential Functions

Let us make an important observation about several of the models constructed in our applications. Each of the models (5), (8), and (24) expresses the derivative as being proportional to some function of the current value of the dependent variable. During much of this text you will study similar models based on submodels assuming proportionality relationships, like the model for a falling body based on the submodels for resistive forces expressed in (16) and (17). The resulting differential equations are characterized by having constant coefficients. Such equations suggest looking for solution functions whose derivatives are "constant multiples" of the function itself. In Appendix B we develop the following important property:

> Whenever a positive function has a derivative equal to a constant multiple of the function itself, then that function is given by an exponential equation
>
> $$y = Ce^{k_1 x}$$
>
> where C is a constant.

This property is fundamental to finding solutions of linear constant coefficient differential equations, as you shall see in Chapter 3. In the preceding examples the exponential consistently appeared as a solution to linear differential equations. This pattern will continue as we solve linear equations.

EXERCISES 1.3

1. If the half-life of radium is 1620 yr, what percentage of a sample of radium remains after 400 yr?

2. If initially there are 100 g of a radioactive substance, and after 3 wk only 20 g remain, how many grams remain after 4 wk?

3. Carbon-14 Dating

In a living organism the ratio between ordinary carbon and radioactive carbon-14 remains roughly constant, equal to about one carbon-14 atom per 10^{12} atoms of regular carbon in the atmosphere. After an organism dies, however, no new carbon can be ingested by eating or breathing, and the proportion of carbon-14 undergoes radioactive decay. The half-life of carbon-14 is approximately 5700 yr. What percentage of carbon-14 is present in the charcoal from a tree killed by a volcanic eruption 1991 yr ago?

4. A humanoid skull is discovered near the remains of an ancient campfire. Archaeologists are convinced the skull is the same age as the original campfire. It is determined from laboratory testing that only 1% of the original amount of carbon-14 remains in the charcoal from the burned wood taken from the campfire. Using carbon-14 dating, estimate the aproximate age of the skull.

5. Interest Compounded Continuously

When an investment is compounded, the interest earned during a period is added to the present amount (principal), so that the interest then earns interest during the next time period.

a) If interest is compounded k times per year with an annual interest rate of r (a percentage per year), establish that the amount in the account at the end of one year is

$$Q = \left(1 + \frac{r}{k}\right)^k Q_0$$

where Q_0 is the original amount invested. Assume that no money is withdrawn from the account during the year.

b) Many banks now offer to compound interest continuously. That is, assume the instantaneous rate of change of the principal Q is proportional to the amount present:

$$\frac{dQ}{dt} = rQ$$

where r is the annual interest rate. If initially \$2000 is invested in an individual retirement account (IRA) earning 7.5% interest annually, how much money will have accumulated in the account at the end of 5 yr?

6. Assuming exponential growth, if a given population doubles in 25 years, how many years will it take to triple? Quadruple?

7. Suppose the population of trout in a large lake is growing exponentially. From 1976 to 1980 the population of trout was found to double to an estimated 5000. Estimate the size of the trout population in 1985. During what month and year will the trout population reach 15,100 fish?

8. Suppose the number of bacteria in a certain culture of yeast grows at a rate proportional to the number present. If the population of a colony of yeast bacteria doubles in 45 min, find the number of bacteria present at the end of 2.5 hr.

9. Elimination of a Drug

Suppose a drug dose of D_0 (in milligrams) is injected directly into the bloodstream. Assume the drug leaves the blood and enters the urine at a rate proportional to the amount of drug present in the blood at any time. If half the original drug dose has entered the urine after 1½ hr, find the time when the amount of drug in the blood is 25% of the original dose.

10. Braking Distance

Assume the braking system of a car is designed in such a way that the maximum braking force increases in proportion to the mass of the car. Basically, this means that if the force per unit area applied by the braking system remains constant, then the surface area in contact with the brakes increases in proportion to the mass of the car. Assume that under a panic stop the maximum braking force is applied continuously. Assume further that no other resistance forces tend to oppose the motion of the car.

a) Find a differential equation modeling the velocity of a car traveling along a flat straight road.

b) If v_0 denotes the velocity at time $t = 0$ when the brakes are initially applied, show that the total distance d_b traveled by the car before it comes to a complete stop is proportional to v_0^2.

c) Estimate the constant of proportionality in (b) from the following table based on tests conducted by the U. S. Bureau of Public Roads. The given braking distances represent average values for 85% of the drivers tested.

Speed (mph)	20	25	30	35	40	45	50	55	60	65	70
Braking Distance (ft)	20	28	41	53	72	93	118	149	182	221	266

11. Inclined Plane

An object sliding down an inclined plane encounters a resistance force called *friction*. The magnitude of the frictional force is a constant μ times the magnitude N of the normal force the surface applies to the object. The constant μ is called the *coefficient of friction* and depends on the surface. The friction tends to oppose the motion of the sliding object. Assuming there are no other resisting forces and that gravitational force is constant, formulate a differential equation model for the velocity of an object as it slides down the plane. Assume the initial velocity is $v(0) = v_0$ (see Fig. 1.25).

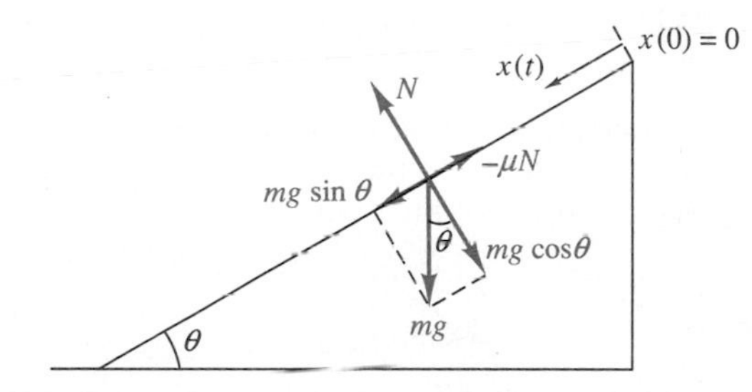

FIGURE 1.25
Forces on an object sliding down an inclined plane.

12. An object of mass 0.5 kg is thrown vertically upward with an initial velocity of 15 m/sec from the roof of a building 40 m high. Neglect all resisting forces and assume only gravity acts on the object.

a) What is the maximum height it reaches above the ground?
b) Assuming the object does not strike the building on its way down, how long does it take to reach the ground?

In problems 13–22 formulate a mathematical model expressed as a differential equation to describe the situation or answer the question. Identify the variables that affect the behavior described in the problem, and formulate any submodels you may wish to consider. What are the assumptions of your model? Be specific. What variables might you initially consider to be constant? What submodels might initially be given by simple proportionalities? Remember that there are no single right answers. You need not solve your differential equation.

13. A 125-lb parachutist opens her parachute when her downward velocity is 100 ft/sec. What is the velocity of the parachutist at any time $t \geqslant 0$?

14. In a chemical reaction two substances A and B combine in equal amounts to produce compound C. Suppose a and b are the initial concentrations of A and B respectively. What is the concentration of C at any time $t \geqslant 0$?

15. Two people are riding in a motorboat, and the combined weight of the people, motor, boat, and equipment is 600 lb. Assuming the boat starts from rest, what is its velocity at any time $t \geqslant 0$?

16. Motion of a Pendulum

A pendulum is made by attaching a weight of mass m to a very light and rigid rod of length L mounted on a hinge so that the system can swing in a vertical plane. Describe the motion of the pendulum.

17. A substance S flows into a certain mixture in a container at a constant rate, and the mixture is kept uniform by stirring. Suppose the uniform mixture simultaneously flows out of the container at another rate (usually different from the first one). How much of S is present in the mixture at any time $t \geqslant 0$?

18. Vibrating Spring

A 12-lb weight is placed upon the lower end of a coil spring suspended from the ceiling. The weight comes to rest in its equilibrium position, stretching the spring 2 in. The weight is then pulled down 1.5 in below its equilibrium position and released from rest at $t = 0$. Find the displacement of the weight away from the equilibrium position at any time $t \geqslant 0$.

19. A balloon is ascending at the rate of 12 m/sec at a height 80 m above the ground when a package is dropped. How long does it take the package to reach the ground?

20. An elevator ascends with an upward acceleration of 4 ft/sec^2. At the instant when its upward speed is 8 ft/sec, a loose bolt drops from the ceiling of the elevator 9 ft from the floor. How long does it take the bolt to hit the floor?

21. Absorption of Light

The German physicist and mathematician Johann Lambert (1728–1777) asserted that the absorption of light in a very thin transparent layer is proportional to the thickness of the layer times the amount of light incident on that layer. Formulate this assertion mathematically. How would you test the model?

22. Spread of a Disease

An individual in a population of N people becomes infected with a virus. Assume that the rate at which the virus is transmitted throughout the population is proportional to the number of infected individuals, y, and also to the number of persons not infected. Assume further that the population of the community remains constant. Predict the number of people infected after t days.

1.4 GRAPHING SOLUTIONS TO FIRST-ORDER DIFFERENTIAL EQUATIONS

We have used the model construction process to create a number of important differential equations models. We are also interested in solving the models we create. However, sometimes we can predict outcomes of models graphically without actually solving them. Before commencing with the study of solution methods for differential equations, let us investigate graphical representations of solutions.

On many occasions in calculus you computed the derivative of some given function to gain more information about it. For instance, the derivatives would indicate regions where the function is increasing or decreasing, locations of relative extrema and inflection points, the nature of the concavity and curvature, and so forth. For each model constructed in Section 1.3 note that a calculus problem still remains, but now the problem is different: information about the derivative of a function is hypothesized and now we desire to find out about the function itself. What are its properties? How does it behave over a specified interval? And so on.

Interpreting the derivative as the slope of the line tangent to the graph of the function is useful for gaining information about the solution to a differential equation. From this information it is possible to sketch qualitatively the solution curves to the equation. These graphs provide considerable insight into the solution, and the graphical techniques are themselves of practical benefit. Since most differential equation models that people construct cannot be solved analytically, other techniques are needed to investigate the behavior of solutions.

Direction Fields

Suppose the first-order differential equation

$$\frac{dy}{dx} = g(x, y)$$

is given. Each time an initial value $y(x_0) = y_0$ is specified, the solution curve is required to pass through the point (x_0, y_0) and have the slope value $g(x_0, y_0)$ there. Therefore, graphically it is possible to draw a short line segment with the correct slope through each point (x, y) in the plane. The resulting configuration is known as the **direction field** or **slope field** of the first-order differential equation. Direction fields for several first-order equations are illustrated in Fig. 1.26. A **solution curve** is then tangent to the

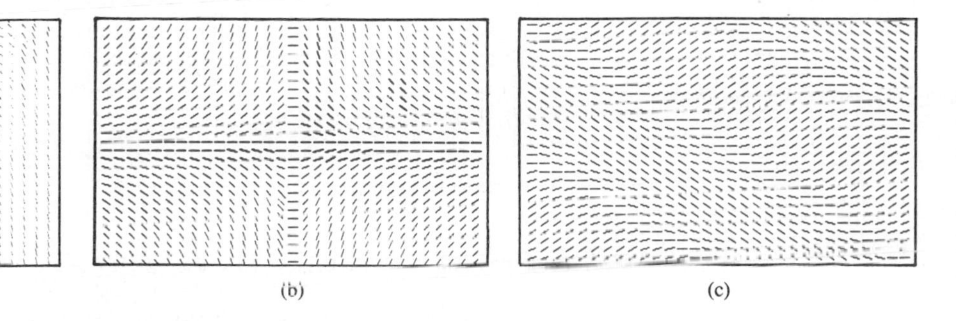

(a) (b) (c)

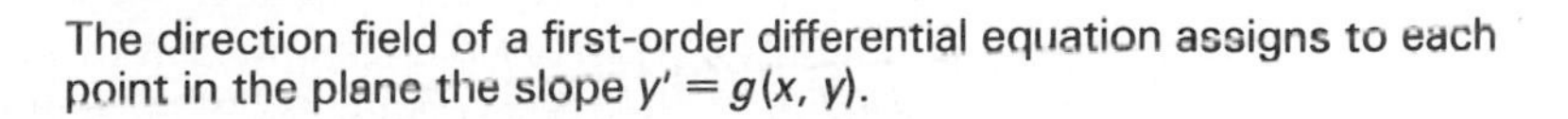

FIGURE 1.26
The direction field of a first-order differential equation assigns to each point in the plane the slope $y' = g(x, y)$.

direction line at each point through which the curve passes. Thus, the direction field gives a visual representation of what the family of possible solution curves to the differential equation looks like. Several solution curves are depicted in Fig. 1.27 for the direction fields presented in Fig. 1.26.

The construction of the direction field for a specific first-order equation can be quite tedious to carry out with pencil and paper. Computer software is available for constructing direction fields; all of our examples were computer-generated.* One method for constructing a direction field is to consider all the points in the xy-plane that are associated with a constant slope c. In other words, consider all points (x, y) satisfying

$$g(x, y) = c.$$

This locus of points is called an **isocline.**

For the logistic growth model (see Section 1.3, Application 4)

$$y' = (2 - y)y,$$

the derivative is constant whenever the dependent variable y has a constant value. Since the isoclines are the family of curves defined as

$$(2 - y)y = \text{constant},$$

these isoclines are simply horizontal lines in the xy-plane. Along each horizontal line the slopes in the direction field have exactly the same value. Figure 1.28 (a) shows several isoclines for the logistic model, and Fig. 1.28 (b) shows the corresponding direction field in which are sketched five solution curves. No two of these solution curves can cross. To cross, two solution curves have to meet at a point of intersection (x_0, y_0), so that both curves could satisfy the same initial condition $y(x_0) = y_0$. However, for the logistic differential equation it turns out that *only one* function can satisfy the equation together with a given initial condition. This issue of uniqueness of solutions is discussed again in Chapter 2.

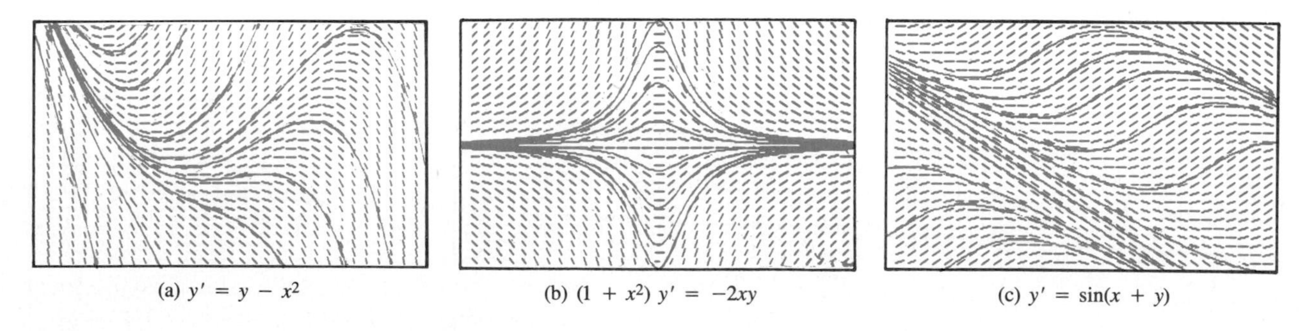

(a) $y' = y - x^2$ (b) $(1 + x^2)\, y' = -2xy$ (c) $y' = \sin(x + y)$

FIGURE 1.27
Direction fields with several solution curves. Each solution curve in the direction field satisfies some specific initial condition $y(x_0) = y_0$.

* *The Calculus Toolkit,* by Ross L. Finney, Dale T. Hoffman, Judah L. Schwartz, and Carroll O. Wilde (Reading, Mass.: Addison-Wesley, 1990), draws direction fields.

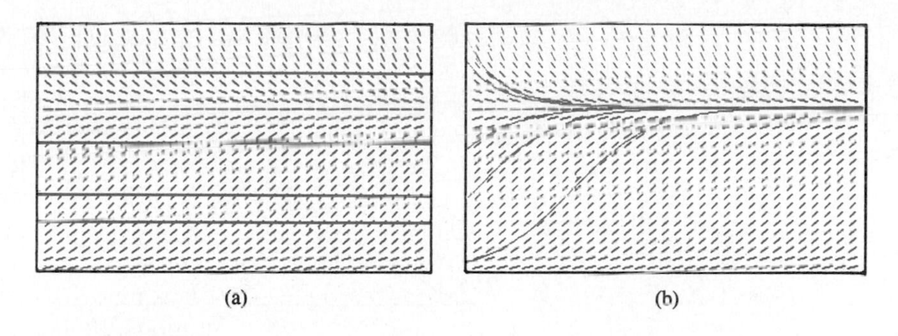

FIGURE 1.28

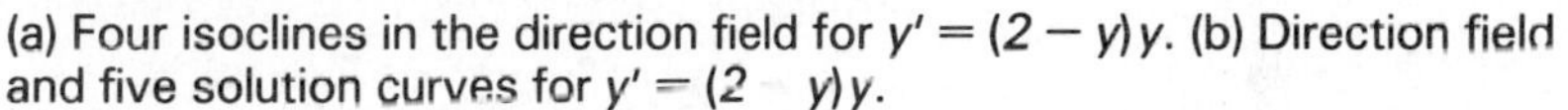
(a) Four isoclines in the direction field for $y' = (2 - y)y$. (b) Direction field and five solution curves for $y' = (2 - y)y$.

Graphing the Derivative

In many modeling applications the instantaneous rate of change dy/dx is assumed to be proportional to some function of the dependent variable y alone. Thus $dy/dx = f(y)$. In such cases it is possible to obtain qualitative information concerning the nature of a solution curve $y(x)$ by investigating the graph of $f(y)$, that is, by plotting dy/dx versus y. This idea is new. In calculus you were accustomed to plotting dy/dx versus the *independent variable* x. We now explore this new idea in several examples.

EXAMPLE 1 EXPONENTIAL GROWTH

Consider the differential equation describing the Malthusian model of population growth discussed in Section 1.3:

$$\frac{dP}{dt} = kP, \quad k > 0. \tag{1}$$

Thus, the Malthusian model assumes a simple proportionality between growth rate and population (see Fig. 1.29a). Since $P > 0$ and $k > 0$, you can see from relation (1) that $dP/dt > 0$. It follows that the population curve $P(t)$ is everywhere increasing. Moreover, the smaller the value of k, the less rapid is the growth in the population over time. For a fixed value of k, as P increases, so does its rate of change dP/dt.

Figure 1.29(b) depicts the direction field for exponential growth model (1). Note that dP/dt is constant whenever the population level P is constant. Thus the isocline curves are horizontal lines. It follows that all the solution curves consist of any one solution curve shifted horizontally; that is, they are horizontal translates of one another. Several such curves are sketched in the direction field in Fig. 1.29(b). The initial population level $P(0) = P_0$ distinguishes these various solution curves.

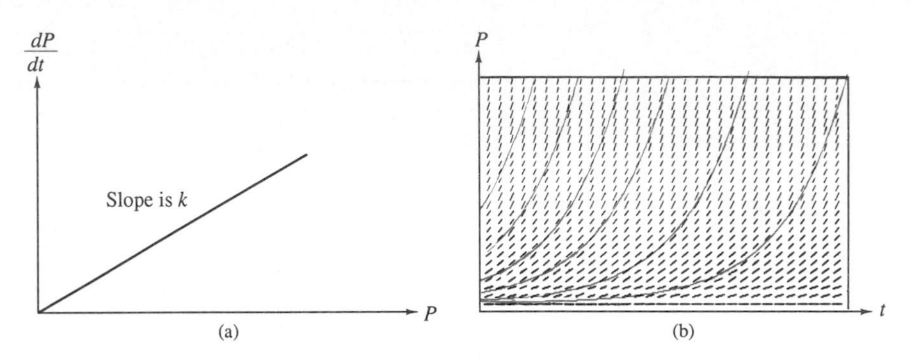

FIGURE 1.29

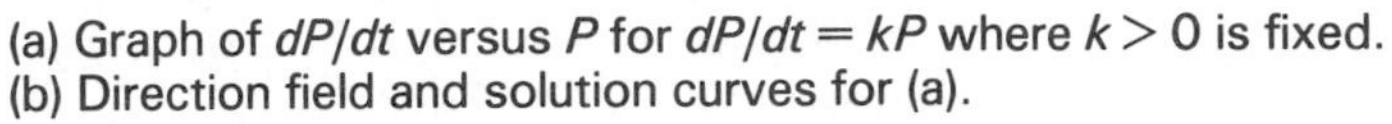
(a) Graph of dP/dt versus P for $dP/dt = kP$ where $k > 0$ is fixed.
(b) Direction field and solution curves for (a).

Rest Points and Autonomous Equations

Additional qualitative information can be obtained from the differential equation itself. For example, we might want to know what happens to the solution curve as the independent variable grows arbitrarily large. Does the solution tend toward some fixed value in that case? We begin with an example.

EXAMPLE 2 THE LOGISTIC GROWTH MODEL

Recall the logistic population growth model constructed in Section 1.3:

$$\frac{dP}{dt} = r(M - P)P. \tag{2}$$

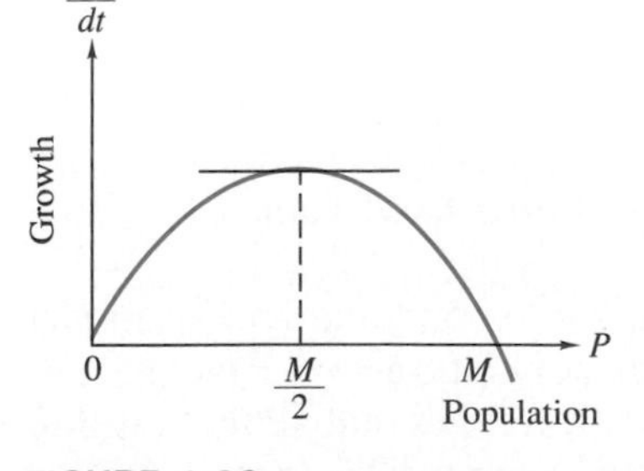

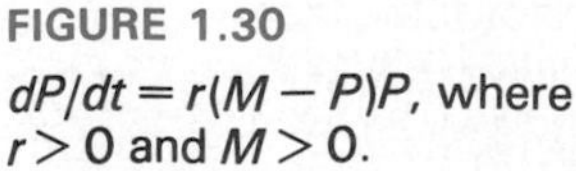
FIGURE 1.30
$dP/dt = r(M - P)P$, where $r > 0$ and $M > 0$.

The graph of dP/dt versus the dependent variable P is the parabola shown in Fig. 1.30. Differentiating equation (2) implicitly with respect to t gives the second derivative

$$P'' = rP'(M - 2P). \tag{3}$$

Since $P'' = 0$ when $P = M/2$, dP/dt is at a maximum at that population level. Note that for $0 < P < M$, the derivative dP/dt is positive and the population $P(t)$ is increasing; for $P > M$, dP/dt is negative and $P(t)$ is decreasing. Moreover, if $P < M/2$, the second derivative P'' is positive and the population curve is concave upward; for $M > P > M/2$, P'' is negative and the population curve is concave downward.

The parabola in Fig. 1.30 has zeros at the population levels $P = 0$ and $P = M$. At those levels $dP/dt = 0$, so no change in the population P can occur. That is, if the population level is at 0, it will remain there forever; if it is at the level M, it will remain there for all time. These observations prompt the following definition.

DEFINITION 1.6

In the graph of dP/dt versus P, points P for which the derivative dP/dt is zero are called **rest points, equilibrium points,** or **critical points** of the differential equation.

Stable and Unstable Rest Points The behavior of solutions near rest points is of significant interest. Let us examine what happens to the population when P is near one of the rest points $P = 0$ or $P = M$.

Suppose the population in model (2) satisfies $P < M$. Then $dP/dt > 0$, and the population increases, getting closer to the maximum M. On the other hand, if $P > M$, then $dP/dt < 0$, and the population will decrease toward M (see Fig. 1.31). Thus no matter what positive value is assigned to it at the start, the population will approach the limiting value M as time tends toward infinity. We say that M is an *asymptotically stable* rest point because whenever the population level is perturbed away from the level, it tends to return there again.*

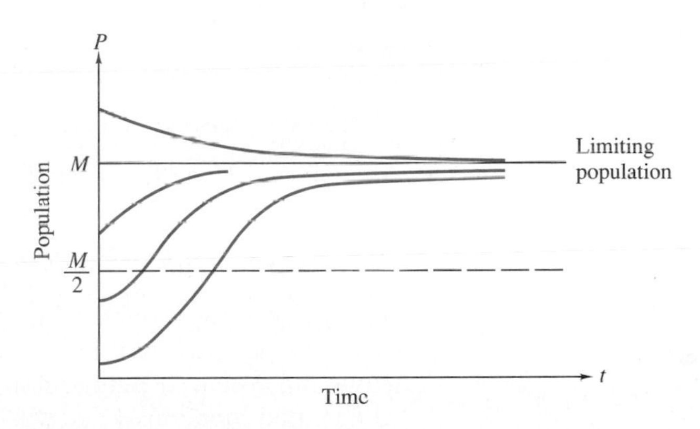

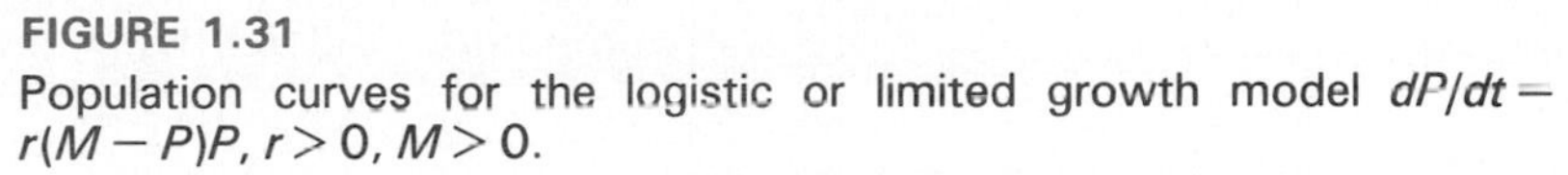

FIGURE 1.31
Population curves for the logistic or limited growth model $dP/dt = r(M - P)P$, $r > 0$, $M > 0$.

* However, if the starting population is not at the level M, the population $P(t)$ cannot reach M in a finite amount of time. This fact follows from the properties that $P = M$ is a solution to the limited growth equation (2) and that two solutions cannot cross.

Next, consider the rest point $P = 0$ in Fig. 1.30. If P is perturbed slightly away from 0 so that $M > P > 0$, then $dP/dt > 0$ and the population increases toward M. In this situation we say that $P = 0$ is an *unstable* rest point because any population not starting at that level tends to move away from it.

In general, equilibrium solutions to differential equations are classified as **stable** or **unstable** according to whether, graphically, nearby solutions stay close to or converge to the equilibrium, or diverge away from the equilibrium, respectively, as the independent variable tends to infinity.

At $P = M/2$, $P'' = 0$ and the derivative dP/dt is at a maximum (see Fig. 1.30). Thus the population P is increasing most rapidly when $P = M/2$, and a point of inflection occurs there in the graph. These features give each solution curve its characteristic *sigmoid* ("S") shape. From this information the family of solutions to the limited growth model must appear qualitatively as shown in Fig. 1.31. The shape of the curves agrees with those obtained from the direction field in Figure 1.28(b)

Notice again that at each population level P, the derivative dP/dt is constant, so the solution curves are horizontal translates of one another. The curves are distinguished by the initial population level $P(0) = P_0$. In particular, note the solution curve when $M/2 < P_0 < M$ in Fig. 1.31.

The procedure illustrated in Examples 1 and 2 is called a **graphical stability analysis.** Note that in the differential equation

$$\frac{dP}{dt} = kP$$

the independent variable t does not appear explicitly on the righthand side. This characteristic makes it an **autonomous equation.** Autonomous equations and systems of autonomous equations constitute an important class of differential equations. (You will learn how to conduct a stability analysis for systems of autonomous differential equations in Section 7.2.)

The Phase Line

Before considering another example, let's consider Eq. (2) from yet another point of view. In Fig. 1.32 we present a **phase line** depicting the limited growth model. On the phase line the two equilibrium points $P = 0$ and $P = M$ are circled. These are the points where $dP/dt = 0$. Also on the phase

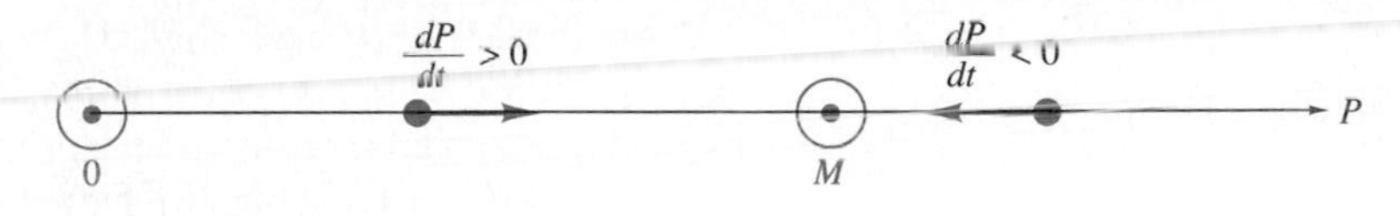

FIGURE 1.32
A phase line for logistic growth.

line we identify the sign of dP/dt for all values of P. In this case dP/dt is positive for $0 < P < M$ and negative for $P > M$. (If $P < 0$ were possible, then we would extend the line to the left of $P = 0$, and dP/dt for $P < 0$ would be negative.) From the sign of dP/dt, we then indicate whether P is increasing or decreasing by an appropriate arrow. If the arrow points toward the right, P is increasing; toward the left it indicates P is decreasing. By knowing the sign of the derivative for all values of P, together with the starting value $P_0 = P(0)$, we can determine what happens as $t \to \infty$. A curve approached asympotically as $t \to \infty$ is called a **steady-state** outcome. The phase line allows us to predict the steady-state outcome of a model from knowing the sign of the derivative and the initial condition.

For the logistic growth model we note that if $P_0 > 0$, then as $t \to \infty$, the population $P(t)$ approaches M, the maximum sustainable population. However, if $P_0 = 0$, then $P(t) \equiv 0$. (See Fig. 1.31.) In Section 2.1 an analytical solution to (2) is obtained.

EXAMPLE 3 NEWTON'S LAW OF COOLING

Consider again the model for the cooling of a space capsule, developed in Section 1.3:

$$\frac{dT_m}{dt} = -k(T_m - \beta), \quad T_m(0) = \alpha, \quad t > 0 \tag{4}$$

where T_m represents the temperature of the space capsule, t represents the amount of time elapsed after splashdown, β represents the temperature of the ocean (which is assumed to be constant), and k is a positive constant of proportionality. We begin our analysis with the phase line for Eq. (4) shown in Fig. 1.33.

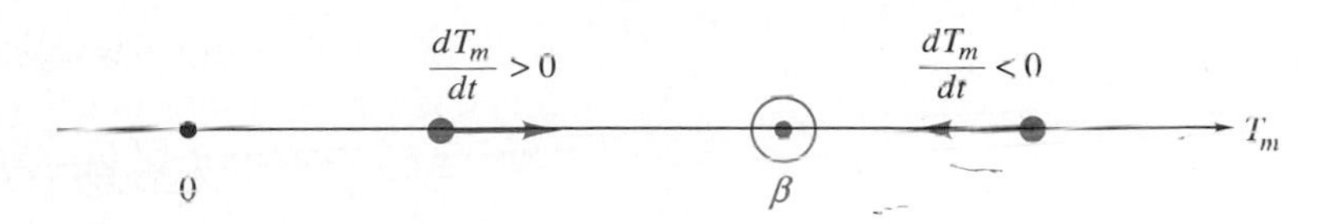

FIGURE 1.33
A phase-line analysis for Eq. (4) reveals that the steady-state outcome is $T_m = \beta$.

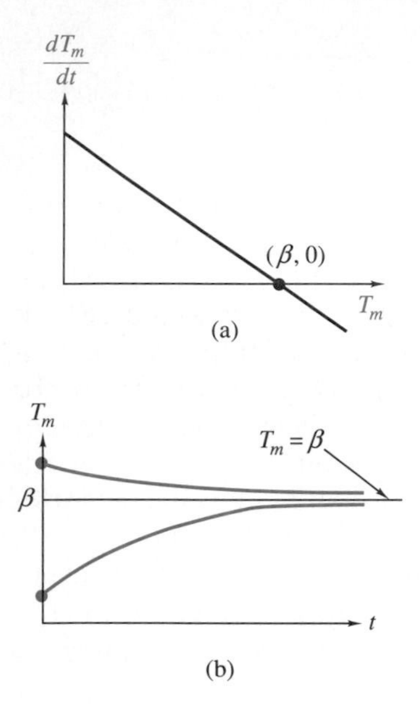

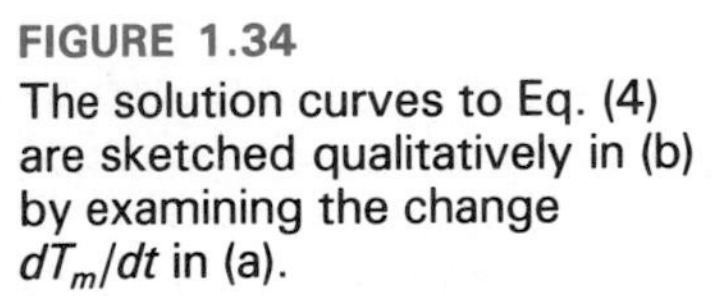

FIGURE 1.34
The solution curves to Eq. (4) are sketched qualitatively in (b) by examining the change dT_m/dt in (a).

Note that $dT_m/dt = 0$ when $T_m = \beta$. Thus β is an equilibrium point. If $T_m < \beta$, then dT_m/dt is positive; for $T_m > \beta$, dT_m/dt is negative. This information is recorded using arrows on the phase line. Thus β is the steady-state outcome for Eq. (4). From the figure we conclude the following:

1. If $T_m(0) < \beta$, the capsule will warm and approach the ocean temperature β.
2. If $T_m(0) = \beta$, then $T_m(t) = \beta$ always.
3. If $T_m(0) > \beta$, the capsule will cool and approach the ocean temperature β.

A phase-line analysis reveals the nature of the equilibrium points for the model under study. In the case at hand, the equilibrium point $T_m = \beta$ is stable.

Now consider the graph of dT_m/dt versus T_m, as depicted in Fig. 1.34(a). From that graph we can see that dT_m/dt has a zero at $T_m = \beta$, so no change in temperature T_m can occur there. If $T_m < \beta$, dT_m/dt is positive, so the temperature of the capsule must increase toward the limiting value β. On the other hand, if $T_m > \beta$, dT_m/dt is negative, so the temperature of the capsule will decrease toward β. Notice that the second derivative $T''_m = -kT'_m$ is negative for $T_m < \beta$ and positive for $T_m > \beta$. From these observations we know that the solution curves for $T_m(t)$ must appear qualitatively as shown in Fig. 1.34(b). In Section 2.1 we will obtain an analytical solution to Eq. (4).

EXAMPLE 4 MOTION IN A GRAVITATIONAL FIELD

We now explore the falling-body problem with a graphical analysis.

Case 1: A Falling Body Encountering No Resistance

For this situation the differential equation

$$v' = g \tag{5}$$

was constructed in Section 1.3. Recall that g represents the acceleration due to gravity in the earth's gravitational field and that we assumed g to be constant.

The graph of dv/dt versus v is the horizontal straight line shown in Fig. 1.35(a). Since dv/dt is always positive, the velocity v increases indefinitely with increasing t. Integration of Eq. (5), combined with the assumption that $v(0) = 0$, confirms this result:

$$v = gt.$$

The graph of $v(t)$ appears in Fig. 1.35(b).

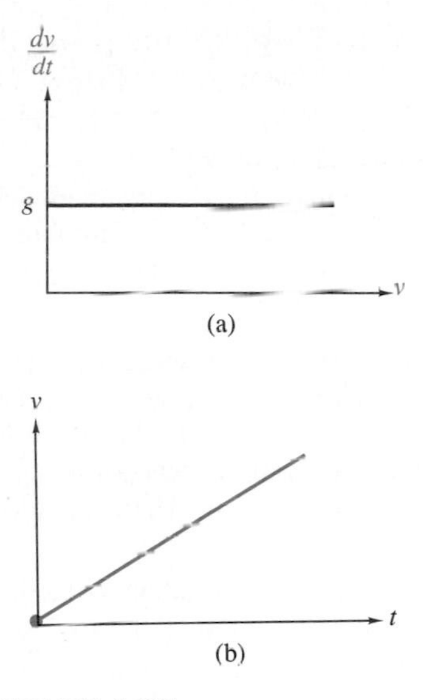

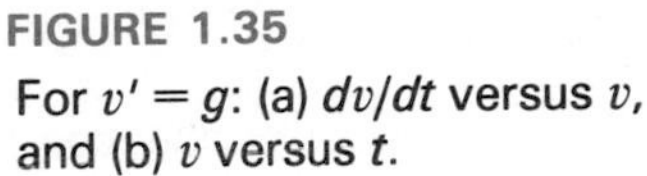

FIGURE 1.35
For $v' = g$: (a) dv/dt versus v, and (b) v versus t.

Case 2: Falling Body Encountering Only Viscous Friction

The model obtained in Section 1.3 was

$$mv' + kv = mg \tag{6}$$

where m is the mass of the object and k is a proportionality constant associated with the frictional resistance. Figure 1.36(a) reveals the graph of dv/dt versus v for model (6). From the graph you can see that $v = mg/k$ is a stable equilibrium point. The limiting value mg/k (the steady state) is called the **terminal velocity** of the falling body. Thus for a body falling subject to a resisting force, the velocity does not increase indefinitely as was the case for the body falling in a vacuum. The solution curves $v(t)$ must appear qualitatively as shown in Fig. 1.36(b).

Case 3: A Falling Body Encountering Viscous Friction and a Resistive Buoyant Force

This refinement produced the model

$$mv' + kv = (m - m_f)g \tag{7}$$

where m_f denotes the mass of the fluid displaced by the object. A graphical analysis of Eq. (7) reveals curves exactly like those depicted in Fig. 1.36(b) except that the terminal velocity is reduced from mg/k to $(m - m_f)g/k$.

In Section 2.8 we will consider a further refinement where an object with variable mass is moving in the earth's gravitational field.

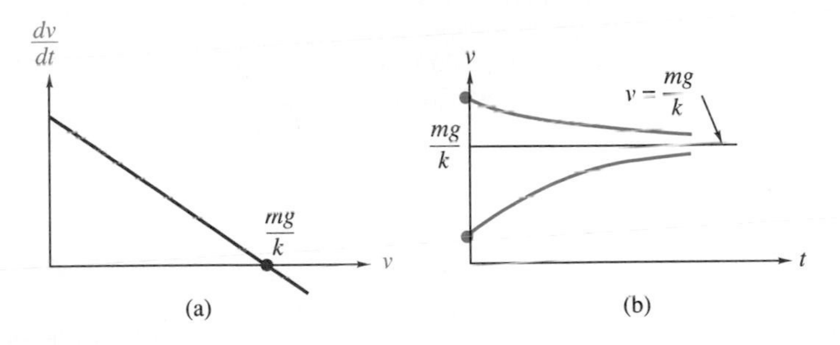

FIGURE 1.36
The solution curves to Eq. (6) are sketched qualitatively in (b) by examining the change dv/dt in (a). Notice in (b) that v aproaches the terminal velocity mg/k as t tends toward infinity.

In Chapter 2 we will consider analytical and numerical methods for solving first-order differential equations.

EXERCISES 1.4

In problems 1–6, construct a direction field for the given differential equation by first finding the isoclines. Then sketch in several solution curves.

1. $\dfrac{dy}{dx} = y$

2. $\dfrac{dy}{dx} = x$

3. $\dfrac{dy}{dx} = x + y$

4. $\dfrac{dy}{dx} = x - y$

5. $\dfrac{dy}{dx} = xy$

6. $\dfrac{dy}{dx} = \dfrac{1}{y}$

Sketch a number of solution curves to the equations in problems 7–10. Show the correct slope, concavity, and any points of inflection. Use graphical analysis.

7. $\dfrac{dy}{dx} = (y+2)(y-3)$

8. $\dfrac{dy}{dx} = y^2 - 4$

9. $\dfrac{dy}{dx} = y^3 - y$

10. $\dfrac{dy}{dx} = x - 2y$

11. Analyze graphically the equation $dy/dt = ry$ when $r < 0$. What happens to any solution curve as t becomes large?

Develop graphically the models in problems 12–15. First graph dP/dt versus P. Then obtain various graphs of P versus t by selecting different initial values of $P(0)$ (as done in the population example in the text). Identify and discuss the nature of the equilibrium points in each model.

12. $\dfrac{dP}{dt} = a - bP, \quad a,b > 0$

13. $\dfrac{dP}{dt} = P(a - bP), \quad a,b > 0$

14. $\dfrac{dP}{dt} = k(M - P)(P - m), \quad k,M,m > 0$

15. $\dfrac{dP}{dt} = kP(M - P)(P - m), \quad k,M,m > 0$

16. Another equation for modeling population growth is the **Gompertz equation:**

$$\frac{dP}{dt} = rP(a - \ln P)$$

where r and a are positive constants.

a) Sketch the graph of dP/dt versus P.
b) Find the equilibrium points of the Gompertz model and determine whether each is stable or unstable.
c) Sketch various solution curves of P versus t.
d) Describe the behavior of $P(t)$ as $t \to \infty$

17. The fish and game department in a certain state is planning to issue deer hunting permits. It is known that if the deer population falls below a certain level m, the deer will become extinct. It is also known that if the deer population goes above the maximum carrying capacity M, the population will decrease to M.

a) Discuss the reasonableness of the following model for the growth rate of the deer population as a function of time:

$$\frac{dP}{dt} = kP(M - P)(P - m)$$

where P is the population of the deer and k is a constant of proportionality. Include a graph of dP/dt versus P as part of your discussion.
b) Explain how this growth-rate model differs from the logistic model $dP/dt = kP(M - P)$. Is it better or worse than the logistic model?
c) Show that when $P > M$ for all t, the limit of $P(t)$ as $t \to \infty$ is M.
d) Discuss what happens if $P < m$ for all t.
e) Graphically discuss the solutions to the differential equation. What are the equilibrium points of the model? Explain the dependence of the steady-state value of P on the initial conditions. How many deer hunting permits should be issued?

18. Sociologists recognize a phenomenon called **social diffusion,** which is the spreading of a piece of information, technological innovation, or cultural fad among a population. The members of the population can be divided into two classes: those who have the information and those who do not. In a fixed population whose size is known, it is reasonable to assume that the rate of diffusion is proportonal to the number who have the information times the number yet to receive it. If X denotes the number of individuals who have the information in a population of N people, then a mathematical model for social diffusion is given by

$$\frac{dX}{dt} = kX(N - X)$$

where t represents time and k is a positive constant.

a) Discuss the reasonableness of the model.
b) Graph dX/dt versus X.
c) From the graph predict the value of X for which the information is spreading most rapidly. How many people eventually receive the information?

19. From a graphical analysis find the terminal velocity of a falling body. In your model assume that the resisting force is proportional to the square of the magnitude of the velocity and that there is no other force opposing the motion.

20. A body of mass m is projected vertically downward with initial velocity v_0. Assume the resisting force is proportional to the square root of the magnitude of the velocity, and find the terminal velocity from a graphical analysis.

21. A falling body of mass m encounters a resisting force proportional to $|v|^r$, where r is a positive constant and $|v|$ is the magnitude of the velocity. Find the terminal velocity from a graphical analysis.

22. A sailboat together with its occupants weighs 3200 lb. The sailboat is running along a straight course with the wind providing a constant force of 50 lb in the direction of motion. The only other force acting on the boat is resistance as the boat moves through the water. The resisting force is numerically equal to five times the boat's speed and the initial velocity is 1 ft/sec. What is the maximum velocity in ft/sec of the boat under this wind?

TOOLKIT PROGRAM	
Antiderivatives and Slope Fields	Graphs solutions of the initial value problem $y' = f(x, y)$, $y(x_0) = y_0$, for your choice of f, x_0, and y_0

CHAPTER 1 REVIEW EXERCISES

For each differential equation in problems 1–21, a) determine whether it is linear or nonlinear, and b) state the order of the differential equation. Assume y is a function of x.

1. $y'' + 3y' + y = x^2 + y^2$

2. $(x - y)\,dx + x^2\,dy = 0$

3. $\dfrac{d^4y}{dx^4} + \dfrac{d^2y}{dx^2} + y^2 = x^2$

4. $(7 - x^2)\,dy + (x^2 + 4)\,dx = 0$

5. $\dfrac{d^3y}{dx^3} + \dfrac{d^2y}{dx^2} + 2y\dfrac{dy}{dx} = x^4$

6. $(6 + y)\,dx + (7 + x)\,dy = 0$

7. $2y'' + x^3y = \sin y$

8. $x^3\left(\dfrac{dy}{dx}\right)^2 + xy = 12\cos x$

9. $2\dfrac{d^2y}{dx^2} + x^4y = 0$

10. $2xy''' + 3y = 3x^2$

11. $x^2yy' + 2y = 7$

12. $\dfrac{d^2y}{dx^2} + x\sqrt{y} = 0$

13. $2yy'' - y = 1$

14. $x^2\dfrac{dy}{dx} + xy = \cos y$

15. $\left(\dfrac{dy}{dx}\right)^2 + 3xy = 5$

16. $2x^2y' + x^3y = \cos x$

17. $\dfrac{d^2y}{dx^2} + x^3y = \sin y$

18. $\dfrac{dy}{dx} + x^2y^2 = 3$

19. $(\sin x)y' + x^2y = e^{x^2}$

20. $2xy' + (\sin x)y = e^{x^2}$

21. $x^2\left(\dfrac{dy}{dx}\right)^2 + 2y = e^x$

22. In the data below, x represents the diameter of a ponderosa pine in inches measured at waist height and y is the number of board feet of lumber. Lumbercutters wish to use a readily measureable dimension (diameter at waist height) to estimate the number of board feet of lumber in the tree. Test the proportionality $y \propto x^3$ by graphing y versus x^3. If the proportionality seems reasonable, estimate the constant of proportionality from your graph.

x	17	19	20	23	25	28	32	38	39	41
y	19	25	32	57	71	113	123	252	259	294

23. In the data below, L represents the length of a black bass in inches and W is its weight in ounces. Test the proportionality $W \propto L^3$ and estimate the constant of proportionality if the assumed proportionality seems reasonable.

L(in.)	14.5	12.5	17.25	14.5	12.63	17.75	14.13	12.63
W(oz)	27	17	41	26	17	49	23	16

24. In the table below, d_r represents the average distance a car travels (in feet) while a driver reacts to a stimulus (reaction distance). Let v be the velocity of the car in mph.

a) Test the assumed proportionality $d_r \propto v$ by plotting d_r versus v.

b) If $d_r \propto v$ seems reasonable, estimate the constant of proportionality from your graph and suggest a submodel $d_r = k_1 v$.

c) Test the proportionality $d_b \propto v^2$ by graphing d_b versus v^2, where d_b is the distance the car travels after applying the brakes.

d) If $d_b \propto v^2$ seems reasonable, estimate the constant of proportionality from your graph and suggest a submodel $d_b = k_2 v^2$.

Reactions and braking distances under typical driving conditions at various speeds.

v	20	25	30	35	40	45	50	55	60	65	70	75	80	(velocity of car)
d_r	22	28	33	39	44	50	55	61	66	72	77	83	88	(reaction distance)
d_b	20	28	41	53	72	93	118	149	182	221	266	318	376	(braking distance)

25. Use the results of problems 23 and 24 to suggest a model based on the empirical data that predicts the total distance traveled by a car while the driver is reacting to a stimulus and braking. Use your model to predict the total stopping distance under average conditions when traveling at 100 mph. (In problem 26 you are asked to formulate a differential equation model based on Newton's laws.)

26. Develop a differential equation model for the total distance required to stop a vehicle after the driver reacts to a stimulus. Assume the distance traveled during reaction is proportional to the velocity v. For the submodel predicting the distance traveled while braking, assume:

a) during deceleration, all forces are negligible except the braking force F_b;

b) vehicles are designed so that the maximum braking force F_b increases in proportion to the mass of the vehicle, that is, $F_b \propto m$;

c) the maximum braking force is applied continuously throughout the stop.

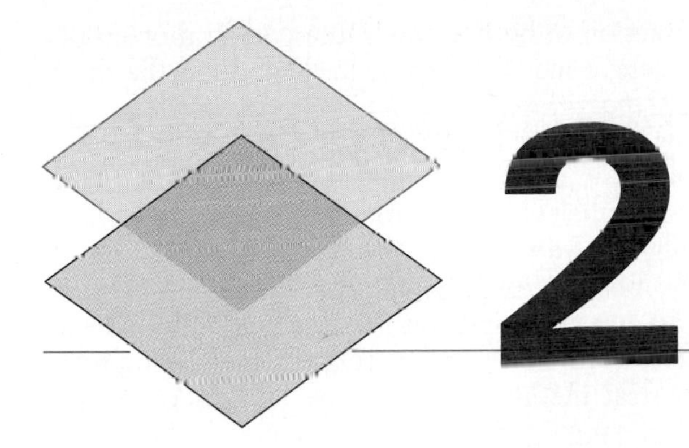

2

First-Order Differential Equations

In Chapter 1 we used the physical interpretation of the derivative as an instantaneous rate of change to model continuous behavior and to approximate the average rate of change for behavior that may take place in discrete steps. In this chapter we apply the tools of calculus directly to solve or approximate solutions to first-order equations. These solution techniques are then applied to solve models of importance to space technology in the final (optional) sections of the chapter.

For a first-order differential equation of the form

$$y' = f(x) \tag{1}$$

in which the derivative is simply a continuous function of the independent variable x, we can always integrate both sides directly to find a solution. However, an antiderivative for the specific function $f(x)$ may not be known or easy to find. However, if additionally an initial condition $y(x_0) = y_0$ is given, then for any specified value $x = x_1$ of the independent variable it is possible to numerically approximate the integral

$$y_1 = y_0 + \int_{x_0}^{x_1} f(x)\, dx \tag{2}$$

using Simpson's rule or another approximation technique to obtain selected values of the dependent variable y.

On the other hand, if the differential equation has the form

$$y' = f(x, y) \tag{3}$$

where the right side is a function of both x and y, then we cannot simply integrate directly. Nevertheless, it may be possible to manipulate the equation algebraically to *separate the variables* in such a way as to permit antidifferentiation. This technique is discussed in Section 2.1.

For first-order equation (3) no satisfactory method exists for writing a general analytic solution. Nevertheless solution methods do exist for special types of first-order equations, and we will present several of these methods in this chapter. Unfortunately, most first-order equations that arise from modeling real-world behavior are not amenable to these analytic solution techniques. If an initial value is known, however, the solution may be approximated numerically. Numerical methods will also be introduced in this chapter. In Chapter 3 we will simplify the differential equation models in order to permit analytic solutions.

2.1 SEPARATION OF VARIABLES

Suppose an object is dropped from a tall building. Initially the object has zero velocity ($v(0) = 0$), but we wish to know the object's velocity at any time $t > 0$. If we neglect all resistance, Newton's second law leads to

$$m \frac{dv}{dt} = mg$$

where m is the mass of the object and g is the acceleration due to gravity. In this formulation positive displacement is being measured downward from the top of the building. The differential equation simply integrates directly to yield

$$v = gt + C.$$

Knowing $v(0) = 0$ allows us to evaluate C:

$$v(0) = 0 = g \cdot 0 + C$$

or

$$C = 0.$$

Thus the solution function $v = gt$ predicts the velocity of an object dropped with zero initial velocity if gravity is the only force considered.

This very simple falling-body example illustrates two features typical of first-order differential equations. First, an integration process is required to obtain y from its derivative y'. Second, the integration process introduces a single arbitrary constant of integration, which can then be evaluated if an initial condition is known.

We can always integrate (at least theoretically) a first-order differential equation of the form

$$y' = f(x)$$

whenever f is a continuous function. However, consider the differential equation

$$\frac{dy}{dx} = f(x, y) \tag{1}$$

where the derivative is a function of both the variables x and y. We may be able to factor $f(x, y)$ into factors containing only x or y terms, but not both:

$$f(x, y) = p(x)q(y).$$

We allow for the possibility that either p or q may be constant functions. When the variables are separable in this way, differential equation (1) becomes

$$\frac{dy}{dx} = p(x)q(y),$$

which can be rewritten as

$$\frac{dy}{q(y)} = p(x)\, dx. \tag{2}$$

> To solve separable equation (2), simply integrate both sides (with respect to the same variable x).

Integration of both sides is permitted because we are assuming y is a function of x. Thus the left side of Eq. (2) is

$$\frac{dy}{q(y)} = \frac{y'(x)}{q(y(x))}\, dx.$$

Substitution of this expression into Eq. (2) gives

$$\frac{y'(x)}{q(y(x))}\, dx = p(x)\, dx.$$

If we set $u = y(x)$ and $du = y'(x)\, dx$, then integration of both sides leads to the solution

$$\int \frac{du}{q(u)} = \int p(x)\, dx + C. \tag{3}$$

Of course, we have to be concerned about any values of x where $q(y(x))$ is zero. Let us consider several examples.

EXAMPLE 1 Solve $y' = 3x^2e^{-y}$.

Solution. We separate the variables and write

$$e^y\,dy = 3x^2\,dx.$$

Integration of each side yields

$$e^y = x^3 + C.$$

Applying the natural logarithm to each side results in

$$y = \ln(x^3 + C). \tag{4}$$

Let's verify that y does solve the given differential equation. Differentiating Eq. (4) we find

$$y' = \frac{3x^2}{(x^3 + C)}.$$

Substitution of y and y' into the original equation then gives

$$\frac{3x^2}{x^3 + C} = 3x^2\,e^{-\ln(x^3+C)}. \tag{5}$$

Since

$$\begin{aligned} e^{-\ln(x^3+C)} &= e^{\ln(x^3+C)^{-1}} \\ &= \frac{1}{x^3 + C}, \end{aligned}$$

Eq. (5) is valid for all values of x satisfying $x^3 + C > 0$, and the differential equation is satisfied.

EXAMPLE 2 Solve $y' = 2(x + y^2x)$.

Solution. The differential equation can be written as

$$\frac{dy}{dx} = 2x(1 + y^2),$$

and separating the variables gives

$$\frac{dy}{1 + y^2} = 2x\,dx.$$

Integration of both sides leads to

$$\tan^{-1} y = x^2 + C$$

or

$$y = \tan(x^2 + C). \tag{6}$$

To verify that Eq. (6) is a solution to the given equation, we differentiate y:

$$\begin{aligned} y' &= 2x \sec^2(x^2 + C) \\ &= 2x[1 + \tan^2(x^2 + C)] \\ &= 2x(1 + y^2). \end{aligned}$$

Substitution into the original differential equation gives the identity

$$2x(1 + y^2) = 2(x + y^2 x).$$

From now on to save space we will not always verify that the function we find by a solution method is in fact a solution to the differential equation (as we have done in the examples so far). However, it is good practice to do so, especially if the solution method is fairly involved.

The Differential Form

In many cases the first-order differential equation appears in **differential form** as

$$M(x, y)\, dx + N(x, y)\, dy = 0. \tag{7}$$

For example, the equations

$$ye^{-x}\, dy + x\, dx = 0, \tag{8}$$

$$\sec x\, dy - x \cot y\, dx = 0, \tag{9}$$

$$(xe^y - e^{2y})\, dy + (e^y + x)\, dx = 0 \tag{10}$$

all have differential form. If separation of variables is to apply to Eq. (7), first write the equation in the form

$$\frac{dy}{dx} = -\frac{M(x, y)}{N(x, y)}.$$

Next look for cancellation of common terms in the numerator and denominator, and then separate the variables, if possible. Finally, integrate each side as before. Observe that Eqs. (8) and (9) are indeed separable, whereas Eq. (10) is not.

EXAMPLE 3 Solve $\sec x\, dy - x \cot y\, dx = 0$.

Solution. After dividing by $\sec x \cot y$ the equation becomes

$$\tan y\, dy - x \cos x\, dx = 0.$$

Integration then gives

$$\ln|\cos y| + x \sin x + \cos x = C.$$

Using separation of variables involves no new ideas, only the ability to recognize factors and integrate. If the correct factoring can be found, the solution technique is simply to integrate. You studied a variety of integration techniques in integral calculus and may have forgotten some of the more important ones. Appendix A provides a brief review of the main integration techniques needed for this text. These techniques are simple u-substitution, integration by parts, and integration of rational functions requiring partial fraction decomposition. We call your attention especially to the convenient *tableau method* for integration by parts and to the *Heaviside method,* which is convenient for certain partial fraction decompositions. You are urged to review these latter two methods since we employ them regularly throughout the text. The following examples illustrate these integration techniques in the context of solving separable differential equations.

EXAMPLE 4 Solve $e^{-x}y' = x$.

Solution. Separating the variables yields

$$dy = xe^x\, dx.$$

Integrating the right side by the tableau method, we have

Sign	Derivatives	Integrals
+	→ x	e^x
−	→ 1	e^x
+	→ 0	e^x

Thus, interpreting the tableau, we get

$$\int xe^x\, dx = +xe^x - 1 \cdot e^x + \int 0 \cdot e^x dx + C$$
$$= (x-1)e^x + C.$$

Therefore,

$$y = (x-1)e^x + C.$$

EXAMPLE 5 Solve $e^{x+y}y' = x$.

Solution. The equation can be written as

$$e^x e^y\,dy = x\,dx.$$

Separating the variables leads to

$$e^y\,dy = xe^{-x}\,dx.$$

Integrating the right side by parts with e^x replaced by e^{-x} in the tableau in Example 4 gives us

$$\int xe^{-x}\,dx = -xe^{-x} - e^{-x} + \int 0 \cdot e^{-x}\,dx.$$

Thus the separable equation integrates to

$$\begin{aligned} e^y &= -xe^{-x} - e^{-x} + C \\ &= -(x+1)e^{-x} + C. \end{aligned}$$

We now solve for y by taking the logarithm of each side to obtain

$$y = \ln[C - (x+1)e^{-x}]. \tag{11}$$

Let us verify that y does solve the original differential equation. Differentiating Eq. (11) yields

$$\begin{aligned} y' &= [C - (x+1)e^{-x}]^{-1} \cdot [-e^{-x} + (x+1)e^{-x}] \\ &= e^{-y} \cdot xe^{-x} \\ &= xe^{-(x+y)}. \end{aligned}$$

Substituting this result into the differential equation yields the identity

$$e^{(x+y)}xe^{-(x+y)} = x.$$

EXAMPLE 6 Solve the differential equation $dy/dx = \ln x$, where $x > 0$.

Solution. Integration of the right side gives

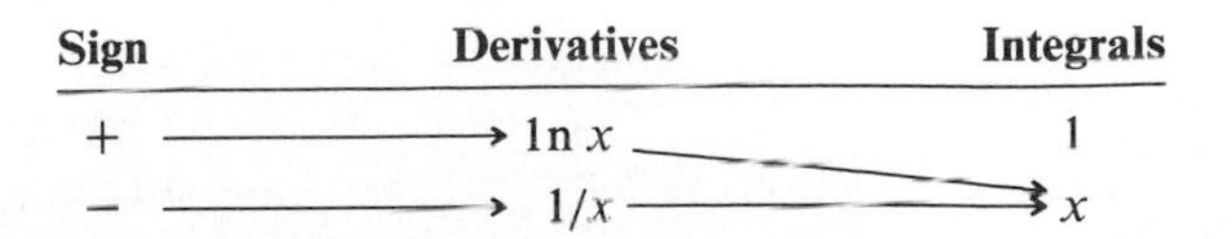

Thus,

$$y = \int \ln x \, dx$$
$$= x \ln x - \int \left(\frac{1}{x}\right) x \, dx + C$$
$$= x \ln x - x + C.$$

EXAMPLE 7 Solve the initial value problem $x^2 y y' = e^y$, where $y(2) = 0$.

Solution. Separating the variables and integrating each side leads to

$$y e^{-y} \, dy = x^{-2} \, dx$$
$$-(y+1)e^{-y} = -\frac{1}{x} + C.$$

To evaluate the arbitrary constant, substitute $y = 0$ and $x = 2$, to obtain

$$-(0+1)1 = -\frac{1}{2} + C$$

or

$$C = -\frac{1}{2}.$$

Thus

$$-(y+1)e^{-y} = -\frac{1}{x} - \frac{1}{2}$$

or

$$2x(y+1) = (2+x)e^y.$$

EXAMPLE 8 Solve $y' = x(1 - y^2)$, where $-1 < y < 1$.

Solution. Separating the variables we have

$$\frac{dy}{1-y^2} = x \, dx.$$

Partial fraction decomposition yields

$$\frac{1}{1-y^2} = \frac{1}{(1+y)(1-y)}$$
$$= \frac{1/2}{1+y} + \frac{1/2}{1-y}.$$

Integration then results in

$$\frac{1}{2} \ln|1+y| - \frac{1}{2} \ln|1-y| = \frac{x^2}{2} + C$$

or

$$\ln\left(\frac{1+y}{1-y}\right) = x^2 + C_1$$

where $C_1 = 2C$. Exponentiating each side leads to

$$\left(\frac{1+y}{1-y}\right) = e^{x^2} e^{C_1}.$$

Setting $e^{C_1} = C_2$ and solving algebraically for y, we have

$$1 + y = C_2 e^{x^2} - y C_2 e^{x^2}$$

or

$$y = \frac{C_2 e^{x^2} - 1}{C_2 e^{x^2} + 1}$$

where $C_2 = e^{C_1}$ is an arbitrary constant.

EXAMPLE 9 NEWTON'S LAW OF COOLING (REVISITED)

In Section 1.3 we developed the following model for the cooling of a space capsule having uniform temperature in all directions:

$$\frac{dT_m}{dt} = -k(T_m - \beta), \quad k > 0$$

where $T_m(0) = \alpha$. Here T_m is the temperature of the mass at any time $t > 0$, β is the constant temperature of the surrounding medium, α is the initial temperature of the mass, and k is a constant of proportionality depending on the thermal properties of the mass. We will solve the above differential equation for T_m.

Solution. After separating the variables we obtain,

$$\frac{dT_m}{T_m - \beta} = -k\,dt.$$

Integration yields

$$\ln|T_m - \beta| = -kt + C.$$

Exponentiating both sides, we get

$$|T_m - \beta| = e^{-kt+C} = e^{-kt} e^C.$$

Since e^C is a constant, we substitute $C_1 = e^C$ into the above equation:

$$|T_m - \beta| = C_1 e^{-kt}.$$

From the initial condition $T_m(0) = \alpha$ we evaluate C_1:

$$|\alpha - \beta| = C_1.$$

Substitution of this result into the solution produces

$$|T_m - \beta| = |\alpha - \beta| e^{-kt}.$$

Assuming the object is initially warmer than the surrounding medium, we have

$$T_m = \beta + (\alpha - \beta)e^{-kt}.$$

The graph of $T_m(t)$ is shown in Fig. 2.1. As $t \to \infty$ you can see that $T_m \to \beta$, in agreement with the graphical analysis presented in Section 1.4.

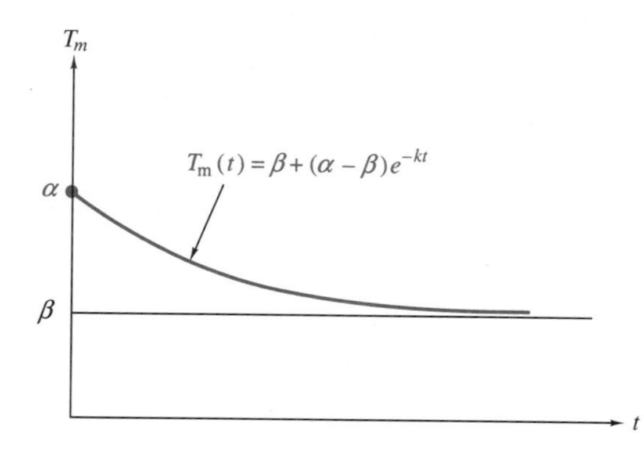

FIGURE 2.1
This graph of $T_m(t)$ assumes that the initial temperature $T_m(0) = \alpha$ is greater than the temperature β of the surrounding medium.

EXAMPLE 10 POPULATION GROWTH WITH LIMITED RESOURCES (REVISITED)

In Chapter 1 we developed the following model for population growth in a limited environment:

$$\frac{dP}{dt} = r(M - P)P = rMP - rP^2$$

where $P(t_0) = P_0$. Here P denotes the population at any time $t > 0$, M is the carrying capacity of the environment, and r is a proportionality constant. Let us solve this model.

Solution. Separating the variables we get

$$\frac{dP}{P(M - P)} = r\,dt.$$

Using partial fraction decomposition on the left side gives us

$$\frac{1}{M}\left(\frac{dP}{P}+\frac{dP}{M-P}\right)=r\,dt.$$

Multiplying by M and integrating then yields

$$\ln P-\ln|M-P|=rMt+C$$

for some arbitrary constant C. Note that since $P>0$, the absolute value symbol in the expression $\ln|P|$ is not necessary. Using the initial condition, we evaluate C in the case where $0<P_0<M$:

$$C=\ln\left(\frac{P_0}{M-P_0}\right)-rMt_0.$$

Substitution for C into the solution and algebraic simplification gives us

$$\ln\left[\frac{P(M-P_0)}{P_0(M-P)}\right]=rM(t-t_0).$$

Exponentiating both sides of this equation we obtain

$$\frac{P(M-P_0)}{P_0(M-P)}=e^{rM(t-t_0)}$$

or

$$P_0(M-P)e^{rM(t-t_0)}=P(M-P_0).$$

Finally, solving this equation algebraically for population P yields the **logistic curve**

$$P(t)=\frac{P_0Me^{rM(t-t_0)}}{M-P_0+P_0e^{rM(t-t_0)}}.$$

If you divide the numerator and denominator of this last expression by $e^{rM(t-t_0)}$ and then take the limit as $t\to\infty$, you will find that $P(t)\to M$. That is, the population tends toward the maximum sustainable population. Our analytic result agrees with our graphical analysis for the case $0<P_0<M$, which was depicted in Fig. 1.31.

Uniqueness of Solutions

Often the solutions to a first-order differential equation can be expressed in the explicit form $y=f(x)$, where each solution is distinguished by a different value of the arbitrary constant of integration. For instance, if we separate the variables in the equation

$$\frac{dy}{dx}=\frac{2y}{x},\quad x\neq 0, \tag{12}$$

we obtain

$$\frac{dy}{y} = 2\frac{dx}{x}.$$

Notice, however, that the algebra is not valid when $y = 0$. Next, integration of both sides of this last result gives us

$$\ln|y| = 2\ln|x| + C_1. \tag{13}$$

Exponentiating both sides of (13) yields

$$|y| = C_2 x^2$$

where $C_2 = e^{C_1} > 0$ is a constant. Finally, by applying the definition of absolute value we have

$$y = Cx^2 \tag{14}$$

where $C = \pm C_2$ is positive or negative according to whether y is positive or negative. However, there are still other solutions not given by Eq. (14). Equation (14) represents a *family of parabolas,* each parabola distinguished by a different value of the constant C. If $C > 0$, the curves $y = Cx^2$ open upward, and if $C < 0$, they open downward (see Fig. 2.2). Exactly one of these parabolas passes through each point in the plane excluding the origin. Thus, by specifying an initial condition $y(x_0) = y_0$, a unique solution curve from the family denoted by Eq. (14) is selected that passes through the point (x_0, y_0).

The question of whether more than one solution curve can pass through a specific point (x_0, y_0) in the plane is an important one. For uniqueness to occur, certain conditions must be met by the function $f(x, y)$ defining the differential equation $y' = f(x, y)$. This question of uniqueness is discussed more fully in Section 2.7. Geometrically, the condition of uniqueness means that two solution curves cannot cross at the point in question. In Fig. 2.2

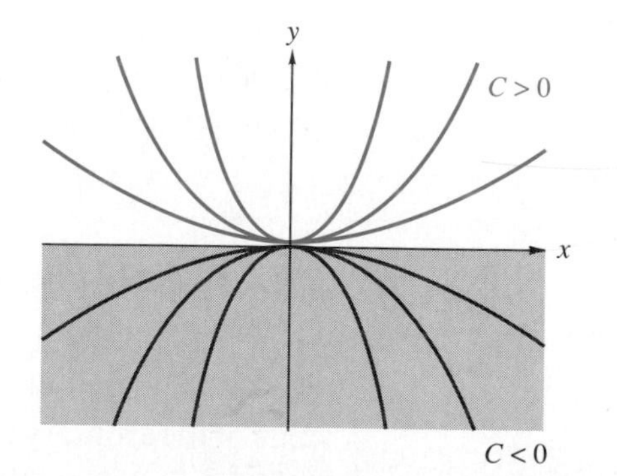

FIGURE 2.2
The family of parabolas $y = Cx^2$.

there is a unique solution parabola passing through each point in the plane with the exception of the origin. Through the origin pass infinitely many solution parabolas, along with other solution curves as well. For instance, notice that the constant function $y \equiv 0$ is a solution to the differential equation. This fact is readily seen when Eq. (12) is written in the form

$$xy' = 2y, \tag{15}$$

which is clearly satisfied when $y \equiv 0$. However, the solution $y \equiv 0$ is *not* a member of the family of solutions represented by Eq. (14), since C_2 is always positive. The difficulty lies with the lack of continuity of the function $f(x, y) = 2y/x$ at the origin. Nevertheless, if we let $C = 0$ in Eq. (14), we do pick up the solution $y \equiv 0$. Even so, there are still other solutions not given by Eq. (14) for any value of C.

The issue of uniqueness of solutions is rather involved for nonlinear first-order equations. However, in the very important case of the linear first-order equation the matter is easily settled. This points up one of the significant differences between linear and nonlinear differential equations. In the next section we will investigate the nature of the linear first-order equation.

EXERCISES 2.1

In problems 1–8 solve the separable differential equation using u-substitution.

1. $\dfrac{dy}{dx} = y^2 - 2y + 1$

2. $\dfrac{dy}{dx} = \sqrt{y} \cos^2 \sqrt{y}$

3. $y' = \dfrac{3y(x+1)^2}{y-1}$

4. $yy' = \sec y^2 \sec^2 x$

5. $y \cos^2 x \, dy + \sin x \, dx = 0$

6. $y' = \left(\dfrac{y}{x}\right)^2$

7. $y' = xe^y \sqrt{x-2}$

8. $y' = xye^{x^2}$

In problems 9–16 solve the separable differential equation using integration by parts.

9. $\sec x \, dy + x \cos^2 y \, dx = 0$

10. $2x^2 \, dx - 3\sqrt{y} \csc x \, dy = 0$

11. $y' = \dfrac{e^y}{xy}$

12. $y' = xe^{x-y} \csc y$

13. $y' = e^{-y} \ln\left(\dfrac{1}{x}\right)$

14. $y' = y^2 \tan^{-1} x$

15. $y' = y \sin^{-1} x$

16. $\sec(2x+1) \, dy + 2xy^{-1} \, dx = 0$

In problems 17–24 solve the separable differential equation using partial fractions.

17. $(x^2 + x - 2) \, dy - 3y \, dx = 0$

18. $y' = (y^2 - 1)x^{-1}$

19. $x(x-1) \, dy - y \, dx = 0$

20. $y' = \dfrac{(y+1)^2}{x^2 + x - 2}$

21. $9y\,dx - (x-1)^2(x+2)\,dy = 0$

22. $e^x\,dy + (y^3 - y^2)\,dx = 0$

23. $\sqrt{1-y^2}\,dx + (x^2 - 2x + 2)\,dy = 0$

24. $(2x - x^2)\,dy + e^{-y}\,dx = 0$

In problems 25–32 solve the separable differential equation.

25. $\sqrt{2xy}\,\dfrac{dy}{dx} = 1$

26. $(\ln x)\,\dfrac{dx}{dy} = xy$

27. $x^2\,dy + y(x-1)\,dx = 0$

28. $ye^x\,dy - (e^{-y} + e^{2x-y})\,dx = 0$

29. $(x \ln y)y' = \left(\dfrac{x+1}{y}\right)^2$

30. $y' = \dfrac{\sin^{-1}x}{2y \ln y}$

31. $y' = e^{-y} - xe^{-y}\cos x^2$

32. $(1 + x + xy^2 + y^2)\,dy = (1-x)^{-1}\,dx$

In problems 33–39 solve the initial value problem.

33. $y^{-2}\dfrac{dx}{dy} = \dfrac{e^x}{e^{2x}+1}, \quad y(0) = 1$

34. $\dfrac{dy}{dx} + xy = x, \quad y(1) = 2$

35. $y' - 2y = 1, \quad y(2) = 0$

36. $2(y^2 - 1)\,dx + \sec x \csc x\,dy = 0, \quad y\left(\dfrac{\pi}{4}\right) = 0$

37. $\dfrac{dP}{dt} + P = Pte^t, \quad P(0) = 1$

38. $\dfrac{dP}{dt} = (P^2 - P)t^{-1}, \quad P(1) = 2$

39. $x\,dy - (y + \sqrt{y})\,dx = 0, \quad y(1) = 1$

2.2

LINEAR EQUATIONS

Serious problems of water pollution face the world's industrialized nations. If the polluted body of water is a river, it can clean itself fairly rapidly once the pollution is stopped, provided excessive damage has not already occurred. But the problem of pollution in a lake or reservoir is not so easily overcome. A polluted lake, such as one of the Great Lakes, contains a large amount of water that must somehow be cleaned. Presently government and industry still rely on natural processes for this cleanup. In this section we shall model how long a given lake might take to return to an acceptable level of pollution through natural processes alone; modeling this problem will allow us to examine some characteristics of linear equations.

Let's make some simplifying assumptions to model the situation. Imagine the lake to be a large container or tank holding a volume $V(t)$ of water at any time t. Assume that when water enters the lake, perfect mixing occurs, so that the pollutants are uniformly distributed throughout the lake at any time. Assume also that the pollutants are not removed from the lake by sedimentation, decay, or any other natural mechanism except the outflow of water from the lake. Moreover, the pollutants flow freely from the lake (unlike DDT, which tends to concentrate in the fatty tissues of animals and thus be retained in biological systems). Let $p(t)$ denote the amount of the pollutant in the lake at time t. Then the **concentration** of the pollutant is the ratio $c(t) = p(t)/V(t)$.

Over the time interval $[t, t + \Delta t]$, the change in the amount of pollutant Δp is the amount of pollutant that enters the lake minus the amount that leaves:

$$\Delta p = \text{amount input} - \text{amount output}.$$

If water enters the lake with a constant concentration of c_{in} in grams per liter at a rate of r_{in} in liters per second, then

$$\text{amount input} \approx r_{in}c_{in}\,\Delta t = \alpha\,\Delta t$$

where $\alpha = r_{in}c_{in}$ is constant. For example, if polluted water with a concentration of 9 g/L of pollutant enters the lake at 10 L/sec, then $\alpha = 90$ g of pollutant enters the lake each second.

If water leaves the lake at a constant rate of r_{out} L/sec, then, since the concentration of pollutant in the lake is given by p/V, we have

$$\text{amount output} \approx r_{out}\frac{p}{V}\,\Delta t.$$

Thus

$$\Delta p \approx \left(\alpha - \frac{pr_{out}}{V}\right)\Delta t.$$

Dividing by Δt and passing to the limit as $\Delta t \to 0$ results in

$$\frac{dp}{dt} = \alpha - \frac{r_{out}}{V}\,p. \tag{1}$$

Suppose $V(0) = V_0$ is the volume of the lake initially. Then $V(t) = V_0 + (r_{in} - r_{out})t$ represents the volume at any time t. Substituting for V in Eq. (1) and rearranging terms gives us

$$\frac{dp}{dt} + \frac{pr_{out}}{V_0 + (r_{in} - r_{out})t} = \alpha \tag{2}$$

where α, r_{out}, V_0, and $(r_{in} - r_{out})$ are all constants. Equation (2) is an example of a *linear first-order differential equation.* Note that if $r_{in} - r_{out} = 0$, Eq. (2) is a separable equation. However, if $r_{in} - r_{out} \neq 0$, Eq. (2) represents a new type of first-order equation.

At the end of this section we will return to the pollution problem to find the amount $p(t)$ of pollutant in the lake at any time t. In order to solve this model we now take up the general question of solving linear first-order equations.

First-Order Linear Equations

The **first-order linear equation** is an equation of the form

$$a_1(x)y' + a_0(x)y = b(x) \tag{3}$$

where $a_1(x)$, $a_0(x)$, and $b(x)$ depend only on the independent variable x, not on y. For example,

$$2xy' - y = xe^{-x}, \tag{4}$$

$$(x^2 + 1)y' + xy = x, \tag{5}$$

$$y' + (\tan x)y = \cos^2 x \tag{6}$$

are all first-order linear equations. The equation

$$y' - x^2 e^{-2y} = 0 \tag{7}$$

is not linear, although it is separable. The equation

$$(2x - y^2)y' + 2y = x \tag{8}$$

is neither linear nor separable.

We assume in Eq. (3) that the functions $a_1(x)$, $a_0(x)$, and $b(x)$ are continuous on an interval and that $a_1(x) \neq 0$ on that interval. Division of both sides of (3) by $a_1(x)$ gives the **standard form** of the linear equation,

$$y' + P(x)y = Q(x) \tag{9}$$

where $P(x)$ and $Q(x)$ are continuous on the interval. The solution method we present for the linear equation proceeds from the standard form.

To provide some insight into the solution form, we are going to solve Eq. (9) in three stages. The first stage is the case when $P(x) \equiv$ constant and $Q(x) \equiv 0$. (The symbol $\equiv$ means "is identically equal to." Thus $P(x) \equiv$ constant says that $P(x)$ is constant for all values of x.) Then the case is considered for $P(x) \equiv$ constant and $Q(x) \not\equiv 0$. Finally, we consider the general case given by Eq. (9).

Case 1: $y' + ky = 0$, $k =$ Constant

The differential equation is separable with

$$\frac{dy}{y} = -k\,dx.$$

Then

$$y = Ce^{-kx}$$

is a solution for any constant C. If we write the last equation as

$$e^{kx}y = C$$

and differentiate implicitly, we obtain

$$e^{kx}y' + ke^{kx}y = 0$$

or

$$e^{kx}(y' + ky) = 0.$$

That is, multiplication of each side of the equation $y' + ky = 0$ by the exponential function e^{kx} results in

$$\frac{d}{dx}(e^{kx}y) = \frac{d}{dx}(C).$$

The solution is now readily obtained by integrating each side. Armed with this insight we consider the second case.

Case 2: $y' + ky = Q(x)$, $k =$ Constant

From our observation in Case 1, if we multiply both sides of Eq. (9) by e^{kx}, we get

$$e^{kx}(y' + ky) = e^{kx}Q(x)$$

or

$$\frac{d}{dx}(e^{kx}y) = e^{kx}Q(x). \tag{10}$$

Integration then gives us

$$e^{kx}y = \int e^{kx}Q(x)\,dx + C \tag{11}$$

where C is an arbitrary constant.

Let us pause to reflect on this procedure. We multiplied the equation $y' + ky = Q(x)$ by a function e^{kx} of the independent variable. The left side was then transformed into the derivative of a product:

$$e^{kx}y' + ke^{kx}y = \frac{d}{dx}(e^{kx}y).$$

Then all that was needed to arrive at a solution was to integrate resulting Eq. (10), obtaining Eq. (11).

Case 3: The General Linear Equation $y' + P(x)y = Q(x)$

Let us try the same idea as in Case 2: multiply both sides of the equation by some function $\mu(x)$ so that the left side is the derivative of the product μy.

That is,

$$\begin{aligned}\mu(x)[y' + P(x)y] &= \frac{d}{dx}[\mu(x)y] \\ &= \mu(x)y' + \mu'(x)y.\end{aligned}$$

The function $\mu(x)$ is not yet known, but from the last equation it must satisfy

$$\mu(x)P(x)y = \mu'(x)y$$

or

$$\frac{\mu'(x)}{\mu(x)} = P(x). \tag{12}$$

We seek only *one* function $\mu(x)$ in our procedure, so assume that $\mu(x)$ is positive over the interval. Then integrating Eq. (12) gives us

$$\ln \mu(x) = \int P(x)\, dx$$

or, exponentiating both sides,

$$\mu(x) = e^{\int P(x)dx}. \tag{13}$$

That is, Eq. (13) determines precisely a function $\mu(x)$ that will work for our procedure. Note that $\mu(x)$ defined by Eq. (13) is indeed positive. The function $\mu(x)$ is called an **integrating factor** for linear first-order Eq. (9).

Now, multiplying Eq. (9) through by the integrating factor (13) results in

$$\mu(x)[y' + P(x)y] = \mu(x)Q(x)$$

or

$$\frac{d}{dx}[\mu(x)y] = \mu(x)Q(x). \tag{14}$$

In order to solve Eq. (14), simply integrate both sides:

$$\mu(x)y = \int \mu(x)Q(x)\, dx + C \tag{15}$$

where $\mu(x)$ is given by Eq. (13). We can then solve explicitly for the solution y by dividing each side of Eq. (15) by the integrating factor $\mu(x)$.

The method of solution presented for linear first-order equations requires two integrations. The first integration in Eq. (13) produces the integrating factor $\mu(x)$; the second leads to the general solution y from Eq. (15). Both integrations are possible because $P(x)$ and $Q(x)$ are assumed to be continuous over the interval. Using the Fundamental Theorem of Calculus, we could easily verify that the function y defined by Eq. (15) does satisfy the original linear Eq. (9) (see Exercises 2.2, problem 23). We will now summarize the solution method.

SOLVING A LINEAR FIRST-ORDER EQUATION

Step 1. Write the linear first-order equation in standard form:

$$y' + P(x)y = Q(x). \tag{16}$$

Step 2. Calculate the integrating factor:

$$\mu(x) = e^{\int P(x)dx}. \tag{17}$$

Step 3. Multiply the right side of equation (16) by μ and integrate:

$$\int \mu(x)Q(x)\,dx + C. \tag{18}$$

Step 4. Write the general solution:

$$\underbrace{\mu(x)}_{\substack{\text{integrating factor} \\ \text{from Step 2}}} y = \underbrace{\int \mu(x)Q(x)\,dx + C}_{\text{result of Step 3}}. \tag{19}$$

Observe in Step 2 that no arbitrary constant of integration is introduced when determining the integrating factor μ. The reason for this is that only a *single* function is sought as an integrating factor, not an entire family of functions. Now let us apply the method to several examples.

EXAMPLE 1 Find the general solution of

$$xy' + y = e^x, \quad x > 0.$$

Solution. *Step 1.* We write the linear equation in standard form:

$$y' + \left(\frac{1}{x}\right)y = \left(\frac{1}{x}\right)e^x.$$

Thus $P(x) = 1/x$ and $Q(x) = e^x/x$.

Step 2. The integrating factor is

$$\begin{aligned} \mu(x) &= e^{\int P(x)dx} = e^{\int dx/x} \\ &= e^{\ln x} = x. \end{aligned}$$

Step 3. We multiply the right side of the equation in Step 1 by $\mu = x$ and integrate the results to get

$$\int \mu(x)Q(x)\,dx = \int x \cdot \left(\frac{1}{x}\right) e^x\,dx$$
$$= \int e^x\,dx$$
$$= e^x + C.$$

Step 4. The general solution is given by Eq. (19):

$$xy = e^x + C$$

or

$$y = \frac{e^x + C}{x}, \quad x > 0.$$

Let us verify that y does indeed solve the original equation. Differentiation of y gives us

$$y' = \frac{-1}{x^2}(e^x + C) + \frac{1}{x}e^x.$$

Then

$$xy' + y = \left[\frac{-1}{x}(e^x + C) + e^x\right] + \left(\frac{1}{x}\right)(e^x + C)$$
$$= e^x$$

so the differential equation is satisfied.

EXAMPLE 2 Find the general solution of

$$y' + (\tan x)y = \cos^2 x$$

over the interval $-\pi/2 < x < \pi/2$.

Solution. *Step 1.* The equation is in standard form with $P(x) = \tan x$ and $Q(x) = \cos^2 x$.

Step 2. The integrating factor is

$$\mu(x) = e^{\int P(x)dx} = e^{\int \tan x dx} = e^{-\ln|\cos x|} = \sec x,$$

since $\cos x > 0$ over the interval $-\pi/2 < x < \pi/2$.

Step 3. Next we integrate the product $\mu(x)Q(x)$:

$$\int \sec x \cos^2 x\,dx = \int \cos x\,dx = \sin x + C.$$

Step 4. The general solution is given by

$$(\sec x)y = \sin x + C$$

or

$$y = \sin x \cos x + C \cos x.$$

EXAMPLE 3 Find the solution of

$$3xy' - y = \ln x + 1, \quad x > 0,$$

satisfying $y(1) = -2$.

Solution. In this example we shall omit the designation of the steps. With $x > 0$, we rewrite the equation in the standard form as

$$y' - \frac{1}{3x}y = \frac{\ln x + 1}{3x}.$$

Then the integrating factor is given by

$$\mu = e^{\int -dx/3x} = e^{(-1/3)\ln x} = x^{-1/3}.$$

Thus

$$x^{-1/3}y = \frac{1}{3}\int (\ln x + 1)x^{-4/3}\,dx.$$

Integration by parts results in the following (with the details left to you to figure out):

$$x^{-1/3}y = -x^{-1/3}(\ln x + 1) + \int x^{-4/3}\,dx + C.$$

Therefore

$$x^{-1/3}y = -x^{-1/3}(\ln x + 1) - 3x^{-1/3} + C$$

or

$$y = -(\ln x + 4) + Cx^{1/3}.$$

When $x = 1$ and $y = -2$ are substituted into the general solution, the arbitrary constant C is evaluated:

$$-2 = -(0 + 4) + C$$

or

$$C = 2.$$

Thus

$$y = 2x^{1/3} - \ln x - 4$$

is the particular solution we seek.

EXAMPLE 4 WATER POLLUTION

We now return to the problem of water pollution of a large lake introduced at the beginning of this section. Suppose a large lake formed by damming a river holds initially 100 million gallons of water. Because a nearby agricultural field was sprayed with a pesticide, the water has become contaminated. The concentration of the pesticide has been measured and is equal to 35 ppm (parts per million) or 35×10^{-6}. The river continues to flow into the lake at a rate of 300 gal/min. The river is only slightly contaminated with pesticide and has a concentration of 5 ppm. The flow of water over the dam can be controlled and is set at 400 gal/min. Assume that no additional spraying causes the lake to become even more contaminated. How long will it be before the water reaches an acceptable level of concentration equal to 15 ppm?

Solution. From the opening discussion of Section 2.2, recall that

$$V(t) = V_0 + (r_{in} - r_{out})t.$$

For the particular lake at hand, we are given that $V_0 = 100 \times 10^6$ and $r_{in} - r_{out} = 300 - 400 = -100$ gal/min. Thus,

$$V(t) = 100 \times 10^6 - 100t$$

represents the volume of the lake at time t. Since $r_{in} - r_{out} = -100$, note that the lake will be empty when $V(t) = 0$ or $t = 10^6$ min $\approx$ 1.9 yr. Hopefully, the contamination in the lake can be reduced to the acceptable level of $15/10^6$ before the lake is empty.

Using the notation introduced in the opening discussion, we have $\alpha = r_{in}c_{in} = 300(5/10^6)$. Thus, from Eq. (2) the differential equation governing the change in pollution is given by

$$\frac{dp}{dt} + \frac{400p}{100 \times 10^6 - 100t} = 15 \times 10^{-4}. \tag{20}$$

The integrating factor for Eq. (20) is

$$\mu = e^{\int 4dt/(10^6 - t)} = e^{-4\ln(10^6 - t)} = (10^6 - t)^{-4}$$

assuming $t < 10^6$. Thus the solution satisfies

$$\begin{aligned}(10^6 - t)^{-4}p(t) &= \int 15 \times 10^{-4}(10^6 - t)^{-4}\,dt \\ &= 5 \times 10^{-4}(10^6 - t)^{-3} + C.\end{aligned}$$

Therefore

$$p(t) = 5 \times 10^{-4}(10^6 - t) + C(10^6 - t)^4. \tag{21}$$

From the initial condition, when $t = 0$ the concentration is $c_0 = p(0)/V_0 = 35 \times 10^{-6}$. Hence

$$p(0) = (35 \times 10^{-6}) \times 100 \times 10^6 = 3500.$$

By substituting this result into Eq. (21) we evaluate the constant of integration C:

$$3500 = 5 \times 10^{-4} \times 10^6 + C \times 10^{24}$$

or $C = 3 \times 10^{-21}$. The particular solution for the level of pollution at any time $t < 10^6$ is therefore

$$p(t) = 5 \times 10^{-4}(10^6 - t) + 3 \times 10^{-21}(10^6 - t)^4. \tag{22}$$

The problem asks for the time t when the concentration level $c(t) = p(t)/V(t) = 15 \times 10^{-6}$. Here t is measured in minutes. Division of Eq. (22) by $V(t)$ and application of this condition yields

$$15 \times 10^{-6} = \frac{5 \times 10^{-4}(10^6 - t) + 3 \times 10^{-21}(10^6 - t)^4}{100(10^6 - t)}.$$

Simplifying algebraically, we get

$$3 \times 10^{-18}(10^6 - t)^3 - 1 = 0.$$

Using a calculator or computer, we find that the solution of this last equation for t gives

$$t \approx 306{,}650 \text{ min} \approx 7 \text{ months}.$$

The linear equation (of any order) is very important to the study of differential equations and constitutes the primary focus of this text. Nevertheless, there are several classes of nonlinear first-order equations that do occur in applications to the physical sciences and that are amenable to analytical solutions. We will take up these special first-order equations in Sections 2.3 and 2.4. We end this section with a brief discussion of the initial value problem for the linear first-order equation.

Uniqueness of Solutions

Unlike the difficulties we can encounter when solving (nonlinear) separable equations, the linear first-order equation always has one and only one solution satisfying a specified initial condition. This result is stated precisely in the following theorem.

THEOREM 2.1

Suppose that $P(x)$ and $Q(x)$ are continuous functions over the interval $\alpha < x < \beta$. Then there is one and only one function $y = y(x)$ satisfying the first-order linear equation

$$y' + P(x)y = Q(x)$$

on the interval and the initial condition

$$y(x_0) = y_0$$

at the specified point x_0 in the interval.

Theorem 2.1 is known as the **existence and uniqueness theorem** for the linear first-order equation. Any real value whatsoever may be assigned to y_0 and the theorem will be satisfied. Thus the particular solution found in Example 3 is the only function satisfying the differential equation and the initial condition specified there. Problems 23 and 24 in the exercises outline a proof of the existence and uniqueness theorem based on the Fundamental Theorem of Calculus.

EXERCISES 2.2

In problems 1–15 find the general solution of the given first-order linear differential equation. State an interval over which the general solution is valid.

1. $y' + 2xy = x$

2. $y' - 3y = e^x$

3. $2y' - y = xe^{x/2}$

4. $\dfrac{y'}{2} + y = e^{-x} \sin x$

5. $xy' + 2y = 1 - x^{-1}$

6. $xy' - y = 2x \ln x$

7. $y' = y - e^{2x}$

8. $y' = \dfrac{2y}{x} + x^3e^x - 1$

9. $x^2 \dfrac{dy}{dx} + xy = 2$

10. $(1 + x) \dfrac{dy}{dx} + y = \sqrt{x}$

11. $x^2\, dy + xy\, dx = (x - 1)^2\, dx$

12. $(1 + e^x)\, dy + (ye^x + e^{-x})\, dx = 0$

13. $e^{-y}\, dx + (e^{-y}x - 4y)\, dy = 0$

14. $(x + 3y^2)\, dy + y\, dx = 0$

15. $y\, dx + (3x - y^{-2} \cos y)\, dy = 0, \quad y > 0$

In problems 16–20 solve the initial value problem.

16. $y' + 4y = 1, \quad y(0) = 1$

17. $\dfrac{dy}{dx} + 3x^2y = x^2, \quad y(0) = -1$

18. $x\, dy + (y - \cos x)\, dx = 0, \quad y\left(\dfrac{\pi}{2}\right) = 0$

19. $xy' + (x - 2)y = 3x^3e^{-x}, \quad y(1) = 0$

20. $y\, dx + (3x - xy + 2)\, dy = 0, \quad y(2) = -1, y < 0$

21. Oxygen flows through one tube into a liter flask filled with air, and the mixture of oxygen and air (considered well stirred) escapes through another tube. Assuming that air contains 21% oxygen, what percentage of oxygen will the flask contain after 5 L have passed through the intake tube?

22. If the average person breathes 20 times per minute, exhaling each time 100 in^3 of air containing 4% carbon dioxide, find the percentage of carbon dioxide in the air of a 10,000 ft^3 closed room 1 hr after a class of 30 students enters. Assume that the air is fresh at the start, that the ventilators admit 1000 ft^3 of fresh air per minute, and that the fresh air contains 0.04% carbon dioxide.

23. Existence

Assume the hypothesis of Theorem 2.1.

a) From the Fundamental Theorem of Calculus we have

$$\frac{d}{dx}\left[\int \mu(x)Q(x)\,dx\right] = \mu(x)Q(x).$$

Use this fact to show that any function y given by Eq. (15) solves linear first-order Eq. (9). *Hint:* differentiate both sides of Eq. (15).

b) If the constant C is given by

$$C = y_0\mu(x_0) - \int_{x_0}^{x} \mu(t)Q(t)\,dt$$

in Eq. (15), show that the resulting function y defined by Eq. (15) satisfies the initial condition $y(x_0) = y_0$.

24 Uniqueness

Assume the hypothesis of Theorem 2.1 and assume that $y_1(x)$ and $y_2(x)$ are both solutions to the linear first-order equation satisfying the initial condition $y(x_0) = y_0$.

a) Verify that $y(x) = y_1(x) - y_2(x)$ satisfies the initial value problem

$$y' + P(x)y = 0, \quad y(x_0) = 0.$$

b) For the integrating factor $\mu(x)$ defined by Eq. (17), show that

$$\frac{d}{dx}\left(\mu(x)[\,y_1(x) - y_2(x)]\right) = 0.$$

Conclude that $\mu(x)[\,y_1(x) - y_2(x)] \equiv$ constant.

c) From part (a), we have $y_1(x_0) - y_2(x_0) = 0$. Since $\mu(x) > 0$ for $\alpha < x < \beta$, use part (b) to establish that $y_1(x) - y_2(x) \equiv 0$ on the interval (α, β). Thus $y_1(x) = y_2(x)$ for all $\alpha < x < \beta$.

2.3 EXACT EQUATIONS

In multivariable calculus you studied a function $\phi(x, y)$ of two independent variables called a potential function. Recall that a **potential function** represents the potential for work an object possesses because of its location in an energy field (such as a gravitational or electromagnetic field). For instance, a 1-lb object 5 ft above the ground resting on a table has a potential difference of 5 ft-lb relative to the ground.

For a fixed constant C, the level curve $\phi(x, y) = C$ represents the points in the x, y plane with exactly the same potential C. Contour lines on a map are examples of **level curves** (or **equipotential curves**) connecting points with the same potential in the earth's gravitational field (because the points lie at the same height above sea level). Figure 2.3 displays a potential function and its level curves. In Section 2.9 (optional) we will develop the potential function associated with an orbiting satellite acted on by the earth's gravitational field.

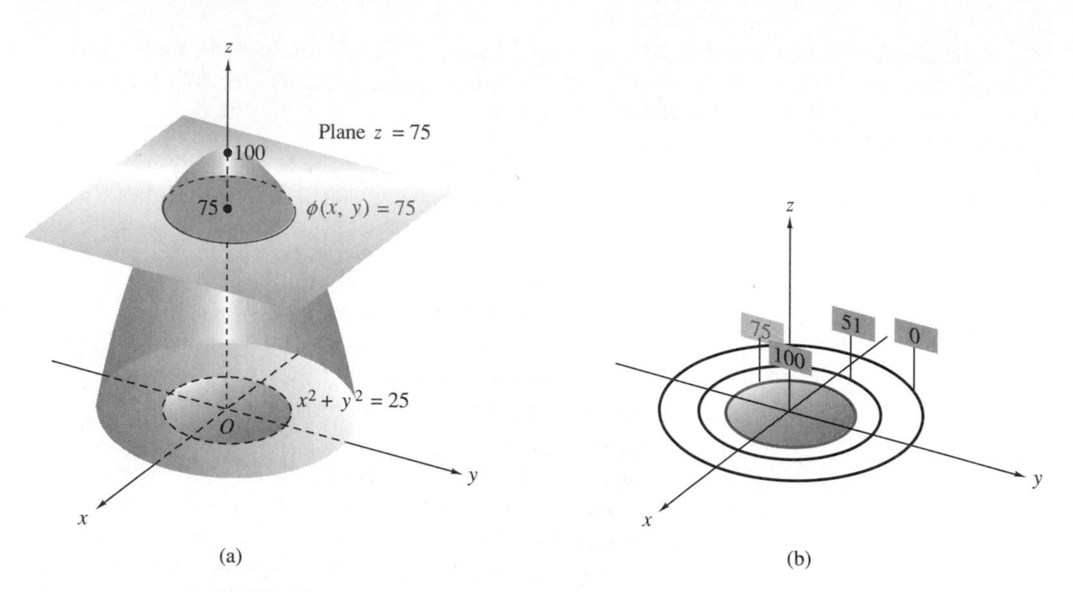

FIGURE 2.3
(a) The potential function $\phi(x, y) = 100 - x^2 - y^2$ measures the potential in a field. (b) Each level curve $\phi(x, y) = C$ (C = constant) connects points with equal potential (equipotential curves). The equipotential curves are described by the differential equation $y' = -x/y$.

The following concept is useful in studying differential equations whose solutions are given by potential functions.

DEFINITION 2.1

The **total differential** of a function $\phi(x, y)$ is

$$d\phi = \frac{\partial \phi}{\partial x}\, dx + \frac{\partial \phi}{\partial y}\, dy. \tag{1}$$

Assume we have a level curve of a potential function $\phi(x, y)$ represented by

$$\phi(x, y) = C. \tag{2}$$

Then the total differential of Eq. (2) gives $d\phi = 0$, or

$$\frac{\partial \phi}{\partial x}\, dx + \frac{\partial \phi}{\partial y}\, dy = 0.$$

We assume that near the point (x, y) Eq. (2) defines y as a function of x. Then, by rearranging terms in the last equation above, we can find the derivative of this function:

$$\frac{dy}{dx} = -\frac{\partial \phi/\partial x}{\partial \phi/\partial y} \tag{3}$$

Of course, Eq. (3) requires that the partial derivative $\partial\phi/\partial y \neq 0$. Assuming the partial derivatives $\partial\phi/\partial x$ and $\partial\phi/\partial y$ exist, they are themselves functions of x and y, so that Eq. (3) can be rewritten as

$$\frac{dy}{dx} = g(x,y) \tag{4}$$

provided that $\partial\phi/\partial y \neq 0$. Notice that under suitable conditions Eq. (2), which describes a family of level curves of a potential function (one curve for each value of C), leads directly to first-order differential Eq. (4). As discussed in Chapter 1, we often start with some information about the rate of change of a function and then seek to find the function. Thus we can ask, given differential Eq. (4), does a method exist to find an implicit solution of the form of Eq. (2) to that differential equation (Eq. 4)? Before proceeding further, let us illustrate the ideas presented so far in this section with a specific example.

EXAMPLE 1 Consider the level curves of the potential function

$$\phi(x, y) = x^3 + 2x^2y + y^2 = C.$$

The total differential is given by

$$d\phi = (3x^2 + 4xy)\,dx + (2x^2 + 2y)\,dy = 0,$$

which leads to the differential equation

$$\frac{dy}{dx} = -\frac{3x^2 + 4xy}{2x^2 + 2y}$$

describing the equipotential curves of $\phi(x, y)$. Working from the reverse direction, that is, given the differential equation

$$\frac{dy}{dx} = -\frac{3x^2 + 4xy}{2x^2 + 2y},$$

we know an implicit solution is defined by the equation

$$x^3 + 2x^2y + y^2 = C.$$

Then, an initial condition such as $y(0) = 3$ allows us to evaluate the arbitrary constant $C = 9$. The particular solution

$$x^3 + 2x^2y + y^2 = 9$$

then corresponds to a specific level curve connecting all the points in the x,y-plane having the same potential value of 9.

Solving Exact Equations

Let us take a first-order differential equation $dy/dx = g(x, y)$ and express it in the differential form

$$M(x, y)\,dx + N(x, y)\,dy = 0. \tag{5}$$

We want to develop answers to the following questions:

1. Does there exist a potential function $\phi(x, y)$ whose total differential (1) is identical to the lefthand side of Eq. (5)?
2. If the potential function ϕ does exist, how do we find it?

DEFINITION 2.2

If a function $\phi(x, y)$ exists such that $\partial\phi/\partial x = M(x, y)$ and $\partial\phi/\partial y = N(x, y)$, then $M(x, y)\,dx + N(x, y)\,dy = 0$ is said to be an **exact differential equation.**

To answer the first question above, we first compare total differential (1) of the level curves of Eq. (2) to the lefthand side of Eq. (5). By matching the coefficients of dx and dy in Eqs. (1) and (5), we see that in order for $\phi(x, y)$ to exist, it must satisfy the conditions

$$\frac{\partial\phi}{\partial x} = M(x, y) \tag{6}$$

and

$$\frac{\partial\phi}{\partial y} = N(x, y). \tag{7}$$

Differentiating Eq. (6) with respect to y and Eq. (7) with respect to x gives us

$$\frac{\partial^2\phi}{\partial y\partial x} = \frac{\partial M}{\partial y} \tag{8}$$

and

$$\frac{\partial^2\phi}{\partial x\partial y} = \frac{\partial N}{\partial x}. \tag{9}$$

Now provided that partial derivative Eqs. (6–9) are continuous, the second partial derivatives $\partial^2\phi/\partial y\partial x$ and $\partial^2\phi/\partial x\partial y$ are equal. Therefore Eqs. (8) and (9) imply the equality

$$\frac{\partial M}{\partial y} = \frac{\partial N}{\partial x}.$$

We summarize these conclusions with the following test for exactness of a first-order differential equation.*

TEST FOR EXACTNESS

The differential equation

$$M(x, y)\,dx + N(x, y)\,dy = 0$$

is exact if and only if

$$\frac{\partial M}{\partial y} = \frac{\partial N}{\partial x}. \tag{10}$$

A Method for Solving the Exact Equation

Now let us address the second question: given that a differential equation is exact, how do we find the potential function $\phi(x, y)$? In this method we take advantage of the knowledge that $M = \partial\phi/\partial x$ and $N = \partial\phi/\partial y$ from Eqs. (6) and (7). First we choose either M or N to integrate, whichever is easier. For purposes of illustration, we choose to integrate M. Thus

$$\phi = \int M\,dx + g(y). \tag{11}$$

Note that the "constant of integration" is an *arbitrary function* $g(y)$, since it would be zero when Eq. (11) was differentiated with respect to x. Next we determine $g(y)$. Since $\partial\phi/\partial y = N$, we differentiate Eq. (11) and equate the result to N:

$$N = \frac{\partial\phi}{\partial y} = \frac{\partial}{\partial y}\left[\int M\,dx + g(y)\right]$$

or

$$N = \frac{\partial}{\partial y}\left(\int M\,dx\right) + g'(y) \tag{12}$$

We match the terms on both sides of Eq. (12) to isolate $g'(y)$ and integrate with respect to y to find $g(y)$. In isolating $g'(y)$ in Eq. (12), it is crucial that

* There are some technical conditions needed for the functions M and N if the test for exactness is to be valid. Namely, the functions together with their first partial derivatives M_x, M_y, N_x, and N_y must be continuous over a simply connected region in the xy-plane. We can think of a simply connected region as one that has "no holes and no boundary points"—like an open interval. Moreover, our argument has shown only that condition (10) is necessary for exactness. The proof of the sufficiency of Eq. (10) can be found in any standard text on multivariable calculus.

$g'(y)$ *be a function of only the variable* y. Otherwise the differential equation will not meet test (10) for exactness (or we will have made an error somewhere in our previous calculations when solving a specific problem). Let us summarize the method and illustrate it with some examples. (An alternate method is presented in problem 39 of Exercises 2.3.)

A METHOD FOR SOLVING EXACT EQUATIONS

Step 1. Write the first-order equation in differential form:

$$M(x, y)\,dx + N(x, y)\,dy = 0.$$

Step 2. Test for exactness: does

$$\frac{\partial M}{\partial y} = \frac{\partial N}{\partial x}?$$

Step 3. If exact, integrate the function M with respect to x or N with respect to y. For illustration we choose M:

$$\phi(x, y) = \int M\,dx + g(y).$$

Step 4. To find the unknown function $g(y)$:

a) Differentiate ϕ with respect to y and equate the result to N:

$$N = \frac{\partial}{\partial y}\left(\int M\,dx\right) + g'(y).$$

b) Integrate $g'(y)$ to find g.

Step 5. Write the implicit solution to the first-order equation:

$$\phi(x, y) = C.$$

Step 6. Evaluate the constant C if an initial condition is specified.

EXAMPLE 2 Solve the initial value problem

$$\frac{dy}{dx} = -\frac{3x^2 + 4xy}{2x^2 + 2y}, \quad y(0) = 3.$$

Solution. We carry out the steps in the method described above.

Step 1. The differential form of the above equation is

$$(3x^2 + 4xy)\,dx + (2x^2 + 2y)\,dy = 0.$$

Step 2. We know that the differential equation is exact because $\partial M/\partial y = 4x = \partial N/\partial x$.

Step 3. We choose to integrate M:

$$\begin{aligned}\phi &= \int M\,dx + g(y)\\ &= \int (3x^2 + 4xy)\,dx + g(y)\\ &= x^3 + 2x^2y + g(y).\end{aligned}$$

Step 4. Then $\partial\phi/\partial y = N$, so

$$0 + 2x^2 + \frac{dg}{dy} = 2x^2 + 2y$$

or

$$g'(y) = 2y.$$

Integration then gives us

$$g(y) = y^2.$$

We omit the constant of integration here because it would simply be absorbed into the arbitrary constant for the general solution in the next step.

Step 5. Therefore $\phi(x, y) = x^3 + 2x^2y + y^2$, so the implicit general solution is

$$x^3 + 2x^2y + y^2 = C.$$

Step 6. The initial condition $y(0) = 3$ yields $C = 9$. Thus the particular solution to the initial value problem is

$$x^3 + 2x^2y + y^2 = 9.$$

EXAMPLE 3 Solve the differential equation

$$(xe^y - e^{2y})\,dy + (e^y + x)\,dx = 0.$$

Solution. We will follow the steps in the above method without itemizing each step separately. The equation is already in differential form, so we move on to test for exactness:

$$\frac{\partial}{\partial x}(xe^y - e^{2y}) = e^y$$

and

$$\frac{\partial}{\partial y}(e^y + x) = e^y,$$

so the equation is exact. Then

$$\frac{\partial\phi}{\partial x} = e^y + x$$

implies that

$$\phi(x, y) = xe^y + \frac{x^2}{2} + g(y).$$

To determine $g(y)$ we use the fact that $\partial\phi/\partial y = N$:

$$\frac{\partial\phi}{\partial y} = xe^y + g'(y) = xe^y - e^{2y}.$$

Hence $g'(y) = -e^{2y}$, and integration gives us

$$g(y) = \frac{-e^{2y}}{2}.$$

The implicit general solution is then $\phi(x, y) = C$, or

$$xe^y + \frac{x^2}{2} - \frac{e^{2y}}{2} = C.$$

EXAMPLE 4 Show that

$$y\,dx + (2x - ye^y)\,dy = 0$$

is not exact, but that by multiplying the equation by the factor $\mu = y$ it becomes exact. Then use that fact to solve the equation.

Solution. To test for exactness we differentiate

$$\frac{\partial}{\partial y}(y) = 1 \quad \text{and} \quad \frac{\partial}{\partial x}(2x - ye^y) = 2.$$

Thus the differential equation fails to be exact. Multiplying the equation by the factor y results in

$$y^2\,dx + (2xy - y^2e^y)\,dy = 0.$$

Testing this last equation for exactness, we find that

$$\frac{\partial}{\partial y}(y^2) = 2y = \frac{\partial}{\partial x}(2xy - y^2e^y).$$

Since the exactness test is satisfied, this last differential equation can be solved by the exactness method:

$$\frac{\partial\phi}{\partial x} = y^2,$$

so that

$$\phi(x, y) = y^2x + g(y).$$

Next we have

$$\begin{aligned}\frac{\partial\phi}{\partial y} &= 2yx + g'(y) \\ &= 2xy - y^2e^y \qquad \text{(by the exactness condition).}\end{aligned}$$

Therefore

$$g'(y) = -y^2e^y.$$

Integration by parts yields

$$\begin{aligned}g(y) &= -y^2e^y + 2ye^y - 2e^y - \int 0 \cdot e^y\, dy \\ &= -e^y(y^2 - 2y + 2).\end{aligned}$$

The implicit general solution is then $\phi(x, y) = C$, or

$$y^2x - e^y(y^2 - 2y + 2) = C.$$

Let us check that our methodology has indeed produced a true solution. Implicit differentiation of the solution equation results in

$$2yy'x + y^2 - y'e^y(y^2 - 2y + 2) - e^y(2yy' - 2y') = 0.$$

Simplifying this equation algebraically leads to

$$y'(2x - ye^y) + y = 0,$$

which in differential form is

$$(2x - ye^y)\, dy + y\, dx = 0.$$

Since this is the original differential equation, our implicit solution does indeed check.

Integrating Factors

In Example 4 the function $\mu = y$ is called an **integrating factor** for the original nonexact differential equation. Generally any factor $\mu(x, y)$ that transforms a nonexact equation into an exact equation is an integrating factor. In fact, the integrating factor

$$\mu(x) = e^{\int P(x)dx}$$

makes the linear first-order equation

$$y' + P(x)y = Q(x)$$

exact. To see this, write the linear equation in differential form, multiply the result by μ, and apply the test for exactness. Thus

$$\mu\, dy + \mu P y\, dx = \mu Q\, dx$$

or

$$\mu\, dy + (\mu P y - \mu Q)\, dx = 0.$$

Applying the test for exactness we have

$$\frac{\partial}{\partial y}(\mu P y - \mu Q) = \mu P$$

and

$$\begin{aligned}\frac{\partial \mu}{\partial x} &= \frac{\partial}{\partial x}\left(e^{\int P(x)\,dx}\right)\\ &= e^{\int P(x)\,dx}\frac{d}{dx}\left[\int P(x)\,dx\right].\end{aligned}$$

From the Fundamental Theorem of Calculus,

$$\frac{d}{dx}\left[\int P(x)\,dx\right] = P(x),$$

so that

$$\frac{\partial \mu}{\partial x} = e^{\int P(x)\,dx}\,P = \mu P.$$

Thus the test for exactness is satisfied. The idea of multiplying the linear equation by a function $\mu(x)$ to transform it into an exact equation was first proposed in the 17th century by Gottfried Leibniz.

EXERCISES 2.3

In problems 1–10 find a differential equation $y' = g(x, y)$ that describes the slopes of the equipotential curves for the given function. No arbitrary constant C should appear in your differential equation.

1. $x^2 + 4y^2 = C$

2. $x^2 y = 1 + Cx$

3. $y = 4 + Ce^{2x}$

4. $y = Cx + 1 + C^2$

5. $y(Cx + y) = 1$

6. $xy = e^{Cx}$

7. $y = (C + \sin x)^2$

8. $y = \tan^{-1}(x + C)$

9. $y = x \sin(C - x)$

10. $x^2 = Cy + C^2$

In problems 11–20 place the differential equation in differential form and test for exactness. Do not solve the equation.

11. $y' = -\dfrac{x+2y}{y^2+2x}$

12. $y' = \dfrac{2y(y-x)}{x(x-4y)}$

13. $(x \sin y - y^2)y' = \cos y$

14. $(x - e^{x+y} \cos y)y' = e^{x+y} \sin y + 1$

15. $x\sqrt{x^2+y^2}\,\dfrac{dx}{dy} = \dfrac{x^2 y}{y - \sqrt{x^2+y^2}}$

16. $(x^2 y + y)y' = 1 + y^2$

17. $(3 + y + 2y^2 \sin^2 x)\dfrac{dx}{dy} = y \sin 2x - 2xy - x$

18. $y' = -\dfrac{2x + y\cos(xy)}{x \cos(xy)}$

19. $[\cos(x^2+y) - 3xy^2]y' + 2x\cos(x^2+y) - y^3 = 0$

20. $xe^y - e^{2y} + (e^y - x)\dfrac{dx}{dy} = 0$

In problems 21–30 the given differential equation is exact. Use the method presented in the text to find an implicit solution.

21. $(x - 2y)\,dx + (y^2 - 2x)\,dy = 0$

22. $(x^2 - 2xy)\,dy - (y^2 - 2xy + 1)\,dx = 0$

23. $(x \cos y - x^2)\,dy + (\sin y - 2xy + x^2)\,dx = 0$

24. $[y \sin(xy)]\,dx + [x \sin(xy)]\,dy = 0$

25. $(e^y \sin x + ye^{-x})\,dx - (e^y + e^{-x} + e^y \cos x)\,dy = 0$

26. $\left(\dfrac{y}{x} + \ln y\right)dx + \left(\dfrac{x}{y} + \ln x + e^y\right)dy = 0$

27. $(x + e^{-y} + x \ln y)\,dy + (y \ln y + e^x)\,dx = 0$

28. $[e^{2x} + x\sin(xy)]\,dy$
$$+\left[2ye^{2x} + y\sin(xy) + \frac{1}{x}\right]dx = 0$$

29. $x(3xy + 8y^2 + 2)\,dx + (x^3 + 8x^2 y + 9y^2)\,dy = 0$

30. $(y \cos x + \sin y - \cos x)\,dx$
$$+ (\sin x + x\cos y - e^y \sin y)\,dy = 0$$

In problems 31–35 the specified differential equation is exact. Solve the given initial value problem.

31. $(xy - y^2 + x)\,dx + \left(\dfrac{x^2}{2} - 2xy + y\right)dy = 0,$
$y(1) = 1$

32. $(ye^x + y)\,dx + (e^x + x + y)\,dy = 0, \quad y(0) = 1$

33. $(2y + e^{-x}\sin y)\,dy + (e^x + x + e^{-x}\cos y)\,dx = 0,$
$y(0) = \dfrac{\pi}{2}$

34. $(y^2 \cos x - 3x^2 y - e^x)$
$$+ (2y \sin x - x^3 + \ln y)y' = 0, \quad y(0) = 1$$

35. $(\cos x \cos y + y)y' + \tan x = \sin x \sin y, \quad y(0) = \dfrac{\pi}{2}$

36. Show that all first-order separable equations are exact. Is the converse true?

37. Determine a function $M(x, y)$ so that the following differential equation is exact:
$$M(x, y)\,dx + (x \sin y + \ln y - ye^y)\,dy = 0.$$

38. a) Show that the following differential equation is not exact:
$$(2y^2 + 3x)\,dx + 2xy\,dy = 0.$$
b) Show that if you multiply the differential equation by x the resulting equation is exact; then solve the equation.

39. Another method for solving the exact equation is as follows. From the differential form of the equation, integrate the functions M and N to find the potential function ϕ:
$$\phi = \int M\,dx + g(y) \tag{a}$$
and
$$\phi = \int N\,dy + h(x) \tag{b}$$
Compare expressions (a) and (b) to determine the functions $g(y)$ and $h(x)$. Then write the implicit solution to the first-order equation
$$\phi(x, y) = C.$$
Use this method to solve problems 21–35.

SUBSTITUTION METHODS

In this section we will present solution techniques for obtaining analytical solutions to three types of special nonlinear first-order equations. These techniques are substitution techniques that transform the special equation into a type of equation studied in earlier sections.

Homogeneous-type Equations

Consider the first-order equation

$$y' = \frac{y}{x} - \cos^2\left(\frac{y}{x}\right), \quad x > 0.$$

A simple check reveals that the equation is not separable, linear (in either variable x or y), or exact. Nevertheless, the form of the equation suggests that the substitution $v = y/x$ might work. Thus substituting $y = vx$ and

$$\frac{dy}{dx} = v + x\frac{dv}{dx}$$

into the equation yields

$$v + x\frac{dv}{dx} = v - \cos^2 v.$$

This last equation is separable in the variables v and x:

$$\sec^2 v \, dv = -\frac{dx}{x}.$$

Integration gives us

$$\tan v = -\ln x + C.$$

Finally, after replacing v with y/x we have the solution

$$\tan\left(\frac{y}{x}\right) = -\ln x + C.$$

For what type or class of differential equations might the substitution $v = y/x$ be fruitful? From the example above we might guess that, if the derivative dy/dx is a function of y/x, the substitution will lead to an equation we know how to solve. Let us see if that conjecture holds true.

Suppose then that the derivative satisfies

$$\frac{dy}{dx} = g\left(\frac{y}{x}\right); \tag{1}$$

that is, the derivative is a function of y/x. The form of Eq. (1) suggests the substitution

$$v = \frac{y}{x} \tag{2}$$

or

$$y = vx.$$

Differentiation gives us

$$\frac{dy}{dx} = v + x\frac{dv}{dx}. \tag{3}$$

Substitution of Eqs. (2) and (3) into Eq. (1) produces

$$v + x\frac{dv}{dx} = g(v). \tag{4}$$

Notice that Eq. (4) is separable no matter what the function g happens to be:

$$\frac{dv}{g(v) - v} = \frac{dx}{x}. \tag{5}$$

To solve Eq. (5) we simply integrate each side:

$$\int \frac{dv}{g(v) - v} = \int \frac{dx}{x}. \tag{6}$$

All that remains to be done to obtain the implicit solution in terms of the original variables x and y is to substitute y/x for v after performing the integrations in Eq. (6).

The class of first-order differential equations for which the derivative equals a function of y/x has a special name given by the following definition.

DEFINITION 2.3

A differential equation

$$\frac{dy}{dx} = f(x, y)$$

is said to be **homogeneous-type** (or simply **homogeneous**) if the function $f(x, y)$ can be expressed as a function $g(y/x)$ of the ratio y/x.

For example, the equations

$$\frac{dy}{dx} = \frac{y}{x} + 1, \tag{7}$$

$$\frac{dy}{dx} = \frac{y}{x} - \cos^2\left(\frac{y}{x}\right), \tag{8}$$

$$\frac{dy}{dx} = \frac{4y^3 - x^3}{3xy^2} = \frac{1}{3}\left[4\left(\frac{y}{x}\right) - \left(\frac{y}{x}\right)^{-2}\right] \tag{9}$$

are all homogeneous-type. However, the equation

$$\frac{dy}{dx} = \frac{y + 2xy}{x^2} = x^{-1}\left(\frac{y}{x}\right) + 2\left(\frac{y}{x}\right) \tag{10}$$

is *not* homogeneous-type because it cannot be written as a function only of y/x.

From the preceding discussion, any homogeneous-type Eq. (1) can be transformed to separable Eq. (5) by the substitutions $y = vx$ and $y' = v + xv'$ from Eqs. (2) and (3). The resulting separable equation can then be solved by integration. Finally, the substitution $v = y/x$ converts the general solution so obtained into an expression involving the original variables x and y.* Following is another example of the method.

EXAMPLE 1 Solve

$$2xy\,dy - (x^2 + 3y^2)\,dx = 0, \quad x > 0.$$

Solution. A quick check reveals that the equation is not separable, linear, or exact. It may not at first appear to be homogeneous-type, but dividing the equation through by x^2 yields

$$2\left(\frac{y}{x}\right) dy - \left(1 + \frac{3y^2}{x^2}\right) dx = 0$$

or

$$\frac{dy}{dx} = \frac{1 + 3y^2/x^2}{2y/x}.$$

Substituting $y = vx$ and $dy/dx = v + x\,dv/dx$ results in

$$v + x\frac{dv}{dx} = \frac{1 + 3v^2}{2v} = \frac{1}{2v} + \frac{3v}{2}.$$

Thus

$$x\frac{dv}{dx} = \frac{1}{2v} + \frac{v}{2} = \frac{1 + v^2}{2v}$$

or

$$\frac{2v\,dv}{1 + v^2} = \frac{dx}{x}.$$

Integration gives us

$$\ln(1 + v^2) = \ln x + C$$

* In 1691 Leibniz proposed the solution method for the homogeneous-type equation as described here.

since $x > 0$. Combining logarithms, we get

$$\ln\left(\frac{1+v^2}{x}\right) = C.$$

Taking the exponential of both sides and substituting $v = y/x$ then yields

$$1 + \left(\frac{y}{x}\right)^2 = C_1 x$$

or

$$x^2 + y^2 = C_1 x^3$$

where $C_1 = e^C$ is an arbitrary (positive) constant.

Bernoulli Equations

An equation of the form

$$y' + P(x)y = Q(x)y^n \tag{11}$$

is called a **Bernoulli equation.*** The Bernoulli equation is very similar in form to the linear first-order equation except that the righthand side contains the factor y^n. If $n = 1$, the Bernoulli equation is separable; if $n = 0$, it is linear. In general, however, the Bernoulli equation is not linear in the dependent variable y.

When $n \neq 0$ and $n \neq 1$, the substitutions

$$v = y^{1-n} \tag{12}$$

and

$$\frac{dv}{dx} = (1-n)y^{-n}\frac{dy}{dx} \tag{13}$$

transform the Bernoulli equation

$$y^{-n}\frac{dy}{dx} + P(x)y^{1-n} = Q(x)$$

into the form

$$\frac{dv}{dx} + (1-n)P(x)v = (1-n)Q(x). \tag{14}$$

* This form was first proposed by Jakob Bernoulli in 1695. In 1696 Leibniz showed that it could be reduced to a linear equation by substituting Eq. (12). Johann Bernoulli (Jakob's younger brother) gave another method of solution, and Jakob solved the equation in 1696 by reducing it to a separable equation.

Equation (14) is linear in the variable v and can be solved by the method presented in Section 2.2. Further substitution of $v = y^{1-n}$ into the solution produces the solution in terms of the original variables x and y.

EXAMPLE 2 Solve

$$\frac{dy}{dx} - y = e^{-x}y^2.$$

Solution. Substitute $v = y^{1-2} = y^{-1}$ and $dv/dx = -y^{-2}\,dy/dx$. Thus $dy/dx = -y^2\,dv/dx = -v^{-2}\,dv/dx$, and the equation becomes

$$-v^{-2}\frac{dv}{dx} - v^{-1} = e^{-x}v^{-2}$$

or

$$\frac{dv}{dx} + v = -e^{-x}.$$

This last equation is linear in v. The integrating factor is

$$\mu = e^{\int dx} = e^x.$$

Hence

$$\begin{aligned} e^x v &= -\int e^x e^{-x}\,dx + C \\ &= -x + C. \end{aligned}$$

Since $v = y^{-1}$, we obtain $y = 1/v$, and substitution results in

$$y = \frac{e^x}{C - x}.$$

We now check to see that y does indeed solve the original equation:

$$\frac{dy}{dx} = \frac{(C-x)e^x + e^x}{(C-x)^2}.$$

Thus

$$\begin{aligned} \frac{dy}{dx} - y &= \frac{e^x}{(C-x)^2} \\ &= e^{-x}\left(\frac{e^x}{C-x}\right)^2 \\ &= e^{-x}y^2. \end{aligned}$$

Riccati Equations

The nonlinear equation

$$\frac{dy}{dx} + P(x)y = Q(x)y^2 + R(x) \tag{15}$$

is known as the **Riccati equation.*** Notice that if $R(x) \equiv 0$ then the Riccati equation becomes a special case of the Bernoulli equation; if $Q(x) \equiv 0$ then Eq. (15) is a linear first-order equation. In applications a Riccati equation occurs in multidimensional control processes using dynamic programming techniques.†

Depending on the functions $P(x)$, $Q(x)$, and $R(x)$, the general solution to Eq. (15) may not be expressible analytically in terms of elementary functions. However, if y_1 is a *known* function that satisfies the Riccati equation, then the substitution

$$y = y_1 + \frac{1}{u} \tag{16}$$

transforms Eq. (15) into a first-order equation that is linear in u. Here $u = u(x)$ is an unknown function to be determined. The argument proceeds as follows.

Differentiation of Eq. (16) leads to

$$y' = y_1' - \frac{1}{u^2}u'. \tag{17}$$

Substituting Eqs. (16) and (17) into Eq. (15) results in

$$y_1' - \left(\frac{1}{u^2}\right)u' + P \cdot \left(y_1 + \frac{1}{u}\right) = Q \cdot \left(y_1^2 + \frac{2y_1}{u} + \frac{1}{u^2}\right) + R. \tag{18}$$

Since y_1 is a solution to Eq. (15),

$$y_1' + Py_1 = Qy_1^2 + R,$$

Eq. (18) reduces to the linear first-order equation

$$\frac{du}{dx} + [2y_1(x)Q(x) - P(x)]u = -Q(x). \tag{19}$$

* This equation is named after Count Jacopo Francesco Riccati (1676–1754), an Italian mathematician and philosopher. He studied this equation extensively when investigating curves whose curvature depends only on the dependent variable y (and not on x). Euler in 1760 proposed the substitution given by Eq. (16) to obtain a linear equation. Riccati's work was also important because he had the idea of reducing second-order equations to first order. We will present this idea in the treatment of higher-order differential equations in subsequent chapters.

† See R. Bellman, *Introduction to the Mathematical Theory of Control Processes* (New York: Academic Press, 1968).

Equation (19) can be solved by the method given in Section 2.2. Substituting $u^{-1} = y - y_1$ into the solution then expresses it in terms of the original variables x and y.

EXAMPLE 3 Solve

$$y' - \left(\frac{1}{x}\right)y = 1 - \left(\frac{1}{x^2}\right)y^2, \quad x > 0.$$

Solution. This is a Riccati equation in which $P(x) = -1/x$, $Q(x) = -1/x^2$, and $R(x) = 1$. It is easy to see that the function $y_1 = x$ is a solution to the equation. Then, from Eq. (19), substituting

$$y = x + \frac{1}{u}$$

leads to the linear first-order equation

$$\frac{du}{dx} + \left[2x\left(-\frac{1}{x^2}\right) - \left(-\frac{1}{x}\right)\right]u = \frac{1}{x^2}$$

or

$$\frac{du}{dx} - \left(\frac{1}{x}\right)u = \frac{1}{x^2}.$$

The integrating factor is

$$\mu = e^{\int -dx/x} = e^{-\ln x} = x^{-1}.$$

Thus

$$x^{-1}u = \int x^{-1}\left(\frac{1}{x^2}\right) dx$$
$$= -\frac{1}{2x^2} + C.$$

Therefore, after algebraic manipulation we find that

$$\frac{1}{u} = \frac{2x}{2Cx^2 - 1}.$$

Finally, substituting $u^{-1} = y - x$ into this last equation gives the solution in terms of the original variables

$$y = x + \frac{2x}{2Cx^2 - 1}$$

where C is an arbitrary constant.

EXERCISES 2.4

In problems 1–10 identify the equation as homogeneous-type, Bernoulli, linear, or Riccati. Do not solve the equation.

1. $\left(\dfrac{x}{y}\right)\dfrac{dy}{dx} + 1 = 3y^2$

2. $y' = \dfrac{xy}{x^2 - xy + y^2}$

3. $y' + x \sin x = \dfrac{y}{x}$

4. $y' = x^3 + \dfrac{2y}{x} + \dfrac{y^2}{x}$

5. $xy' = y(\ln x - \ln y)$

6. $(\cos x)\dfrac{dy}{dx} - y \sin x + y^2 = 0$

7. $y' + 2xy = e^{-x} + y^2$

8. $\dfrac{dy}{dx} + \dfrac{\tan y}{x \sin y} = 0$

9. $\dfrac{y}{x}\dfrac{dx}{dy} = xy^3 + y^2 - 1$

10. $(x + \sqrt{y^2 - xy})\,dy - y\,dx = 0$

In problems 11–20 solve the homogeneous-type equation using the method discussed in the text.

11. $2xy\dfrac{dy}{dx} = -(x^2 + y^2)$

12. $(x - y)\,dy + (x + y)\,dx = 0$

13. $x(y - x)\dfrac{dy}{dx} = x^2 + y^2$

14. $\dfrac{dy}{dx} = \dfrac{2y}{x - 2y}$

15. $(x - y)\,dx + x\,dy = 0$

16. $y' - \dfrac{y}{x} = \cos\left(\dfrac{y - x}{x}\right), \quad x > 0$

17. $xy' = y \ln y - y \ln x$

18. $xy' - y = xe^{y/x}$

19. $(y + \sqrt{x^2 - xy})\,dx - x\,dy = 0$

20. $(x - y \ln y + y \ln x)\,dx + x(\ln y - \ln x)\,dy = 0$

In problems 21–32 solve the given Bernoulli equation.

21. $y' - y = -y^2$

22. $y' - y = xy^2$

23. $x^2y' + 2xy = y^3, \quad x > 0$

24. $xy' + y = y^{-2}, \quad x > 0$

25. $y' = 2y - e^x y^2$

26. $xy' = x^2y^2 - y, \quad x > 0$

27. $xy\,dx + (x^2 - 3y)\,dy = 0$

28. $y' + xy^3e^{-2x} = y$

29. $xy' + y + x^2y^2e^x = 0, \quad x > 0$

30. $3xy^2y' - 3y^3 = x^4 \cos x, \quad x > 0$

31. $x^3y^2 - y' + xy = 0$

32. $x\,dy = y(xy - 1)\,dx, \quad x > 0$

In problems 33–38, solve the given Riccati equation. First verify that the given function is a particular solution to the differential equation. Then use the method presented in the text to find a more general solution containing one arbitrary constant.

33. $y' + 2y + y^2 = 0, \quad y_1 = -2$

34. $y' = xy^2 + (1 - 2x)y + x - 1, \quad y_1 = 1$

35. $y' = x^3(y - x)^2 + \dfrac{y}{x}, \quad y_1 = x, \quad x > 0$

36. $y' = x^3 + \dfrac{2y}{x} - \dfrac{y^2}{x}, \quad y_1 = -x^2, \quad x > 0$

37. $y' = 1 + \dfrac{y}{x} - \dfrac{y^2}{x^2}, \quad y_1 = x, \quad x > 0$

38. $y' = -\dfrac{2}{x} + \left(\dfrac{1}{x} - 2\right)y + y^2, \quad y_1 = 2, \quad x > 0$

39. Use the substitution $x + y = v$ to solve the following nonlinear equations.

a) $y' = (x + y)^2$

b) $y' = \sin(x + y)$

c) $y' = \dfrac{x + y + 4}{2x + 2y - 1}$

40. **Clairaut Equations**

The equation

$$y = xy' + \frac{1}{2}(y')^2$$

is an example of a **Clairaut equation.***

a) Show that the family of straight lines

$$y = Cx + \frac{1}{2}C^2$$

solves the differential equation where C is an arbitrary constant.

b) Show that $y = -x^2/2$ also solves the given equation. Note that this function is not part of the family of straight lines. It is called a singular solution (see part d).

c) More generally, show that the family of straight lines

$$y = Cx + f(C)$$

solves Clairaut's equation

$$y = xy' + f(y').$$

d) Show that another solution to the equation in part (c) may be given parametrically by

$$x = -f'(t),$$
$$y = f(t) - tf'(t)$$

where $t = y'$ and $f''(t) \neq 0$. This solution is the **singular solution.**

* Named after the French mathematician Alexis Claude Clairaut (1713–1765).

2.5 NUMERICAL APPROXIMATION: EULER'S METHOD

In this section we explore the geometric basis for constructing numerical solutions to first-order differential equations.* Recall from Section 1.1 that a numerical solution to a differential equation consists of a table of values of the dependent variable y for preselected values of the independent variable x. The y values are necessarily approximations to the true values of the solution function. The x values are customarily assigned by selecting an increment value Δx and "stepping forward" from some initial value of x to $x + \Delta x$, then to $x + 2\Delta x$, and so forth; for each step in x value we obtain the corresponding numerical approximation to the value of y. One method of this type that is basic to understanding all such numerical methods is Euler's method, which is revealing because of its geometric simplicity. To set a context for developing this method, we will first take a closer look at the information provided in the statement of an initial value problem.

Initial Value Problems

For the models developed in Chapter 1, we found an equation relating a derivative to some function of the independent and dependent variables.

* This section is adapted from UMAP Unit 625, which was written by the authors and David H. Cameron. The adaptation is presented with the permission of COMAP, Inc./UMAP, 60 Lowell Street, Arlington, MA 02174.

That is,

$$\frac{dy}{dx} = g(x, y)$$

where g is some function in which either x or y may not appear explicitly. Moreover, some starting value was given; that is, $y(x_0) = y_0$. Finally, we were interested in the values of y for a specific interval of x values, say $x_0 \leq x \leq b$. In summary, we determined **first-order initial value problems** of the general form

$$\frac{dy}{dx} = g(x, y), \qquad y(x_0) = y_0, \quad x_0 \leq x \leq b. \tag{1}$$

As we have seen from our previously developed models, initial value problems constitute an important class of problems. Let us examine the three parts of the model.

The Differential Equation $dy/dx = g(x, y)$ We are interested in finding a function $y = f(x)$ whose derivative satisfies the equation $dy/dx = g(x, y)$. Although we do not know f, we can compute its derivative given particular values of x and y. As a result we can describe the direction field associated with the differential equation, as discussed in Section 1.4.

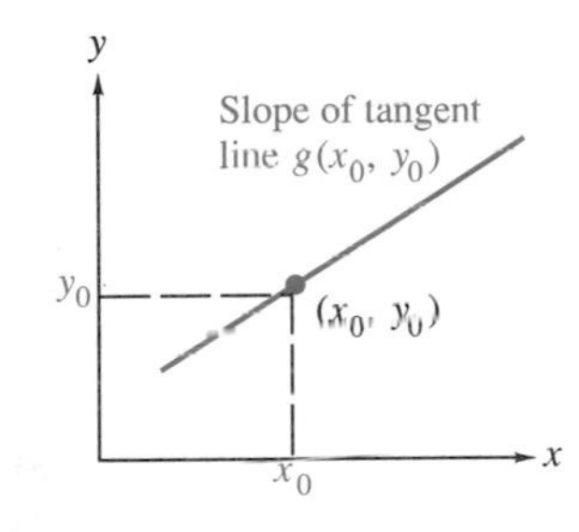

FIGURE 2.4

The solution curve passes through the point (x_0, y_0) and has a slope equal to $g(x_0, y_0)$.

The Initial Value $y(x_0) = y_0$ The initial value equation states that, for the initial value x_0 of the independent variable, we know the corresponding y value is $f(x_0) = y_0$. Geometrically this means that the point (x_0, y_0) lies on the solution curve (see Fig. 2.4). So we know where our solution curve begins. Moreover, from the differential equation $dy/dx = g(x, y)$ we know that the slope of the solution curve at (x_0, y_0) is the number $g(x_0, y_0)$. This is also depicted in Fig. 2.4.

The Interval $x_0 \leq x \leq b$ The condition $x_0 \leq x \leq b$ gives the particular interval of the x-axis with which we are concerned. Thus, we would like to relate y with x over the interval $x_0 \leq x \leq b$ by finding the solution function $y = f(x)$ passing through the point (x_0, y_0) with a slope of $g(x_0, y_0)$ at that point (Fig. 2.5). Note that the function $y = f(x)$ is continuous over $x_0 \leq x \leq b$ because its derivative exists there (recall Theorem 1.1).

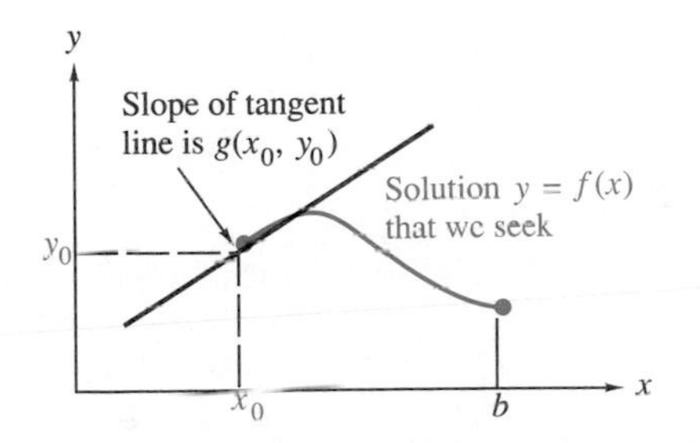

FIGURE 2.5

The solution $y = f(x)$ to the initial value problem is a continuous function over the interval from x_0 to b.

Approximating Solutions to Initial Value Problems

We shall now study a method that utilizes the three parts of the initial value problem, together with the geometrical interpretation of the derivative, to construct a sequence of discrete points in the plane that numerically approximate the points on the actual solution curve $y = f(x)$. We begin with an example we can solve analytically in order to compare the exact with the approximate results.

EXAMPLE 1 INTEREST COMPOUNDED CONTINUOUSLY

Consider the problem of determining the present or future value of a principal amount of money deposited in an account earning continuously compounded interest. (For background information, see Exercises 1.3, problem 5.) Let us investigate a numerical solution. Suppose at time $t = 0$, \$1000 is deposited at 7% annual interest compounded continuously. We want to know how much money will be in the account when $t = 20$ yr. If $Q(t)$ represents the amount of money in the account at any time t, then the (instantaneous) rate of change of Q is the interest rate times Q. Thus

$$\frac{dQ}{dt} = 0.07Q, \qquad Q(0) = 1000, \quad 0 \leq t \leq 20. \tag{2}$$

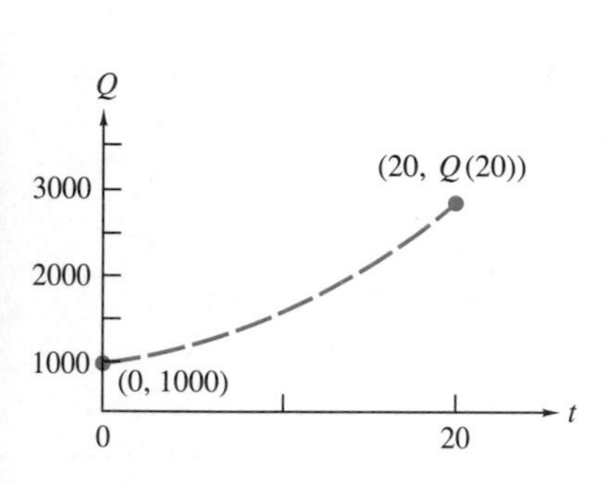

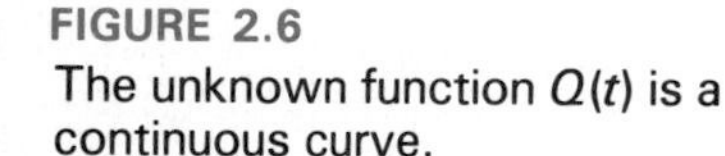

FIGURE 2.6
The unknown function $Q(t)$ is a continuous curve.

We want to find the function $Q(t)$ that solves this initial value problem. Figure 2.6 sketches in the above information with a dashed curve to represent the unknown function Q.

We know the derivative of $Q(t)$ for any known values of Q and t. In particular, we know that at the point (0, 1000) the derivative is $dQ/dt = 0.07(1000) = 70$. Since the derivative can be interpreted as the slope of the line tangent to the curve, we sketch a tangent line with slope 70 to our unknown function Q at $t = 0$ (Fig. 2.7).

Since we do not know $Q(t)$, we cannot determine exactly the value $Q(20)$. However, we can approximate it by the *value on the tangent line* when $t = 20$. Now the equation of the tangent line T in point-slope form is given by

$$T - 1000 = 70(t - 0).$$

In other words,

$$T = Q_0 + \left.\frac{dQ}{dt}\right|_{t=t_0} \Delta t$$

where $t_0 = 0$, $Q_0 = 1000$, $dQ/dt|_{t=0} = 70$, and $\Delta t = 20 - 0 = 20$. Thus

$$T(20) = 1000 + 70(20) = 2400.$$

Then we make the approximation

$$Q(20) \approx 2400 = T(20) = Q_1 \tag{3}$$

to the value of the unknown function at $t = 20$. Thus, starting with \$1000, we estimate that there will be \$2400 in the account when $t = 20$ yr if interest is compounded continuously at an annual rate of 7%.

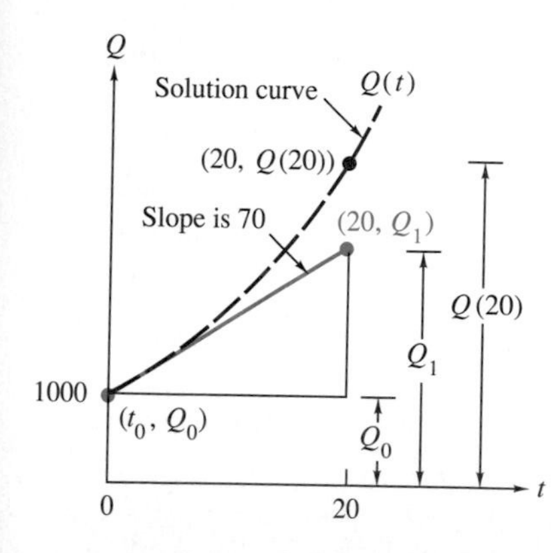

FIGURE 2.7
The point $(20, Q_1)$ on the tangent line T approximates the actual solution point $(20, Q(20))$.

Using Two Steps to Improve the Estimate

It should be emphasized that we have used the known starting value $Q(0) = 1000$ to calculate the estimate $Q(20) \approx 2400$. How can the approximation be improved to get a more accurate picture of the solution curve? The tangent-

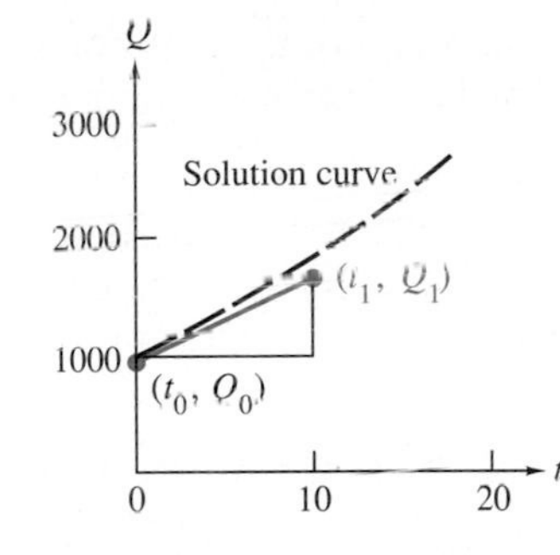

FIGURE 2.8
The point (t_1, Q_1) estimates the solution curve at the halfway value $t = 10$.

line procedure assumed that the derivative Q' is the constant 70 over the entire interval $0 \leq t \leq 20$, but actually the slope of the solution curve changes as Q and t change. It would be reasonable to expect the estimate of $Q(20)$ to be more accurate if another estimate were made at an intermediate point. Setting $\Delta t = 20/2 = 10$ in the same problem, we obtain

$$
\begin{aligned}
Q(10) \approx Q_1 = Q_0 + \left.\frac{dQ}{dt}\right|_{t=0} \Delta t \\
= 1000 + 70(10) = 1700.
\end{aligned}
$$

This intermediate estimate is depicted in Fig. 2.8.

Next we use the estimate $Q(10) \approx 1700$ to approximate the derivative at $t = 10$ from the formula

$$
\left.\frac{dQ}{dt}\right|_{t=10} = 0.07Q(10).
$$

Note an important difference between our calculations for $Q'(0)$ and $Q'(10)$:

> We know the value of the derivative at $t = 0$ *exactly* because we know $Q(0) = 1000$, but we must *estimate* the derivative at $t = 10$ because $Q(10) \approx 1700$ is only an estimate.

Now we calculate our estimate $Q(20)$:

$$
\begin{aligned}
Q(20) \approx Q_2 = Q_1 + \left.\frac{dQ}{dt}\right|_{t=10} \Delta t \\
= Q_1 + 0.07Q(10)\Delta t \\
\approx 1700 + 0.07(1700)(10) \\
= 2890.
\end{aligned}
$$

This two-step process is shown in Fig. 2.9. You will see later on that the

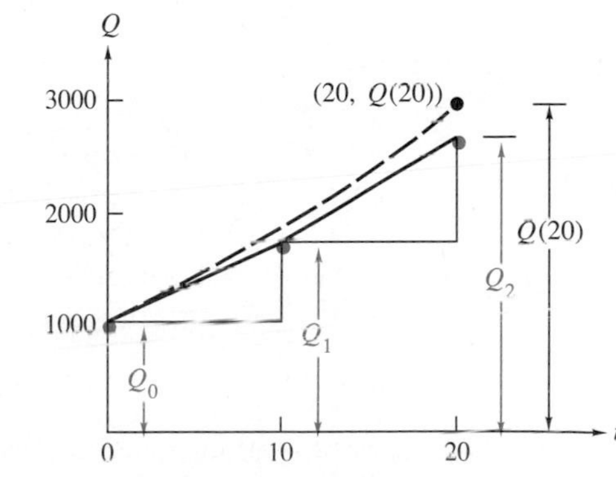

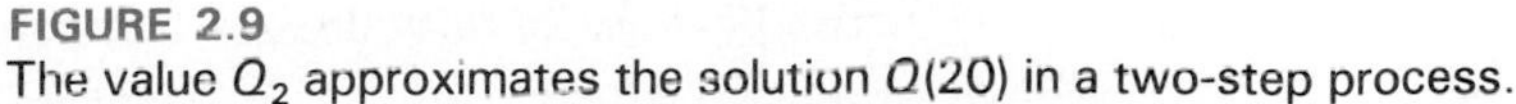

FIGURE 2.9
The value Q_2 approximates the solution $Q(20)$ in a two-step process.

approximation $Q(20) \approx 2890$ is closer to the actual value of the solution at $t = 20$. With more subdivisions we could expect further improvement of the approximation.

Euler's Method

Let us generalize the procedure given in Example 1 to the initial value problem

$$\frac{dy}{dx} = g(x, y), \qquad y(x_0) = y_0, \quad x_0 \leqslant x \leqslant b.$$

We approximate the solution through **Euler's method,** which involves building a **table of approximate values** to the solution in a step-by-step fashion as indicated in Fig. 2.10 and outlined below. Notice that an error is produced at each step.

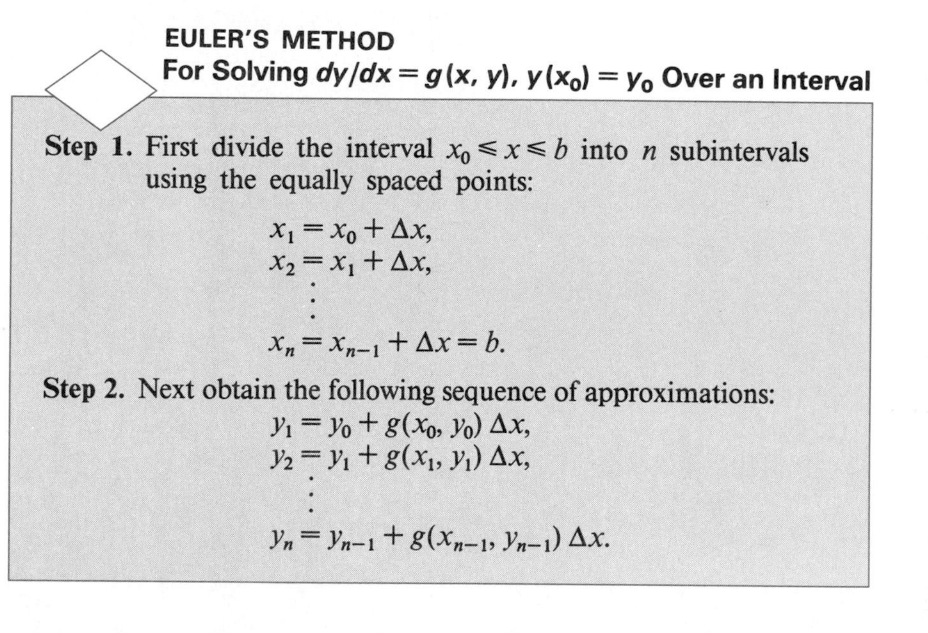

EULER'S METHOD
For Solving $dy/dx = g(x, y)$, $y(x_0) = y_0$ Over an Interval

Step 1. First divide the interval $x_0 \leqslant x \leqslant b$ into n subintervals using the equally spaced points:

$$\begin{aligned} x_1 &= x_0 + \Delta x, \\ x_2 &= x_1 + \Delta x, \\ &\vdots \\ x_n &= x_{n-1} + \Delta x = b. \end{aligned}$$

Step 2. Next obtain the following sequence of approximations:

$$\begin{aligned} y_1 &= y_0 + g(x_0, y_0)\, \Delta x, \\ y_2 &= y_1 + g(x_1, y_1)\, \Delta x, \\ &\vdots \\ y_n &= y_{n-1} + g(x_{n-1}, y_{n-1})\, \Delta x. \end{aligned}$$

Improving the Estimate of the Graph

Example 1 showed us that step size is important in making approximations. Since there was a significant change in the estimate of $Q(20)$ (from 2400 to 2890) when we reduced the step size Δt in the compound interest example, we should not be too confident in our results at this point. However, Euler's method can easily be coded for computer implementation to facilitate re-

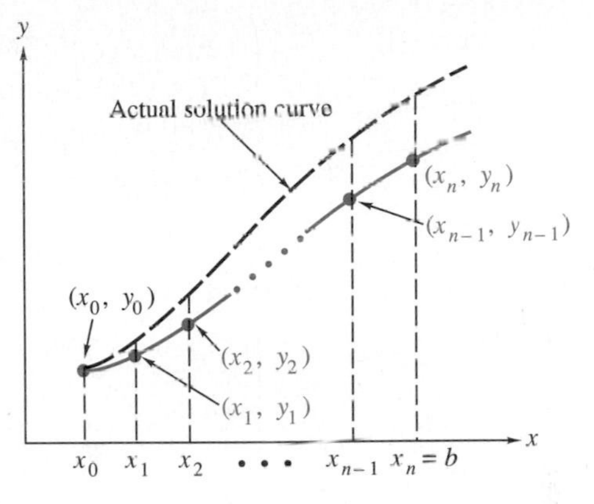

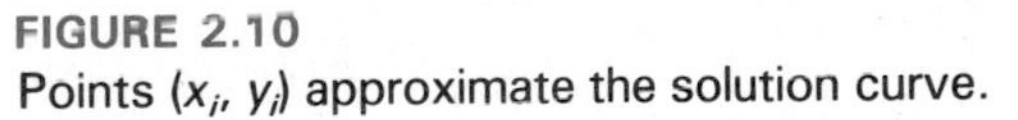
FIGURE 2.10
Points (x_i, y_i) approximate the solution curve.

ductions in step size. Using a calculator program, we apply Euler's method using step sizes of $\Delta t = 20$, $\Delta t = 10$, and $\Delta t = 1$ to obtain approximations for Q at every integer value of t between 0 and 20. To get an idea of what the unknown solution function looks like, we plot the points obtained from the various step sizes on a single graph (Fig. 2.11).

It might be tempting to reduce the step size even further in order to obtain greater accuracy, as suggested by Fig. 2.11. However, each additional calculation not only requires additional computer time but more importantly

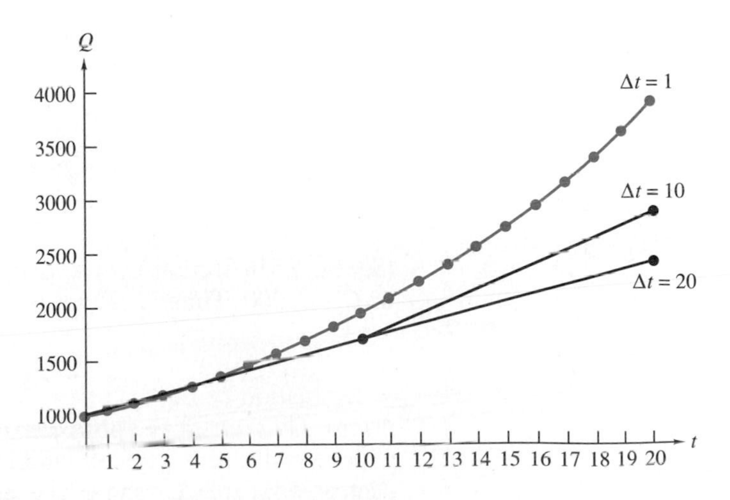

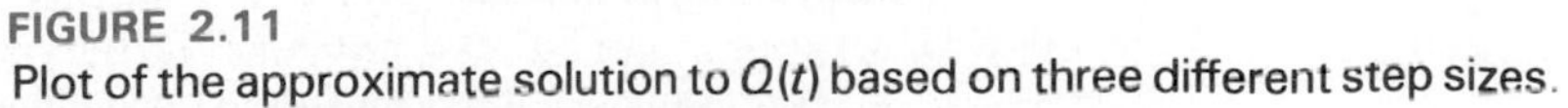
FIGURE 2.11
Plot of the approximate solution to $Q(t)$ based on three different step sizes.

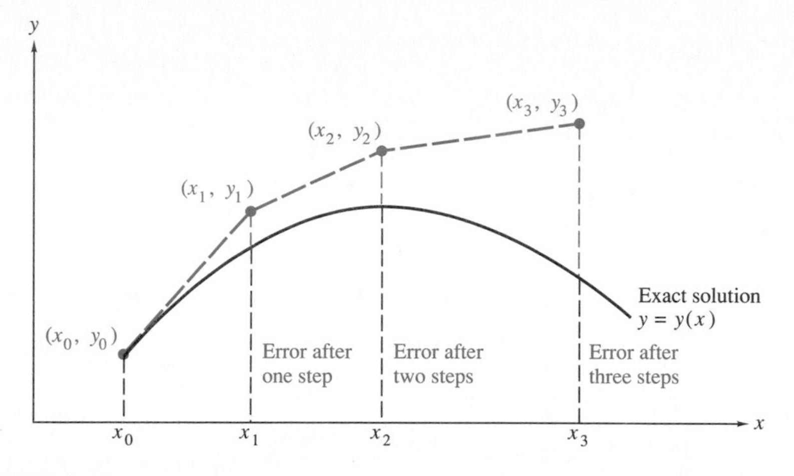

FIGURE 2.12
Errors arise at each step of Euler's method, and the approximations y_k may move further away from the exact solution.

introduces roundoff error (see Fig. 2.12). Since these errors tend to accumulate, an ideal method would improve the theoretical accuracy of the approximation yet minimize the number of calculations. We will briefly discuss errors later in this section. In order to evaluate the accuracy of the method in this example, we will now find the analytic solution. Of course, normally a numerical approximation technique is used only when the solution cannot be found analytically.

Finding the Solution Analytically

After separating the variables and integrating the initial value problem

$$\frac{dQ}{dt} = 0.07Q, \qquad Q(0) = 1000, \quad 0 \leq t \leq 20,$$

we obtain the general solution

$$Q = C_1 e^{0.07t}. \tag{4}$$

(See Eq. 25 in Section 1.3.) Applying the initial condition $Q(0) = 1000$ gives us $C_1 = 1000$. Thus

$$Q = 1000e^{0.07t}. \tag{5}$$

Evaluation of Eq. (5) at $t = 20$ yields $Q(20) = 4055.20$ to two decimal places. Thus our three approximations for $Q(20)$ (2400 with $\Delta t = 20$, 2890 with $\Delta t = 10$, and 3869.68 with $\Delta t = 1$) do become more accurate as Δt decreases.* Table 2.1 shows the numerical approximation of values com-

* Calculations for this example were done with 13 digits and rounded to two decimal places for presentation.

TABLE 2.1

Comparison of the Numerical Solution by Euler's Method with Actual Solution Curve

$$\frac{dQ}{dt} = 0.07Q, \quad Q(0) = 1000, \quad \Delta t = 1$$

t	Euler $Q(t)$	Actual $Q(t)$	Absolute Error	t	Euler $Q(t)$	Actual $Q(t)$	Absolute Error
0	1000.00	1000.00	0.00	10	1967.15	2013.75	46.60
1	1070.00	1072.51	2.51	11	2104.85	2159.77	54.92
2	1144.90	1150.27	5.37	12	2252.19	2316.37	64.18
3	1225.04	1233.68	8.64	13	2409.85	2484.32	74.47
4	1310.80	1323.13	12.33	14	2578.53	2664.46	85.93
5	1402.55	1419.07	16.52	15	2759.03	2857.65	98.62
6	1500.73	1521.96	21.23	16	2952.16	3064.85	112.69
7	1605.78	1632.32	26.54	17	3158.82	3287.08	128.26
8	1718.19	1750.67	32.48	18	3379.93	3525.42	145.49
9	1838.46	1877.61	39.15	19	3616.53	3781.04	164.51
				20	3869.68	4055.20	185.52

puted by Euler's method when $\Delta t = 1$ compared with the actual solution values computed from Eq. (5). Notice that the errors in this example accumulate. Fig. 2.11 shows the approximation to the solution curve produced by our three step sizes.

Error

How does your calculator or computer (which performs only arithmetic operations) compute sin x, and how accurate is the result? From your previous studies of calculus, recall the Maclaurin series representation for sin x:

$$\sin x = x - \frac{x^3}{3!} + \frac{x^5}{5!} - \frac{x^7}{7!} + \cdots + \frac{(-1)^n x^{2n+1}}{(2n+1)!} + \cdots. \tag{6}$$

Let us use the series to estimate sin $\pi/6 = 0.5$ and compute the absolute value of the error by consecutively considering the first term, the sum of the first two terms, the sum of the first three terms, and finally the sum of the first four terms. (Try it on your computational device and compare the results.)

Number of terms	1	2	3	4
Absolute error (approximately)	0.0236	0.0004	10^{-5}	10^{-6}

What are the sources of error in these calculations? First, error is introduced in terminating the infinite series of Eq. (6) after only a finite number of terms.

This is called **truncation error.** For a convergent series, truncation error equals the sum of the discarded terms and it can sometimes be accurately estimated. More important here is the fact that the theoretical truncation error decreases as more and more terms are considered in the sum.

A second type of error is introduced by the computational device itself. A machine has only a finite number of available digit positions to represent all possible numbers. When a machine performs arithmetic operations, often using approximate representations of numbers, machine or **roundoff error** is introduced. For example, consider a calculator or computer that uses eight-digit arithmetic. Then the number ⅓ is represented by the decimal .33333333, and $3 \times$ ⅓ is the number .99999999 instead of the actual value 1. The error of 10^{-8} is due to rounding off. The ideal real number for ⅓ is an *infinite* string of decimal digits .33333 . . ., but any calculator or computer can do arithmetic only with numbers having finite precision. When many arithmetic operations are performed in succession, each with its own roundoff error, the accumulated effect of rounding off can be significant and alter the results of the actual numbers that are supposed to be the answer. The problem of dealing with roundoff error is a central issue when using numerical solution methods. To reduce roundoff error, mathematicians develop techniques to minimize the number of arithmetic operations required to calculate a desired result.

Let us explore how combined truncation and roundoff errors work in Euler's method. From Taylor's theorem in elementary calculus, the function $f(x)$ satisfies the series

$$f(x_0 + \Delta x) = f(x_0) + f'(x_0)\,\Delta x + \left[\frac{f''(x_0)}{2}(\Delta x)^2 + \cdots\right]$$

under suitable conditions on f.* If we disregard the infinitely many terms in the bracket, we obtain the approximation used in Euler's method. To reduce this truncation error, we reduce the size of Δx. However, as we reduce Δx, the number of steps required to reach a fixed abscissa $b > x_0$ increases, requiring more computations. Thus, while the truncation errors decrease as Δx is reduced, roundoff errors may eventually cause the total error to increase. This idea is shown in Fig. 2.13 where $n = (b - x_0)/\Delta x$ is the minimum number of steps required to reach b from the starting value x_0.

Euler's method has a relatively high truncation error requiring a large number of steps to reduce those effects. But the larger number of steps means that roundoff error becomes more significant. In Section 2.6 we will discuss two other methods that produce better results.

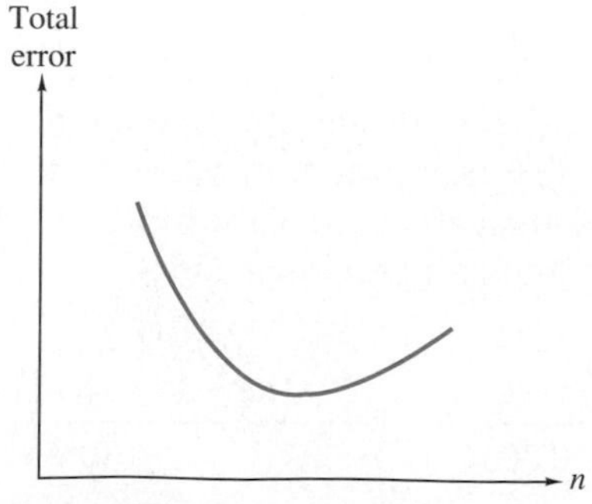

Total error

n

FIGURE 2.13

Roundoff error eventually increases as the number of calculations increases.

* The function f must be differentiable at x_0 for all orders; moreover, the remainder term $f^{(n)}(\zeta)(\Delta x)^n/n!$, where ζ is some point satisfying $x_0 < \zeta < x_0 + \Delta x$, must go to zero as $n \to \infty$.

EXERCISES 2.5

1. When interest is compounded, the interest earned is added to the principal amount so that it may also earn interest. For a one-year period, the principal amount P is given by

$$P = \left(1 + \frac{i}{n}\right)^n P(0)$$

where i is the annual interest rate (given as a decimal), and n is the number of times during the year that the interest is compounded.

To lure depositors banks offer to compound interest at different intervals: semiannually, quarterly, or daily. A certain bank advertises that it compounds interest *continuously*. If \$100 is deposited initially, formulate a mathematical model that describes the growth of the initial deposit during the first year. Assume an annual interest rate of 10%.

2. Use the differential equation model formulated in problem 1 to answer the following.

a) From the derivative evaluated at $t = 0$, determine an equation of the tangent line T passing through the point (0, 100).

b) Estimate $Q(1)$ by finding $T(1)$, where $Q(t)$ denotes the amount of money in the bank at time t (assuming no withdrawals).

c) Estimate $Q(1)$ using a step size of $\Delta t = 0.5$.

d) Estimate $Q(1)$ using a step size of $\Delta t = 0.25$.

e) Plot the estimates you obtained for $\Delta t = 1.0$, 0.5, and 0.25 to approximate the graph of $Q(t)$.

3. a) For the differential equation model obtained in problem 1, find $Q(t)$ by separating the variables and integrating.

b) Evaluate $Q(1)$.

c) Compare your previous estimates of $Q(1)$ with its actual value.

d) Find the effective annual interest rate computed in part (c) with interest compounded:

i) Semiannually: $\left(1 + \frac{0.10}{2}\right)^2$

ii) Quarterly: $\left(1 + \frac{0.10}{4}\right)^4$

iii) Daily: $\left(1 + \frac{0.10}{365}\right)^{365}$

e) Estimate the limit of $(1 + 0.10/n)^n$ as $n \to \infty$ by evaluating the expression for $n = 1000$, 10,000, 100,000.

f) What is $\lim_{n\to\infty} (1 + 1/n)^n$?

4. Using Euler's method with a step size of $\Delta x = 0.1$, determine a table of approximate values for the solution to the initial value problem

$$y' = y + e^x - 2, \quad y(0) = 2,$$

when $x = 0.1$, $x = 0.2$, $x = 0.3$, and $x = 0.4$. Solve the linear first-order equation analytically, and compare your results.

5. Using Euler's method with a step size of $\Delta x = 0.2$, estimate the solution value $y(1)$ to the initial value problem

$$y' = e^{x+y}, \quad y(0) = 0.$$

Solve the equation analytically using one of the methods presented in this chapter, and compare your results.

6. Using Euler's method with a step size of $\Delta x = 0.5$, estimate the solution value $y(3)$ to the initial value problem

$$(2x + y)\,dx + (x + 2y)\,dy = 0, \quad y(1) = 0.$$

Solve the equation analytically. Do you have much confidence in your numerical solution? Explain your conclusion.

TOOLKIT PROGRAMS

First Order Initial Value Problem

2.6

NUMERICAL APPROXIMATION: IMPROVED EULER'S AND RUNGE–KUTTA METHODS (Optional)

In Section 2.5 the interpretation of the derivative as the slope of the line tangent to the curve was used to formulate Euler's method for numerically approximating solutions to first-order initial value problems. The method converges slowly, and reductions in the step size produce undesirable round-off errors. In this section we will again use the interpretation of the derivative as the slope of the tangent line to discover techniques that produce better approximations for a given number of calculations.

Finding an Average Value for the Derivative

In Euler's method we estimated the "rise" in the solution function being approximated by estimating the slope (derivative) at the left endpoint and multiplying by the "run" (the step size, or length of the interval over which the approximation is taking place). For example, suppose we know $f(a)$ and are attempting to approximate the differentiable function f over $[a, b]$. In the one-step method we estimate $f(b)$ by

$$f(b) \approx f(a) + f'(a)(b - a). \tag{1}$$

Let us analyze this estimate in view of the Mean Value Theorem for differential calculus.

> **MEAN VALUE THEOREM**
>
> If a function f is continuous on the closed interval $[a, b]$ and differentiable on the open interval (a, b), then there is at least one point c in (a, b) such that
>
> $$f(b) = f(a) + f'(c)(b - a). \tag{2}$$

The geometric interpretation of the Mean Value Theorem is illustrated in Fig. 2.14. From the figure and from Eq. (2) you can see that the slope of the curve $y = f(x)$ at $x = c$ equals the slope of the secant line joining the endpoints A and B on the curve. That is, there exists an *average derivative* that, if found, could be used to precisely calculate $f(b)$. We may interpret $f'(a)$ in Eq. (1) as an *estimate* of the average derivative in the interval $[a, b]$. We would not expect it to be a very good estimate since it is computed at the left endpoint and the derivative is unlikely to be constant throughout the entire interval. Thus we seek to improve our estimate of the average derivative over the interval $[a, b]$. We will do this in the Improved Euler's method by estimating the derivative at more points in the interval $[a, b]$, and in the Runge–Kutta methods by using a weighted averaging process.

Limitations with Euler's Method

Consider again the continuous-interest problem explored in Section 2.5. We are interested in determining how much money will be in the account after 20 yr starting with an initial investment of \$1000 continuously compounded

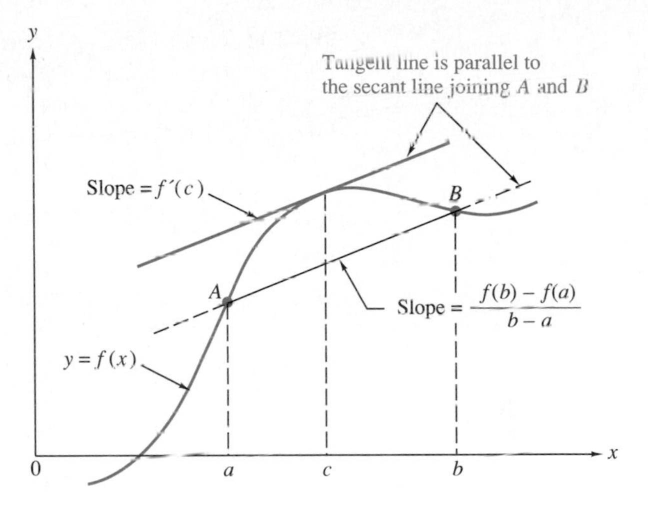

FIGURE 2.14
The Mean Value Theorem states that there is at least one point c between a and b for which the slope of the curve $f'(c)$ equals the slope of the secant line joining the points A and B.

at an annual interest rate of 7%. Letting $Q(t)$ represent the amount of money in the account at any time t yields the model

$$\frac{dQ}{dt} = 0.07Q, \qquad Q(0) = 1000, \quad 0 \leq t \leq 20. \tag{3}$$

In using Euler's method in Section 2.5, we found the value of the derivative at the point (0, 1000) to be 70. We then assumed that the derivative dQ/dt was constant over the interval [0, 20] and used the point-slope form of a line to estimate the value at $t = 20$ yr. The next improvement we presented reduced the step size Δt by one-half (from 20 to 10) and computed the estimate of $Q(20)$ in two steps. Finally, we reduced Δt to 1 and estimated $Q(20)$ using 20 steps. We noticed in all cases that an error was produced at each step and that the errors accumulated with each step.

Improved Euler's Method with One Step

The Improved Euler's method estimates the average derivative over an interval by averaging the estimates of the derivative computed at the left and right endpoints.

As we saw with the "unimproved" Euler's method (which we refer to throughout simply as "Euler's method"), we can calculate the value of the slope at the point (0, 1000) for the initial value problem. We then use that slope value to determine an equation of the tangent line T in point-slope form:

$$T - 1000 = 70(t - 0).$$

Thus

$$T = Q_0 + m_0 \, \Delta t$$

where $Q_0 = 1000$, $m_0 = dQ/dt|_{t=0} = 70$, and $\Delta t = t - 0$. When $t = 20$, we have,

$$\begin{aligned} T(20) &= 1000 + 70(20) \\ &= 2400. \end{aligned}$$

Then we make the approximation

$$T(20) = 2400 \approx Q(20)$$

to the value of the unknown function Q at $t = 20$. See Fig. 2.15 for a geometrical interpretation.

Averaging the Slopes at the Endpoints At this stage we have approximated $Q(20)$ using Euler's method. We can use the approximation to estimate the slope at $t = 20$. That is,

$$m_1 = \left.\frac{dQ}{dt}\right|_{t=20} = 0.07Q(20) \approx 0.07\,(2400) = 168.$$

Now we have values for the slope at both endpoints of the interval [0, 20]: $m_0 = 70$ and $m_1 = 168$ for the left and right endpoints, respectively. Euler's method uses only the left endpoint value. Can we get a better approximation of the **average derivative** in the interval [0, 20]? Intuitively, if we use both of our slopes the approximation should be better. Let us incorporate both slope

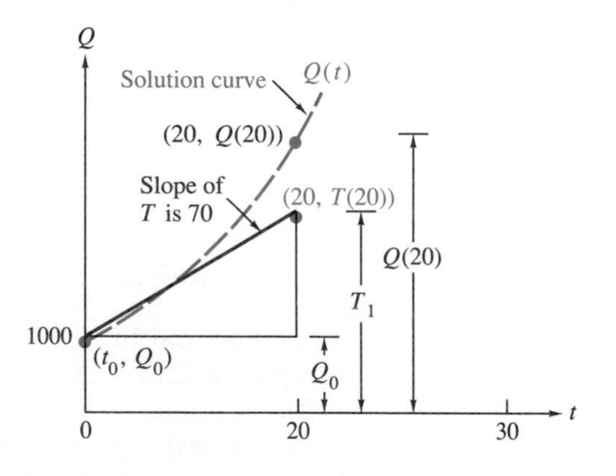

FIGURE 2.15
The point $(20, T(20))$ on the tangent line approximates the actual solution point $(20, Q(20))$.

values by taking their simple average:

$$\begin{aligned} m_{\text{aver}} &= \frac{m_0 + m_1}{2} \\ &= \frac{70 + 168}{2} \\ &= 119. \end{aligned}$$

Predicting the Next Q Value We now use this average as the slope of a line passing through the point (0, 1000). Then we use this new line to predict the value of $Q(20)$. We call this approximate value Q_1, so for a one-step approximation, we have

$$Q_1 = Q_0 + m_{\text{aver}}\, \Delta t \tag{4}$$

where $Q_0 = 1000$, $m_{\text{aver}} = (m_0 + m_1)/2$, and $\Delta t = t - 0$. When $t = 20$ we have

$$Q_1 = 1000 + 119(20) = 3380. \tag{5}$$

Now we can make the approximation

$$Q(20) \approx Q_1 = 3380.$$

Figure 2.16 presents a geometrical interpretation of the Improved Euler's method.

Comparing Euler's and the Improved Euler's Methods Let us interpret result (5): the Improved Euler's method with one step estimates that, at the end of year 20, $3380 will be in the account when interest is compounded continuously at an annual interest rate of 7%. Recall that the

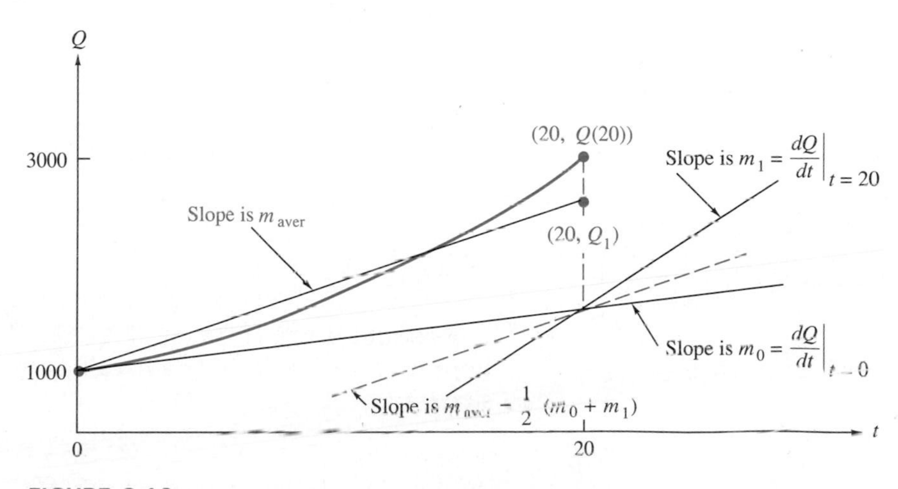

FIGURE 2.16

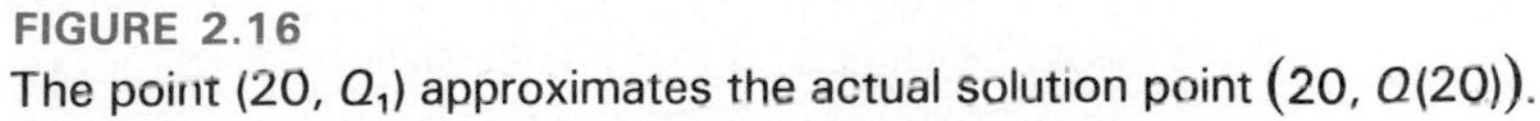
The point $(20, Q_1)$ approximates the actual solution point $(20, Q(20))$.

approximation for $Q(20)$ was \$2400 using Euler's method with a step size of $\Delta t = 20$. In Section 2.5 we calculated the solution analytically and found the exact value to be \$4055.20 rounded to two decimal places. Thus, comparing the approximations for Euler's and the Improved Euler's methods you see there is a significant improvement in using the Improved Euler's method.

With the Improved Euler's method notice that we estimated $Q(20)$ twice. First we *predicted* the value for $Q(20)$ using Euler's method. Then we used that value to *estimate* the righthand slope at $Q(20)$. Averaging the right- and lefthand slopes, we improved our estimate of the average derivative to *correct* the original estimate of $Q(20)$.

Improved Euler's Method with Two Steps

In Section 2.5 we improved our estimate for $Q(20)$ in Euler's method by reducing the step size Δt. This observation suggests we try the same idea for the Improved Euler's method. Let us first make an estimate at the halfway point in our example when $t = 10$. For $\Delta t = 20/2 = 10$ we compute the Improved Euler's estimate for $Q(10)$:

$$Q(10) \approx Q_1 = Q_0 + m_{\text{aver}}\,\Delta t.$$

Calculation of m_{aver} We know the value of the slope of the tangent line at $t = 0$ is 70. As before, we can use the equation of the tangent line at $t = 0$ to compute the Euler's method estimate of $Q(10)$:

$$\begin{aligned} Q(10) &\approx Q_0 + \left.\frac{dQ}{dt}\right|_{t=0} \Delta t \\ &= 1000 + 70(10) \\ &= 1700. \end{aligned}$$

We then use this estimate to approximate the slope at the right endpoint when $t = 10$:

$$m_1 = 0.07(1700) = 119.$$

Next we compute the average slope:

$$m_{\text{aver}} = \frac{m_0 + m_1}{2} = \frac{70 + 119}{2} = 94.5$$

Predicting the Halfway Q Value Using this average slope we now improve our estimate of $Q(10)$, calling the improved value Q_1 for the two-step Improved Euler's method:

$$\begin{aligned} Q(10) \approx Q_1 &= Q_0 + m_{\text{aver}}\,\Delta t \\ &= 1000 + 94.5(10) \\ &= 1945. \end{aligned}$$

We are going to estimate $Q(20)$ in a similar manner. That is,

$$Q(20) \approx Q_2 = Q_1 + m_{\text{aver}}\,\Delta t \tag{6}$$

where m_{aver} now represents the average of the estimates to the slopes at $t = 10$ and $t = 20$.

Calculation of the Next m_{aver} For our estimate of the slope at the left endpoint $t = 10$, we use our estimate of $Q(10)$ and the differential equation (3):

$$m_0 = \left.\frac{dQ}{dt}\right|_{t=10} \approx 0.07(1945) = 136.15.$$

To estimate the slope at $t = 20$, we first compute the Euler's method estimate of $Q(20)$:

$$\begin{aligned} Q(20) &\approx Q_1 + \left.\frac{dQ}{dt}\right|_{t=10} \Delta t \\ &= 1945 + 136.15(10) \\ &= 3306.50. \end{aligned}$$

We then use this approximation to estimate the value of the slope at $t = 20$:

$$m_1 = 0.07(3306.50) = 231.455.$$

We improve our estimate by averaging the left- and righthand estimates for the slope to obtain

$$m_{\text{aver}} = \frac{m_0 + m_1}{2} = \frac{136.15 + 231.455}{2} = 183.803$$

Predicting the Next Q Value Finally, we compute the Improved Euler's estimate by correcting our approximation to $Q(20)$:

$$\begin{aligned} Q(20) \approx Q_2 &= Q_1 + m_{\text{aver}}\,\Delta t \\ &= 1945 + 183.803(10) \\ &= 3783.03. \end{aligned}$$

We compare the two-step estimate of \$3783.03 and the one-step estimate of \$3380 with the analytic solution of \$4055.20. As expected, the two-step Improved Euler's method is an improvement over the one-step Improved Euler's method.

Improved Euler's Method with n Steps

Now we will generalize the two-step procedure to an n-step procedure and place it in algorithmic form to facilitate computer coding. Given the initial value problem

$$\frac{dy}{dx} = g(x, y), \qquad y(x_0) = y_0, \quad x_0 \leq x \leq b,$$

we approximate a solution numerically using the following procedure. In this way a table of approximate values to the solution is built up in a step-by-step fashion. The Improved Euler's method is easily coded for computer implementation to facilitate reductions in the step size.

IMPROVED EULER'S METHOD
For Solving $dy/dx = g(x, y)$, $y(x_0) = y_0$ Over an Interval

Step 1. First divide the interval $x_0 \leq x \leq b$ into n subintervals using equally spaced points:

$$x_1 = x_0 + \Delta x,$$
$$x_2 = x_1 + \Delta x,$$
$$\vdots$$
$$x_n = x_{n-1} + \Delta x = b.$$

Step 2. Next obtain the following sequence of approximations:

$$y_1 = y_0 + [g(x_0, y_0) + g(x_1, y_1^*)] \frac{\Delta x}{2},$$

$$y_2 = y_1 + [g(x_1, y_1) + g(x_2, y_2^*)] \frac{\Delta x}{2},$$

$$\vdots$$

$$y_n = y_{n-1} + [g(x_{n-1}, y_{n-1}) + g(x_n, y_n^*)] \frac{\Delta x}{2},$$

where $y_n^* = y_{n-1} + [g(x_{n-1}, y_{n-1})] \Delta x$.

Applying the Improved Euler's Method

Table 2.2 shows the numerical approximation values computed by the Improved Euler's method for our continuous interest example. Note that, although the absolute error increases with each step, it does so much more slowly than with Euler's method (see Table 2.1), thus representing a significant improvement.

We now present the numerical method for solving first-order equations that is most often used in practice. For our continuous interest example the approximations this method produces will equal the actual solution values rounded to the nearest penny with a step size of $\Delta t = 1$.

TABLE 2.2

Comparison of the Numerical Solution by Improved Euler's Method with Actual Solution Curve

$\frac{dQ}{dt} = 0.07Q, \quad Q(0) = 1000, \quad \Delta t = 1$

t	Improved Euler $Q(t)$	Actual $Q(t)$	Absolute Error	t	Improved Euler $Q(t)$	Actual $Q(t)$	Absolute Error
0	1000.00	1000.00	0.00	10	2012.66	2013.75	1.09
1	1072.45	1072.51	0.06	11	2158.48	2159.77	1.29
2	1150.15	1150.27	0.12	12	2314.86	2316.37	1.51
3	1233.48	1233.68	0.20	13	2482.57	2484.32	1.75
4	1322.84	1323.13	0.29	14	2662.43	2664.46	2.03
5	1418.68	1419.07	0.39	15	2855.33	2857.65	2.32
6	1521.47	1521.96	0.49	16	3062.20	3064.85	2.65
7	1631.70	1632.32	0.62	17	3284.05	3287.08	3.03
8	1749.91	1750.67	0.76	18	3521.98	3525.42	3.44
9	1876.69	1877.61	0.92	19	3777.15	3781.04	3.89
				20	4050.80	4055.20	4.40

Runge-Kutta Methods

When we used the average of the estimates of the derivatives at both endpoints of the interval rather than only the estimate at the left endpoint, we improved the approximation to the solution. How can we make further improvements? Perhaps if we obtain additional estimates of the derivative at one or more *interior points* of the interval, we will see even further improvement. A class of approximation techniques that estimate the derivatives at various points in the interval and then compute a **weighted-average derivative** is the **Runge–Kutta methods**, named after the two German mathematicians who developed them, Carl D. T. Runge (1856–1927) and W. Kutta (1867–1944).

The Runge–Kutta methods are classified by order, where the order depends on the number of slope estimates used at each step. We begin by examining a second-order method. Again assuming that $y' = g(x, y)$, we set the nth approximation equal to

$$y_n = y_{n-1} + aK_1 + bK_2, \tag{7}$$

where

$$K_1 = g(x_{n-1}, y_{n-1})\,\Delta x, \tag{8}$$
$$K_2 = g(x_{n-1} + \alpha\Delta x, y_{n-1} + \beta K_1)\,\Delta x. \tag{9}$$

Here a, b, α, and β are constants satisfying the restrictions

$$a + b = 1, \quad b\alpha = ½, \quad \text{and} \quad b\beta = ½. \tag{10}$$

Note that K_1 and K_2 are estimates of the rise since they equal slope estimates multiplied by the step size. The constants a and b allow for assigning various weights to those estimates. If you substitute $a = ½$, $b = ½$, $\alpha = 1$, and $\beta = 1$, you will notice that the Improved Euler's method is a particular second-order Runge–Kutta method.

A very popular method for approximating solutions to first-order initial value problems is the **fourth-order Runge–Kutta method.** The equations for this procedure are derived in any standard numerical analysis text. Here we will illustrate the methods geometrically. The resulting procedure is summarized in the following algorithm. Note that K_1 and K_4 are estimates to the rise using slope estimates computed at the endpoints of the interval $[x_{n-1}, x_n]$. The rises K_2 and K_3 are computed using slope estimates at the midpoint of the interval. Thus the method computes a **weighted average** of four estimates of the rise in the interval, weighting the estimates at the midpoint more heavily.

FOURTH-ORDER RUNGE–KUTTA METHOD

For Solving $\frac{dy}{dx} = g(x, y)$, $y(x_0) = y_0$ Over an Interval

Step 1. First divide the interval $x_0 \leq x \leq b$ into p subintervals using equally spaced points:

$$x_1 = x_0 + \Delta x,$$
$$x_2 = x_1 + \Delta x,$$
$$\vdots$$
$$x_p = x_{p-1} + \Delta x = b.$$

Step 2. For $n = 1, 2, 3, \ldots, p$ obtain the following sequence of approximations:

$$y_n = y_{n-1} + \frac{K_1 + 2K_2 + 2K_3 + K_4}{6}$$

where

$$K_1 = g(x_{n-1}, y_{n-1})\,\Delta x,$$
$$K_2 = g\left(x_{n-1} + \frac{\Delta x}{2}, y_{n-1} + \frac{K_1}{2}\right)\Delta x,$$
$$K_3 = g\left(x_{n-1} + \frac{\Delta x}{2}, y_{n-1} + \frac{K_2}{2}\right)\Delta x,$$
$$K_4 = g(x_{n-1} + \Delta x, y_{n-1} + K_3)\,\Delta x.$$

Geometric Interpretation of the Fourth-Order Runge–Kutta Method

Let us now interpret the fourth-order Runge–Kutta method geometrically.

First Estimate of the Rise Consider point A in Fig. 2.17. Its coordinates are (x_{n-1}, y_{n-1}). An Euler's method estimate of the slope at A is $g(x_{n-1}, y_{n-1})$. Multiplying the slope $g(x_{n-1}, y_{n-1})$ by the run Δx, we see that K_1 is an Euler's method estimate of the corresponding rise in the approximating line.

Second Estimate of the Rise In Fig. 2.17 the point B is the midpoint of the line segment AA'. Thus the coordinates of B must be $(x_{n-1} + \Delta x/2, y_{n-1} + K_1/2)$. An *estimate* of the slope at B is $g(x_{n-1} + \Delta x/2, y_{n-1} + K_1/2)$. If this estimate is multiplied by Δx, we will have a second estimate K_2 of the rise for an approximating line, as depicted in Fig. 2.18.

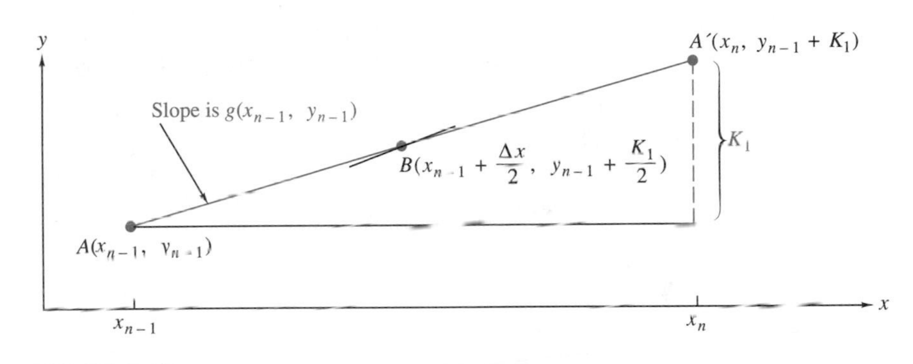

FIGURE 2.17
K_1 is an Euler's method estimate of the rise computed using an estimate of the slope at the left endpoint.

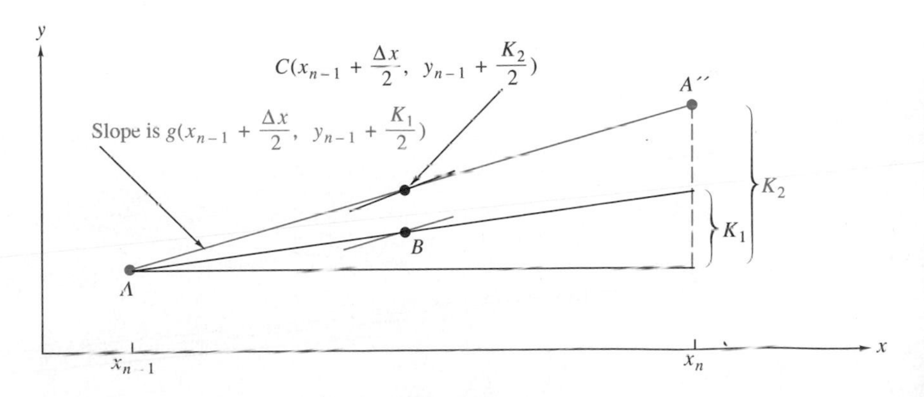

FIGURE 2.18
K_2 is an Euler's method estimate of the rise computed from point A using the slope estimated at B.

Third Estimate of the Rise In a similar fashion, the midpoint C of the line segment AA'' has coordinates $(x_{n-1} + \Delta x/2, y_{n-1} + K_2/2)$. An estimate of the slope at C is $g(x_{n-1} + \Delta x/2, y_{n-1} + K_2/2)$, and K_3 yields another estimate to the rise of the approximating line over the interval $[x_{n-1}, x_n]$. Here K_3 is computed by multiplying the slope $g(x_{n-1} + \Delta x/2, y_{n-1} + K_2/2)$ by the run Δx, as depicted in Fig. 2.19.

Fourth Estimate of the Rise In Fig. 2.19 note that the coordinates of point D are $(x_{n-1} + \Delta x, y_{n-1} + K_3)$. An estimate of the slope at point D is $g(x_{n-1} + \Delta x, y_{n-1} + K_3)$ yielding K_4 as an estimate of the rise (see Fig. 2.20).

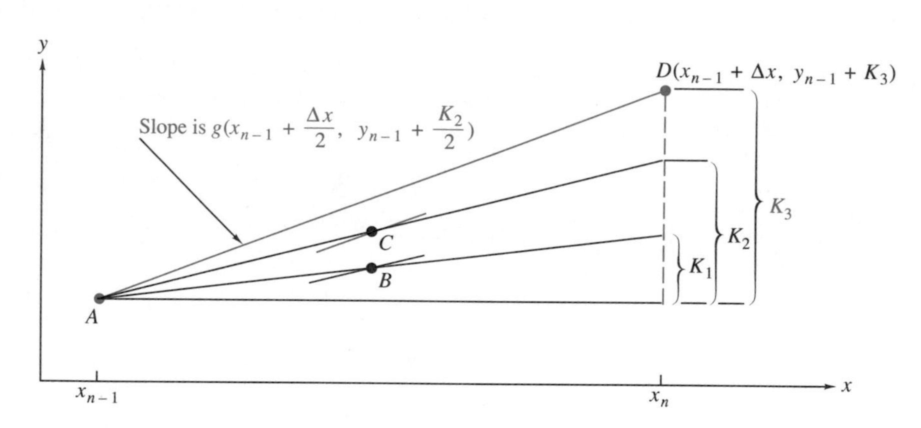

FIGURE 2.19
K_3 is an estimate of the rise based on a slope estimated at point C.

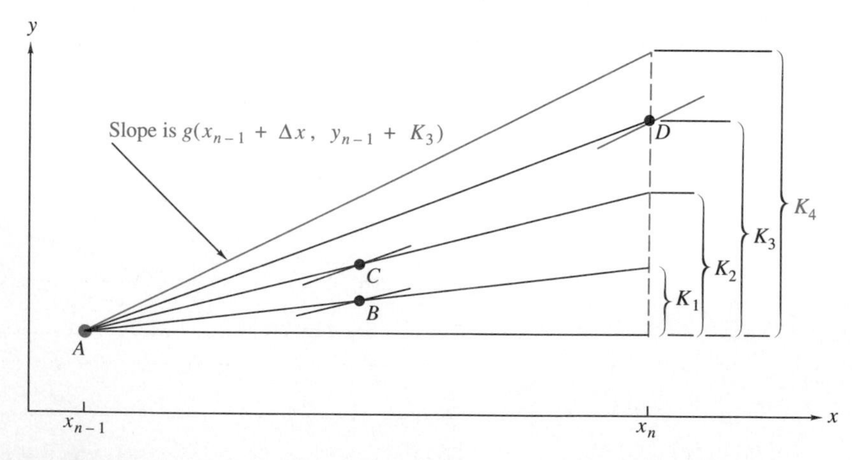

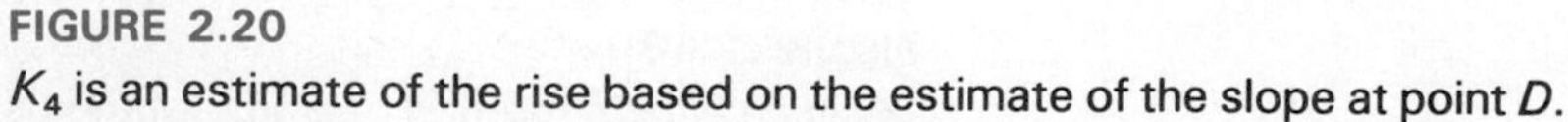
FIGURE 2.20
K_4 is an estimate of the rise based on the estimate of the slope at point D.

Let us summarize our procedure. We have computed four estimates for the rise of the approximating line over the interval $[x_{n-1}, x_n]$: K_1 based on an estimate of the slope at the left end of the interval (point A); K_2 and K_3 based on estimates of the slope at the center of the interval (points B and C); and K_4 based on an estimate of the slope at the right end of the interval (point D).

Weighted Estimate of the Rise If we weight estimates K_2 and K_3 twice as heavily as K_1 and K_4, we have the **weighted estimate** of the rise for an approximating line:

$$\text{rise} = \frac{K_1 + 2K_2 + 2K_3 + K_4}{6}.$$

Different weighting schemes are possible, each giving a distinct Runge–Kutta method.

Calculating the Value of the Dependent Variable Finally, we estimate y_n on the approximating line:

$$y_n = y_{n-1} + \frac{K_1 + 2K_2 + 2K_3 + K_4}{6}.$$

Figure 2.21 displays the final approximating line for obtaining y_n.

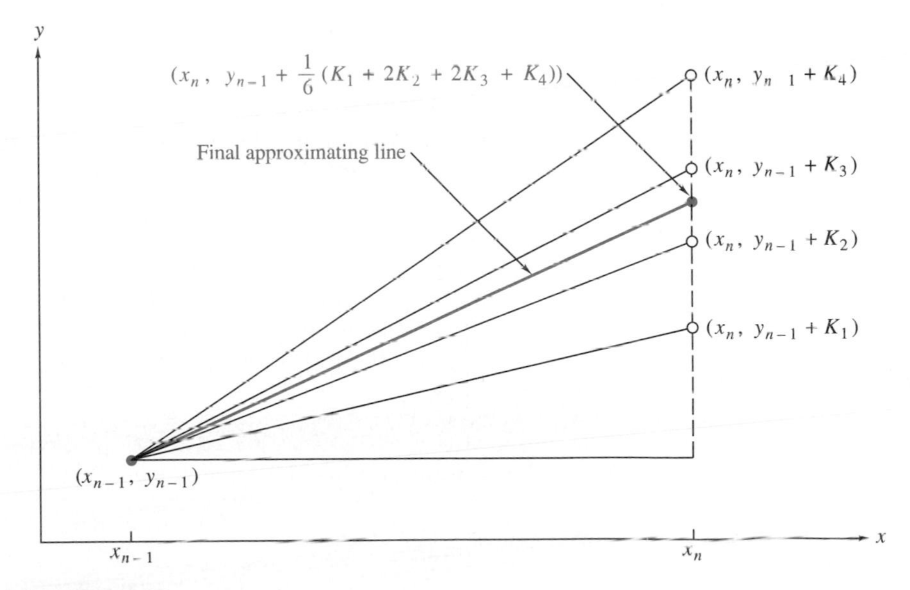

FIGURE 2.21

The final estimate to y_n reflects a weighted estimate of the rise based on slopes estimated at the left, center (twice), and right endpoints of the interval.

Applying the Runge–Kutta Method and Comparing Methods for Accuracy

We now return to the illustrative example concerning the amount of money accrued at the end of year 20 with 7% annual interest compounded continuously. We obtained a value of $4055.20 with a step size of $\Delta t = 1$ using the fourth-order Runge–Kutta method presented above. This result agrees with the analytic solution value to the nearest cent.

Table 2.3 compares the results of using Euler's method, the Improved Euler's method, and a fourth-order Runge–Kutta method for various step sizes to calculate the continuous-interest model shown in Eq. (3). The rounded-off analytic solution is $4055.20.

Roundoff Error As discussed in Section 2.5, each operation performed on a computer potentially produces roundoff error. Thus, in judging the usefulness of a particular numerical procedure, we need to be concerned with the number of operations required.

> To compare different methods, *we consider step sizes that produce an equivalent number of operations.*

Each time a derivative is approximated, a function is evaluated. Potentially, a function evaluation involves a large number of arithmetic operations on a computer. (Consider how a computer estimates $\sin x$ using the Taylor series.) Since Euler's method requires one derivative calculation at each step, the Improved Euler's method requires two derivatives, and the fourth-order Runge–Kutta method requires four, the approximations listed in Table 2.4 require roughly the same number of function evaluations. That is, since each step of the Runge–Kutta method requires four times as many derivative calculations as Euler's method, for instance, the total number of steps used in Euler's method can be four times as many as in the Runge–Kutta method to achieve the same number of function evaluations when comparing the accuracy of the results of the two methods. For a comparable step size we see the superiority of the Runge–Kutta method for this example.

TABLE 2.3

The Estimates of $Q(20)$ for $dQ/dt = 0.07Q$, $Q(0) = 1000$

	Euler		Improved Euler		Runge–Kutta	
Step Size	*Estimate*	*Error*	*Estimate*	*Error*	*Estimate*	*Error*
1.0	3869.68	185.52	4050.80	4.40	4055.20	<0.002
0.5	3959.26	95.94	4054.07	1.13	4055.20	$<10^{-4}$
0.25	4006.39	48.81	4054.91	0.34	4055.20	$<10^{-5}$

TABLE 2.4

Estimating $Q(20)$ for the Same Number of Operations

Method	Step Size	Estimate	Error
Euler	0.25	4006.39	48.81
Improved Euler	0.50	4054.07	1.13
Runge–Kutta	1.00	4055.20	<0.002

Summary

In this section we improved on Euler's method of approximating a solution to a first-order initial value problem presented in Section 2.5. We first considered the Improved Euler's method, which simply averages the estimated slopes at the endpoints of the interval. Then we studied Runge–Kutta methods, which employ weighted averages of various estimates to the slopes at interior points in the interval. Of the methods investigated here, the fourth-order Runge–Kutta method usually provides the most accurate approximation and is most frequently used in practice.

EXERCISES 2.6

The symbol ☐ indicates problems that are optional and require a calculator or computer.

1. A certain bank advertises that it compounds interest continuously. If \$100 is deposited initially with an annual interest rate of 10%, then the following equation can be used to model the problem:

$$\frac{dQ}{dt} = 0.10Q, \qquad Q(0) = 100, \quad 0 \leq t \leq 1.$$

   a) Find the value of m_{aver}.
   b) Estimate $Q(1)$ by finding Q_1 using the Improved Euler's method with one step.

2. Consider again the model of problem 1 and estimate $Q(1)$ using the Improved Euler's method with two steps.

3. Given the same model as in problem 1:
   a) Estimate $Q(1)$ using a step size $\Delta t = 0.25$ in the Improved Euler's method.
   ☐ b) Use a computer or calculator program to estimate $Q(1)$ using a step size of $\Delta t = 0.10$.

4. Again consider the model described in problem 1 and the fourth-order Runge–Kutta method presented in the text.
   a) Estimate $Q(1)$ using a step size of $\Delta t = 0.25$.
   ☐ b) Use a computer or calculator program to estimate $Q(1)$ with a step size $\Delta t = 0.10$.

Problems 5–7 require a calculator.

5. Solve problem 4 in Exercises 2.5 using the fourth-order Runge–Kutta method.

6. Solve problem 5 in Exercises 2.5 using the fourth-order Runge–Kutta method.

7. Solve problem 6 in Exercises 2.5 using the fourth-order Runge–Kutta method.

TOOLKIT PROGRAMS

First Order Initial Value Problem

EXISTENCE AND UNIQUENESS OF SOLUTIONS (Optional)

In Sections 2.5 and 2.6 we approximated a solution to the initial value problem

$$y' = g(x, y), \qquad y(x_0) = y_0, \quad x_0 \leqslant x \leqslant b, \tag{1}$$

by constructing a table of approximate values for selected values $x_1, x_2, \ldots, x_n$ of the independent variable over the interval. Each value in the table was produced individually, in succession, from the previous point. However, there are times when simply having a table of approximate function values for the solution is inadequate. For instance, we may need to investigate the qualitative nature of the solution function or use the solution for theoretical purposes. In cases where it is difficult or impossible to obtain an analytical solution, it may be possible to obtain a *function* that approximates the solution.

In this section we will show how you can construct a sequence of *approximate solution functions.* Each function is defined over the entire interval $x_0 \leqslant x \leqslant b$ and is obtained from the previous approximating function in a certain way. As we repeat the procedure, the new approximating function gets closer to the solution function than was the case with the previous approximation. Although the procedure is not practical from a computational point of view, it is important to the theoretical issues of existence and uniqueness of solutions to first-order initial value problems.

Approximating the Exponential Function

To get a clearer picture of a sequence of functions approximating a fixed function, recall the Maclaurin polynomials for the function $f(x) = e^x$ from calculus. The polynomials are all defined over the entire real line and are given by

$$\begin{aligned}
P_0(x) &= 1,\\
P_1(x) &= 1 + x,\\
P_2(x) &= 1 + x + \frac{x^2}{2},\\
P_3(x) &= 1 + x + \frac{x^2}{2} + \frac{x^3}{6},\\
&\vdots\\
P_n(x) &= 1 + x + \frac{x^2}{2} + \frac{x^3}{6} + \cdots + \frac{x^n}{n!}.
\end{aligned}$$

For a specified interval, say $-1 \leqslant x \leqslant 1$, if we neglect any numerical round-off errors, the Maclaurin polynomials get closer and closer to e^x as the degree

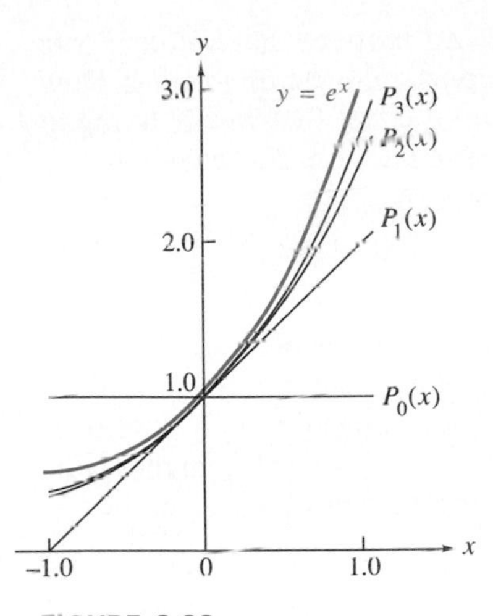

FIGURE 2.22
The graph of the function $f(x) = e^x$ and its first four Maclaurin polynomials over the interval $[-1, 1]$.

n increases (see Fig. 2.22). In fact, if you choose some positive error tolerance $\epsilon > 0$, a degree N exists such that the error between $P_N(x)$ and e^x is less than ϵ no matter what value of x in the interval you select for your evaluation. Symbolically we can express this as

$$|P_N(x) - e^x| < \epsilon \qquad \text{for all } x \text{ in } [-1, 1] \tag{2}$$

for some N that depends on ϵ. Thus we can substitute the polynomial

$$P_N(x) = 1 + x + \frac{x^2}{2} + \cdots + \frac{x^N}{N!}$$

for the function $f(x) = e^x$ over the interval $[-1, 1]$. Of course we would need to determine the actual value of N so that Eq. (2) is satisfied *before* the substitution is made. Useful information for a more thorough investigation of the error involved in Eq. (2) is contained in the following important result from calculus.

TAYLOR'S THEOREM WITH REMAINDER

If the function f has derivatives of all orders in an open interval I containing $x = a$, then for each positive integer n and for each x in I,

$$f(x) = f(a) + f'(a)(x-a) + f''(a)\frac{(x-a)^2}{2!} + \cdots$$

$$+ f^{(n)}(a)\frac{(x-a)^n}{n!} + R_n(x, a) \tag{3}$$

where

$$R_n(x, a) = f^{(n+1)}(c)\frac{(x-a)^{n+1}}{(n+1)!} \tag{4}$$

for some c between a and x.

Formula (4) can now be used to estimate the error between $P_N(x)$ and e^x in Eq. (2). For the function $f(x) = e^x$,

$$|R_N(x, 0)| = \frac{e^c|x|^{N+1}}{(N+1)!} < \frac{3}{(N+1)!}$$

since $e^c < 3$ for $-1 < c < 1$. Thus, for instance, if we choose $\epsilon = 10^{-3}$, it suffices that $N \geq 7$ in order to satisfy Eq. (2).

We are going to employ a similar idea in approximating the solution to the initial value problem given in Eq. (1). That is, we will construct a sequence of functions (not necessarily polynomials) $y_1(x), y_2(x), \ldots, y_n(x), \ldots$ such that for some N the function $y_N(x)$ differs from the actual solution function

$y(x)$ by no more than some chosen $\epsilon > 0$, no matter what value of x is selected from the interval $x_0 \leq x \leq b$. An important question to ask is, How do we know a solution $y(x)$ exists (aside from the fact that we are trying to approximate it)? We discuss the existence issue later on in the section.

Recursive Approximation

The procedure we will use in constructing the sequence of approximations $y_1, y_2, \ldots, y_n$ is **recursive** in nature; that is, at each stage of the procedure the approximation y_k is obtained from a previous one. You may have seen recursive approximation used before in your study of mathematics. For instance, you know that for a positive integer n its **factorial** is given by

$$n! = 1 \cdot 2 \cdot 3 \cdots n,$$

the product of the positive integers up to and including n. The number $n!$ can also be defined recursively:

$$1! = 1 \qquad \text{and} \qquad n! = n \cdot (n-1)! \tag{5}$$

Thus, to calculate $n!$ you multiply the previous calculation $(n-1)!$ by n. Of course, you have to start someplace, so $1! = 1$ is specified. Using definition (5), we have

$$\begin{aligned} 2! &= 2 \cdot 1! = 2 \cdot 1 &&= 2 \\ 3! &= 3 \cdot 2! = 3 \cdot 2 &&= 6 \\ 4! &= 4 \cdot 3! = 4 \cdot 6 &&= 24 \\ 5! &= 5 \cdot 4! = 5 \cdot 24 &&= 120 \end{aligned}$$

and so forth. Notice how the calculations in Eq. (5) are built up in a successive fashion.

In order to get an idea of how we might come up with a recursive method for finding a sequence of function approximations to the initial value problem (1), let us first examine a recursive procedure you studied in calculus to solve an algebraic equation.

Newton's Method

For a given continuous function, recall Newton's method for solving the equation

$$f(x) = 0. \tag{6}$$

That is, we are interested in finding a value $x = r$ so that $f(r)$ has the value 0. We call r a **root** or **zero** of the function f. Geometrically $x = r$ is the location where the graph of f crosses the x-axis. In Newton's method we construct a sequence of approximations to the root by starting with an initial guess value x_1 and then obtaining successive approximations according to the following recursion formula:

$$x_n = x_{n-1} - \frac{f(x_{n-1})}{f'(x_{n-1})}. \tag{7}$$

Notice how the nth approximation x_n is obtained from the immediately preceding approximation x_{n-1} as illustrated in Fig. 2.23.

Let us consider Eq. (7) from another point of view. Suppose one of the approximations turns out to be *exactly* the root value r. For example, suppose $x_{n-1} = r$. Then we calculate the next approximation from Eq. (7):

$$x_n = r - \frac{f(r)}{f'(r)} = r - 0.$$

Thus $x_n = r$ also, and using Eq. (7) we can see that the root r must then satisfy the condition

$$r = r - \frac{f(r)}{f'(r)}.$$

In other words, r solves the equation $f(x) = 0$ if and only if it solves the equation

$$x = x - \frac{f(x)}{f'(x)}. \tag{8}$$

In summary, to solve Eq. (6) we produce the equivalent Eq. (8) that has exactly the same solutions. Newton's method then solves Eq. (8) recursively using the procedure defined by Eq. (7) with an initial guess. We will now apply a similar methodology to solve the initial value problem of Eq. (1). First we need to come up with an equation that has the same solutions as Eq. (1).

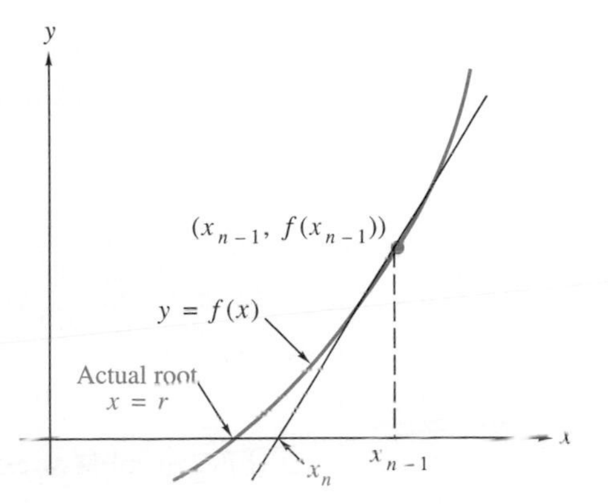

FIGURE 2.23

The approximation x_n is the point where the tangent line to the graph at the point $(x_{n-1}, f(x_{n-1}))$ crosses the x-axis. The approximation x_n is closer to the actual root r than is the previous approximation x_{n-1}.

The Fundamental Integral Equation

Suppose the function $y(x)$ solves the initial value problem of Eq. (1). Then

$$y'(x) = g(x, y(x)) \tag{9}$$

over $x_0 \leq x \leq b$, and $y(x_0) = y_0$. Let us integrate both sides of Eq. (9) over the subinterval $[x_0, x]$:

$$\int_{x_0}^{x} y'(t)\, dt = \int_{x_0}^{x} g(t, y(t))\, dt \tag{10}$$

(We introduce the dummy variable t during the integration since x is the true independent variable over our original interval $x_0 \leq x \leq b$.) By applying the Fundamental Theorem of Calculus to the left side of Eq. (10), we obtain

$$y(x) - y(x_0) = \int_{x_0}^{x} g(t, y(t))\, dt$$

or

$$y(x) = y_0 + \int_{x_0}^{x} g(t, y(t))\, dt. \tag{11}$$

That is, if $y(x)$ solves the initial value problem of Eq. (1), then it must also solve **integral equation** (11).

Conversely, suppose $y(x)$ solves integral equation (11). Then

$$\frac{d[y(x)]}{dx} = \frac{d}{dx}\left[y_0 + \int_{x_0}^{x} g(t, y(t))\, dt\right]$$

or

$$y'(x) = \frac{d}{dx}\left[\int_{x_0}^{x} g(t, y(t))\, dt\right] \tag{12}$$

To differentiate the right side of Eq. (12) we need to apply the following theorem from integral calculus.

SECOND FUNDAMENTAL THEOREM OF CALCULUS

If f is continuous on $[x_0, b]$, then

$$F(x) = \int_{x_0}^{x} f(t)\, dt$$

is differentiable at every point x in (x_0, b) and

$$F'(x) = f(x).$$

In other words, the derivative of an integral with respect to the upper limit of integration is simply the integrand evaluated at that upper limit. By applying this result to the righthand side of Eq. (12), we obtain

$$y'(x) = g(x, y(x)),$$

so $y(x)$ solves Eq. (1). Moreover, after substituting $x = x_0$ into Eq. (11), we find

$$y(x_0) = y_0 + \int_{x_0}^{x_0} g(t, y(t))\, dt = y_0 + 0.$$

Therefore, the solutions to Eqs. (1) and (11) are exactly the same (under suitable restrictions, such as the continuity of g).

Imitating Newton's method, we now solve Eq. (11) recursively.

Successive Approximations

To solve integral Eq. (11) we start with a guess, an initial function $y_1(x)$. We can choose $y_1(x)$ to be almost any function that is convenient, as long as it is differentiable; this requirement is necessary since the function is supposed to approximate the solution to a differential equation. Typically we choose $y_1(x)$ to be the constant function $y_1(x) \equiv y_0$. Then we obtain the successive approximations $y_2(x), y_3(x), y_4(x), \ldots$ recursively according to the following formula:

$$\underbrace{y_n(x)}_{\substack{\text{new}\\ \text{approximation}}} = y_0 + \int_{x_0}^{x} g(t, \underbrace{y_{n-1}(t)}_{\substack{\text{previous}\\ \text{approximation}}})\, dt. \tag{13}$$

This method is known as **Picard's method of successive approximations,** named after the French mathematician Émile Picard (1856–1941). In a more advanced course, such as advanced calculus, you will be shown that the sequence $\{y_n(x)\}$ converges to one and only one function $y(x)$ that solves the initial value problem. We will discuss this issue of uniqueness in more detail below, but first let us consider a specific example.

EXAMPLE 1 Solve the initial value problem

$$y' = y + e^x, \quad y(0) = 1. \tag{14}$$

Solution. We begin with the initial guess $y_1(x) \equiv 1$ and use the method of successive approximations to find the first approximation:

$$\begin{aligned} y_2(x) &= y_0 + \int_{x_0}^{x} [y_1(t) + e^t]\, dt \\ &= 1 + \int_0^x (1 + e^t)\, dt \\ &= 1 + \left[t + e^t \right]_0^x \\ &= 1 + x + e^x - (0 + 1) \end{aligned}$$

or

$$y_2(x) = x + e^x.$$

We calculate the next approximation:

$$\begin{aligned} y_3(x) &= y_0 + \int_{x_0}^{x} [y_2(t) + e^t]\, dt \\ &= 1 + \int_0^x (t + 2e^t)\, dt \\ &= 1 + \left[\frac{t^2}{2} + 2e^t \right]_0^x \end{aligned}$$

or

$$y_3(x) = -1 + \frac{x^2}{2} + 2e^x.$$

And the next approximation is

$$\begin{aligned} y_4(x) &= y_0 + \int_{x_0}^{x} [y_3(t) + e^t]\, dt \\ &= 1 + \int_0^x \left(-1 + \frac{t^2}{2} + 3e^t\right) dt \\ &= 1 + \left[-t + \frac{t^3}{6} + 3e^t \right]_0^x \end{aligned}$$

or

$$y_4(x) = -2 - x + \frac{x^3}{6} + 3e^x. \tag{15}$$

We could continue to obtain further approximations, but let us stop here to compare our approximation $y_4(x)$ with an analytical solution to see how good the approximation Eq. (15) is.

Comparison with the Analytical Solution Differential Eq. (14) is linear and can be written as

$$y' - y = e^x.$$

An integrating factor is

$$\mu(x) = e^{\int -dx} = e^{-x}.$$

Thus

$$e^{-x}y = \int e^{-x} \cdot e^x \, dx = x + C.$$

When $x = 0$, $y = 1$, so $C = 1$. Therefore the solution to the initial value problem is

$$y = (x + 1)e^x. \tag{16}$$

As a basis for comparing Eq. (16) with our approximation in Eq. (15), we first replace e^x by its third-degree Maclaurin polynomial approximation. Then we substitute that expression into both the approximation $y_4(x)$ and the analytical solution $(x + 1)e^x$ and compare results. Thus

$$\begin{aligned} y_4(x) &\approx -2 - x + \frac{x^3}{6} + 3\left(1 + x + \frac{x^2}{2} + \frac{x^3}{6}\right) \\ &= 1 + 2x + \frac{3x^2}{2} + \frac{2x^3}{3} \end{aligned} \tag{17}$$

and note that

$$\begin{aligned} (x + 1)e^x &\approx (x + 1)\left(1 + x + \frac{x^2}{2} + \frac{x^3}{6}\right) \\ &= 1 + 2x + \frac{3x^2}{2} + \frac{2x^3}{3} + \frac{x^4}{6}. \end{aligned} \tag{18}$$

Hence Eqs. (17) and (18) differ by only the single term $x^4/6$. For the interval $-1 \leq x \leq 1$ (which contains the initial value $x_0 = 0$), for instance, the term $x^4/6$ never gets larger than $1/6$.

As we continue to calculate successive approximations $y_5(x)$, $y_6(x)$, . . ., the resulting functions get closer and closer to the analytical solution $y = (x + 1)e^x$. Of course the power of the method is to solve first-order differential equations we cannot solve by any (known) analytical method. We presented an elementary example here so you could see how the successive approximations actually compare to the true analytical solution. In practice the integrations in Eq. (13) may be very difficult or impossible to carry out.

Existence and Uniqueness of Solutions

Consider again the general first-order initial value problem:

$$y' = g(x, y), \quad y(x_0) = y_0. \tag{19}$$

In our presentation and discussions of the solution methods given in this chapter, we have simply *assumed a solution does exist* and then proceeded to find it. However, there are two fundamental issues that cannot be ignored.

1. **Existence:** Does a solution $y = y(x)$ to Eq. (19) actually exist?
2. **Uniqueness:** Is there more than one function that satisfies Eq. (19)?

A complete and thorough discussion of the existence and uniqueness questions is more appropriate for an advanced course that includes the theory of differential equations. Nevertheless, we would like to indicate some of the pitfalls here and state the main result on existence and uniqueness that we will use throughout the remainder of the text. We begin with an example.

EXAMPLE 2 Given the initial value problem

$$y' = 3y^{2/3}, \quad y(0) = 0, \tag{20}$$

we might choose to solve it by the method of separable variables:

$$\frac{dy}{3y^{2/3}} = dx, \quad y \neq 0,$$

and integration of both sides gives us

$$y^{1/3} = x + C$$

or

$$y = (x + C)^3. \tag{21}$$

Substitution of the initial condition $y(0) = 0$ yields $C = 0$ and the particular solution

$$y = x^3. \tag{22}$$

You can check that function (22) does indeed satisfy Eq. (20), but so does the constant function $y(x) \equiv 0$:

$$0 = \frac{d}{dx}(0) = 3(0)^{2/3},$$

and the initial condition is also satisfied. However, *there is no value of the constant C for which the family of solutions given by Eq. (21) yields the function $y(x) \equiv 0$.* In fact, the situation is even more severe than presently appears. To see why, consider Eq. (21) as graphed in Fig. 2.24. Select one of the curves with a positive value of C, say $C = 1$ for definiteness, and travel upward along the negative branch of this curve (below the x-axis) until you reach the x-axis; then (smoothly) latch onto the x-axis and travel through the origin, continuing until you decide to transfer to the upper branch (above the x-axis) of a curve with a negative value of C, say $C = -2$. The resulting continuously differentiable curve you have just traced out is yet another solution to initial value problem (20) (see Fig. 2.24). Thus *there are infinitely many solutions to Eq. (20).* Obviously Eq. (21) is not the general solution in the sense of expressing all possible solutions for suitable selections of values for the parameter C.

In this example, the difficulty occurs because separation of the variables lead to the expression $dy/3y^{2/3}$ where division by zero takes place whenever we reach the x-axis. That is, Eq. (21) results in unique solutions provided we never cross the x-axis.

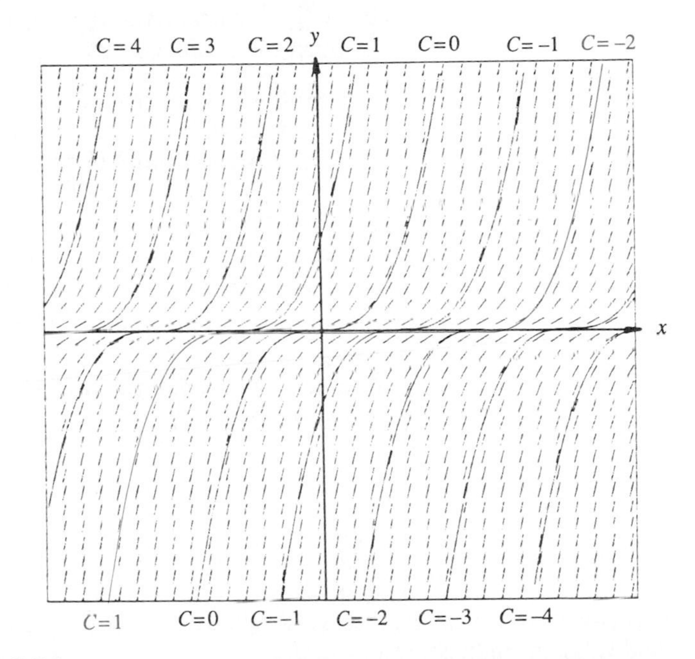

FIGURE 2.24
The family of integral curves $y = (x + C)^3$ solves differential Eq. (20). However, the family does not contain the solution curve $y(x) \equiv 0$ nor any of the infinitely many solutions like the colored curve.

Here is the main result on existence and uniqueness we use throughout the text. (There are stronger versions of this theorem, but we will have no occasion to use them.)

EXISTENCE AND UNIQUENESS THEOREM

For the initial value problem

$$y' = g(x, y), \quad y(x_0) = y_0,$$

assume that $g(x, y)$ and $\partial g/\partial y$ are continuous functions in a region of the x,y-plane containing the point (x_0, y_0). Then there exists an interval $[a, b]$ containing x_0 for which there is a unique solution $y(x)$ to the initial value problem.

In Example 2, Eq. (20), the partial derivative $\partial g/\partial y = 2y^{-1/3}$ fails to be continuous in any region containing the x-axis. However, provided we stay away from the x-axis the theorem guarantees that a unique solution will exist (namely, the branch of the appropriate curve in the family (21) that passes through the initial condition point (x_0, y_0)).

Remarks One proof of this theorem shows that the sequence of successive approximations $y_1(x), y_2(x), \ldots, y_n(x)$, defined by Eq. (13) converges to the unique solution. Also, there are versions of the existence and uniqueness theorem for higher-order differential equations, but we will omit stating those results here.

The analytical techniques presented in this chapter do not apply to solving all first-order equations. To obtain analytical solutions for more equations, including those of higher order, we will simplify our models to *linear differential equations only.* The conditions guaranteeing the existence and uniqueness of solutions for linear equations are easily satisfied. They are stated in Theorem 2.1.

EXERCISES 2.7

In problems 1–6, use the method of successive approximations to find y_1, y_2, y_3, and y_4.

1. $y' = y, \quad y(0) = 1$

2. $y' = xy, \quad y(0) = 1$

3. $y' = x - y, \quad y(0) = 0$

4. $y' = x + y, \quad y(0) = 1$

5. $y' = x^2 + y, \quad y(0) = 0$

6. $y' = 1 + y^2, \quad y(0) = 0$

In problems 7–12, determine if a unique solution to the initial value problem is guaranteed to exist on an interval containing x_0.

7. $y' = x^2 + y^2, \quad y(0) = 1$

8. $y' - y = x, \quad y(1) = 1$

9. $y' - x\sqrt{y} = 0, \quad y(0) = 0$

10. $y' = \sqrt{xy}, \quad y(1) = 1$

11. $y' - y^3 = 0, \quad y(0) = 0$

12. $xy' = y, \quad y(0) = 0$

TOOLKIT PROGRAMS

Antiderivatives and Direction Fields

2.8

LAUNCHING A ROCKET INTO SPACE (Optional)

In this section we will consider the problem of launching a rocket into space. How fast is the rocket moving when its fuel is burned up? Will the rocket escape the gravitational pull of the earth or fall back to its surface? We will examine these questions in the following two examples.

EXAMPLE 1 THE SPEED OF A SINGLE-STAGE ROCKET

After launch, the rocket is continuously losing mass in its initial stage of flight as fuel is being burned. Moreover, that mass is being propelled away from the rocket at a significant speed. Thus we cannot simply apply Newton's second law ($F = ma$) as we did in Section 1.3, but we can use the more general formulation of his laws, which asserts that the net force acting on the rocket–fuel–exhaust system is the instantaneous rate of change of the system's momentum. Let us formulate more precisely the problem of determining the rocket's speed.

Assumptions Let $m_r(t)$ denote the mass of the rocket and fuel at any time t, and let $m_e(t)$ represent the mass of all the ejected exhaust gases. Both m_r and m_e are changing over time, but we assume their sum is always constant:

$$m_r(t) + m_e(t) = m_s \tag{1}$$

where m_s is the initial mass of the rocket–fuel system at time $t = 0$. Note from Eq. (1) that $dm_e/dt = -dm_r/dt$.

Suppose the rocket is moving vertically upward with velocity v relative to the earth. Let u denote the velocity of the aggregate of exhaust gases *relative to the rocket;* that is, u is the mean value of the velocities of all the individual exhaust gas molecules that have been ejected. Thus we consider m_e to be the mass of a single entity moving with velocity u relative to the rocket at any instant.

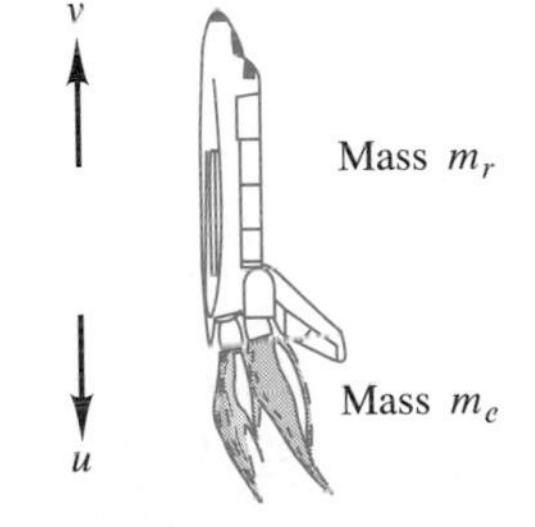

FIGURE 2.25
The rocket–fuel system of mass m_r is traveling upward with velocity v, and the exhaust gases of mass m_e are moving in the opposite direction with velocity u relative to the rocket.

All of the variables m_r, m_e, and v are changing continuously over time. The velocity u is in the direction opposite to v, and we assume that the exhaust gases are being ejected from the rocket at a constant speed. So we assume that $|u| = -u$ is constant and equal to the speed s at which the gases are ejected from the rocket. This situation is depicted in Fig. 2.25.

Forces on the Exhaust Gases First we will consider only the exhaust gases that have already been ejected as a single point mass at time t. Since u is the mean velocity of the ejected gases relative to the rocket and v is the velocity of the rocket relative to the earth, the sum $u + v$ is the mean velocity of the ejected exhaust gases *relative to the earth.* If we assume that gravity is the only appreciable force acting on the gases after leaving the rocket, then at time t the instantaneous rate of change of the momentum of the already ejected exhaust gases is given by

$$m_e \frac{d}{dt}(u+v) = -m_e g \tag{2}$$

where the negative sign appears because the gravitational attraction is opposite to the direction of motion.

Total Force Acting on the Rocket–Fuel–Exhaust System What forces are acting on the rocket–fuel–exhaust system as it leaves the earth's surface? Certainly gravitational attraction is the dominant force. During the early stages of flight, frictional drag is also important, but as a first approximation we will neglect this factor. Eventually fluid compressibility becomes a factor, but we will neglect that also. Therefore the net external force acting on the entire system is

$$F = F_p - F_r = -m_s g. \tag{3}$$

where F_p is the propulsive force and F_r is the resisting force.

Interrelationships among the Variables According to Newton's second law of motion,

$$F = \frac{d}{dt}\,(\text{momentum of the system}).$$

Under our assumptions, at any time t we have

$$\text{momentum of the system} = m_r v + m_e(u+v).$$

Thus

$$F = \frac{d}{dt}\,[m_r v + m_e(u+v)],$$

or, differentiating and substituting from Eq. (3), we get

$$m_r \frac{dv}{dt} + v\frac{dm_r}{dt} + m_e \frac{d}{dt}(u+v) + (u+v)\frac{dm_e}{dt} = -m_s g. \tag{4}$$

We know that $m_s = m_r + m_e$ and $dm_e/dt = -dm_r/dt$ from Eq. (1) and that $m_e d(u+v)/dt = -m_e g$ from Eq. (2). After substituting these expressions into Eq. (4), we have

$$m_r v' + v\frac{dm_r}{dt} - m_e g - (u+v)\frac{dm_r}{dt} = -m_r g - m_e g,$$

which yields the equation

$$m_r v' - u\frac{dm_r}{dt} = -m_r g. \tag{5}$$

Now assume the rocket fuel is being consumed at a constant rate r during the time interval $0 \leq t \leq T$. Then

$$m_r(t) = m_s - rt, \quad 0 \leq t \leq T. \tag{6}$$

Recall our assumption that the speed $|u| = -u$ of the aggregate of exhaust gases relative to the rocket is a constant s (say the constant speed at which the gases are being ejected from the rocket at full throttle until the fuel is depleted). Substitution of Eq. (6) into Eq. (5) gives us the model

$$(m_s - rt)v' - sr = -(m_s - rt)g. \tag{7}$$

Solving the Model Let us solve the model given by Eq. (7). Dividing by $m_s - rt$ and writing the equation in differential form, we have

$$dv - \frac{sr}{m_s - rt}\, dt = -g\, dt.$$

Integration then gives us

$$v + s \ln |m_s - rt| = -gt + C$$

From the initial condition $v(0) = v_0$ we evaluate C:

$$v_0 + s \ln m_s = C.$$

Substitution of this result into our solution yields

$$v = v_0 - gt + s \ln \left(\frac{m_s}{m_s - rt}\right). \tag{8}$$

Here v_0 is the velocity of the rocket at time $t = 0$.

Velocity at Burnout Equation (8) gives the velocity of the rocket at any time t. How fast is the rocket moving when the fuel is completely used up? At burnout time $t = T$ the mass of the rocket is simply m_1, the mass of the casing and instruments. From Eq. (6) we know that $T = (m_s - m_1)/r$, and substitution of this value into Eq. (8) gives the velocity

$$\begin{aligned} v_1 &= v_0 - gT + s \ln\left(\frac{m_s}{m_s - rT}\right) \\ &= v_0 - g\left[\frac{(m_s - m_1)}{r}\right] + s \ln\left(\frac{m_s}{m_1}\right) \end{aligned}$$

or

$$v_1 = v_0 - \frac{gm_s}{r}\left(1 - \frac{m_1}{m_s}\right) + s \ln\left(\frac{m_s}{m_1}\right). \tag{9}$$

Equation (9) expresses the velocity v_1 at burnout in terms of the physical "design" constants m_1 (the mass of the rocket casing and instruments), m_s (the mass of the entire system), and s (the "thrust" capability of the rocket engines), as well as the initial velocity v_0. An interesting question, which we will consider next, is whether a single-stage rocket can be designed so that the burnout velocity v_1 given by Eq. (9) is sufficient to propel the rocket out of the earth's gravitational field and into space.

EXAMPLE 2 ESCAPING THE EARTH'S GRAVITATIONAL FIELD

Let us investigate the velocity-of-escape problem. That is, *when the rocket's fuel supply has been exhausted, what burnout velocity v_1 must the rocket have attained at time $t = T$ in order to escape the earth's gravitational attraction and continue its journey into space?*

Assumptions In our previous discussion we assumed the rocket fuel burnout takes place well within the earth's atmosphere; that is, during the burn phase we assumed the distance r between the rocket and the center of the earth to be the constant radius R of the earth. This situation is depicted in Fig. 2.26(a). However, as the rocket leaves the vicinity of the earth, the distance r increases appreciably relative to R, as shown in Fig. 2.26(b). How does the force of gravity when the rocket is in outer space compare with the force of gravity when the rocket is within the earth's thin membrane of atmosphere? From Newton's law of gravitational attraction (Section 1.3, Eq. 12) the ratio of these two forces is given by

$$\frac{Gm_1m_E/r^2}{Gm_1m_E/R^2} = \frac{R^2}{r^2}. \tag{10}$$

Note that mass m_1, which the rocket attained at burnout, no longer changes because all of the rocket fuel has been depleted. At the point of burnout the

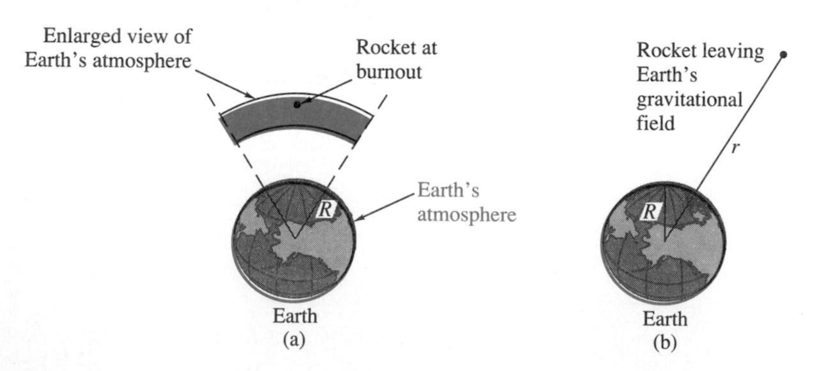

FIGURE 2.26
(a) During the burn phase the rocket remains within the thin membrane of the earth's atmosphere, where we assume its distance r from the center of the earth is the constant radius R of the earth. (b) As the rocket leaves the vicinity of the earth, the distance r increases appreciably relative to R.

earth's gravitational attraction on the rocket is

$$F_p = -\frac{Gm_1 m_E}{R^2} = -m_1 g. \tag{11}$$

Interrelationships among the Variables Equation (11) is equivalent to assuming that at burnout, $r - R \approx 0$, and gravity is the constant g. After the burnout point we assume from Eq. (10) that the force of gravity diminishes as the ratio R^2/r^2. Thus, after burnout the submodel for the propulsion force is given by

$$\begin{aligned} F_p &= -m_1 g \frac{R^2}{r^2} \\ m_1 \frac{dv}{dt} &= -\frac{m_1 g R^2}{r^2} \end{aligned} \tag{12}$$

or

$$\frac{dv}{dt} = -\frac{gR^2}{r^2}. \tag{13}$$

Solving the Model We cannot simply integrate Eq. (13) directly for v because r is an unknown function of t. However, using the chain rule from calculus we observe that

$$\frac{dv}{dt} = \frac{dv}{dr}\frac{dr}{dt} = \frac{dv}{dr} v.$$

Substitution of this result into Eq. (13) gives us

$$v \frac{dv}{dr} = -\frac{gR^2}{r^2}$$

or

$$v\, dv = gR^2 \left(-\frac{dr}{r^2}\right). \tag{14}$$

Integrating both sides of Eq. (14) then yields

$$\frac{v^2}{2} = \frac{gR^2}{r} + C. \tag{15}$$

When $v = v_1$ at burnout time $t = T$, we assume $r \approx R$. Thus we calculate the constant C as

$$C = \frac{v_1^2}{2} - gR.$$

Substituting this value of C into Eq. (15) and simplifying algebraically, we obtain

$$v^2 = \frac{2gR^2}{r} + v_1^2 - 2gR \tag{16}$$

for the velocity of the rocket at any time $t > T$. Let us interpret the solution (16).

Determining the Escape Velocity Note that the first term $2gR^2/r$ in Eq. (16) is positive and approaches zero as r becomes large. Suppose that $v_1^2 - 2gR$ is nonnegative. Then from Eq. (16), v^2 can never be zero. Since $v = v_1$ is positive at the instant $t = T$ when the fuel supply is exhausted, and since v is a continuous function, it follows that v must remain positive forever. That is, in this case the rocket leaves the earth's gravitational field and continues its journey into space.

On the other hand, suppose that $v_1^2 - 2gR$ is negative. Then as r becomes larger and larger (as the rocket gets farther away from the earth), a critical point is reached when the right side of Eq. (16) becomes zero:

$$\frac{2gR^2}{r} + v_1^2 - 2gR = 0.$$

At that point the velocity v is zero. Since no fuel remains in the rocket to provide additional propulsion, the earth's gravitational field continues to act on the vehicle and pulls it back to earth. Thus *the minimum value for the rocket to escape occurs when*

$$v_1^2 - 2gR = 0$$

or

$$v_1 = \sqrt{2gR}. \tag{17}$$

Assuming that $g = 32.2$ ft/sec^2 and $R = 3964$ mi, the **escape velocity** required of the rocket is

$$\begin{aligned} v_1 &= \sqrt{2(32.2)\left(\frac{1}{5280}\right)(3964)} \\ &\approx 6.95 \text{ mi/sec} \\ &\approx 25{,}000 \text{ mi/hr}. \end{aligned}$$

Escape Velocity and Burnout Let us compare this last result with the burnout speed v_1 in Eq. (9) attained when the fuel is exhausted at time $t = T$. We assume that the rocket is launched from the ground (at rest), so $v_0 = 0$. The ratio m_1/m_s of the mass of the rocket casing and instruments to the mass of the entire system depends on the materials, but a typical value is about 0.2. A typical value for the relative exhaust speed s, for both liquid and solid fuel rockets, is about 1.9 mi/sec. Thus, from Eq. (9) with $r = \rho > 0$ we have

$$\begin{aligned} v_1 &= 0 - \frac{gm_s}{\rho}(1 - 0.2) + 1.9 \ln\left(\frac{1}{0.2}\right) \\ &< 1.9 \ln(5) \approx 3 \text{ mi/sec}. \end{aligned}$$

Therefore, under current technologies (such as $m_1/m_s \approx 0.2$, $s \approx 1.9$, and so forth), it is impossible for a single-stage rocket to escape the earth's gravitational field. After the initial or first stage of rocket fuel is depleted, part of the rocket mass m_s (namely, the part that was carrying the fuel and is now empty) must be discarded. Then a second-stage rocket engine must be fired in order to boost the speed of the remaining mass toward the required escape velocity of 6.95 mi/sec.*

EXERCISES 2.8

1. Show that if an object of mass m is placed nine-tenths of the distance from the center of the earth to the center of the moon, then the gravitational attractions of the earth and moon on the mass are equal (and the object is at rest if there are no other forces of propulsion). Assume the mass of the earth is 81 times the mass of the moon.

2. Consider a rocket that prior to launch weighs 2000 lb, including 1600 lb of fuel. Assume that throughout the launch the fuel is burned at a constant rate and that the exhaust gases are being ejected at the constant speed of 5000 ft/sec. If the rocket is fired from rest and it takes 60 sec to reach burnout, what is the velocity of the rocket at burnout? Neglect air resistance and assume the earth's gravitational attraction is constant during the burn phase.

3. The volume of a spherical raindrop increases as it falls through the atmosphere due to the adhesion of mist particles on its surface. Assume that the raindrop retains its spherical shape thoughout the fall and that the instantaneous rate of change of its volume with respect to the distance fallen is proportional to the surface area at that distance. What is the radius of the raindrop as a function of the distance fallen?

4. An object initially at rest is pulled on a sled by a force of 15 lb across a frozen pond. The object and sled together weigh 160 lb. Assuming that friction is negligible and the air resistance is proportional to the velocity of the sled, find the velocity of the sled at any time $t \geqslant 0$.

2.9 MOTION OF A SATELLITE (Optional)

In this section we will derive the function expressing the potential energy associated with a satellite in orbit around the earth. Our purpose is to connect the concept of a potential function as a solution to an exact first-order differential equation (see Section 2.3) with a real-world example illustrating how a potential function arises in describing potential energy. We employ vector notation in our discussion and use the dot and cross products you studied in calculus. We denote vectors by boldface lowercase letters and scalars by ordinary lowercase letters. Thus, for instance, $\mathbf{r}$ is a vector and $|\mathbf{r}| = r$ denotes its (scalar) magnitude. We use only very elementary results

* Actually a three-stage rocket is required. An interesting and complete development of this problem is contained in "Lagrange Multipliers and the Design of Multistage Rockets." Undergraduate Mathematics and Its Applications Project, (Arlington, Mass.: COMAP, Inc., 1981), module 517.

concerning vectors, such as *the cross product of two parallel vectors is the zero vector* ***0***. We assume you know that the derivative of a vector function is obtained by differentiating each of its scalar component functions.

In this application we show that the equipotential curves for a satellite under the influence of a gravitational field are concentric circles. Suppose a small artificial satellite is placed into orbit above the earth's surface and acted on by the earth's gravitational field. According to Newton's law of gravitational attraction, the force **F** of attraction on the satellite of mass m has magnitude

$$|\mathbf{F}| = \frac{GMm}{r^2} \tag{1}$$

where r is the distance of the satellite from the center of the earth, M is the mass of the earth, and $G = 6.670 \times 10^{-8}$ dyne-cm^2/gm^2 is the universal gravitational constant. The force is directed toward the earth along the line joining the earth's center with the center of mass of the satellite. Since the mass of the satellite is very small compared with the earth's mass, we assume that $M + m \approx M$ and consider the satellite to be a single point moving in orbit. Assuming no forces act on the satellite other than the earth's gravitational attraction, *we are interested in describing the orbit of the satellite; that is, its path of motion.* This central force problem is often referred to in celestial mechanics as the **two-body problem.** Specifically, we would like answers to the following questions:

1. Does the satellite tend to "wander" in its orbit, with the result that its path is not in a single plane?
2. What kind of paths are possible for the satellite? That is, can the satellite follow any kind of path when its force of propulsion is only gravity?
3. How are the kinetic and potential energies of the satellite related?

Figure 2.27 displays a path we might think is possible for the satellite's motion.

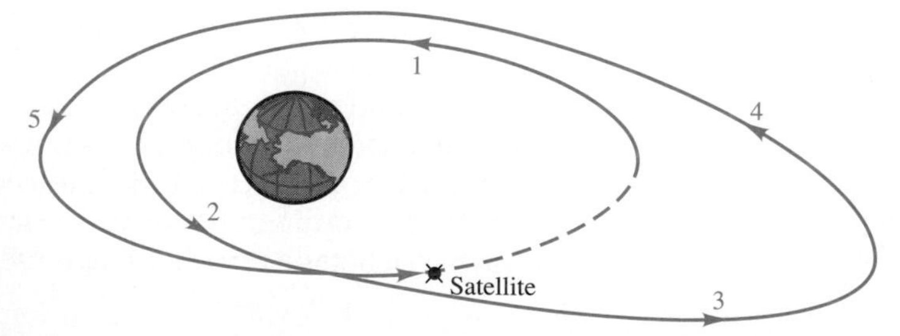

FIGURE 2.27
What kind of paths are possible for a satellite in motion, assuming no forces are acting other than the earth's gravitational attraction?

The Equation of Motion

Let (X, Y, Z) be a fixed nonrotating coordinate system with origin at the center of the earth. Let $\mathbf{r}$ denote the vector from the center of the earth to the center of the satellite. Let (X', Y', Z') denote a second coordinate system, fixed in space, whose axes are parallel to the (X, Y, Z) frame. Let $\mathbf{r}_m$ and $\mathbf{r}_M$ denote the position vectors of the satellite and earth relative to the (X', Y', Z') system, respectively (see Fig. 2.28). Thus, $\mathbf{r} = \mathbf{r}_m - \mathbf{r}_M$. Moreover, from Newton's laws in the frame (X', Y', Z') we have

$$m\mathbf{r}_m'' = \left(-\frac{GMm}{r^2}\right)\frac{\mathbf{r}}{r} \tag{2}$$

and

$$M\mathbf{r}_M'' = \left(\frac{GMm}{r^2}\right)\frac{\mathbf{r}}{r} \tag{3}$$

for the gravitational force acting on the satellite and the earth, respectively. Canceling the masses from each side of Eqs. (2) and (3) and subtracting the results gives us

$$\mathbf{r}_m'' - \mathbf{r}_M'' = \left(-\frac{G(M+m)}{r^2}\right)\frac{\mathbf{r}}{r}.$$

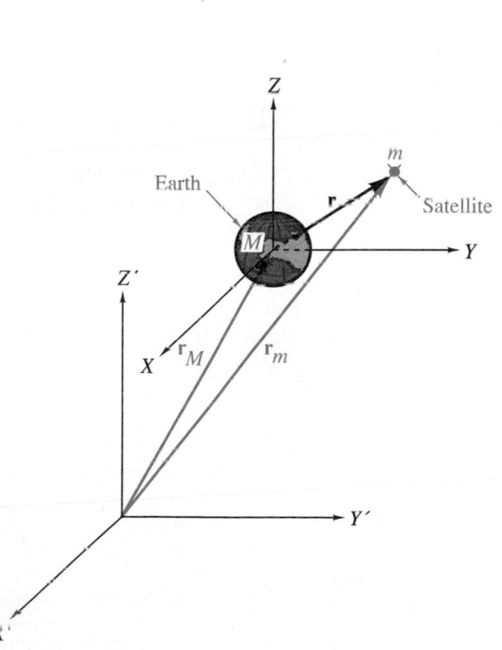

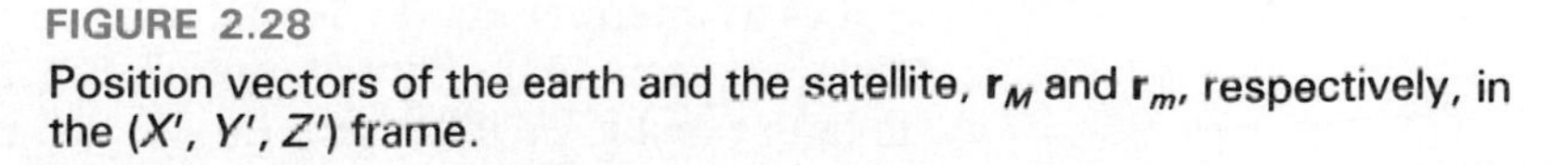

FIGURE 2.28
Position vectors of the earth and the satellite, $\mathbf{r}_M$ and $\mathbf{r}_m$, respectively, in the (X', Y', Z') frame.

Then the assumptions $M + m \approx M$ and $\mathbf{r} = \mathbf{r}_m - \mathbf{r}_M$ give the equation

$$\mathbf{r}'' = \left(-\frac{GM}{r^3}\right)\mathbf{r} \tag{4}$$

as the **vector equation of motion for the satellite.** Notice that we only needed the fixed frame (X', Y', Z') to derive Eq. (4) and so can now discard that frame. Thus we can measure position, velocity, and acceleration in any coordinate system we choose, such as the convenient (X, Y, Z) earth frame. Equation (4) is typical of a **central force** in which the force vector always points toward a fixed center.

It is convenient to define a parameter γ, called the **gravitational parameter,** as

$$\gamma = GM.$$

Then Eq. (4) simplifies to

$$\mathbf{r}'' + \frac{\gamma}{r^3}\mathbf{r} = \mathbf{0}. \tag{5}$$

We can also write Eq. (5) as a *system of first-order equations:*

$$\mathbf{r}' = \mathbf{v}, \tag{6}$$

$$\mathbf{v}' = -\frac{\gamma}{r^3}\mathbf{r}, \tag{7}$$

Conservation of Angular Momentum

The position vector $\mathbf{r}$ and the velocity vector $\mathbf{v}$ of the satellite depend on time; that is, $\mathbf{r} = \mathbf{r}(t)$ and $\mathbf{v} = \mathbf{v}(t)$. From Eq. (7) the vectors $\mathbf{r}$ and $\mathbf{v}'$ are parallel. Thus,

$$\mathbf{r} \times \mathbf{v}' = \mathbf{0}.$$

However,

$$\begin{aligned} \frac{d}{dt}(\mathbf{r} \times \mathbf{v}) &= (\mathbf{r} \times \mathbf{v}') + (\mathbf{r}' \times \mathbf{v}) \\ &= (\mathbf{r} \times \mathbf{v}') + (\mathbf{v} \times \mathbf{v}) \\ &= \mathbf{0} + \mathbf{0} = \mathbf{0}. \end{aligned}$$

Therefore

$$\mathbf{r} \times \mathbf{v} = \mathbf{h} \tag{8}$$

where $\mathbf{h}$ is a constant vector.

Let us interpret Eq. (8). Recall from calculus that the velocity vector $\mathbf{v}$ is always tangent to the path of motion. Since $\mathbf{r} \times \mathbf{v}$ is **normal** (perpendicular) to both $\mathbf{r}$ and $\mathbf{v}$, we infer from Eq. (8) that $\mathbf{r}$ and $\mathbf{v}$ always lie in a fixed plane that has the normal vector **h**. Therefore *the orbit of the satellite lies in a plane.* We conclude that there is no "wandering" in the path of the satellite and that the orbit must appear like that depicted in Fig. 2.29. It turns out that the orbit is an ellipse with one focus located at the center of the earth, but we are not going to derive this result. The constant vector **h** in Eq. (8) is called the **specific angular momentum** of the satellite. The plane of the orbit, which is constant and fixed in space, is called the **orbital plane.** Next we will calculate the potential energy associated with the satellite.

Conservation of Mechanical Energy

First, let us determine the work done in moving the satellite from one point in space to another against the force of gravity:

$$W = \int_{r_1}^{r_2} \mathbf{F} \cdot d\mathbf{r} = \int_{r_1}^{r_2} m\mathbf{v}' \cdot d\mathbf{r}$$

or, from Eq. (7)

$$\begin{aligned} W &= \int_{r_1}^{r_2} -\frac{m\gamma}{r^3}\mathbf{r} \cdot d\mathbf{r} \\ &= \int_{t_1}^{t_2} -\frac{m\gamma}{r^3}\mathbf{r} \cdot \mathbf{r}'\, dt \end{aligned} \tag{9}$$

where $r(t_1) = r_1$ and $r(t_2) = r_2$. We know that $d(r^2)/dt = d(\mathbf{r} \cdot \mathbf{r})/dt = 2\mathbf{r} \cdot \mathbf{r}'$ and $d(r^2)/dt = 2r\, dr/dt = 2rr'$. Thus $\mathbf{r} \cdot \mathbf{r}' = rr'$. Substitution of this result into Eq. (9) gives us

$$\begin{aligned} W &= \int_{t_1}^{t_2} -\frac{m\gamma}{r^3} rr'\, dt \\ &= \int_{r_1}^{r_2} -\frac{m\gamma}{r^2}\, dr \\ &= \frac{m\gamma}{r_2} - \frac{m\gamma}{r_1}. \end{aligned} \tag{10}$$

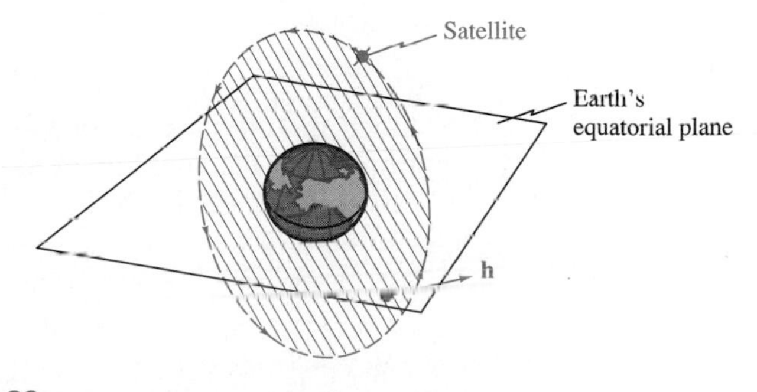

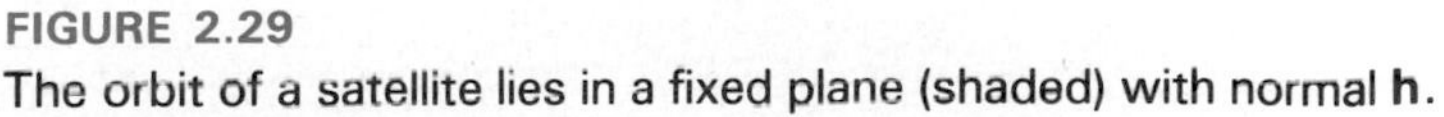

FIGURE 2.29
The orbit of a satellite lies in a fixed plane (shaded) with normal **h**.

Since the work done equals the change in potential energy, we can interpret the expression $-\gamma/r$ as the **potential energy per unit mass** of the satellite. Note that the potential energy is always negative.

Next, we take the dot product of $\mathbf{v}$ with Eq. (7):

$$\mathbf{v} \cdot \mathbf{v}' = -\frac{\gamma}{r^3} \mathbf{v} \cdot \mathbf{r},$$

$$vv' = -\frac{\gamma}{r^3} r'r,$$

or

$$v \frac{dv}{dt} = -\frac{\gamma}{r^2} \frac{dr}{dt}.$$

Integrating both sides of this last equation gives

$$\frac{v^2}{2} = \frac{\gamma}{r} + \Xi$$

or

$$\Xi = \frac{v^2}{2} - \frac{\gamma}{r} \tag{11}$$

where Ξ is a constant. Equation (11) tells us that *the sum of the kinetic energy per unit mass and the potential energy per unit mass of the satellite is always constant.* This principle is called the **conservation of mechanical energy.** The constant Ξ in Eq. (11) is called the **specific mechanical energy.** Thus the mechanical energy of the satellite remains constant throughout its orbital flight. Since the orbit is an ellipse, the distance r of the satellite from the center of the earth varies along the orbital path. As, the satellite gets *closer* to the earth, the potential energy $-\gamma/r$ decreases; at the same time the kinetic energy $v^2/2$ must *increase,* so the satellite speeds up. Finally, let us express the potential energy in terms of coordinates.

Exact Equation for Potential Energy

Since the orbit of the satellite is an ellipse in the orbital plane, we will introduce a rectangular coordinate system in this plane, with the origin at the center of the earth (see Fig. 2.30). The potential energy expressed in (x, y) coordinates is the function $\phi(x, y)$ given by

$$\begin{aligned} \phi(x, y) &= -\frac{\gamma}{r} \\ &= -\frac{\gamma}{\sqrt{x^2 + y^2}}. \end{aligned} \tag{12}$$

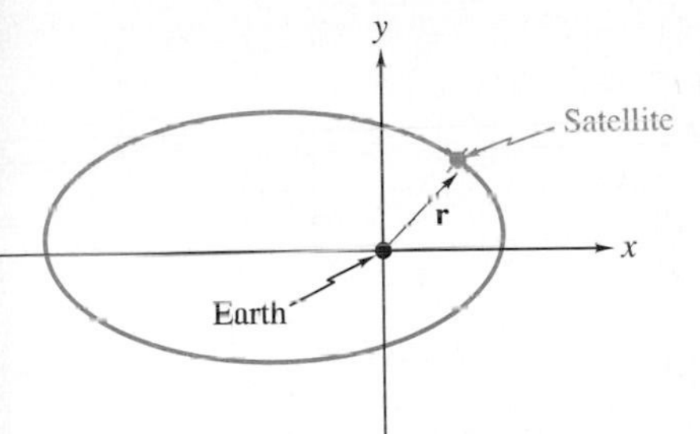

FIGURE 2.30
A rectangular coordinate system with the origin at the center of the earth is introduced in the orbital plane of the satellite.

Now *we assume further that the satellite is in a circular orbit.* This assumption means that the distance r of the satellite from the center of the earth is constant, so $\phi(x, y) \equiv$ constant. In that case,

$$0 = d\phi = \frac{\partial\phi}{\partial x}\,dx + \frac{\partial\phi}{\partial y}\,dy$$

$$= \frac{\gamma x}{(x^2 + y^2)^{3/2}}\,dx + \frac{\gamma y}{(x^2 + y^2)^{3/2}}\,dy.$$

That is, in the case of a circular orbit the potential energy of the satellite is the potential function $\phi(x, y)$ for which $\phi(x, y) = C$ is the general solution to the exact differential equation

$$x(x^2 + y^2)^{-3/2}\,dx + y(x^2 + y^2)^{-3/2}\,dy = 0. \tag{13}$$

Therefore the equipotential curves associated with the potential energy of the satellite are precisely those circles in the orbital plane whose centers lie at the center of the earth (see Fig. 2.31). This fact should come as no surprise given our development. Notice that for a satellite in circular orbit, the potential energy is constant, so the kinetic energy must also be constant by the principle of conservation of energy (11). Therefore a satellite in circular orbit moves at constant speed v.*

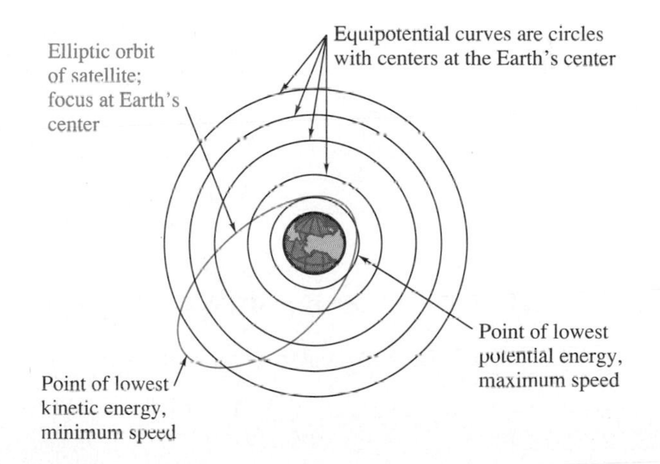

FIGURE 2.31
The equipotential curves of the satellite are circles concentric about the earth in the orbital plane. When the satellite is closest to the earth, it has minimum potential energy and therefore is traveling at maximum speed by Eq. (11).

* For further information regarding satellite motion, see *Fundamentals of Astrodynamics* by Roger R. Bate, Donald D. Mueller, and Jerry E. White (New York: Dover Publications, 1971).
A classic text in this area is *An Introduction to Celestial Mechanics* by Forest Ray Moulton (orig. ed., New York: Macmillan Company, 1914; 2d rev. ed., New York: Dover Publications, 1970). This text also introduces the more complicated three-body problem.

EXERCISES 2.9

1. If a satellite moves along a curve, its position, velocity, and acceleration can be expressed in terms of the moving unit vectors

$$\mathbf{u}_r = (\cos\theta)\mathbf{i} + (\sin\theta)\mathbf{j}$$
$$\mathbf{u}_\theta = -(\sin\theta)\mathbf{i} + (\cos\theta)\mathbf{j}.$$

a) Draw a diagram showing the vectors $\mathbf{u}_r$ and $\mathbf{u}_\theta$.

b) Show that $\mathbf{u}_r$ and $\mathbf{u}_\theta$ are perpendicular unit vectors and that the position vector of the satellite is $\mathbf{r} = r\mathbf{u}_r$.

c) Show that the velocity $\mathbf{v}$ of the satellite is given by

$$\mathbf{v} = \frac{dr}{dt}\mathbf{u}_r + \left(r\frac{d\theta}{dt}\right)\mathbf{u}_\theta.$$

d) Show that the acceleration $\mathbf{a}$ of the satellite is given by

$$\mathbf{a} = \left[\frac{d^2r}{dt^2} - r\left(\frac{d\theta}{dt}\right)^2\right]\mathbf{u}_r + \left[r\frac{d^2\theta}{dt^2} + 2\frac{dr}{dt}\frac{d\theta}{dt}\right]\mathbf{u}_\theta.$$

2. Kepler's Second Law

Use the equation for velocity in problem 1(c) and Eq. (8) for the specific angular momentum $\mathbf{h}$ to derive

$$\mathbf{h} = \left(r^2\frac{d\theta}{dt}\right)\mathbf{k}.$$

a) Assume that t is zero when the satellite is closest to the center of the earth. Then $r(0) = r_0$. Show that

$$|\mathbf{v}(0)| = v_0 = \left[r\frac{d\theta}{dt}\right]_{t=0}.$$

b) Show that the constant vector $\mathbf{h}$ satisfies

$$\mathbf{h} = r_0v_0\mathbf{k}.$$

c) Recall from calculus that the area element in polar coordinates is

$$dA = \frac{1}{2}r^2\,d\theta.$$

From this and part (b) conclude that

$$\frac{dA}{dt} = \frac{r_0v_0}{2}.$$

In other words, dA/dt is constant. This is Kepler's second law.

For the earth, $r_0 \approx 150{,}000{,}000$ km, $v_0 \approx 30$ km/sec, and $dA/dt \approx 2{,}250{,}000{,}000$ km²/sec.

3. Orbital Period

Recall from calculus that the area of an ellipse is given by $A = \pi ab$, where a is the length of the semimajor axis and b is the length of the semiminor axis.

a) If T denotes the **orbital period** of a satellite (the time it takes to complete one orbital revolution), use problem 2(c) to establish that

$$A = \frac{1}{2}Tr_0v_0$$

by integrating dA/dt from $t = 0$ to $t = T$.

b) Since for any ellipse the eccentricity e satisfies the relation

$$b = a\sqrt{1 - e^2},$$

show that

$$T = \frac{2\pi a^2}{r_0v_0}\sqrt{1 - e^2}.$$

CHAPTER 2 REVIEW EXERCISES

In problems 1–34 find the general or particular solution, as appropriate.

1. $y' = -\dfrac{x^2 + 6xy + y^2}{y + 2xy + 3x^2}$

2. $y' = \dfrac{xe^{7x}}{(y + 4)(y - 3)}$

3. $y' = \dfrac{x^2 + 6}{x - 2}$

4. $\dfrac{dy}{dx} = \dfrac{xe^{-3x}}{(y - 2)(y + 4)},\quad y(1) = 3$

5. $\dfrac{dy}{dx} = -\dfrac{2xy + 3x + y^2}{x^2 + 2xy},\quad y(1) = 2$

6. $\dfrac{dy}{dx} = -\dfrac{3yx^2 + y + x}{x^3 + x + y^2},\quad y(0) = 3$

7. $(y\cos x + 2xe^y)\,dx + (\sin x + x^2e^y + 2)\,dy = 0,\quad y(\pi) = 0$

8. $(\ln x)\,dx + (xy - x)\,dy = 0$

9. $(2x^3 - xy^2 - 2y + 3)\,dx + (-x^2y - 2x)\,dy = 0,\quad y(2) = 1$

10. $2x(y^2 + 1)\,dx - x^2y\,dy = 0$

11. $(3x^2y^2 - y^3 - 2x)\,dx + (1 - 3xy^2 + 2x^3y)\,dy = 0,\quad y(1) = 2$

12. $(y + 1)\,dy + (y - yx)\,dx = 0$

13. $(x^2 + y)\,dy + (2xy + x^2)\,dx = 0, \quad y(3) = 0$

14. $2x(y - 1)\,dx - y\,dy = 0$

15. $x\,dy + (3y - 6x^3)\,dx = 0$

16. $x\dfrac{dy}{dx} + y = x \sin x^2$

17. $x^3\dfrac{dy}{dx} - 3x^2y = x^3$

18. $xy' + y = \sin x$

19. $(4x^3y^3 - 2xy)\,dx + (3x^4y^2 - x^2 + 3)\,dy = 0$

20. $(x^2 - 1)(y + 1)^2\,dy + x\,dx = 0$

21. $x\dfrac{dy}{dx} + 2x^2y = 4x^2$

22. $(2x^3 - xy^2 - 2y + 3)\,dx + (-x^2y - 2x)\,dy = 0$

23. $xy' + y = xe^{x^2}$

24. $x^2dy + y^2dx = 0$

25. $x\dfrac{dy}{dx} + y = \sin x$

26. $(x^3 + x^2y + \sin y)\,dy + (3x^2y + xy^2 + e^x)\,dx = 0$

27. $(y + 2)\,dx + y(x + 4)\,dy = 0$

28. $(6x^2 + 4xy + y^2)\,dx + (2x^2 + 2xy - 3y^2)\,dy = 0$

29. $x^3y' - 3x^2y = x^3$

30. $(4 + 2y^3)\,dx + (2 + 12y^2x)\,dy = 0$

31. $(3xy + 3x - 4)\,dy + (y + 1)^2\,dx = 0$

32. $y^4\,dx + (2xy^3 - 1)\,dy = 0$

33. $(2r \sin\theta + \cos\theta)\,dr - (r \sin\theta - r^2 \cos\theta)\,d\theta = 0$

34. $\dfrac{dx}{dt} = 6 - \dfrac{3x}{50}, \quad x(0) = 0$

35. A small amount A_0 of a radioactive substance is placed in a lead container. After 24 hr it is observed that $6/7$ of the original amount is remaining. If the rate of decay is proportional to the amount of substance present at any time, what is the half-life of this substance? How long will it take for the substance to be reduced to $1/5$ of its original amount?

36. The rate of increase of the gypsy moth caterpillar population is directly proportional to the number P of the caterpillars present at any time t. The results of a survey show that the caterpillars increased from 2 million in 1979 to 3 million in 1981. Predict the gypsy moth caterpillar population in 1985.

37. It is claimed that a painting for sale is 400 years old. The pigment in the painting contains white lead ($_{210}$Pb), a radioactive isotope with a half-life of 22 years. Careful measurements indicate that 97.5% of the original amount of $_{210}$Pb in the pigment has disintegrated. Assume that the rate of decay is proportional to the amount present at any time. Predict the actual age of the painting.

38. Recall Newton's law of cooling from Chapter 1. In a room at a constant temperature of 70°F, the temperature of a cup of coffee changes from 200° at noon to 190° at 1 PM. Predict the temperature of the coffee at 3:30 PM.

39. The body of a murder victim was discovered at 11 PM one evening. The police doctor on call arrived at 11:30 PM and immediately took the temperature of the body, which was 94.6°F. He again took the temperature after one hour, when it was 93.4°F, and he noted that the temperature of the room was a constant 70°F. Use Newton's law of cooling to estimate the time of death, assuming the victim's normal body temperature was 98.6°F.

40. A frozen turkey is removed from a freezer kept at 20°F and placed in a refrigerator at 40°F to thaw. After 3 hr, the temperature of the turkey is 22°F. You know that the instantaneous rate of change of turkey temperature with respect to time is proportional to the difference in temperature of the turkey and its surroundings.

  a) Write a boundary-value differential equation problem that models this situation.
  b) When will the temperature of the turkey reach 35°F, assuming uniform warming?

41. In an oil refinery a storage tank contains 2000 gal of gasoline that initially has 100 lb of an additive dissolved in it. In preparation for winter weather, gasoline containing 2 lb of additive per gallon is pumped into the tank at a rate of 40 gal/min. The well-mixed solution is pumped out at the same rate. Find the amount of additive in the tank at time $t = 20$.

42. The rate of change of the concentration of pollution in Lake Michigan is equal to the difference between the concentration of polluted water entering the lake and the concentration of water leaving. Assume that water containing a constant concentration of k kg/km³ of pollutants enters the lake at a rate of 158 km³/yr, and water leaves the lake at the same rate. Also assume that the volume of Lake Michigan remains constant at 4900 km³.

a) Write a differential equation that models the rate of change of concentration of pollution in Lake Michigan. Solve the equation.

b) If the initial concentration of pollution is 50 kg/km^3, find the particular solution to the problem.

c) The fastest possible cleanup of the lake will occur if all pollution inflow ceases. This is represented by $k = 0$. If all pollution into the lake was stopped immediately, how long would it take to reduce pollution to 50% of its current level?

d) Graph your solution for the first 70 yr after pollution stops. What happens to the concentration as time goes on?

In problems 43–48 use the numerical method specified to make the required approximation. Show intermediate calculations, carrying four decimal places.

43. $y' = (x - y)^2, \quad y(0) = 0.5.$
Use Euler's method and step size $h = 0.1$ to estimate $y(0.5)$.

44. $y^{-2}\,dy = \dfrac{dx}{2 - x^{1/2}}, \quad y(0) = 1.$
Use Euler's method and step size $h = 0.3$ to estimate $y(0.6)$.

45. $y' = \dfrac{x + y}{2y + 3}, \quad y(1) = 3.$
Use the Improved Euler's method and step size $h = 0.2$ to estimate $y(1.6)$.

46. $y' = 2xy + 3y, \quad y(3) = 14.$
Use the Improved Euler's method and step size $h = 0.2$ to estimate $y(3.6)$.

47. $y' = 2xy + 2y, \quad y(3) = 2.$
Use a fourth-order Runge–Kutta method and step size $h = 0.4$ to estimate $y(3.8)$.

48. $y' = \dfrac{x + 1}{3y + 1}, \quad y(1) = 2.$
Use a fourth-order Runge–Kutta method and step size $h = 0.4$ to estimate $y(1.8)$.

49. A projectile of mass 0.10 kg shot vertically upward with an initial velocity of 8 m/sec is slowed due to the forces exerted by gravity and air resistance.

a) If the force due to air resistance equals 0.002 times the square of the projectile's instantaneous velocity acting opposite the direction of the velocity, write an initial-value differential equation problem that models this situation. Use velocity as the dependent variable.

b) Apply a fourth-order Runge–Kutta method with $h = 0.05$ to estimate the projectile's instantaneous velocity at 0.05, 0.10, and 0.15 sec.

50. An automobile headlight mirror is designed to reflect the light given off by the headlamp in rays parallel to the road surface (horizontal) (Fig. 2.32). By using the principle of optics that the angle of light incidence equals the angle of light reflection, we can derive a first-order differential equation that models the desired shape of the mirror:

$$\frac{dy}{dx} = \frac{-x + (x^2 + y^2)^{1/2}}{y}, \quad y > 0.$$

The mirror is designed so that the distance of the mirror directly above the lamp is 1 in, so $y(0) = 1$. Use the fourth-order Runge–Kutta method with $h = 1$ to find the vertical distance of the mirror from the horizontal line running through the lamp for distances of 1 and 2 in in front of the lamp.

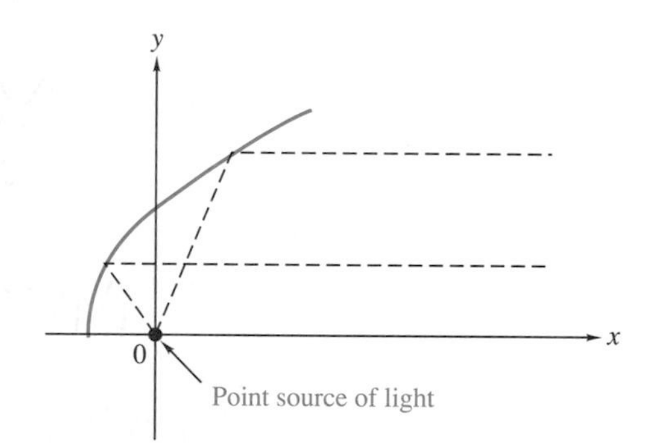

FIGURE 2.32
By using the principle of optics that the angle of light incidence equals the angle of light reflection, we can derive a first-order differential equation that models the desired shape of the mirror:

51. A rocket with an initial mass of 25,000 kg is launched from the earth's surface. It expels exhaust gas at a constant rate of 1000 kg/sec at a constant velocity of 400 m/sec relative to the rocket. If there are no external forces acting on the rocket, the time rate of change of the rocket's displacement from earth can be modeled by the following differential equation:

$$\frac{dx}{dt} = -400 \ln \left| \frac{25{,}000 - 1000t}{25{,}000} \right| - 9.8t.$$

Using the fourth-order Runge–Kutta method and $h = 1$, estimate the displacement of the rocket 1 and 2 sec after launch.

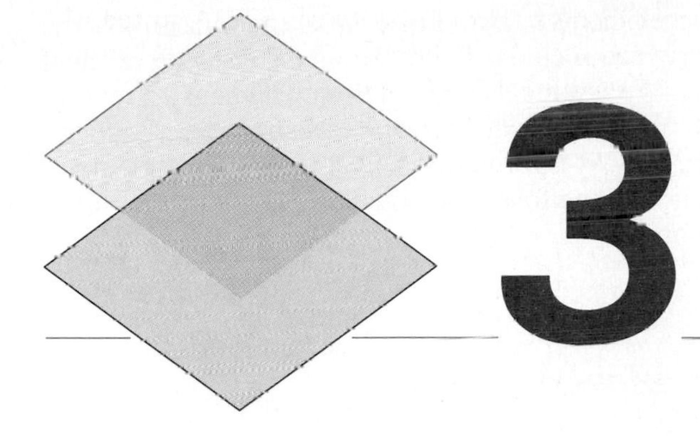

3

Homogeneous Linear Equations with Constant Coefficients: Modeling without Forcing Functions

In this part of the text we construct and solve models of the form

$$a_n y^{(n)} + a_{n-1} y^{(n-1)} + \cdots + a_1 y' + a_0 y = g(x) \tag{1}$$

where the coefficients a_i are constants (some of which may be zero), but $a_n \neq 0$. The positive integer n is called the **order** of the differential equation. That is, the order of a differential equation is the order of the highest-appearing derivative.

Differential Eq. (1) is an example of a linear equation.

DEFINITION 3.1

An ***n*th-order linear differential equation** has the form

$$a_n(x) y^{(n)} + a_{n-1}(x) y^{(n-1)} + \cdots + a_1(x) y' + a_0(x) y = g(x) \tag{2}$$

The function $g(x)$ is called the **forcing function.** If $g(x) \equiv 0$, differential Eq. (2) is said to be **homogeneous**; otherwise it is **nonhomogeneous.***

* These terms are used in the sense of homogeneous linear algebraic equations, like $ax + by + cz = 0$. They are not related to equations that are *homogeneous-type*, a term used to describe a special class of first-order differential equations discussed in Section 2.4.

Notice that all of the coefficients $a_i(x)$ in Eq. (2) are functions only of the independent variable x (which includes the possibility that some of them may be constants or zero). We assume that the leading coefficient $a_n(x)$ is not identically equal to zero. Values of x where $a_n(x) = 0$ cause special difficulties in solving linear equations. We will address this issue in Chapter 9 in our discussion of series solutions. The following equations are all examples of linear differential equations:

$$\frac{dy}{dx} = ky,$$

$$2y'' + 5y' - 3y = 0,$$

$$y' + \frac{1}{x} y = 2,$$

$$x^2y'' + xy' - 2y = \sin x.$$

The first two equations are homogeneous and have constant coefficients; the last two are nonhomogeneous and have at least one variable coefficient.

Notice that in a linear differential equation the dependent variable y or any of its derivatives appear to at most the first power. Moreover, there are neither products of y with any of its derivatives nor products of the derivatives. Also, no functional expressions involving y or any of its derivatives can appear. The following are examples of *nonlinear* differential equations:

$$y\frac{dy}{dx} = x,$$

$$y'' + x \sin y = 0,$$

$$y'' + 2(y')^2 = e^x.$$

The forcing function for nonhomogeneous linear equations may be **continuous** over the interval of interest, or it may be **discontinuous** (such as a step function or square wave). We also distinguish those forcing functions that are **periodic,** like a sine wave for instance. Examples of these types of forcing functions are depicted in Fig. 3.1.

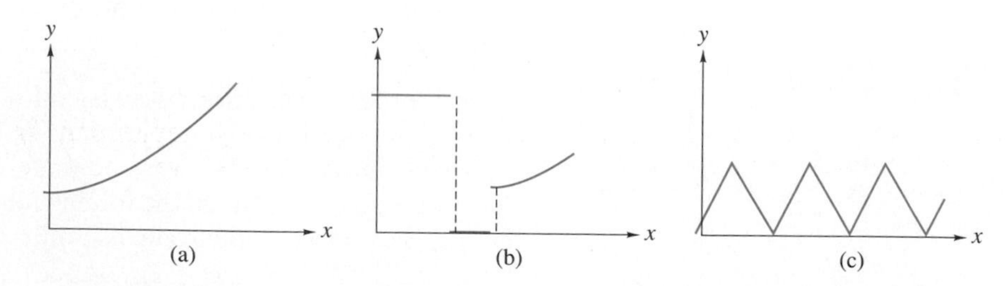

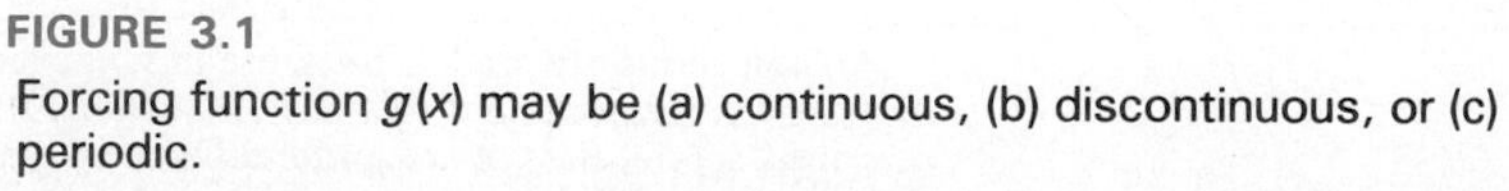

FIGURE 3.1

Forcing function $g(x)$ may be (a) continuous, (b) discontinuous, or (c) periodic.

Homogeneous and nonhomogeneous linear equations with constant coefficients provide good first approximations to much of the physical behavior studied at the undergraduate level. Section 3.3 and later sections will investigate solutions to homogeneous equations with constant coefficients a_i. Solutions to nonhomogeneous equations will be the subject of later chapters, presented according to continuous forcing functions (Chapter 4), discontinuous forcing functions (Chapter 5), and periodic and impulse forcing functions (Chapter 6).

Before discussing solutions to homogeneous equations, Section 3.1 will develop some models for classical behaviors you will study in depth in your science and engineering courses. These models motivate the investigation of the class of linear differential equations with constant coefficients and their solution techniques.

3.1 MODELING WITH LINEAR DIFFERENTIAL EQUATIONS HAVING CONSTANT COEFFICIENTS

Recall the discussion of proportionality in Section 1.3. The quantity y is proportional to x if and only if $y = kx$ for some constant k. Note that whenever we assume a proportionality behavior for the dependent variable or any of its derivatives, a submodel having a constant coefficient must result. The graph of the proportionality relationship is then a straight line through the origin, so the use of a proportionality as a simplifying assumption can be thought of as a **linearization** assumption.

EXAMPLE 1

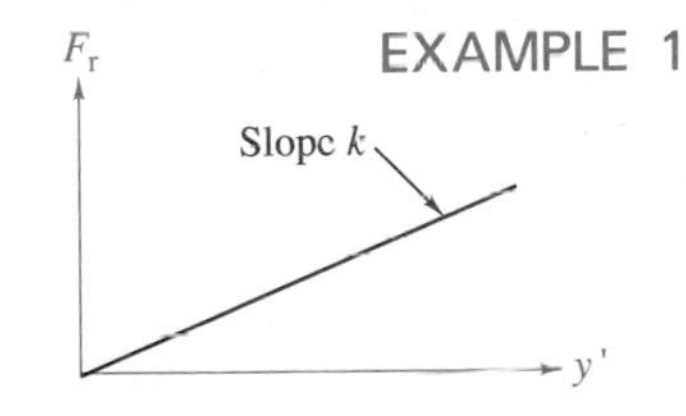

FIGURE 3.2
Submodel $F_r = ky'$ for $k > 0$ is a constant-coefficient submodel relating the force of friction and velocity.

Suppose we are interested in the distance $y(t)$ traveled by a falling object, and we are developing a submodel for the frictional force of drag F_r caused by air resistance. If we assume the air resistance is proportional to the velocity of the falling object, then we obtain the following constant-coefficient submodel

$$F_r = kv = ky'.$$

A graph of this submodel is depicted in Fig. 3.2. Using a linear submodel as a simplifying assumption is a very powerful and useful tool in mathematical modeling.

EXAMPLE 2

Consider again the problem of describing the motion of an object that is initially at rest and dropped from a hovering helicopter. Assuming no resistive forces we constructed the following model in Section 1.3:

$$F = F_p - F_r,$$
$$ma = mg - 0,$$

or, since $a = y''$,

$$y'' = g. \tag{1}$$

Although this differential equation is nonhomogeneous, it can readily be solved by integrating twice to yield

$$y = \frac{gt^2}{2} + c_1 t + c_2. \tag{2}$$

The constants of integration c_1 and c_2 can be evaluated using the initial conditions $y'(0) = v_0$ and $y(0) = y_0$, to yield the particular solution

$$y = \frac{gt^2}{2} + v_0 t + y_0. \tag{3}$$

Note that solution (2) to second-order differential Eq. (1) has *two* arbitrary parameters c_1 and c_2 resulting from two integrations. In Section 3.2 we will establish that the general solution to a linear second-order equation always contains two arbitrary constants.

EXAMPLE 3 DYNAMIC EQUILIBRIUM OR STEADY STATE MOTION

When an object is in dynamic equilibrium, the forces of propulsion F_p and resistance F_r are in perfect balance. Thus if we sum the forces acting on the body as depicted in Fig. 3.3, we have from Newton's second law that

$$F = F_p - F_r,$$
$$ma = 0,$$

or

$$my'' = 0. \tag{4}$$

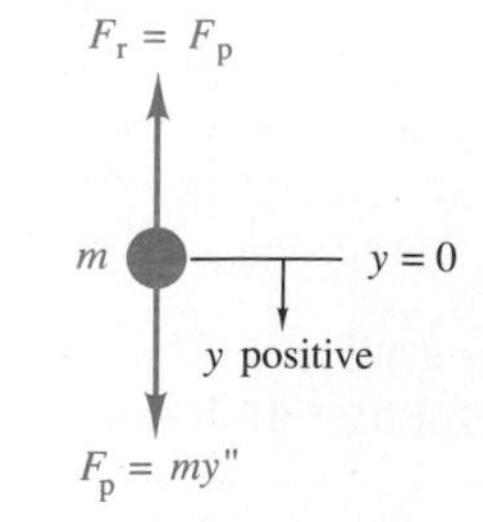

FIGURE 3.3
The forces of propulsion F_p and resistance F_r are in perfect balance.

This second-order linear homogeneous equation can be integrated twice to yield

$$y = c_3 t + c_4. \tag{5}$$

Again the two constants of integration c_3 and c_4 can be evaluated from the information $y(0) = y_0$ and $y'(0) = v_0$ at time $t = 0$ to yield

$$y = v_0 t + y_0.$$

Let us compare solution (5) with solution (2) obtained previously for the falling body with no resistance. The function

$$y = \frac{gt^2}{2} + c_1 t + c_2$$

turns out to be the general solution to the *nonhomogeneous equation* $y'' = g$. On the other hand, the function

$$y = c_3 t + c_4$$

is the general solution to the *homogeneous equation* $y'' = 0$. (These facts will be established more generally in Sections 3.2 and 4.1.) Observe that the general solution to the nonhomogeneous equation differs from the solution to the homogeneous equation by only the single term $gt^2/2$. Note further that this term $y = gt^2/2$ is a function that solves the nonhomogeneous equation $y'' = g$. This behavior for solutions is always exhibited for linear equations (as you will see in Section 4.1). Thus, for this example the general solution to the nonhomogeneous equation is given by

$$y = c_1 t + c_2 + \frac{gt^2}{2}.$$

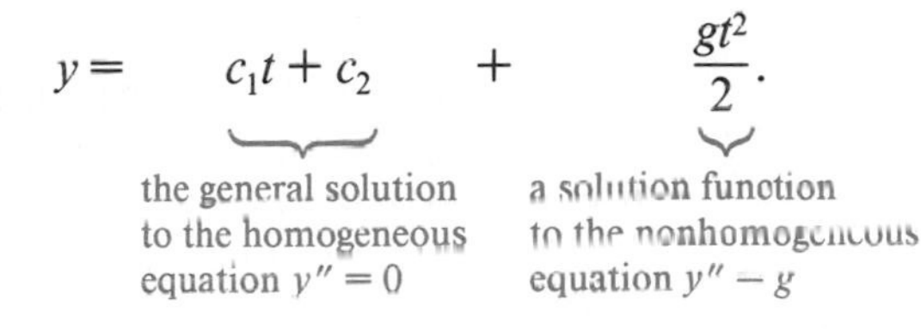

This important property for linear equations is not shared by nonlinear equations.

EXAMPLE 4 AN AUTOMOBILE SUSPENSION SYSTEM

How stiff should the springs in an automobile be and how heavy duty should the shocks be for a safe and comfortable ride? What does "tuning" a suspension system mean? To answer these questions we will now construct a differential equations model for an idealized automobile suspension system. Assume that the automobile is resting on a hydraulic lift and that the wheels are hanging freely. Let us develop submodels to approximate the effect of the spring and shock absorber system (Fig. 3.4).

Submodel for the Shock Absorber As the plunger in the shock absorber causes the fluid to move, a certain resistance F_f to this motion is exhibited—a sort of internal molecular friction. This **viscous drag** is caused by short-range molecular cohesive forces, and it opposes the upward (or downward) motion of the suspended wheel. In both liquids and gases, viscous drag is proportional to the velocity as long as the velocity is slow enough

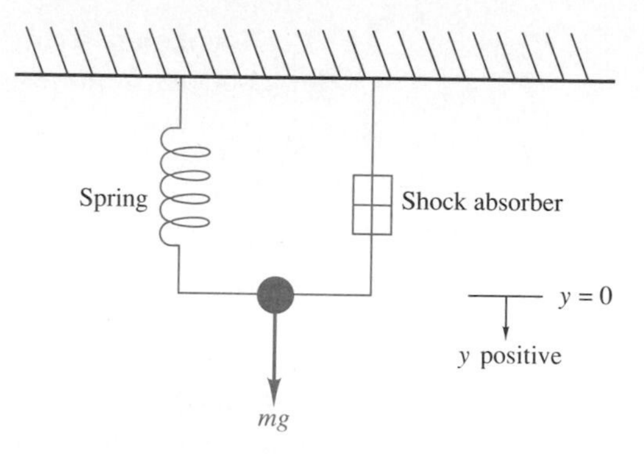

FIGURE 3.4
An idealization of an automobile wheel supported by a spring and a shock absorber.

to move the molecules of the fluid from point to point without any rotational motion or turbulence. Thus

$$F_f = \delta y' \tag{6}$$

for some positive constant δ (called the **damping coefficient**). Of course, this approximation can be reasonable only over a restricted range of values for the velocity y'.

Submodel for the Spring Now consider the spring, suspended as diagrammed in Fig. 3.5(a). With no weight attached, the spring assumes a natural equilibrium of length ℓ_0. Next suppose a weight mg (such as a wheel) is attached to the spring (see Fig. 3.5b). In this situation the spring stretches a distance ℓ and achieves a new equilibrium in which the propulsion and restoring forces of the spring are equal. The propulsion force is the weight mg, but how shall we model the restoring force of the spring, F_r? Again we assume a simple proportionality:

> *The restoring force is proportional to the distance the spring is stretched or squeezed beyond its natural equilibrium point.*

This result is known as **Hooke's law.** We know it is not a precise relationship; consider what happens when you try to stretch a coil far beyond the equilibrium point. Nevertheless, Hooke's law has been empirically shown to be reasonably accurate over a restricted range of values. Thus when the propulsion and restoring forces are balanced, the spring is in an equilibrium position as characterized by

$$F_p = F_r$$

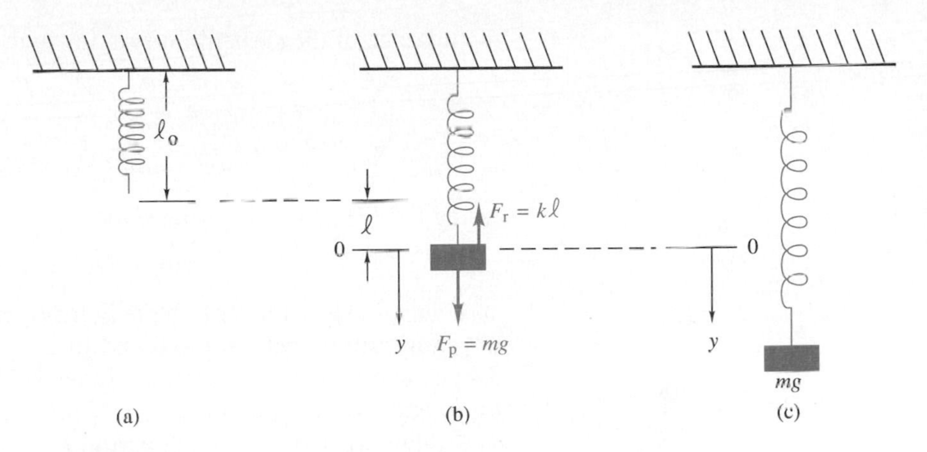

FIGURE 3.5
A spring seeks an equilibrium position where the restoring force of the spring equals the forces of propulsion. Hooke's law hypothesizes a simple proportionality between the distance the spring is stretched beyond its natural equilibrium and the restoring force.

or

$$mg = k\ell. \tag{7}$$

Suppose we stretch the spring a distance y beyond the previous equilibrium position of $y = 0$ shown in Fig. 3.5(b); Fig. 3.5(c) diagrams the new situation. What is the motion of the end of the spring where the weight is attached? That is, we want to know y at any time as the spring moves up and down. From Newton's second law, we know that

$$F = F_p - F_r,$$
$$ma = mg - k(\ell + y),$$
$$my'' = mg - k(\ell + y).$$

Using Eq. (7) we have

$$my'' = -ky. \tag{8}$$

Since we are not modeling any damping forces, we expect model (8) to predict vertical periodic oscillations about the equilibrium position $y = 0$. In Section 3.4 we will analyze the solutions to Eq. (8) and note the effect that the constants k and m have on the nature of the oscillations.

Modeling the Suspended Wheel Let us incorporate our submodels for the spring and the shock absorber to model the motion of the suspended wheel depicted in Fig. 3.4. Setting the origin $y = 0$ at the natural equilibrium position as above, we will now model the motion $y(t)$ resulting from some disturbance of the suspension system (such as an initial displacement):

$$F = F_p - F_f - F_r,$$
$$ma = mg - \delta y' - k(\ell + y),$$
$$my'' = mg - \delta y' - k(\ell + y).$$

Using Eq. (7) and simplifying gives us

$$my'' + \delta y' + ky = 0. \tag{9}$$

You would expect the motion predicted by model (9) to be oscillatory about the equilibrium position $y = 0$ and to eventually damp to zero. In Section 3.4 you will see not only that this is indeed the case but also how the constants m, δ, and k determine the nature of the damping.

Finally, we add a forcing function $f(t)$ to the system. This function may approximate the effect of an external disturbance on the system, such as periodic bumps or potholes in the road. [Several different types of functions for $f(t)$ are developed in Chapters 4 and 5. The more realistic $f(t)$, the more difficult is the solution to the model.] Inclusion of the forcing function results in a second-order linear nonhomogeneous differential equation with constant coefficients:

$$my'' + \delta y' + ky = f(t). \tag{10}$$

A solution to Eq. (10) is developed in Section 4.5. For simplicity in our presentation here, we assume the car is on a hydraulic lift with the wheel hanging freely. In the exercises at the end of this section you will be asked to model the action of the wheel when it is in contact with the ground.

EXAMPLE 5 CIRCULAR MOTION UNDER GRAVITY; UNDAMPED AND DAMPED PENDULUMS

We now investigate the motion of a pendulum. Suppose a mass m is suspended from a ceiling by a long thin rod. Let ℓ denote the length of the pendulum rod and θ denote the angle of displacement from the vertical, as illustrated in Fig. 3.6. We are interested in describing the motion of the pendulum. If we know the angle θ at any time, we can then predict the pendulum's position, amplitude, and period. The amplitude is the maximum angle of displacement away from the vertical, and the period is the time required for the pendulum bob to swing through one complete cycle. We can set the pendulum in motion simply by releasing it from some initial raised position $\theta = \theta_0$ with zero initial velocity, by giving it an initial velocity at the moment of release, or even by providing some kind of periodic forcing system, such as a motor. Among the forces to consider in our model are gravity, friction at the hinge, and resistance due to the fluid in which the pendulum resides. We will begin with a "simple" pendulum, ignoring fric-

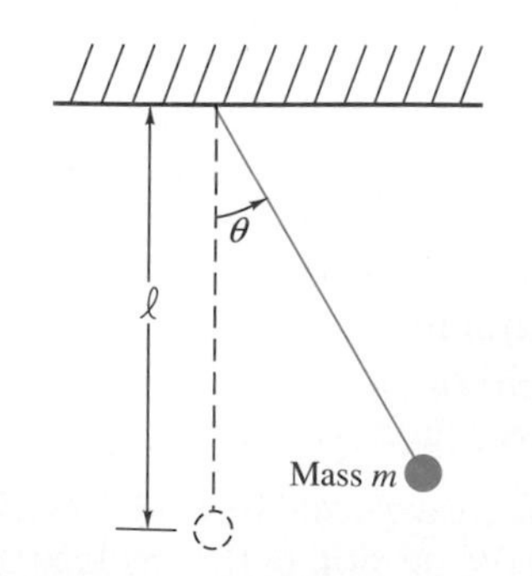

FIGURE 3.6
A simple pendulum.

tion at the hinge and any damping due to fluid resistance; that is, we will treat the pendulum as if it is moving in a vacuum.

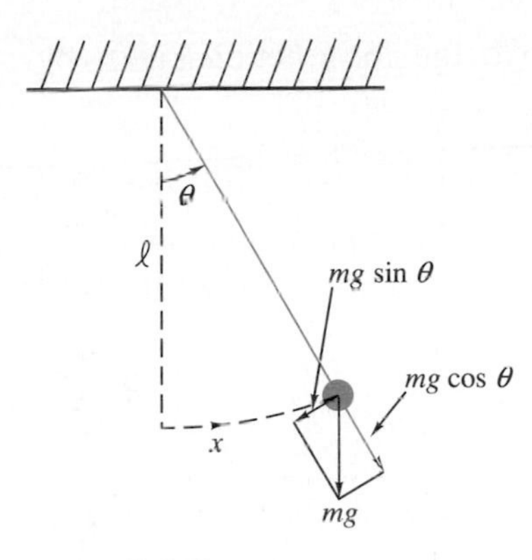

FIGURE 3.7
Assume the force due to gravity is the only force acting on the simple pendulum in this free-body diagram valid at any time.

Assumptions

Let us make some simplifying assumptions. We assume the pendulum rod to be "massless" and the (entire) mass m of the pendulum to be concentrated at a single point. Neglecting buoyancy, friction at the hinge, and fluid resistance, we have the free-body diagram shown in Fig. 3.7. Note that the angle of displacement θ and the arc length x are measured starting from the vertical position, and moving positive counterclockwise to the right.

Undamped Pendulums

The force of gravity mg can be resolved into a tangential component, $mg \sin \theta$, and a component normal or perpendicular to the path of the pendulum, $mg \cos \theta$. Using the coordinate system indicated in Fig. 3.7 together with Newton's second law, we sum the forces in the tangential direction of motion to obtain the force of propulsion:

$$F = F_p,$$
$$ma = -mg \sin \theta,$$
$$mx'' = -mg \sin \theta,$$

or

$$x'' + g \sin \theta = 0. \tag{11}$$

The negative sign for the force $mg \sin \theta$ occurs because it is a *restoring force* that is always directed toward the equilibrium (vertical) position. Thus, when θ is positive and the pendulum is to the right of the vertical position, the restoring force is negative and directed tangentially towards the left; when θ is negative and the pendulum is to the left of the vertical, the restoring force is positive and directed tangentially towards the right (see Fig. 3.7).

Nonlinear Model To simplify expression (11), we use the relation $x = \ell\theta$ for the length of arc subtended by an angle θ measured in radians. Substituting $x'' = \ell\theta''$ into Eq. (11) yields

$$\ell\theta'' + g \sin \theta = 0. \tag{12}$$

Model (12) describes an idealization of a simple undamped, unforced pendulum. Although Eq. (12) models an extremely simple behavior (since resistive forces have been neglected), the differential equation cannot be solved analytically by elementary methods. It is another example of a nonlinear equation. The nonlinearity in this case is due to the presence of the sin θ term.

Linear Model The Maclaurin expansion for the sine function is given by

$$\sin \theta = \theta - \frac{\theta^3}{3!} + \frac{\theta^5}{5!} - \quad . . . + \frac{(-1)^n \theta^{2n+1}}{(2n+1)!} + \quad . . .$$

where the angle θ is measured in radians. Thus, for very small angles (just how "small" needs to be investigated), we can make the further approximation that

$$\sin \theta \approx \theta$$

to obtain the linear model

$$\ell \theta'' + g\theta = 0. \tag{13}$$

Differential Eq. (13) is a *linearization* of Eq. (12).

Damped Pendulums

Next we consider the effect air or fluid resistance has on the pendulum. The force of resistance F_d acts in opposition to the motion. For low velocities we can reasonably assume a "viscous damping" submodel in which F_d is proportional to the instantaneous change in arc length; that is, the resistance is proportional to the velocity along the path traced out by the pendulum bob. Thus $F_d = -\beta x'$ where β is a positive constant. Our force equation then becomes

$$F = F_p + F_d,$$
$$ma = -mg \sin \theta - \beta x',$$

or

$$mx'' = -mg \sin \theta - \beta x'.$$

Substituting $x' = \ell \theta'$ and $x'' = \ell \theta''$ as before yields the equation

$$m\ell \theta'' + \beta \ell \theta' + mg \sin \theta = 0 \tag{14}$$

for the damped pendulum. Finally, if a forcing function $f(t)$ is incorporated to model any additional external force imposed on the pendulum system, we obtain

$$m\ell \theta'' + \beta \ell \theta' + mg \sin \theta = f(t). \tag{15}$$

Again using $\sin \theta \approx \theta$ we obtain the linearized approximation

$$m\ell \theta'' + \beta \ell \theta' + mg\theta = f(t). \tag{16}$$

Later in this chapter we will analyze the effect that different values of the constants m, ℓ, and β have on the resulting motion of the pendulum when $f(t) \equiv 0$.

For the next example we turn our attention from the physics of motion to electricity. As a basis for comparison, first note that the mass of an object together with its position are fundamental to a description of motion. According to Newton's law of gravitational attraction the force of mutual attraction between two masses m_1 and m_2 separated by a distance r is proportional to $m_1 m_2/r^2$. **Potential energy** is the energy (the capacity for doing work) of a system of masses due to the relative position of the parts of the system. It takes at least two masses to have potential energy. For instance, under the earth's gravity an object has potential energy when raised above some "zero" level such as the ground. *The gravitational potential equals the amount of work necessary to raise the object to that height above the ground.* Newton's second law proposes a proportionality to describe the "flow of masses," or motion in a gravitational field. Now let us study what happens with electricity.

EXAMPLE 6 AN ELECTRIC CIRCUIT

The basic quantity in electricity is the **charge** q. Analogous to the law of gravitational attraction is **Coulomb's law:**

> The force E of mutual attraction between two charges q_1 and q_2 is given by the proportionality
>
> $$E \propto \frac{q_1 q_2}{r^2}. \tag{17}$$

Hence the **electric force** at a point in an electric field is the force that would be exerted on a unit positive charge placed there. The **difference in electric potential** between two points in an electric field is the amount of work necessary to move a unit positive charge from one point to the other.

To describe motion in a gravitational field we use displacement (or position) x and velocity $v = dx/dt$. Analogously, in an electric field we use charge q and its flow or **current** $i = dq/dt$. As you might guess, there are interesting similarities between motion in a gravitational field and the flow of electrons (the carriers of charge) in an electric field. These concepts are developed in great detail in physics and electricity courses. Here we simply develop submodels for some components of electrical networks involving **resistors, inductors,** and **capacitors** using simple proportionalities between the electric potential, charge, and current.

Consider the electric circuit shown in Fig. 3.8. It consists of four components: voltage source, resistor, inductor, and capacitor. We introduce the following notation:

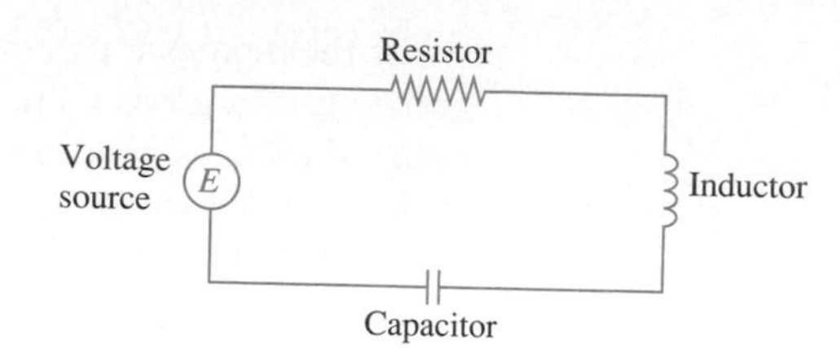

FIGURE 3.8
An electric circuit.

q: charge at a cross section of a conductor measured in **coulombs** (abbreviated c);
i: current or rate of change of charge dq/dt (flow of electrons) at a cross section of a conductor measured in **amperes** (abbreviated A);
E: electric (potential) source measured in **volts** (abbreviated V);
V: difference in potential between two points along the conductor measured in **volts** (V).

You can think of electrical flow as being like fluid flow, where the voltage source is the pump and the resistor, inductor, and capacitor tend to block the flow. A battery or generator is an example of a source, producing a voltage that causes the current to flow through the circuit when the switch is closed. An electric light bulb or appliance would provide resistance. The inductance is due to a magnetic field that opposes any change in the current as it flows through a coil. The capacitance is normally created by two metal plates that alternate charges and thus reverse the current flow.

Submodel for Resistance Let us begin by considering a submodel for the resistor. Ohm observed that the current i, flowing through a resistor caused by a potential difference across it, is (approximately) proportional to the potential difference. He named this constant of proportionality $1/R$ and called R the **resistance.** Thus

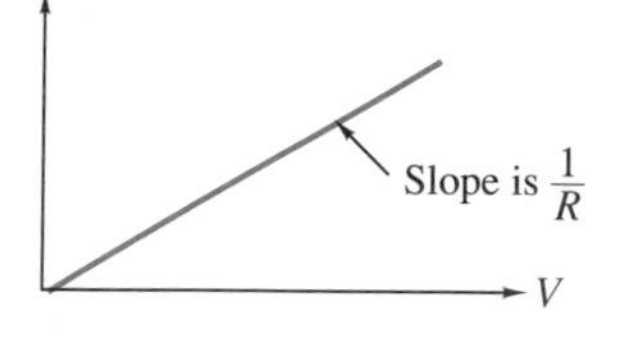

FIGURE 3.9
The current induced across a resistor is proportional to the potential difference across the resistor.

$$i = \frac{1}{R} V. \tag{18}$$

This idea is illustrated in Fig. 3.9. In actuality the behavior is much more complex than that modeled by Eq. (18) and depends on several other factors (temperature, for instance).

In a similar experimental manner, approximations to the voltage drops across the inductor and capacitor have been observed to satisfy proportionalities.

Submodel for Inductance The voltage drop across the inductor is proportional to the change in the current:

$$V = L \frac{di}{dt}. \tag{19}$$

where the constant of proportionality L is the **inductance.**

Submodel for Capacitance The voltage drop across the capacitor is proportional to the charge:

$$V = \frac{1}{C} q. \tag{20}$$

where C is the **capacitance** and q is the charge on the capacitor.

Modeling an Electric Circuit Let us use these three simple submodels to develop a model describing the current flow in an electric circuit. The German physicist Gustav R. Kirchhoff (1824–1887) observed the following result through experimentation:

KIRCHHOFF'S SECOND LAW

The sum of the voltage drops in a closed circuit is equal to the impressed voltage.

In Fig. 3.8 the "impressed" or applied voltage is the voltage source $E(t)$. Thus, from Kirchhoff's second law we have

$$L\frac{di}{dt} + Ri + \frac{1}{C} q = E(t). \tag{21}$$

Using the relation $i = dq/dt$, we can rewrite Eq. (21) as

$$Lq'' + Rq' + \frac{1}{C} q = E(t). \tag{22}$$

Summary

Note that Eq. (22) is a second-order linear *nonhomogeneous* differential equation with constant coefficients modeling the *LRC* electric circuit shown in Fig. 3.8. Thus it has exactly the same mathematical form as the model for motion of a physical object when damping, viscous friction, and a forcing function are all present (for example, damped motion under gravity, damped spring–mass systems, and damped pendulums). *Only the interpretations for the constants of proportionality change.* Hence, the differential equation models we are going to investigate provide good first approximations to classical behaviors of mechanics and electromagnetism. We summarize our analogies for the physics of motion versus electricity in the following table.

LINEAR SECOND-ORDER CONSTANT-COEFFICIENT MODELS

Mechanical System	Electrical System
$my'' + \delta y' + ky = f(t)$	$Lq'' + Rq' + \frac{1}{C}q = E(t)$
y: displacement	q: charge
y': velocity	q': current
y'': acceleration	q'': change in current
m: mass	L: inductance
δ: damping constant	R: resistance
k: spring constant	$1/C$: where C is the capacitance
$f(t)$: forcing function	$E(t)$: voltage source
Potential field: $F \propto \frac{m_1 m_2}{r^2}$	Potential field: $E \propto \frac{q_1 q_2}{r^2}$

EXERCISES 3.1

1. An object initially at rest and weighing 80 lb falls from a great height and encounters air resistance 3 times its instantaneous velocity. Assuming no other forces act on the object, formulate a second-order initial value problem for the distance the object falls. Assume that the acceleration due to gravity is the constant 32 ft/sec^2.

2. An object of mass m is thrown vertically downward with an initial velocity of v_0 ft/sec in a medium resisting the motion with a force proportional to the velocity. Formulate a second-order initial value problem for the vertical distance the object travels as a function of time.

3. If a spring is stretched 0.37 in. by a 14-lb force, what stretch will be produced by a 9-lb force? by a 22-lb force? Assume Hooke's law.

4. A man weighing 192 lb jumps off a building that is 2000 ft high. Assume that as he falls he is subject to a drag force that numerically is equal to 12 times his instantaneous velocity. His initial velocity is assumed to be zero. Formulate an initial value problem expressing the man's vertical displacement as a function of time.

5. A 16-lb weight is attached to the lower end of a coil spring suspended from the ceiling and having a spring constant of 1 lb/ft. The resistance in the spring–mass system is numerically equal to the instantaneous velocity. At $t = 0$ the weight is set in motion from a position 2 ft below its equilibrium position by giving it a downward velocity of 2 ft/sec. Write an initial value problem that models the given situation.

6. An 8-lb weight stretches a spring 4 ft. The spring–mass system resides in a medium offering a resistance to the motion that is numerically equal to 1.5 times the instantaneous velocity. If the weight is released at a position 2 ft above its equilibrium position with a downward velocity of 3 ft/sec, write an initial value problem modeling the given situation.

7. A 20-lb weight is hung on an 18-in. spring and stretches it 6 in. The weight is pulled down 5 in. and 5 lb are added to the weight. If the weight is now released with a downward velocity of v_0 in/sec, write an initial value problem modeling the vertical displacement.

8. A 10-lb weight is suspended by a spring that is stretched 2 in. by the weight. Assume a resistance whose magnitude is $20/\sqrt{g}$ lb times the instantaneous velocity v in feet per second. If the weight is pulled down 3 in. below its equilibrium position and released, formulate an initial value problem modeling the behavior of the spring–mass system.

9. A mass weighing 8 lb stretches a spring 3 in. The spring–mass system resides in a medium with a damping constant of 2 lb-sec/ft. If the mass is released from its equilibrium position with a velocity of 4 in/sec in the downward direction, write an initial value problem modeling the vertical displacement.

10. An *LRC* circuit is set up with an inductance of 1/5 henry, a resistance of 1 ohm and a capacitance of 5/6 farad. Assuming the initial charge is 2 coulombs and the initial current is 4 amperes, write a second-order initial value problem for the charge on the capacitor.

11. An (open) electrical circuit consists of an inductor, a resistor, and a capacitor. There is an initial charge of 2 coulombs on the capacitor. At the instant the circuit is closed, a current of 3 amperes is present and a voltage of $E(t) = 20 \cos t$ is applied. In this circuit the voltage drop across the resistor is 4 times the instantaneous change in the charge, the voltage drop across the capacitor is 10 times the charge, and the voltage drop across the inductor is 2 times the instantaneous change in the current. Write an initial value problem to model the circuit.

12. An inductor of 2 henrys is connected in series with a resistor of 12 ohms, a capacitor of 1/16 farad, and a 300 volt battery. Initially, the charge on the capacitor is zero and the current is zero. Formulate an initial value problem modeling this electrical circuit.

13. A spring–mass system consists of a spring with a spring constant of 3 lb/ft suspending a 16-lb weight. The system resides in a medium with a damping constant of 1 lb-sec/ft. There is an external force of $f(t) = 10 \cos 8t$ lb on the system. Using a resistance of 10 ohms, find an *LRC* circuit that is governed by exactly the same differential equation as that modeling the mechanical spring–mass system.

14. A spring–mass system resides in a medium with a damping constant of 4 lb-sec/ft. The spring constant is 16 lb/ft, and a mass weighing 20 lb is attached to the spring. Assuming an inductance of 0.02 henrys, find an *LRC* circuit that is the analogue of the mechanical system in the sense that both systems are governed by the same differential equation.

3.2

LINEAR SECOND-ORDER EQUATIONS: BASIC THEORY

In modeling the motion of a falling object with no resistive forces we obtained the second-order equation

$$y'' = g$$

having the solution

$$y = \frac{gt^2}{2} + c_1 t + c_2$$

where c_1 and c_2 are arbitrary constants. It would seem reasonable to expect the general solution of a second-order equation to contain two arbitrary constants. In this section we investigate the nature of the general solution to an important class of second-order equations that arise frequently in modeling real-world behaviors. This is the class of linear equations.

A linear second-order equation has the form

$$a_2(x)\frac{d^2y}{dx^2} + a_1(x)\frac{dy}{dx} + a_0(x)y = b(x) \tag{1}$$

where $a_2(x)$, $a_1(x)$, $a_0(x)$, and $b(x)$ are assumed to be continuous functions of x over the open interval $\alpha < x < \beta$.

EXAMPLE 1 Let us examine a few second-order equations for linearity.

a) The equation

$$y'' + 3y' + 2y = 0$$

is linear and homogeneous because $b(x)$, the term on the right side of Eq. (1) here equals zero.

b) The equation

$$x^2y'' - 2xy' + y = 0$$

is also linear and homogeneous.

c) The equation

$$x^2y'' - 3xy' + 3y = x^4e^x$$

is linear and nonhomogeneous because $b(x) \not\equiv 0$.

d) The equation

$$y'' - y^2 = 0$$

is nonlinear because it fails to be in the form of Eq. (1).

In a linear equation no terms involving the dependent variable y or any of its derivatives can appear in any way but to the first power, nor can they appear within function expressions like $\sin y$ or e^y. Also mixed products of y or its derivatives, like $y\, dy/dx$ and $(dy/dx)(d^2y/dx^2)$ for instance, cannot appear in a linear equation. Of course, y or some of its derivatives may be absent and not appear at all.

In Chapter 4 you will see that to solve a nonhomogeneous linear differential equation we first solve an associated homogeneous equation. In this chapter we focus our attention on finding the general solution to the homogeneous linear second-order equation.

Solutions to Homogeneous Second-Order Equations

For the homogeneous second-order equation

$$a_2(x)y'' + a_1(x)y' + a_0(x)y = 0, \tag{2}$$

we assume that $a_2(x)$ is never zero for any x over the interval $\alpha < x < \beta$. (In Chapter 9 we will take up the question of $a_2(x)$ having zero values.) Thus we can divide by $a_2(x)$ and write Eq. (2) in the **standard form**

$$y'' + p(x)y' + q(x)y = 0 \tag{3}$$

where $p(x)$ and $q(x)$ are continuous over the interval. The following example illustrates a general principle that holds for linear homogeneous equations.

EXAMPLE 2 By direct substitution it is easy to show that the two exponential functions

$$y_1(x) = e^{-x} \quad \text{and} \quad y_2(x) = e^{-2x}$$

solve the homogeneous linear equation

$$y'' + 3y' + 2y = 0.$$

Moreover, any *linear combination*

$$y = c_1e^{-x} + c_2e^{-2x},$$

where c_1 and c_2 are arbitrary constants, also solves the homogeneous equation. By direct substitution we have

$$\begin{aligned}(c_1e^{-x} + c_2e^{-2x})'' &+ 3(c_1e^{-x} + c_2e^{-2x})' + 2(c_1e^{-x} + c_2e^{-2x}) \\ &= (c_1e^{-x} + 4c_2e^{-2x}) + 3(-c_1e^{-x} - 2c_2e^{-2x}) + 2(c_1e^{-x} + c_2e^{-2x}) \\ &= (c_1 - 3c_1 + 2c_1)e^{-x} + (4c_2 - 6c_2 + 2c_2)e^{-2x} \\ &= 0.\end{aligned}$$

We now state a more general result.

SUPERPOSITION PRINCIPLE

> If $y_1(x)$ and $y_2(x)$ are two solutions to homogeneous linear Eq. (2), then for any constants c_1 and c_2 the function
>
> $$y(x) = c_1y_1(x) + c_2y_2(x) \tag{4}$$
>
> is also a solution. The function $y(x)$ is said to be a **linear combination** of the functions y_1 and y_2.

As in Example 2, it is an easy matter to verify the principle of superposition: substitute the first and second derivatives of the righthand side of Eq. (4) into homogeneous linear differential Eq. (2) and you will see readily that it is satisfied (see problem 16 in Exercises 3.2). It is important for you to realize that the principle of superposition does not hold for *nonlinear* differential equations. It also does not hold for *nonhomogeneous linear* equations (see problem 17).

From the superposition principle note that the following are true.

a) A sum of two solutions $y_1 + y_2$ of a homogeneous linear differential equation is also a solution. (Choose $c_1 = c_2 = 1$ in Eq. 4.)

b) Any constant multiple $c_1 y_1$ of a solution $y_1(x)$ of a homogeneous linear differential equation is also a solution. (Choose $c_2 = 0$ in Eq. 4.)

c) The **trivial solution** $y(x) \equiv 0$ *always* satisfies the homogeneous linear differential equation. (Choose $c_1 = c_2 = 0$ in Eq. 4.) However, in applied problems the trivial solution is generally of no interest because the zero function as a solution is usually not a practical result.

A key result for homogeneous linear second-order Eq. (2) is that two solutions y_1 and y_2 do exist, such that every solution y can be written as some linear combination (4) by choosing suitable values for the constants c_1 and c_2. It is important to realize that exactly *two* solutions are needed for a *second-order* linear equation. However, not just any pair of solutions will do. A special relationship between the solutions is required. We are going to explore this relationship and develop these ideas more fully (although we will not prove them in this text). Let us begin the discussion with a useful definition and an example.

DEFINITION 3.2

If every solution to homogeneous linear Eq. (2) can be written as some linear combination of the two solutions y_1 and y_2, then

$$y = c_1 y_1(x) + c_2 y_2(x),$$

written with arbitrary constants c_1 and c_2, is called the **general solution** to Eq. (2).

EXAMPLE 3

The constant functions $y_1 = 5$ and $y_2 = -10$ are both solutions to the equation $y'' = 0$. However, the linear combination

$$y = c_1(5) + c_2(-10)$$

does not provide the solution $y = x$ for any choice of the real constants c_1 and c_2. In fact, if c_1 and c_2 are arbitrary constants, then the expression $5c_1 - 10c_2$ is simply another arbitrary constant (which we could name c_3). The difficulty here lies with the fact that the two solutions $y_1 = 5$ and $y_2 = -10$ are *not independent* of each other because y_2 is a constant multiple of y_1. We need to avoid this kind of relationship if the solutions are to produce the general solution.

If we choose the two solutions $y_1(x) = x$ and $y_2(x) = 1$, then the expression

$$y = c_1 x + c_2$$

does give the general solution to $y'' = 0$. The reason for this is that the solution functions are *independent* in a way we will now make precise. Later on we will state a theorem validating this assertion.

Linear Dependence and Independence

Although we are here primarily concerned with two functions, we state the following definition valid for any finite set of functions for purposes of generality and completeness in our presentation.

DEFINITION 3.3

The set of functions $f_1(x), f_2(x), \ldots, f_k(x)$ is **linearly dependent** on an interval I if and only if there exist constants $b_1, b_2, \ldots, b_k$, *not all zero,* such that

$$b_1 f_1(x) + b_2 f_2(x) + \cdots + b_k f_k(x) = 0 \tag{5}$$

for every x in I. If the set of functions is *not* linearly dependent we say that it is **linearly independent** on I.

EXAMPLE 4 The functions $f_1(x) = \cos x$, $f_2(x) = \sin(x - \pi/2)$, and $f_3(x) = x$ form a linearly dependent set over any interval I. The equation

$$b_1 \cos x + b_2 \sin(x - \pi/2) + b_3 x = 0$$

is valid for all x if $b_1 = b_2 = 1$ and $b_3 = 0$.

EXAMPLE 5 The functions $y_1 = e^{2x}$ and $y_2 = e^{-x}$ are linearly independent over any interval I. If

$$b_1 e^{2x} + b_2 e^{-x} = 0 \tag{6}$$

where b_1 and b_2 are not both zero, then assuming that $b_1 \neq 0$ we have

$$e^{2x} = ce^{-x}$$

where c is the constant $-b_2/b_1$. But then $c = e^{3x}$ for each x in I contrary to the assumption that c is *constant.* A similar contradiction is produced if we assume that $b_2 \neq 0$. Thus Eq. (6) is impossible unless $b_1 = b_2 = 0$. This establishes the claim that the two functions are linearly independent.

The argument used in Example 5 is quite general. Two functions $f_1(x)$ and $f_2(x)$ are linearly dependent if and only if there exist constants b_1 and b_2, not both zero, such that

$$b_1 f_1(x) + b_2 f_2(x) = 0 \tag{7}$$

for every x in I. Since at least one constant is not zero, we may assume without loss of generality that $b_1 \neq 0$. Then we can solve Eq. (7) for $f_1(x)$ to obtain the relation

$$f_1(x) = -\frac{b_2}{b_1} f_2(x).$$

This leads to the following result, which is often convenient to apply:

> Two functions are linearly dependent on I if and only if one of the functions is a constant multiple of the other function.

EXAMPLE 6 The functions $y_1 = e^x$ and $y_2 = xe^x$ are linearly independent over any interval I. If they were not, then one of the functions would be a constant multiple of the other, say

$$xe^x = ce^x.$$

But this last equation implies that $x = c$, which is contrary to c being a constant. A similar contradiction occurs if e^x is a constant multiple of xe^x.

The next example shows that whether a set of functions is linearly independent (or dependent) results not only from the functions but also the interval on which they are defined. That is, the same set of functions may be linearly independent over one interval and linearly dependent over another.

EXAMPLE 7 **a)** The functions $y_1 = x$ and $y_2 = |x|$ are linearly independent over any interval containing both positive and negative values of x. To see why, suppose that

$$b_1 x + b_2|x| = 0. \tag{8}$$

If $x > 0$, then $|x| = x$ and Eq. (8) implies that $b_1 + b_2 = 0$. On the other hand, if $x < 0$, then $|x| = -x$ and Eq. (8) implies that $b_1 - b_2 = 0$. The only solution is $b_1 = b_2 = 0$, so the functions are linearly independent.

b) The functions $y_1 = x$ and $y_2 = |x|$ are linearly dependent over any interval containing only positive or only negative values for x. For instance, if the interval consists only of negative values, then

$$y_1 = x = -1(-x) = -1 \cdot |x| = -y_2.$$

Thus the functions are linearly dependent.

EXAMPLE 8 If $r_1 \neq r_2$ are constants, then the two functions

$$y_1 = e^{r_1 x} \quad \text{and} \quad y_2 = e^{r_2 x}$$

form a linearly independent set over any interval I. For if

$$e^{r_1 x} = ce^{r_2 x},$$

then

$$c = e^{(r_1 - r_2)x}$$

for every x in I. Thus c fails to be a constant. Likewise we cannot write y_2 as a constant multiple of y_1.

EXAMPLE 9 The three functions e^x, $\cos x$, and $\sin x$ form a linearly independent set on any interval I. For suppose there exist constants b_1, b_2, and b_3 satisfying the relation

$$b_1 e^x + b_2 \cos x + b_3 \sin x = 0. \tag{9}$$

Differentiating this equation twice yields the following two equations, which must also be valid:

$$b_1 e^x - b_2 \sin x + b_3 \cos x = 0 \tag{10}$$

and

$$b_1 e^x - b_2 \cos x - b_3 \sin x = 0. \tag{11}$$

Solving the system of Eqs. (9–11) by the method of determinants, we find that the determinant of the coefficients is

$$\begin{vmatrix} e^x & \cos x & \sin x \\ e^x & -\sin x & \cos x \\ e^x & -\cos x & -\sin x \end{vmatrix} = 2e^x.$$

Since the exponential function is never zero, the only simultaneous solution to Eqs. (9–11) is $b_1 = b_2 = b_3 = 0$.

General Solution to the Homogeneous Linear Equation

We now have the necessary elements to state the result establishing the form of the general solution to the homogeneous linear second-order differential equation.

THEOREM 3.1

If $a_2(x)$, $a_1(x)$, and $a_0(x)$ are functions continuous over the interval I, and if $a_2(x) \neq 0$ on I, then the second-order equation

$$a_2(x)y'' + a_1(x)y' + a_0(x)y = 0 \tag{12}$$

has two linearly independent solutions $y_1(x)$ and $y_2(x)$ on I. Furthermore, if $y_1(x)$ and $y_2(x)$ are *any* two linearly independent solutions, and if $y(x)$ is a solution to Eq. (12) on I, then there exist unique constants c_1 and c_2 such that

$$y(x) = c_1 y_1(x) + c_2 y_2(x). \tag{13}$$

Theorem 3.1 tells us that the general solution to second-order linear Eq. (12) is given by linear combination (13), where c_1 and c_2 are arbitrary constants, whenever y_1 and y_2 are two linearly independent solutions. This observation motivates the following definition.

DEFINITION 3.4

Two solutions y_1 and y_2 of Eq. (12) are said to form a **fundamental set** of solutions if every solution to Eq. (12) can be expressed as a linear combination of y_1 and y_2.

From Theorem 3.1 we conclude that two solutions form a fundamental set if and only if they form a linearly independent set. In summary:

Whenever $\{y_1, y_2\}$ is a fundamental set of solutions, or equivalently a linearly independent set of solutions, then

$$y = c_1 y_1(x) + c_2 y_2(x)$$

is the general solution to Eq. (12), where c_1 and c_2 are arbitrary constants.

Following is an example.

EXAMPLE 10 The functions $y_1 = e^x$ and $y_2 = xe^x$ form a fundamental set of solutions to the differential equation

$$y'' - 2y' + y = 0.$$

To see this, we substitute the solution functions directly into the differential equation:

$$(e^x)'' - 2(e^x)' + e^x = e^x - 2e^x + e^x = 0$$

and

$$(xe^x)'' - 2(xe^x)' + xe^x = (x+2)e^x - 2(x+1)e^x + xe^x = 0.$$

Thus both functions satisfy the differential equation. Since they are linearly independent (which we showed in Example 6), they form a fundamental set of solutions. (In the next section you will learn how to find y_1 and y_2.) Therefore

$$y = c_1 e^x + c_2 xe^x$$

is the general solution to the given second-order equation.

Second-Order Initial Value Problems

To determine a unique solution to a first-order linear equation it was sufficient to specify the value of the solution at a single point. Since the general solution to a second-order equation contains two arbitrary constants (which is also true in the nonhomogeneous case), it is necessary to specify two conditions. One way of doing this is to specify the value of the solution and the value of the derivative at a single point: $y(x_0) = y_0$ and $y'(x_0) = y_0'$. These are called **initial conditions.** Thus, to determine a unique solution of a second-order equation it is necessary to specify a point that lies on the solution curve as well as the slope of the curve at that point. The following result, which we state without proof, guarantees the existence of a unique solution for both homogeneous and nonhomogeneous linear second-order equations.

THEOREM 3.2

Suppose the functions $p(x)$, $q(x)$, and $g(x)$ are continuous on the open interval $\alpha < x < \beta$. Then there exists one and only one function $y = y(x)$ satisfying both the differential equation

$$y'' + p(x)y' + q(x)y = g(x) \tag{14}$$

on the interval, and the initial conditions

$$y(x_0) = y_0 \qquad \text{and} \qquad y'(x_0) = y_0'$$

at the specified point x_0 in the interval.

Theorem 3.2 is known as the **existence and uniqueness theorem** for the linear second-order equation. It is important to realize that any real values can be assigned to y_0 and y_0' and the theorem applies.

EXAMPLE 11 Let us find the solution to the differential equation introduced in Example 10:

$$y'' - 2y' + y = 0$$

satisfying the initial conditions $y(0) = 1$ and $y'(0) = -1$.

Solution. The general solution is

$$y = c_1 e^x + c_2 x e^x.$$

Then

$$y' = c_1 e^x + c_2(x+1)e^x.$$

From the initial conditions we have

$$1 = c_1 + c_2 \cdot 0 \qquad \text{and} \qquad -1 = c_1 + c_2 \cdot 1.$$

Thus $c_1 = 1$ and $c_2 = -2$. The unique solution satisfying the initial conditions is

$$y = e^x - 2xe^x.$$

In Theorem 3.2 the initial conditions that determine a unique solution to Eq. (14) are conditions on the value of the solution and its derivative at a *single point* x_0 in the interval. Another approach to determine the values of the two arbitrary constants in the general solution is to specify the solution values at *two different points* in the interval, say $y(x_1) = A$ and $y(x_2) = B$. Such problems, known as **boundary value problems,** do not always possess a solution. These problems are studied in more advanced mathematics texts.* (See also problem 45 in Exercises 3.2.)

In the next section we will discuss a method for finding a fundamental set of solutions to linear homogeneous second-order equations with *constant coefficients;* that is, when $a_2(x)$, $a_1(x)$, and $a_0(x)$ are all constants in linear Eq. (12).

EXERCISES 3.2

In problems 1–10 show that both y_1 and y_2 satisfy the given differential equation. Form the linear combination $y_3 = c_1 y_1 + c_2 y_2$ and demonstrate that y_3 is also a solution.

1. $y'' - 7y' + 12y = 0, \quad y_1 = e^{3x}, y_2 = e^{4x}$
2. $2y'' + y' - y = 0, \quad y_1 = e^{-x}, y_2 = e^{x/2}$

* See H. Sagan, *Boundary and Eigenvalue Problems in Mathematical Physics* (New York: Wiley, 1961).

3. $y'' + y' - 6y = 0, \quad y_1 = e^{-3x}, y_2 = e^{2x} - e^{-3x}$

4. $y'' + y = 0, \quad y_1 = \sin x, y_2 = \cos x$

5. $2y'' + y = 0, \quad y_1 = \sin\left(\frac{x}{\sqrt{2}}\right), y_2 = \cos\left(\frac{x}{\sqrt{2}}\right)$

6. $y'' + 2y' + 2y = 0, \quad y_1 = e^{-x}\sin x, y_2 = e^{-x}\cos x$

7. $y'' + 2y' + y = 0, \quad y_1 = e^{-x}, y_2 = xe^{-x}$

8. $y'' - 4y' + 4y = 0, \quad y_1 = e^{2x}, y_2 = xe^{2x}$

9. $16y'' + 8y' + y = 0, \quad y_1 = e^{-x/4}, y_2 = xe^{-x/4}$

10. $y'' - 2y' + y = 0, \quad y_1 = e^x, y_2 = (1 + x)e^x$

In problems 11–15 classify the given second-order equation as linear or nonlinear. If it is linear, is it homogeneous or nonhomogeneous?

11. $x^2y'' + xy' - y + 2x^2e^x = 0, \quad x > 0$

12. $y'' + (y')^2 + xy = 0$

13. $y'' + 2y' = \sqrt{xy}, \quad xy > 0$

14. $xy'' + (1 - 2x)y' = (1 - x)y, \quad x > 0$

15. $x'' - e^tx' + tx = \sqrt{t}e^{-t}$

16. Verify the principle of superposition.

17. Show that the superposition principle fails to hold in the following cases of nonlinear or nonhomogeneous equations.

a) $y'' - y = 2, \quad y_1 = e^x - 2, y_2 = e^{-x} - 2$

b) $y'' - y' = x, \quad y_1 = -x - \frac{x^2}{2}, y_2 = e^x - x - \frac{x^2}{2}$

c) $y'' + (y')^2 = 0, \quad y_1 = 1, \quad y_2 = \ln x$

In problems 18–35 determine whether the given set of functions is linearly independent or linearly dependent on the specified interval.

18. $\{x, x^3\}, \quad -\infty < x < \infty$

19. $\{x^2, x^4\}, \quad -\infty < x < \infty$

20. $\{x - 1, x + 3\}, \quad -\infty < x < \infty$

21. $\{x + 2, |x| + 2\}, \quad -1 < x < 1$

22. $\{0, x\}, \quad 0 < x < \infty$

23. $\{\sin x, \sin^2 x\}, \quad 0 \leq x < 2\pi$

24. $\{x, e^{-x}\}, \quad -\infty < x < \infty$

25. $\{1 + e^x, e^x\}, -\infty < x < \infty$

26. $\{e^{3x}, \sin 3x\}, \quad -\infty < x < \infty$

27. $\{x, \sin x\}, \quad -\infty < x < \infty$

28. $\{\sin x + \cos x, \cos x\}, \quad -\infty < x < \infty$

29. $\{x, \sqrt{x}\}, \quad 0 < x < \infty$

30. $\{x|x|, x^2\}, \quad -\infty < x < \infty$

31. $\{x, xe^{-x}, x^2e^{-x}\}, \quad -\infty < x < \infty$

32. $\{x, e^x, \cos x\}, \quad -\infty < x < \infty$

33. $\{1, \cos x, x\cos x\}, \quad -\infty < x < \infty$

34. $\{e^x, e^{2x}, e^{3x}\}, \quad -\infty < x < \infty$

35. $\{\sin x, \sin 2x, \sin 3x\}, \quad -\pi/2 \leq x \leq \pi/2$

In problems 36–44 show that y_1 and y_2 solve the differential equation. Find the unique solution satisfying the initial conditions.

36. $y'' + 6y' + 5y = 0, \quad y(0) = 0, \quad y'(0) = 3, \quad y_1 = e^{-5x}, y_2 = e^{-x}$

37. $y'' + y' - 6y = 0, \quad y(0) = 1, \quad y'(0) = 0, \quad y_1 = e^{2x}, y_2 = e^{-3x}$

38. $y'' + 4y = 0, \quad y(0) = 1, \quad y'(0) = -1, \quad y_1 = \cos 2x, y_2 = \sin 2x$

39. $y'' + 2y' + 5y = 0, \quad y(0) = 1, y'(0) = 1, y_1 = e^{-x}\sin 2x, y_2 = e^{-x}\cos 2x$

40. $y'' - 6y' + 9y = 0, \quad y(0) = 0, \quad y'(0) = 1, \quad y_1 = e^{3x}, y_2 = xe^{3x}$

41. $y'' + 4y' + 4y = 0, \quad y(0) = 1, \quad y'(0) = 0, \quad y_1 = e^{-2x}, y_2 = xe^{-2x}$

42. $9y'' - 12y' + 4y = 0, \quad y(0) = 1, y'(0) = 1, y_1 = e^{2x/3}, y_2 = xe^{2x/3}$

43. $x^2y'' - xy' + y = 0, \quad y(1) = 1, \quad y'(1) = -1, \quad y_1 = x, y_2 = x\ln x$

44. $xy'' - y' = 0, \quad y(1) = 1, y'(1) = -2, y_1 = 1, y_2 = x^2$

45. **Boundary Value Problems**

a) Show that $y_1 = \cos 2x$ and $y_2 = \sin 2x$ is a fundamental set of solutions to the differential equation $y'' + 4y = 0$.

b) Find a *unique solution* satisfying the boundary conditions $y(0) = 0$ and $y(\pi/12) = 1$.
c) Show there is *no solution* satisfying the boundary conditions $y(0) = 0$ and $y(\pi) = 1$.
d) Show there are *infinitely many solutions* satisfying the boundary conditions $y(0) = 0$ and $y(\pi) = 0$.

46. Prove that if $y_1(x)$ and $y_2(x)$ are differentiable and linearly dependent on the interval I, then

$$y_1(x)y_2'(x) - y_1'(x)y_2(x) = 0$$

for every x in the interval.

47. Use the result of problem 46 to establish that $\{\cos^2 x, 1 + \cos 2x\}$ is a linearly dependent set on $-\infty < x < \infty$.

48. Prove that if y_1 and y_2 form a fundamental set of solutions for

$$y'' + p(x)y' + q(x)y = 0,$$

then $y_3 = y_1 + y_2$ and $y_4 = y_1 - y_2$ also form a fundamental set of solutions.

49. Suppose $y_1(x)$ is a solution of the homogeneous equation

$$y'' + p(x)y' + q(x)y = 0$$

on the interval $\alpha < x < \beta$. If for some point x_0 in the interval, $y_1(x_0) = y_1'(x_0) = 0$, prove that $y_1(x) \equiv 0$. That is, if $y_1(x)$ is tangent to the x-axis at some point of the interval, then y_1 is identically zero.

50. Prove that if y_1 and y_2 assume a maximum or minimum value at the same point x_0 in the interval $\alpha < x < \beta$, then they cannot form a fundamental set of solutions to the homogeneous equation

$$y'' + p(x)y' + q(x)y = 0$$

on that interval.

3.3

CONSTANT-COEFFICIENT HOMOGENEOUS LINEAR EQUATIONS

We now turn our attention to finding two linearly independent solutions (that is, a fundamental set of solutions) y_1 and y_2 to the homogeneous linear second-order equation

$$a_2 y'' + a_1 y' + a_0 y = 0 \tag{1}$$

where a_2, a_1, and a_0 are assumed to be constants. In Section 3.1 we used proportionality arguments to construct models of this type.

As an example, recall the model for the motion of a suspended wheel supported by a shock absorber with no forcing function

$$my'' + \delta y' + ky = 0, \qquad y(0) = y_0,\ y'(0) = y_0'.$$

The submodels leading to this equation used three proportionality arguments: the acceleration is proportional to the net external force, the shock absorber damping force is proportional to the velocity, and the spring restoring force is proportional to the displacement from the equilibrium position.

What might the general solution to a constant-coefficient homogeneous linear second-order equation look like? In the case of the homogeneous linear first-order equation

$$a_1 y' + a_0 y = 0,$$

we have

$$y' = ky \tag{2}$$

where $k = -a_0/a_1$ is constant. Thus we seek a solution whose first derivative is proportional to the function itself. In Chapter 2 we found that the general solution to Eq. (2) is the exponential function

$$y = Ce^{kx} \tag{3}$$

for some arbitrary constant C.

To solve constant-coefficient second-order Eq. (1) we seek a function that, when multiplied by a constant and added to a constant times its first derivative plus a constant times its second derivative, sums identically to zero. Considering that we used proportionality arguments to construct constant-coefficient models, it would not be surprising to again find the exponential function as a solution. (Note that we verified exponential functions to be solutions to the constant-coefficient equations in Examples 2 and 10 of Section 3.2.)

Can you think of functions other than exponentials having the property that some derivatives are proportional to the function itself? After a little thought you might recall that the sine and cosine functions behave in that way. If you differentiate the function $y = \sin kx$, for instance, its second derivative is $y'' = -k^2 \sin kx$. Thus y'' is proportional to y. Hence, given the differential equation

$$y'' + k^2 y = 0, \tag{4}$$

you should not be surprised to find the two functions

$$y_1 = \sin kx \qquad \text{and} \qquad y_2 = \cos kx$$

as solutions to Eq. (4). Since they are linearly independent, the general solution is

$$y = c_1 \sin kx + c_2 \cos kx. \tag{5}$$

(Later in the discussion you will see that sine and cosine functions are actually related to the exponential function.)

Nature of Solutions to Homogeneous Linear Second-Order Equations with Constant Coefficients

Consider second-order Eq. (1). Since we assume the leading coefficient a_2 to be nonzero, we divide Eq. (1) through by a_2 to obtain the **standard form**

$$y'' + py' + qy = 0 \tag{6}$$

where p and q are constants. Based on our preceding discussion of the exponential functions, we assume a solution to Eq. (6) of the form $y = e^{rx}$. Computing the derivatives and substituting the results into Eq. (6) yields

$$r^2 e^{rx} + pre^{rx} + qe^{rx} = 0.$$

Since the exponential function is never zero, we can divide through by e^{rx} to obtain the quadratic equation

$$r^2 + pr + q = 0. \tag{7}$$

Equation (7) is called the **auxiliary equation** or **characteristic equation** associated with differential Eq. (6). Let us summarize this result.

> The exponential function $y = e^{rx}$ is a solution to second-order Eq. (6) if and only if the constant r is a solution to auxiliary Eq. (7).

The solution to quadratic Eq. (7) produces the two roots

$$r = \frac{-p \pm \sqrt{p^2 - 4q}}{2}. \tag{8}$$

The nature of the two linearly independent solutions to differential Eq. (6) depends on whether the discriminant associated with Eq. (8), $p^2 - 4q$, is positive, negative, or zero. We are going to discuss the three possibilities in turn. Since we seek two linearly independent solutions to Eq. (6), the following result is useful. (You will be asked to establish the result in Exercises 3.3.)

THEOREM 3.3

Let r_1 and r_2 denote real or complex numbers, and let α and β denote real numbers. Then each pair of the following functions forms a linearly independent set on every interval I:

a) e^{r_1x} and e^{r_2x} $(r_1 \neq r_2)$;
b) e^{r_1x} and xe^{r_1x};
c) $e^{\alpha x} \cos \beta x$ and $e^{\alpha x} \sin \beta x$.

Case 1: General Solution for Distinct Real Roots

If the discriminant $p^2 - 4q > 0$, then auxiliary Eq. (7) has two real and distinct roots, r_1 and r_2. Thus $y_1 = e^{r_1x}$ and $y_2 = e^{r_2x}$ are solutions to Eq. (6). Furthermore, from Theorem 3.3(a) we know that y_1 and y_2 are linearly independent. We conclude the following result.

THEOREM 3.4

If r_1 and r_2 are distinct real roots to auxiliary Eq. (7), then the general solution to the second-order constant-coefficient linear Eq. (6) is

$$y = c_1e^{r_1x} + c_2e^{r_2x}.$$

EXAMPLE 1 Find the general solution of the differential equation

$$y'' - y' - 6y = 0.$$

Solution. Substitution of $y = e^{rx}$ into the differential equation yields the auxiliary equation

$$r^2 - r - 6 = 0,$$

which factors as

$$(r - 3)(r + 2) = 0.$$

The roots are $r_1 = 3$ and $r_2 = -2$. Thus the general solution is

$$y = c_1 e^{3x} + c_2 e^{-2x}.$$

EXAMPLE 2 Find the unique solution of the second-order initial value problem

$$2y'' + y' - y = 0, \qquad y(0) = 1,\ y'(0) = 0.$$

Solution. The auxiliary equation is

$$2r^2 + r - 1 = 0$$

or

$$(2r - 1)(r + 1) = 0.$$

The roots are $r_1 = 1/2$ and $r_2 = -1$. Thus the general solution is

$$y = c_1 e^{x/2} + c_2 e^{-x}.$$

To use the initial conditions we need the derivative

$$y' = \frac{1}{2} c_1 e^{x/2} - c_2 e^{-x}.$$

Then $y(0) = 1$ and $y'(0) = 0$ produce the equations

$$c_1 + c_2 = 1 \qquad \text{and} \qquad \frac{1}{2} c_1 - c_2 = 0.$$

Solving we find that $c_1 = 2/3$ and $c_2 = 1/3$. Thus, the unique solution satisfying the initial conditions is

$$y = \frac{2}{3} e^{x/2} + \frac{1}{3} e^{-x}.$$

Notice from Example 2 that we do not need to write the second-order constant-coefficient equation in standard form to find the correct roots to the auxiliary equation.

Case 2: General Solution for Distinct Complex Roots

If the discriminant $p^2 - 4q < 0$, then the roots to auxiliary Eq. (7) are the complex numbers

$$r_1 = \frac{-p + \sqrt{p^2 - 4q}}{2} \quad \text{and} \quad r_2 = \frac{-p - \sqrt{p^2 - 4q}}{2}.$$

These roots have the general form $r_1 = \alpha + i\beta$ and $r_2 = \alpha - i\beta$, where the real part is $\alpha = -p/2$, the imaginary part is $\beta = \sqrt{|p^2 - 4q|}/2$, and $i^2 = -1$. *Note that complex roots always occur in pairs where the sign of the imaginary part changes.* Thus, from Theorem 3.3(a), $y_1 = e^{(\alpha + i\beta)x}$ and $y_2 = e^{(\alpha - i\beta)x}$ are linearly independent solutions to Eq. (6). However, these solutions are *complex valued,* whereas differential Eq. (6) is *real valued.* In most physical applications it is desirable to express the general solution in terms of real-valued functions.

How can we find two linearly independent real-valued solutions to Eq. (6) from our complex-valued solutions? The following result will aid us. It expresses the complex exponential in terms of the trigonometric functions.

EULER'S IDENTITY*

For the complex number $i\beta$,

$$e^{i\beta} = \cos\beta + i\sin\beta. \tag{9}$$

Thus our two complex solutions y_1 and y_2 can be expressed as

$$y_1 = e^{(\alpha + i\beta)x} = e^{\alpha x}e^{i\beta x} = e^{\alpha x}(\cos\beta x + i\sin\beta x) \tag{10}$$

and

$$y_2 = e^{(\alpha - i\beta)x} = e^{\alpha x}e^{-i\beta x} = e^{\alpha x}(\cos\beta x - i\sin\beta x). \tag{11}$$

By the superposition principle, all possible linear combinations of y_1 and y_2 must also be solutions to Eq. (6). For instance,

$$y_3 = \frac{1}{2}y_1 + \frac{1}{2}y_2 = e^{\alpha x}\cos\beta x \tag{12}$$

and

$$y_4 = \left(\frac{1}{2i}\right)y_1 + \left(-\frac{1}{2i}\right)y_2 = e^{\alpha x}\sin\beta x \tag{13}$$

* Leonhard Euler (1707–1783) is considered by many to be the greatest 18th-century mathematician. His mathematical interests ranged across many areas of pure and applied mathematics, and he made enormous contributions to the mathematical literature. Euler gave the general solution to constant-coefficient homogeneous linear differential equations in 1743. Later he extended his results to solve nonhomogeneous equations.

are solutions to Eq. (6). From Theorem 3.3(c), we know that $e^{\alpha x} \cos \beta x$ and $e^{\alpha x} \sin \beta x$ are linearly independent solutions.

In our derivation of real solution y_4 we multiplied by the complex constant $1/2i$. You may find that procedure unsatisfactory if linear combinations are to be formed using only real constants. Let us discuss an alternate procedure where multiplication by a complex number can be avoided. The following result from the theory of complex variables guarantees that the real and imaginary parts of our complex-valued solutions y_1 and y_2 are themselves *real solutions* to second-order differential Eq. (6).

THEOREM 3.5

The complex function $u(x) + iv(x)$ is a solution to differential Eq. (6) if and only if each of the real-valued functions $u(x)$ and $v(x)$ is a solution.

Thus, knowing that y_1 and y_2 are solutions to Eq. (6), the preceding theorem implies that $e^{\alpha x} \cos \beta x$, $e^{\alpha x} \sin \beta x$, and $-e^{\alpha x} \sin \beta x$ are real-valued solutions to Eq. (6). Using Theorem 3.3(c) we can choose $e^{\alpha x} \cos \beta x$ and $e^{\alpha x} \sin \beta x$ as a fundamental set of solutions. Theorem 3.1 in Section 3.2 then gives the following general solution to a constant-coefficient second-order equation when the roots to the auxiliary equation are complex.

THEOREM 3.6

If $r_1 = \alpha + i\beta$ and $r_2 = \alpha - i\beta$ are complex roots to auxiliary Eq. (7), then the general solution to second-order constant-coefficient Eq. (6) is

$$y = e^{\alpha x}(c_1 \cos \beta x + c_2 \sin \beta x).$$

EXAMPLE 3 Find the general solution to the differential equation

$$y'' - 4y' + 5y = 0.$$

Solution. The auxiliary equation is

$$r^2 - 4r + 5 = 0.$$

The roots are the complex pair $r = (4 \pm \sqrt{16 - 20})/2$ or $r_1 = 2 + i$ and $r_2 = 2 - i$. Thus $\alpha = 2$ and $\beta = 1$ give the general solution

$$y = e^{2x}(c_1 \cos x + c_2 \sin x).$$

EXAMPLE 4 Find the unique solution of the second-order initial value problem

$$y'' - 2y' + 5y = 0, \qquad y(0) = 0,\ y'(0) = 1.$$

Solution. The auxiliary equation is

$$r^2 - 2r + 5 = 0,$$

which has the complex roots $r_1 = 1 + 2i$ and $r_2 = 1 - 2i$. Therefore the general solution is

$$y = e^x(c_1 \cos 2x + c_2 \sin 2x).$$

To use the initial conditions we need the derivative

$$y' = e^x[(c_1 + 2c_2) \cos 2x + (c_2 - 2c_1) \sin 2x].$$

Then $y(0) = 0$ and $y'(0) = 1$ yield the equations

$$c_1(1) + c_2(0) = 0,$$
$$(c_1 + 2c_2)(1) + (c_2 - 2c_1)(0) = 1.$$

By solving we find that $c_1 = 0$ and $c_2 = \frac{1}{2}$. Thus the unique solution is

$$y = \frac{1}{2} e^x \sin 2x.$$

Case 3: General Solution for Repeated Real Roots

Consider now the situation when the coefficients of second-order Eq. (6) satisfy $p^2 - 4q = 0$. In this case auxiliary Eq. (7) has the double root $r = m$. In fact, from quadratic formula (8), $m = -p/2$. From our previous development we know that $y_1 = e^{mx}$ is a solution to Eq. (6). We need a second, linearly independent solution in order to write the general solution.

Form of the Second Solution Since multiplication of e^{mx} by a constant fails to produce a linearly independent solution, suppose we try to determine a second solution by multiplying e^{mx} by a *function.* That is, let us try to determine a second solution of the form

$$y_2 = u(x)e^{mx} \tag{14}$$

where $u(x)$ is an unknown function we wish to determine. Thus we need to find a function $u(x)$ such that y_2 is a solution of Eq. (6) and y_2 and y_1 are linearly independent functions.

Finding the Unknown Function $u(x)$ We find the function $u(x)$ by requiring that y_2 in Eq. (14) satisfy the differential equation. To expedite our calculations we first rewrite the equation from our knowledge that $p^2 -$

$4q = 0$ and $m = -p/2$ is the value of the double root. Then $p = -2m$ and $q = p^2/4 = m^2$, so second-order Eq. (6) becomes

$$y'' - 2my' + m^2y = 0. \tag{15}$$

Computing the first and second derivatives of y_2 in Eq. (14) gives us

$$y_2' = u'e^{mx} + mue^{mx}$$

and

$$y_2'' = u''e^{mx} + 2mu'e^{mx} + m^2ue^{mx}.$$

Using the form of Eq. (15), we then demand that y_2 satisfy

$$y_2'' - 2my_2' + m^2y_2 = 0. \tag{16}$$

Substituting for y_2 and its derivatives into Eq. (16) yields

$$(u'' + 2mu' + m^2u)e^{mx} - 2m(u' + mu)e^{mx} + m^2ue^{mx} = 0.$$

After algebraically combining like terms on the lefthand side of this equation we have

$$u''e^{mx} = 0.$$

Since the exponential function is never zero, it follows that

$$u'' = 0.$$

Now *any* function of the form $u = c_2x + c_1$, for arbitrary constants c_1 and c_2, satisfies $u'' = 0$. We seek only one function u so that $y_2 = ue^{mx}$ is independent of y_1. For simplicity we choose $c_2 = 1$ and $c_1 = 0$. (Note that if we choose $c_2 = 0$ the resulting function y_2 is only a constant multiple of y_1. In that case the functions would not be linearly independent.) Thus

$$u = x. \tag{17}$$

General Solution for Repeated Roots Substitution of $u = x$ from Eq. (17) into the form of Eq. (14) yields the linearly independent solution

$$y_2 = xe^{mx}. \tag{18}$$

We know that the functions $y_1 = e^{mx}$ and $y_2 = xe^{mx}$ are linearly independent from Theorem 3.3(b). Thus we have the following result.

THEOREM 3.7

If m is a double root of auxiliary Eq. (7), then the general solution to second-order constant-coefficient Eq. (6) is

$$y = c_1e^{mx} + c_2xe^{mx}.$$

Remark The technique we employed above is a powerful one. We obtained our second independent solution by changing the arbitrary *constant* or *parameter* c in the solution $y = ce^{mx}$ to a *function* $u(x)$ *to be determined* in such a way that $y_2 = ue^{mx}$ is an independent solution. Later on in the sequel (Section 4.4) you will see how we employ this same idea when we solve nonhomogeneous equations. Later still (in Section 9.1) we will use the idea to construct a second independent solution from a known solution for linear equations with *variable coefficients.*

EXAMPLE 5 Find the general solution to

$$y'' + 4y' + 4y = 0.$$

Solution. The auxiliary equation is

$$r^2 + 4r + 4 = 0,$$

which factors into

$$(r + 2)^2 = 0.$$

Thus $r = -2$ is a double root. From Theorem 3.7 we have that $y_1 = e^{-2x}$ and $y_2 = xe^{-2x}$ are linearly independent solutions. Therefore the general solution is

$$y = c_1e^{-2x} + c_2xe^{-2x}.$$

EXAMPLE 6 Find the general solution to

$$9y'' - 6y' + y = 0.$$

Solution. The auxiliary equation is

$$9r^2 - 6r + 1 = 0.$$

The roots are $r = (6 \pm \sqrt{36 - 36})/18$, so $r = 1/3$ is a double root. Therefore the general solution is given by

$$y = c_1e^{x/3} + c_2xe^{x/3}.$$

EXAMPLE 7 Solve the second-order initial value problem

$$y'' - 10y' + 25y = 0, \qquad y(0) = 1,\ y'(0) = -1.$$

Solution. The auxiliary equation is

$$r^2 - 10r + 25 = 0$$

or

$$(r-5)^2 = 0.$$

Thus $r = 5$ is a double root giving rise to the general solution

$$y = c_1 e^{5x} + c_2 x e^{5x}.$$

The derivative is

$$y' = 5c_1 e^{5x} + c_2 e^{5x} + 5c_2 x e^{5x}.$$

The initial conditions $y(0) = 1$ and $y'(0) = -1$ yield

$$c_1 + c_2 \cdot 0 = 1$$

and

$$5c_1 + c_2 + 5c_2 \cdot 0 = -1.$$

Solving this system of equations gives us $c_1 = 1$ and $c_2 = -6$. Therefore the unique solution is

$$y = e^{5x} - 6xe^{5x}.$$

We now summarize our results for solving the homogeneous second-order linear differential equation with constant coefficients.

GENERAL SOLUTION TO THE HOMOGENEOUS SECOND-ORDER LINEAR EQUATION WITH CONSTANT COEFFICIENTS

$$a_2 y'' + a_1 y' + a_0 y = 0$$

1. If the roots r_1 and r_2 of the auxiliary equation are real and distinct, the general solution is

$$y = c_1 e^{r_1 x} + c_2 e^{r_2 x}.$$

2. If the roots of the auxiliary equation are complex, $r_1 = \alpha + i\beta$ and $r_2 = \alpha - i\beta$, the general solution is

$$y = e^{\alpha x}(c_1 \cos \beta x + c_2 \sin \beta x).$$

3. If $r = m$ is a double root of the auxiliary equation, the general solution is

$$y = c_1 e^{mx} + c_2 x e^{mx}.$$

EXERCISES 3.3

In problems 1–30 find the general solution of the given equation.

1. $y'' - y' - 12y = 0$
2. $3y'' - y' = 0$
3. $y'' + 3y' - 4y = 0$
4. $y'' - 9y = 0$
5. $y'' - 4y = 0$
6. $y'' - 64y = 0$
7. $2y'' - y' - 3y = 0$
8. $9y'' - y = 0$
9. $8y'' - 10y' - 3y = 0$
10. $3y'' - 20y' + 12y = 0$
11. $y'' + 9y = 0$
12. $y'' + 4y' + 5y = 0$
13. $y'' + 25y = 0$
14. $y'' + y = 0$
15. $y'' - 2y' + 5y = 0$
16. $y'' + 16y = 0$
17. $y'' + 2y' + 4y = 0$
18. $y'' - 2y' + 3y = 0$
19. $y'' + 4y' + 9y = 0$
20. $4y'' - 4y' + 13y = 0$
21. $y'' = 0$
22. $y'' + 8y' + 16y = 0$
23. $y'' + 4y' + 4y = 0$
24. $y'' - 6y' + 9y = 0$
25. $y'' + 6y' + 9y = 0$
26. $4y'' - 12y' + 9y = 0$
27. $4y'' + 4y' + y = 0$
28. $4y'' - 4y' + y = 0$
29. $9y'' + 6y' + y = 0$
30. $9y'' - 12y' + 4y = 0$

In problems 31–40 find the unique solution of the second-order initial value problem.

31. $y'' + 6y' + 5y = 0, \quad y(0) = 0, y'(0) = 3$
32. $y'' + 16y = 0, \quad y(0) = 2, y'(0) = -2$
33. $y'' + 12y = 0, \quad y(0) = 0, y'(0) = 1$
34. $12y'' + 5y' - 2y = 0, \quad y(0) = 1, y'(0) = -1$
35. $y'' + 8y = 0, \quad y(0) = -1, y'(0) = 2$
36. $y'' + 4y' + 4y = 0, \quad y(0) = 0, y'(0) = 1$
37. $y'' - 4y' + 4y = 0, \quad y(0) = 1, y'(0) = 0$
38. $4y'' - 4y' + y = 0, \quad y(0) = 4, y'(0) = 4$
39. $4y'' + 12y' + 9y = 0, \quad y(0) = 2, y'(0) = 1$
40. $9y'' - 12y' + 4y = 0, \quad y(0) = -1, y'(0) = 1$

In problems 41–55 find the general solution.

41. $y'' - 2y' - 3y = 0$
42. $6y'' - y' - y = 0$
43. $4y'' + 4y' + y = 0$
44. $9y'' + 12y' + 4y = 0$
45. $4y'' + 20y = 0$
46. $y'' + 2y' + 2y = 0$
47. $25y'' + 10y' + y = 0$
48. $6y'' + 13y' - 5y = 0$
49. $4y'' + 4y' + 5y = 0$
50. $y'' + 4y' + 6y = 0$
51. $16y'' - 24y' + 9y = 0$
52. $6y'' - 5y' - 6y = 0$
53. $9y'' + 24y' + 16y = 0$
54. $4y'' + 16y' + 52y = 0$
55. $6y'' - 5y' - 4y = 0$

In problems 56–60 solve the initial value problem.

56. $y'' - 2y' + 2y = 0, \quad y(0) = 0, y'(0) = 2$
57. $y'' + 2y' + y = 0, \quad y(0) = 1, y'(0) = 1$
58. $4y'' - 4y' + y = 0, \quad y(0) = -1, y'(0) = 2$
59. $3y'' + y' - 14y = 0, \quad y(0) = 2, y'(0) = -1$
60. $4y'' + 4y' + 5y = 0, \quad y(\pi) = 1, y'(\pi) = 0$

61. Establish the validity of the three linearly independent sets of functions in Theorem 3.3.

62. Show that the general solution of $y'' - 16y = 0$ is
$$y = c_1 \sinh 4x + c_2 \cosh 4x.$$

63. Show that if a_2, a_1, and a_0 are positive constants, then all solutions of
$$a_2 y'' + a_1 y' + a_0 y = 0$$
approach zero as x tends toward infinity.

3.4 AN APPLICATION: VIBRATIONS IN AN AUTOMOBILE SUSPENSION SYSTEM

Look at the automobile driving simulator in Fig. 3.10. The simulator is designed to recreate almost any totally safe real-world driving experience. Automotive engineers are very concerned with designing cars to provide safe and comfortable rides. Prior to building a prototype car to test in the simulator (which itself is based on thousands of equations), engineers use mathematical models to predict how the automobile will behave. We will now look at a model of an automobile suspension system using what you have learned so far about differential equations.

In Section 3.1 we developed one model,

$$my'' + \delta y' + ky = 0, \tag{1}$$

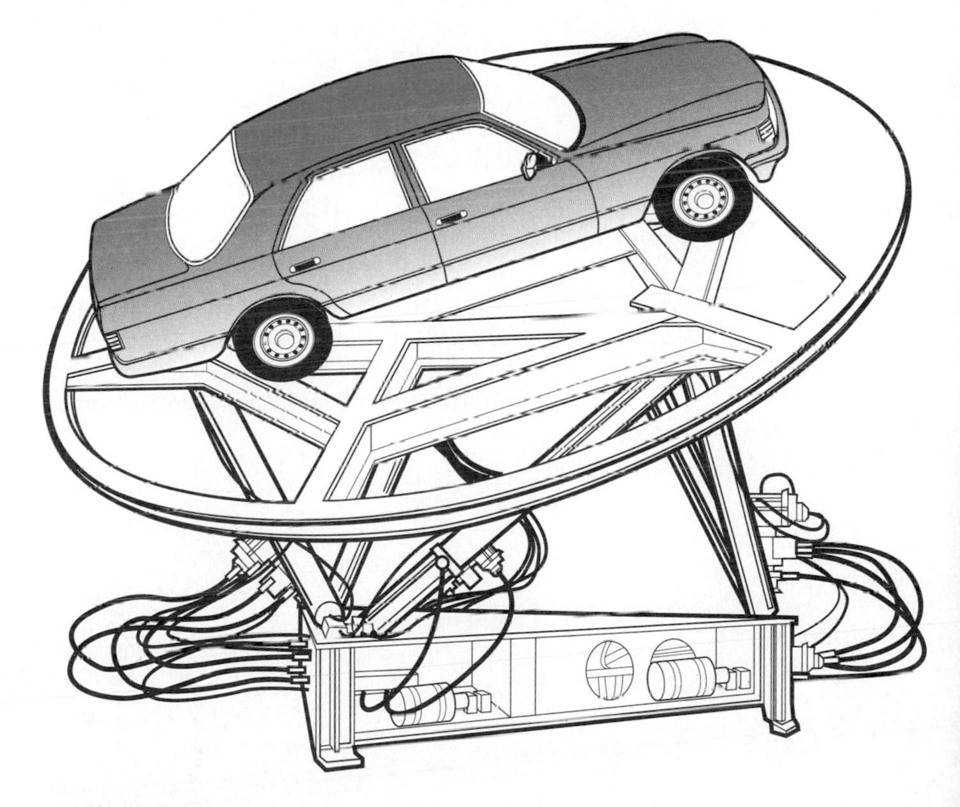

FIGURE 3.10
Driving simulator used to test a prototype automobile. Prior to building the prototype, mathematical models are analyzed to design a car providing a safe, smooth, and comfortable ride.

to approximate the action of a point mass m supported by a shock absorber (with damping coefficient δ) and a coil spring (with spring constant k). In Eq. (1) y represents the downward displacement from the natural equilibrium position of $y = 0$ (see Fig. 3.11). We now assume m to be the portion of the automobile's mass supported by one tire and analyze its motion for different choices of the spring and shock absorber constants, k and δ. [In Chapter 4 we will also consider what happens in the presence of a forcing function $f(t)$ representing various road conditions.]

We will interpret the results of our analysis in terms of an automobile suspension system. However, the analysis applies equally well to any straight-line motion having a viscous drag proportional to speed and a resistive force proportional to displacement. Another example is the motion of a pendulum. The motion is represented by the equation

$$m\ell\theta'' + \beta\ell\theta' + mg\theta = 0 \tag{2}$$

where m is the mass of the pendulum, ℓ is its length, β is a constant of proportionality for the viscous drag, and θ is the angular displacement from the vertical position (measured in radians). Equation (2) is a *linear approximation* for the action of an unforced damped pendulum.

As still another example, the equation

$$Lq'' + Rq' + \frac{1}{C}q = 0 \tag{3}$$

models an electrical circuit. Here L is the inductance, R is the resistance, C is the capacitance, and q is the charge. Equation (3) approximates the variation of the charge in an LRC circuit without an electromotive forcing function. Notice that Eqs. (1–3) all have the same fundamental mathematical form. That is, they are constant-coefficient homogeneous linear second-order equations. The analysis of each of these equations is the same; only the interpretation of the coefficients changes.

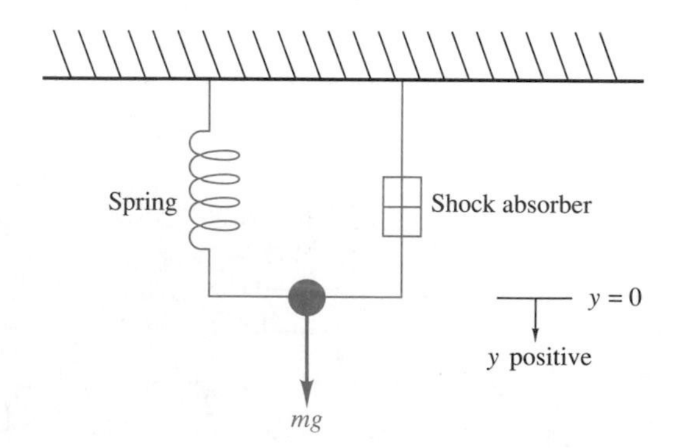

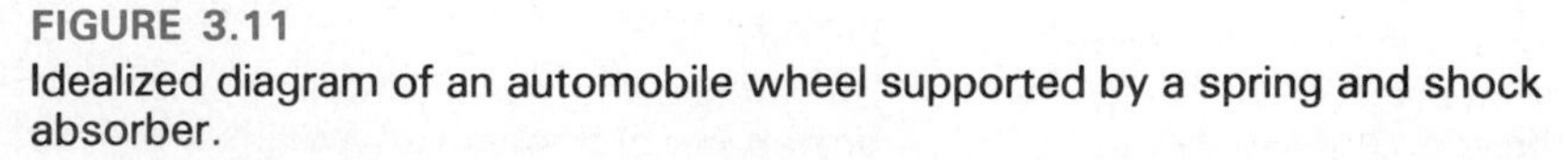

FIGURE 3.11
Idealized diagram of an automobile wheel supported by a spring and shock absorber.

Model of the Suspension System

We begin our analysis by dividing Eq. (1) through by the nonzero mass m and specifying initial conditions for position and velocity:

$$y'' + \frac{\delta}{m}y' + \frac{k}{m}y = 0, \quad y(0) = y_0, \; y'(0) = v_0. \tag{4}$$

Here y_0 represents an initial vertical displacement from the equilibrium, and v_0 is the initial velocity of the center of mass. The solution to Eq. (4) will describe the vertical motion of the center of mass as the system reacts to the discharge of the initial conditions.

Problem Identification The automotive engineer wants to choose values for the *design ratios* δ/m and k/m, which ensure both safety and comfort. To simplify our analysis, we will make the substitutions $2d = \delta/m$ and $\omega_0^2 = k/m$ so that Eq. (4) becomes

$$y'' + 2dy' + \omega_0^2 y = 0, \quad y(0) = y_0, \; y'(0) = v_0. \tag{5}$$

Let us first analyze the action of the mass with no resistive force. Then we will study the effects on the system when we add first a coil spring and then a shock absorber as well. In each situation we will graph the solution and interpret it in terms of the automobile. We will also determine the sensitivity of the results and their engineering implications as the design ratios δ/m and k/m are increased (if an increase is appropriate).

System with no Resistive Forces

To understand the effects of adding a coil spring and a shock absorber to the system, we assume initially that there is no resistive force whatsoever. This assumption produces the model

$$y'' = 0, \qquad y(0) = y_0, \; y'(0) = v_0. \tag{6}$$

The particular solution, with arbitrary constants evaluated from the initial conditions, is

$$y = v_0 t + y_0.$$

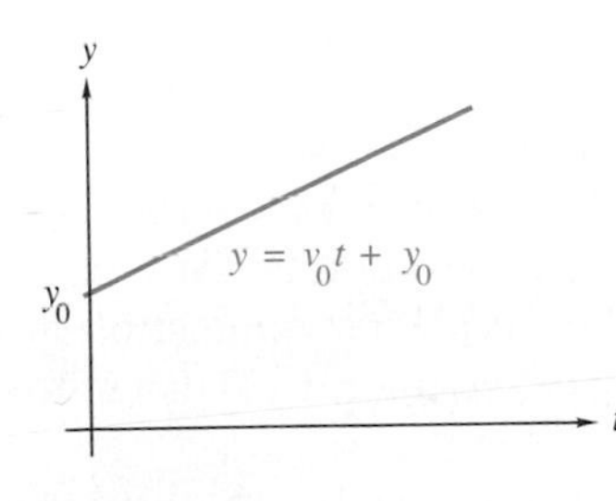

FIGURE 3.12
In the absence of any resistive force, displacement increases at the constant rate v_0.

The graph of the solution (Fig. 3.12) reveals what you might expect: the displacement y increases at the constant rate v_0. This behavior occurs because no resistance is present to counter the initial motion. The solution indicates that the displacement increases linearly with time. Clearly this model is too simple to describe the behavior of the system.

System with Coil Spring

Next we add to the system a coil spring with spring constant k. This assumption and the substitution of ω_0^2 for k/m yield the model

$$y'' + \omega_0^2 y = 0, \qquad y(0) = y_0,\ y'(0) = v_0. \tag{7}$$

The auxiliary equation is

$$r^2 + \omega_0^2 = 0,$$

which has the purely imaginary roots $\pm\omega_0 i$. The particular solution is then

$$y = y_0 \cos \omega_0 t + \frac{v_0}{\omega_0} \sin \omega_0 t \tag{8}$$

after the arbitrary constants are evaluated from the initial conditions. Interpreting solution (8), you can see that the motion is *oscillatory*. In general, the presence of nonzero imaginary parts in the roots to the auxiliary equation leads to oscillatory motion. The motion expressed by Eq. (8) is called **simple harmonic motion.** Let us investigate the nature of this motion.

Simple Harmonic Motion The oscillatory motion represented by solution (8) has several distinct features. Because of the presence of the sine and cosine terms, simple harmonic motion is *periodic.*

> The **period** of the simple harmonic motion described by Eq. (8) is
>
> $$\tau = \frac{2\pi}{\omega_0}.$$

In other words, $y(t + n\tau) = y(t)$ for all integers n where $\tau = 2\pi/\omega_0$. The period τ is the time it takes the mass to complete one cycle in its motion. Thus the mass m is moving up and down in a repetitive fashion through a lowest point to a highest point, then back to the lowest point, and so forth. The motion does not necessarily start at the highest or lowest point, but it reaches both within each complete cycle. Moreover, the largest positive displacement (lowest point) of m away from the equilibrium position $y = 0$ always equals the largest negative displacement (highest point). This maximum displacement is called the **amplitude** of the simple harmonic motion. To find the value of the amplitude, we are going to rewrite Eq. (8) in a more revealing form.

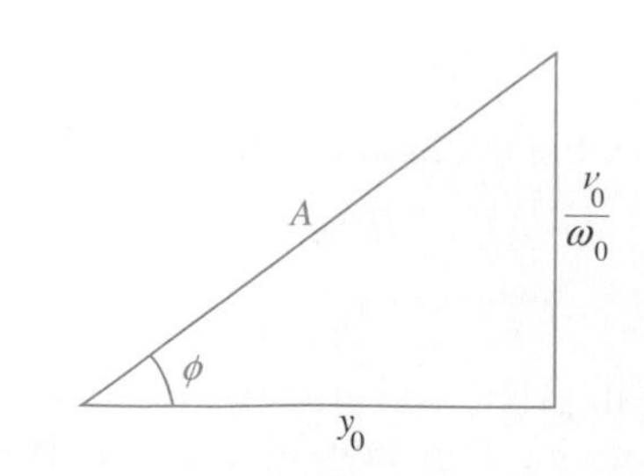

FIGURE 3.13
The coefficients y_0 and v_0/ω_0 form the legs of a right triangle.

We will let the coefficients of the trigonometric terms in Eq. (8) form the legs of the right triangle displayed in Fig. 3.13. Then

$$A = \sqrt{y_0^2 + \left(\frac{v_0}{\omega_0}\right)^2}.$$

Then the following relations are readily seen from Fig. 3.13:

$$\cos \phi = \frac{y_0}{A} \tag{9}$$

and

$$\sin \phi = \frac{v_0}{\omega_0 A}. \tag{10}$$

Notice that $\tan \phi = v_0/\omega_0 y_0$.

Dividing Eq. (8) by A and substituting from Eqs. (9) and (10) leads to

$$\frac{y}{A} = \cos \phi \cos \omega_0 t + \sin \phi \sin \omega_0 t. \tag{11}$$

If you compare Eq. (11) with the trigonometric difference formula

$$\cos (\alpha - \beta) = \cos \alpha \cos \beta + \sin \alpha \sin \beta,$$

you can see that

$$\frac{y}{A} = \cos (\omega_0 t - \phi)$$

or

$$y = A \cos (\omega_0 t - \phi). \tag{12}$$

In Eq. (12), $\phi = \tan^{-1} v_0/\omega_0 y_0$ is a radian angle called the **phase shift** of the simple harmonic motion. Its geometric interpretation is shown in Fig. 3.14, which graphs the displacement y versus $\omega_0 t$.

Since $|\cos (\omega_0 t - \phi)| \leq 1$, the motion in Eq. (12) always lies between $y = A$ and $y = -A$. Thus A represents the amplitude of the oscillatory motion.

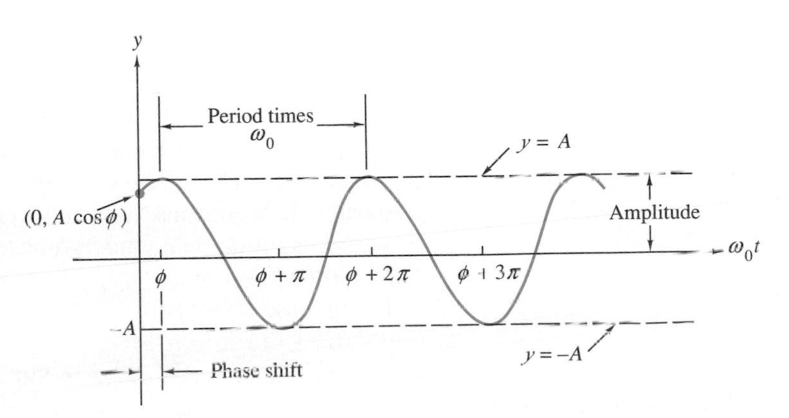

FIGURE 3.14
Simple harmonic motion.

The **amplitude** of the simple harmonic motion of Eqs. (8) or (12) is

$$A = \sqrt{y_0^2 + \left(\frac{v_0}{\omega_0}\right)^2}.$$

For a particular spring–mass system (where k and m have fixed constant values), you can see that the amplitude of the simple harmonic motion increases if you increase either the initial displacement y_0 or the initial velocity v_0. This result probably comes as no surprise. Since we assume there is no energy-dissipating device in our system, the simple harmonic motion continues indefinitely. The presence of the coil spring simply *changes the direction of the motion* in a periodic fashion.

Physical Interpretation of Simple Harmonic Motion Let us substitute the design ratio k/m for ω_0^2 and interpret the results for our automobile suspension system. The amplitude becomes

$$A = \sqrt{y_0^2 + v_0^2\left(\frac{m}{k}\right)}.$$

It depends on the initial conditions and the value of k/m. The period becomes

$$\tau = \frac{2\pi}{\omega_0} = \frac{2\pi\sqrt{m}}{\sqrt{k}}.$$

It depends only on the ratio k/m and not on the initial conditions.

Since most automobile passengers are more sensitive to the number of oscillations for a given unit of time, it is useful to define the reciprocal of the period, or **frequency,** as

$$f = \frac{1}{\tau} = \frac{\sqrt{k}}{2\pi\sqrt{m}}.$$

For instance, if the period is ½ sec, then the frequency is 2 cycles/sec. Notice that the frequency of the vibrations depends only on the natural properties of the materials used, as specified by k and m. We call f the system's **natural frequency.** It becomes a very important consideration when there is a periodic forcing function having its own frequency. We will take up this problem in Section 4.5.

Finally, the phase shift

$$\phi = \tan^{-1}\left(\frac{v_0\sqrt{m}}{y_0\sqrt{k}}\right)$$

depends on both the initial conditions and the ratio k/m. We summarize these results for simple harmonic motion in Table 3.1.

TABLE 3.1

Simple Harmonic Motion

$y'' + \left(\frac{k}{m}\right)y = 0,\quad y(0) = y_0,\quad y'(0) = v_0 \quad y = A\cos\left(\sqrt{\frac{k}{m}}t - \phi\right)$

Amplitude	$A = \sqrt{y_0^2 + \frac{mv_0^2}{k}}$
Period	$\tau = 2\pi\sqrt{\frac{m}{k}}$
Frequency	$f = \frac{1}{2\pi}\sqrt{\frac{k}{m}}$
Phase shift	$\phi = \tan^{-1}\left(\frac{v_0\sqrt{m}}{y_0\sqrt{k}}\right)$

What are the engineering implications of our analysis? Remember, an engineer basically controls the ratio k/m in the design of a suspension system. For a given mass m, if the spring is made stiffer (so that the ratio k/m is increased), then the amplitude of the vibration *decreases*. Likewise, the phase shift decreases. However, the frequency *increases*. By conducting experiments for various values of the ratio k/m, an engineer can determine a range of acceptable values based on some criteria defining comfort and safety.

Remark The analysis that led to simple harmonic motion applies as well to the following analogous systems:

Undamped pendulum:

$$m\ell\theta'' + mg\theta = 0.$$

***LC* circuit:**

$$Lq'' + \frac{1}{C}q = 0.$$

System with Spring and Shock Absorber

Now let us incorporate a shock absorber into the suspension system. The model is

$$y'' + 2dy' + \omega_0^2 y = 0, \qquad y(0) = y_0,\ y'(0) = v_0 \tag{13}$$

TABLE 3.2

Possible Solutions for Damped Motion $2d=\frac{\delta}{m}$, $\omega_0^2=\frac{k}{m}$

$d^2-\omega_0^2$	Parameter Relations	Solution	Nature of Damping
Positive	$d>\omega_0$ or $\delta^2>4km$	$y=c_1e^{(-d+\sqrt{d^2-\omega_0^2})t}+c_2e^{(-d-\sqrt{d^2-\omega_0^2})t}$	Overdamped
Zero	$d=\omega_0$ or $\delta^2=4km$	$y=(c_1t+c_2)e^{-dt}$	Critically damped
Negative	$d<\omega_0$ or $\delta^2<4km$	$y=e^{-dt}[c_1\cos\sqrt{\omega_0^2-d^2}\,t+c_2\sin\sqrt{\omega_0^2-d^2}\,t]$	Underdamped

where $2d=\delta/m$ and $\omega_0^2=k/m$ are the design ratios the automotive engineer will eventually control.

The auxiliary equation is

$$r^2+2dr+\omega_0^2=0, \tag{14}$$

which has two roots:

$$r_1=-d+\sqrt{d^2-\omega_0^2} \quad \text{and} \quad r_2=-d-\sqrt{d^2-\omega_0^2}.$$

The discriminant $d^2-\omega_0^2$ depends only on the ratios δ/m and k/m. The discriminant determines the nature of the solution, and there are three possibilities. They are summarized in Table 3.2 and we will discuss each case in turn.

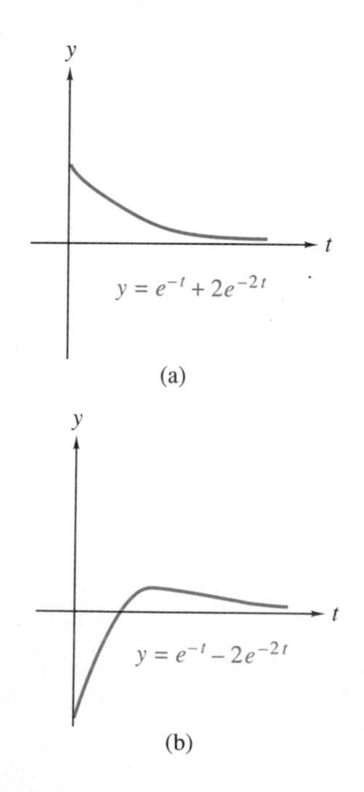

FIGURE 3.15
Overdamped motion occurs when the discriminant $d^2-\omega_0^2$ is positive.

Case 1: $d^2-\omega_0^2>0$ for Overdamped Motion

Both roots to the auxiliary equation are real numbers. Since $d>\sqrt{d^2-\omega_0^2}$, both roots are also negative. The general solution to the differential equation in Eq. (13) is

$$y=c_1e^{(-d+\sqrt{d^2-\omega_0^2})t}+c_2e^{(-d-\sqrt{d^2-\omega_0^2})t}. \tag{15}$$

Thus, as t approaches infinity, the solution $y(t)$ approaches zero asymptotically. This situation is referred to as **overdamped motion.** No oscillation occurs in the system. Examples illustrating overdamped motion are graphed in Fig. 3.15.

Case 2: $d^2-\omega_0^2=0$ for Critically Damped Motion

This case is not likely to occur for an automobile suspension system. That is, given the small errors inherent in the engineering and manufacturing processes, it is highly improbable that d^2 will equal exactly ω_0^2. Nevertheless, we include a discussion of this situation for the purpose of completeness in our mathematical analysis. In this case auxiliary Eq. (14) has the double negative root $-d$. The general solution is

$$y = (c_1 t + c_2)e^{-dt}. \tag{16}$$

To examine what happens to the solution as t approaches infinity, we need the following result, which you studied in calculus.

L'HÔPITAL'S RULE

If $\lim_{t\to\infty} f(t) = \infty$ and $\lim_{t\to\infty} g(t) = \infty$, then

$$\lim_{t\to\infty} \frac{f(t)}{g(t)} = \lim_{t\to\infty} \frac{f'(t)}{g'(t)}.$$

Applying l'Hôpital's rule to solution (16) we find that

$$\lim_{t\to\infty} \frac{c_1 t + c_2}{e^{dt}} = \lim_{t\to\infty} \frac{c_1}{de^{dt}} = 0.$$

The initial conditions determine the values of c_1 and c_2 in Eq. (16). If both c_1 and c_2 are positive, then the factor $c_1 t + c_2$, and hence the solution, is always positive (since $t \geq 0$ in our analysis). Likewise, if both c_1 and c_2 are negative, the solution is always negative. However, if either c_1 or c_2 is negative (but not both), then the factor $c_1 t + c_2 = 0$ when t has the positive value $-c_2/c_1$. In this case the solution *must cross* the equilibrium position $y = 0$. A typical example of each of these two situations is displayed in Fig. 3.16. There is no oscillation in the system about the equilibrium position.

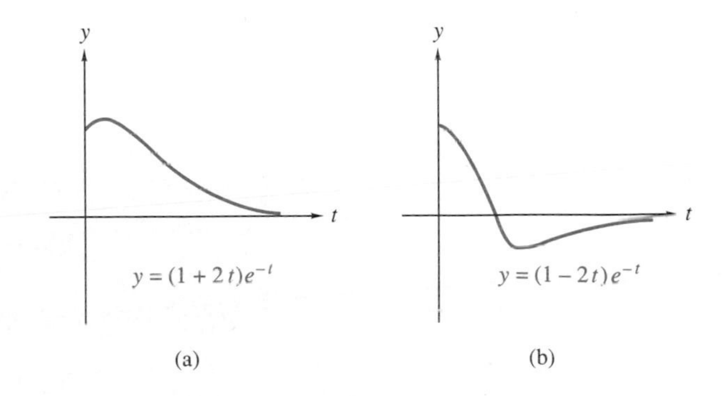

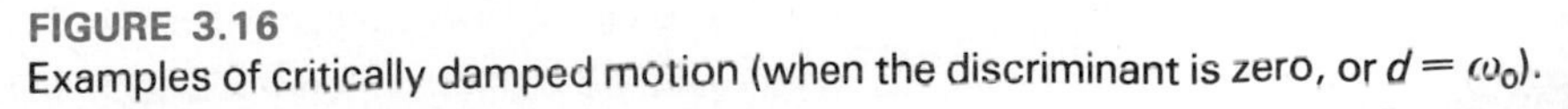
FIGURE 3.16
Examples of critically damped motion (when the discriminant is zero, or $d = \omega_0$).

Case 3: $d^2 - \omega_0^2 < 0$ for Underdamped Motion

In this case the general solution to Eq. (13) is

$$y = e^{-dt}(c_1 \cos \sqrt{\omega_0^2 - d^2}\, t + c_2 \sin \sqrt{\omega_0^2 - d^2}\, t). \tag{17}$$

The same procedure we employed to convert the solution equation for simple harmonic motion into a more revealing form can help us here. Solution (17) can be rewritten as

$$y = Ae^{-dt} \cos(\sqrt{\omega_0^2 - d^2}\, t - \phi) \tag{18}$$

where $A = \sqrt{c_1^2 + c_2^2}$ and $\phi = \tan^{-1}(c_2/c_1)$.

Since

$$|\cos(\sqrt{\omega_0^2 - d^2}\, t - \phi)| \leq 1,$$

the displacement y must lie between the curves $y = Ae^{-dt}$ and $y = -Ae^{-dt}$. These curves act as an "envelope" that contains the sinusoidal oscillations. The term e^{-dt} is a decaying exponential, and the resulting oscillatory motion is referred to as **underdamped motion.** A typical solution curve is sketched in Fig. 3.17. Although the motion is not truly periodic, the time between consecutive maximum peaks in the oscillation is called the **quasi period.** The frequency is again the reciprocal of the period and has the value $(1/2\pi)\sqrt{\omega_0^2 - d^2}$. Thus the frequency depends only on the design ratios k/m and δ/m.

Physical Interpretation of Damped Vibrations

Let us interpret damped motion in terms of the design ratios δ/m and k/m. In all three cases the effect of adding a shock absorber to the system dampens the displacement from the equilibrum position to zero. If the shock absorber

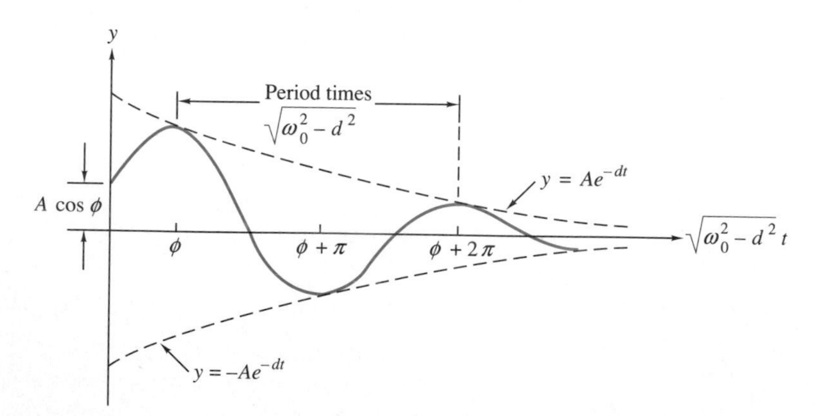

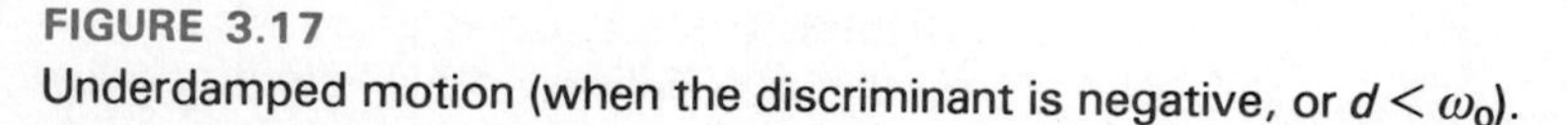

FIGURE 3.17
Underdamped motion (when the discriminant is negative, or $d < \omega_0$).

is weak relative to the mass–spring system (that is, if $\delta^2 < 4km$), the motion is underdamped and the mass oscillates about the equilibrium point $y = 0$ as the displacement is being damped out. If the shock absorber is strong relative to the mass–spring system ($\delta^2 > 4km$), the motion is overdamped and the mass does not oscillate in periodic fashion.

Let us consider further the case of underdamped motion. To determine the effect of varying the ratio δ/m relative to k/m, we need to find the arbitrary constants c_1 and c_2 in Eq. (17) from the initial conditions $y(0) = y_0$ and $y'(0) = v_0$. In Exercises 3.4 you will be asked to derive the following particular solution:

$$y = e^{-\delta t/2m}\left[y_0 \cos\left(\frac{\sqrt{4km - \delta^2}}{2m}\right)t + \frac{2mv_0 + y_0\delta}{\sqrt{4km - \delta^2}} \sin\left(\frac{\sqrt{4km - \delta^2}}{2m}\right)t \right] \tag{19}$$

Analysis of (19) reveals that the *exponential envelope* $y = \pm Ae^{-dt}$ (shown in Fig. 3.17) depends on the design ratio δ/m, the initial conditions y_0 and v_0, and the ratio k/m. However, frequency in vibration depends only on the constants m, k, and δ, independent of the initial condition values. Thus underdamped motion with no forcing function is like simple harmonic motion to the extent that the frequency of the damped vibrations is an intrinsic property of the automobile suspension system.

Remark The analysis performed for the automobile suspension system also applies to the following behaviors:

Linearized damped unforced pendulum:

$$m\ell\theta'' + \beta\ell\theta' + mg\theta = 0.$$

***LRC* circuit without a forcing function:**

$$Lq'' + Rq' + \frac{1}{C}q = 0.$$

These two applications are addressed in Exercises 3.4.

Incorporating a Forcing Function

What happens when there is a nonzero forcing function? In driving down a concrete highway, with bumps where the concrete sections are unevenly joined, you may have noticed that your automobile suspension vibrations increase or decrease as you vary the speed of the car. That is, the resulting vibrations are worse at a particular speed. We will address this important refinement in Chapter 4, which considers the nonhomogeneous equation. As might be expected, when the forcing function is itself periodic (like the bumps in the concrete highway), the relationship between the frequency of vibrations in the forcing function and the natural frequency of the car plays an essential role in the overall effects. Inadequate attention paid to the combined effects of these frequencies can lead to disastrous consequences.

EXERCISES 3.4

Mechanical units in the British and metric systems may be helpful in doing the problems.

Unit	British System	MKS System
Distance	Feet (ft)	Meters (m)
Mass	Slugs	Kilograms (kg)
Time	Seconds (sec)	Seconds (sec)
Force	Pounds (lb)	Newtons (N)
g(earth)	32 ft/sec^2	9.81 m/sec^2

1. A 16-lb weight is attached to the lower end of a coil spring suspended from the ceiling and having a spring constant of 1 lb/ft. The resistance in the spring–mass system is numerically equal to the instantaneous velocity. At $t=0$ the weight is set in motion, from a position 2 ft below its equilibrium position, by giving it a downward velocity of 2 ft/sec. At the end of π sec determine whether the mass is above or below the equilibrium position and if so by what distance.

2. An 8-lb weight stretches a spring 4 ft. The spring–mass system resides in a medium offering a resistance to the motion equal to 1.5 times the instantaneous velocity. If the weight is released at a position 2 ft above its equilibrium position with a downward velocity of 3 ft/sec, find its position relative to the equilibrium position 2 sec later.

3. A 20-lb weight is hung on an 18-in. spring, stretching it 6 in. The weight is pulled down 5 in. and 5 lb are added to the weight. If the weight is now released with a downward velocity of v_0 in/sec, find the position of the mass relative to the equilibrium in terms of v_0 and valid for any time $t \geqslant 0$.

4. A mass of 1 slug is attached to a spring whose constant is 25/4 lb/ft. Initially the mass is released 1 ft above the equilibrium position with a downward velocity of 3 ft/sec, and the subsequent motion takes place in a medium that offers a damping force numerically equal to 3 times the instantaneous velocity. An external force $f(t)$ is driving the system, but assume that initially $f(t) \equiv 0$. Formulate and solve an initial value problem that models the given system. Interpret your results.

5. A 10-lb weight is suspended by a spring that is stretched 2 in. by the weight. Assume a resistance whose magnitude is $40/\sqrt{g}$ lb times the instantaneous velocity in feet per second. If the weight is pulled down 3 in. below its equilibrium position and released, find the time required to reach the equilibrium position for the first time.

6. A weight stretches a spring 6 in. It is set in motion at a point 2 in. below its equilibrium position with a downward velocity of 2 in/sec.

a) When does the weight return to its starting position?
b) When does it reach its highest point?
c) Show that the maximum velocity is $2\sqrt{2g+1}$ in/sec.

7. A weight of 10 lb stretches a spring 10 in. The weight is drawn down 2 in. below its equilibrium position and given an initial velocity of 4 in/sec. An identical spring has a different weight attached to it. This second weight is drawn down from its equilibrium position a distance equal to the amplitude of the first motion and then given an initial velocity of 2 ft/sec. If the amplitude of the second motion is twice that of the first, what weight is attached to the second spring?

8. A weight stretches one spring 3 in. and a second weight stretches another spring 9 in. If both weights are simultaneously pulled down 1 in. below their respective equilibrium positions and then released, find the first time after $t=0$ when their velocities are equal.

9. A weight of 16 lb stretches a spring 4 ft. The weight is pulled down 5 ft below the equilibrium position and then released. What initial velocity v_0 given to the weight would have the effect of doubling the amplitude of the vibration?

10. A mass weighing 8 lb stretches a spring 3 in. The spring–mass system resides in a medium with a damping constant of 2 lb-sec/ft. If the mass is released from its equilibrium position with a velocity of 4 in/sec in the downward direction, find the time required for the mass to return to its equilibrium position for the first time.

11. A weight suspended from a spring executes damped vibrations with a period of 2 sec. If the damping factor decreases by 90% in 10 sec, find the acceleration of the weight when it is 3 in. below its equilibrium position and is moving upward with a speed of 2 ft/sec.

12. A 10-lb weight stretches a spring 2 ft. If the weight is pulled down 6 in. below its equilibrium position and released, find the highest point reached by the weight. Assume the spring–mass system resides in a medium offering a resistance of $10/\sqrt{g}$ lb times the instantaneous velocity in feet per second.

13. An *LRC* circuit is set up with an inductance of 1/5 henry, a resistance of 1 ohm, and a capacitance of 5/6 farad. Assuming the initial charge is 2 coulombs and the initial current is 4 amperes, find the solution function describing the charge on the capacitor at any time. What is the charge on the capacitor after a long period of time?

14. An (open) electrical circuit consists of an inductor, a resistor, and a capacitor. There is an initial charge of 2 coulombs on the capacitor. At the instant the circuit is closed, a current of 3 amperes is present but no external voltage is being applied. In this circuit the voltage drops at three points are numerically related as follows: across the capacitor, 10 times the charge; across the resistor, 4 times the instantaneous change in the charge; and across the inductor, 2 times the instantaneous change in the current. Find the charge on the capacitor as a function of time.

15. An inductor of 2 henrys is connected in series with a resistor of 12 ohms, a capacitor of 1/16 farads, and a battery initially assumed to be 0 volts. The initial instantaneous charge q on the capacitor is 10 coulombs and the initial current q' is 0. Formulate and solve an initial value problem that models the circuit described. Interpret your results.

16. A **seconds pendulum** is one that swings through its complete arc in 1 sec; that is, it has a period of 2 sec. Using the linearized undamped model, find the length of a seconds pendulum with small amplitude.

17. A pendulum with no resistance has a period equal to 1.5 sec. It is pulled aside 5° from its vertical equilibrium position. Assuming the linearized undamped pendulum model, find the angular velocity $d\theta/dt$ of the pendulum as it passes through the vertical position for the first time.

18. A particle moves along the x axis in accordance with the law

$$x'' + 10x' + 9x = 0.$$

From a point 2 ft to the right of the origin the particle at time $t = 0$ is projected toward the left with a velocity of 20 ft/sec.

a) Find the time when the particle reaches its leftmost position.

b) Find the distance traveled and the velocity at the end of 1 sec.

19. A mathematical model representing the angular displacement $\theta(t)$ of a swinging door is

$$I\theta'' = -k_1\theta - k_2\theta'$$

where θ is the angle measured from the equilibrium position of the door, I is the moment of inertia of the door with respect to the hinge axis, k_1 is the constant of proportionality of a spring that acts to close the door, and k_2 is the constant of proportionality of a hydraulic mechanism that acts as a damper opposing the movement of the door (see the accompanying figure). For the system shown, $I = 1$ lb-sec^2/ft, $k_1 = 4$ lb/ft, and $k_2 = 5$ lb-sec/ft. The initial conditions are $\theta(0) = \pi/3$ and $\theta'(0) = 0$.

a) Formulate and solve an initial value problem that models the system shown in the figure and determine the angle $\theta(t)$ as a function of time ($t > 0$) for the given system.

b) What does your model predict as t becomes large? If a computer is available, graph your results.

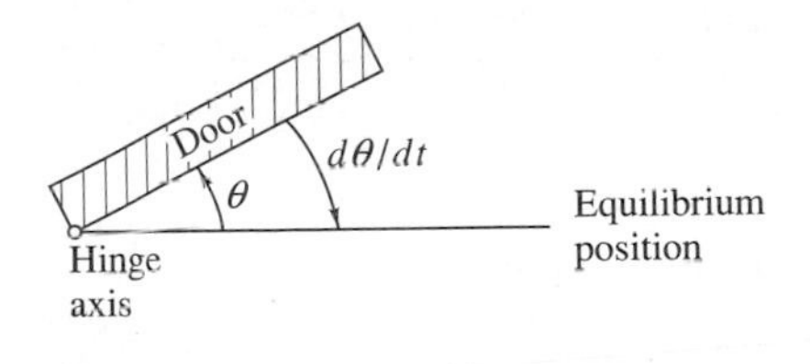

FIGURE 3.18
A swinging door.

20. Derive particular solution (19) in the text.

21. A mass m is free to move along a plane and is subject to a spring force $-k^2x$, a friction force $-bx'$, and an external force $F(t)$ as indicated in the accompanying figure. For this system, $m = 1$, $b = 4$, $k^2 = 20$, $F(t) = 0$, $x(0) = 5$, and

$x'(0) = 10$. Formulate and solve an initial value problem to model the given system. Interpret your results.

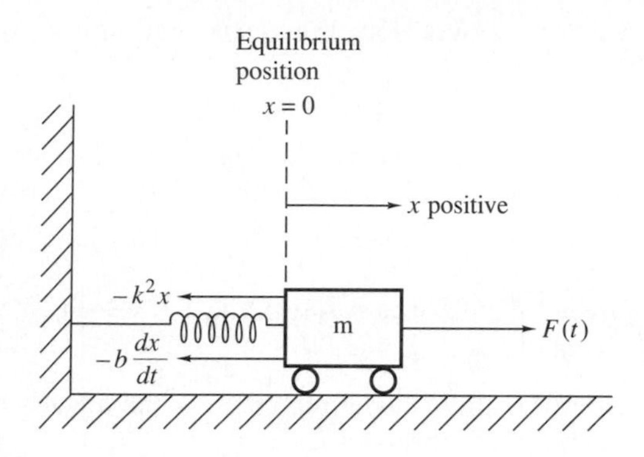

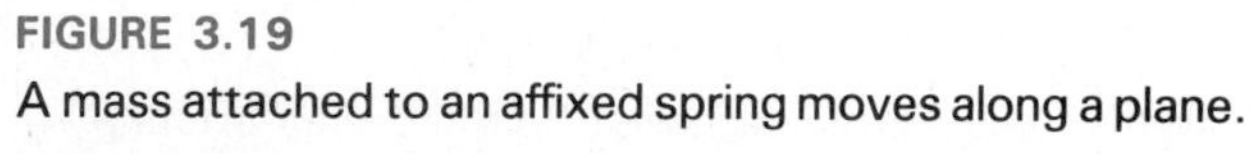

FIGURE 3.19
A mass attached to an affixed spring moves along a plane.

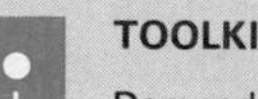

TOOLKIT PROGRAM

Damped Oscillator — Enables you to explore solutions of the equation $y'' + 2dy' + \omega_0^2 y = 0$. The program can help you to learn more about amplitude, frequency, angular frequency (ω_0), period, and conservation of energy.

3.5 HIGHER-ORDER LINEAR EQUATIONS

In this section we will study higher-order linear equations that have the form

$$a_n(x)y^{(n)} + a_{n-1}(x)y^{(n-1)} + \cdots + a_1(x)y' + a_0(x)y = 0 \tag{1}$$

where it is assumed that each $a_i(x)$ is continuous over an interval I. Higher-order equations are of interest for two reasons. First, they often arise when modeling behavior in the physical world. Second, *nonhomogeneous second-order constant-coefficient* equations,

$$a_2y'' + a_1y' + a_0y = g(x), \tag{2}$$

can sometimes be converted into higher-order *homogeneous* equations with constant coefficients. For this conversion to be possible, $g(x)$ must belong to a special class of functions discussed in this section. For instance, the nonhomogeneous equation

$$y'' = g, \tag{3}$$

where g is the constant acceleration of gravity, can be differentiated to yield the homogeneous third-order equation

$$y''' = 0. \tag{4}$$

Likewise, the nonhomogeneous equation

$$y'' + 2y' + 3y = x^2 + x + 1 \tag{5}$$

can be thrice differentiated to yield the fifth-order equation

$$y^{(5)} + 2y^{(4)} + 3y''' = 0. \tag{6}$$

In Sections 4.1 and 4.2 we will present the technique of *undetermined coefficients* for building the general solution to Eq. (5) based on the general solution to Eq. (6). Thus we will solve certain nonhomogeneous equations by first solving a higher-order homogeneous equation. The reason why differentiating nonhomogeneous Eqs. (3) and (5) leads to homogeneous equations is that their forcing functions $g(x) = g$ and $g(x) = x^2 + x + 1$ themselves solve constant-coefficient homogeneous equations.

Operator Notation

It is convenient to represent the lefthand side of Eq. (1) using the following **operator notation:**

$$L[y] = a_n(x)y^{(n)} + a_{n-1}(x)y^{(n-1)} + \cdots + a_1(x)y' + a_0(x)y. \qquad (7)$$

That is, if $y = y(x)$ and its first n derivatives exist over an interval I, then the **differential operator L** transforms $y(x)$ into a new function $L[y]$. The function $L[y]$ is formed by taking the first n derivatives of y (where the *zero derivative* simply refers to the function y itself), multiplying each derivative by some function of x, and forming the sum according to Eq. (7).*

For example, the differential operator

$$L[y] = 2xy'' + (1 - 4x)y' + (2x - 1)y \qquad (8)$$

transforms the function $y = e^{2x}$ as follows:

$$\begin{aligned} L[e^{2x}] &= 2x(4e^{2x}) + (1 - 4x)(2e^{2x}) + (2x - 1)e^{2x} \\ &= (2x + 1)e^{2x}. \end{aligned}$$

You can think of L as a "machine" that processes functions. You input an nth-order differentiable function $y = y(x)$ into the machine L, and it outputs a new function denoted $L[y]$. (We will use a similar idea later in the text when discussing the Laplace transform.)

A special feature of the operator L is its *linearity*. This property means that for any two functions $y_1(x)$ and $y_2(x)$ and any constants c_1 and c_2,

$$L[c_1y_1 + c_2y_2] = c_1L[y_1] + c_2L[y_2]. \qquad (9)$$

The linearity property of Eq. (9) follows immediately from two results you studied in elementary calculus:

1. The derivative of a sum of functions is the sum of the derivatives.
2. The derivative of a constant times a function is the constant times the derivative of the function.

* A more precise statement of this relation is as follows:

$$L[y] = a_ny^{(n)} + a_{n-1}y^{(n-1)} + \cdots + a_1y' + a_0y.$$

The value of this function at x is then

$$L[y](x) = a_n(x)y^{(n)}(x) + a_{n-1}(x)y^{(n-1)}(x) + \cdots + a_1(x)y'(x) + a_0(x)y(x).$$

We are using the abbreviated notation $L[y]$ to refer to $L[y](x)$ in Eq. (7).

Solutions to Higher-Order Linear Equations

Using the operator notation, the nth-order linear equation

$$a_n(x)y^{(n)} + a_{n-1}(x)y^{(n-1)} + \cdots + a_1(x)y' + a_0(x)y = g(x) \tag{10}$$

can be written in the form

$$L[y] = g(x). \tag{11}$$

A **solution** to Eq. (11) is then a continuous function $y = \phi(x)$, defined over the interval I, which the operator L transforms into the function $g(x)$: $L[\phi] = g(x)$. For example, the differential operator

$$L[y] = y'' - 3y' + 2y$$

transforms the function $y = e^x$ into the zero function $g(x) \equiv 0$:

$$L[e^x] = e^x - 3e^x + 2e^x = 0.$$

This means that $y = e^x$ solves the homogeneous equation $L[y] = 0$. In Chapter 4 you will study methods for solving the nonhomogeneous equation $L[y] = g(x)$ for constant-coefficient differential operators when $g(x) \not\equiv 0$.

In Section 3.2 you learned that the general solution to a second-order linear homogeneous equation is formed by taking an arbitrary linear combination of any two linearly independent solutions. That result extends to higher-order linear equations.

THEOREM 3.8

If $a_n(x), a_{n-1}(x), \ldots, a_1(x)$, and $a_0(x)$ are continuous over the interval I, and if $a_n(x) \neq 0$ in I, then the nth-order linear homogeneous equation

$$L[y] = 0 \tag{12}$$

has n linearly independent solutions $y_1(x), y_2(x), \ldots, y_n(x)$ on I. Furthermore, if $y_1(x), y_2(x), \ldots, y_n(x)$ is any set of n linearly independent solutions and if $y(x)$ is a solution to Eq. (12), then there exist unique constants $c_1, c_2, \ldots, c_n$ such that

$$y(x) = c_1y_1 + c_2y_2 + \cdots + c_ny_n. \tag{13}$$

In other words, any solution to Eq. (12) is given by a linear combination of n linearly independent solutions. The n linearly independent solutions y_1, $y_2, \ldots, y_n$ then form a **fundamental set of solutions** to Eq. (12). Note that Theorem 3.8 applies to linear equations with variable coefficients as well as to equations having constant coefficients.

Linear Independence and the Wronskian

Definition 3.3 stated the condition for linear independence of a set of k functions over an interval I. Subsequently we saw that two functions are linearly independent over I if and only if neither function is a constant multiple of the other. In Example 9 of Section 3.2 the definition was applied to establish that the three functions e^x, $\cos x$, and $\sin x$ form a linearly independent set over any interval I. The result followed from the fact that the determinant

$$\begin{vmatrix} e^x & \cos x & \sin x \\ e^x & -\sin x & \cos x \\ e^x & -\cos x & -\sin x \end{vmatrix} = 2e^x$$

is never zero on any interval I. The reasoning used in that example applies to any set of three functions, as we shall now demonstrate.

The three functions $f_1(x)$, $f_2(x)$, and $f_3(x)$ are linearly independent over I if and only if

$$c_1 f_1(x) + c_2 f_2(x) + c_3 f_3(x) = 0 \tag{14}$$

for all x in I implies that $c_1 = c_2 = c_3 = 0$. Differentiation of Eq. (14) twice results in

$$c_1 f_1'(x) + c_2 f_2'(x) + c_3 f_3'(x) = 0 \tag{15}$$

and

$$c_1 f_1''(x) + c_2 f_2''(x) + c_3 f_3''(x) = 0. \tag{16}$$

The homogeneous system of algebraic Eqs. (14–16) has only the trivial solution $c_1 = c_2 = c_3 = 0$ if the determinant

$$\begin{vmatrix} f_1(x) & f_2(x) & f_3(x) \\ f_1'(x) & f_2'(x) & f_3'(x) \\ f_1''(x) & f_2''(x) & f_3''(x) \end{vmatrix}$$

is not zero for at least one x in the interval I. This result holds true for any number of functions, as the following theorem now states. Each function is assumed to be differentiable at least $n - 1$ times.

THEOREM 3.9

If the determinant

$$\begin{vmatrix} f_1(x) & f_2(x) & \cdots & f_n(x) \\ f_1'(x) & f_2'(x) & \cdots & f_n'(x) \\ \vdots & \vdots & & \vdots \\ f_1^{(n-1)}(x) & f_2^{(n-1)}(x) & \cdots & f_n^{(n-1)}(x) \end{vmatrix} \tag{17}$$

is not zero for at least one x in the interval I, then the set of functions $\{f_1(x), f_2(x), \ldots, f_n(x)\}$ is linearly independent over I.

Determinant (17) is called the **Wronskian*** of the n functions and is denoted by $W[f_1(x), f_2(x), \ldots, f_n(x)]$. We also write $W[f_1, f_2, \ldots, f_n]$ or $W[f_1, f_2, \ldots, f_n](x)$ for the Wronskian.

EXAMPLE 1 If a, b, and c are distinct real numbers, the functions e^{ax}, e^{bx}, e^{cx} form a linearly independent set over any interval. This follows from an application of Theorem 3.9:

$$W[e^{ax}, e^{bx}, e^{cx}] = \begin{vmatrix} e^{ax} & e^{bx} & e^{cx} \\ ae^{ax} & be^{bx} & ce^{cx} \\ a^2e^{ax} & b^2e^{bx} & c^2e^{cx} \end{vmatrix}$$

$$= e^{ax}e^{bx}e^{cx} \begin{vmatrix} 1 & 1 & 1 \\ a & b & c \\ a^2 & b^2 & c^2 \end{vmatrix}$$

$$= e^{ax}e^{bx}e^{cx} \begin{vmatrix} 1 & 1 & 1 \\ 0 & b-a & c-a \\ 0 & 0 & (c^2-a^2)-(b+a)(c-a) \end{vmatrix}$$

$$= e^{ax}e^{bx}e^{cx}(b-a)(c-a)(c-b) \neq 0$$

In Example 1 the Wronskian is never zero on any interval. The condition that the Wronskian is never zero is sufficient to show that a set of functions is linearly independent, but it is not in general a necessary condition (see Exercises 3.5, problem 47). However, the situation is different when the functions are in fact solutions to a homogeneous linear differential equation.

THEOREM 3.10

Let $y_1, y_2, \ldots, y_n$ be solutions of

$$a_n(x)y^{(n)} + a_{n-1}(x)y^{(n-1)} + \cdots + a_1(x)y' + a_0(x)y = 0 \quad (18)$$

where each term $a_i(x)$ is continuous on the interval I. Then the set $\{y_1, y_2, \ldots, y_n\}$ is linearly independent on I if and only if the Wronskian $W[y_1, y_2, \ldots, y_n] \neq 0$ for all x in I.

The next result reveals a powerful property of the Wronskian for a set of solutions to a homogeneous linear differential equation. We again omit the proof.

* Named after the Polish mathematician Jozef Wronski (1778–1853).

THEOREM 3.11

If the functions $y_1, y_2, \ldots, y_n$ are solutions to homogeneous linear differential Eq. (18) on the interval I, then the Wronskian $W[y_1, y_2, \ldots, y_n]$ is either never zero on I or it is identically zero on I.

Theorems 3.10 and 3.11 are useful in determining whether a set $\{y_1, y_2, \ldots, y_n\}$ forms a fundamental set of solutions to homogeneous linear nth-order Eq. (18). Simply evaluate $W[y_1, y_2, \ldots, y_n]$ at some conveniently selected point x_0 in I. If the value of the Wronskian is not zero at x_0, then the set is a fundamental set of solutions.

Let us summarize the facts about linear independence and fundamental sets of solutions. Assume that $y_1, y_2, \ldots, y_n$ are solutions to homogeneous Eq. (18), where each $a_i(x)$ is continuous on the interval I. Then the following four statements are equivalent; that is, each one implies the other three.

1. $\{y_1, y_2, \ldots, y_n\}$ is a fundamental set of solutions on I.
2. $\{y_1, y_2, \ldots, y_n\}$ is linearly independent on I.
3. $W[y_1, y_2, \ldots, y_n] \neq 0$ for at least one x_0 in I.
4. $W[y_1, y_2, \ldots, y_n] \neq 0$ for every x in I.

For the remainder of this section we are concerned with linear differential equations having constant coefficients.

Polynomial Operator

In the special case that the coefficient functions in linear nth-order differential Eq. (10) are constants, the differential operator L takes a particularly revealing form. We introduce the **derivative operator** notation:

$$D = \frac{d}{dx}.$$

That is, D *denotes differentiation with respect to the independent variable* x. Thus, Dy means dy/dx. Likewise, D^2 denotes second-order differentiation d^2/dx^2, D^3 denotes d^3/dx^3, and so forth. Thus the differential operator

$$L[y] = a_n y^{(n)} + a_{n-1} y^{(n-1)} + \cdots + a_1 y' + a_0 y \tag{19}$$

can be written in the form

$$(a_n D^n + a_{n-1} D^{n-1} + \cdots + a_1 D + a_0) y.$$

Since this expression looks like a polynomial in powers of D, we adopt polynomial notation and write

$$P(D)y = (a_n D^n + a_{n-1} D^{n-1} + \cdots + a_1 D + a_0)y \tag{20}$$

where $a_n \neq 0$ and each a_i is a constant. In this polynomial operator notation the homogeneous nth-order equation with constant coefficients takes the form

$$P(D)y = 0. \tag{21}$$

The advantages of this notation will become apparent as we investigate solutions for constant-coefficient linear equations.

Auxiliary Equation

Assuming that $a_0 \neq 0$ and solving Eq. (21) for the function $y(x)$ yields us

$$y = -\frac{a_n}{a_0} y^{(n)} - \frac{a_{n-1}}{a_0} y^{(n-1)} - \cdots - \frac{a_1}{a_0} y'.$$

Thus we seek a solution function $y = y(x)$ the multiples of whose various derivatives sum to the function itself. From our knowledge of the exponential function we try $y = e^{rx}$. Substitution into Eq. (21) leads to

$$e^{rx}(a_n r^n + a_{n-1} r^{n-1} + \cdots + a_1 r + a_0) = 0.$$

Division by $e^{rx} \neq 0$ then yields the *auxiliary equation*

$$a_n r^n + a_{n-1} r^{n-1} + \cdots + a_1 r + a_0 = 0. \tag{22}$$

Just as we found for the second-order constant-coefficient linear equation, a root $r = m$ of the auxiliary equation leads to a solution $y = e^{mx}$.

An important fact from algebra is that a polynomial of degree n with real coefficients has exactly n roots in the complex number system. Some of these roots may be real (that is, have a zero imaginary part). Whenever $\alpha + \beta i$ is a complex root, its complex conjugate $\alpha - \beta i$ is also a root. Thus complex roots having a nonzero imaginary part always occur in conjugate pairs.

Recall from algebra that whenever $r = m$ is a real root of the auxiliary equation, the linear term $r - m$ is a factor of the equation:

$$a_n r^n + a_{n-1} r^{n-1} + \cdots + a_1 r + a_0 = (r - m)Q(r) \tag{23}$$

where $Q(r)$ is a polynomial of degree $n - 1$.

In the case that $r = \alpha + \beta i$ is a complex root with a nonzero imaginary part, then both $r - \alpha - \beta i$ and $r - \alpha + \beta i$ are factors of the auxiliary equation because complex roots occur in conjugate pairs. From the expansion

$$(r - \alpha - \beta i)(r - \alpha + \beta i) = r^2 - 2\alpha r + \alpha^2 + \beta^2,$$

you can see that the complex roots $\alpha \pm \beta i$ give rise to the factorization

$$a_n r^n + a_{n-1} r^{n-1} + \cdots + a_1 r + a_0 = (r^2 - 2\alpha r + \alpha^2 + \beta^2)Q(r) \tag{24}$$

where $Q(r)$ is a polynomial of degree $n - 2$.

Factorizations (23) and (24) of the auxiliary equation correspond exactly to factorizations of the polynomial operator $P(D)$ in Eq. (20). Thus, for instance, if $r = m$ is a root of the auxiliary equation, then

$$P(D) = (D - m)Q(D)$$

is a corresponding factorization of the polynomial operator. This correspondence is one of the advantages of the polynomial operator notation.

Roots of Polynomials

The quadratic formula gives the real or complex roots to a quadratic equation. A procedure also exists for finding the real or complex roots of a cubic polynomial, but it involves changing the variable and using the cube roots of unity, which is not nearly as simple as applying the quadratic formula. Finding the roots of a polynomial of higher degree is generally nontrivial, and no algebraic algorithm can be produced that works in every case.

Nevertheless, we can sometimes recognize a perfect square and identify factors. For example,

$$r^4 - 16 = 0$$

factors into

$$(r^2 - 4)(r^2 + 4) = 0$$

or

$$(r + 2)(r - 2)(r^2 + 4) = 0.$$

We readily determine the four roots as $r = \pm 2, \pm 2i$.

If the **leading coefficient** (part of the term having the highest power of the polynomial) is 1, then the constant term is equal to $(-1)^n$ times the product of all the roots. For instance, the polynomial

$$r^3 - 7r + 6 = (r - 1)(r - 2)(r + 3),$$

and you see that the constant term 6 is the product of the three roots 1, 2, and -3 multiplied by $(-1)^3$. Thus in some instances dividing by the highest-order term a_n and testing the possible integer factors of the new constant term may lead to a root. The knowledge of a root can then be used to reduce the order of the polynomial through polynomial division. Following is an example of this procedure.

EXAMPLE 2 Division of the polynomial

$$2r^3 - 4r^2 - 2r + 4 = 0 \tag{25}$$

by the leading coefficient $a_3 = 2$ yields

$$r^3 - 2r^2 - r + 2 = 0.$$

The possible integer factors of the resulting constant term 2 are $1, -1, 2$, and -2. By testing $r = 1$ we see that it is a root of the polynomial. Then we can divide the factor $r - 1$ out of the polynomial (using synthetic division if you wish):

$$\begin{array}{r|l} & r^2 - \ \ r - 2 \\ \hline r - 1 & r^3 - 2r^2 - \ \ r + 2 \\ & \underline{r^3 - \ \ r^2} \\ & \quad - r^2 - r \\ & \quad \underline{- r^2 + r} \\ & \qquad\quad -2r + 2 \\ & \qquad\quad \underline{-2r + 2} \\ & \qquad\qquad\quad 0 \end{array}$$

Since the remainder is 0, $r - 1$ and $r^2 - r - 2$ are factors of polynomial (25). Moreover, we can factor the quadratic term: $r^2 - r - 2 = (r + 1)(r - 2)$. Then

$$r^3 - 2r^2 - r + 2 = (r - 1)(r + 1)(r - 2)$$

gives a complete factorization of Eq. (25) leading to the roots -1, 1, and 2.

It may happen that the roots to a polynomial are irrational numbers or otherwise difficult to find. In such cases a numerical scheme like Newton's method for approximating roots might be used. Newton's method is easy to program on a computer or programmable calculator. On the other hand, it may be just as convenient and revealing to approximate the solution to the differential equation by using a numerical solution technique. Several numerical approximation methods for second-order equations are discussed in Section 4.6. Let us now investigate how the nature of the roots to the auxiliary equation governs the form of the solution to the corresponding homogeneous equation.

Distinct Real Roots

If all n roots of auxiliary Eq. (22) are real and distinct, we have $r = r_1, r_2, \ldots, r_n$. Then the exponential functions

$$e^{r_1 x}, e^{r_2 x}, \ldots, e^{r_n x}$$

form a fundamental set of solutions to the homogeneous equation $P(D)y = 0$. The general solution to $P(D)y = 0$ is given by

$$y = c_1 e^{r_1 x} + c_2 e^{r_2 x} + \cdots + c_n e^{r_n x}. \tag{26}$$

Complex Roots

As noted before, complex roots always occur in conjugate pairs: $\alpha + \beta i$ and $\alpha - \beta i$. Using Euler's identity (see Section 3.3, Eq. (9)), we find that each complex pair yields two linearly independent solutions: $e^{\alpha x} \cos \beta x$ and $e^{\alpha x} \sin \beta x$.

For example, knowing the complex roots $1 \pm 2i$ and $3 \pm 4i$ to an auxiliary fourth-order polynomial produces the general solution

$$y = e^{x}(c_1 \cos 2x + c_2 \sin 2x) + e^{3x}(c_3 \cos 4x + c_4 \sin 4x).$$

Repeated Real Roots

Suppose now that $r = m$ is a root of the auxiliary equation repeated k times. Then from our earlier discussion, the term $(D - m)^k$ is a factor of the polynomial operator:

$$P(D)y = (D - m)^k Q(D)y. \tag{27}$$

Armed with this information we can establish that each of the k functions

$$e^{mx},\ xe^{mx},\ x^2 e^{mx},\ \ldots,\ x^{k-1} e^{mx} \tag{28}$$

solves the differential equation. This result follows from the following property:

$$(D - m)^k x^j e^{mx} = 0 \qquad \text{whenever } j < k. \tag{29}$$

To prove Eq. (29) we need merely apply the operator $D - m$ to the function $x^j e^{mx}$ k times. Consider, for example, $(D - m)^3 x^2 e^{mx}$:

$$\begin{aligned}
(D - m)x^2 e^{mx} &= 2xe^{mx} + mx^2 e^{mx} - mx^2 e^{mx}, \\
(D - m)^2 x^2 e^{mx} &= (D - m)2xe^{mx} = 2e^{mx} + 2mxe^{mx} - 2mxe^{mx}, \\
(D - m)^3 x^2 e^{mx} &= (D - m)2e^{mx} = 2me^{mx} - 2me^{mx} = 0.
\end{aligned}$$

Hence $(D - m)^3 x^2 e^{mx} = 0$, which is in agreement with Eq. (29).

It can be shown that the Wronskian of the functions in list (28) is never zero, so they form a fundamental set of solutions. We present an example.

EXAMPLE 3 The differential equation

$$y''' - 9y'' + 27y' - 27y = 0$$

has the auxiliary equation

$$r^3 - 9r^2 + 27r - 27 = 0 = (r-3)^3.$$

The roots are therefore $r = 3, 3, 3$. The general solution is then given by

$$y = c_1e^{3x} + c_2xe^{3x} + c_3x^2e^{3x}.$$

Repeated Complex Pairs

A repeated root may be complex-valued. For example, suppose the complex roots $r = \alpha \pm \beta i$ are repeated twice. Then the complex functions e^{rx} and xe^{rx} are both solutions. Application of Euler's identity to each complex exponential results in the following four linearly independent solutions:

$$e^{\alpha x}\cos\beta x, \quad e^{\alpha x}\sin\beta x, \quad xe^{\alpha x}\cos\beta x, \quad xe^{\alpha x}\sin\beta x.$$

These solutions form a fundamental set because their Wronskian is never zero. Let us consider some examples.

EXAMPLE 4 The differential equation

$$y^{(4)} + 2y'' + y = 0$$

has the auxiliary equation

$$r^4 + 2r^2 + 1 = (r^2+1)^2 = 0.$$

The roots are $r = \pm i, \pm i$. The general solution is

$$y = c_1\cos x + c_2\sin x + c_3x\cos x + c_4x\sin x.$$

EXAMPLE 5 The differential equation

$$y^{(6)} - y^{(5)} + 2y^{(4)} - 2y''' + y'' - y' = 0$$

has the auxiliary equation

$$r^6 - r^5 + 2r^4 - 2r^3 + r^2 - r = r(r-1)(r^2+1)^2 = 0.$$

The roots are $r = 0, 1, \pm i, \pm i$. The general solution is

$$y = c_1 + c_2e^x + c_3\cos x + c_4\sin x + c_5x\cos x + c_6x\sin x.$$

Summary of Higher-Order Solutions

Table 3.3 summarizes the various solution forms to higher-order linear equations with constant coefficients. You can see that although each situa-

TABLE 3.3

Nature of Solutions to the nth-Order Linear Homogeneous Equation with Constant Coefficients

$$a_n y^{(n)} + a_{n-1} y^{(n-1)} + \cdots + a_1 y' + a_0 y = 0$$

Roots	General Solution	Type of Functions
Distinct and real: $r = r_1, r_2, \ldots, r_n$	$y = c_1 e^{r_1 x} + c_2 e^{r_2 x} + \cdots + c_n e^{r_n x}$	Exponentials
Distinct and complex pairs (purely imaginary): $r = \pm\beta_1 i, \pm\beta_2 i, \ldots, \pm\beta_k i$, where $n = 2k$	$y = c_1 \cos \beta_1 x + c_2 \sin \beta_1 x + c_3 \cos \beta_2 x + c_4 \sin \beta_2 x + \cdots + c_{n-1} \cos \beta_k x + c_n \sin \beta_k x$	Trigonometric functions
Distinct and complex pairs (real and imaginary): $r = \alpha_1 \pm \beta_1 i, \alpha_2 \pm \beta_2 i, \ldots, \alpha_k \pm \beta_k i$, where $n = 2k$	$y = e^{\alpha_1 x}(c_1 \cos \beta_1 x + c_2 \sin \beta_1 x) + e^{\alpha_2 x}(c_3 \cos \beta_2 x + c_4 \sin \beta_2 x) + \cdots + e^{\alpha_k x}(c_{n-1} \cos \beta_k x + c_n \sin \beta_k x)$	Trigonometric with exponential factors
Repeated real: $r = m, m, \ldots, m$ (n times)	$y = (c_1 + c_2 x + \cdots + c_{n-1} x^{n-2} + c_n x^{n-1}) e^{mx}$	Exponential with polynomial factor
Repeated zero: $r = 0, 0, \ldots, 0$ (n times)	$y = c_1 + c_2 x + \cdots + c_{n-1} x^{n-2} + c_n x^{n-1}$	Polynomials
Repeated complex pairs: $r = \alpha \pm \beta i$ (k times), where $n = 2k$	$y = e^{\alpha x}(c_1 \cos \beta x + c_2 \sin \beta x) + x e^{\alpha x}(c_3 \cos \beta x + c_4 \sin \beta x) + \cdots + x^{k-1} e^{\alpha x}(c_{n-1} \cos \beta x + c_n \sin \beta x)$	Trigonometric with exponentials times powers of x factors

tion assumes a solution of the form $y = e^{rx}$, the type of the solution may vary considerably: purely exponential, trigonometric, trigonometric with exponential coefficients, and so forth. Each type has a different implication for the physical behavior being modeled. A trigonometric function indicates the presence of oscillations that are periodic in nature. Trigonometric functions are present whenever at least two roots have imaginary parts. If a complex root $\alpha \pm \beta i$ has the nonzero real part α, the corresponding oscillatory motion is either "amplified" or "damped" by the exponential factor $e^{\alpha x}$, depending on whether α is positive (amplification) or negative (damping). The functions presented in Table 3.3 comprise an important class of functions for many enginering applications. You will find it helpful to refer back to this table when the method of undetermined coefficients is presented in Chapter 4.

EXERCISES 3.5

In problems 1–5 assume that $y = e^{rx}$ and write the auxiliary equation.

1. $y''' - y'' - y' + y = 0$
2. $(2D^3 - 4D^2 - 2D + 4)y = 0$
3. $y^{(4)} - 5y'' + 4y = 0$

4. $(D^4 - 1)y = 0$

5. $(D^4 + 1)y = 0$

In problems 6–15 find the roots of the given algebraic equation.

6. $r^3 - r^2 - r + 1 = 0$

7. $r^3 - 2r^2 - 2r + 1 = 0$

8. $r^4 - 5r^2 + 4 = 0$

9. $r^4 - 1 = 0$

10. $r^4 + 1 = 0$

11. $r^4 - 2r^2 + 1 = 0$

12. $r^4 + r^3 - r^2 - r = 0$

13. $4r^4 + 3r^2 - 1 = 0$

14. $r^4 + r = 0$

15. $r^5 + 3r^3 - 4r = 0$

In problems 16–25 write the general solution corresponding to the given set of roots.

16. $r = 1, 1, -1$

17. $r = -2, -2, -2$

18. $r = 1, 2, -1$

19. $r = 0, 0, 3$

20. $r = 2, -2, -1, 1$

21. $r = 0, 3, 3, -1$

22. $r = 1, -1, \pm i$

23. $r = \pm i, \pm i$

24. $r = -1 \pm 2i, -1 \pm 2i$

25. $r = 0, 0, \pm i, \pm i, \pm i$

In problems 26–40 find the general solution to the given higher-order equation.

26. $y^{(4)} + 2y'' + y = 0$

27. $y''' + 3y'' + 3y' + y = 0$

28. $y''' - y' = 0$

29. $y^{(4)} - y = 0$

30. $(D^3 + 2D^2 - D - 2)y = 0$

31. $(D^3 - D^2 - 4D + 4)y = 0$

32. $(D^3 - D^2 + 4D - 4)y = 0$

33. $(D^3 + 3D^2 + 4D + 2)y = 0$

34. $(D^3 + 3D^2 + D - 5)y = 0$

35. $(4D^3 + 8D^2 + 13D + 9)y = 0$

36. $(D^3 - 2D^2 + 2D - 1)y = 0$

37. $(D^4 - 5D^2 - 6)y = 0$

38. $(D^4 + 24D^2 + 144)y = 0$

39. $(D^4 + 4D^3 + 4D^2 - 4D - 5)y = 0$

40. $(D^4 + 2D^3 - 2D^2 - 8D - 8)y = 0$

In problems 41–45 solve the given initial value problem.

41. $(D^3 + D^2)y = 0, \quad y(0) = 1, y'(0) = -1, y''(0) = 3$

42. $(D^3 + D^2 + D + 1)y = 0, \quad y(0) = 2, y'(0) = y''(0) = 0$

43. $(D^3 - 3D^2 - D + 3)y = 0, \quad y(0) = 2, y'(0) = -3, y''(0) = -1$

44. $(D^4 + D^2)y = 0, \quad y(0) = 1, \quad y'(0) = 1, \quad y''(0) = -1, y'''(0) = 0$

45. $(D^4 + D^2)y = 0, \quad y(0) = 0, y'(0) = 2, y''(0) = 0, y'''(0) = -1$

46. Each of the following sets of functions is known to be a solution set to a linear homogeneous third-order differential equation (not necessarily the same equation for each set). Determine if each forms a fundamental set of solutions. Justify your answer.

a) $y_1 = 8x + 3, y_2 = 7, y_3 = 4x - 2$
b) $y_1 = 2x + 1, y_2 = 3, y_3 = 2x^2 + 1$
c) $y_1 = 5x + 1, y_2 = 7x - 2, y_3 = 15$
d) $y_1 = e^x, y_2 = 5e^x - 1, y_3 = 1 - e^x$

47. a) Use the definition to show that the set $\{x, xe^x\}$ is linearly independent over any interval I.
b) Show that the Wronskian $W[x, xe^x]$ may assume the value zero.

CHAPTER 3 REVIEW EXERCISES

1. Show that $y_1 = x^{-1}$ and $y_2 = x^2$ form a fundamental set of solutions for the following linear differential equation:

$$x^2y'' - 2y = 0, \quad 0 < x < \infty.$$

2. Show that $y_1 = x^{-2}$ and $y_2 = x^{-2} \ln x$ form a fundamental set of solutions for the following linear differential equation:

$$x^2y'' + 5xy' + 4y = 0, \quad 0 < x < \infty.$$

3. Show that $y_1 = x^{-1/2}$ and $y_2 = x^{-1/2} \ln x$ form a fundamental set of solutions for the following linear differential equation:

$$4x^2y'' + 8xy' + y = 0, \quad 0 < x < \infty$$

4. Show that $y_1 = x^{-1}$ and $y_2 = x^4$ form a fundamental set of solutions for the following linear differential equation:

$$x^2y'' - 2xy' - 4y = 0, \quad 0 < x < \infty.$$

In problems 5–15 find the general or particular solution to the given differential equation, whichever is appropriate.

5. $y''' - 3y' - 2y = 0$

6. $y''' - 4y'' + 5y' - 2y = 0$

7. $y''' + 3y'' - 4y = 0$

8. $y''' + 12y'' + 36y' = 0$

9. $y'' + 144y = 0, \quad y(0) = 2, \quad y(\pi/24) = 3$

10. $y''' - 25y' = 0$

11. $y'' - 4y' + 13y = 0, \quad y(0) = 1, y(\pi/6) = 0$

12. $y''' - 4y' = 0$

13. $y'' + 4y' + 4y = 0, \quad y(0) = 1, y(\pi) = 0$

14. $y'' + 4y = 0, \quad y(0) = 1, y(\pi/4) = 1$

15. $y''' - 16y' = 0$

16. A steel spring hanging from a horizontal support has a natural length of 10 cm. A 10-g mass is attached to the spring, causing it to stretch 5 cm. The mass is then set in motion from the equilibrium position with a downward velocity of 10 cm/sec.

a) Model this system with an initial-value differential equation problem, assuming that damping forces are negligible.
b) What is the position of the mass after 1 sec of motion? 3 sec?
c) How long does it take for the mass to return to the equilibrium position for the first time?
d) What is the velocity of the mass after 1 and 3 sec of motion?
e) What is the amplitude of the system?

17. Assume that the spring–mass system described in problem 16 now travels through a fluid that exerts a damping force (in g-cm/sec^2) of 90 times the instantaneous velocity (in centimeters per second) of the mass.

a) Find the positions of the mass after 0.1, 0.4, 1.0, and 3.0 sec of motion.
b) Is this system overdamped, underdamped, or critically damped?

18. An automobile shock absorber–coil spring system is designed to support 768 lb, the portion of the automobile's weight it supports. The spring has a constant of 60 slugs per inch.

a) Assume that the automobile's shock absorber is so worn that it provides no effective damping force. If the system is jarred from equilibrium by an instantaneous force that imparts an initial velocity of 30 in/sec upward, find a particular solution that describes the vertical displacement of the automobile over time. Graph the particular solution for the first three seconds of motion. Describe the system's performance.
b) Now assume that the shock absorber is replaced. The new shock absorber exerts a damping force (in slug-in/sec^2) that is equal to 60 times the instantaneous vertical velocity of the system (in inches per second). Model this improved system with an initial value problem, and solve it subject to the conditions described in part (a). Graph the resulting equation for the first three seconds of motion. How has the system's performance improved? Is this system overdamped, underdamped, or critically damped?

19. A shock absorber–coil spring system for an imported automobile is designed to support 350 kg. The spring has a constant of 140,000 kg/cm. The shock absorber exerts a damping force (in kg-cm/sec^2) that is equal to 3500 times the instantaneous vertical velocity of the system (in centimeters per second). The system is jarred from equilibrium by an instantaneous force that imparts an initial velocity of 75 m/sec upward. Model this system with an initial value problem and solve it. Graph the resulting equation for the first three seconds of motion. What is the maximum displacement from equilibrium that the system experiences? Is this system overdamped, underdamped, or critically damped?

20. Army cannons have been developed with stationary carriages and sliding gun tube–breech block assemblies (see Fig. 3.20). A damping piston is positioned between the gun tube–breech block assembly and the carriage to absorb the recoil of firing. A recoil spring, positioned likewise, pushes the gun tube and breech block back to the in-battery position, the position the gun tube–breech block assembly is in when the cannon is ready to fire. The gun tube–breech block assembly has a mass of 1500 kg. The recoil spring has

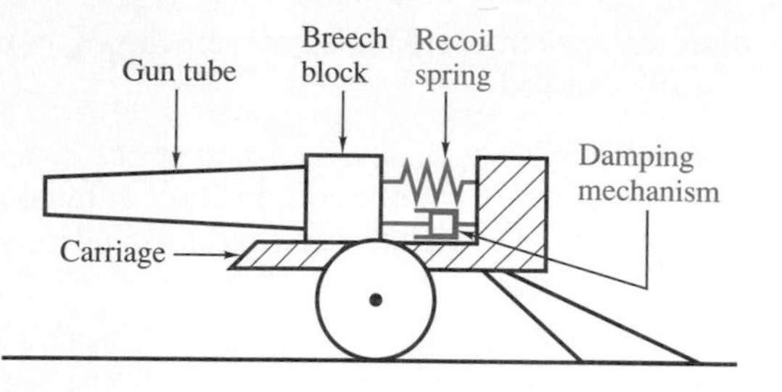

FIGURE 3.20
Recoil system diagram.

a spring constant of 19,500 N/m. The damping mechanism exerts a force (in newtons) numerically equal to 9000 times the instantaneous velocity of the gun tube–breech block assembly (in meters per second). When the cannon is fired, the gun tube is pushed to the rear at an instantaneous velocity of 5 m/sec.

a) Model this system as an initial value problem and find the particular solution that gives the displacement of the gun tube–breech block assembly with respect to time.
b) Graph the solution for the first three seconds of motion.
c) At what time does the gun tube–breech block assembly first return to the in-battery position? What is the assembly's instantaneous velocity at this time?
d) At what time does the assembly reach its maximum displacement? What is its maximum displacement?

21. A 2-g mass is suspended from the end of a rod 2.45 m long to form a pendulum. The pendulum is set in motion by displacing the mass 1 radian from the equilibrium position and releasing it. Assuming that the pendulum encounters no resistance and experiences only small angular displacements, find its angular displacement after ½ sec of motion. Use 9.8 m/sec^2 as the acceleration of an object due to gravity.

22. Assume that the pendulum in problem 21 travels through a fluid that exerts a damping force (in g-m/sec^2) that is 8 times the instantaneous velocity (in meters per second) along the path traced by the mass. Now find the pendulum's angular displacement after ½ second of motion. Is the pendulum system overdamped, underdamped, or critically damped?

23. An open switch is located in a simple *LRC* electrical circuit. The capacitor, with a capacitance 1/60 farads, has an initial charge of 20 coulombs. The resistor has a resistance of 50 ohms. The inductor has an inductance of 10 henrys.

a) Find an equation that gives the charge on the capacitor for any time after the switch is closed.
b) What is the charge on the capacitor after 1 sec?
c) What is the current in the circuit after 1 sec?

24. A mass is suspended from the end of an elastic shaft, as shown in Fig. 3.21. The torsional motion of the mass can be modeled by a differential equation in a manner similar to that of other systems described in this chapter. Let $\theta(t)$ represent the angular displacement of the mass at any time t. The resultant torque on the system will equal the product of the mass's moment of inertia and its angular acceleration, $\theta''(t)$. Twisting the mass out of equilibrium causes the shaft to exert a restoring torque on the mass that is proportional to the angular displacement. The system can also experience a damping torque, which we will assume is proportional to the instantaneous angular velocity $\theta'(t)$ of the mass. The differential equation that describes this system is

$$I\theta'' + c\theta' + k\theta = 0$$

where I is the mass's moment of inertia, c is the damping constant, and k is the shaft constant.

a) Assume that for a given torsion system $I = 2$ kg-m^2, $k = 10$ N-m, and there is no damping torque present. If the mass is twisted 2 radians from equilibrium and released, find the equation that gives the angular displacement of the mass at any time. How much twist has the mass undergone after 1 sec?
b) If a damping torque (in newton-meters) equal to 4 times the instantaneous angular velocity (in radians per meter) is exerted on the system in part (a), find the new angular displacement equation. Is the torsion system overdamped, underdamped, or critically damped?

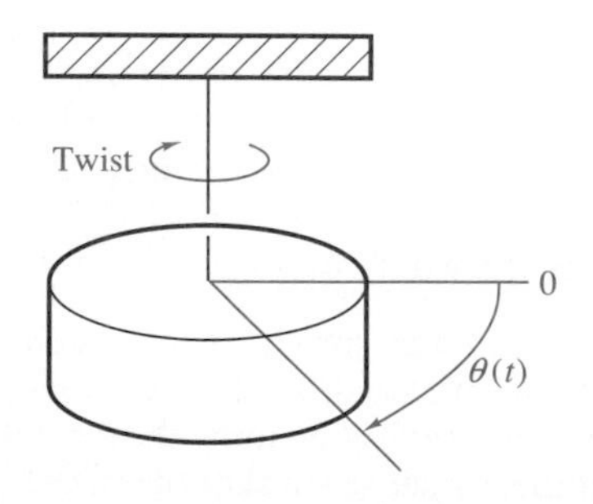

FIGURE 3.21
A mass suspended from an elastic shaft.

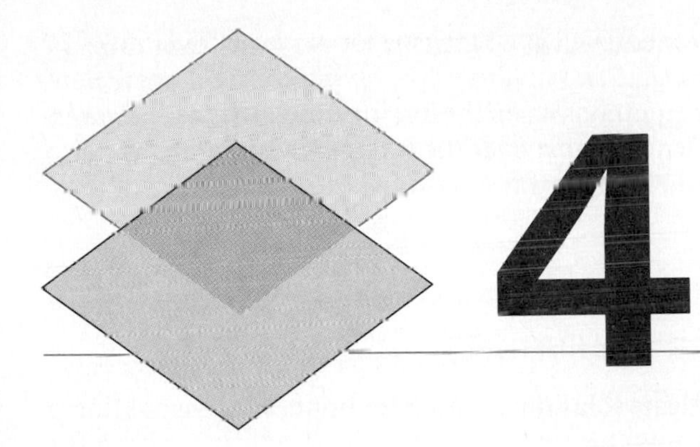

4 Nonhomogeneous Linear Equations: Modeling with Continuous Forcing Functions

In Chapter 3 we investigated solution procedures for homogeneous linear differential equations with constant coefficients. That is, we considered models of the form

$$a_n y^{(n)} + a_{n-1} y^{(n-1)} + \cdots + a_1 y' + a_0 y = g(x)$$

where $g(x) = 0$ and the coefficients a_i are all constants such that $a_n \neq 0$. However, in many behaviors the forcing function $g(x)$ is not identically zero. For example, in modeling the automobile suspension system in Chapter 3 we obtained the second-order differential equation

$$my'' + \delta y' + ky = f(t)$$

where the forcing function $f(t)$ approximates the effect of external disturbances to the system, such as periodic bumps and potholes in the road. In the case of modeling the behavior of an electrical system, we obtained

$$Lq'' + Rq' + \frac{1}{C} q = E(t)$$

where $E(t)$ is a voltage source (such as a battery) that may vary over time. In this chapter we will investigate two methods for solving constant-coefficient nonhomogeneous linear equations when the forcing function is a *continuous function* of the independent variable over the interval in question. We also present numerical solution methods for second-order differential equations.

4.1

SOLUTIONS TO NONHOMOGENEOUS LINEAR EQUATIONS

Before we begin to investigate solution methods for nonhomogeneous linear equations, we need to know the nature of the general solution. Using the operator notation introduced in Chapter 3, we write the nonhomogeneous linear equation as

$$L[y] = g(x) \tag{1}$$

where

$$L[y] = a_n(x)y^{(n)} + a_{n-1}(x)y^{(n-1)} + \cdots + a_1(x)y' + a_0(x)y. \tag{2}$$

Here it is assumed that $a_n(x)$ is never zero and each $a_i(x)$ coefficient is continuous over the interval I. In this chapter we will also assume that the forcing function $g(x)$ is continuous over I. (Discontinuous forcing functions are studied in Chapter 5.) For the nonhomogeneous case, $g(x) \not\equiv 0$.

Form of the General Solution

Suppose a fundamental set of solutions $\{y_1, y_2, \ldots, y_n\}$ is known for the associated homogeneous equation

$$L[y] = 0 \tag{3}$$

where L is the linear operator given by Eq. (2). Then the general solution to Eq. (3) is

$$y_c = c_1y_1 + c_2y_2 + \cdots + c_ny_n \tag{4}$$

where each c_i is an arbitrary constant, as discussed in Chapter 3. Thus $L[y_c] = 0$. Notice that the general solution to homogeneous Eq. (3) is denoted by the subscripted notation y_c. The reason will become clear in the ensuing discussion.

Next, suppose we could somehow come up with a single function y_p that satisfies *nonhomogeneous* Eq. (1); that is,

$$L[y_p] = g(x). \tag{5}$$

Then observe that the sum function

$$y = y_c + y_p \tag{6}$$

also solves nonhomogeneous Eq. (1):

$$\begin{aligned} L[y] &= L[y_c + y_p] \\ &= L[y_c] + L[y_p] \\ &= 0 + g(x) \\ &= g(x). \end{aligned}$$

Moreover, if $y = y(x)$ is the general solution to Eq. (1), it must be in the form of Eq. (6). If we have any function y_p satisfying Eq. (1), then

$$\begin{aligned} L[y - y_p] &= L[y] - L[y_p] \\ &= g(x) - g(x) \\ &= 0. \end{aligned}$$

Thus

$$y_c = y - y_p$$

is the general solution of Eq. (3). It follows that the general solution to nonhomogeneous Eq. (1) satisfies Eq. (6). Let us summarize these results.

THEOREM 4.1

The **general solution** $y = y(x)$ to the nonhomogeneous linear differential equation

$$L[y] = g(x)$$

has the form

$$y = y_c + y_p$$

where the function y_c, called the **complementary solution,** is the general solution to the associated homogeneous equation $L[y] = 0$, and the function y_p, called a **particular solution,** is any function satisfying $L[y_p] = g(x)$. Thus the general solution to the nonhomogeneous equation is the sum of the complementary solution and *any* particular solution.

*n*th-Order Initial Conditions

The general solution

$$\begin{aligned} y &= y_c + y_p \\ &= c_1 y_1 + c_2 y_2 + \cdots + c_n y_n + y_p \end{aligned}$$

to $L[y] = g(x)$ contains n arbitrary constants $c_1, c_2, \ldots, c_n$. To determine a unique solution to this nth-order linear differential equation it is necessary to specify a point that lies on the solution curve as well as the values of its first $n-1$ derivatives there:

$$y(x_0) = y_0,\ y'(x_0) = y_0',\ \ldots,\ y^{(n-1)}(x_0) = y_0^{(n-1)}.$$

These are called **initial conditions.** The following result guarantees the existence and uniqueness of solutions to the nth-order initial value problem.* Note that the theorem generalizes the existence and uniqueness theorem (Theorem 3.2) for linear second-order equations.

THEOREM 4.2

Suppose the functions $a_n(x)$, $a_{n-1}(x)$, $\ldots$, $a_1(x)$, $a_0(x)$, and $g(x)$ are continuous on the open interval I: $\alpha < x < \beta$ and that $a_n(x) \neq 0$ in I. Let x_0 be any point in I. Then there exists one and only one function $y = y(x)$ satisfying the differential equation

$$a_n(x)y^{(n)} + a_{n-1}(x)y^{(n-1)} + \cdots + a_1(x)y' + a_0(x)y = g(x)$$

on the interval with the initial conditions

$$y(x_0) = y_0,\ y'(x_0) = y_0',\ \ldots,\ y^{(n-1)}(x_0) = y_0^{(n-1)}$$

at the specified point x_0 in the interval. Here the constants y_0, y_0', $\ldots$, $y_0^{(n-1)}$ can have any values whatsoever.

In particular, Theorem 4.2 guarantees the existence and uniqueness of solutions to constant-coefficient linear differential equations subject to prescribed initial conditions. For example, the fourth-order equation

$$y^{(4)} - 2y''' + y'' = \sin x,$$

subject to the conditions $y'''(0) = y''(0) = y'(0) = y(0) = 0$, has a unique solution over any interval. Likewise, the nonconstant-coefficient equation

$$(x+1)y''' + e^x y'' + 2y = x^2$$

subject to the conditions $y''(0) = 0$, $y'(0) = -1$, and $y(0) = 1$, has a unique solution for $x > -1$ (where the leading coefficient is never zero). Let us explore the first of these examples further.

* A complete proof of the existence and uniqueness theorem can be found in *An Introduction to Ordinary Differential Equations* by E. A. Coddington (Englewood Cliffs, N.J., Prentice-Hall, 1961), chap. 6.

EXAMPLE 1 By direct substitution it is easy to verify that $y_p = -\frac{1}{2} \cos x$ solves the equation

$$y^{(4)} - 2y''' + y'' = \sin x.$$

Let us find the unique solution satisfying the initial conditions $y'''(0) = y''(0) = y'(0) = y(0) = 0$.

First we solve the associated homogeneous equation

$$y^{(4)} - 2y''' + y'' = 0,$$

which has the auxiliary polynomial

$$\begin{aligned} r^4 - 2r^3 + r^2 &= 0, \\ r^2(r^2 - 2r + 1) &= 0, \\ r^2(r-1)^2 &= 0. \end{aligned}$$

The roots are $m = 0, 0, 1, 1$. Therefore

$$y_c = c_1 + c_2 x + c_3 e^x + c_4 x e^x,$$

and the general solution to the nonhomogeneous equation is

$$\begin{aligned} y &= y_c + y_p \\ &= c_1 + c_2 x + c_3 e^x + c_4 x e^x - \tfrac{1}{2} \cos x. \end{aligned}$$

Next we use the initial conditions to evaluate the arbitrary constants in the general solution. After calculating the derivatives we have

$$\begin{aligned} y' &= c_2 + c_3 e^x + c_4(1 + x)e^x + \tfrac{1}{2} \sin x, \\ y'' &= c_3 e^x + c_4(2 + x)e^x + \tfrac{1}{2} \cos x, \\ y''' &= c_3 e^x + c_4(3 + x)e^x - \tfrac{1}{2} \sin x. \end{aligned}$$

Evaluating y and its derivatives at $x = 0$ results in the equations

$$\begin{aligned} c_1 \quad + c_3 \qquad\quad - \tfrac{1}{2} &= 0, \\ c_2 + c_3 + \ c_4 \qquad &= 0, \\ c_3 + 2c_4 + \tfrac{1}{2} &= 0, \\ c_3 + 3c_4 \qquad &= 0. \end{aligned}$$

Substituting $c_3 = -3c_4$ from the last into the third equation yields

$$-3c_4 + 2c_4 + \tfrac{1}{2} = 0.$$

Thus $c_4 = \frac{1}{2}$ and $c_3 = -\frac{3}{2}$. From the second equation $c_2 = 1$. Finally, from the first equation $c_1 = 2$. Therefore substitution of these values into the general solution produces the unique solution to the initial value problem:

$$y = 2 + x - \frac{3}{2} e^x + \frac{1}{2} x e^x - \frac{1}{2} \cos x.$$

Constant-Coefficient Linear Equations

For the constant-coefficient linear equation

$$L[y] = a_n y^{(n)} + a_{n-1} y^{(n-1)} + \cdots + a_1 y' + a_0 y = g(x) \tag{7}$$

where $a_n \neq 0$, you learned in Chapter 3 how to find the general solution to the associated homogeneous equation

$$a_n y^{(n)} + a_{n-1} y^{(n-1)} + \cdots + a_1 y' + a_0 y = 0 \tag{8}$$

that we now call the complementary solution y_c (so $L[y_c] = 0$). Our task now is to come up with a particular solution y_p to the nonhomogeneous equation $L[y] = g(x)$. We will next investigate one method for finding y_p. The method is algebraic in nature; no integration is involved. However, the method applies only for special cases of the continuous forcing function $g(x)$. (For more general types of continuous forcing functions, another method will be presented in Section 4.4.)

Conversion to a Homogeneous Higher-Order Equation

Assume now that $g(x)$ is one of, or a sum of, the function types listed in Table 3.3. Then $g(x)$ solves a homogeneous constant-coefficient linear differential equation. In this situation you will see that Eq. (1) can be converted to a *homogeneous* higher-order equation with constant coefficients. From this conversion we can determine the *form* of the particular solution y_p to nonhomogeneous Eq. (1). This solution form is then substituted into Eq. (1) to determine certain constants, called *undetermined coefficients,* appearing in the form that arises from the conversion process. The resulting substitution produces the desired particular solution y_p.

From Theorem 4.1 the general solution to Eq. (1) is given by

$$y = y_c + y_p$$

where

$$L[y_c] = 0 \qquad \text{and} \qquad L[y_p] = g(x).$$

The complementary solution y_c is the general solution to Eq. (1) when $g(x) \equiv 0$ (which is the associated homogeneous situation). Assuming $g(x)$ is itself a solution to a homogeneous linear equation with constant coefficients, Eq. (1) can be converted into a homogeneous higher-order equation by the appropriate differentiation operations. We denote this higher-order equation by $L_h[y] = 0$. That is,

$$L[y_c + y_p] = g(x)$$

converts to

$$L_h[y_c + y_p] = 0.$$

Therefore the particular solution y_p must satisfy two conditions:

Condition 1: $L_h[y_c + y_p] = 0,$
Condition 2: $L[y_p] = g(x).$

Condition 1 *determines the form* of y_p only. That is, it determines y_p up to *unknown* constants called undetermined coefficients. Application of Condition 2 then determines these coefficients. We now describe a methodology for using the two conditions to produce a particular solution y_p. Justification for the methodology is provided in Section 4.2.

FINDING A PARTICULAR SOLUTION y_p TO $L[y] = g(x)$

Step 1. Find the complementary solution y_c, where

$$L[y_c] = 0.$$

Step 2. Find the general solution y_h solving the higher-order equation

$$L_h[y_h] = 0.$$

The solution y_h is determined from L and $g(x)$.

Step 3. The form of the particular solution is denoted by y_q and obtained as

$$y_q = y_h - y_c.$$

Step 4. Determine the unknown constants, called **undetermined coefficients** in the solution form y_q by requiring that

$$L[y_q] = g(x).$$

Substitution of these numerical values into the form y_q produces a particular solution y_p.

Step 2 of the procedure is the key to implementing the method correctly. In some cases we are able to find the operator L_h rather easily, and then determining y_h becomes a straightforward process. We now illustrate the above methodology with several examples in which we find the operator L_h giving the homogeneous higher-order equation. But how do we find the general solution y_h without first finding L_h? Later in this section we will show you a simple procedure for finding y_h whenever the function $g(x)$ is of the assumed type. The following examples are intended to give some insight into the process.

EXAMPLE 2 Find a particular solution y_p to the nonhomogeneous second-order equation

$$L[y] = y'' = g$$

where g is the constant acceleration of gravity.

Solution. Differentiation of the given equation produces the higher-order equation

$$L_h[y] = y''' = 0.$$

Step 1. We need to solve the homogeneous equation

$$L[y] = y'' = 0$$

for the complementary solution y_c. In this case integration gives us

$$y_c = c_1 + c_2 x.$$

Step 2. Solution of $L_h[y] = 0$ yields

$$y_h = c_1 + c_2 x + c_3 x^2.$$

Step 3. The form of the particular solution we seek is

$$y_q = y_h - y_c = c_3 x^2.$$

The arbitrary constant c_3 is the undetermined coefficient in this case.

Step 4. Requiring that y_q satisfy the original equation means that

$$y_q'' = g.$$

Since $y_q' = 2c_3 x$ and $y_q'' = 2c_3$, we obtain

$$2c_3 = g$$

or $c_3 = g/2$. Therefore

$$y_p = \frac{gx^2}{2}.$$

EXAMPLE 3 Find a particular solution y_p to the first-order equation

$$L[y] = y' + 3y = x.$$

Solution. We differentiate the equation twice to obtain the higher-order equation

$$L_h[y] = y''' + 3y'' = 0.$$

Step 1. Solving $L[y] = 0$, we find the auxiliary equation

$$r + 3 = 0.$$

The root is -3. Thus the complementary solution is

$$y_c = c_1 e^{-3x}.$$

Step 2. Solving $L_h[y] = 0$ leads to the auxiliary equation

$$r^3 + 3r^2 = 0$$

or

$$r^2(r + 3) = 0.$$

The roots are $-3, 0, 0$. Thus

$$y_h = c_1 e^{-3x} + c_2 + c_3 x.$$

Step 3. The form of the particular solution is

$$y_q = y_h - y_c = c_2 + c_3 x.$$

The arbitrary constants c_2 and c_3 are the undetermined coefficients.

Step 4. To find the undetermined coefficients we require that y_q satisfy the original nonhomogeneous equation: $L[y_q] = x$; that is,

$$y_q' + 3y_q = x.$$

Hence

$$c_3 + 3(c_2 + c_3 x) = x$$

or

$$(c_3 + 3c_2) + 3c_3 x = x.$$

Since this last equation must hold for *all* values of x, the coefficients of like powers of x on each side of the equation must match (be equal). Therefore

$$\begin{aligned} c_3 + 3c_2 &= 0, \\ 3c_3 \qquad &= 1. \end{aligned}$$

Solution of these equations yields $c_3 = 1/3$ and $c_2 = -1/9$. Substitution of these values into the form y_q gives the particular solution we seek:

$$y_p = -\frac{1}{9} + \frac{x}{3}.$$

EXAMPLE 4 Find a particular solution y_p to the nonhomogeneous second-order equation

$$L[y] = y'' + y' = x.$$

Solution. We differentiate the equation twice to obtain the higher-order equation

$$L_h[y] = y^{(4)} + y''' = 0.$$

Step 1. Solving $L[y] = 0$ results in the auxiliary equation

$$r^2 + r = 0.$$

The roots are $-1, 0$. Thus

$$y_c = c_1 e^{-x} + c_2.$$

Step 2. Solving $L_h[y] = 0$ gives us the auxiliary equation

$$r^4 + r^3 = 0.$$

The roots are $-1, 0, 0, 0$. Thus

$$y_h = c_1 e^{-x} + c_2 + c_3 x + c_4 x^2.$$

Step 3. The form of the particular solution is

$$y_q = y_h - y_c = c_3 x + c_4 x^2.$$

The arbitrary constants c_3 and c_4 are the undetermined coefficients.

Step 4. To find the undetermined coefficients we require that y_q satisfy $L[y_q] = x$. Thus

$$y_q'' + y_q' = x.$$

Now

$$y_q' = c_3 + 2c_4 x \qquad \text{and} \qquad y_q'' = 2c_4.$$

Thus

$$2c_4 + (c_3 + 2c_4 x) = x$$

or

$$(2c_4 + c_3) + 2c_4 x = x.$$

Since this last equation must hold for *all* values of x, the coefficients of like powers of x on each side of the equation must match. Therefore

$$\begin{aligned} 2c_4 + c_3 &= 0, \\ 2c_4 \qquad &= 1. \end{aligned}$$

Solution of these equations yields $c_4 = 1/2$ and $c_3 = -1$. Substitution of these values into y_q gives the particular solution

$$y_p = -x + \frac{x^2}{2}.$$

In Examples 2–4 the forcing function $g(x)$ was very elementary. For this reason we were able to produce the higher-order equation $L_h[y] = 0$ by simply differentiating both sides of $L[y] = g(x)$ a few times. The higher-order equation allowed us to find the general solution y_h from which we could then write the solution form y_q in Step 3.

Suppose, however, that $g(x) = xe^x$. In this case, the way to produce the equation $L_h[y] = 0$ may not be so evident. What derivative operations should be performed on $g(x)$ to cause it to become identically zero? Knowing how to solve constant coefficient homogeneous linear equations, as summarized in Table 3.3, tells us that the function $g(x)$ comes from the repeated roots 1, 1. The auxiliary equation corresponding to these roots is

$$(r-1)(r-1)=0$$

or

$$r^2 - 2r + 1 = 0.$$

Therefore $g(x) = xe^x$ is a solution to the homogeneous equation

$$y'' - 2y' + y = 0. \tag{9}$$

Thus the nonhomogeneous equation

$$L[y] = xe^x \tag{10}$$

can be converted to a homogeneous higher-order equation according to the operations suggested by Eq. (9). That is,

$$(L[y])'' - 2(L[y])' + L[y] = (xe^x)'' - 2(xe^x)' + xe^x$$

becomes

$$L_h[y] = 0.$$

From the types of functions possible for $g(x)$ shown in Table 3.3, you can see these calculations could be complicated in practice. Fortunately it is not necessary to know the differential operator L_h that causes $g(x)$ to become zero. *All that is needed are the roots of the auxiliary polynomial associated with constant-coefficient differential Eq. (9) for which* $g(x)$ is a solution. You can obtain these roots quite simply from $g(x)$ by reading Table 3.3 from the rightmost column to the left. When these roots are written as a single list together with all the roots of the original auxiliary equation associated with $L[y] = 0$ in Step 1, the function y_h can be produced immediately. (We will justify our methodology at the end of the presentation in Section 4.2.)

FINDING THE HIGHER-ORDER FUNCTION y_h TO THE CONSTANT COEFFICIENT EQUATION $L[y] = g(x)$

Step 1. Denote the roots of the auxiliary polynomial associated with $L[y] = 0$ by m:

$$m = m_1, m_2, \ldots, m_n. \tag{11}$$

Some of these roots may be repeated, and some of them may be complex (with a nonzero imaginary part).

Step 2. The function $g(x)$ is assumed to solve a constant-coefficient homogeneous linear equation:

$$L_1[y] = 0.$$

Denote the roots of the auxiliary polynomial associated with this equation by m^*:

$$m^* = m_1^*, m_2^*, \ldots, m_k^*. \tag{12}$$

Determine the m^* roots from your knowledge of solutions to constant-coefficient equations summarized in Table 3.3. Some of the m^* roots may be identical to some of the m roots in the list (11). That does not matter—be sure to list *all* the m^* roots.

Step 3. Write the m and m^* roots from the lists in (11) and (12) as a *single* list of roots denoted by M_h:

$$M_h = m_1, m_2, \ldots, m_n, m_1^*, m_2^*, \ldots, m_k^*. \tag{13}$$

Step 4. The function y_h is the general solution to the homogeneous equation

$$L_h[y] = 0$$

whose associated auxiliary polynomial has *all* the roots M_h. You can write the solution y_h from the roots M_h using Table 3.3.

We present several examples.

EXAMPLE 5 Find the function y_h for the nonhomogeneous equation

$$L[y] = y'' - 2y' - 3y = 1 - x^2.$$

Solution. The auxiliary polynomial for $L[y] = 0$ is

$$r^2 - 2r - 3 = 0$$

or

$$(r+1)(r-3)=0.$$

The roots are $m=-1, 3$.

The roots associated with the forcing function $g(x)=1-x^2$ are $m^*=0, 0, 0$. That is, thinking of $g(x)$ as a solution to a constant-coefficient homogeneous equation, the root 0 would have to be repeated three times to obtain the term x^2 as well as the constant 1.

Writing the roots as a single list gives

$$M_h=-1, 3, 0, 0, 0.$$

Thus

$$y_h=c_1e^{-x}+c_2e^{3x}+c_3+c_4x+c_5x^2.$$

EXAMPLE 6 Find the function y_h for the nonhomogeneous equation

$$L[y]=y''+y'-2y=xe^x.$$

Solution. The auxiliary polynomial for $L[y]=0$ is

$$r^2+r-2=(r+2)(r-1)=0.$$

The roots are $m=-2, 1$.

The forcing function xe^x comes from the repeated roots $m^*=1, 1$. Writing the roots as a single list gives us

$$M_h=-2, 1, 1, 1$$

Thus

$$y_h=c_1e^{-2x}+c_2e^x+c_3xe^x+c_4x^2e^x.$$

EXAMPLE 7 Find the function y_h for the nonhomogeneous equation

$$y''+y=\cos x.$$

Solution. The auxiliary polynomial is

$$r^2+1=0.$$

The roots are $m=\pm i$.

The forcing function $\cos x$ comes from the roots $m^*=\pm i$. The combined roots are

$$M_h=\pm i, \pm i.$$

Thus

$$y_h = c_1 \cos x + c_2 \sin x + c_3 x \cos x + c_4 x \sin x.$$

EXAMPLE 8 Find the function y_h for the nonhomogeneous equation

$$y'' - 2y' = 6e^{2x} - 4 \sin x.$$

Solution. The auxiliary polynomial for $L[y] = 0$ is

$$r^2 - 2r = 0.$$

The roots are $m = 0, 2$.

The righthand side of the equation, $6e^{2x} - 4 \sin x$, comes from the roots $m^* = 2, \pm i$.

Writing all the roots as a single list gives us

$$M_h = 0, 2, 2, \pm i.$$

Thus

$$y_h = c_1 + c_2 e^{2x} + c_3 x e^{2x} + c_4 \cos x + c_5 \sin x.$$

In the previous examples we have concentrated on Steps 2 and 3 in the procedure for finding a particular solution to the constant-coefficient nonhomogeneous linear equation. Step 4 in the procedure is to find the undetermined coefficients in the solution form $y_q = y_h - y_c$. This step requires that y_q satisfy the original nonhomogeneous equation. We will focus our attention on this step in Section 4.2. The following exercises give you some practice in the first three steps.

EXERCISES 4.1

In problems 1–6 find a particular solution to the first-order equation using the method illustrated in Examples 2–4 of the text.

1. $y' - y = x + 1$
2. $2y' - 3y = x^2$
3. $y' + 2y = x^2 - x$
4. $3y' + y = 2x^2 - 1$
5. $y' - y = 1 - 3x^2$
6. $y' + 4y = x^3$

In problems 7–16 determine the roots m^* if the given function is interpreted as a solution of a homogeneous higher-order linear differential equation with constant coefficients.

7. $3x^2 - 1$
8. $x^2(1 - 2x)$
9. $2 - 5e^{-x}$
10. $x + 7e^{2x}$
11. $3e^x - 4xe^{-x}$
12. $x - 2 \cos x$
13. $e^x \sin x - x$
14. $xe^{2x} \cos x - 1$
15. $e^{3x} + e^{-x} \sin 2x + x$
16. $e^x \sin 2x + 3e^{-x} \cos 2x$

In problems 17–36 find the higher-order solution y_h as outlined in the text and illustrated in Examples 5–8.

17. $y'' - y = 2e^{-x}$
18. $y'' - y = 2xe^{-x} + 1$
19. $y'' + y' - 2y = e^x + e^{-2x}$

20. $y'' - 4y = x - e^{2x}$

21. $y'' - 2y' + y = e^x + \sin x$

22. $y'' + 2y' + y = 3x + e^{-x}$

23. $y''' - y' = x^2$

24. $y'' - 2y' = x^2 - 3e^{2x}$

25. $y'' + y = 1 - xe^x$

26. $y'' + 4y = \cos 2x$

27. $y'' + 3y = 1 - \sin \sqrt{3}x$

28. $y'' - y' = e^x \cos x$

29. $y'' - 2y' + 2y = 2e^x \sin x$

30. $y'' - 4y' + 5y = e^{2x} \sin x$

31. $y'' + y = x \cos x$

32. $y'' - y = x \sin x - \cos x$

33. $y'' + 2y' + 2y = x^2 + \sin x$

34. $y'' + 2y' + 5y = e^{-x} + \cos 2x$

35. $y'' + 2y' + 5y = x(x + \cos x)$

36. $y'' - 4y' + 5y = x^2 - \sin 2x$

In problems 37–42 verify that the given function is a particular solution to the specified nonhomogeneous equation. Find the general solution and evaluate its arbitrary constants to find the unique solution satisfying the equation and the given initial conditions.

37. $y'' + y' = x, \quad y_p = \dfrac{x^2}{2} - x, \; y(0) = 0, \; y'(0) = 0$

38. $y''' + y = x, \quad y_p = 2 \sin x + x, \; y(0) = 0, \; y'(0) = 0$

39. $\frac{1}{2}y'' + y' + y = 4e^x(\cos x - \sin x), \quad y_p = 2e^x \cos x,$ $y(0) = 0, \; y'(0) = 1$

40. $y'' - y' - 2y = 1 - 2x, \; y_p = x - 1, \; y(0) = 0, \; y'(0) = 1$

41. $y'' - 2y' + y = 2e^x, \quad y_p = x^2 e^x, \; y(0) = 1, \; y'(0) = 0$

42. $y'' - 2y' + y = x^{-1}e^x, \; x > 0, \; y_p = xe^x \ln x, \; y(1) = e,$ $y'(1) = 0$

4.2

FINDING A PARTICULAR SOLUTION: UNDETERMINED COEFFICIENTS

In solving the constant-coefficient linear nonhomogeneous equation

$$L[y] = g(x) \tag{1}$$

where $g(x)$ is itself a solution to a homogeneous equation with constant coefficients, you learned in Section 4.1 how to find the higher-order solution y_h satisfying

$$L_h[y_h] = 0.$$

The *solution form* for a particular solution to Eq. (1) is then given by

$$y_q = y_h - y_c$$

where y_c is the complementary solution satisfying $L[y_c] = 0$.

To find a particular solution y_p we require that y_q satisfy differential Eq. (1):

$$L[y_q] = g(x). \tag{2}$$

This last condition results in a system of linear algebraic equations in the unknown coefficients appearing in y_q. When the system is solved, thereby determining the coefficients, the values of the coefficients are substituted into the form y_q to produce a particular solution y_p. We illustrate this solution procedure by finding y_p in the next several examples.

EXAMPLE 1 Find a particular solution y_p for the nonhomogeneous equation

$$L[y] = y'' - 2y' - 3y = 1 - x^2. \tag{3}$$

Solution. From Example 5 in the previous section, we know that the roots of the auxiliary polynomial are $m = -1, 3$. The complementary solution is

$$y_c = c_1e^{-x} + c_2e^{3x}.$$

We also found in Example 5 that

$$y_h = c_1e^{-x} + c_2e^{3x} + c_3 + c_4x + c_5x^2.$$

Thus

$$y_q = y_h - y_c = c_3 + c_4x + c_5x^2.$$

To avoid the extensive use of subscripts as we proceed with the solution, we rename the constants and write the solution form as

$$y_q = Ax^2 + Bx + C.$$

We need to determine the unknown coefficients A, B, and C. When we substitute the polynomial and its derivatives into differential Eq. (3), we have

$$2A - 2(2Ax + B) - 3(Ax^2 + Bx + C) = 1 - x^2$$

or

$$-3Ax^2 + (-4A - 3B)x + (2A - 2B - 3C) = 1 - x^2.$$

This last equation holds for all values of x if its two sides are identical quadratic polynomials. Thus we equate corresponding powers of x to obtain

$$\begin{aligned} -3A \qquad\qquad &= -1, \\ -4A - 3B \qquad &= 0, \\ 2A - 2B - 3C &= 1. \end{aligned}$$

These equations imply in turn that $A = 1/3$, $B = -4/9$, and $C = 5/27$. Therefore, after substituting these values into the quadratic expression, we obtain the particular solution

$$y_p = \frac{1}{3}x^2 - \frac{4}{9}x + \frac{5}{27}.$$

EXAMPLE 2 Find a particular solution y_p for the nonhomogeneous equation

$$(D^2 + D - 2)y = xe^x. \tag{4}$$

Solution. From Example 6 in Section 4.1 we can deduce that

$$y_c = c_1 e^{-2x} + c_2 e^x$$

and

$$y_h = c_1 e^{-2x} + c_2 e^x + c_3 x e^x + c_4 x^2 e^x.$$

Then

$$y_q = y_h - y_c = Axe^x + Bx^2 e^x,$$

where we renamed c_3 as A and c_4 as B to avoid using a lot of subscripts.

To determine coefficients A and B we demand that y_q satisfy the given differential Eq. (4):

$$(D^2 + D - 2)y_q = xe^x. \tag{5}$$

Calculating the derivatives on the lefthand side of Eq. (5), we have

$$Dy_q = Ae^x + (A + 2B)xe^x + Bx^2 e^x$$

and

$$D^2 y_q = (2A + 2B)e^x + (A + 4B)xe^x + Bx^2 e^x.$$

We then substitute these derivatives into Eq. (5) and collect like terms on the lefthand side:

$$D^2 y_q + Dy_q - 2y_q = xe^x$$

$$(2A + 2B + A)e^x + (A + 4B + A + 2B - 2A)xe^x$$
$$+ (B + B - 2B)x^2 e^x = xe^x$$

or

$$(3A + 2B)e^x + 6Bxe^x = xe^x. \tag{6}$$

Now Eq. (6) is an identity in x, so the coefficients of corresponding terms on each side of this equation must be equal:

$$3A + 2B = 0,$$
$$6B = 1.$$

Solution of this algebraic system gives $B = 1/6$ and $A = (-2/3)\,B = -1/9$. Substitution of these numerical values into solution form (5) then gives our desired particular solution

$$y_p = -\frac{1}{9}xe^x + \frac{1}{6}x^2 e^x.$$

EXAMPLE 3 Find a particular solution y_p to the nonhomogeneous equation

$$y'' + y = \cos x. \tag{7}$$

Solution. From Example 7 in Section 4.1 we can see that

$$y_c = c_1\cos x + c_2\sin x$$

and

$$y_h = c_1\cos x + c_2\sin x + c_3 x\cos x + c_4 x\sin x.$$

Then

$$y_q = y_h - y_c = Ax\cos x + Bx\sin x$$

where we have renamed c_3 as A and c_4 as B.

We require that y_q satisfy the relation $y_q'' + y_q = \cos x$. Differentiating y_q we obtain

$$\begin{aligned} y_q' &= A\cos x + B\sin x - Ax\sin x + Bx\cos x, \\ y_q'' &= 2B\cos x - 2A\sin x - Bx\sin x - Ax\cos x. \end{aligned}$$

Hence

$$y_q'' + y_q = \cos x$$

becomes

$$2B\cos x - 2A\sin x = \cos x. \tag{8}$$

Since Eq. (8) is an identity in x, we equate coefficients of like terms on both sides to obtain the system

$$\begin{aligned} 2B &= 1, \\ -2A &= 0. \end{aligned}$$

Solution of this system gives $A = 0$ and $B = \tfrac{1}{2}$. Hence

$$y_p = \frac{x}{2}\sin x.$$

EXAMPLE 4 Find a particular solution y_p to the nonhomogeneous equation

$$y'' - 2y' = 6e^{2x} - 4\sin x. \tag{9}$$

Solution. From Example 8 in Section 4.1 we know that

$$y_c = c_1 + c_2e^{2x}$$

and

$$y_h = c_1 + c_2e^{2x} + c_3xe^{2x} + c_4\cos x + c_5\sin x.$$

Then

$$y_q = y_h - y_c = Axe^{2x} + B\cos x + C\sin x$$

where $c_3 = A$, $c_4 = B$, and $c_5 = C$. Requiring y_q to satisfy Eq. (9) means that

$$y_q'' - 2y_q' = 6e^{2x} - 4 \sin x.$$

Thus differentiation of y_q gives us

$$\begin{aligned} y_q' &= Ae^{2x} + 2Axe^{2x} - B \sin x + C \cos x, \\ y_q'' &= 4Ae^{2x} + 4Axe^{2x} - C \sin x - B \cos x. \end{aligned}$$

Hence

$$y_q'' - 2y_q' = 6e^{2x} - 4 \sin x$$

translates to

$$2Ae^{2x} + (2B - C) \sin x + (-B - 2C) \cos x = 6e^{2x} - 4 \sin x.$$

Equating coefficients yields

$$\begin{aligned} 2A &= 6, \\ 2B - C &= -4, \\ -B - 2C &= 0. \end{aligned}$$

Solution of the system gives $A = 3$, $B = -8/5$, and $C = 4/5$. Substituting the numerical values of the coefficients into the solution form then yields the particular solution

$$y_p = 3xe^{2x} - \frac{8}{5} \cos x + \frac{4}{5} \sin x.$$

We now summarize the method of undetermined coefficients for solving a nonhomogeneous linear equation with constant coefficients.

METHOD OF UNDETERMINED COEFFICIENTS TO FIND THE GENERAL SOLUTION TO $L[y] = g(x)$

Step 1. Verify the forcing function form. Check that $g(x)$ contains only sums of terms of the form

$$x^k e^{mx}, \qquad x^k e^{mx} \sin bx, \qquad x^k e^{mx} \cos bx$$

(possibly multiplied by constants), where $k \geq 0$ is an integer and m and b are real numbers (possibly 0).

Step 2. Find the complementary solution y_c. Solve the associated homogeneous equation

$$L[y] = 0.$$

Step 3. Find the higher-order solution y_h. Write the m and m^* roots as a single list to find y_h, as described in Section 4.1.

Step 4. Write the general form of a particular solution:

$$y_q = y_h - y_c.$$

Step 5. Solve for the undetermined coefficients. Require the solution form y_q to satisfy the relation

$$L[y_q] = g(x).$$

Substitute the resulting values into y_q thereby converting it into a particular solution y_p.

Step 6. Form the general solution. The general solution to the nonhomogeneous equation is given by the sum

$$y = y_c + y_p.$$

We now present several examples.

EXAMPLE 5 Solve the differential equation

$$(D^2 - 1)y = x + \sin x. \tag{10}$$

Solution. We follow our solution procedure outlined above.

Step 1. The forcing function $g(x) = x + \sin x$ contains only terms of the required form. Thus we can employ the method of undetermined coefficients for determining y_p.

Step 2. By solving the associated homogeneous equation

$$(D^2 - 1)y = 0,$$

we find that the roots to the auxiliary equation $r^2 - 1 = 0$ are $m = -1, 1$. Thus

$$y_c = c_1e^{-x} + c_2e^x.$$

Step 3. Considering the forcing function, the x term comes from the root 0, repeated twice; the $\sin x$ term comes from the complex root $0 \pm i$. Thus the roots associated with the forcing function are $m^* = 0, 0, \pm i$.

Writing the roots as a single list gives us $M_h = -1, 1, 0, 0, \pm i$. Thus

$$y_h = c_1e^{-x} + c_2e^x + c_3 + c_4x + c_5\cos x + c_6\sin x.$$

Step 4. After deleting the complementary solution y_c from the solution in

Step 3 and renaming the coefficients, we have the solution form

$$y_q = A + Bx + C\cos x + E\sin x.$$

Step 5. Calculating $(D^2 - 1)y_q$ we obtain,

$$Dy_q = B - C\sin x + E\cos x,$$
$$D^2y_q = \quad -E\sin x - C\cos x.$$

Then

$$(D^2 - 1)y_q = x + \sin x$$

translates to

$$(-E\sin x - C\cos x) - A - Bx - C\cos x - E\sin x = x + \sin x$$

or

$$-A - Bx - 2C\cos x - 2E\sin x = x + \sin x.$$

Equating coefficients for like terms on both sides of this last equation gives the system

$$-A = 0, \qquad -B = 1, \qquad -2C = 0, \qquad -2E = 1.$$

Solution of the system is easily seen to be $A = 0$, $B = -1$, $C = 0$, and $E = -½$. Substituting these numerical values into y_q yields the particular solution

$$y_p = -x - \frac{1}{2}\sin x.$$

Step 6. The general solution to nonhomogeneous Eq. (10) is

$$y = y_c + y_p = c_1e^{-x} + c_2e^{x} - x - \frac{1}{2}\sin x.$$

We now work another example illustrating the method of undetermined coefficients, this time without naming each step. Make sure you can identify which step is being performed at each stage of the solution procedure.

EXAMPLE 6 Find the general solution to the equation

$$(D^2 - D - 2)y = 6x + 6e^{-x}. \tag{11}$$

Solution. The method of undetermined coefficients applies since the forcing function has the required form. The auxiliary equation of Eq. (11) is

$$r^2 - r - 2 = (r - 2)(r + 1) = 0,$$

which has the roots $m = -1, 2$. The complementary solution is

$$y_c = c_1 e^{-x} + c_2 e^{2x}.$$

The right-hand side of Eq. (11) comes from the roots

$$m^* = 0, 0, -1.$$

Writing these roots as a single list leads to the general solution

$$y_h = c_1 e^{-x} + c_2 e^{2x} + c_3 + c_4 x + c_5 x e^{-x}.$$

Thus the correct solution form is

$$y_q = A + Bx + Cxe^{-x}.$$

Then

$$\begin{aligned} Dy_q &= B \quad - Cxe^{-x} + Ce^{-x}, \\ D^2 y_q &= \quad + Cxe^{-x} - 2Ce^{-x}. \end{aligned}$$

Thus Eq. (11) becomes

$$(-2A - B) - 2Bx + (-2C + C + C)xe^{-x} + (-C - 2C)e^{-x} = 6x + 6e^{-x}.$$

Equating coefficients results in the system of equations

$$-2A - B = 0, \qquad -2B = 6, \qquad -3C = 6.$$

Therefore $B = -3$, $C = -2$, and $A = 3/2$. The particular solution is

$$y_p = \frac{3}{2} - 3x - 2xe^{-x},$$

and the general solution to Eq. (11) is

$$y = c_1 e^{-x} + c_2 e^{2x} + \frac{3}{2} - 3x - 2xe^{-x}. \tag{12}$$

EXAMPLE 7 Solve differential Eq. (11) subject to the conditions $y(0) = 1$ and $y'(0) = -1$.

Solution. We must use the given conditions to evaluate the arbitrary constants c_1 and c_2 in general solution (12). First we need to know y'. From Eq. (12) we determine that

$$y' = -c_1 e^{-x} + 2c_2 e^{2x} - 3 - 2e^{-x} + 2xe^{-x}.$$

Then $y(0) = 1$ and $y'(0) = -1$ yield the algebraic equations

$$\begin{aligned} c_1 + c_2 + \frac{3}{2} &= 1, \\ -c_1 + 2c_2 - 5 &= -1. \end{aligned}$$

Solving this system, we obtain $c_1 = -5/3$ and $c_2 = 7/6$. Therefore

$$y = -\frac{5}{3}e^{-x} + \frac{7}{6}e^{2x} + \frac{3}{2} - 3x - 2xe^{-x}$$

is the solution to Eq. (11) subject to the given conditions.

Justifying the Method of Undetermined Coefficients

Consider the constant-coefficient linear nonhomogeneous differential equation

$$P(D)y = g(x) \tag{13}$$

where it is assumed that $g(x)$ is itself a solution to a constant-coefficient homogeneous equation:

$$Q(D)g(x) = 0. \tag{14}$$

We use the polynomial operator notation [$P(D)$ and $Q(D)$] to remind us that the linear equations have constant coefficients. The notation also facilitates our discussion. From Eqs. (13) and (14) you can see that

$$Q(D)P(D)y = Q(D)g(x);$$

that is,

$$L_h[y] = Q(D)P(D)y = 0. \tag{15}$$

Now let the roots of the auxiliary equation $P(m) = 0$ be

$$m = m_1, m_2, \ldots, m_n. \tag{16}$$

Let the roots of the auxiliary equation $Q(m) = 0$ be

$$m^* = m_1^*, m_2^*, \ldots, m_k^*. \tag{17}$$

Taken altogether, the roots in Eqs. (16) and (17) are the roots of the auxiliary equation $Q(m)P(m) = 0$ associated with the homogeneous higher-order equation $L_h[y] = 0$ in Eq. (15). Therefore the general solution y_h to Eq. (15) contains the complementary solution y_c to nonhomogeneous Eq. (13), in addition to some other terms we call y_q. That is, we have the following result.

> The general solution y_h to
>
> $$Q(D)P(D)y = 0$$
>
> has the form
>
> $$y_h = y_c + y_q. \tag{18}$$

Any function y_p that satisfies the original equation $P(D)y = g(x)$ must also satisfy $Q(D)P(D)y = 0$. Since Eq. (18) represents the general solution to $Q(D)P(D)y = 0$, we must be able to find our particular solution y_p by appropriately determining values of the constants appearing in Eq. (18). That is, the family of solutions y_h contains a particular solution y_p. This observation means that

$$P(D)(y_c + y_q) = g(x)$$

or by linearity that

$$P(D)y_c + P(D)y_q = g(x).$$

Since $P(D)y_c = 0$, we obtain the requirement that

$$P(D)y_q = g(x). \tag{19}$$

In other words, for suitable numerical values of the coefficients appearing in it, the solution form y_q is converted into the particular solution y_p.

EXERCISES 4.2

In problems 1–20 solve the given differential equation by the method of undetermined coefficients.

1. $y'' + y' - 6y = 5e^{2x}$
2. $y'' - 2y' - 8y = 1 - 3e^{-2x}$
3. $y'' - 9y = x + 2e^{3x}$
4. $y'' - 3y' - 10y = x(1 - e^{-2x})$
5. $y'' - 4y' + 4y = 2 - e^x \sin x$
6. $y'' - 6y' + 9y = x + 5e^{3x}$
7. $2y'' + y' - y = 1 + e^x \cos x$
8. $2y'' - 5y' - 12y = e^{4x} + 2 \cos x$
9. $y'' + 2y' + 2y = e^x + \sin x$
10. $y'' + 2y' + 2y = e^{-x} + 3 \cos x$
11. $y'' + 4y = 1 - xe^x$
12. $y'' + y = 2 \sin x$
13. $y'' + 9y = x + \cos 3x$
14. $y'' + 16y = x^2 - 3 \sin 4x$
15. $y''' - y' = 2 + e^x$
16. $y''' - y'' - 2y' = e^{-x} + 3e^{2x}$
17. $y''' - 4y'' + 4y' = x^2 - \sin x$
18. $2y''' - 5y'' - 3y' = 2e^{3x} + e^x$
19. $y^{(4)} - 2y'' + y = x - 2e^{-x}$
20. $y^{(4)} + 2y'' + y = xe^{-x}$

In problems 21–30 solve the given initial value problem.

21. $y'' + y' = x, \quad y(0) = 1, y'(0) = 0$
22. $y'' - 3y' = e^{3x}, \quad y(0) = 1, y'(0) = 0$
23. $y'' - 3y' + 2y = 1 - e^x, \quad y(0) = 0, y'(0) = 1$
24. $y'' - y' - 6y = xe^{3x}, \quad y(0) = -1, y'(0) = 1$
25. $y''' + 2y'' = x^2, \quad y(0) = 1, y'(0) = 1, y''(0) = 0$
26. $y''' - 2y'' + y' = \sin x, \quad y(0) = 0, y'(0) = 0, y''(0) = 0$
27. $y'' + y = x^2, \quad y\left(\frac{\pi}{2}\right) = 1, y'\left(\frac{\pi}{2}\right) = 0$
28. $y'' + 2y' + 2y = xe^{-x}, \quad y(0) = 1, y'(0) = -1$
29. $y^{(4)} - y = x + e^x, \quad y(0) = 0, \ y'(0) = 0, \ y''(0) = 0, \ y'''(0) = 0$
30. $y^{(4)} + y''' = x, \quad y(0) = 1, \ y'(0) = 0, \ y''(0) = -1, \ y'''(0) = 0$

VARIATION OF PARAMETERS: LINEAR FIRST-ORDER EQUATIONS (Optional)

In Section 2.2 you were introduced to the linear first-order equation

$$a_1(x)y' + a_0(x)y = g(x).$$

Assuming that $a_1(x)$ is never zero, we wrote this equation in the standard form

$$y' + P(x)y = Q(x) \tag{1}$$

where $P(x)$ and $Q(x)$ are continuous functions over an interval. Notice that Eq. (1) has the *variable coefficient* $P(x)$, so it is not a constant-coefficient linear equation. You learned how to solve Eq. (1) by the integrating factor method. That is, you found the integrating factor

$$\mu(x) = e^{\int P(x)\,dx}$$

leading to the general solution

$$\mu(x)y = \int \mu(x)Q(x)\,dx + C.$$

As a preview to solving nonhomogeneous higher-order linear equations where the forcing function $g(x)$ is not of the special type suitable for undetermined coefficients, we present in this section another method for solving first-order Eq. (1). This method, called **variation of parameters,** first determines the complementary solution y_c to the associated homogeneous equation

$$y' + P(x)y = 0. \tag{2}$$

From the complementary solution a particular solution y_p will be constructed in a special way. The power of the methodology is that it generalizes to nonhomogeneous higher-order linear equations (even if they possess variable coefficients). We present the more general method for constant coefficient linear equations in the next section.

The associated homogeneous Eq. (2) is easily solved by the method of separable variables studied in Section 2.1. In fact, the complementary solution is

$$y_c = Ce^{-\int P(x)\,dx}. \tag{3}$$

To simplify the notation, let us write

$$y_1(x) = e^{-\int P(x)\,dx}. \tag{4}$$

Then

$$y_c = Cy_1(x) \tag{5}$$

is the complementary solution to Eq. (1) where C is an arbitrary constant. Notice that $y_1(x)$ is never zero. Now we want to find a particular solution y_p to nonhomogeneous Eq. (1).

Constructing a Particular Solution y_p

It is important to note from Eq. (5) that y_c is a *constant* C times the solution $y_1(x)$. We encountered a similar situation in Section 3.3 for repeated roots to the auxiliary equation. In that case the root yielded but a single solution $y_1(x)$, and we sought a second, linearly independent solution. Suppose we emulate the procedure used there and vary the constant C in Eq. (5), replacing it by an unknown *function* $v(x)$. That is, we propose that the particular solution take the form

$$y_p = v(x)\,y_1(x) \tag{6}$$

for some function $v(x)$. In order to determine the function v we demand that y_p satisfy nonhomogeneous Eq. (1). If we substitute Eq. (6) into Eq. (1), we obtain

$$(v'y_1 + vy_1') + Pvy_1 = Q \tag{7}$$

where we have omitted the functional notation for v, y_1, P, and Q in order to simplify the expression. We next rewrite Eq. (7) as

$$v'y_1 + v(y_1' + Py_1) = Q. \tag{8}$$

Now y_1 solves homogeneous Eq. (2), so

$$y_1' + Py_1 = 0.$$

Thus we can simplify Eq. (8) to obtain the following condition on the derivative v':

$$v'y_1 = Q. \tag{9}$$

Then

$$v' = \frac{Q}{y_1}$$

and we simply integrate to obtain the function

$$v = \int \frac{Q(x)}{y_1(x)}\,dx. \tag{10}$$

There is no need to include an arbitrary constant of integration in Eq. (10) since we seek *any* single function y_p in Eq. (6) that serves as a particular solution. The constant would simply get absorbed into C in Eq. (5) anyway, when we form the general solution $y = y_c + y_p$. Thus,

$$\begin{aligned} y &= y_c + y_p \\ &= Cy_1(x) + v(x)y_1(x) \\ &= Cy_1(x) + \left[\int \frac{Q(x)}{y_1(x)}\,dx\right]y_1(x). \end{aligned}$$

After substituting $e^{-\int P(x)\,dx}$ for $y_1(x)$ and rearranging terms we have the following *general solution to first-order linear Eq. (1):*

$$y = e^{-\int P(x)\,dx}\left[\int Q(x)e^{\int P(x)\,dx}\,dx + C\right]. \tag{11}$$

Notice that Eq. (11) contains only one arbitrary constant C. Since the integrating factor method used $\mu(x) = e^{\int P(x)\,dx}$, you can see that Eq. (11) yields the same general solution form as was found previously in Section 2.2.

The form of general solution (11) appears formidable with all those integrals. Fortunately you do not need to memorize the solution formula. We can use the following procedure to find the general solution.

SOLVING THE FIRST-ORDER LINEAR EQUATION $y' + P(x)y = Q(x)$

Step 1. Place the first-order equation in standard form:

$$y' + P(x)y = Q(x).$$

Step 2. Find the complementary solution y_c. By separable variables, solve the associated homogeneous equation

$$y' + P(x)y = 0$$

to find

$$y_c = Cy_1(x)$$

where

$$y_1(x) = e^{-\int P(x)\,dx}.$$

Step 3. Assume the particular solution form $y_p = vy_1$. Write the condition on the derivative of the unknown function v:

$$v'y_1 = Q(x).$$

Step 4. Integrate the derivative v' for the function v:

$$v = \int v'\,dx = \int \frac{Q(x)}{y_1(x)}\,dx.$$

Step 5. Substitute v into the form for y_p and write the general solution:

$$y = y_c + y_p.$$

We present several examples.

EXAMPLE 1 Find the general solution to

$$y' + 2xy = x.$$

Solution. The equation is written in standard form. Solving

$$y' + 2xy = 0$$

for the complementary solution by the method of separable variables yields

$$y_c = Ce^{-\int 2x\,dx} = Ce^{-x^2}.$$

Thus $y_1 = e^{-x^2}$.

The particular solution is assumed to have the form

$$y_p = ve^{-x^2}$$

where the unknown function v satisfies the condition

$$v'e^{-x^2} = x.$$

Integration for v yields

$$v = \int xe^{x^2}\,dx = \frac{1}{2}e^{x^2}.$$

Thus

$$y_p = \frac{1}{2}e^{x^2}e^{-x^2} = \frac{1}{2},$$

and the general solution is

$$y = y_c + y_p = Ce^{-x^2} + \frac{1}{2}.$$

EXAMPLE 2 Find the general solution to

$$2y' - y = xe^{x/2}.$$

Solution. After writing the equation in standard form we have

$$y' - \frac{1}{2}y = \frac{x}{2}e^{x/2}.$$

Solving the associated homogeneous equation

$$y' - \frac{y}{2} = 0$$

for the complementary solution yields

$$y_c = e^{\int dx/2} = Ce^{x/2}.$$

Thus $y_1 = e^{x/2}$.

The particular solution has the form

$$y_p = ve^{x/2}$$

where v satisfies the condition

$$v'e^{x/2} = \frac{x}{2}e^{x/2}.$$

Integration gives

$$v = \int v'\,dx = \int \frac{x}{2}\,dx = \frac{x^2}{4}.$$

Thus

$$y_p = \left(\frac{x^2}{4}\right)e^{x/2},$$

and the general solution is

$$y = y_c + y_p = \left(C + \frac{x^2}{4}\right)e^{x/2}.$$

EXAMPLE 3 Find the general solution to

$$xy' + 2y = 1 - \frac{1}{x}, \quad x > 0.$$

Solution. Writing the equation in standard form gives us

$$y' + \frac{2}{x}y = \frac{1}{x} - \frac{1}{x^2}.$$

Solving the homogeneous equation

$$y' + \frac{2}{x}y = 0$$

for the complementary solution gives

$$y_c = e^{-\int (2/x)\,dx} = Ce^{-2\ln x} = Cx^{-2}.$$

The form of the particular solution is assumed to be

$$y_p = vx^{-2}$$

where v satisfies the condition

$$v'x^{-2} = \frac{1}{x} - \frac{1}{x^2}.$$

Integration gives

$$v = \int (x-1)\,dx = \frac{x^2}{2} - x.$$

Thus

$$y_p = \left(\frac{x^2}{2} - x\right)x^{-2} = \frac{1}{2} - x^{-1}.$$

The general solution is

$$y = y_c + y_p = Cx^{-2} - x^{-1} + \frac{1}{2}.$$

EXAMPLE 4 Find the general solution to

$$y' + (\tan x)y = \cos^2 x$$

over the interval $-\pi/2 < x < \pi/2$.

Solution. The equation is given in standard form. Solving the associated homogeneous equation

$$y' + (\tan x)y = 0$$

for the complementary solution gives us

$$y_c = e^{-\int \tan x\,dx} = Ce^{\ln|\cos x|} = C\cos x$$

where $|\cos x| = \cos x$ since $-\pi/2 < x < \pi/2$.

The form of the particular solution is

$$y_p = v\cos x$$

where v satisfies the condition

$$v'\cos x = \cos^2 x.$$

Integration yields

$$v = \int \cos x\,dx = \sin x.$$

Thus

$$y_p = \sin x\cos x,$$

and the general solution is

$$y = y_c + y_p = (C + \sin x)\cos x.$$

EXAMPLE 5 Find the general solution to

$$3xy' - y = \ln x + 1, \quad x > 0.$$

Solution. Writing the equation in standard form gives us

$$y' - \frac{1}{3x}y = \frac{\ln x + 1}{3x}.$$

Solving the associated homogeneous equation

$$y' - \frac{y}{3x} = 0$$

yields

$$y_c = e^{\int dx/3x} = Ce^{(\ln x)/3} = Cx^{1/3}.$$

Then

$$y_p = vx^{1/3}$$

where v satisfies the condition

$$v'x^{1/3} = \frac{\ln x + 1}{3x}.$$

Thus

$$v = \int (\ln x + 1)\frac{x^{-4/3}}{3}\,dx.$$

Integration by parts gives us

$$\begin{aligned} v &= -x^{-1/3}(\ln x + 1) + \int x^{-4/3}\,dx \\ &= -x^{-1/3}(\ln x + 1) - 3x^{-1/3}. \end{aligned}$$

Hence

$$y_p = vx^{1/3} = -(\ln x + 1) - 3 = -(\ln x + 4).$$

The general solution is

$$y = y_c + y_p = Cx^{1/3} - \ln x - 4.$$

EXAMPLE 6 Solve the initial value problem

$$\frac{dy}{dx} + 3x^2y = x^2, \quad y(0) = -1.$$

Solution. The equation is written in standard form. Solving the associated homogeneous equation for the complementary solution yields

$$y_c = e^{-\int 3x^2 dx} = Ce^{-x^3}.$$

Then

$$y_p = ve^{-x^3}$$

where

$$v'e^{-x^3} = x^2.$$

Integration gives

$$v = \int x^2 e^{x^3}\, dx = \frac{e^{x^3}}{3}.$$

Thus

$$y_p = ve^{-x^3} = \frac{1}{3}.$$

The general solution is

$$y = y_c + y_p = Ce^{-x^3} + \frac{1}{3}.$$

Evaluating the arbitrary constant C from the initial condition leads to

$$-1 = Ce^0 + \frac{1}{3} \qquad \text{or} \qquad C = -\frac{4}{3}.$$

Therefore

$$y = \frac{1 - 4e^{-x^3}}{3}.$$

EXERCISES 4.3

In problems 1–15 find the general solution of the given first-order differential equation.

1. $y' + 2xy = x^3$
2. $y' - 3y = e^x$
3. $\dfrac{y'}{2} + y = e^{-x} \sin x$
4. $2y' - y = x^2 e^{-x/2}$
5. $xy' + 4y = x - x^3, \quad x > 0$
6. $xy' - y = 2x \ln x$
7. $y' = y - e^{2x}$
8. $y' = \dfrac{2y}{x} + x^3 e^x - 1, \quad x > 0$
9. $x^2 \dfrac{dy}{dx} + xy = 2, \quad x > 0$
10. $(1 + x) \dfrac{dy}{dx} + y = \sqrt{x}$
11. $x^2\, dy + xy\, dx = (x - 1)^2 dx, \quad x > 0$
12. $(1 + e^x)\, dy + (ye^x + e^{-x})\, dx = 0$

13. $e^{-y}\,dx + (e^{-y}x - 4y)\,dy = 0$

14. $(x + 3y^2)\,dy + y\,dx = 0, \quad y > 0$

15. $y\,dx + (3x - y^{-2}\cos y)\,dy = 0, \quad y > 0$

In problems 16–20 solve the initial value problem.

16. $y' + 4y = 1, \quad y(0) = 1$

17. $\dfrac{dy}{dx} + 3x^2y = x^5, \quad y(0) = -1$

18. $x\,dy + (y - \cos x)\,dx = 0, \quad y(\pi/2) = 0, x > 0$

19. $xy' + (x - 2)y = 3x^3e^{-x}, \quad y(1) = 0, \; x > 0$

20. $y\,dx + (3x - xy + 2)\,dy = 0, \quad y(2) = -1$

4.4

VARIATION OF PARAMETERS: SECOND- AND HIGHER-ORDER EQUATIONS

In this section we continue our investigation into solving nonhomogeneous second-order differential equations with constant coefficients. The method we develop here applies for any continuous forcing function, not just the special types required in the method of undetermined coefficients in Section 4.2. Our procedure generalizes the idea we used to solve the first-order equation in Section 4.3.

Suppose we have the constant-coefficient second-order equation

$$a_2y'' + a_1y' + a_0y = g(x) \tag{1}$$

where $g(x)$ is a continuous forcing function and $a_2 \neq 0$. Let

$$y_c = c_1y_1(x) + c_2y_2(x) \tag{2}$$

denote the complementary solution to the associated homogeneous equation. We now *vary the parameters c_1 and c_2 and replace them by functions $v_1(x)$ and $v_2(x)$*. We propose that the solution form

$$y_p = v_1(x)y_1(x) + v_2(x)y_2(x) \tag{3}$$

solve nonhomogeneous Eq. (1). So we must determine the functions v_1 and v_2. Since we have two functions to determine, we must impose two conditions on them. Of course the conditions we impose have to be valid for all values of x over the solution interval in question.

Determining the Conditions

The particular solution y_p is required to solve Eq. (1), so we need to compute its derivatives. For the first derivative the product rule gives us

$$y_p' = v_1y_1' + v_2y_2' + v_1'y_1 + v_2'y_2 \tag{4}$$

where we have suppressed the notation for the independent variable x to simplify the expression. At this point we impose the first of our two conditions by requiring in Eq. (4) that

$$v_1'y_1 + v_2'y_2 = 0. \tag{5}$$

Condition (5) is not entirely arbitrary. It prevents any second derivatives of v_1 and v_2 from arising. Thus we can ignore the sum of the last two terms in Eq. (4) when we calculate the second derivative of y_p. Using the product rule again, we differentiate

$$y_p' = v_1 y_1' + v_2 y_2'$$

to obtain

$$y_p'' = v_1 y_1'' + v_2 y_2'' + v_1' y_1' + v_2' y_2'. \tag{6}$$

Next we substitute the expressions for y_p and its derivatives into nonhomogeneous Eq. (1), remembering that condition (5) simplifies the first derivative. Thus from Eqs. (3–6) we get

$$\begin{aligned} a_2(v_1 y_1'' + v_2 y_2'' + v_1' y_1' + v_2' y_2') + a_1(v_1 y_1' + v_2 y_2') \\ + a_0(v_1 y_1 + v_2 y_2) = g. \end{aligned}$$

This equation can be rewritten as

$$\begin{aligned} v_1(a_2 y_1'' + a_1 y_1' + a_0 y_1) + v_2(a_2 y_2'' + a_1 y_2' + a_0 y_2) \\ + a_2(v_1' y_1' + v_2' y_2') = g. \end{aligned} \tag{7}$$

Moreover, since y_1 and y_2 are solutions to the homogeneous equation associated with Eq. (1), the first two terms in Eq. (7) are zero. This observation leaves us with the second condition on our parameter functions:

$$a_2(v_1' y_1' + v_2' y_2') = g$$

or, since $a_2 \neq 0$,

$$v_1' y_1' + v_2' y_2' = \frac{g}{a_2}. \tag{8}$$

Conditions (5) and (8) give two equations in the unknowns v_1' and v_2'. The determinant of the coefficients in these two equations is the Wronskian

$$W[y_1, y_2] = \begin{vmatrix} y_1(x) & y_2(x) \\ y_1'(x) & y_2'(x) \end{vmatrix} \tag{9}$$

of the independent solutions $y_1(x)$ and $y_2(x)$ to the associated homogeneous equation

$$a_2 y'' + a_1 y' + a_0 y = 0. \tag{10}$$

From Theorem 3.10 we know that Wronskian (9) is never zero on the solution interval. Thus from the method of determinants (also known as *Cramer's rule*) it is possible to solve Eqs. (5) and (8) simultaneously for the derivatives $v_1'(x)$ and $v_2'(x)$. Next we integrate the results to find the desired functions v_1 and v_2. Substitution of these functions into Eq. (3) yields a particular solution y_p to nonhomogeneous Eq. (1). Let us summarize this method for finding y_p. Then we will consider several examples.

METHOD OF VARIATION OF PARAMETERS FOR SOLVING $a_2y'' + a_1y' + a_0y = g(x)$

Step 1. Find the complementary solution y_c. Solve the associated homogeneous equation

$$a_2y'' + a_1y' + a_0y = 0.$$

Write the complementary solution as

$$y_c = c_1y_1(x) + c_2y_2(x) \tag{11}$$

where y_1 and y_2 are any two independent solutions of the homogeneous equation, and c_1 and c_2 are arbitrary constants.

Step 2. Vary the parameters. Replace constants c_1 and c_2 by the (unknown) functions $v_1(x)$ and $v_2(x)$, and write the form of the particular solution as

$$y_p = v_1(x)y_1(x) + v_2(x)y_2(x). \tag{12}$$

Step 3. Write the conditions on the derivatives v_1' and v_2':

$$\begin{aligned} v_1'y_1 + v_2'y_2 &= 0, \\ v_1'y_1' + v_2'y_2' &= \frac{g}{a_2}. \end{aligned} \tag{13}$$

The independent variable x has been suppressed in these conditions to simplify the notation.

Step 4. Solve the conditions for v_1' and v_2'. Solve system of equations (13) for the derivatives v_1' and v_2'.

Step 5. Integrate the derivatives v_1' and v_2'. Perform the antidifferentiations to find the parameter functions:

$$v_1(x) = \int v_1'(x)\,dx, \tag{14}$$

$$v_2(x) = \int v_2'(x)\,dx. \tag{15}$$

Omit the arbitrary constants of integration in Eqs. (14) and (15) since you are seeking *any* particular solution of the given equation.

Step 6. Write the particular solution y_p. Substitute the functions $v_1(x)$ and $v_2(x)$ from Step 5 into Eq. (12) for y_p.

Step 7. Form the general solution. The general solution to the nonhomogeneous equation is the sum of the complementary and particular solutions found in Steps 1 and 6:

$$y = y_c + y_p.$$

EXAMPLE 1 Find the general solution to the equation

$$y'' + y = \tan x. \tag{16}$$

Solution. *Step 1.* The solution of the homogeneous equation

$$y'' + y = 0$$

is given by

$$y_c = c_1 \cos x + c_2 \sin x. \tag{17}$$

Step 2. We vary the parameters in Eq. (17) and write

$$y_p = v_1(x) \cos x + v_2(x) \sin x \tag{18}$$

for the form of the particular solution.

Step 3. Since $y_1(x) = \cos x$ and $y_2(x) = \sin x$, the conditions to be satisfied by the parameters are

$$\begin{aligned} v_1' \cos x + v_2' \sin x &= 0, \\ -v_1' \sin x + v_2' \cos x &= \tan x. \end{aligned} \tag{19}$$

Note that the leading coefficient a_2 equals 1 on the left-hand side of Eq. (16). Thus the right-hand side of the second equation in Eq. (19) is really $\tan x$ divided by 1.

Step 4. Solution of system (19) gives

$$v_1' = \frac{\begin{vmatrix} 0 & \sin x \\ \tan x & \cos x \end{vmatrix}}{\begin{vmatrix} \cos x & \sin x \\ -\sin x & \cos x \end{vmatrix}} = \frac{-\tan x \sin x}{\cos^2 x + \sin^2 x} = \frac{-\sin^2 x}{\cos x}.$$

Likewise,

$$v_2' = \frac{\begin{vmatrix} \cos x & 0 \\ -\sin x & \tan x \end{vmatrix}}{\begin{vmatrix} \cos x & \sin x \\ -\sin x & \cos x \end{vmatrix}} = \sin x.$$

Step 5. After integrating v_1' and v_2' we have

$$\begin{aligned} v_1(x) &= \int \frac{-\sin^2 x}{\cos x}\, dx \\ &= -\int (\sec x - \cos x)\, dx \\ &= -\ln|\sec x + \tan x| + \sin x, \end{aligned} \tag{20}$$

$$v_2(x) = \int \sin x\, dx = -\cos x. \tag{21}$$

Note that we have omitted the constants of integration in determining v_1 and v_2. They would merely be absorbed into the arbitrary constants in complementary solution (17).

Step 6. Substituting v_1 and v_2 from Eqs. (20) and (21) into y_p in Eq. (18) yields

$$\begin{aligned} y_p &= [-\ln|\sec x + \tan x| + \sin x] \cos x + (-\cos x) \sin x \\ &= (-\cos x) \ln|\sec x + \tan x|. \end{aligned}$$

Step 7. The general solution to Eq. (16) is

$$y = c_1 \cos x + c_2 \sin x - (\cos x)\ln|\sec x + \tan x|.$$

EXAMPLE 2 Solve the nonhomogeneous equation

$$(D^2 + D - 2)y = xe^x. \tag{22}$$

Solution. We solved this equation by the method of undetermined coefficients in Example 2 of Section 4.2. For purposes of comparing methods, let us now solve the same equation by the variation-of-parameters method.

Step 1. The auxiliary polynomial is

$$r^2 + r - 2 = (r + 2)(r - 1),$$

giving the complementary solution

$$y_c = c_1 e^{-2x} + c_2 e^x.$$

Step 2. We vary the parameters and write a particular solution form:

$$y_p = v_1(x)e^{-2x} + v_2(x)e^x. \tag{23}$$

Step 3. Noting that $a_2 = 1$ on the left-hand side of differential Eq. (22), we see that the conditions are

$$\begin{aligned} v_1' e^{-2x} + v_2' e^x &= 0, \\ -2v_1' e^{-2x} + v_2' e^x &= xe^x. \end{aligned}$$

Step 4. Solving the above system for v_1', v_2' gives us

$$v_1' = \frac{\begin{vmatrix} 0 & e^x \\ xe^x & e^x \end{vmatrix}}{\begin{vmatrix} e^{-2x} & e^x \\ -2e^{-2x} & e^x \end{vmatrix}} = \frac{-xe^{2x}}{3e^{-x}} = -\frac{1}{3} xe^{3x}.$$

Likewise, we have

$$v_2' = \frac{\begin{vmatrix} e^{-2x} & 0 \\ -2e^{-2x} & xe^x \end{vmatrix}}{3e^{-x}} = \frac{xe^{-x}}{3e^{-x}} = \frac{x}{3}.$$

Step 5. Integrating to obtain the parameter functions, we have

$$\begin{aligned} v_1(x) &= \int -\frac{1}{3} xe^{3x}\,dx \\ &= -\frac{1}{3}\left(\frac{xe^{3x}}{3} - \int \frac{e^{3x}}{3}\,dx\right) \\ &= \frac{1}{27}(1-3x)e^{3x} \end{aligned} \tag{24}$$

and

$$v_2(x) = \int \frac{x}{3}\,dx = \frac{x^2}{6}. \tag{25}$$

Step 6. Substituting Eqs. (24) and (25) into Eq. (23) gives us the particular solution:

$$\begin{aligned} y_p &= \left[\frac{(1-3x)e^{3x}}{27}\right]e^{-2x} + \left(\frac{x^2}{6}\right)e^x \\ &= \frac{1}{27}e^x - \frac{1}{9}xe^x + \frac{1}{6}x^2e^x. \end{aligned} \tag{26}$$

Step 7. The general solution to Eq. (22) is

$$y = c_1e^{-2x} + c_2e^x - \frac{1}{9}xe^x + \frac{1}{6}x^2e^x$$

where the term $(1/27)e^x$ in Eq. (26) has been absorbed into the term c_2e^x in the complementary solution. The general solution agrees with our result using the method of undetermined coefficients. Notice, however, that the two methods produce different particular solution functions in this case. The expression for y_p as derived from the method of undetermined coefficients does not contain the term $e^x/27$, which appears in particular solution (26). However, as we have emphasized earlier, y_p is not unique.

Let us work another example, this time omitting explicit designation of the various steps in the procedure.

EXAMPLE 3 Solve the nonhomogeneous equation

$$y'' - 3y' + 2y = \frac{e^{3x}}{e^x + 1}. \tag{27}$$

Solution. The auxiliary polynomial is

$$r^2 - 3r + 2 = (r-1)(r-2),$$

so the complementary solution is

$$y_c = c_1e^x + c_2e^{2x}.$$

We then form the particular solution:

$$y_p = v_1(x)e^x + v_2(x)e^{2x}. \tag{28}$$

The conditions are

$$v_1'e^x + v_2'e^{2x} = 0,$$
$$v_1'e^x + 2v_2'e^{2x} = \frac{e^{3x}}{e^x + 1}.$$

Solving this system for v_1' and v_2', we have:

$$v_1' = \frac{\begin{vmatrix} 0 & e^{2x} \\ \dfrac{e^{3x}}{e^x+1} & 2e^{2x} \end{vmatrix}}{\begin{vmatrix} e^x & e^{2x} \\ e^x & 2e^{2x} \end{vmatrix}} = -\frac{e^{2x}}{e^x + 1}$$

and

$$v_2' = \frac{\begin{vmatrix} e^x & 0 \\ e^x & \dfrac{e^{3x}}{e^x+1} \end{vmatrix}}{e^{3x}} = \frac{e^x}{e^x + 1}.$$

Integration then yields

$$\begin{aligned} v_1(x) &= \int -\frac{e^{2x}}{e^x+1}\,dx \\ &= \int (u^{-1} - 1)\,du, \qquad \text{where } u = e^x + 1 \\ &= \ln(e^x + 1) - (e^x + 1) \end{aligned}$$

and

$$v_2(x) = \int \frac{e^x}{e^x+1}\,dx = \ln(e^x + 1).$$

Substituting these expressions into Eq. (28) gives us the particular solution

$$\begin{aligned} y_p &= [\ln(e^x + 1) - e^x - 1]e^x + [\ln(e^x + 1)]e^{2x} \\ &= (e^x + e^{2x})\ln(e^x + 1) - e^{2x} - e^x. \end{aligned}$$

Finally, we arrive at the general solution:

$$\begin{aligned} y &= y_c + y_p \\ &= c_1e^x + c_2e^{2x} + (e^x + e^{2x})\ln(e^x + 1) \end{aligned}$$

where the terms $-e^{2x}$ and $-e^x$ in y_p have been absorbed into the complementary solution y_c.

Higher-Order Equations

The method of variation of parameters we have described easily extends to equations of order greater than 2. For the linear nth-order equation

$$a_n y^{(n)} + a_{n-1} y^{(n-1)} + \cdots + a_1 y' + a_0 y = g(x) \tag{29}$$

where $a_n \neq 0$, we first find the complementary solution

$$y_c = c_1 y_1(x) + c_2 y_2(x) + \cdots + c_n y_n(x). \tag{30}$$

Then we *vary the parameters* to form the particular solution

$$y_p = v_1(x) y_1(x) + v_2(x) y_2(x) + \cdots + v_n(x) y_n(x). \tag{31}$$

Next we impose the following ***n* conditions** on the derivatives of the parameter functions:

$$\begin{aligned}
v_1' y_1 + v_2' y_2 + \cdots + v_n' y_n &= 0, \\
v_1' y_1' + v_2' y_2' + \cdots + v_n' y_n' &= 0, \\
&\ \,\vdots \\
v_1' y_1^{(n-1)} + v_2' y_2^{(n-1)} + \cdots + v_n' y_n^{(n-1)} &= \frac{g}{a_n}.
\end{aligned} \tag{32}$$

Because the terms $y_i(x)$ are linearly independent solutions of the associated homogeneous equation, Theorem 3.10 guarantees that the determinant of the coefficients for system (32) is *never* zero. That is, the Wronskian

$$W[y_1, y_2, \ldots, y_n] = \begin{vmatrix} y_1(x) & y_2(x) & \cdots & y_n(x) \\ y_1'(x) & y_2'(x) & \cdots & y_n'(x) \\ \vdots & \vdots & & \vdots \\ y_1^{(n-1)} & y_2^{(n-1)} & \cdots & y_n^{(n-1)} \end{vmatrix} \tag{33}$$

is never zero over the solution interval. Thus we can solve system (32) for the derivatives v_i'. We then integrate these derivatives to find the desired functions $v_i(x)$ for $i = 1, \ldots, n$. Substitution of the results into form (31) then gives a particular solution y_p. As before, the general solution is $y = y_c + y_p$. Following is an example.

EXAMPLE 4 Find the general solution of

$$(D^3 - D^2 + D - 1)y = x. \tag{34}$$

Solution. The auxiliary polynomial

$$r^3 - r^2 + r - 1 = (r - 1)(r^2 + 1)$$

yields the complementary solution

$$y_c = c_1 e^x + c_2 \cos x + c_3 \sin x.$$

After varying the parameters we have the particular solution form

$$y_p = v_1(x)e^x + v_2(x)\cos x + v_3(x)\sin x. \tag{35}$$

The conditions are as follows:

$$\begin{aligned} v_1' e^x + v_2' \cos x + v_3' \sin x &= 0, \\ v_1' e^x - v_2' \sin x + v_3' \cos x &= 0, \\ v_1' e^x - v_2' \cos x - v_3' \sin x &= x. \end{aligned}$$

Solving this system by the method of determinants yields the derivatives

$$\begin{aligned} v_1'(x) &= \frac{x}{2} e^{-x}, \\ v_2'(x) &= -\frac{x}{2}(\cos x - \sin x), \\ v_3'(x) &= -\frac{x}{2}(\cos x + \sin x). \end{aligned}$$

Integration by parts gives

$$\begin{aligned} v_1(x) &= -\frac{1}{2}(x+1)e^{-x}, \\ v_2(x) &= -\frac{1}{2}[x(\sin x + \cos x) - \sin x + \cos x], \\ v_3(x) &= \frac{1}{2}[x(\cos x - \sin x) - \sin x - \cos x]. \end{aligned}$$

After substituting these functions into form (35) and simplifying the result algebraically, we find the particular solution to be

$$y_p = -x - 1.$$

We leave it to you to work out the details of the solution in problem 26 of Exercises 4.4. The general solution is

$$y = c_1 e^x + c_2 \cos x + c_3 \sin x - x - 1.$$

We conclude this section with several observations and remarks.

Remark The method of variation of parameters applies to *all linear nonhomogeneous differential equations,* including those with *variable coefficients.* In order to implement the method, you have to start with the complementary solution y_c to the associated homogeneous equation. At this stage in your study of differential equations, you only know how to solve *homogeneous higher-order equations* with *constant coefficients.* Note, however, that you can solve *first-order linear equations* with *variable coefficients.* In Chapter 9 we will investigate solving certain linear higher-order equations with variable coefficients.

Remark The method of variation of parameters requires n integrations to find the functions $v_1(x), v_2(x), \ldots, v_n(x)$ from their derivatives. It is not always possible using antidifferentiation, to find closed-form analytical solutions for the $v_i(x)$ terms in terms of elementary functions. In such cases we may express the terms $v_i(x)$ as integrals or seek a numerical approximation to the solution (see Section 4.6).

Remark If $n > 3$, it may be very tedious or impractical to solve system (32) giving the conditions for the derivatives $v_1'(x), v_2'(x), \ldots, v_n'(x)$. Thus, although the method of variation of parameters is very powerful from a theoretical point of view, it does have practical limitations.

EXERCISES 4.4

In problems 1–20 find the general solution of the given second-order equation using the method of variation of parameters.

1. $y'' + y = \sec x, \quad -\frac{\pi}{2} < x < \frac{\pi}{2}$
2. $y'' + y = \sin^2 x$
3. $y'' - 3y' + 2y = \dfrac{1}{1 + e^{-x}}$
4. $y'' + 2y' + y = \dfrac{e^{-x}}{x}, \quad x > 0$
5. $y'' + 2y' + y = 4e^{-x} \ln x, \quad x > 0$
6. $y'' + 3y' + 2y = \cos e^x$
7. $y'' - 3y' + 2y = \sin e^{-x}$
8. $y'' - y = e^{-2x} \sin e^{-x}$
9. $y'' + 8y' + 16y = \dfrac{e^{-4x}}{x^2}, \quad x > 0$
10. $y'' + 2y' + 2y = 2e^{-x} \tan x, \quad -\frac{\pi}{2} < x < \frac{\pi}{2}$
11. $y'' - y = \sin^2 x$
12. $y'' - y = (1 + e^{-x})^2$
13. $y'' + 2y' + 2y = 3e^{-x} \sin^2 x$
14. $y'' + y = \sec^3 x, \quad -\frac{\pi}{2} < x < \frac{\pi}{2}$
15. $y'' - y = 2(1 - e^{-2x})^{-1/2}, \quad x > 0$
16. $y'' - y' - 2y = \dfrac{3e^{2x}}{1 + e^x}$
17. $y'' + 2y' + 2y = 2e^{-x} \tan^2 x, \quad -\frac{\pi}{2} < x < \frac{\pi}{2}$
18. $y'' - 2y' + y = e^x \sec^2 x, \quad -\frac{\pi}{2} < x < \frac{\pi}{2}$
19. $y'' + y = \sec^2 x \csc x, \quad 0 < x < \frac{\pi}{2}$
20. $y'' + 2y' + y = e^{-x} \tan^{-1} x$

In problems 21–25 find a particular solution y_p to the given third-order equation using the method of variation of parameters.

21. $y''' + 4y' = 3 \csc 2x, \quad 0 < x < \frac{\pi}{2}$
22. $y''' + y'' = x$
23. $y''' - y'' + y' - y = e^x$
24. $y''' - y' = 2e^{-x} \sin 2x$
25. $y''' + y' = \sec^2 x, \quad -\frac{\pi}{2} < x < \frac{\pi}{2}$
26. Work out the details of the solution to Example 4 in the text.

In problems 27–31 solve the given initial value problem.

27. $y'' + 9y = 9 \sec^2 3x, \quad y(0) = y'(0) = 0$
28. $y'' - 6y' + 9y = e^{3x} \ln x, \quad y(1) = y'(1) = 0$

29. $y'' + y = \cos^2 x, \quad y(0) = 0, y'(0) = 1$

30. $y'' + 2y' + y = \frac{e^{-x}}{x^2}, \quad y(1) = y'(1) = 0$

31. $y'' + y = \sec^3 x, \quad y(0) = \frac{1}{2}, y'(0) = 0$

In problems 32–34 two linearly independent solutions y_1 and y_2 are given to the associated homogeneous equation of the variable-coefficient nonhomogeneous equation. (You will learn how to find these solutions in Chapter 9.) Use the method of variation of parameters to find a particular solution to the nonhomogeneous equation. Assume $x > 0$ in each problem.

32. $x^2y'' + xy' - y = x, \quad y_1 = x^{-1}, y_2 = x$

33. $x^2y'' + 2xy' - 2y = x^2, \quad y_1 = x^{-2}, y_2 = x$

34. $x^2y'' - 3xy' + 4y = x^3, \quad y_1 = x^2, y_2 = x^2 \ln x$

4.5

AN APPLICATION: VIBRATIONS IN AN AUTOMOBILE SUSPENSION SYSTEM WITH A CONTINUOUS FORCING FUNCTION

In Section 3.4 we studied the motion of an automobile with a suspension system consisting of a coil spring and a shock absorber. We developed the following model.

Free (unforced) motion:

$$my'' + \delta y' + ky = 0 \qquad (1)$$

where

- y: function of time representing the vertical displacement from the equilibrium position;
- m: portion of the mass of the car supported by the suspension system;
- δ: proportionality constant representing the damping of the shock absorber;
- k: proportionality constant representing the "stiffness" of the spring.

We investigated the nature of the solutions to Eq. (1) using the model refinement process. Specifically, we studied each of the following models in turn.

Undamped free motion (spring only):

$$y'' + \omega_0^2 y = 0. \tag{2}$$

Damped free motion (spring and shock absorber):

$$y'' + 2dy' + \omega_0^2 y = 0 \tag{3}$$

where

$$\omega_0^2 = \frac{k}{m} \quad \text{and} \quad 2d = \frac{\delta}{m}.$$

Let us examine the effect that incorporating a forcing function has on the resulting motion. The model is

$$my'' + \delta y' + ky = f(t). \tag{4}$$

How shall we model the forcing function? Two questions need to be addressed. First, for various road conditions likely to be encountered, how does the external or "excitation" force actually vary with time? That is, what is the amplitude and the time of application? Second, can the assumed forcing function be approximated with familiar or known mathematical functions? Unfortunately, for purposes of convenience and simplicity, these two steps are often combined into one step coupled with the further restriction that only "nice" elementary mathematical functions be used. These are generally continuous functions known to produce "closed-form" analytical solutions (no infinite series) applying the techniques of undetermined coefficients or variation of parameters.

In Fig. 4.1 we present several representations of different types of possible forcing functions $f(t)$. Figure 4.1(a) shows a force that varies sinusoidally with amplitude F_0 and period $2\pi/\omega$. Such a force might approximate the effect of driving at a particular speed over a section of an old country dirt or gravel road with a washboardlike surface. The force shown in Fig. 4.1(a) could also be used to approximate a force that is actually applied in a periodic fashion by a continuous function. This periodic force might represent the "wake-up strips" placed in front of toll booths on a modern highway. On the other hand, the continuous periodic function shown in Fig. 4.1(b) or the discontinuous periodic function shown in Fig. 4.1(c) might be better representations of these wake-up strips. They might also approximate other road conditions that are present. (Discontinuous forcing functions will be studied in Chapter 5.) Finally, in Fig. 4.1(d) we present a forcing function of very large amplitude applied for a very short duration of time. This type of function might model the force experienced when driving over a New York pothole. The force is called an *impulse force.* An ability to model impulse forces and solve the resulting differential equation adds significantly to the realism that can be captured in the mathematical model. (Impulse forces will be treated in Section 6.2.)

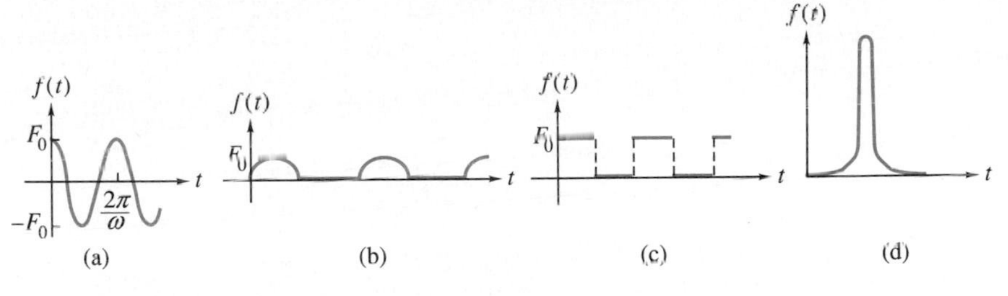

FIGURE 4.1
Approximations of various road conditions. (a) A washboard effect; (b, c) wake-up strips; (d) a New York pothole effect.

We begin our analysis by assuming a simple road condition (or a mathematical approximation to a more complicated condition) modeled by an elementary continuous forcing function of the form

$$f(t) = F_0 \cos \omega t$$

where the constant F_0 is the amplitude of this forcing function and $\omega/2\pi$ is its frequency. We now rewrite Eq. (4) using the constants $\omega_0^2 = k/m$ and $2d = \delta/m$.

Forced motion:

$$y'' + 2dy' + \omega_0^2 y = \frac{F_0}{m} \cos \omega t. \tag{5}$$

First we will investigate this forced-motion model when only the coil spring is present. Then we will study the case where both the spring and the shock absorber are present.

Undamped Forced Motion

The differential equation in this case is

$$y'' + \omega_0^2 y = \frac{F_0}{m} \cos \omega t. \tag{6}$$

The general solution to Eq. (6) has the form

$$y = y_c + y_p$$

where y_c is the complementary solution and y_p is any particular solution. In Section 3.4 we derived

$$y_c = \sqrt{y_0^2 + \left(\frac{v_0}{\omega_0}\right)^2} \cos(\omega_0 t - \phi) \tag{7}$$

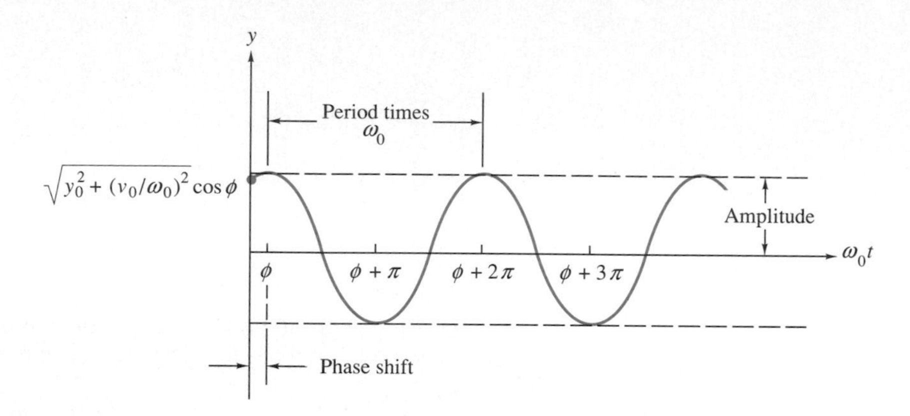

FIGURE 4.2
Simple harmonic motion.

where $\tan \phi = v_0/y_0\omega_0$. In Eq. (7) y_0 and v_0 are the initial conditions for the homogeneous equation $y'' + \omega_0^2 y = 0$ associated with Eq. (6). The graph of y_c versus $\omega_0 t$ is presented again in Fig. 4.2.

The simple harmonic motion described by Eq. (7) is periodic and has the **natural frequency** of $\omega_0/2\pi$. Thus when a particular solution is added to y_c, the curve in Fig. 4.2 might be translated, damped out, or possibly reinforced, among other possibilities. You will see that the result actually depends on the frequency $\omega/2\pi$ of the forcing function.

A question of interest to the automotive engineer is, How does the nature of the solution to Eq. (7) vary for different values of ω? Specifically, how does the relationship between the automobile's natural frequency $\omega_0/2\pi$ and the frequency $\omega/2\pi$ of the forcing function affect the motion? For nonhomogeneous Eq. (6), we impose the initial conditions

$$y(0) = 0 \qquad \text{and} \qquad y'(0) = 0$$

and study two cases.

Case 1: $\omega_0 \neq \omega$

Equation (6) is solved by the method of undetermined coefficients. The roots to the auxiliary equation are $m = \pm\omega_0 i$ yielding the complementary solution

$$y_c = c_1 \cos \omega_0 t + c_2 \sin \omega_0 t.$$

The roots to the auxiliary polynomial for the forcing function are $m^* = \pm\omega i$. Since $\omega_0 \neq \omega$, the correct trial solution is

$$y_q = A \cos \omega t + B \sin \omega t$$

where A and B are the undetermined coefficients. The derivatives are

$$Dy_q = B\omega \cos \omega t - A\omega \sin \omega t$$

and

$$D^2y_q = -A\omega^2 \cos \omega t - B\omega^2 \sin \omega t.$$

Requiring y_q to satisfy Eq. (6) means that

$$(D^2 + \omega_0^2)y_q = \frac{F_0}{m} \cos \omega t$$

or

$$A(\omega_0^2 - \omega^2) \cos \omega t + B(\omega_0^2 - \omega^2) \sin \omega t = \frac{F_0}{m} \cos \omega t.$$

Solving the previous equation for the undetermined coefficients gives $A = F_0/m(\omega_0^2 - \omega^2)$ and $B = 0$. The particular solution is

$$y_p = \frac{F_0}{m(\omega_0^2 - \omega^2)} \cos \omega t.$$

The general solution is then

$$y = c_1 \cos \omega_0 t + c_2 \sin \omega_0 t + \frac{F_0}{m(\omega_0^2 - \omega^2)} \cos \omega t. \tag{8}$$

Substituting the initial conditions $y(0) = y'(0) = 0$ into Eq. (8) leads to $c_2 = 0$ and $c_1 = -F_0/m(\omega_0^2 - \omega^2)$. Therefore the solution to the initial value problem is

$$y = \frac{F_0}{m(\omega_0^2 - \omega^2)} (\cos \omega t - \cos \omega_0 t). \tag{9}$$

Interpreting the Solution We now interpret solution (9). To facilitate our analysis of the solution we make the following substitutions:

$$\phi = \frac{\omega_0 + \omega}{2} \quad \text{and} \quad \psi = \frac{\omega_0 - \omega}{2}.$$

Then $\phi + \psi = \omega_0$ and $\phi - \psi = \omega$. From the identity

$$2 \sin \phi \sin \psi = \cos(\phi - \psi) - \cos(\phi + \psi),$$

we can rewrite Eq. (9) as follows:

$$\begin{aligned} y &= \frac{F_0}{m(\omega_0^2 - \omega^2)} (\cos \omega t - \cos \omega_0 t) \\ &= \frac{F_0}{m(\omega_0^2 - \omega^2)} [\cos(\phi - \psi)t - \cos(\phi + \psi)t] \\ &= \frac{F_0}{m(\omega_0^2 - \omega^2)} 2 \sin \phi t \sin \psi t \end{aligned}$$

or

$$y = \frac{2F_0}{m(\omega_0^2 - \omega^2)} \sin \frac{(\omega_0 + \omega)t}{2} \sin \frac{(\omega_0 - \omega)t}{2}. \tag{10}$$

In Eq. (10) notice that $|\sin(\omega_0 + \omega)t/2| \leq 1$. Thus the motion described by solution (10) is contained within the **sinusoidal envelope**

$$y_e = \pm \left[\frac{2F_0}{m(\omega_0^2 - \omega^2)} \right] \sin \frac{(\omega_0 - \omega)t}{2}. \tag{11}$$

This envelope has the frequency $(\omega_0 - \omega)/4\pi$ and is graphed in Fig. 4.3.

Note also in Eq. (10) that $\omega_0 + \omega > \omega_0 - \omega$ because both ω_0 and ω are positive for the physical problem under consideration. Thus the oscillation $y = \sin[(\omega_0 + \omega)t/2]$ is contained within the envelope, is also sinusoidal, and has a frequency of $(\omega_0 + \omega)/4\pi$. A sinusoidal motion whose amplitude also varies in a sinusoidal fashion (as just described) is called a **beat.** Figure 4.4 illustrates a specific situation, where $\omega_0 + \omega$ is several times larger than $\omega_0 - \omega$.

Closeness of Frequencies The previous analysis is valid only when $\omega \neq \omega_0$. However, before considering the case when $\omega = \omega_0$, let us see what happens to the envelope as the frequency $\omega/2\pi$ of the forcing function approaches the natural frequency $\omega_0/2\pi$ of the suspension system.

Taking the limit of the amplitude in Eq. (10) we have

$$\lim_{\omega \to \omega_0} \frac{2F_0}{m(\omega_0^2 - \omega^2)} = \infty.$$

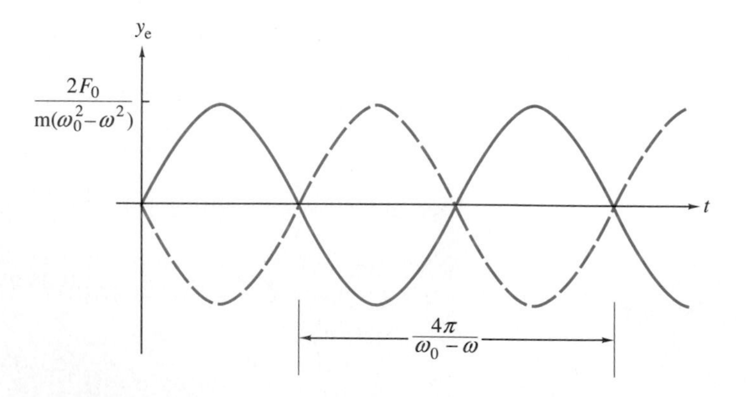

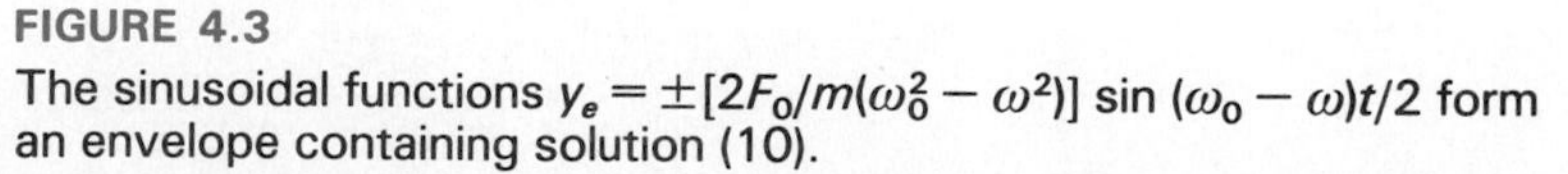
FIGURE 4.3
The sinusoidal functions $y_e = \pm[2F_0/m(\omega_0^2 - \omega^2)] \sin(\omega_0 - \omega)t/2$ form an envelope containing solution (10).

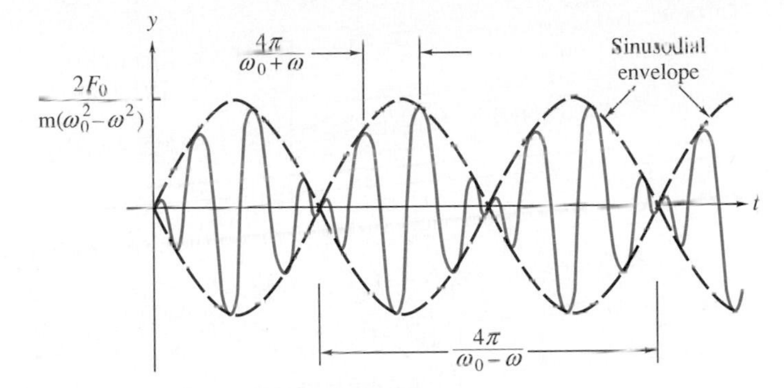

FIGURE 4.4
Beat vibrations represented by Eq. (10).

Thus the amplitude of the vibrations becomes arbitrarily large as the natural and forcing-function frequencies approach each other. In summary:

> The periodic forcing function $F_0 \cos \omega t$ causes arbitrarily large vibrations as its frequency $\omega/2\pi$ approaches the natural frequency $\omega_0/2\pi$ of the suspension system.

When the two frequencies actually coincide, we call the resulting behavior **resonance.** Resonance can cause tremendous damage in automobile suspension systems, aircraft, suspension bridges, ships, electrical circuits, and other systems.

Case 2: $\omega = \omega_0$

Satisfaction of this model requires the two frequencies to be equal exactly. Since the model is only an approximation, this force situation is highly unlikely to occur in practice. Nevertheless, it is important to study this case for mathematical completeness and understanding of the problem. Thus in this case Eq. (6) can be rewritten as

$$y'' + \omega^2 y = \frac{F_0}{m} \cos \omega t. \tag{12}$$

The roots of the auxiliary equation are $m = \pm \omega i$. Likewise, the roots of the auxiliary polynomial for the forcing function are $m^* = \pm \omega i$. The complementary solution is

$$y_c = c_1 \cos \omega t + c_2 \sin \omega t.$$

The correct trial solution is

$$y_q = At \cos \omega t + Bt \sin \omega t.$$

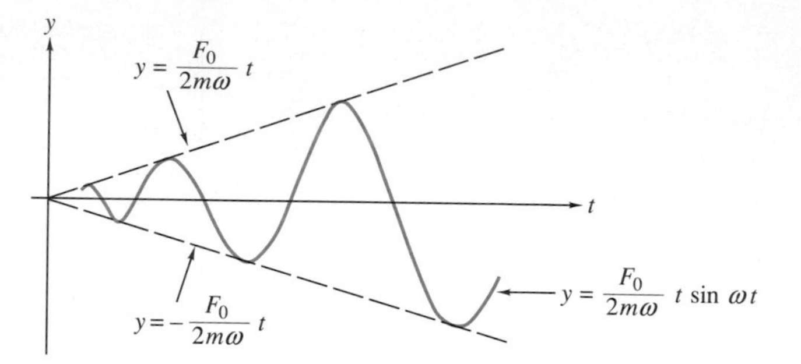

FIGURE 4.5
Solution (14) oscillates between the lines $y = \pm(F_0/2m\omega)t$. In this behavior of resonance, the vibrations grow larger and larger as $t \to \infty$.

Requiring that y_q satisfy the equation

$$(D^2 + \omega^2)y_q = \frac{F_0}{m}\cos\omega t$$

yields the values $A = 0$ and $B = F_0/2m\omega$. Therefore

$$y_p = \frac{F_0}{2m\omega}t\sin\omega t.$$

The general solution to Eq. (12) is

$$y = c_1\cos\omega t + c_2\sin\omega t + \frac{F_0}{2m\omega}t\sin\omega t. \tag{13}$$

The initial conditions $y(0) = y'(0) = 0$ yield $c_1 = c_2 = 0$. Thus the solution to the initial value problem in this case is

$$y = \frac{F_0}{2m\omega}t\sin\omega t. \tag{14}$$

Since $|\sin\omega t| \leqslant 1$, the motion represented by Eq. (14) oscillates sinusoidally between the lines $y = \pm(F_0/2m\omega)t$. The situation is shown in Fig. 4.5.

Our model and Fig. 4.5 predict that when $\omega = \omega_0$ the amplitude of the vibrations grows larger and larger as t increases. This resonant behavior causes great concern for an engineer because it implies ultimate failure of the suspension system.

Damped Forced Motion

In Section 3.4 we studied the solutions to the homogeneous model

$$y'' + 2dy' + \omega_0^2 y = 0. \tag{15}$$

The nature of the solutions depends on the value of the discriminant $d^2 - \omega_0^2$ as summarized in Table 4.1.

TABLE 4.1

Solutions for Damped Motion Where $2d = \frac{\delta}{m}$, $\omega_0^2 = \frac{k}{m}$

$d^2 - \omega_0^2$	Parameter Relations	Solution	Nature of Damping
Positive	$d > \omega_0$ or $\delta^2 > 4km$	$y = c_1 e^{(-d+\sqrt{d^2-\omega_0^2})t} + c_2 e^{(-d-\sqrt{d^2-\omega_0^2})t}$	Overdamped
Zero	$d = \omega_0$ or $\delta^2 = 4km$	$y = (c_1 t + c_2)e^{-dt}$	Critically damped
Negative	$d < \omega_0$ or $\delta^2 < 4km$	$y = e^{-dt}[c_1 \cos(\sqrt{\omega_0^2 - d^2}\, t) + c_2 \sin(\sqrt{\omega_0^2 - d^2}\, t)]$	Underdamped

Note in all three cases that the solution y_c approaches zero as t tends toward infinity. This result makes sense because no new energy is being added to the system, while the energy provided by the initial conditions is being dissipated though the damping factor e^{-dt}. Think of y_c as representing a **transient solution.** It permits satisfaction of the initial conditions but gradually disappears as the initial energy is dissipated.

Now suppose a periodic forcing function is present. The differential equation governing the motion is

$$y'' + 2dy' + \omega_0^2 y = \frac{F_0}{m} \cos \omega t. \tag{16}$$

The particular solution y_p we now seek provides the **steady-state solution.** That is, since the general solution to nonhomogeneous system (16) is the sum of the complementary solution y_c and the particular solution y_p, with y_c approaching zero as t becomes large, the solution to the nonhomogeneous system "settles down" to

$$y = y_p.$$

Solution of the Damped Model The correct trial solution to Eq. (16) is

$$y_q = A \cos \omega t + B \sin \omega t.$$

Substituting y_q into Eq. (16) and solving for the undetermined coefficients yields

$$A = \frac{F_0(\omega_0^2 - \omega^2)}{m[4d^2\omega^2 + (\omega_0^2 - \omega^2)^2]}$$

and

$$B = \frac{2F_0 d\omega}{m[4d^2\omega^2 + (\omega_0^2 - \omega^2)^2]}.$$

Therefore the steady-state solution is

$$y_p = \frac{F_0}{m[4d^2\omega^2 + (\omega_0^2 - \omega^2)^2]}[(\omega_0^2 - \omega^2)\cos\omega t + 2d\omega \sin\omega t] \quad (17)$$

In a manner similar to rewriting the equation for simple harmonic motion, we use the substitutions

$$\cos\phi = \frac{\omega_0^2 - \omega^2}{\sqrt{4d^2\omega^2 + (\omega_0^2 - \omega^2)^2}} \quad (18)$$

and

$$\sin\phi = \frac{2d\omega}{\sqrt{4d^2\omega^2 + (\omega_0^2 - \omega^2)^2}}. \quad (19)$$

After rewriting Eq. (17) we have

$$y_p \frac{m\sqrt{4d^2\omega^2 + (\omega_0^2 - \omega^2)^2}}{F_0} = \frac{(\omega_0^2 - \omega^2)\cos\omega t + 2d\omega\sin\omega t}{\sqrt{4d^2\omega^2 + (\omega_0^2 - \omega^2)^2}}.$$

Substituting Eqs. (18) and (19) into the previous equation yields

$$y_p \frac{m\sqrt{4d^2\omega^2 + (\omega_0^2 - \omega^2)^2}}{F_0} = \cos\phi\cos\omega t + \sin\phi\sin\omega t.$$

Recalling the identity $\cos(\alpha - \beta) = \cos\alpha\cos\beta + \sin\alpha\sin\beta$ and solving for y_p yields

$$y_p = \frac{F_0}{m\sqrt{4d^2\omega^2 + (\omega_0^2 - \omega^2)^2}}\cos(\omega t - \phi) \quad (20)$$

where $\tan\phi = 2d\omega/(\omega_0^2 - \omega^2)$. Finally, substituting $k/m = \omega_0^2$ and $\delta/m = 2d$ into Eq. (20) gives the steady-state solution in terms of the design parameters of the suspension system:

$$y_p = \frac{F_0}{\sqrt{\delta^2\omega^2 + (k - m\omega^2)^2}}\cos(\omega t - \phi) \quad (21)$$

where $\tan\phi = \delta\omega/(k - m\omega^2)$.

Interpreting the Solution As we would expect, the steady-state solution y_p has the same frequency as the forcing function since the effects of the initial conditions disappear as t becomes large.

It is important to note that steady-state solution (21) represents simple harmonic motion with an amplitude of $F_0/\sqrt{\delta^2\omega^2 + (k - m\omega^2)^2}$. Note that the denominator in Eqs. (20) or (21) cannot equal zero for $\delta > 0$, but it does equal $\delta\omega$ $(= 2md\omega)$ when $\omega_0 = \omega$. Thus the resonant behavior discussed previously for forced undamped motion cannot occur in the presence of damping due to the shock absorber. Nevertheless, the oscillations can become quite large as $\omega \to \omega_0$ if $\delta\omega$ is small enough.

Remarks The above analysis for periodic forcing functions of the form $F_0 \cos \omega t$ applies to the following analogous physical systems.

Linearized damped forced pendulum:

$$m\ell\theta'' + \beta\ell\theta' + mg\theta = f(t).$$

***LRC* circuit with forced voltage:**

$$Lq'' + Rq' + \frac{1}{C}q = E(t).$$

EXERCISES 4.5

Mechanical Units in the British and metric systems may be helpful in doing the problems.

Unit	British System	MKS System
Distance	Feet (ft)	Meters (m)
Mass	Slugs	Kilograms (kg)
Time	Second (sec)	Seconds (sec)
Force	Pounds (lb)	Newtons (N)
g(earth)	32 ft/sec^2	9.81 m/sec^2

1. A package weighing 192 lb falls off a building that is 2000 ft high. Assume that as the package falls, it is subject to a drag force that is numerically equal to 12 times the instantaneous velocity. The initial velocity is assumed to be zero. Solve an initial value problem determining the vertical displacement of the package as a function of time. At which of the following times does the package hit the ground, $t = 112.87$, $t = 125.5$, or $t = 147.11$? What is its speed when it hits the ground?

2. A 16-lb weight is attached to the lower end of a coil spring suspended from the ceiling and having a spring constant of 1 lb/ft. The resistance in the spring–mass system is numerically equal to the instantaneous velocity. At $t = 0$ the weight is set in motion from a position 2 ft below its equilibrium position by giving it a downward velocity of 2 ft/sec. At the same instant an external force given by $f(t) = -2.5 \sin t$ (in pounds) is applied to the system. At the end of π sec determine whether the mass is above or below the equilibrium position and by how much.

3. A 2-kg mass is attached to the lower end of a coil spring suspended from the ceiling. The mass comes to rest in its equilibrium position thereby stretching the spring 1.96 m. The mass is in a viscous medium that offers a resistance in newtons numerically equal to 4 times the instantaneous velocity measured in meters per second. The mass is then pulled down 2 m below its equilibrium position and released with a downward velocity of 3 m/sec. At this same instant an external force given by $f(t) = 20 \cos t$ (in newtons) is applied to the system. At the end of π sec determine if the mass is above or below its equilibrium position and by how much.

4. An 8-lb weight stretches a spring 4 ft. The spring–mass system resides in a medium offering a resistance to the motion equal to 1.5 times the instantaneous velocity, and an external force given by $f(t) = 6 + e^{-t}$ (in pounds) is being applied. If the weight is released at a position 2 ft above its equilibrium position with downward velocity of 3 ft/sec, find its position relative to the equilibrium after 2 sec have elapsed.

5. A 16-lb weight stretches a spring 4 ft. This spring–mass system is in a medium with a damping constant of 4.5 lb-sec/ft, and an external force given by $f(t) = 4 + e^{-2t}$ (in pounds) is being applied. What is the solution function describing the position of the mass at any time if the mass is released from 2 ft below the equilibrium position with an initial velocity of 4 ft/sec downward?

6. A 10-kg mass is attached to a spring having a spring constant of 140 N/m. The mass is started in motion from the equilibrium position with an initial velocity of 1 m/sec in the upward direction and with an applied external force given by $f(t) = 5 \sin t$ (in newtons). The mass is in a viscous medium with a coefficient of resistance equal to 90 N-sec/m. Formulate an initial value problem that models the given system; solve the model and interpret the results.

7. A mass of 1 slug is attached to a spring whose constant is 25/4 lb/ft. Initially the mass is released 1 ft above the equilibrium position with a downward veocity of 3 ft/sec, and the subsequent motion takes place in a medium that offers a damping force numerically equal to 3 times the instantaneous velocity. An external force given by $f(t) = 10$ (in pounds) is driving the system. Formulate and solve an initial value problem that models the given system. Interpret your results.

8. A mass of 4 kg stretches a spring 40 cm. The spring–mass system resides in a viscous medium with a coefficient of resistance equal to 25 N-sec/m. At time $t = 0$ an external force given by $f(t) = \cos 10t$ (in newtons) is applied to the system. Determine the steady-state solution to the system.

9. An 8-lb weight is attached to a spring having a spring constant of 16 lb/ft. This spring–mass system is in a medium with a damping constant of 0.25 lb-sec/ft, and an external force given by $f(t) = \cos 2t$ (in pounds) is being applied. If initially the mass is set in motion at the equilibrium position with a downward velocity of 2 ft/sec, what is the steady-state solution?

10. A mass of 0.2 kg is attached to a spring having a spring constant of 80 N/m. The spring–mass system is in a medium with a damping constant of 4 N-sec/m. The mass is subjected to an external sinusoidal force given by $f(t) = 2 \sin 30t$ (in newtons). In the steady-state solution, what is the amplitude of the forced oscillation?

11. A mass m is subject to a resistive force of $-\delta y'$ but *no* springlike restoring force.

a) Show that the displacement y as a function of time is

$$y = C - \left(\frac{mv_0}{\delta}\right) e^{-\delta t/m}.$$

b) At time $t = 0$ an external force $f(t) = F_0 \cos \omega t$ is switched on. Find the values of A and ϕ in the steady-state solution $y(t) = A \cos(\omega t - \phi)$.

12. A mass weighing 4 lb stretches a spring 3 in. The spring–mass system resides in a viscous medium with a damping constant of 2 lb-sec/ft. The spring system is acted on by an external force given by $f(t) = 4 \sin 4t$ (in pounds). If the mass is released from its equilibrium position with a velocity of 4 in/sec in the downward direction, find the position of the mass at any time t.

13. Suppose $L = 10$ henrys, $R = 10$ ohms, $C = 1/500$ farads, $E = 100$ volts, $q(0) = 10$ coulombs, and $q'(0) = i(0) = 0$. Formulate and solve an initial value problem that models the given LRC circuit. Interpret your results.

14. An inductor of 2 henrys is connected in series with a resistor of 12 ohms, a capacitor of 1/16 farads, and a battery assumed to be 300 volts. The initial instantaneous charge q on the capacitor is 10 coulombs and the initial current is $q'(0) = 0$. Formulate and solve an initial value problem that models the circuit described. Interpret your results.

15. An LRC circuit is set up with an inductance of 1/5 henry, resistance of 1 ohm, and capacitance of 5/6 farads. Assuming the initial charge is 2 coulombs, the initial current is 4 amperes, and the impressed voltage is given by $E(t) = 1 + e^{-t}$ (in volts), find the solution function describing the charge on the capacitor at any time. What is the charge on the capacitor over a long period of time?

16. A series circuit consisting of an inductor, a resistor, and a capacitor is open. There is an initial charge of 2 coulombs on the capacitor, and 3 amperes of current is present in the circuit at the instant the circuit is closed. A voltage given by $E(t) = 20 \cos t$ is applied. In this circuit the voltage drops are numerically equal to the following: across the resistor to 4 times the instantaneous change in the charge, across the capacitor to 10 times the charge, and across the inductor to 2 times the instantaneous change in the current. Find the charge in the capacitor as a function of time. Determine the charge on the capacitor and the current at time $t = 10$.

17. A mass m is free to move along a plane and is subject to a spring force $-k^2x$, a frictional force $-bx'$, and an external

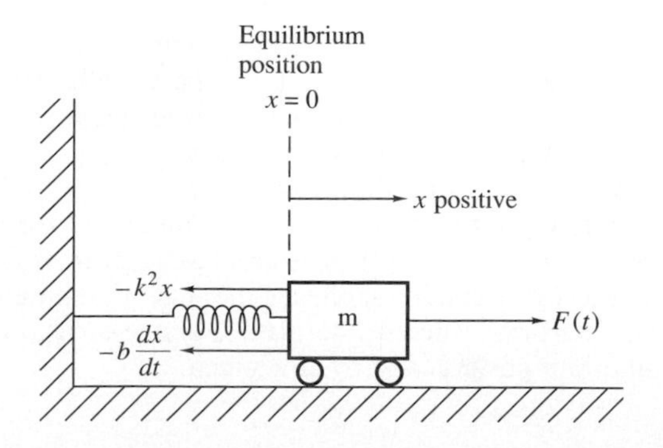

force $F(t)$, as indicated in the accompanying figure. For the system shown, $m=1$, $b=4$, $k^2=20$, $F(t)=10\cos 2t$, $x(0)=5$, and $x'(0)=10$. Formulate and solve an initial value problem to model the given system. Interpret your results.

18. A 3200-lb vehicle rests on a spring–shock absorber system on each of four wheels. The basic data yield the following model:

$$\frac{800}{32}y'' + 50y' + 425y = f(t)$$

where $f(t) = 25e^{-t}\cos 4t$ represents the force in pounds resulting from a bumpy road. Use the technique of variation of parameters to find the general solution.

19. For certain conditions the vibrations transmitted by the road to the car seem negligible. At other times the vibrations seem to add more and more to the instability of the car, causing large oscillations that in theory grow without bound. What is the term used to describe this situation? Give an example of a forcing function $f(t)$ that would cause unbounded oscillations. Solve the model using any method, and graph the results if a computer is available.

20. The opening and closing of small, high-performance hydraulic valves used to control guidance jets in spacecraft are accomplished by the use of a torque motor as illustrated in the accompanying figure. If the mass m of the rotor is 32 slugs, the coefficient of viscous damping is 128 lb-sec/ft, the spring constant is 96 lb/ft, the radius r is 1 ft, and the motor is driven by a torque M equal to 32 cos t (in foot-pounds), then the initial value problem describing the rotation of the motor shaft is

$$\theta'' + 4\theta' + 3\theta = \cos t$$

where $\theta(0) = \theta'(0) = 0$. Find the angular displacement of the motor shaft using the technique of variation of parameters.

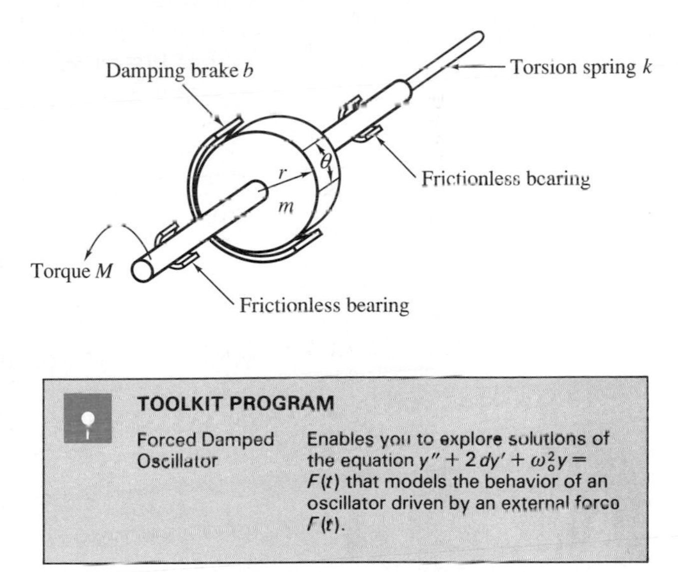

TOOLKIT PROGRAM

Forced Damped Oscillator	Enables you to explore solutions of the equation $y'' + 2dy' + \omega_0^2 y = F(t)$ that models the behavior of an oscillator driven by an external force $F(t)$.

4.6 NUMERICAL SOLUTIONS TO SECOND-ORDER DIFFERENTIAL EQUATIONS (Optional)

In Chapter 2 three numerical methods for approximating the solution to a first-order initial value problem were presented:

- Euler's method
- Improved Euler's method
- Fourth-order Runge–Kutta method.

In this section we will investigate numerical approximations to second-order initial value problems.

Assume that the second-order differential equation can be solved for the second derivative y'' and written in **normal form** as

$$y'' = g(x, y, y'). \tag{1}$$

The initial conditions

$$y(x_0) = y_0 \qquad \text{and} \qquad y'(x_0) = v_0 \tag{2}$$

are also imposed. Thus we seek a solution to Eq. (1) satisfying initial conditions (2). To expedite the numerical solution process, we introduce the new variable

$$v = y'(x). \tag{3}$$

Then $v' = y''$, and the second-order initial value problem can be written as a pair of first-order equations:

$$\begin{aligned} y' &= v, \\ v' &= g(x, y, v), \end{aligned} \tag{4}$$

subject to the initial conditions

$$y(x_0) = y_0 \qquad \text{and} \qquad v(x_0) = v_0. \tag{5}$$

Any one of the three numerical approximation methods discussed previously can be used to solve system (4) subject to conditions (5). However, whichever numerical method is used must be applied to *both* first-order Eqs. (4) at the same time. For instructional purposes we first illustrate the procedure using Euler's method.

Euler's Method

First we subdivide the interval $x_0 \leqslant x \leqslant b$ for which the solution is desired into subintervals of equal length $\Delta x = h$. The Euler approximation moves along the tangent line to each curve. Thus the $k + 1$st approximations are computed from the immediately preceding approximations according to the following formulas.

$$\begin{aligned} y_{k+1} &= y_k + hy'(x_k), \\ v_{k+1} &= v_k + hv'(x_k). \end{aligned}$$

From Eqs. (4) these approximations reduce to

$$\begin{aligned} y_{k+1} &= y_k + hv_k, \\ v_{k+1} &= v_k + hg(x_k, y_k, v_k). \end{aligned} \tag{6}$$

The initial conditions $y_0 = y(x_0)$ and $v_0 = y'(x_0)$ start the approximation procedure. In a step-by-step fashion we then build a table of values to the solution $y(x)$ as well as its derivative $y'(x)$. We now summarize the method, which is easy to program on a computer or programmable calculator.

EULER'S METHOD FOR SOLVING $y'' = g(x, y, y')$, $y(x_0) = y_0$, $y'(x_0) = v_0$ OVER AN INTERVAL

Step 1. Divide the interval $x_0 \leq x \leq b$ into n subintervals using the equally spaced points:

$$x_1 = x_0 + h, \quad x_2 = x_1 + h, \quad \ldots, \quad x_n = x_{n-1} + h = b.$$

Step 2. Obtain the sequence of approximations in the order written:

$$\begin{aligned} y_1 &= y_0 + hv_0, \\ v_1 &= v_0 + hg(x_0, y_0, v_0); \\ y_2 &= y_1 + hv_1, \\ v_2 &= v_1 + hg(x_1, y_1, v_1); \\ &\vdots \\ y_n &= y_{n-1} + hv_{n-1}, \\ v_n &= v_{n-1} + hg(x_{n-1}, y_{n-1}, v_{n-1}). \end{aligned}$$

EXAMPLE 1 Using Euler's method, solve the second-order initial value problem

$$y'' + y' - 2y = 0, \qquad y(0) = \frac{3}{2}, \quad y'(0) = 0 \tag{7}$$

over the interval $0 \leq x \leq 1$.

Solution. We choose the step size $h = \Delta x = 0.1$. Rewriting the second-order equation as a pair of first-order equations gives us

$$\begin{aligned} v &= y', \\ v' &= y'' = 2y - y'. \end{aligned}$$

Substitution yields

$$\begin{aligned} y' &= v, \\ v' &= 2y - v. \end{aligned} \tag{8}$$

We carry out the first approximations starting with the initial values $y_0 = 3/2$ and $v_0 = 0$. Then

$$y_1 = y_0 + hv_0 = \frac{3}{2} + 0.1(0) = 1.5$$

and

$$v_1 = v_0 + h(2y_0 - v_0) = 0 + 0.1(3 - 0) = 0.3.$$

The process is repeated for the second approximations:

$$y_2 = y_1 + hv_1 = 1.5 + 0.1(0.3) = 1.53$$

and

$$v_2 = v_1 + h(2y_1 - v_1) = 0.3 + 0.1(3 - 0.3) = 0.57.$$

TABLE 4.2

Estimates for $y'' + y' - 2y = 0$ where $y(0) = 1.5$, $y'(0) = 0$

x	Euler $y(x)$	Actual $y(x)$	Euler $v = y'(x)$	Actual $y'(x)$
0.0	1.500	1.500	0.000	0.000
0.1	1.500	1.515	0.300	0.286
0.2	1.530	1.557	0.570	0.551
0.3	1.587	1.624	0.819	0.801
0.4	1.669	1.716	1.055	1.042
0.5	1.774	1.833	1.283	1.281
0.6	1.903	1.973	1.509	1.521
0.7	2.054	2.137	1.739	1.767
0.8	2.227	2.326	1.976	2.024
0.9	2.425	2.542	2.224	2.294
1.0	2.647	2.786	2.486	2.583

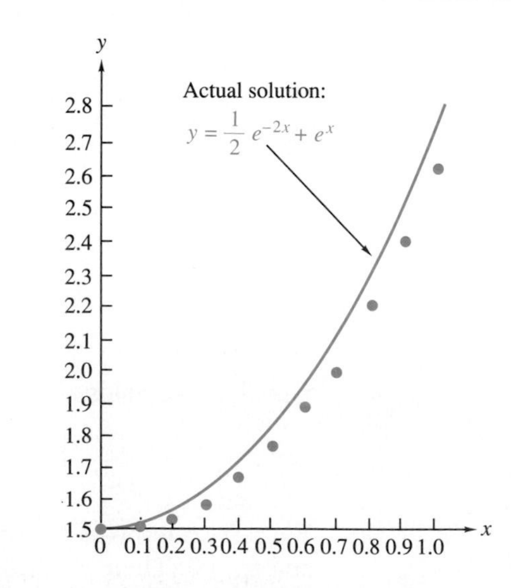

FIGURE 4.6
Scatterplot of numerical approximations versus the actual solution curve for the initial value problem $y'' + y' - 2y = 0$ where $y(0) = 3/2$, $y'(0) = 0$.

Continuing in this way allows us to produce a table of values (Table 4.2). The approximations are compared against the actual solution $y = e^{-2x}/2 + e^x$ obtained by the methods discussed in Section 3.3. A plot of the solution curve superimposed on a scatterplot of the Euler estimates to the solution is shown in Fig. 4.6.

As was the case for the first-order equation, values derived from Euler's method for second-order initial value problems converge slowly to the actual values, and reductions in step size produce undesirable numerical errors. The Improved Euler's method generally yields better results.

Improved Euler's Method

The Improved Euler's method calculates the average of the values for the derivative at the endpoints of the step interval. Assuming the kth approximations y_k and v_k to Eq. (4) are known, the next approximations are computed according to the formulas

$$y_{k+1} = y_k + \frac{h}{2}(v_k + v_{k+1})$$

and

$$v_{k+1} = v_k + \frac{h}{2}[g(x_k, y_k, v_k) + g(x_{k+1}, y_{k+1}, v_{k+1})].$$

However, these last equations contain the terms y_{k+1} and v_{k+1} on *both* sides of the equation. Therefore we first predict the following values using Euler's method:

$$\begin{aligned} v^*_{k+1} &= v_k + hg(x_k, y_k, v_k), \\ y^*_{k+1} &= y_k + \frac{h}{2}[v_k + v^*_{k+1}]. \end{aligned} \tag{9}$$

Then we improve these values according to the formulas

$$\begin{aligned} v_{k+1} &= v_k + \frac{h}{2}[g(x_k, y_k, v_k) + g(x_{k+1}, y^*_{k+1}, v^*_{k+1})], \\ y_{k+1} &= y_k + \frac{h}{2}[v_k + v_{k+1}]. \end{aligned} \tag{10}$$

Rather than pursuing the Improved Euler's method, let us make additional estimates to the derivatives and average them for a Runge–Kutta method. It is this method that is most often used in practice.

Fourth-Order Runge–Kutta Method

Recall from Section 2.6 that the Runge–Kutta methods obtained a weighted average of additional estimates to the derivative over the step interval. Application of the fourth-order Runge-Kutta method to each of first-order Eqs. (4) leads to the following algorithm for approximating the solution to the second-order initial value problem. The estimates in the algorithm must be calculated in the same order in which they are listed.

FOURTH-ORDER RUNGE–KUTTA METHOD FOR SOLVING $y'' = g(x, y, y')$, $y(x_0) = y_0$, $y'(x_0) = v_0$ OVER AN INTERVAL

Step 1. Divide the interval $x_0 \leq x \leq b$ into n subintervals using the equally spaced points

$$x_1 = x_0 + h,\ x_2 = x_1 + h,\ \ldots,\ x_n = x_{n-1} + h = b.$$

Step 2. Introduce the variable v to represent the derivative y'. Then, assuming the kth estimates y_k and v_k are known, calculate the following slope estimates in the order listed:

$$M_1 = v_k$$

$$L_1 = g(x_k, y_k, v_k),$$

$$M_2 = v_k + \frac{hL_1}{2},$$

$$L_2 = g\left(x_k + \frac{h}{2}, y_k + \frac{hM_1}{2}, v_k + \frac{hL_1}{2}\right),$$

$$M_3 = v_k + \frac{hL_2}{2},$$

$$L_3 = g\left(x_k + \frac{h}{2}, y_k + \frac{hM_2}{2}, v_k + \frac{hL_2}{2}\right),$$

$$M_4 = v_k + hL_3,$$

$$L_4 = g(x_k + h, y_k + hM_3, v_k + hL_3).$$

Step 3. Calculate the next estimates to the solution and its slope:

$$y_{k+1} = y_k + \frac{h}{6}(M_1 + 2M_2 + 2M_3 + M_4),$$

$$v_{k+1} = v_k + \frac{h}{6}(L_1 + 2L_2 + 2L_3 + L_4).$$

Step 4. Repeat Steps 2 and 3 for $k = 0, 1, 2, \ldots, n - 1$.

EXAMPLE 2 Using the fourth-order Runge–Kutta method, solve the second-order initial value problem in Example 1.

Solution. Choose the step size $h = 0.1$ and use the equations

$$y' = v,$$
$$v' = 2y - v.$$

The slope estimates for Step 2 are as follows:

$$M_1 = v_0 = 0,$$
$$L_1 = 2y_0 - v_0 = 3 - 0 = 3,$$
$$M_2 = v_0 + \frac{hL_1}{2} = 0 + \frac{0.3}{2} = 0.15,$$
$$L_2 = 2\left(y_0 + \frac{hM_1}{2}\right) - \left(v_0 + \frac{hL_1}{2}\right) = 2\left(\frac{3}{2}\right) - \frac{0.3}{2} = 2.85,$$
$$M_3 = v_0 + \frac{hL_2}{2} = 0 + \frac{0.285}{2} = 0.1425,$$
$$L_3 = 2\left(y_0 + \frac{hM_2}{2}\right) - \left(v_0 + \frac{hL_2}{2}\right)$$
$$= 2\left(\frac{3}{2} + \frac{0.015}{2}\right) - \left(0 + \frac{0.285}{2}\right) = 2.8625,$$
$$M_4 = v_0 + hL_3 = 0 + 0.28625 = 0.28625,$$
$$L_4 = 2(y_0 + hM_3) - (v_0 + hL_3) = 2\left(\frac{3}{2} + 0.01425\right) - (0 + 0.28625)$$
$$= 2.74225.$$

Thus Step 3 gives us the next estimates to the solution

$$y_1 = y_0 + \frac{h}{6}(M_1 + 2M_2 + 2M_3 + M_4) = 1.5 + \frac{0.087125}{6}$$
$$\approx 1.51452$$

and its derivative

$$v_1 = v_0 + \frac{h}{6}(L_1 + 2L_2 + 2L_3 + L_4) = 0 + \frac{1.716725}{6}$$
$$\approx 0.28612.$$

The slope estimates for the next iteration in Step 2 are calculated as follows:

$$M_1 = v_1 = 0.28612,$$
$$L_1 = 2y_1 - v_1 = 2.74292,$$

$$M_2 = v_1 + \frac{hL_1}{2} = 0.423266,$$
$$L_2 = 2\left(y_1 + \frac{hM_1}{2}\right) - \left(v_1 + \frac{hL_1}{2}\right) = 2.634386,$$
$$M_3 = v_1 + \frac{hL_2}{2} = 0.4178393,$$
$$L_3 = 2\left(y_1 + \frac{hM_2}{2}\right) - \left(v_1 + \frac{hL_2}{2}\right) = 2.6535273,$$
$$M_4 = v_1 + hL_3 = 0.55147273,$$
$$L_4 = 2(y_1 + hM_3) - (v_1 + hL_3) = 2.56113513.$$

Thus Step 3 gives us

$$y_2 = y_1 + \frac{h}{6}(M_1 + 2M_2 + 2M_3 + M_4) \approx 1.55652$$

and

$$v_2 = v_1 + \frac{h}{6}(L_1 + 2L_2 + 2L_3 + L_4) \approx 0.55078.$$

Continuing in this fashion yields a table of approximations (Table 4.3). Notice that the approximations agree to three decimal places with the actual analytical solution $y = e^{-2x}/2 + e^x$.

TABLE 4.3

Estimates for $y'' + y' - 2y = 0$ where $y(0) = 1.5$, $y'(0) = 0$

x	Runge–Kutta $y(x)$	Actual $y(x)$	Runge–Kutta $b = y'(x)$	Actual $y'(x)$
0.0	1.500	1.500	0.000	0.000
0.1	1.515	1.515	0.286	0.286
0.2	1.557	1.557	0.551	0.551
0.3	1.624	1.624	0.801	0.801
0.4	1.716	1.716	1.042	1.042
0.5	1.833	1.833	1.281	1.281
0.6	1.973	1.973	1.521	1.521
0.7	2.137	2.137	1.767	1.767
0.8	2.326	2.326	2.024	2.024
0.9	2.542	2.542	2.294	2.294
1.0	2.786	2.786	2.583	2.583

EXAMPLE 3 Use the fourth-order Runge–Kutta method to solve the second-order initial value problem

$$y'' + 3y' + 2y = \sin e^x, \qquad y(0) = y'(0) = 0$$

over the interval $0 \leq x \leq 1$.

Solution. We choose the step size $h = 0.1$ and write the differential equation as a pair of first-order equations:

$$\begin{aligned} y' &= v, \\ v' &= \sin e^x - 3v - 2y. \end{aligned}$$

The initial conditions mean that $y_0 = v_0 = 0$. We then execute Step 2 in the Runge–Kutta procedure:

$$\begin{aligned}
M_1 &= v_0 = 0, \\
L_1 &= \sin e^0 - 3v_0 - 2y_0 = \sin 1 \approx 0.84147, \\
M_2 &= v_0 + \frac{hL_1}{2} \approx 0.04207, \\
L_2 &= \sin e^{0.05} - 3\left(v_0 + \frac{hL_1}{2}\right) - 2\left(y_0 + \frac{hM_1}{2}\right) \approx 0.74183, \\
M_3 &= v_0 + \frac{hL_2}{2} \approx 0.037092, \\
L_3 &= \sin e^{0.05} - 3\left(v_0 + \frac{hL_2}{2}\right) - 2\left(y_0 + \frac{hM_2}{2}\right) \approx 0.75257, \\
M_4 &= v_0 + hL_3 \approx 0.075257, \\
L_4 &= \sin e^{0.1} - 3(v_0 + hL_3) - 2(y_0 + hM_3) \approx 0.66035.
\end{aligned}$$

Thus Step 3 gives us the next estimate

$$y_1 = y_0 + \frac{h}{6}(M_1 + 2M_2 + 2M_3 + M_4) \approx 0.00389$$

and

$$v_1 = v_0 + \frac{h}{6}(L_1 + 2L_2 + 2L_3 + L_4) \approx 0.07484.$$

Continuing this procedure gives us the values in Table 4.4, rounded to three decimal places. The approximations are compared against the actual analytical solution $y = c_1e^{-2x} + c_2e^{-x} - e^{-2x}\sin e^x$ obtained by the method of

TABLE 4.4

Estimates for $y'' + 3y' + 2y = \sin e^x$ where $y(0) = y'(0) = 0$

x	Runge–Kutta $y(x)$	Actual $y(x)$	Runge–Kutta $v = y'(x)$	Actual $y'(x)$
0.0	1.000	1.000	0.000	0.000
0.1	0.004	0.004	0.075	0.075
0.2	0.014	0.014	0.133	0.133
0.3	0.030	0.030	0.178	0.178
0.4	0.050	0.050	0.210	0.210
0.5	0.072	0.072	0.231	0.231
0.6	0.096	0.096	0.242	0.242
0.7	0.120	0.120	0.242	0.242
0.8	0.143	0.143	0.230	0.230
0.9	0.165	0.165	0.205	0.205
1.0	0.184	0.184	0.165	0.166

variation of parameters. The initial conditions evaluate the constants: $c_1 = \sin 1 - \cos 1$ and $c_2 = \cos 1$.

Variable Coefficients

At this stage in our study of differential equations we have investigated solution methods of linear equations with variable coefficients only in the first-order case (see Section 4.3). In Chapter 9 we will consider some analytical methods involving infinite series to solve some fairly elementary linear second-order equations with variable coefficients. However, we can now easily find numerical approximations to second-order equations with variable coefficients. In fact, the equation need not even be linear. Following is an example of solving an equation with variable coefficients.

EXAMPLE 4 Use the fourth-order Runge–Kutta method to solve the second-order initial value problem

$$x^2y'' - xy' + y = 4x \ln x, \qquad y(1) = 1, \quad y'(1) = 0$$

over the interval $1 \leq x \leq 2$.

Solution. We choose the step size $h = 0.1$ and write the second-order differential equation as a pair of first-order equations:

$$y' = v,$$
$$v' = \frac{4 \ln x + v}{x} - \frac{y}{x^2}.$$

From $x_0 = 1$, $y_0 = 1$, and $v_0 = 0$ we calculate the estimates for Step 2:

$$M_1 = v_0 = 0,$$

$$L_1 = \frac{4 \ln x_0 + v_0}{x_0} - \frac{y_0}{x_0^2} = -1,$$

$$M_2 = v_0 + \frac{hL_1}{2} = -0.05,$$

$$L_2 = \frac{4 \ln\left(x_0 + \frac{h}{2}\right) + v_0 + \frac{hL_1}{2}}{x_0 + \frac{h}{2}} - \frac{y_0 + \frac{hM_1}{2}}{\left(x_0 + \frac{h}{2}\right)^2} \approx -0.76878,$$

$$M_3 = v_0 + \frac{hL_2}{2} \approx -0.038439,$$

$$L_3 = \frac{4 \ln\left(x_0 + \frac{h}{2}\right) + v_0 + \frac{hL_2}{2}}{x_0 + \frac{h}{2}} - \frac{y_0 + \frac{hM_2}{2}}{\left(x_0 + \frac{h}{2}\right)^2} \approx -0.755503,$$

$$M_4 = v_0 + hL_3 \approx -0.07555,$$

$$L_4 = \frac{4 \ln (x_0 + h) + v_0 + hL_3}{x_0 + h} - \frac{y_0 + hM_3}{(x_0 + h)^2} \approx -0.545369.$$

Thus

$$y_1 = y_0 + \frac{h}{6}(M_1 + 2M_2 + 2M_3 + M_4) \approx 0.99579$$

and

$$v_1 = v_0 + \frac{h}{6}(L_1 + 2L_2 + 2L_3 + L_4) \approx -0.076566.$$

TABLE 4.5

Estimates for $x^2y'' - xy' + y = 4x \ln x$ where $y(1) = 1$, $y'(1) = 0$

x	Runge–Kutta $y(x)$	Actual $y(x)$	Runge–Kutta $v = y'(x)$	Actual $y'(x)$
1.0	1.000	1.000	0.000	0.000
1.1	0.996	0.996	−0.077	−0.077
1.2	0.986	0.986	−0.112	−0.112
1.3	0.975	0.975	−0.113	−0.113
1.4	0.964	0.964	−0.085	−0.085
1.5	0.958	0.958	−0.032	−0.032
1.6	0.959	0.959	0.041	0.041
1.7	0.967	0.967	0.132	0.132
1.8	0.986	0.986	0.239	0.239
1.9	1.015	1.015	0.358	0.358
2.0	1.058	1.058	0.490	0.490

Continuing in this fashion, we build the table of estimates shown as Table 4.5. The true solution turns out to be

$$y = c_1 x + c_2 x \ln x + \frac{2x}{3} (\ln x)^3$$

where $c_1 = 1$ and $c_2 = -1$ are evaluated from the initial conditions. Table 4.5 reveals that the Runge–Kutta estimates agree with the actual solution values to three decimal places.

Remark Whenever a numerical method is used to approximate the solution to a differential equation, the accuracy of the results must be questioned. We explored this difficult issue briefly in Section 2.5. However, a thorough discussion of error is most appropriate for a course in numerical analysis. A beginner should start with a fairly small value for h, since decreasing values of h generally reduce truncation error and improve the accuracy, at least until computational roundoff error becomes a serious problem. One procedure you can employ to check your results is to work with one value of h and then repeat the process using $h/2$.

EXERCISES 4.6

I. In problems 1–10 use Euler's method to calculate the first three approximations y_1, y_2, y_3, and v_1, v_2, v_3 for the specified step size. Then repeat your calculations for $h/2$. Solve the initial value problem using an appropriate analytical method and compare your numerical results against the actual solution values.

1. $y'' = y, \quad y(0) = 2, y'(0) = 0, h = 0.1$
2. $y'' = -y, \quad y(0) = 0, y'(0) = 1, h = 0.2$
3. $y'' = 2(y' - y), \quad y(0) = y'(0) = 1, h = 0.1$
4. $y'' + y' = x, \quad y(0) = 1, y'(0) = 0, h = 0.2$
5. $y'' - 3y' = e^x, \quad y(0) = 1, y'(0) = 0, h = 0.1$
6. $y'' - 3y' = e^{3x}, \quad y(0) = 1, y'(0) = 0, h = 0.1$
7. $y'' + y = x^2, \quad y(0) = 0, y'(0) = 1, h = 0.2$
8. $y'' + y = \tan x, \quad y(0) = 1, y'(0) = -1, h = 0.1$
9. $y'' + y = \sec x, \quad y(0) = 1, y'(0) = 0, h = 0.1$
10. $y'' - y' - 2y = 5 \sin x, \quad y(0) = y'(0) = -1, h = 0.2$

II. Repeat problems 1–10 using the fourth-order Runge–Kutta method. Compare your results against those obtained by Euler's method as well as against the actual solution obtained analytically.

III. In problems 11–12, use the fourth-order Runge–Kutta method to calculate the first three approximations for the specified step size. Then repeat your calculations for $h/2$.

11. $xy'' + (1 - x)y' - y = 0, \quad y(1) = y'(1) = 1, h = 0.1$
12. $xy'' + y' = x + 1, \quad y(1) = \frac{1}{4}, y'(1) = -\frac{1}{2}, h = 0.2$

CHAPTER 4 REVIEW EXERCISES

In problems 1–17 solve the differential equation for the general or particular solution, as appropriate.

1. $y'' + 2y' - 3y = 12e^x, \quad y(0) = 0, y'(0) = 1$
2. $y'' + y' - 2y = 3e^x, \quad y(0) = 0, y'(0) = -2$
3. $y'' + 4y' + 4y = xe^{-2x}, \quad y(0) = 0, y'(0) = 1$
4. $y'' - y' - 2y = 6e^{2x}, \quad y(0) = 0, y'(0) = 1$
5. $y'' - 2y' + y = 4xe^{-x}$

6. $y'' - 4y' + 4y = xe^{3x}$

7. $y'' + 2y' + y = 25xe^{4x}$

8. $y'' + 6y' + 9y = 25xe^{2x}$

9. $y'' - 3y' - 4y = 2 \sin x$

10. $y'' + 4y' + 4y = 8 \sin 2x, \quad y(0) = -1, y'(\pi) = 1$

11. $\theta'' + 16\theta = t^2, \quad \theta(0) = 1, \theta'(0) = 5$

12. $y'' + y = 3xe^{2x}, \quad y(\pi) = 0, y'(0) = 1$

13. $x'' + 4x = 3 \sin 2t$

14. $x'' + 2x' = 3t + 2 + e^t$

15. $y'' + 16y = -\sec 4x$

16. $y'' + 25y = -\csc 5x$

17. $y'' + 4y = \sec 2x$

18. A series circuit consisting of an inductor, a resistor, and a capacitor is open, and there is a charge of 2 C on the capacitor. A current of 2 A is present in the circuit at the instant the circuit is closed and a voltage given by $E(t) = -5 \sin t$ is applied. In this circuit the voltage drop across the resistor is numerically equal to 2 times the instantaneous change in the charge, the inductance is 1 henry, and the voltage drop across the capacitor is equal to 2 times the charge. Find the charge in the capacitor as a function of time. What is the current in the circuit at the end of $\pi/2$ sec?

19. A buoyant cubical box 1 ft on a side weighs 32 lb. If we place the box in a perfectly still pool of water, it will come to rest partially submerged in a buoyant equilibrium position. We then use a machine to produce waves that exert a force (in pounds) equal to $\cos t + e^{-t}$ on the box. By using Archimedes' principle (that an object submerged in a fluid is buoyed up by a force equal to the weight of the fluid displaced) and assuming no damping, we can derive the following differential equation to model the vertical motion of the box:

$$x'' + 64x = \cos t + e^{-t}$$

where x represents the vertical displacement of the box from its equilibrium position in still water, and the density of the water is taken as 64 lb/ft³. Find a particular solution to this problem that describes the vertical displacement of the box from equilibrium over time.

20. An automobile shock absorber–coil spring system is designed to support 768 lb, the portion of the automobile's weight it supports. The spring has a constant of 60 slugs/in. The effect of a bumpy road on the system can be described by the function $f(t) = 250 \sin 4t$ (in slug-in/sec²), which acts upward on the tire. The system is initially in equilibrium at rest.

a) Assume that the automobile's shock absorber is so worn that it provides no effective damping force. Find a particular solution that describes the vertical displacement of the automobile over time. Graph the particular solution for the first three seconds of motion. Describe the system's performance.

b) Now assume that the shock absorber is replaced. The new shock absorber exerts a damping force (in pounds) that is equal to 60 times the instantaneous vertical velocity of the system (in inches per second). Model this improved system with an initial value problem, and solve it subject to the conditions described in part (a) above. Graph the resulting equation for the first three seconds of motion. How has the system's performance improved? Is this system overdamped, underdamped, or critically damped?

21. A shock absorber–coil spring system for an imported automobile is designed to support 350 kg. The spring has a constant of 140,000 kg/cm. The shock absorber exerts a damping force (in kg-cm/sec²) that is equal to 3500 times the instantaneous vertical velocity of the system (in centimeters per second). The system is jarred from equilibrium by a force, $f(t) = 1750\, e^{-2t} \sin 3t$ (in kg-cm/sec²), acting upward on the tire that represents a bumpy road. Model this system with an initial value problem and solve it. Graph the resulting equation for the first three seconds of motion. What is the maximum displacement from equilibrium that the system experiences? Is this system overdamped, underdamped, or critically damped?

22. We wish to determine information about a cannon system (see Chapter 3 Review Exercises, problem 20). The gun tube–breech block assembly has a mass of 1500 kg. The recoil spring has a spring constant of 19,500 N/m. The damping mechanism exerts a force (in newtons) numerically equal to 9000 times the instantaneous velocity of the gun tube–breech block assembly (in meters per second). When the cannon is fired, the gas pressure force from the firing of the round exerts a force of 40,500 N on the gun tube–breech block assembly for ⅕ sec. The force is zero after ⅕ sec.

a) Model this system as an initial value problem and find the particular solution that gives the displacement of the gun tube–breech block assembly with respect to time.

b) Graph the solution for the first three seconds of motion.
c) At what time does the gun tube–breech block assembly first return to the in-battery position? What is the assembly's instantaneous velocity at this time?
d) At what time does the assembly reach its maximum displacement? What is its maximum displacement?

23. A 2-g mass is suspended from the end of a rod 2.45 m long to form a pendulum. The pendulum system travels through a fluid that exerts a damping force (in g-m/sec^2) that is 8 times the instantaneous velocity (in meters per second) along the path traced by the mass. A variable forcing function of $4.9\, e^{-5t} \cos 3t$ (in g-m/sec^2) acts tangentially to the path of motion of the mass. The pendulum is set in motion by displacing the mass 1 radian from the equilibrium position and releasing it from rest. The pendulum only experiences small angular displacements. Find its angular displacement after ½ sec of motion. Use 9.8 m/sec^2 as the acceleration of an object due to gravity. Is the pendulum system overdamped, underdamped, or critically damped?

24. A 3-kg mass is suspended from the end of a rod 4.9 m long to form a pendulum. The pendulum system travels through a fluid that exerts a damping force (in newtons) that is numerically equal to 4 times the instantaneous velocity (in meters per second) along the path traced by the mass. A variable forcing function $f(t) = 58.8\, e^{-t}$ (in newtons) acts tangentially to the path of motion of the mass. The pendulum initially begins in the equilibrium position with an angular velocity of 7 radians per second. The pendulum experiences only small angular displacements. Find its angular displacement after ½ sec of motion. Use 9.8 m/sec^2 as the acceleration of an object due to gravity. Is the pendulum system overdamped, underdamped, or critically damped?

25. A series electrical circuit consists of a capacitor with a capacitance of 1/12 farads, a resistor with a resistance of 7 ohms, and an inductor with an inductance of 1 henry. The circuit is connected to a voltage source that exerts a variable voltage of $200 \sin t$ (in volts). Initially there is no charge on the capacitor. The circuit has an initial current of 26 amperes.

a) Find an equation that gives the charge on the capacitor for any time after the switch is closed.
b) What is the charge on the capacitor after 1 sec?
c) What is the current in the circuit after 1 sec?

26. A simple *LRC* electrical circuit consists of a capacitor with a capacitance of 1/60 farads, a resistor with a resistance of 50 ohms, and an inductor with an inductance of 10 henrys. The circuit is connected to a 24-volt battery. Initially there is no charge on the capacitor and no current in the circuit.

a) Find an equation that gives the charge on the capacitor for any time after the switch is closed.
b) What is the charge on the capacitor after 1 sec?
c) What is the current in the circuit after 1 sec?

27. The *LRC* circuit in problem 26 is connected to an alternating current source that applies a voltage $E(t) = 104 \cos 2t$ (in volts). There is no initial charge on the capacitor or current in the circuit.

a) Find an equation that gives the charge on the capacitor for any time after the switch is closed.
b) What is the charge on the capacitor after 1 sec?
c) What is the current in the circuit after 1 sec?
d) Would a 5-ampere fuse have its capacity exceeded in this circuit?
e) Graph the transient response of the circuit (the complementary function of the differential equation that models the circuit).
f) Graph the general solution to the differential equation that models the circuit. What happens to the transient response as time increases? What happens to the steady-state solution (the particular solution of the general solution)?

28. In problem 24 of the Chapter 3 Review Exercises, it was shown that a torsion system with no external force can be modeled by the differential equation $I\theta'' + c\theta' + k\theta = 0$. If an external torque $T(t)$ is applied to the system, the model becomes $I\theta'' + c\theta' + k\theta = T(t)$. Assume that for a given torsion system $I = 1$ kg-m^2, $k = 4$ N-m, and that a damping torque (in newton-meters) numerically equal to 4 times the instantaneous angular velocity (in radians per meter) is exerted. A variable external torque $T(t) = \cos t$ (in newton-meters) is applied to the system. The system begins turning from the equilibrium position with no initial angular velocity. Find the equation that gives the angular displacement of the mass at any time. What is θ after 1 sec? Is the torsion system overdamped, underdamped, or critically damped?

29. Assume that for a given torsion system $I = 2$ kg-m^2, $k = 10$ N-m, and a damping torque (in newton-meters) numerically equal to 4 times the instantaneous angular velocity (in radians per meter) is exerted. A variable external torque $T(t) = 100\, te^{-4t} \cos 2t$ (in newton-meters) is applied to the system. If the mass is twisted 2 radians from equilibrium and released from rest, find the equation that gives the angular displacement of the mass at any time. How much twist has the mass undergone after 1 sec? Is the torsion system overdamped, underdamped, or critically damped?

30. A researcher is investigating the vibration of the human heart by having a patient rest on a horizontal table containing springs that allow the table to vibrate horizontally but not vertically. Let x denote the horizontal displacement, in centimeters, of the table from its equilibrium position. Let M represent the combined mass, in grams, of the patient and the portion of the table that is set in motion. If the table is subject to a damping force proportional to its horizontal instantaneous velocity, then the motion of the table can be modeled by the differential equation:

$$M\frac{d^2x}{dt^2} + \beta\frac{dx}{dt} + \gamma x = F(t),$$

where β and γ are proportionality constants characteristic of the table and its springs, respectively, and $F(t)$ is the force on the system due to the pumping action of the heart. Suppose that m is the mass of blood pumped out of the heart during each vibration, and y is the instantaneous center of mass of the pumped quantity of blood, as measured from the center of the heart. The force on the system is then $F(t) = m\, d^2y/dt^2$. We will approximate y as the function $y(t) = A \sin \omega t$, where $\omega = 2\pi f$ rad/sec, f is the patient's heart rate in beats per second, and A is a constant. Using all of this information, we obtain the following differential equation:

$$\frac{d^2x}{dt^2} + \frac{\beta}{M}\frac{dx}{dt} + \frac{\gamma}{M}x = -\frac{m\omega^2 A}{M}\sin \omega t.$$

a) Assume that $\beta/M = 1/10\ \text{sec}^{-1}$, $\gamma/M = 9/1600\ \text{sec}^{-2}$, and $A/M = 0.01$ cm/g. Find the general solution of the resulting differential equation in terms of the parameters m and ω.
b) In your solution, you should have obtained the fractions $9/1600$ and $1/1600$. Since these numbers are both negligible with respect to ω^2, simplify the steady-state solution (the particular solution of the general solution) by assuming they are approximately zero. Using this simplification, does the maximum displacement of the table depend on the patient's heart rate?
c) Using only the simplified steady-state solution obtained in part (b), determine the mass of blood, m, pumped by the heart with each heartbeat if the maximum displacement of the table is 0.02 cm.
d) Is the table and spring system underdamped or overdamped? For what value of γ/M will the system be critically damped if M and β do not change?

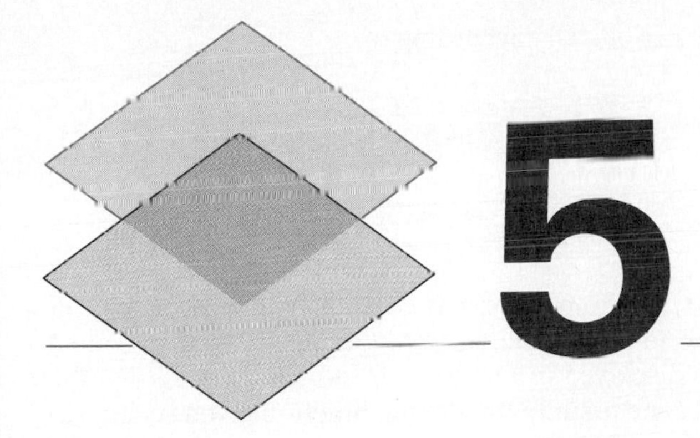

5

Laplace Transform I: Modeling with Discontinuous Forcing Functions

In Chapters 3 and 4 we have been studying constant-coefficient differential equations of the form

$$a_n y^{(n)} + a_{n-1} y^{(n-1)} + \cdots + a_1 y' + a_0 y = g(x)$$

where the forcing function $g(x)$ is assumed to be *continuous* over the solution interval. However, in many situations modeling real-world behavior the forcing function fails to be continuous. For example, in studying the behavior of an electrical circuit modeled by

$$Lq'' + Rq' + \frac{1}{C} q = E(t) \tag{1}$$

(see Example 6 in Section 3.1), the voltage source $E(t)$ may vary in a discontinuous fashion. One possibility is

$$E(t) = \begin{cases} t, & 0 \leq t < T, \\ 0, & T \leq t, \end{cases} \tag{2}$$

in which the voltage source is turned on and increases linearly according to $E = t$ until time T, when it is turned off. Another possibility is that the voltage source may vary periodically according to

$$E(t) = \begin{cases} 1, & 0 \leqslant t < \dfrac{T}{2}, \\ 0, & \dfrac{T}{2} \leqslant t < T, \end{cases} \tag{3}$$

such that $E(t + T) = E(t)$. Here the source is alternately turned on and off for equal lengths of time in a periodic fashion. Graphs of voltage source functions (2) and (3) are displayed in Fig. 5.1.

Let us examine Eq. (1) and assume the forcing function given by Eq. (2) and the initial conditions $q(0) = q_0$ and $i(0) = i_0$, where $i = dq/dt$ is the current. What would be required to solve it? First, we would solve the second-order initial value problem

$$Lq'' + Rq' + \frac{1}{C}q = t, \qquad q(0) = q_0, \quad i(0) = i_0, \tag{4}$$

over the interval $0 \leqslant t < T$. Then we would calculate $q_1 = q(T)$ and $i_1 = q'(T) = i(T)$ from the solution. Finally we would solve the initial value problem

$$Lq'' + Rq' + \frac{1}{C}q = 0, \qquad q(T) = q_1, \quad i(T) = i_1, \tag{5}$$

using the new forcing function $E(t) \equiv 0$ for $t \geqslant T$. Note that this procedure involves solving *two* initial value problems, each corresponding to one of the two continuous pieces of the forcing function. Calculating the new initial values q_1 and i_1 serves to piece together the two solutions.

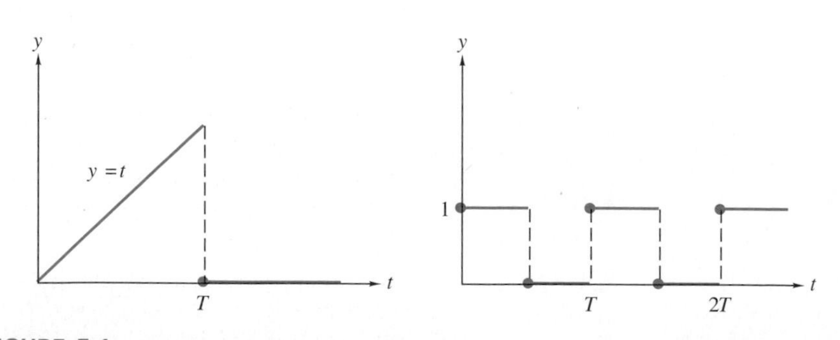

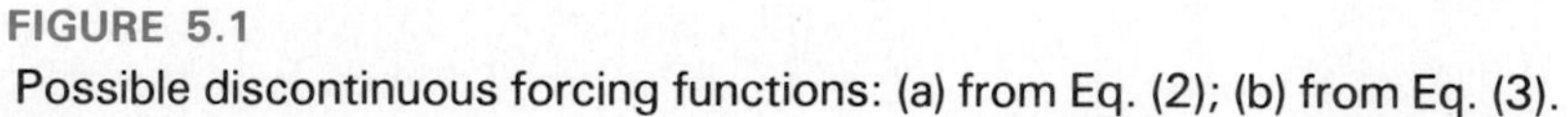
FIGURE 5.1
Possible discontinuous forcing functions: (a) from Eq. (2); (b) from Eq. (3).

In the case of forcing function (3) there would be *infinitely many* initial value problems—one corresponding to each continuous portion of the periodic forcing function. Moreover, at each point of discontinuity it would be required to calculate the new initial condition for the next continuous piece. In this chapter we introduce a new solution procedure that allows for the solution of initial value problems with a discontinuous or periodic forcing function in a single step. The method also applies to the situation when the forcing function injects an "instantaneous impulse" into the system, such as a pothole in the road or a baseball striking a bat. The method is known as the *Laplace transform technique.** We will see that it converts the initial value problem in differential equations into an algebraic equation—an astonishing result.

We will begin by discussing the transform idea in general terms. Then we will illustrate how the Laplace transform is used to solve a first-order initial value problem with a continuous forcing function (so you can observe the technique with a simple example you already know how to solve). This preliminary discussion in Section 5.1 will be followed by a deeper investigation of the Laplace transform, its fundamental properties, and its application to solving linear constant-coefficient initial value problems with discontinuous and continuous forcing functions in Sections 5.2–5.5.

5.1 INTRODUCTION TO THE LAPLACE TRANSFORM

One way to think of a real-valued function f of a single real variable is as a machine, much like a computer or calculator. You select a value of the independent variable t from the domain of the function and feed the value as input into the function "machine." As output the machine produces the unique number $f(t)$, the value of the function at t. This familiar idea is illustrated in Fig. 5.2.

For example, the function $f(t) = t^2$ can be called the squaring function: you input a number t and the function squares it to produce the value t^2. In this case the value of t can be any real number. Likewise, the function $f(t) = \sqrt{t}$ is the square root function: input t and $\sqrt{t}$ results as output. In this situation the input value for t can only be a nonnegative number if f is to be real valued. The trigonometric, logarithmic, and exponential keys on a calculator behave in a similar fashion. You input a value of t, and the calculator produces the corresponding unique function value when the selected func-

* Pierre Simon de Laplace (1749–1827) was a French mathematician who made many important discoveries in mathematical physics. He worked on the problem of measuring heat, as well as the wave propagation of sound, the tension in the surface layer of water, and celestial mechanics. His last words were reported to be, "What we know is very slight; what we don't know is immense."

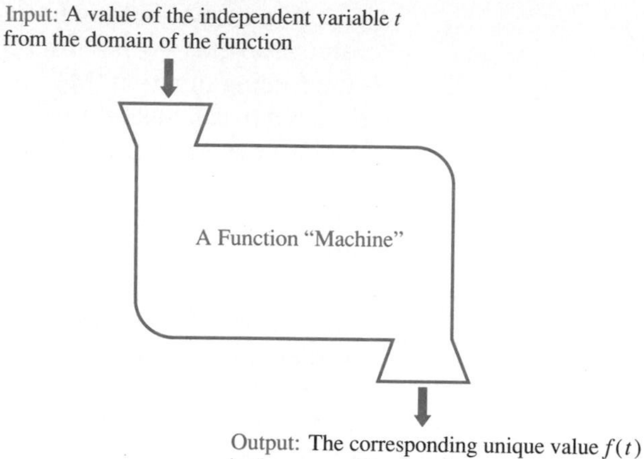

FIGURE 5.2
The function $f(t)$ is a machine with input t that outputs the numerical value $f(t)$.

tion key is pressed or activated. You must be careful to use a value of t appropriate to the domain of the function in order for the calculator to produce a correct output. Otherwise most calculators will signal that you have made an error.

Laplace Transform as a Machine

In an analogous way we are going to create a machine called the **Laplace transform.** We are not going to describe now exactly how the machine operates; we will define the transform in Section 5.2. For the moment think of the machine as a "black box" like a new computer you have not used before. The properties of the Laplace transform presented in this section will be developed fully in the next several sections.

The Laplace transform machine differs from the function machine shown in Fig. 5.2 in that this new machine receives *a function* as input instead of a number. As output it produces another function. These two functions—the input and the output—must have domains (*every* function has a domain). We refer to the domain of the input function as the ***t* domain,** and to the domain of the output function as the ***s* domain.** That is, if you input the function $f(t)$ into the Laplace transform machine, it produces the output function $F(s)$. Our notation will denote the input functions using lowercase letters such as f, g, and y, and the corresponding output functions by such uppercase letters as F, G, and Y. Thus we write $f(t) \to F(s)$, for instance, to signify that the function $f(t)$ is transformed into the function $F(s)$. This idea is illustrated in Fig. 5.3.

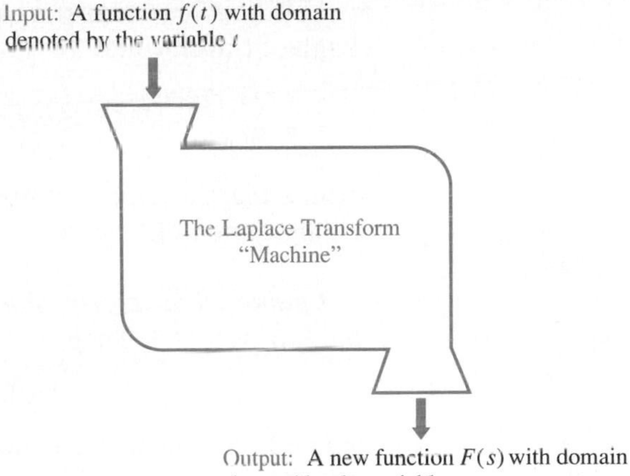

FIGURE 5.3
Laplace transform as a machine that operates on functions rather than numbers.

We are going to state several basic properties of the Laplace transform and tell you what it does to some familiar elementary functions. These results will be validated in Section 5.2 when we will define the Laplace transform. The notation

$$\mathscr{L}\{f(t)\} = F(s) \tag{1}$$

means that the Laplace transform of the function $f(t)$ is the function $F(s)$.

We are going to show you how the Laplace transform works to solve the first-order initial value problem

$$y' - 4y = e^{3t}, \quad y(0) = 1.$$

This elementary example will help you understand the transform process. Of course you can already solve this problem using variation of parameters or the integrating factor method. Nevertheless, this example will illustrate the principles of the Laplace transform in a very clear way. In order to use the transform to solve the problem, we need to know what the transform does to a derivative y' and to the exponential function e^{3t}.

Properties of the Laplace Transform

The following properties of the Laplace transform are important in understanding its usefulness.

Linearity of the Laplace Transform If $f_1(t)$ and $f_2(t)$ are functions having Laplace transforms, and c_1 and c_2 are any constants, then

$$\mathscr{L}\{c_1 f_1(t) + c_2 f_2(t)\} = c_1 \mathscr{L}\{f_1(t)\} + c_2 \mathscr{L}\{f_2(t)\}. \tag{2}$$

Laplace Transform of an Exponential For the real constant a, the Laplace transform of the exponential function $f(t) = e^{at}$ is given by

$$\mathscr{L}\{e^{at}\} = \frac{1}{s-a}, \qquad s > a. \tag{3}$$

Notice that the Laplace transform converts an exponential function of t into an algebraic function of s.

Laplace Transform of a Derivative Let $y'(t)$ denote the derivative of the function $y(t)$. Then

$$\mathscr{L}\{y'(t)\} = sY(s) - y(0) \tag{4}$$

where $Y(s)$ is the Laplace transform $\mathscr{L}\{y(t)\}$ of the original function. Again notice that the Laplace transform converts the operation of differentiation into the algebraic operation of multiplying the transform by s and subtracting a constant.

Armed with the information in Eqs. (2–4) we can solve the first-order initial value problem stated earlier.

EXAMPLE 1 SOLVING A FIRST-ORDER INITIAL VALUE PROBLEM

Consider the initial value problem

$$y' - 4y = e^{3t} \tag{5}$$

where $y(0) = 1$. Using the Laplace transform, let us operate on both sides of Eq. (5) to obtain

$$\mathscr{L}\{y' - 4y\} = \mathscr{L}\{e^{3t}\}. \tag{6}$$

Applying the linearity of the transform to the lefthand side of Eq. (6) and exponential rule (3) to the righthand side results in

$$\mathscr{L}\{y'\} - 4\mathscr{L}\{y\} = \frac{1}{s-3}.$$

Next we use derivative rule (4) and obtain

$$sY(s) - y(0) - 4Y(s) = \frac{1}{s-3}. \tag{7}$$

Substituting the initial condition $y(0) = 1$ into the lefthand side of Eq. (7) and solving for $Y(s)$ yields

$$Y(s) = \frac{1}{(s-3)(s-4)} + \frac{1}{s-4}.$$

From partial fraction decomposition we get

$$\frac{1}{(s-3)(s-4)} = -\frac{1}{s-3} + \frac{1}{s-4}.$$

(You may wish to review partial fraction decomposition in Appendix A. The Heaviside technique discussed there is especially easy to use in this example.) Substitution of the decomposition result into $Y(s)$ gives us

$$Y(s) = -\frac{1}{s-3} + \frac{2}{s-4}. \tag{8}$$

Now $Y(s)$ is the Laplace transform of the solution function $y(t)$ we seek. The final task is to obtain $y(t)$ from $Y(s)$. That is, we must undo or *invert* the Laplace transform operation. To perform the inversion in this example, we observe that the two terms of $Y(s)$ in Eq. (8) come from exponential functions according to rule (3):

$$\mathscr{L}\{e^{3t}\} = \frac{1}{s-3} \quad \text{and} \quad \mathscr{L}\{e^{4t}\} = \frac{1}{s-4}.$$

Thus we conclude that

$$\begin{aligned}
\mathscr{L}\{y(t)\} &= Y(s) \\
&= -\frac{1}{s-3} + \frac{2}{s-4} \\
&= -\mathscr{L}\{e^{3t}\} + 2\mathscr{L}\{e^{4t}\} \\
&= \mathscr{L}\{-e^{3t} + 2e^{4t}\}
\end{aligned}$$

where the last equality follows from the linearity property of the Laplace transform. Finally, since the transforms are the same,

$$y(t) = -e^{3t} + 2e^{4t} \tag{9}$$

is the solution of the initial value problem.

To check our solution we verify Eq. (5). From solution (9)

$$y' = -3e^{3t} + 8e^{4t}$$

and

$$y' - 4y = -3e^{3t} + 8e^{4t} + 4e^{3t} - 8e^{4t} = e^{3t},$$

which agrees with Eq. (5). Also, $y(0) = -e^0 + 2e^0 = 1$, so the initial condition is satisfied.

Laplace Transforms of Some Elementary Functions

We list a few more Laplace transforms of some familiar functions so you can practice using the transform to solve first-order initial value problems in the exercises:

$$\mathcal{L}\{1\} = \frac{1}{s}, \qquad s > 0. \tag{10}$$

$$\mathcal{L}\{t^n\} = \frac{n!}{s^{n+1}}, \qquad s > 0, \quad n = 1, 2, 3, \ldots. \tag{11}$$

$$\mathcal{L}\{\sin kt\} = \frac{k}{s^2 + k^2}, \qquad s > 0. \tag{12}$$

$$\mathcal{L}\{\cos kt\} = \frac{s}{s^2 + k^2}, \qquad s > 0. \tag{13}$$

Results (10–13) are derived in Section 5.2.

Remark In using the Laplace transform technique, notice how differential Eq. (5) was converted into an *algebraic equation* for $Y(s)$ in Eq. (7). We then solved this equation algebraically for $Y(s)$ and used partial fraction decomposition to write $Y(s)$ as a sum of terms each of which is easily recognizable as the Laplace transform of some known function. Note also that the solution procedure incorporated the initial conditions. Thus the solution obtained through the Laplace transform technique automatically satisfied the initial conditions and no arbitrary constants were involved.

EXERCISES 5.1

In problems 1–16 use rules (2), (3), and (10–13) to find the Laplace transform of the given function.

1. $f(t) = 5$
2. $f(t) = -3t$
3. $f(t) = 2t - 1$
4. $f(t) = 2 - t^2$
5. $f(t) = e^{t+1}$
6. $f(t) = -4e^{-t}$
7. $f(t) = t + 2e^t$
8. $f(t) = e^{t-1} + 3t$
9. $f(t) = e^{-2t} + t^2$
10. $f(t) = e^{t/2} + \sin t$
11. $f(t) = \cos t + 2 \sin t$
12. $f(t) = t^2 - \cos 3t$
13. $f(t) = (t + 1)^2$
14. $f(t) = \sin (2t - \pi)$
15. $f(t) = \sin 3t \cos 3t$
16. $f(t) = \sin^2 t$

In problems 17–25 use the Laplace transform method to solve the first-order initial value problem.

17. $y' + 2y = 0, \quad y(0) = 1$
18. $y' + 2y = 3, \quad y(0) = 1$
19. $y' - 2y = t, \quad y(0) = 1$
20. $y' - 2y = t + 1, \quad y(0) = 1$
21. $y' + y = e^{2t} + t, \quad y(0) = -1$
22. $y' + y = \sin 2t, \quad y(0) = 0$
23. $y' + 4y = \cos 2t, \quad y(0) = 2$
24. $y' + 4y = 1 - \sin t, \quad y(0) = -1$

25. $2y' + y = e^t + 2 \sin t, \quad y(0) = 0$

26. An object is pulled on a sled by a force of 15 lb across a smooth frozen pond. Assume that there are no frictional forces but that there is a force of air resistance due to wind conditions that numerically equals 3 times the velocity of the sled. If object and sled together weigh 160 lb and the sled starts from rest, what is the velocity v of the sled at the end of 5 sec? Use the method of Laplace transforms to find v as a function of t.

27. An electrical circuit with an inductor and a resistor has a variable voltage source $E(t) = \sin 2t$ that forces a current i through the circuit. The inductance is $L = 10$ henrys, and the resistance is $R = 0.2$ ohms. Assuming that the initial current is $i(0) = 0$ amps, use the method of Laplace transforms to find i as a function of t.

28. An electrical circuit is connected with a resistance R of 3 ohms, an inductance L of 1 henry, and a variable voltage source $E(t) = 10 \sin 10t$ volts. If the current is initially 0, use the method of Laplace transforms to find i as a function of t. What is the current when $t = 3$ sec?

29. A large vat contains 500 gal of fresh water. An acid solution containing 10% sulfuric acid runs into the vat at the rate of 4 gal/min. The mixture is kept uniform by stirring and runs out of the vat at the same rate of 4 gal/min. Find the amount of acid in the vat at the end of 1 hr. Use the method of Laplace transforms.

30. A tank contains 200 gal of brine with 50 lb of salt in solution. Brine containing 1 lb/gal of salt runs into the tank at the rate of 3 gal/min. The mixture, kept uniform by stirring, runs out of the tank at the same rate of 3 gal/min. Use the method of Laplace transforms to find the time when the concentration of salt in the tank reaches twice the original concentration.

31. Suppose you have a laser light beam in water, as indicated in Fig. 5.4. Let $y = N_r$ denote the **beam brightness** or **radiance**, which is the number of watts per meter squared per steradian. The beam starts at the source with brightness N_0. We wish to predict the brightness N_r at a distance r from the source. From experimentation it is known that the rate of decrease of N_r (by absorption and scattering out of the beam) is proportional to N_r. It is also known that the rate of increase of N_r (by scattering of ambient brightness, such as daylight) into the beam is known to be N_* per unit path length, where N_* is some function of r.

a) Justify the model

$$\frac{dN_r}{dr} + \alpha N_r = N_*.$$

This equation is the **transfer of radiance** (**brightness**) in a general medium that scatters and absorbs light. In many applications α is itself a function of the distance r.

b) Solve the differential equation by the Laplace transform method if $\alpha = 2$, $N_* = 1 - e^{-r}$, and $N_0 = 25$.

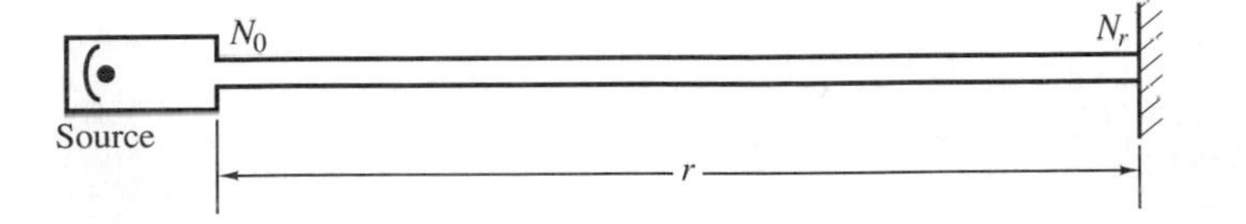

FIGURE 5.4

5.2 DEFINITION AND BASIC PROPERTIES OF THE LAPLACE TRANSFORM

In Section 5.1 we considered the Laplace transform as a black-box machine and explored how it could be used to solve first-order initial value problems. We are interested in solving higher-order initial value problems as well, especially those that have discontinuous forcing functions. If we are going to use the Laplace transform productively in our solution efforts, it is necessary to investigate what the transform actually does and how it behaves.

DEFINITION 5.1

The **Laplace transform** of the function $f(t)$, $0 \leq t < \infty$ is the function $F(s) = \mathcal{L}\{f(t)\}$ defined by

$$F(s) = \int_0^\infty e^{-st} f(t)\, dt = \lim_{b \to \infty} \int_0^b e^{-st} f(t)\, dt \tag{1}$$

provided the limit exists.

Let us use the definition to calculate the Laplace transform of some elementary functions.

EXAMPLE 1 Find the Laplace transform of $f(t) = e^{2t}$.

Solution. We apply Definition 5.1 to the function in question:

$$\begin{aligned}
\mathcal{L}\{e^{2t}\} &= \lim_{b \to \infty} \int_0^b e^{-st} e^{2t}\, dt \\
&= \lim_{b \to \infty} \int_0^b e^{(2-s)t}\, dt \\
&= \lim_{b \to \infty} \frac{1}{2-s} [e^{(2-s)t}]_0^b, \qquad s \neq 2 \\
&= \lim_{b \to \infty} \frac{1}{2-s} (e^{(2-s)b} - 1).
\end{aligned}$$

If $s > 2$, then the exponent $(2 - s)b$ is negative and $e^{(2-s)b} \to 0$ as $b \to \infty$. Therefore

$$\mathcal{L}\{e^{2t}\} = \frac{1}{s-2}, \qquad s > 2. \tag{2}$$

If $s < 2$, then the integral does not converge and the Laplace transform is not defined.

EXAMPLE 2 Find the Laplace transform of $f(t) = k$ for any constant k.

Solution. Definition 5.1 tells us that

$$\begin{aligned}\mathscr{L}\{k\} &= \lim_{b\to\infty} \int_0^b e^{-st}k\,dt \\ &= \lim_{b\to\infty} -\frac{k}{s}\,[e^{-st}]_0^b, \qquad s \neq 0 \\ &= \lim_{b\to\infty} -\frac{k}{s}\,(e^{-sb} - 1) \\ &= -\frac{k}{s}\,(0 - 1), \qquad s > 0.\end{aligned}$$

Therefore

$$\mathscr{L}\{k\} = \frac{k}{s}, \qquad s > 0. \tag{3}$$

If $s < 0$, then the integral does not converge and the Laplace transform is not defined.

EXAMPLE 3 Find the Laplace transform of $f(t) = t$.

Solution. From Definition 5.1 we have

$$\mathscr{L}\{t\} = \lim_{b\to\infty} \int_0^b e^{-st}t\,dt.$$

Integration by parts gives us

$$\begin{aligned}\mathscr{L}\{t\} &= \lim_{b\to\infty} \left[-\frac{t}{s}\,e^{-st} - \frac{1}{s^2}\,e^{-st}\right]_0^b, \qquad s \neq 0 \\ &= \lim_{b\to\infty} \left(-\frac{b}{s}\,e^{-sb} - \frac{1}{s^2}\,e^{-sb} + 0 + \frac{1}{s^2}\right).\end{aligned}$$

The limit

$$\lim_{b\to\infty} \frac{b}{s}\,e^{-sb} = \lim_{b\to\infty} \frac{b}{se^{sb}}$$

is an indeterminate (∞/∞) form. According to l'Hôpital's rule we calculate the limit of the derivative of the numerator divided by the derivative of the denominator (differentiating with respect to b):

$$\lim_{b\to\infty} \frac{b}{se^{sb}} = \lim_{b\to\infty} \frac{1}{s^2 e^{sb}} = 0, \qquad s > 0.$$

Thus

$$\mathscr{L}\{t\} = -0 - 0 + 0 + \frac{1}{s^2}$$

or

$$\mathscr{L}\{t\} = \frac{1}{s^2}, \qquad s > 0. \tag{4}$$

If $s < 0$, then the integral does not converge and the Laplace transform is not defined.

EXAMPLE 4 Find the Laplace transform of $f(t) = \sin 3t$.

Solution. Using Definition 5.1 we find that

$$\mathscr{L}\{\sin 3t\} = \lim_{b\to\infty} \int_0^b e^{-st} \sin 3t \, dt.$$

Integration by parts gives us

$$\begin{aligned}
\mathscr{L}\{\sin 3t\} &= \lim_{b\to\infty}\left[\left(-\frac{1}{s} e^{-st} \sin 3t - \frac{3}{s^2} e^{-st} \cos 3t\right)\Bigg|_0^b - \frac{9}{s^2}\int_0^b e^{-st} \sin 3t \, dt\right], \qquad s \neq 0 \\
&= \lim_{b\to\infty}\left(-\frac{1}{s} e^{-sb} \sin 3b - \frac{3}{s^2} e^{-sb} \cos 3b + \frac{3}{s^2} - \frac{9}{s^2}\int_0^b e^{-st} \sin 3t \, dt\right) \\
&= \frac{3}{s^2} - \frac{9}{s^2} \lim_{b\to\infty} \int_0^b e^{-st} \sin 3t \, dt, \qquad s > 0.
\end{aligned}$$

Combining the integral on the righthand side with the lefthand side gives us

$$\left(1 + \frac{9}{s^2}\right) \lim_{b\to\infty} \int_0^b e^{-st} \sin 3t \, dt = \frac{3}{s^2}.$$

After dividing through by $1 + \frac{9}{s^2}$ we find that

$$\mathscr{L}\{\sin 3t\} = \lim_{b\to\infty} \int_0^b e^{-st} \sin 3t \, dt$$
$$= \frac{\frac{3}{s^2}}{1 + \frac{9}{s^2}},$$

or, simplifying algebraically, that

$$\mathscr{L}\{\sin 3t\} = \frac{3}{s^2+9}, \qquad s > 0. \tag{5}$$

EXAMPLE 5 Find $\mathscr{L}\{te^{-3t}\}$.

Solution. We apply Definition 5.1:

$$\mathscr{L}\{te^{-3t}\} = \int_0^\infty e^{-st} te^{-3t} \, dt$$
$$= \int_0^\infty te^{-(s+3)t} \, dt.$$

Integration by parts gives us

$$\mathscr{L}\{te^{-3t}\} = \left[-\frac{t}{s+3} e^{-(s+3)t} - \frac{1}{(s+3)^2} e^{-(s+3)t} \right]_0^\infty$$
$$= \lim_{b\to\infty} \left[-\frac{b}{s+3} - \frac{1}{(s+3)^2} \right] e^{-(s+3)b} + \frac{1}{(s+3)^2}$$
$$= \frac{1}{(s+3)^2}, \qquad s > -3.$$

Linearity of the Laplace Transform

If a and b are real constants, and $f(t)$ and $g(t)$ are functions that have Laplace transforms, then

$$\int_0^\infty [af(t) + bg(t)]e^{-st} \, dt = a \int_0^\infty f(t)e^{-st} \, dt + b \int_0^\infty g(t)e^{-st} \, dt$$

whenever both integrals converge. Therefore whenever $f(t)$ and $g(t)$ have Laplace transforms,

$$\mathscr{L}\{af(t) + bg(t)\} = a\mathscr{L}\{f(t)\} + b\mathscr{L}\{g(t)\}. \tag{6}$$

Equation (6) states that the Laplace transform is a *linear operator* on the set of functions possessing Laplace transforms.

EXAMPLE 6 Find the Laplace transform of the function

$$f(t) = 3 - 2 \sin 3t + 7e^{2t}.$$

Solution. From the linearity of the Laplace transform we have

$$\begin{aligned}\mathscr{L}\{3 - 2 \sin 3t + 7e^{2t}\} &= \mathscr{L}\{3\} - 2\mathscr{L}\{\sin 3t\} + 7\mathscr{L}\{e^{2t}\} \\ &= \frac{3}{s} - 2\frac{3}{s^2+9} + 7\frac{1}{s-2} \\ &= \frac{10s^3 - 12s^2 + 102s - 54}{s(s^2+9)(s-2)}\end{aligned}$$

provided that $s > 2$. Note that the domain of the transform is defined so that *each* Laplace transform of the individual functions in the linear expression exists.

Functions of Exponential Order

The improper integral defining the Laplace transform in Eq. (1) need not converge. To understand the situation, suppose $f(t)$ is a positive function for $t > 0$. Then the integral

$$\int_0^\infty f(t)e^{-st}\, dt \tag{7}$$

can be thought of as representing the "area" under the positive curve $f(t)e^{-st}$ and above the positive t-axis. But if this area fails to be finite (so it is undefined), then integral (7) does not converge; we say it **diverges.** The difficulty in the divergent situation is that the integrand function $f(t)e^{-st}$ simply gets too large or fails to approach zero fast enough to give a finite area. However, if the original function $f(t)$ is sufficiently "damped," this difficulty does not occur. The following definition is useful.

DEFINITION 5.2

The function $f(t)$ is said to be **of exponential order** if there exist positive constants M and c and a number $T > 0$ such that

$$|f(t)| \leq Me^{ct}, \qquad t \geq T. \tag{8}$$

EXAMPLE 7 The functions $f(t) = t$, $f(t) = 3 \sin t$, and $f(t) = 5e^{-2t}$ are all of exponential order. To see this, note the values of M, c, and T from Eq. (8) for each equation:

$$\begin{aligned} |t| &\leq e^t, &\quad (M = 1,\quad c = 1,\quad T > 0),\\ |3 \sin t| &\leq 3e^t, &\quad (M = 3,\quad c = 1,\quad T > 0),\\ |5e^{-2t}| &\leq 5e^t, &\quad (M = 5,\quad c = 1,\quad T > 0). \end{aligned}$$

The functions are shown in Fig. 5.5. The graphs reveal that each function is bounded by an exponential in accordance with inequality (8). In each case the number T can be *any* positive constant whatsoever because each function is bounded by an exponential for all $t \geq 0$.

A function like $f(t) = e^{t^2}$ is not of exponential order. It grows faster than e^{ct} for any positive constant c as soon as the variable t satisfies $t > c > 0$ (see Fig. 5.6). However, all positive integral powers t^n are of exponential order. This last result follows from n applications of l'Hôpital's rule. For example,

$$\lim_{t\to\infty} \left|\frac{t^2}{e^t}\right| = \lim_{t\to\infty} \frac{2t}{e^t} = \lim_{t\to\infty} \frac{2}{e^t} = 0.$$

Therefore for some constant $T > 0$,

$$\frac{t^2}{e^t} < 1, \qquad t > T,$$

which implies that

$$|t^2| < e^t, \qquad t > T.$$

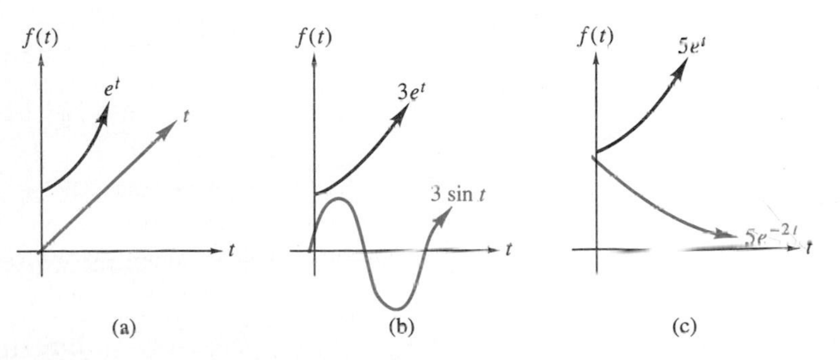

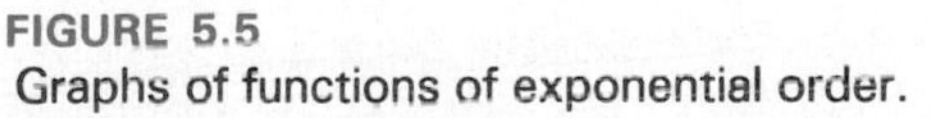

FIGURE 5.5
Graphs of functions of exponential order.

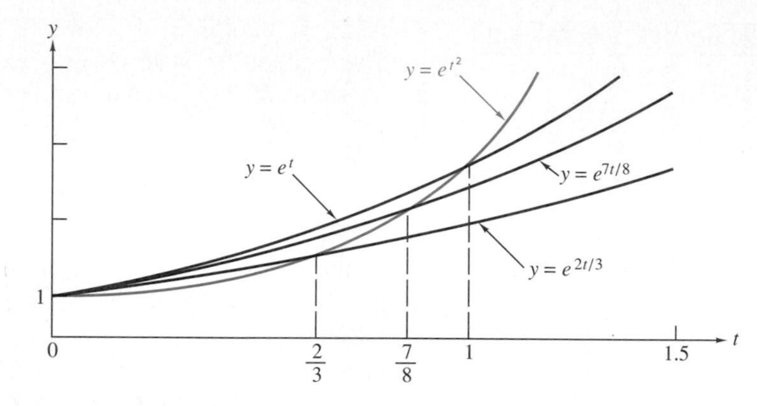

FIGURE 5.6
The function $f(t) = e^{t^2}$ is not of exponential order.

Piecewise Continuous Functions

Other difficulties for the convergence of the integral defining the Laplace transform are that the function $f(t)$ may have too many points of discontinuity, or it may become unbounded near a point of discontinuity and tend toward $+\infty$ or $-\infty$. These problems are avoided by functions satisfying the following property.

DEFINITION 5.3

The function $f(t)$ is said to be **piecewise continuous** for $t \geq 0$ if all of the following are true:

1. There are at most finitely many points $0 = t_0 < t_1 < t_2 < \cdots < t_k$ at which f has discontinuities.
2. f is continuous on each of the open intervals

$$(0, t_1), (t_1, t_2), \ldots, (t_{k-1}, t_k), (t_k, \infty).$$

3. The left- and righthand limits of f at each point t_i,

$$\lim_{t \to t_{i+}} f(t) \quad \text{and} \quad \lim_{t \to t_{i-}} f(t),$$

are finite.

Figure 5.7 shows a typical piecewise continuous function. In contrast, the function shown in Fig. 5.8 is not piecewise continuous because

$$\lim_{t \to t_{2+}} f(t) = +\infty.$$

The following result, given without proof, states sufficient conditions to guarantee the existence of the Laplace transform of a function.

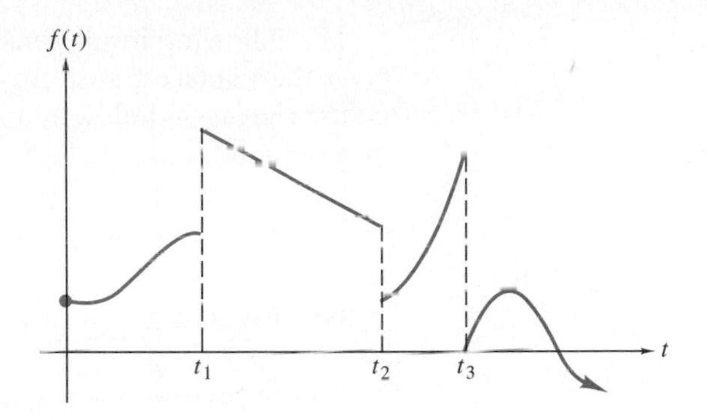

FIGURE 5.7

A typical piecewise continuous function.

> **THEOREM 5.1**
>
> If $f(t)$ is piecewise continuous for $t \geqslant 0$ and of exponential order for some positive constant c and for $t \geqslant T$, then $\mathscr{L}\{f(t)\}$ exists for $s > c$.

We remark that the conditions of the theorem are sufficient but not necessary for the existence of the Laplace transform. The function $f(t) = t^{-1/2}$ is not piecewise continuous over the domain $t \geqslant 0$ but does have a Laplace transform given in terms of the **Gamma function*** (see Exercises 5.2, problem 32).

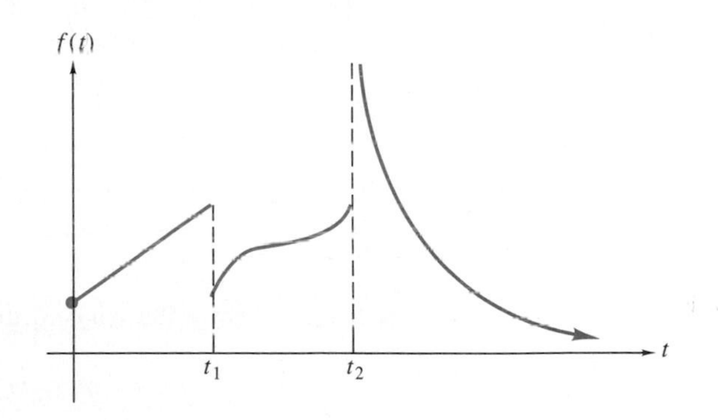

FIGURE 5.8
Function that fails to be piecewise continuous because it becomes unbounded as t approaches t_2 from the right.

* The Gamma function is defined by the integral

$$\Gamma(x) = \int_0^\infty t^{x-1}e^{-t}\,dt, \qquad x > 0.$$

The following result generalizes some of our preceding examples and gives the Laplace transform for a number of familiar functions from calculus. The proofs follow by applying Definition 5.1 of the Laplace transform, as we did in Examples 1–6.

THEOREM 5.2

a) $\mathscr{L}\{k\} = \dfrac{k}{s}, \qquad s > 0.$

b) $\mathscr{L}\{e^{at}\} = \dfrac{1}{s-a}, \qquad s > a.$

c) $\mathscr{L}\{t^n\} = \dfrac{n!}{s^{n+1}}, \qquad s > 0, \qquad n = 1, 2, 3, \ldots .$

d) $\mathscr{L}\{\sin kt\} = \dfrac{k}{s^2 + k^2}, \qquad s > 0.$

e) $\mathscr{L}\{\cos kt\} = \dfrac{s}{s^2 + k^2}, \qquad s > 0.$

f) $\mathscr{L}\{\sinh kt\} = \dfrac{k}{s^2 - k^2}, \qquad s > k.$

g) $\mathscr{L}\{\cosh kt\} = \dfrac{s}{s^2 - k^2}, \qquad s > k.$

The next example illustrates how the use of algebraic or trigonometric identities may simplify the transformation process. The s domain is usually not needed explicitly in applications, so we forgo stating it unless necessary.

EXAMPLE 8 Find $\mathscr{L}\{\sin t \cos t\}$.

Solution. Using the trigonometric identity $\sin 2t = 2 \sin t \cos t$, we have

$$\begin{aligned}\mathscr{L}\{\sin t \cos t\} &= \mathscr{L}\left\{\frac{1}{2}\sin 2t\right\}\\ &= \frac{1}{2}\left(\frac{2}{s^2+4}\right)\\ &= \frac{1}{s^2+4}.\end{aligned}$$

First Translation Theorem

We have seen that evaluating the integral defining the Laplace transform often involves integration by parts. It would be tedious, for example, to find $\mathscr{L}\{e^{-3t}\sin t \cos t\}$ directly. The following result, sometimes called the **first translation theorem**, presents a useful labor-saving device.

> **THEOREM 5.3**
>
> If a is any real number, then
>
> $$\mathscr{L}\{e^{at}f(t)\} = F(s-a)$$
>
> where $F(s) = \mathscr{L}\{f(t)\}$.

Proof. The proof follows directly from Definition 5.1.

$$\begin{aligned}\mathscr{L}\{e^{at}f(t)\} &= \int_0^\infty e^{-st}e^{at}f(t)\,dt \\ &= \int_0^\infty e^{-(s-a)t}f(t)\,dt \\ &= F(s-a)\end{aligned}$$

Applying Theorem 5.3 we can easily evaluate the Laplace transform of the function $e^{at}f(t)$ by finding the transform of the function $f(t)$ and then translating, or shifting, the result $F(s)$ to $F(s-a)$. For notational convenience in the examples to follow, we employ the symbolism

$$\mathscr{L}\{e^{at}f(t)\} = \mathscr{L}\{f(t)\}|_{s\to s-a}. \tag{9}$$

EXAMPLE 9 Find $\mathscr{L}\{e^{-2t}t^3\}$.

Solution. Using Theorem 5.3 we get

$$\begin{aligned}\mathscr{L}\{e^{-2t}t^3\} &= \mathscr{L}\{t^3\}|_{s\to s-(-2)} \\ &= \frac{3!}{s^4}\bigg|_{s\to s+2} \\ &= \frac{3!}{(s+2)^4}.\end{aligned}$$

EXAMPLE 10 Find $\mathscr{L}\{e^{-3t}\sin t \cos t\}$.

Solution. Theorem 5.3 tells us that

$$\begin{aligned}\mathscr{L}\{e^{-3t}\sin t\cos t\} &= \mathscr{L}\{\sin t\cos t\}|_{s\to s+3}\\ &= \frac{1}{s^2+4}\bigg|_{s\to s+3} \qquad \text{(from Example 8)}\\ &= \frac{1}{s^2+6s+13}.\end{aligned}$$

EXERCISES 5.2

In problems 1–13 use Definition 5.1 to find the Laplace transform of the given function.

1. $f(t) = e^{-4t}$
2. $f(t) = \cos 2t$
3. $f(t) = \sinh 3t$
4. $f(t) = t^2$
5. $f(t) = t^4$
6. $f(t) = te^{3t}$
7. $f(t) = t^2e^{-t}$
8. $f(t) = e^t \sin t$
9. $f(t) = e^{-t}\cos t$
10. $f(t) = \cosh kt$
11. $f(t) = \begin{cases} -1, & 0 \le t < 2 \\ 1, & t \ge 2 \end{cases}$
12. $f(t) = \begin{cases} t, & 0 \le t < 1 \\ -1, & t \ge 1 \end{cases}$
13. $f(t) = \begin{cases} \cos t, & 0 \le t < \pi \\ t, & t \ge \pi \end{cases}$

In problems 14–29, use Theorem 5.2 to find the Laplace transform of the given function.

14. $f(t) = 1 - e^t$
15. $f(t) = 2 + 3t - t^2$
16. $f(t) = (1+t)^2$
17. $f(t) = (1+t)(3-t)$
18. $f(t) = (1+e^t)^2$
19. $f(t) = (1+e^t)(1-e^{-t})$
20. $f(t) = 2e^t - \sin 3t$
21. $f(t) = 5t^2 + \cos 2t$
22. $f(t) = e^{-t} + 3\cos 4t$
23. $f(t) = 1 - 2\cosh t$
24. $f(t) = 2\sinh 3t - t$
25. $f(t) = 3e^{-2t} + 2t^3 - 5$
26. $f(t) = \sin 2t - 3\cos t$
27. $f(t) = e^{5t-1} + 4\sin t$
28. $f(t) = \cos^2 t$
29. $f(t) = \sin^2 t$

30. Show directly that the function $f(t) = t^3$ is of exponential order.

31. Show directly that the function $f(t) = t^n$ is of exponential order for any positive integer n.

32. a) Show that the integral

$$\Gamma(x) = \int_0^\infty t^{x-1}e^{-t}\,dt$$

converges if $x > 0$.

b) Using integration by parts, show that $\Gamma(x+1) = x\Gamma(x)$ for $x > 0$.

c) Show that

$$\mathscr{L}\{t^\alpha\} = \frac{\Gamma(\alpha+1)}{s^{\alpha+1}}, \quad \alpha > -1.$$

d) Find $\mathscr{L}\{t^{-1/2}\}$. *Note:* $\Gamma(\tfrac{1}{2}) = \sqrt{\pi}$.

e) Apply part (b) to conclude that $\Gamma(n+1) = n!$ when n is a nonnegative integer.

33. Using Definition 5.1 prove Theorem 5.1.

In problems 34–43 use the first translation theorem to find the indicated Laplace transform.

34. $\mathscr{L}\{e^{2t}\sin t\}$
35. $\mathscr{L}\{e^{3t}\cos t\}$
36. $\mathscr{L}\{e^{-t}\sinh 2t\}$
37. $\mathscr{L}\{e^{-2t}\cosh t\}$
38. $\mathscr{L}\{t^2e^{3t}\}$
39. $\mathscr{L}\{t^5e^{-t}\}$
40. $\mathscr{L}\{e^t(1+t)\}$
41. $\mathscr{L}\{e^{-t}(1+t)^2\}$
42. $\mathscr{L}\{e^{2t}(1-\cos t)\}$
43. $\mathscr{L}\{e^{-t}\sin^2 t\}$

5.3

INVERSE LAPLACE TRANSFORM

In Section 5.1 the Laplace transform was used to solve first-order differential equations. To apply that procedure successfully, the last step required us to find the function $y(t)$ whose Laplace transform $Y(s)$ had been derived through the transformation process. Stated more succinctly, given $Y(s)$ it is required to find the function $y(t)$ corresponding to this transform. This idea prompts the following definition.

DEFINITION 5.4

> If $\mathscr{L}\{f(t)\} = F(s)$, then $f(t)$ is the **inverse Laplace transform** of $F(s)$ and is written
>
> $$f(t) = \mathscr{L}^{-1}\{F(s)\}.$$

We can think of using the inverse transform as running the Laplace transform machine in reverse: we *input* the function $F(s)$ and obtain as *output* the original function $f(t)$. This idea is expressed in Fig. 5.9. When given a function $F(s)$ and required to find a corresponding function $f(t)$, we first need to answer two questions:

1. Does a function $f(t)$ exist for the given function $F(s)$?
2. If $f(t)$ does exist, is it unique?

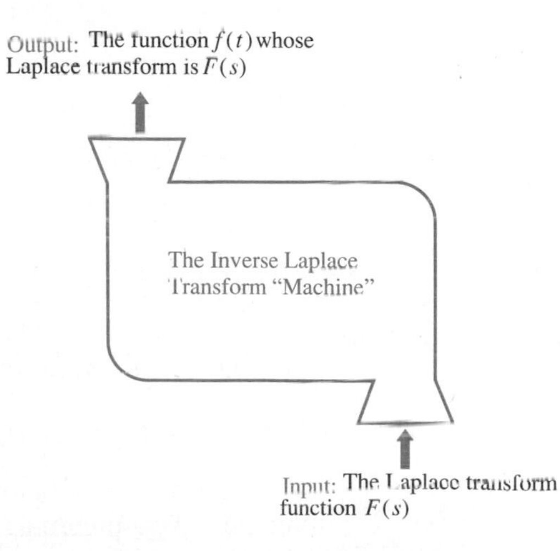

FIGURE 5.9

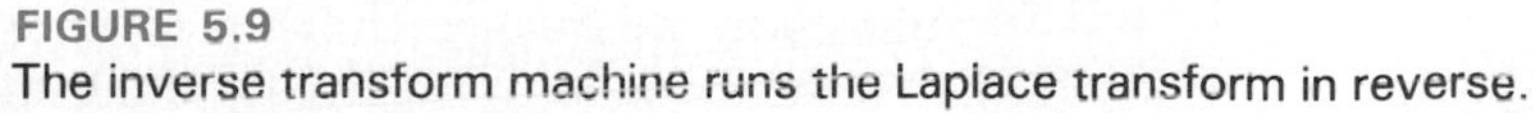
The inverse transform machine runs the Laplace transform in reverse.

At the end of this section we will establish that there can be *only one* continuous function $f(t)$ for which $\mathscr{L}\{f(t)\} = F(s)$. The inverse transform then produces this unique function when $F(s)$ is known: $\mathscr{L}^{-1}\{F(s)\} = f(t)$. This conclusion holds only for continuous functions. In general, a given transform may correspond to many piecewise continuous functions.

Although it is beyond the scope of this text to pursue in depth the first question posed above, the following result establishes a property that must be possessed by $F(s)$ if it is to be the Laplace transform of some piecewise continuous function of exponential order.

THEOREM 5.4

If $f(t)$ is piecewise continuous for $t \geq 0$ and of exponential order for some positive constants M and c and $t > T$, then

$$\lim_{s \to \infty} F(s) = \lim_{s \to \infty} \mathscr{L}\{f(t)\} = 0.$$

Proof. We apply the definitions of the Laplace transform and exponential order:

$$\begin{aligned}
0 \leq |\mathscr{L}\{f(t)\}| &\leq \int_0^\infty e^{-st}|f(t)|dt \\
&\leq M \int_0^\infty e^{-st}e^{ct}\, dt \\
&= -M\left[\frac{e^{-(s-c)t}}{s-c}\right]_0^\infty \\
&= \lim_{b \to \infty}\left(\frac{M}{c-s}e^{-(s-c)b}\right) + \frac{M}{s-c} \\
&= \frac{M}{s-c}, \qquad s > c.
\end{aligned}$$

Thus as $s \to \infty$ it follows that $M/(s-c) \to 0$ and $|\mathscr{L}\{f(t)\}| \to 0$.

As a result of Theorem 5.4 we can determine that certain functions, such as $F(s) = s$, cannot be the Laplace transform of any piecewise continuous function of exponential order. When a function $F(s)$ does possess an inverse transform, we can sometimes read a Laplace transform table in reverse to find the function $f(t)$. This idea is illustrated by the next example.

EXAMPLE 1 Evaluate $\mathscr{L}^{-1}\left\{\frac{1}{s^4}\right\}$.

Solution. We multiply and divide by 3! and read Theorem 5.2(c) in reverse:

$$\mathscr{L}^{-1}\left\{\frac{1}{s^4}\right\} = \left(\frac{1}{3!}\right)\mathscr{L}^{-1}\left\{\frac{3!}{s^4}\right\} = \frac{t^3}{3!}.$$

The following property of the inverse transform is very useful when finding the inverse transform through the use of tables.

TABLE 5.1

Laplace Transforms of Some Basic Functions

$f(t)$	$\mathscr{L}\{f(t)\} = F(s)$
1	$\frac{1}{s}$
e^{at}	$\frac{1}{s-a}$
t^n	$\frac{n!}{s^{n+1}}$
$\sin kt$	$\frac{k}{s^2+k^2}$
$\cos kt$	$\frac{s}{s^2+k^2}$
$\sinh kt$	$\frac{k}{s^2-k^2}$
$\cosh kt$	$\frac{s}{s^2-k^2}$

Linearity of the Inverse Laplace Transform

The inverse Laplace transform is also linear. That is, for constants α and β,

$$\mathscr{L}^{-1}\{\alpha F(s) + \beta G(s)\} = \alpha\mathscr{L}^{-1}\{F(s)\} + \beta\mathscr{L}^{-1}\{G(s)\}$$

where F and G are themselves transforms of some functions f and g.

The inverse Laplace transform of some functions is itself defined by an integral, but the integral involves complex variables.* Evaluation of such an integral is beyond the scope of this course, so we will use our knowledge of the Laplace transforms of elementary functions to find inverse transforms. It is convenient to collect the transforms of basic functions into a table for quick reference (see Table 5.1). To find the Laplace transform of a function given in the left column, read the table from left to right:

$$f(t) \xrightarrow{\mathscr{L}} F(s)$$

The inverse transform reverses the procedure. Start with a function of s in the righthand column and read the table from right to left:

$$f(t) \xleftarrow{\mathscr{L}^{-1}} F(s)$$

The next several examples illustrate finding the inverse transform by using Table 5.1.

EXAMPLE 2 Evaluate $\mathscr{L}^{-1}\left\{\frac{3s-1}{s(s-1)}\right\}$.

FACTORIAL

$5! = 5 \cdot 4 \cdot 3 \cdot 2 \cdot 1$

Solution. Using partial fraction decomposition, we write

$$\frac{3s-1}{s(s-1)} = \frac{A}{s} + \frac{B}{s-1}.$$

* Further information may be obtained in *Advanced Engineering Mathematics* by Michael D. Greenberg (Englewood Cliffs, N.J.: Prentice-Hall, 1988), pp. 534–537.

Then $A = -1/-1 = 1$ and $B = 2/1 = 2$ by the Heaviside technique (see Appendix A). Thus

$$\mathscr{L}^{-1}\left\{\frac{3s-1}{s(s-1)}\right\} = \mathscr{L}^{-1}\left\{\frac{1}{s}\right\} + \mathscr{L}^{-1}\left\{\frac{2}{s-1}\right\}.$$

From Table 5.1 we see that

$$\mathscr{L}^{-1}\left\{\frac{3s-1}{s(s-1)}\right\} = 1 + 2e^t.$$

Partial fraction decomposition plays a major role in evaluating inverse Laplace transforms, as Examples 3–6 illustrate.

EXAMPLE 3 Evaluate $\mathscr{L}^{-1}\left\{\dfrac{1}{(s+1)(s+3)(2s-1)}\right\}$.

Solution. From partial fraction decomposition, we assume that

$$\frac{1}{(s+1)(s+3)(2s-1)} = \frac{A}{s+1} + \frac{B}{s+3} + \frac{C}{2s-1}.$$

By the Heaviside technique (which applies when distinct linear factors appear in the denominator), we have

$$A = \frac{1}{2(-3)} = -\frac{1}{6}$$

$$B = \frac{1}{-2(-7)} = \frac{1}{14}$$

$$C = \frac{1}{\tfrac{3}{2}(\tfrac{7}{2})} = \frac{4}{21}.$$

Thus

$$\begin{aligned}
\mathscr{L}^{-1}\left\{\frac{1}{(s+1)(s+3)(2s-1)}\right\} &= \mathscr{L}^{-1}\left\{-\frac{1}{6(s+1)} + \frac{1}{14(s+3)} + \frac{4}{21(2s-1)}\right\} \\
&= -\frac{1}{6}\mathscr{L}^{-1}\left\{\frac{1}{s+1}\right\} + \frac{1}{14}\mathscr{L}^{-1}\left\{\frac{1}{s+3}\right\} \\
&\quad + \frac{4}{21}\mathscr{L}^{-1}\left\{\frac{1}{2s-1}\right\} \\
&= -\frac{1}{6}e^{-t} + \frac{1}{14}e^{-3t} + \frac{2}{21}\mathscr{L}^{-1}\left\{\frac{1}{s-\tfrac{1}{2}}\right\} \\
&= -\frac{1}{6}e^{-t} + \frac{1}{14}e^{-3t} + \frac{2}{21}e^{t/2}.
\end{aligned}$$

EXAMPLE 4 Evaluate $\mathscr{L}^{-1}\left\{\dfrac{s-4}{(s+1)(s^2+4)}\right\}$.

Solution. We write

$$\frac{s-4}{(s+1)(s^2+4)} = \frac{A}{s+1} + \frac{Bs+C}{s^2+4}.$$

Then

$$\begin{aligned} s-4 &= A(s^2+4) + (Bs+C)(s+1) \\ &= (A+B)s^2 + (B+C)s + (4A+C). \end{aligned}$$

The polynomials on both sides of this last equation are equal if and only if corresponding powers of s are equal, yielding

$$\begin{aligned} A + B \quad &= 0, \\ B + C &= 1, \\ 4A \quad + C &= -4. \end{aligned}$$

Solution of the system gives us $A = -1$, $B = 1$, and $C = 0$. Therefore

$$\begin{aligned} \mathscr{L}^{-1}\left\{\frac{s-4}{(s+1)(s^2+4)}\right\} &= \mathscr{L}^{-1}\left\{\frac{s}{s^2+4} - \frac{1}{s+1}\right\} \\ &= \mathscr{L}^{-1}\left\{\frac{s}{s^2+4}\right\} - \mathscr{L}^{-1}\left\{\frac{1}{s+1}\right\} \\ &= \cos 2t - e^{-t}. \end{aligned}$$

The first translation theorem (Theorem 5.3) can be interpreted in terms of the inverse transform. From Theorem 5.3, $\mathscr{L}\{e^{at}f(t)\} = F(s-a)$. Thus inverting the transform gives us

$$e^{at}f(t) = \mathscr{L}^{-1}\{F(s-a)\}.$$

But $f(t) = \mathscr{L}^{-1}\{F(s)\}$, so after substitution we have

$$e^{at}\mathscr{L}^{-1}\{F(s)\} = \mathscr{L}^{-1}\{F(s-a)\}$$

or, rearranging terms,

$$\mathscr{L}^{-1}\{F(s-a)\} = e^{at}\mathscr{L}^{-1}\{F(s)\}. \tag{1}$$

Examples 5–7 illustrate the use of this result.

EXAMPLE 5 Evaluate $\mathscr{L}^{-1}\left\{\frac{5}{(s-1)^3}\right\}$.

Solution. Using Eq. (1) we have

$$\begin{aligned}\mathscr{L}^{-1}\left\{\frac{5}{(s-1)^3}\right\} &= 5e^t\mathscr{L}^{-1}\left\{\frac{1}{s^3}\right\}\\ &= \frac{5e^t}{2!}\mathscr{L}^{-1}\left\{\frac{2!}{s^3}\right\}\\ &= \frac{5}{2}t^2e^t.\end{aligned}$$

EXAMPLE 6 Evaluate $\mathscr{L}^{-1}\left\{\frac{4-2s}{(s^2+1)(s-1)^2}\right\}$.

Solution. Using partial fraction decomposition results in

$$\frac{4-2s}{(s^2+1)(s-1)^2} = \frac{As+B}{s^2+1} + \frac{C}{s-1} + \frac{D}{(s-1)^2}.$$

Thus

$$\begin{aligned}4-2s &= (As+B)(s-1)^2 + C(s^2+1)(s-1) + D(s^2+1)\\ &= (A+C)s^3 + (-2A+B-C+D)s^2 + (A-2B+C)s\\ &\quad +(B-C+D).\end{aligned}$$

Equating like powers of s on both sides results in the following system of equations:

$$\begin{aligned}A \qquad + C \qquad &= 0,\\ -2A + B - C + D &= 0,\\ A - 2B + C \qquad &= -2,\\ B - C + D &= 4.\end{aligned}$$

Summing these equations gives us $2D = 2$ or $D = 1$. Substitution of the last equation of the system into the second yields

$$0 = -2A + (B - C + D) = -2A + 4.$$

Thus $A = 2$. Then from the first equation we have $C = -A = -2$. Finally, the fourth equation gives us $B = 4 + C - D = 1$. Therefore

$$\mathscr{L}^{-1}\left\{\frac{4-2s}{(s^2+1)(s-1)^2}\right\} = \mathscr{L}^{-1}\left\{\frac{2s+1}{s^2+1} - \frac{2}{s-1} + \frac{1}{(s-1)^2}\right\}$$

$$= 2\mathscr{L}^{-1}\left\{\frac{s}{s^2+1}\right\} + \mathscr{L}^{-1}\left\{\frac{1}{s^2+1}\right\}$$

$$- 2\mathscr{L}^{-1}\left\{\frac{1}{s-1}\right\} + \mathscr{L}^{-1}\left\{\frac{1}{(s-1)^2}\right\}$$

$$= 2\cos t + \sin t - 2e^t + e^t\,\mathscr{L}^{-1}\left\{\frac{1}{s^2}\right\}$$

$$= 2\cos t + \sin t + e^t(t-2).$$

EXAMPLE 7 Evaluate $\mathscr{L}^{-1}\left\{\dfrac{s}{s^2-6s+13}\right\}$.

Solution. Completing the square in the denominator gives us

$$\frac{s}{s^2-6s+13} = \frac{s}{(s-3)^2+4}$$

$$= \frac{s-3}{(s-3)^2+4} + \frac{3}{(s-3)^2+4}.$$

By Eq. (1),

$$\mathscr{L}^{-1}\left\{\frac{s}{s^2-6s+13}\right\} = \mathscr{L}^{-1}\left\{\frac{s-3}{(s-3)^2+4}\right\} + \frac{3}{2}\mathscr{L}^{-1}\left\{\frac{2}{(s-3)^2+4}\right\}$$

$$= e^{3t}\mathscr{L}^{-1}\left\{\frac{s}{s^2+4}\right\} + \frac{3}{2}e^{3t}\mathscr{L}^{-1}\left\{\frac{2}{s^2+4}\right\}$$

$$= e^{3t}\cos 2t + \frac{3}{2}e^{3t}\sin 2t.$$

Uniqueness of the Inverse Transform

We now address the second of the two questions posed in our opening discussion: If $f(t)$ does exist for a given function $F(s)$, is it unique? In general, different piecewise continuous functions can produce the same Laplace transform. To see this, suppose $f(t)$ and $g(t)$ are two piecewise continuous functions for $t > 0$ such that $f(t) = g(t)$ at all points of continuity for both functions, but $f(t) \neq g(t)$ for at least one point of discontinuity. Then we know from calculus that, for every finite interval $[0, b]$ where $b > 0$, the following equality holds:

$$\int_0^b e^{-st}f(t)\,dt = \int_0^b e^{-st}g(t)\,dt.$$

It follows that

$$\int_0^\infty e^{-st}f(t)\,dt = \int_0^\infty e^{-st}g(t)\,dt.$$

Therefore $F(s) = G(s)$. However, distinct *continuous* functions do produce distinct Laplace transforms. This result is stated (without proof) in the following theorem.*

THEOREM 5.5

Suppose f and g are continuous functions of exponential order for $t \geqslant 0$. If $F(s) = G(s)$, then

$$f(t) \equiv g(t), \qquad t \geqslant 0.$$

In other words, given the function $F(s)$, there is *only one* continuous function $f(t)$ for which $\mathscr{L}\{f(t)\} = F(s)$; that is, the inverse transform $f(t) = \mathscr{L}^{-1}\{F(s)\}$ is unique. As an example, the only continuous function of exponential order whose Laplace transform is zero is the identically zero function $f(t) \equiv 0$.

In Section 5.4 we will employ the Laplace transform to solve linear first- and second-order differential equations with constant coefficients and initial conditions.

EXERCISES 5.3

For problems 1–12 use Table 5.1 together with the linearity property to find the inverse transform.

1. $\mathscr{L}^{-1}\left\{\dfrac{2}{s+4}\right\}$
2. $\mathscr{L}^{-1}\left\{\dfrac{3}{s^5}\right\}$
3. $\mathscr{L}^{-1}\left\{\dfrac{3s}{s^2+2}\right\}$
4. $\mathscr{L}^{-1}\left\{\dfrac{5}{s^2+3}\right\}$
5. $\mathscr{L}^{-1}\left\{\dfrac{1}{s^2+4} - \dfrac{1}{3s^2}\right\}$
6. $\mathscr{L}^{-1}\left\{\dfrac{s}{s^2-2} - \dfrac{2}{s-4}\right\}$
7. $\mathscr{L}^{-1}\left\{\dfrac{2}{s^2-1} - \dfrac{3s}{s^2+1}\right\}$
8. $\mathscr{L}^{-1}\left\{\dfrac{1}{s-2} + \dfrac{3}{s+2} - \dfrac{4}{s^3}\right\}$
9. $\mathscr{L}^{-1}\left\{\dfrac{1}{5s+2} - \dfrac{1}{2s^3}\right\}$
10. $\mathscr{L}^{-1}\left\{\dfrac{2}{s^4-1}\right\}$
11. $\mathscr{L}^{-1}\left\{-\dfrac{2}{s^4+2s^2}\right\}$
12. $\mathscr{L}^{-1}\left\{\dfrac{s-1}{(s+2)(s-2)}\right\}$

* For a proof see D. V. Widder, *The Laplace Transform* (Princeton, N.J.: Princeton University Press, 1946). Theorem 5.5 is a special case of a slightly more general result proved by M. Lerch (1860–1922).

In problems 13–26 find the inverse transform.

13. $\mathscr{L}^{-1}\left\{\dfrac{1}{(s+2)(s-4)(2s+1)}\right\}$

14. $\mathscr{L}^{-1}\left\{\dfrac{4}{s(3-2s)(s-1)}\right\}$

15. $\mathscr{L}^{-1}\left\{\dfrac{4-s^2}{s(s^2+2)}\right\}$

16. $\mathscr{L}^{-1}\left\{\dfrac{5+s-3s^2}{(s+2)(s^2-1)}\right\}$

17. $\mathscr{L}^{-1}\left\{\dfrac{s^2-8s-2}{(s-4)(s^2+2)}\right\}$

18. $\mathscr{L}^{-1}\left\{\dfrac{s(3s+4)}{2(s+1)(s^2-2)}\right\}$

19. $\mathscr{L}^{-1}\left\{\dfrac{5s+4}{2s^2-s-1}\right\}$

20. $\mathscr{L}^{-1}\left\{\dfrac{6-s}{3s^2+4s-4}\right\}$

21. $\mathscr{L}^{-1}\left\{\dfrac{s+3}{6s^2+s-1}\right\}$

22. $\mathscr{L}^{-1}\left\{\dfrac{1}{(s^2+4)(s^2+9)}\right\}$

23. $\mathscr{L}^{-1}\left\{\dfrac{5-12s+4s^2}{1-3s+4s^2-12s^3}\right\}$

24. $\mathscr{L}^{-1}\left\{\dfrac{2s^2+s-4}{s^3-s^2-2s}\right\}$

25. $\mathscr{L}^{-1}\left\{\dfrac{1+s+3s^3-3s^4}{(s+1)(2s^4+s^2)}\right\}$

26. $\mathscr{L}^{-1}\left\{\dfrac{6s^5-2s^4-2s^2-s+1}{(s-1)(2s^5-s^3)}\right\}$

In problems 27–36 use the first translation theorem and Table 5.1 to find the inverse transform.

27. $\mathscr{L}^{-1}\left\{\dfrac{3}{(s+2)^2}\right\}$

28. $\mathscr{L}^{-1}\left\{\dfrac{2}{s^2-2s+5}\right\}$

29. $\mathscr{L}^{-1}\left\{\dfrac{s-1}{s^2+2s+2}\right\}$

30. $\mathscr{L}^{-1}\left\{\dfrac{s-1}{s^2-4s+8}\right\}$

31. $\mathscr{L}^{-1}\left\{\dfrac{s^2+3s+4}{s(s+2)^2}\right\}$

32. $\mathscr{L}^{-1}\left\{\dfrac{4s-2}{4s^2+4s+9}\right\}$

33. $\mathscr{L}^{-1}\left\{\dfrac{4s}{4s^2-4s-15}\right\}$

34. $\mathscr{L}^{-1}\left\{\dfrac{4s}{4s^2-4s+5}\right\}$

35. $\mathscr{L}^{-1}\left\{\dfrac{2s^3-2s^2-2s+1}{s^2(s-1)^2}\right\}$

36. $\mathscr{L}^{-1}\left\{\dfrac{3s+4}{(s+1)(s+2)^2}\right\}$

37. Show that

$$\mathscr{L}^{-1}\left\{\frac{s}{(s+a)^2+b^2}\right\} = e^{-at}\frac{(b\cos bt - a\sin bt)}{b}.$$

38. If $a > 0$, show that

$$\mathscr{L}^{-1}\{F(as)\} = \frac{1}{a}f\left(\frac{t}{a}\right)$$

where $\mathscr{L}\{f(t)\} = F(s)$.

5.4

LAPLACE TRANSFORM AND THE DERIVATIVE

Our purpose in introducing the Laplace transform is to provide a method for solving constant-coefficient linear differential equations, especially when the forcing function is piecewise continuous or periodic. In this section we will pursue the methodology already introduced in Section 5.1 for solving equations with continuous forcing functions. In Section 5.5 we will take up the problem of piecewise continuous forcing functions. To begin we need the Laplace transform of the derivative of a function to evaluate such expressions as $\mathscr{L}\{y'\}$ and $\mathscr{L}\{y''\}$.

Laplace Transform of a Derivative

Let us calculate the Laplace transform of the derivative of a function. Suppose $y = f(t)$ is continuous and of exponential order and that $dy/dt = f'(t)$ is piecewise continuous for $t \geqslant 0$. Assume that $t_1 < t_2 < \cdots < t_k$ are the points of discontinuity of f'. For any real number $b > t_k$ we have

$$\int_0^b e^{-st} f'(t)\, dt = \int_0^{t_1} e^{-st} f'(t)\, dt + \int_{t_1}^{t_2} e^{-st} f'(t)\, dt + \cdots + \int_{t_k}^{b} e^{-st} f'(t)\, dt.$$

Integration by parts of each term on the righthand side gives us

$$\int_0^b e^{-st} f'(t)\, dt = e^{-st} f(t)\Big|_0^{t_1} + e^{-st} f(t)\Big|_{t_1}^{t_2} + \cdots + e^{-st} f(t)\Big|_{t_k}^{b} + s\int_0^{t_1} e^{-st} f(t)\, dt + s\int_{t_1}^{t_2} e^{-st} f(t)\, dt + \cdots + s\int_{t_k}^{b} e^{-st} f(t)\, dt.$$

Since f is continuous, the righthand side simplifies to yield

$$\int_0^b e^{-st} f'(t)\, dt = e^{-sb} f(b) - f(0) + s\int_0^b e^{-st} f(t)\, dt.$$

Taking the limit as $b \to \infty$, we find that $e^{-sb} f(b) \to 0$ (because f is of exponential order), and the preceding equality becomes

$$\int_0^\infty e^{-st} f'(t)\, dt = s\int_0^\infty e^{-st} f(t)\, dt - f(0).$$

Let us summarize this result.

THEOREM 5.6

If $f(t)$ is continuous and of exponential order for $t \geqslant 0$, and if $f'(t)$ is piecewise continuous for $t \geqslant 0$, then

$$\mathscr{L}\{f'(t)\} = sF(s) - f(0) \tag{1}$$

where $\mathscr{L}\{f(t)\} = F(s)$.

Since the second derivative is the derivative of $f'(t)$, we can apply Eq. (1) to determine the transform of $f''(t)$:

$$\begin{aligned}\mathscr{L}\{f''(t)\} &= s\mathscr{L}\{f'(t)\} - f'(0)\\ &= s[sF(s) - f(0)] - f'(0)\end{aligned}$$

or

$$\mathscr{L}\{f''(t)\} = s^2F(s) - sf(0) - f'(0). \tag{2}$$

For the third derivative we can use Eqs. (1) and (2):

$$\begin{aligned}\mathscr{L}\{f'''(t)\} &= s\mathscr{L}\{f''(t)\} - f''(0), &&\text{since } f'''(t) = \frac{d}{dt}f''(t)\\ &= s[s^2F(s) - sf(0) - f'(0)] - f''(0), &&\text{from Eq. (2)}\end{aligned}$$

or

$$\mathscr{L}\{f'''(t)\} = s^3F(s) - s^2f(0) - sf'(0) - f''(0). \tag{3}$$

Of course, to ensure that Eq. (3) is valid, the conditions of Theorem 5.6 require that $f(t)$, $f'(t)$, and $f''(t)$ all be continuous and of exponential order for $t \geq 0$ and that $f'''(t)$ be piecewise continuous for $t \geq 0$.

The next several examples illustrate the use of Eqs. (1) and (2) in solving constant-coefficient linear second-order differential equations.

EXAMPLE 1 Solve $y'' - 5y' + 6y = 2e^{-t}$ subject to the initial conditions $y(0) = 0$ and $y'(0) = 1$.

Solution. Transforming both sides gives us

$$\mathscr{L}\{y''\} - 5\mathscr{L}\{y'\} + 6\mathscr{L}\{y\} = 2\mathscr{L}\{e^{-t}\},$$

$$s^2Y(s) - sy(0) - y'(0) - 5[sY(s) - y(0)] + 6Y(s) = \frac{2}{s+1}.$$

Using the initial conditions and simplifying results in

$$(s^2 - 5s + 6)Y(s) - 1 = \frac{2}{s+1}$$

or

$$Y(s) = \frac{2}{(s+1)(s-3)(s-2)} + \frac{1}{(s-3)(s-2)}.$$

After applying the Heaviside technique to complete the partial fraction decomposition of the righthand side, we find

$$\begin{aligned}Y(s) &= \frac{2/12}{s+1} + \frac{2/4}{s-3} + \frac{-2/3}{s-2} + \frac{1}{s-3} + \frac{-1}{s-2}\\ &= \frac{1/6}{s+1} + \frac{3/2}{s-3} - \frac{5/3}{s-2}.\end{aligned}$$

After taking inverse transforms, the solution is

$$y(t) = \frac{e^{-t}}{6} + \frac{3e^{3t}}{2} - \frac{5e^{2t}}{3}.$$

EXAMPLE 2 Solve $y'' + 4y = -5e^{-t}$ subject to the initial conditions $y(0) = 0$ and $y'(0) = 1$.

Solution. Transforming both sides gives us

$$\mathscr{L}\{y''\} + 4\mathscr{L}\{y\} = -5\mathscr{L}\{e^{-t}\},$$

$$s^2Y(s) - sy(0) - y'(0) + 4Y(s) = -\frac{5}{s+1}.$$

Using the initial conditions and simplifying results in

$$(s^2 + 4)Y(s) - 1 = -\frac{5}{s+1},$$

$$(s^2 + 4)Y(s) = \frac{s-4}{s+1},$$

or

$$Y(s) = \frac{s-4}{(s+1)(s^2+4)}.$$

We note from Example 4 in Section 5.3 that the inverse transform of each side gives the solution

$$y(t) = \cos 2t - e^{-t}.$$

EXAMPLE 3 Solve $y'' + y = 4te^t$ subject to the initial conditions $y(0) = -2$ and $y'(0) = 0$.

Solution. Transforming both sides gives us

$$\mathscr{L}\{y''\} + \mathscr{L}\{y\} = \mathscr{L}\{4te^t\},$$

$$s^2Y(s) - sy(0) - y'(0) + Y(s) = \frac{4}{(s-1)^2},$$

where the righthand side is obtained from the first translation theorem. Using the initial conditions and simplifying yields

$$(s^2 + 1)Y(s) + 2s = \frac{4}{(s-1)^2}$$

or

$$Y(s) = \frac{4}{(s^2+1)(s-1)^2} - \frac{2s}{s^2+1}.$$

We decompose the first term on the righthand side into partial fractions:

$$\frac{4}{(s^2+1)(s-1)^2} = \frac{As+B}{s^2+1} + \frac{C}{s-1} + \frac{D}{(s-1)^2}.$$

Thus

$$\begin{aligned} 4 &= (As+B)(s-1)^2 + C(s^2+1)(s-1) + D(s^2+1) \\ &= (A+C)s^3 + (-2A+B-C+D)s^2 + (A-2B+C)s \\ &\quad + (B-C+D). \end{aligned}$$

Equating like powers of s on both sides results in the system

$$\begin{aligned} A \quad\quad + C \quad\quad &= 0, \\ -2A + B - C + D &= 0, \\ A - 2B + C \quad\quad &= 0, \\ B - C + D &= 4. \end{aligned}$$

After summing the four equations we find $2D = 4$, so $D = 2$. Subtracting the second equation in the system from the fourth results in $2A = 4$ or $A = 2$. Then $C = -2$ from the first equation, and $B = 0$ from the fourth. Hence

$$\begin{aligned} Y(s) &= \left[\frac{2s}{s^2+1} - \frac{2}{s-1} + \frac{2}{(s-1)^2}\right] - \frac{2s}{s^2+1} \\ &= \frac{2}{(s-1)^2} - \frac{2}{s-1}. \end{aligned}$$

After taking inverse transforms, the solution is

$$y(t) = 2te^t - 2e^t.$$

Derivative of the Laplace Transform

The next theorem gives the derivative of the Laplace transform. This result is extremely useful in calculating transforms of functions that have the term t as a multiplicative factor.

THEOREM 5.7

If $f(t)$ is piecewise continuous for $t \geq 0$ and of exponential order for some $c > 0$, and if $F(s) = \mathscr{L}\{f(t)\}$, then

$$\frac{d}{ds}F(s) = -\mathscr{L}\{tf(t)\}, \qquad s > c.$$

Proof. In advanced calculus it is possible to show that as long as $f(t)$ is piecewise continuous and of exponential order, we may differentiate under the integral sign when finding the derivative of the Laplace transform of $f(t)$. Thus

$$\begin{aligned}\frac{d}{ds}F(s) &= \frac{d}{ds}\int_0^\infty e^{-st}f(t)\,dt \\ &= \int_0^\infty \frac{\partial}{\partial s}[e^{-st}f(t)]\,dt \\ &= -\int_0^\infty e^{-st}tf(t)\,dt \\ &= -\mathscr{L}\{tf(t)\}.\end{aligned}$$

Theorem 5.7 can be used to calculate Laplace transforms of additional functions. Examples 4–8 illustrate this application.

EXAMPLE 4 Evaluate $\mathscr{L}\{te^{at}\}$.

Solution. From Theorem 5.7 we see that

$$\begin{aligned}\mathscr{L}\{te^{at}\} &= -\frac{d}{ds}\mathscr{L}\{e^{at}\} \\ &= -\frac{d}{ds}\left(\frac{1}{s-a}\right) \\ &= \frac{1}{(s-a)^2}.\end{aligned}$$

EXAMPLE 5 Evaluate $\mathscr{L}\{t\cos kt\}$.

Solution. Theorem 5.7 tells us that

$$\begin{aligned}\mathscr{L}\{t\cos kt\} &= -\frac{d}{ds}\mathscr{L}\{\cos kt\} \\ &= -\frac{d}{ds}\left(\frac{s}{s^2+k^2}\right) \\ &= \frac{s^2-k^2}{(s^2+k^2)^2}.\end{aligned}$$

EXAMPLE 6 Evaluate $\mathscr{L}\{t^2e^{at}\}$.

Solution. From Theorem 5.7 we have

$$\begin{aligned}\mathscr{L}\{t^2e^{at}\} &= -\frac{d}{ds}\mathscr{L}\{te^{at}\}\\ &= \frac{d^2}{ds^2}\mathscr{L}\{e^{at}\}\\ &= \frac{d^2}{ds^2}\left(\frac{1}{s-a}\right)\\ &= \frac{2}{(s-a)^3}.\end{aligned}$$

EXAMPLE 7 Evaluate $\mathscr{L}\{\sin t + t\cos t\}$.

Solution. Noting that $\sin t + t\cos t$ is the derivative of $t\sin t$, we have

$$\begin{aligned}\mathscr{L}\{\sin t + t\cos t\} &= \mathscr{L}\left\{\frac{d}{dt}(t\sin t)\right\}\\ &= s\mathscr{L}\{t\sin t\}, \qquad \text{(by Theorem 5.6)}\\ &= -s\frac{d}{ds}\mathscr{L}\{\sin t\}, \qquad \text{(by Theorem 5.7)}\\ &= -s\left[-\frac{2s}{(s^2+1)^2}\right]\\ &= \frac{2s^2}{(s^2+1)^2}.\end{aligned}$$

EXAMPLE 8 Evaluate $\mathscr{L}\{te^{2t}\cos 3t\}$.

Solution. Theorem 5.7 yields

$$\mathscr{L}\{te^{2t}\cos 3t\} = -\frac{d}{ds}\mathscr{L}\{e^{2t}\cos 3t\}.$$

From the first translation theorem we see that

$$\begin{aligned}\mathscr{L}\{e^{2t}\cos 3t\} &= \mathscr{L}\{\cos 3t\}|_{s\to s-2}\\ &= \left.\frac{s}{(s^2+9)}\right|_{s\to s-2}\\ &= \frac{s-2}{s^2-4s+13}.\end{aligned}$$

After combining these two results we get

$$\begin{aligned}\mathscr{L}\{te^{2t}\cos 3t\} &= -\frac{d}{ds}\left(\frac{s-2}{s^2-4s+13}\right)\\ &= -\frac{(s^2-4s+13)-(s-2)(2s-4)}{(s^2-4s+13)^2}\\ &= \frac{s^2-4s-5}{(s^2-4s+13)^2}.\end{aligned}$$

EXERCISES 5.4

In problems 1–25 solve the given initial value problem using the Laplace transform method.

1. $y'' + 4y = 1, \quad y(0) = y'(0) = 0$
2. $y'' + y = 2t, \quad y(0) = 0, y'(0) = 1$
3. $y'' + y' = t, \quad y(0) = 1, y'(0) = 0$
4. $y'' + 2y' + y = 2e^{-t}, \quad y(0) = y'(0) = 0$
5. $y'' + y' - 6y = 5e^{2t}, \quad y(0) = y'(0) = 0$
6. $y'' - 3y' = e^{3t}, \quad y(0) = 1, y'(0) = 0$
7. $y'' - y' - 6y = e^{3t}, \quad y(0) = -1, y'(0) = 1$
8. $y'' + 2y' + 2y = te^{-t}, \quad y(0) = 1, y'(0) = -1$
9. $y'' + y' = \cos t, \quad y(0) = 0, y'(0) = -1$
10. $y'' - y' = e^t \cos t, \quad y(0) = y'(0) = 0$
11. $y'' - 2y' + y = \sin t, \quad y(0) = 1, y'(0) = 0$
12. $y'' - 3y' + 2y = \sin t, \quad y(0) = 0, y'(0) = 1$
13. $y'' - y' - 6y = \cos 2t, \quad y(0) = y'(0) = 0$
14. $y'' + 3y' + 2y = e^t \sin t, \quad y(0) = y'(0) = 0$
15. $y'' - 4y' + 4y = t^2e^t, \quad y(0) = y'(0) = 0$
16. $y'' - 2y' = e^t \sinh t, \quad y(0) = y'(0) = 0$
17. $y'' - 5y' - 6y = e^{3t} + t, \quad y(0) = 2, y'(0) = 1$
18. $y'' + y' + y = t^2, \quad y(0) = y'(0) = 1$
19. $y'' - 4y' + 4y = 2e^{2t} + \cos t, \quad y(0) = 1, y'(0) = -1$
20. $y'' - 2y' + y = te^t \sin t, \quad y(0) = y'(0) = 0$
21. $y''' - y'' - 4y' + 4y = e^{-t}, \quad y(0) = 0, y'(0) = 0, y''(0) = 0$
22. $2y''' + 3y'' - 3y' - 2y = \sin t, \quad y(0) = y'(0) = y''(0) = 0$
23. $y''' - y'' - 4y' + 4y = te^{-t}, \quad y(0) = y'(0) = 0, y''(0) = 1$
24. $y''' + 2y'' - y' - 2y = \cos t, \quad y(0) = y'(0) = 1, y''(0) = 0$
25. $y''' - y'' + y' - y = t^2, \quad y(0) = 0, y'(0) = -1, y''(0) = 1$

In problems 26–32 find the indicated Laplace transform.

26. $\mathscr{L}\{t^3e^{at}\}$
27. $\mathscr{L}\{t \sin kt\}$
28. $\mathscr{L}\{t^2 \cos kt\}$
29. $\mathscr{L}\{\cosh t + t \sinh t\}$
30. $\mathscr{L}\{kt \sin kt - \cos kt\}$
31. $\mathscr{L}\{te^t \sin t\}$
32. $\mathscr{L}\{te^{-2t} \cos 5t\}$
33. Find $\mathscr{L}\{y^{(n)}\}$ by mathematical induction.
34. The Laplace transforms of some functions can be obtained from their Taylor series expansions. For example, use the Taylor series expansion

$$\cos t = \sum_{n=0}^{\infty} \frac{(-1)^n t^{2n}}{(2n)!}$$

and assume the Laplace transform of this series can be computed term by term to verify that

$$\mathscr{L}\{\cos t\} = \frac{s}{s^2+1}, \quad s > 1.$$

FINIST

35. Let $f(t)$ be piecewise continuous and of exponential order for some positive constant c and $t \geq 0$. Assume the limit of $f(t)/t$ is finite as $t \to 0^+$. Using Theorem 5.7, show that

$$\mathcal{L}\left\{\frac{f(t)}{t}\right\} = \int_s^\infty F(u)\,du$$

where $F(s) = \mathcal{L}\{f(t)\}$.

36. Use Theorem 5.6 to find $\mathcal{L}\{\cos^2 t\}$.

37. A 10-kg mass is attached to a spring having a spring constant of 140 N/m. The mass is started in motion from the equilibrium position with an initial velocity of 1 m/sec in the upward direction and with an applied external force of $f(t) = 5 \sin t$ (in newtons). The mass is in a viscous medium with a coefficient of resistance equal to 90 N-sec/m. Formulate an initial value problem that models the given system, and solve the model using the method of Laplace transforms.

38. An 8-lb weight is attached to a spring having a spring constant of 16 lb/ft. This spring–mass system is in a medium with a damping constant of 0.25 lb-sec/ft, and an external force of $f(t) = \cos 2t$ (in pounds) is being applied. If initially the mass is set in motion at the equilibrium position with a downward velocity of 2 ft/sec, what is the steady-state solution? Solve the problem using the method of Laplace transforms.

39. An inductor of 2 henrys is connected in series with a resistor of 12 ohms, a capacitor of 1/16 farads, and a battery assumed to be 300 volts. The initial instantaneous charge q on the capacitor is 10 coulombs, and the initial current is $q'(0) = 0$. Formulate an initial value problem that models the circuit described, and solve the model using the method of Laplace transforms.

40. A series circuit consisting of an inductor, a resistor, and a capacitor is open. There is an initial charge of 2 coulombs on the capacitor, and 3 amperes of current is present in the circuit at the instant the circuit is closed. A voltage given by $E(t) = 20 \cos t$ is applied. In this circuit the voltage drop across the resistor is numerically equal to 4 times the instantaneous change in the charge, the voltage drop across the capacitor is numerically equal to 10 times the charge, and the voltage drop across the inductor is numerically equal to 2 times the instantaneous change in the current. Find the charge in the capacitor as a function of time using the method of Laplace transforms. Determine the charge on the capacitor and the current at time $t = 10$ sec.

41. A 3200-lb vehicle rests on a spring–shock absorber system on each of four wheels. The basic data yield the following model:

$$\frac{800}{32}y'' + 50y' + 425y = f(t),$$

where $f(t) = 25e^{-t}\cos 4t$ represents the force resulting from a bumpy road. Assuming the initial conditions are $y(0) = y'(0) = 0$, use the method of Laplace transforms to find the solution.

5.5

SOLVING DIFFERENTIAL EQUATIONS WITH PIECEWISE CONTINUOUS FORCING FUNCTIONS

We now turn our attention to the situation in which the forcing function is piecewise continuous for $t \geq 0$. For our purposes throughout this section, we also assume that $f(t) \equiv 0$ for negative values of t. For example, consider the function

$$f(t) = \begin{cases} 2, & 0 \leq t < 2, \\ -1, & 2 \leq t < 3, \\ 0, & 3 \leq t, \end{cases} \tag{1}$$

presented in Fig. 5.10.

There are two steps involved when working with piecewise continuous functions in connection with the Laplace transform:

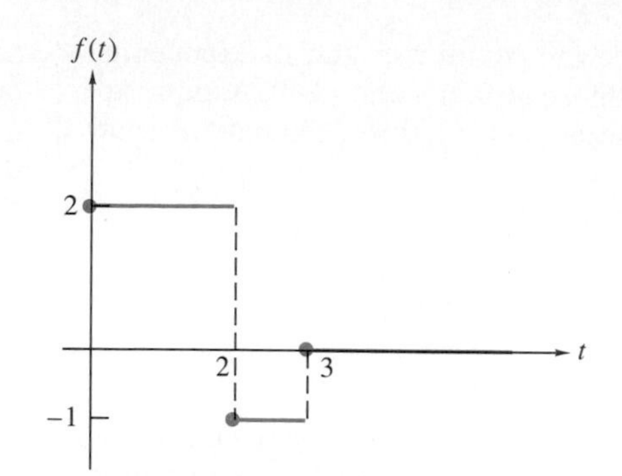

FIGURE 5.10
The piecewise continuous function given in Eq. (1).

Step 1. Write the piecewise continuous function in terms of a *single* formula.

Step 2. Determine the Laplace transform of the expression resulting from the first step.

To implement Step 1 the following special function is useful.

DEFINITION 5.5

The **unit step function** is defined to be

$$U(t) = \begin{cases} 0, & t < 0, \\ 1, & 0 \leq t. \end{cases}$$

The functions $U(t-a)$ and $U(a-t)$ are useful in expressing a piecewise continuous function that has a jump discontinuity at $t = a$. The functions $U(t)$, $U(t-a)$, and $U(a-t)$ are graphed in Fig. 5.11.

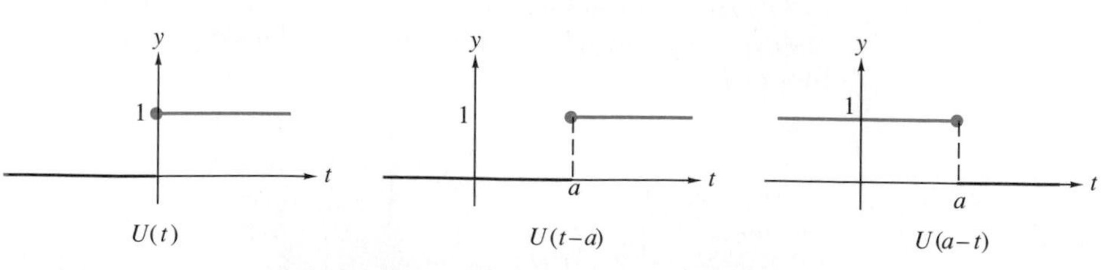

FIGURE 5.11
The unit step functions.

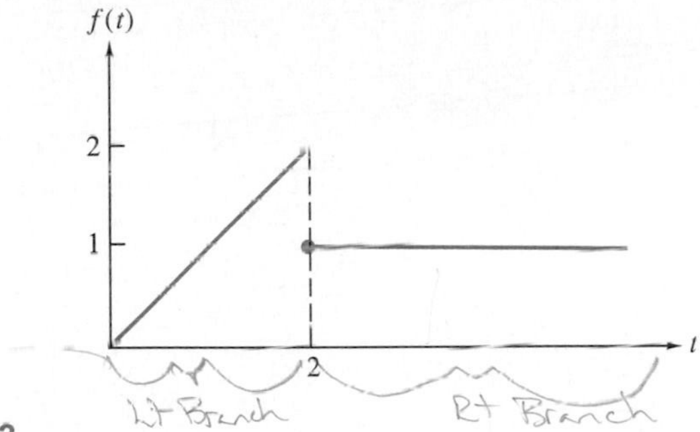

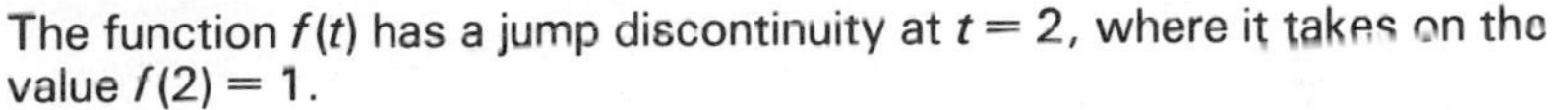

FIGURE 5.12
The function $f(t)$ has a jump discontinuity at $t = 2$, where it takes on the value $f(2) = 1$.

The unit step functions $U(t - a)$ and $U(a - t)$ are essentially off-on functions. They are used to turn one function off and a second function on as t moves across the point $t = a$ along the t-axis. Let us see how this is accomplished using a specific example.

The function

$$f(t) = \begin{cases} t, & 0 \leq t < 2, \\ 1, & 2 \leq t, \end{cases}$$

is shown in Fig. 5.12. This function has a jump discontinuity at $t = 2$ defining two different branches of f. To the left of $t = 2$ the function is defined by $f(t) = t$. We refer to this branch as the **left branch** associated with the point $t = 2$. Likewise, to the right of $t = 2$ we have the **right branch** $f(t) \equiv 1$. The unit step function $U(t - 2)$ can be used to turn off the left branch and turn on the right branch as t moves across the point $t = 2$ in the following way:

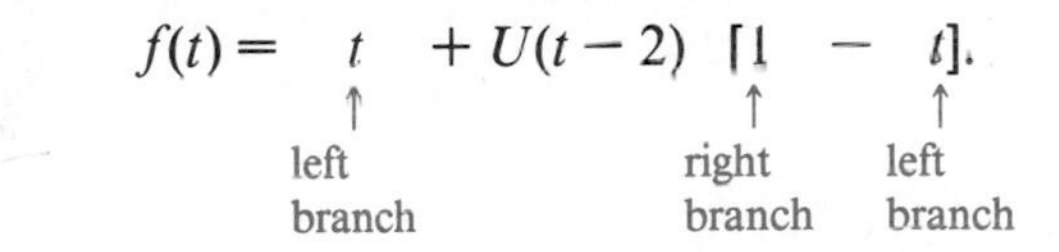

$$f(t) = \underset{\substack{\uparrow \\ \text{left branch}}}{t} + U(t-2)\,[\underset{\substack{\uparrow \\ \text{right branch}}}{1} - \underset{\substack{\uparrow \\ \text{left branch}}}{t}].$$

To understand why the formula is valid, notice that if $t < 2$ then $U(t - 2) = 0$ and from the formula we have $f(t) = t + 0 = t$. On the other hand, when $t \geq 2$, then $U(t - 2) = 1$, and from the formula $f(t) = t + [1 - t] = 1$. These values agree with the definition of the specified function $f(t)$.

To generalize this application of the step function, suppose at the point $t = a$ the function f consists of a left branch and a right branch, as in our example. The situation is depicted in Fig. 5.13. We also assume that, at the point of discontinuity $t = a$ (if the function has a jump discontinuity there), the function takes on the value associated with the *right* branch. Then the unit step function $U(t - a)$ can be used to express $f(t)$ in terms of the two branches as t moves across $t = a$ along the t-axis. This procedure is effected according to the following rule:

$$\text{Left branch} + U(t - a)[\text{right branch} - \text{left branch}]. \tag{2}$$

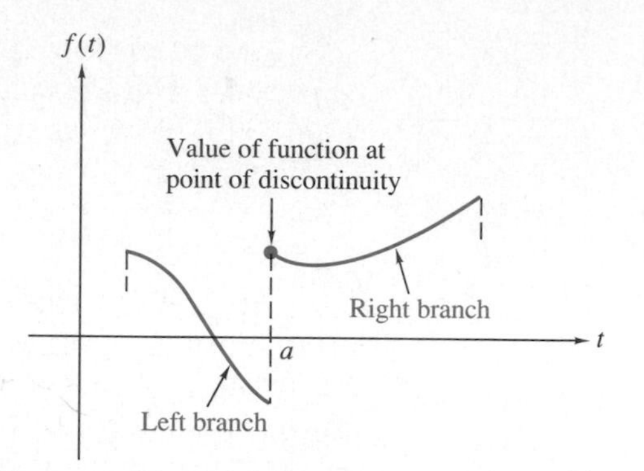

FIGURE 5.13
The unit step function $U(t-a)$ is used to "turn off" the left branch and turn on the right branch as t crosses $t=a$. At $t=a$ the function $f(t)$ has its value along the right branch.

The left branch refers to the curve *immediately to the left* of the point $t=a$. Similarly, the right branch lies immediately to the right of $t=a$. These curves are depicted and labeled in Figs. 5.13 and 5.14; which figure applies depends on how the function f is defined if there is a point of discontinuity at $t=a$.

If instead the piecewise continuous function assumes the value along the left branch at a jump discontinuity $t=a$ (as in Fig. 5.14), the function $U(a-t)$ is used to express the function according to the following rule:

$$\text{Right branch} + U(a-t)[\text{left branch} - \text{right branch}]. \tag{3}$$

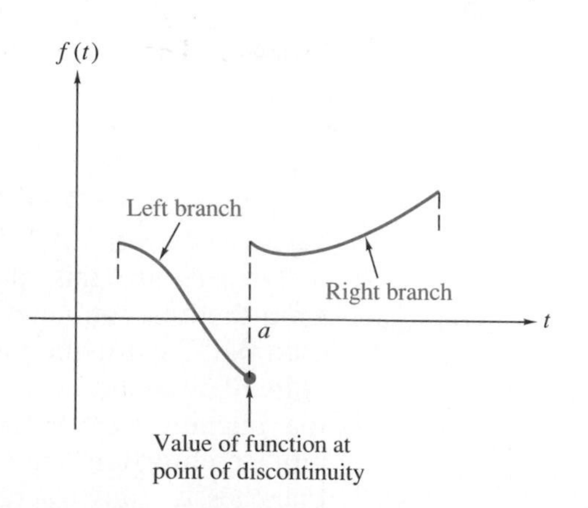

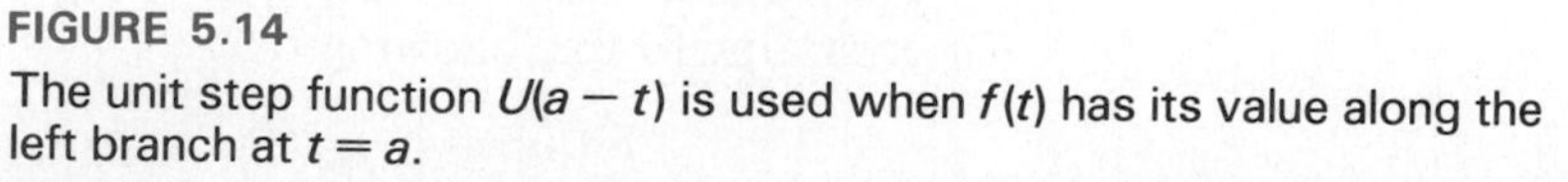

FIGURE 5.14
The unit step function $U(a-t)$ is used when $f(t)$ has its value along the left branch at $t=a$.

In the development to follow we assume the function values will always be defined on the right branch at each jump discontinuity, as shown in Fig. 5.13. Thus we will always be using the unit step function $U(t-a)$ for our representations.

EXAMPLE 1 The piecewise continuous function $f(t)$ given in Eq. (1) can be written as a single formula by using a unit step function and rule (2) at each point of discontinuity. At the jump discontinuity $t=2$, the left branch is $f(t) \equiv 2$ and the right branch is $f(t) \equiv -1$; at the jump discontinuity $t=3$, the left branch is $f(t) \equiv -1$ and the right branch is $f(t) \equiv 0$ (see Fig. 5.10). Thus,

$$f(t) = 2 + U(t-2)[-1-2] + U(t-3)[0-(-1)]$$

or

$$f(t) = 2 - 3U(t-2) + U(t-3). \tag{4}$$

To see this relationship, note that for $t < 2$, both $U(t-2)$ and $U(t-3)$ equal 0. Thus $f(t) = 2$ for $0 \leq t < 2$. When $2 \leq t < 3$, $U(t-2) = 1$ but $U(t-3) = 0$. Then $f(t) = 2 - 3 = -1$ for $2 \leq t < 3$. Finally, when $t \geq 3$, both $U(t-2)$ and $U(t-3)$ equal 1, yielding $f(t) = 2 - 3 + 1 = 0$ as desired.

EXAMPLE 2 Express the function

$$f(t) = \begin{cases} t, & 0 \leq t < \pi, \\ 0, & \pi \leq t < 2\pi, \\ \sin t, & 2\pi \leq t, \end{cases}$$

in terms of unit step functions.

Solution. The points $t = \pi$ and $t = 2\pi$ separate the three branches of this function, as depicted in Fig. 5.15. Thus,

$$\begin{aligned} f(t) &= t + U(t-\pi)[0-t] + U(t-2\pi)[\sin t - 0] \\ &= [1 - U(t-\pi)]t + U(t-2\pi)\sin t. \end{aligned}$$

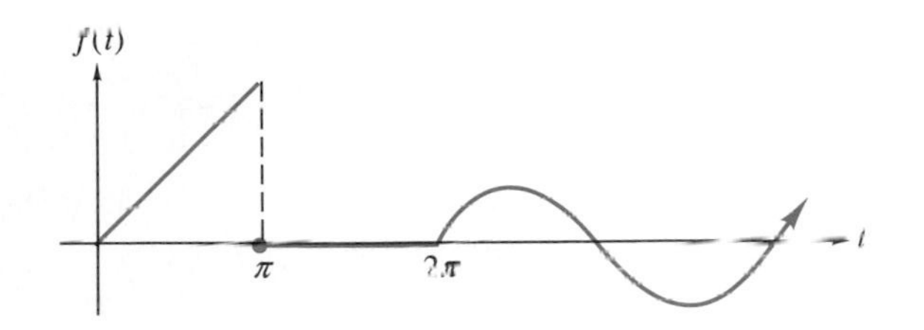

FIGURE 5.15
$f(t)$ in Example 2.

Second Translation Theorem

Our next task is to determine the Laplace transform of a piecewise continuous function written in terms of unit step functions. The next theorem, known as the **second translation theorem,** gives the needed result.

> **THEOREM 5.8**
>
> If $F(s) = \mathscr{L}\{f(t)\}$ exists for $s > c \geq 0$ and if $a > 0$ is constant, then
>
> $$\mathscr{L}\{f(t-a)U(t-a)\} = e^{-as}F(s), \qquad s > c.$$

Proof. Applying the definition of the Laplace transform, we have

$$\mathscr{L}\{f(t-a)U(t-a)\} = \int_0^\infty e^{-st}f(t-a)U(t-a)\,dt.$$

The definition of the unit step function $U(t-a)$ tells us that

$$\int_0^\infty e^{-st}f(t-a)U(t-a)\,dt = \int_a^\infty e^{-st}f(t-a)\,dt.$$

Substituting $u = t - a$ and $du = dt$ into the integral on the righthand side results in

$$\begin{aligned}\mathscr{L}\{f(t-a)U(t-a)\} &= \int_a^\infty e^{-st}f(t-a)\,dt \\ &= \int_0^\infty e^{-s(u+a)}f(u)\,du \\ &= e^{-sa}\int_0^\infty e^{-su}f(u)\,du \\ &= e^{-as}\mathscr{L}\{f(t)\}.\end{aligned}$$

EXAMPLE 3 Evaluate $\mathscr{L}\{U(t-\pi)\}$.

Solution. Making the identification $a = \pi$ and $f(t) \equiv 1$ in the second translation theorem gives us

$$\begin{aligned}\mathscr{L}\{U(t-\pi)\} &= e^{-\pi s}\mathscr{L}\{1\} \\ &= \frac{e^{-\pi s}}{s}.\end{aligned}$$

EXAMPLE 4 Evaluate the Laplace transform of piecewise continuous function (1) given as the single formula (4) in Example 1.

Solution. After applying the linearity of the Laplace transform and the second translation theorem we have

$$\begin{aligned}\mathscr{L}\{2-3U(t-2)+U(t-3)\} &= \mathscr{L}\{2\}-3\mathscr{L}\{U(t-2)\}+\mathscr{L}\{U(t-3)\}\\ &= \frac{2}{s}-3e^{-2s}\mathscr{L}\{1\}+e^{-3s}\mathscr{L}\{1\}\\ &= \frac{2}{s}-\frac{3e^{-2s}}{s}+\frac{e^{-3s}}{s}.\end{aligned}$$

EXAMPLE 5 Evaluate the Laplace transform of the piecewise continuous function presented in Fig. 5.14.

Solution. We use the expression for the function found in Example 2. Thus

$$\begin{aligned}&\mathscr{L}\{t-tU(t-\pi)+(\sin t)U(t-2\pi)\}\\ &\quad= \mathscr{L}\{t\}-\mathscr{L}\{tU(t-\pi)\}+\mathscr{L}\{(\sin t)U(t-2\pi)\}\\ &\quad= \mathscr{L}\{t\}-\mathscr{L}\{(\pi+t-\pi)U(t-\pi)\}+\mathscr{L}\{[\sin(t-2\pi)]U(t-2\pi)\}\\ &\quad= \mathscr{L}\{t\}-\pi\mathscr{L}\{U(t-\pi)\}-\mathscr{L}\{(t-\pi)U(t-\pi)\}\\ &\qquad+\mathscr{L}\{[\sin(t-2\pi)]U(t-2\pi)\}\\ &\quad= \mathscr{L}\{t\}-\pi e^{-\pi s}\mathscr{L}\{1\}-e^{-\pi s}\mathscr{L}\{t\}+e^{-2\pi s}\mathscr{L}\{\sin t\}\\ &\quad= \frac{1}{s^2}-\frac{\pi e^{-\pi s}}{s}-\frac{e^{-\pi s}}{s^2}+\frac{e^{-2\pi s}}{s^2+1}.\end{aligned}$$

Notice in Example 5 that we could not compute the transform $\mathscr{L}\{tU(t-\pi)\}$ directly. The function $f(t)=t$ (in this example) must first be translated from t to $t-\pi$. Thus we wrote $t=\pi+(t-\pi)$ and calculated the transforms

$$\mathscr{L}\{\pi U(t-\pi)\} \quad \text{and} \quad \mathscr{L}\{(t-\pi)U(t-\pi)\}$$

in accordance with the second translation theorem. A similar remark applies to the function $\sin t$.

Inverse Form

The inverse form of the second translation theorem is

$$f(t-a)U(t-a)=\mathscr{L}^{-1}\{e^{-as}F(s)\} \tag{5}$$

where $a>0$ is constant and $f(t)=\mathscr{L}^{-1}\{F(s)\}$.

Interpreting Eq. (5) we see that the second translation theorem is called for whenever we want to find the inverse transform of a function containing a multiplicative exponential factor of the form e^{-as}. For instance, to find $\mathcal{L}^{-1}\{e^{-2s}/s^2\}$ we first find $\mathcal{L}^{-1}\{1/s^2\} = t$ and note that $a = 2$. Then, according to inverse form (5) of the second translation theorem,

$$\mathcal{L}^{-1}\{e^{-2s}/s^2\} = (t-2)U(t-2).$$

In applications it is convenient to use the translation notation

$$f(t)|_{t\to t-a} \qquad \text{or} \qquad \mathcal{L}^{-1}\{F(s)\}|_{t\to t-a}$$

to express $f(t-a)$ when $f(t) = \mathcal{L}^{-1}\{F(s)\}$. (Note that this is similar to the notation introduced by Eq. (9) at the end of Section 5.2.)

EXAMPLE 6 Find $\mathcal{L}^{-1}\left\{\dfrac{e^{-s}}{s(s-1)}\right\}$.

Solution. The presence of the factor e^{-s} signals us that we must invoke the inverse form of the second translation theorem. From partial fraction decomposition we get

$$\frac{1}{s(s-1)} = \frac{1}{s-1} - \frac{1}{s}.$$

Then

$$\begin{aligned}
\mathcal{L}^{-1}\left\{\frac{e^{-s}}{s(s-1)}\right\} &= \mathcal{L}^{-1}\left\{\frac{e^{-s}}{s-1}\right\} - \mathcal{L}^{-1}\left\{\frac{e^{-s}}{s}\right\} \\
&= \mathcal{L}^{-1}\left\{\frac{1}{s-1}\right\}\Bigg|_{t\to t-1} U(t-1) \\
&\quad - \mathcal{L}^{-1}\left\{\frac{1}{s}\right\}\Bigg|_{t\to t-1} U(t-1) \\
&= e^t|_{t\to t-1}\, U(t-1) - 1|_{t\to t-1}\, U(t-1) \\
&= e^{t-1}\, U(t-1) - U(t-1) \\
&= (e^{t-1}-1)U(t-1).
\end{aligned}$$

In order to apply the second translation theorem we *must* have an expression of the form

$$f(t-a)U(t-a)$$

in which the function f is translated along the t-axis. However, as was the case in Example 5, we are required often to find the Laplace transform of an expression

$$g(t)U(t-a). \tag{6}$$

Here the unit step function $U(t-a)$ is multiplied by a *function of t* rather than by the required translated function of $t-a$. Then, to apply the second translation theorem, we must find a function f so that

$$g(t) = f(t-a). \tag{7}$$

(We already know the function g.) Once f is known, the second translation theorem can be applied to find $\mathscr{L}\{g(t)U(t-a)\}$ by calculating $\mathscr{L}\{f(t-a)U(t-a)\}$.

Let us look at Example 5 again. If we let

$$g(t) = t,$$

then $f(u) = \pi + u$ satisfies the equation

$$f(t-\pi) = \pi + (t-\pi) = g(t).$$

Likewise, if

$$g(t) = \sin t,$$

then $f(u) = \sin u$ satisfies the equation

$$f(t-2\pi) = \sin(t-2\pi) = g(t)$$

because the sine function has a period of 2π.

We should realize that it may be algebraically nontrivial to find f knowing g so that Eq. (7) is satisfied. However, in the case where $g(t)$ is a polynomial function, there are simple procedures for the conversion process. We next discuss one such procedure. A second procedure, known as Horner's algorithm, is discussed in Appendix C.

Converting a Polynomial from Powers of t to Powers of $t-a$

Suppose we desire to compute the Laplace transform of

$$(t^2 - t + 2)U(t-2).$$

Here $g(t) = t^2 - t + 2$ is a polynomial in powers of t. We need to express that polynomial in powers of $t-2$ in order to apply the second translation theorem. One method is to use algebraic substitution. The following example illustrates this method.

EXAMPLE 7 Convert $g(t) = t^2 - t + 2$ to a polynomial in powers of $t-2$.

Solution. We begin the process of converting the polynomial from powers of t to powers of $t-2$:

$$t^2 - t + 2 = [(t-2)^2 - (-4t+4)] - t + 2.$$

To obtain the righthand side, $(t-2)^2$ was substituted for t^2 and then the additional terms $-4t+4$ were subtracted from the result in order to maintain the original value t^2. Collecting constants and t terms in the result gives us

$$t^2 - t + 2 = (t-2)^2 + 3t - 2.$$

Next we substitute $t-2$ for the t term on the right and add 6 to maintain equality:

$$\begin{aligned} t^2 - t + 2 &= (t-2)^2 + [3(t-2) + 6] - 2 \\ &= (t-2)^2 + 3(t-2) + 4. \end{aligned}$$

In summary, for

$$g(t) = t^2 - t + 2,$$

the function

$$f(u) = u^2 + 3u + 4$$

satisfies the relation

$$f(t-2) = (t-2)^2 + 3(t-2) + 4 = g(t).$$

EXAMPLE 8 Find $\mathscr{L}\{(t^2 - t + 2)U(t-2)\}$.

Solution. We note that the expression inside the braces takes the form of Eq. (6). We can apply the second translation theorem if we can convert $g(t) = t^2 - t + 2$ into the form of Eq. (7). From the result of Example 7,

$$\begin{aligned} \mathscr{L}\{(t^2 - t + 2)U(t-2)\} &= \mathscr{L}\{[(t-2)^2 + 3(t-2) + 4]U(t-2)\} \\ &= \mathscr{L}\{(t-2)^2 U(t-2)\} + 3\mathscr{L}\{(t-2)U(t-2)\} + 4\mathscr{L}\{U(t-2)\} \\ &= e^{-2s}\mathscr{L}\{t^2\} + 3e^{-2s}\mathscr{L}\{t\} + 4e^{-2s}\mathscr{L}\{1\} \\ &= e^{-2s}\left(\frac{2}{s^3} + \frac{3}{s^2} + \frac{4}{s}\right). \end{aligned}$$

The algebraic substitution method used in Example 7 was easy to implement because the polynomial $g(t)$ is only a quadratic. However, for a polynomial of degree 5 we would begin by substituting $(t-2)^5$ for t^5. Then we have to determine the additional terms to subtract out to maintain the relationship

$$t^5 = (t-2)^5 - \text{additional terms.}$$

Then we have to treat the lower-order terms t^4, t^3, and so forth in turn, so that the resulting polynomial in powers of $t-2$ is equal to the original polynomial in powers of t. Algebraically this procedure would be very tedious.

Fortunately other methods exist requiring much less effort. For instance, the polynomial could be expanded in a Taylor series about $t = a$ ($a = 2$ in Example 7), or we could use synthetic division. (Synthetic division is explained in Appendix C.)

Our motivation for studying the Laplace transform is to solve constant-coefficient linear differential equations with discontinuous forcing functions. We have now developed the appropriate tools. Let us consider several examples.

EXAMPLE 9 Solve $y' - y = f(t)$ where

$$f(t) = \begin{cases} 1, & 0 \leq t < 1, \\ -1, & 1 \leq t, \end{cases}$$

subject to the initial condition $y(0) = 0$.

Solution. After writing the forcing function in terms of unit step functions, we get

$$f(t) = 1 + U(t-1)[-1-1] = 1 - 2U(t-1).$$

Thus

$$\mathcal{L}\{y'\} - \mathcal{L}\{y\} = \mathcal{L}\{1\} - 2\mathcal{L}\{U(t-1)\},$$

$$sY(s) - y(0) - Y(s) = \frac{1}{s} - \frac{2e^{-s}}{s}.$$

Using the initial condition and simplifying algebraically, we have

$$Y(s) = \frac{1}{s(s-1)} - \frac{2e^{-s}}{s(s-1)}.$$

Then we take inverse transforms to get

$$\begin{aligned} y(t) &= \mathcal{L}^{-1}\left\{\frac{1}{s(s-1)}\right\} - 2\mathcal{L}^{-1}\left\{\frac{e^{-s}}{s(s-1)}\right\} \\ &= \mathcal{L}^{-1}\left\{\frac{1}{s-1}\right\} - \mathcal{L}^{-1}\left\{\frac{1}{s}\right\} - 2\mathcal{L}^{-1}\left\{\frac{e^{-s}}{s-1}\right\} + 2\mathcal{L}^{-1}\left\{\frac{e^{-s}}{s}\right\} \\ &= e^t - 1 - [2e^t|_{t\to t-1} U(t-1)] + [2|_{t\to t-1} U(t-1)] \\ &= e^t - 1 - 2(e^{t-1} - 1)U(t-1). \end{aligned}$$

EXAMPLE 10 Assume an electrical circuit is governed by the initial value problem

$$q'' - q' - 6q = E(t)$$

where

$$E(t) = \begin{cases} t, & 0 \leq t < 1, \\ 1, & 1 \leq t, \end{cases}$$

subject to the initial conditions $q(0) = 0$ and $q'(0) = 0$. Find the charge $q(t)$.

Solution. Writing E in terms of a unit step function gives us

$$\begin{aligned} E(t) &= t + U(t-1)(1-t) \\ &= t - (t-1)U(t-1). \end{aligned}$$

Thus

$$\begin{aligned} \mathscr{L}\{q''\} - \mathscr{L}\{q'\} - 6\mathscr{L}\{q\} &= \mathscr{L}\{t\} - \mathscr{L}\{(t-1)U(t-1)\}, \\ s^2Q(s) - sq(0) - q'(0) - sQ(s) + q(0) - 6Q(s) &= \frac{1}{s^2} - \frac{e^{-s}}{s^2}, \\ (s^2 - s - 6)Q(s) &= \frac{1}{s^2} - \frac{e^{-s}}{s^2}. \end{aligned}$$

After solving for the transform, we get

$$Q(s) = \frac{1}{s^2(s-3)(s+2)} - \frac{e^{-s}}{s^2(s-3)(s+2)}.$$

We then use partial fraction decomposition (omitting the details here):

$$\frac{1}{s^2(s-3)(s+2)} = \frac{1/36}{s} - \frac{1/6}{s^2} + \frac{1/45}{s-3} - \frac{1/20}{s+2}.$$

Taking inverse transforms results in

$$\begin{aligned} q(t) &= \frac{1}{36} - \frac{t}{6} + \frac{e^{3t}}{45} - \frac{e^{-2t}}{20} - U(t-1)\left[\frac{1}{36} - \frac{t}{6} + \frac{e^{3t}}{45} - \frac{e^{-2t}}{20}\right]_{t \to t-1} \\ &= \frac{1}{36} - \frac{t}{6} + \frac{e^{3t}}{45} - \frac{e^{-2t}}{20} - \left[\frac{1}{36} - \frac{t-1}{6} + \frac{e^{3(t-1)}}{45} - \frac{e^{-2(t-1)}}{20}\right]U(t-1). \end{aligned}$$

EXAMPLE 11 Solve $y'' - y' = f(t)$ where

$$f(t) = \begin{cases} e^{-t}, & 0 \leq t < 1, \\ 0, & 1 \leq t, \end{cases}$$

subject to the initial conditions $y(0) = y'(0) = 0$.

Solution. Writing the forcing function in terms of unit step functions we get

$$f(t) = e^{-t} - e^{-t}U(t-1).$$

Taking the Laplace transform of the differential equation gives us

$$\mathscr{L}\{y''\} - \mathscr{L}\{y'\} = \mathscr{L}\{e^{-t}\} - \mathscr{L}\{e^{-t}U(t-1)\}$$

or

$$s^2Y(s) - sy(0) - y'(0) - sY(s) + y(0) = \frac{1}{s+1} - \frac{1}{e}\mathscr{L}\{e^{-(t-1)}U(t-1)\}.$$

Applying the initial conditions and collecting like terms leads to

$$(s^2 - s)Y(s) = \frac{1}{s+1} - \frac{e^{-s}}{e(s+1)}.$$

Solving for $Y(s)$ we find

$$Y(s) = \frac{1}{s(s-1)(s+1)} - \frac{e^{-s}}{es(s-1)(s+1)}.$$

Using the Heaviside technique results in

$$\frac{1}{s(s-1)(s+1)} = -\frac{1}{s} + \frac{½}{s-1} + \frac{½}{s+1}.$$

Therefore

$$\begin{aligned} Y(s) = &-\frac{1}{s} + \frac{1}{2(s-1)} + \frac{1}{2(s+1)} \\ &- \frac{1}{e}\left[-\frac{1}{s} + \frac{1}{2(s-1)} + \frac{1}{2(s+1)}\right]e^{-s}. \end{aligned}$$

Taking inverse transforms gives us

$$\begin{aligned} y(t) &= \left[-1 + \frac{e^t + e^{-t}}{2}\right] - e^{-1}\left[-1 + \frac{e^t + e^{-t}}{2}\right]_{t \to t-1} U(t-1) \\ &= \cosh t - 1 + e^{-1}[1 - \cosh(t-1)]\, U(t-1). \end{aligned}$$

EXERCISES 5.5

In problems 1–10 write the given piecewise continuous function in terms of unit step functions. Graph each function.

1. $f(t) = \begin{cases} 0, & 0 \le t < 2 \\ 1, & 2 \le t \end{cases}$

2. $f(t) = \begin{cases} -1, & 0 \le t < 1 \\ 2, & 1 \le t \end{cases}$

3. $f(t) = \begin{cases} 0, & 0 \le t < \frac{1}{2} \\ t, & \frac{1}{2} \le t \end{cases}$

4. $f(t) = \begin{cases} 2t, & 0 \le t < 1 \\ 1 - t, & 1 \le t \end{cases}$

5. $f(t) = \begin{cases} t, & 0 \le t < 1 \\ 2t^2, & 1 \le t \end{cases}$

6. $f(t) = \begin{cases} \sin t, & 0 \le t < \frac{\pi}{2} \\ 0, & \frac{\pi}{2} \le t \end{cases}$

7. $f(t)=\begin{cases} 3e^{-t}, & 0 \leq t < 1 \\ -1, & 1 \leq t \end{cases}$

8. $f(t)=\begin{cases} 1, & 0 \leq t < 1 \\ t-1, & 1 \leq t < 2 \\ t^2-1, & 2 \leq t \end{cases}$

9. $f(t)=\begin{cases} \cos t, & 0 \leq t < \frac{\pi}{2} \\ 1, & \frac{\pi}{2} \leq t < \pi \\ -\sin t, & \pi \leq t \end{cases}$

10. $f(t)=\begin{cases} 1-t^2, & 0 \leq t < 1 \\ 1, & 1 \leq t < 2 \\ t^3, & 2 \leq t \end{cases}$

In problems 11–20 determine the Laplace transform of the given function.

11. $f(t)=\begin{cases} 2, & 0 \leq t < 1 \\ -1, & 1 \leq t \end{cases}$

12. $f(t)=\begin{cases} 1, & 0 \leq t < \frac{1}{2} \\ 0, & \frac{1}{2} \leq t < 2 \\ -1, & 2 \leq t \end{cases}$

13. $f(t)=\begin{cases} 3, & 0 \leq t < 2 \\ t, & 2 \leq t \end{cases}$

14. $f(t)=\begin{cases} 1+t, & 0 \leq t < 1 \\ 1-t, & 1 \leq t \end{cases}$

15. $f(t)=\begin{cases} 1-t^2, & 0 \leq t < 1 \\ 1, & 1 \leq t \end{cases}$

16. $f(t)=\begin{cases} 1+t^2, & 0 \leq t < 1 \\ t, & 1 \leq t \end{cases}$

17. $f(t)=\begin{cases} \sin t, & 0 \leq t < 2\pi \\ 1, & 2\pi \leq t \end{cases}$

18. $f(t)=\begin{cases} -1, & 0 \leq t < 2 \\ t, & 2 \leq t < 3 \\ t^2, & 3 \leq t \end{cases}$

19. $f(t)=\begin{cases} 0, & 0 \leq t < 1 \\ e^{t-1}, & 1 \leq t \end{cases}$

20. $f(t)=\begin{cases} \cos t, & 0 \leq t < \pi \\ 0, & \pi \leq t < 2\pi \\ 1, & 2\pi \leq t \end{cases}$

In problems 21–38 solve the given initial value problem using Laplace transforms.

21. $y'+2y=\begin{cases} 2, & 0 \leq t < 3 \\ 1, & 3 \leq t \end{cases}$

$y(0)=0$

22. $y'-y=\begin{cases} t, & 0 \leq t < 5 \\ 1, & 5 \leq t \end{cases}$

$y(0)=1$

23. $y'+y=\begin{cases} 1, & 0 \leq t < 2 \\ t, & 2 \leq t < 4 \\ 0, & 4 \leq t \end{cases}$

$y(0)=0$

24. $y'+3y=e^t U(t-1)+U(t-3)$

$y(0)=0$

25. $y'-5y=\begin{cases} t, & 0 \leq t < 1 \\ t^3-1, & 1 \leq t \end{cases}$

$y(0)=1$

26. $y''-3y'+2y=\begin{cases} 1, & 0 \leq t < 4 \\ 0, & 4 \leq t \end{cases}$

$y(0)=y'(0)=0$

27. $y''-3y'+2y=\begin{cases} 1, & 0 \leq t < 2 \\ -1, & 2 \leq t \end{cases}$

$y(0)=1,\ y'(0)=0$

28. $y''-5y'+4y=\begin{cases} 0, & 0 \leq t < 3 \\ 2, & 3 \leq t \end{cases}$

$y(0)=0,\ y'(0)=1$

29. $y''+2y'-3y=\begin{cases} t, & 0 \leq t < 1 \\ 0, & 1 \leq t \end{cases}$

$y(0)=0,\ y'(0)=-1$

30. $y''-2y'-3y=\begin{cases} 1, & 0 \leq t < 2 \\ 0, & 2 \leq t < 3 \\ 1, & 3 \leq t \end{cases}$

$y(0)=1,\ y'(0)=-1$

31. $y''+2y'+y=\begin{cases} e^t, & 0 \leq t < 1 \\ 0, & 1 \leq t \end{cases}$

$y(0)=y'(0)=0$

32. $y''+2y'+y=\begin{cases} 1-t, & 0 \leq t < 1 \\ 1, & 1 \leq t \end{cases}$

$y(0)=1,\ y'(0)=0$

33. $y'' - y = tU(t-1) - t$
$y(0) = y'(0) = 0$

34. $y'' - 4y = e^{3t}U(t-2)$
$y(0) = 1,\ y'(0) = 0$

35. $y'' - y' = e^{-t} + e^{t}U(t-3)$,
$y(0) = -1,\ y'(0) = 1$

36. $y'' + y = \begin{cases} 0, & 0 \leq t < 1 \\ t, & 1 \leq t \end{cases}$

$y(0) = 1,\ y'(0) = 0$

37. $y'' + 4y = \begin{cases} \sin 2t, & 0 \leq t < \pi \\ 0, & \pi \leq t \end{cases}$

$y(0) = 0,\ y'(0) = 1$

38. $y'' + 2y' = 4 - 3U(t-1)$
$y(0) = 0,\ y'(0) = 2$

In problems 39–44 write the given polynomial in powers of $t - a$ for the specified value of a. Use Taylor series expansions or the synthetic division method presented in Appendix C.

39. $t^3 - 2t^2 + t - 7,\quad a = -1$

40. $t^4 + t^3 + t^2 + t + 1,\quad a = 1$

41. $t^4 + t^3 + t^2 + t + 1,\quad a = -2$

42. $t^5 - 1,\quad a = 1$

43. $t^5,\quad a = -1$

44. $t^4 - 2t^3 + t,\quad a = \frac{1}{2}$

45. Derive the transform $\mathscr{L}\{U(a-t)\} = \dfrac{1 - e^{-as}}{s}$.

46. Let $g(t-a) = f(t)$. Show that
$\mathscr{L}\{U(t-a)f(t)\} = e^{-as}\mathscr{L}\{g(t)\} = e^{-as}\mathscr{L}\{f(t+a)\}$.

CHAPTER 5 REVIEW EXERCISES

In problems 1–7, find the Laplace transform of the given function using the definition.

1. $f(t) = \begin{cases} 2, & 0 \leq t < 7 \\ 6, & t \geq 7 \end{cases}$

2. $f(t) = \begin{cases} t, & 0 \leq t < 3 \\ 1, & t \geq 3 \end{cases}$

3. $f(t) = \begin{cases} 1 + t, & 0 \leq t < 1 \\ 0, & t \geq 1 \end{cases}$

4. $f(t) = \begin{cases} \sin t, & 0 \leq t < \pi \\ 0, & \pi \leq t \end{cases}$

5. $f(t) = \begin{cases} 0, & 0 \leq t < 3 \\ \dfrac{e^t}{3}, & t \geq 3 \end{cases}$

6. $f(t) = \begin{cases} te^t, & 0 \leq t < 3 \\ 1, & t \geq 3 \end{cases}$

7. $f(t) = \begin{cases} 1, & 0 \leq t < 1 \\ -1, & 1 \leq t < 2 \\ t, & t \geq 2 \end{cases}$

In problems 8–10, graph the functions in the xy-plane.

8. $y(x) = -3 + (\cos x)U(x - 2\pi) + 3U(x - 2\pi)$

9. $y(x) = 5 + xU(x-4) - 5U(x-4)$

10. $y(x) = 4 + (\sin x)U(x - 2\pi) - 4U(x - 2\pi)$

In problems 11–14, write the graphed functions in terms of unit step functions and find their Laplace transforms.

11.

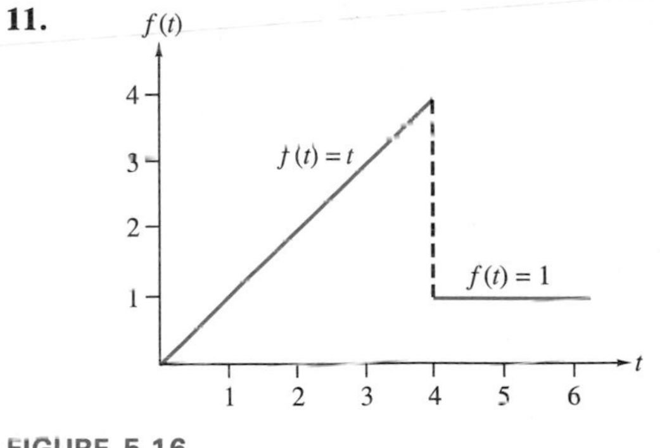

FIGURE 5.16

12.

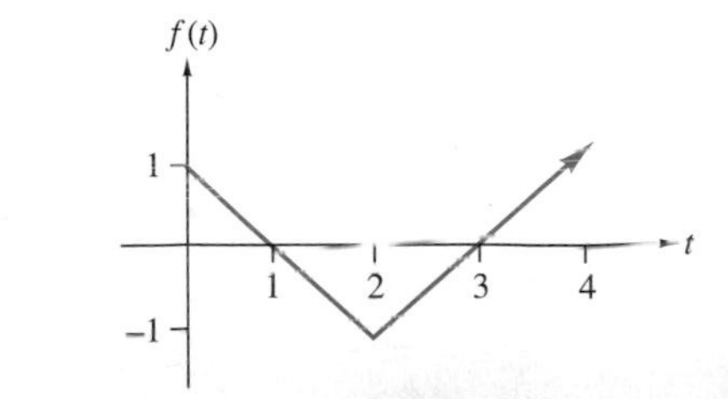

FIGURE 5.17

13.

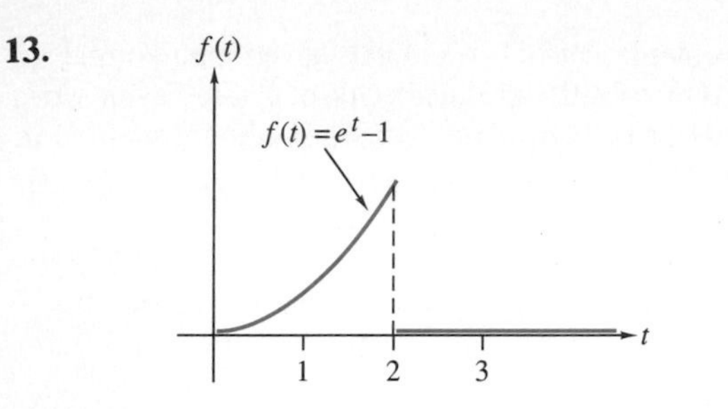

FIGURE 5.18

14.

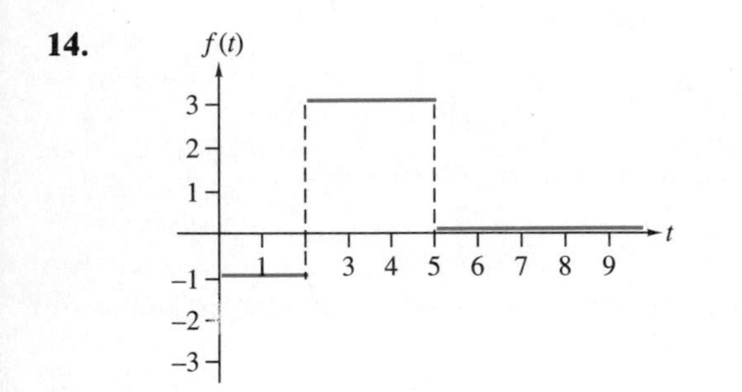

FIGURE 5.19

In problems 15 and 16, write the given functions in terms of the unit step function and find their Laplace transforms.

15. $f(t)=\begin{cases} 2, & 0 \leqslant t < 1 \\ -2, & t \geqslant 1 \end{cases}$

16. $f(t)=\begin{cases} 0, & 0 \leqslant t < 3 \\ 2, & 3 \leqslant t < 5 \\ 0, & t \geqslant 5 \end{cases}$

In problems 17–20, find $\mathscr{L}\{f(t)\}$ for the given functions.

17. $f(t) = te^{-4t} \sin 7t$

18. $f(t) = te^{3t} \cos 5t$

19. $f(t) = te^{-3t} \cos 3t$

20. $f(t) = t^2 U(t-1)$

In problems 21–31, find $\mathscr{L}^{-1}\{F(s)\}$ for the given functions.

21. $F(s) = \dfrac{3s^2 + 8}{s^3 + 4s}$

22. $F(s) = \dfrac{e^{-5s}}{s+3}$

23. $F(s) = \dfrac{s^2 + s + 3}{s^3 + 3s^2}$

24. $F(s) = \dfrac{2e^{-3s}}{s^2 + 4}$

25. $F(s) = \dfrac{se^{-5s}}{s^2 + 9}$

26. $F(s) = \dfrac{24e^{-6s}}{s^5}$

27. $F(s) = \dfrac{s+1}{(s^2 + 2s + 2)^2}$

28. $F(s) = \dfrac{-e^{-3s}}{s(s+1)}$

29. $F(s) = \dfrac{2s + 10}{s^2 + 12s + 40}$

30. $F(s) = \dfrac{e^{-2\pi s}}{s^2 + 1}$

31. $F(s) = \dfrac{1}{s^2 + 3s - 4}$

In problems 32–36, solve the initial value problem.

32. $y'' - 8y' - 20y = \begin{cases} 1, & 0 \leqslant t < 4 \\ 3, & t \geqslant 4 \end{cases}$

$y(0) = 0,\ y'(0) = 1.$

33. $y'' - y' - 6y = \begin{cases} 1, & 0 \leqslant t < 5 \\ 4, & t \geqslant 5 \end{cases}$

$y(0) = 0,\ y'(0) = -1.$

34. $y'' - 16y = \begin{cases} 1, & 0 \leqslant t < \pi \\ -1, & \pi \leqslant t < 2\pi \\ 0, & t \geqslant 2\pi \end{cases}$

$y(0) = 0,\ y'(0) = 0.$

35. $y'' + 4y' + 8y = \begin{cases} t, & 0 \leqslant t < 2 \\ 0, & t \geqslant 2 \end{cases}$

$y(0) = 0,\ y'(0) = 1.$

36. $y'' + 4y = \begin{cases} 2t, & 0 \leqslant t \leqslant 3 \\ 6, & t > 3 \end{cases}$

$y(0) = 1,\ y'(0) = 0.$

37. The displacement $x(t)$ of a mass on a spring can be modeled by the differential equation

$$mx'' + \beta x' + kx = f(t).$$

In a given system the mass $m = 1$ slug, the damping constant $\beta = 6$ lb-sec/ft, the spring constant $k = 5$ lb/ft, and the external force driving the system is $f(t) = \cos t$. Initially the mass is released from rest at the equilibrium position. Solve for $x(t)$ using Laplace transforms.

38. In a spring–mass system (see problem 37), a mass of 1 slug is attached to a spring with a spring constant of 30 lb/ft and placed in a medium with a damping constant of 11 lb-sec/ft. An external force of 5 lb is applied to the system for the first 2 sec and is reduced to 2 lb thereafter. Find a function describing the position of the mass at any time if the mass is released from the equilibrium position with an initial velocity of 2 ft/sec downward.

39. An automobile shock absorber–coil spring system is designed to support 768 lb, the portion of the automobile's weight it supports. The spring has a constant of 60 slugs/in. The system is tested by running it on a bumpy section of road followed by a flat section. The bumpy section of road exerts a force $f(t) = 250 \sin 4t$ (in slug-in/sec^2) upward on the system for $3\pi/8$ sec. The system is initially in equilibrium at rest.

a) Assume that the automobile's shock absorber is so worn that it provides no effective damping force. Express the force exerted by the road on the system in terms of unit step functions. Find a particular solution that describes the vertical displacement of the automobile over time using the method of Laplace transforms. Graph the particular solution for the first 3 sec of motion. Describe the system's performance.

b) Now assume that the shock absorber is replaced. The new shock absorber exerts a damping force (in pounds) that is numerically equal to 60 times the instantaneous vertical velocity of the system (in inches per second). Model this improved system with an initial value problem and solve it using Laplace transforms subject to the conditions described in part (a). Graph the resulting equation for the first 3 sec of motion.

40. A shock absorber–coil spring system for an imported automobile is designed to support 350 kg. The spring has a constant of 122,150 kg/cm. The shock absorber exerts a damping force (in kilogram centimeters per second squared) that is numerically equal to 3500 times the instantaneous vertical velocity of the system (in centimeters per second). Beginning in equilibrium at rest, the system is tested by running it over a bumpy road surface that exerts a force of $3500\, e^{-5t} \cos 2t$ (in kilogram centimeters per second squared) upward on the system for $3\pi/2$ sec. After $3\pi/2$ sec, the system encounters a flat surface.

a) Express the force exerted by the two road surfaces in terms of unit step functions.

b) Model this system with an initial value problem and solve it.

c) Graph the resulting equation for the first 3 sec of motion.

d) What is the maximum displacement from equilibrium that the system experiences?

41. We wish to determine information about a cannon system (see Chapter 3 Review Exercises, problem 20). The gun tube–breech block assembly has a mass of 1500 kg. The recoil spring has a spring constant of 19,500 N/m. The damping mechanism exerts a force (in newtons) numerically equal to 9000 times the instantaneous velocity of the gun tube–breech block assembly (in meters per second). When the cannon is fired, the gas pressure force from the firing of the round exerts a force of 40,500 N on the gun tube–breech block assembly for ⅕ sec. The force is zero after ⅕ sec. Model this system as an initial value problem, and find the particular solution that gives the displacement of the gun tube–breech block assembly with respect to time by using the method of Laplace transforms.

42. A 2-g mass is suspended from the end of a rod 2.45 m long to form a pendulum. The pendulum system travels through a fluid that exerts a damping force (in gram meters per second squared) that is numerically 4 times the instantaneous velocity (in meters per second) along the path traced by the mass. The pendulum is set in motion by displacing the mass 1 radian from the equilibrium position and releasing it from rest. Initially, no forcing function acts on the system. After 1 sec of motion, a variable forcing function given by $10e^{-t}$ (in gram meters per second squared) acts tangentially to the path of motion of the mass. The pendulum only experiences small angular displacements. Use 9.8 m/sec^2 as the acceleration of an object due to gravity.

a) Write the forcing function in terms of unit step functions.

b) Find the equation that gives the angular displacement of the pendulum at any time using the method of Laplace transforms.

c) Find the angular displacement of the pendulum after ½ sec of motion.

43. The charge $q(t)$ in an *LRC* series circuit can be modeled by the differential equation

$$Lq'' + Rq' + \frac{1}{C}q = E(t).$$

In a given series circuit, $L = 1$ henry, $R = 6$ ohms, $C = 1/8$ farads, and the impressed voltage $E(t) = \cos t$ volts. Initially both the charge and the current are zero. Find $q(t)$ using the method of Laplace transforms.

44. In a given *LRC* series circuit (see problem 43), $L = 1$ henry, $R = 11$ ohms, and $C = 1/28$ farads. The impressed voltage is 2 volt for the first 3 sec and 9 volt thereafter. If the initial charge is zero and the initial current is 3 amperes, find a function $q(t)$ describing the charge on the capacitor at any time.

45. An *LRC* circuit has an inductance of 1 henry, a capacitance of 0.04 farads, and no resistance. The external voltage applied to the circuit is $25t$ volts for the first 3 sec and a constant 75 volts thereafter. Initially there is no charge on the capacitor and no current in the circuit.

a) Write the external voltage function in terms of unit step functions.

b) Find an equation that gives the charge on the capacitor for any time using the method of Laplace transforms.

46. A simple *LRC* circuit consists of a capacitor with a capacitance of $1/60$ farads, a resistor with a resistance of 50 ohms, and an inductor with an inductance of 10 henries. The circuit is connected to an alternating current source that applies a voltage of $104 \cos 2t$ volts for $\pi/4$ sec. After $\pi/4$ sec, the circuit is disconnected from the voltage source. Initially, there is no charge on the capacitor and no current in the circuit.

a) Write the external voltage function in terms of unit step functions.

b) Find an equation that gives the charge on the capacitor for any time using the method of Laplace transforms.

c) Graph the equation for charge.

d) Find the charge on the capacitor after $\pi/8$ sec and $3\pi/8$ sec.

e) Find the current in the circuit after $\pi/8$ sec. Would this value for current exceed the capacity of a 1-ampere fuse?

47. The same *LRC* circuit described in problem 46 is connected to a switch that will allow the circuit to receive voltage from either a battery (direct current) or an alternating current source. This is similar to electric alarm clocks that have a battery backup. Initially, the circuit is connected to a 24-volt battery. After 2π sec the circuit is switched to the alternating current source that exerts a voltage of $104 \cos 2t$ volts. Initially there is no charge on the capacitor and no current in the circuit.

a) Write the external voltage function in terms of unit step functions.

b) Find an equation that gives the charge on the capacitor for any time using the method of Laplace transforms.

c) Find the charge on the capacitor after 8 sec.

48. Assume you are given the torsion system described in problem 28 of the Chapter 4 Review Exercises except that $I = 2$ kg-m^2, $k = 10$ N-m, and a damping torque (in newton meters) equal to 4 times the instantaneous angular velocity (in radians per meter) is exerted. A variable external torque of $10t$ (in newton meters) is applied to the system for the first second of motion, and a torque of $100\, te^{-4t}$ (in newton meters) is applied thereafter. The mass is initially twisted 2 radians from equilibrium and released from rest.

a) Express the external torque in terms of unit step functions.

b) Find the equation that gives the angular displacement of the mass at any time using the method of Laplace transforms.

c) How much twist has the mass undergone after 1 sec?

49. We wish to investigate the shear stress on a structural girder of a tall building that is being designed to withstand the effects of earthquakes. A simple model that approximates the crosswise forces acting on a building girder is the familiar equation $mx'' + bx' + kx = f(t)$, where $x(t)$ is the crosswise displacement of the girder, $f(t)$ is the net applied shear stress, and m, b, and k are positive constants dependent on the characteristics of the girder. The girder is normally subject to a time dependent stress induced by the swaying of the building. Assume that this stress is described by the function $A \cos t$. Assume that this stress acts on the girder, which begins in equilibrium at rest, for $\pi/2$ seconds when an earthquake strikes, changing the shear stress to $Be^{-t} \cos t$. The constants of the girder that we are investigating are $m = 1$, $b = 5$, and $k = 6$.

a) Express the external shear stress on the girder in terms of unit step functions.

b) Using Laplace transforms, find the equation for the crosswise displacement of the girder over time.

c) What is the crosswise displacement of the girder after $\pi/4$ and π seconds?

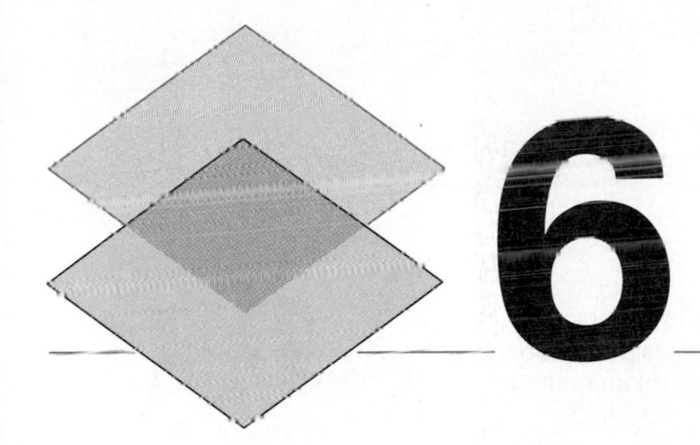

6

Laplace Transform II: Modeling with Periodic or Impulse Forcing Functions

In Chapter 5 we presented the elementary concepts of the Laplace transform and its use in solving constant-coefficient differential equations with piecewise continuous forcing functions. In this chapter we present additional topics associated with the Laplace transform.

In Section 4.5 we modeled the vibrations of an automobile suspension system consisting of a coil spring and a shock absorber with an external forcing function. There we investigated the equation

$$my'' + \delta y' + ky = f(t).$$

Recall that m is the portion of the mass of the car supported by the suspension system, δ is the damping constant for the shock absorber, and k is the stiffness of the spring. Several models were discussed to account for various road conditions represented by the forcing function $f(t)$. The graphs of these models are presented again as Fig. 6.1. In Section 4.5 we studied the solution for the case where the forcing function takes the continuous form

$$f(t) = F_0 \cos \omega t$$

as depicted in Fig. 6.1(a). In this chapter we investigate solution techniques for forcing functions shown in the remaining graphs. These forcing functions

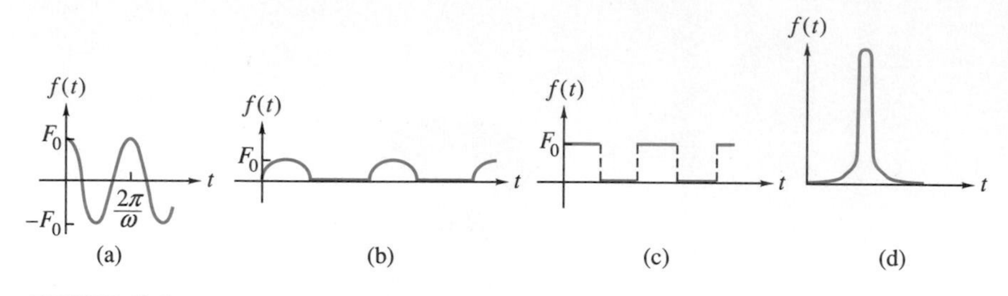

FIGURE 6.1
Approximations of various road conditions. (a) A washboard effect; (b, c) wakeup strips; (d) approximation of a New York pothole effect.

are of a periodic nature, as in Figs. 6.1(b, c), or of an impulse nature, as in Fig. 6.1(d).

In Section 6.1 we will investigate the problem of solving differential equations when the forcing function is periodic, like the sine and cosine functions. However, in the more general case the periodic function may have *infinitely many discontinuities.* Then in Section 6.2 we will consider the situation where the forcing function is of an "impulse" nature, like the force applied when a baseball is struck by a bat or the force that might be experienced when driving through a New York pothole.

Another important topic is *convolution,* which is concerned with taking the inverse transform of a product and has applications in solving certain *integral equations.* We will present an introduction to those ideas in Section 6.3. Finally, in Section 6.4 we will develop an application of the Laplace transform to control systems. These concepts play a significant role in modern engineering.

6.1 PERIODIC FORCING FUNCTIONS

Numerous situations occur in engineering when the forcing function for a constant-coefficient linear second-order differential equation is periodic in nature. For instance, in the equation

$$Lq'' + Rq' + \frac{1}{C}q = E(t)$$

modeling an electrical circuit, it may be the case that the impressed voltage $E(t)$ is periodic, as depicted in Fig. 6.2. Figure 6.2(a) represents a situation in which the impressed voltage $E(t)$ is turned on and off repeatedly for short, equal time intervals, and while the voltage is on it is constant. In Fig. 6.2(b) the voltage increases linearly to a maximum value, at which instant it is turned off and then immediately on again, building linearly to the same

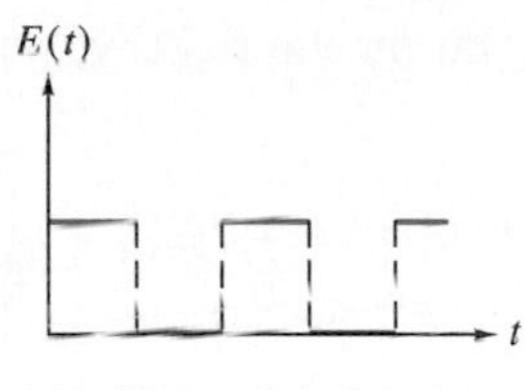

(a) Square wave function

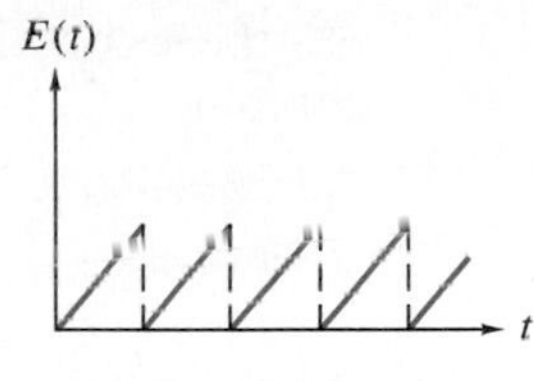

(b) Sawtooth function

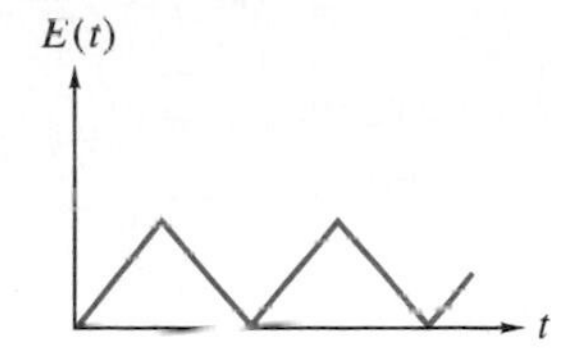

(c) Triangular wave function

FIGURE 6.2
Examples of periodic forcing functions.

maximum as before; it repeats this pattern at short time intervals. In Fig. 6.2(c), the voltage starts at zero, increases linearly to a maximum, and decreases linearly to zero, repeating this pattern indefinitely. These functions are commonly referred to as the **square wave, sawtooth wave,** and **triangular wave** functions, respectively.

DEFINITION 6.1

A function $f(t)$ defined for $t \geqslant 0$ is **periodic** with period $T > 0$ if

$$f(t+T) = f(t).$$

The Laplace transform technique is very useful for solving differential equations with periodic forcing functions. The following theorem establishes that the transform of a periodic function is obtained by integrating over one period.

THEOREM 6.1

Let $f(t)$ be piecewise continuous and of exponential order for $t \geqslant 0$. If $f(t)$ is periodic with period T, then

$$\mathscr{L}\{f(t)\} = \frac{1}{1-e^{-sT}} \int_0^T e^{-st} f(t)\, dt. \tag{1}$$

Proof. Applying the definition of the Laplace transform gives us

$$\begin{aligned}\mathscr{L}\{f(t)\} &= \int_0^\infty e^{-st} f(t)\, dt \\ &= \int_0^T e^{-st} f(t)\, dt + \int_T^\infty e^{-st} f(t)\, dt.\end{aligned}$$

In the second integral on the right we substitute $u = t - T$ and $du = dt$ to obtain

$$\begin{aligned}\mathcal{L}\{f(t)\} &= \int_0^T e^{-st}f(t)\,dt + \int_0^\infty e^{-s(u+T)}f(u+T)\,du \\ &= \int_0^T e^{-st}f(t)\,dt + e^{-sT}\int_0^\infty e^{-su}f(u)\,du \\ &= \int_0^T e^{-st}f(t)\,dt + e^{-sT}\mathcal{L}\{f(t)\}.\end{aligned}$$

Combining the last term on the righthand side with the lefthand side yields

$$(1 - e^{-sT})\mathcal{L}\{f(t)\} = \int_0^T e^{-st}f(t)\,dt.$$

Solving this last equation for $\mathcal{L}\{f(t)\}$ immediately yields the desired result, Eq. (1).

EXAMPLE 1 Determine the Laplace transform of the sawtooth function $f(t) = t$, $0 \leq t < 1, f(t+1) = f(t)$, depicted in Fig. 6.2(b).

Solution. From Eq. (1) we get

$$\mathcal{L}\{f(t)\} = \frac{1}{1 - e^{-s}}\int_0^1 te^{-st}\,dt.$$

Integration by parts yields

$$\begin{aligned}\mathcal{L}\{f(t)\} &= \frac{1}{1 - e^{-s}}\left[-\frac{te^{-st}}{s} - \frac{e^{-st}}{s^2}\right]_0^1 \\ &= \frac{1}{1 - e^{-s}}\left(-\frac{e^{-s}}{s} - \frac{e^{-s}}{s^2} + \frac{1}{s^2}\right) \\ &= \frac{1}{1 - e^{-s}}\left(\frac{1 - e^{-s}}{s^2} - \frac{e^{-s}}{s}\right) \\ &= \frac{1 - (s+1)e^{-s}}{s^2(1 - e^{-s})} \\ &= \frac{1 + s - (s+1)e^{-s} - s}{s^2(1 - e^{-s})} \\ &= \frac{1+s}{s^2} - \frac{1}{s(1 - e^{-s})}.\end{aligned}$$

EXAMPLE 2 Find the inverse transform $\mathscr{L}^{-1}\left\{\dfrac{1}{(s+1)(1-e^{-s})}\right\}$.

Solution. Recall from calculus the geometric series

$$\frac{1}{1-r} = 1 + r + r^2 + r^3 + \cdots, \qquad |r| < 1.$$

Thus if $r = e^{-s}$, we have

$$\frac{1}{1-e^{-s}} = 1 + e^{-s} + e^{-2s} + e^{-3s} + \cdots, \qquad s > 0.$$

The Laplace transform under investigation can then be written as

$$\frac{1}{(s+1)(1-e^{-s})} = \frac{1}{s+1}(1 + e^{-s} + e^{-2s} + \cdots).$$

For each positive integer k the second translation theorem gives

$$\begin{aligned}\mathscr{L}^{-1}\left\{\frac{e^{-ks}}{s+1}\right\} &= e^{-t}\big|_{t\to t-k}U(t-k)\\ &= e^{-(t-k)}U(t-k).\end{aligned}$$

Therefore applying this result term by term to the problem at hand yields

$$\begin{aligned}f(t) &= \mathscr{L}^{-1}\left\{\frac{1}{(s+1)(1-e^{-s})}\right\}\\ &= e^{-t} + e^{-(t-1)}U(t-1) + e^{-(t-2)}U(t-2) + \cdots.\end{aligned}$$

Let us interpret $f(t)$ geometrically.

Consider the interval $n < t < n + 1$ for some fixed positive integer n. Over this interval we have

$$\begin{aligned}f(t) &= e^{-t} + e^{-(t-1)} + \cdots + e^{-(t-n)}\\ &= e^{-t}(1 + e + e^2 + \cdots + e^n)\\ &= e^{-t}\left(\frac{e^{n+1}-1}{e-1}\right)\\ &= \frac{e^{-t+n+1}}{e-1} - \frac{e^{-t}}{e-1}, \qquad n < t < n+1.\end{aligned}$$

The second term on the right is a continuous exponential function decaying rapidly to zero as t increases. The first term ranges from $e/(e-1)$ at $t = n$ to $1/(e-1)$ at $t = n + 1$, *regardless of the value of* n. Therefore, at each integer value for t this first term has a jump discontinuity with jump value

$$\frac{e}{e-1} - \frac{1}{e-1} = 1.$$

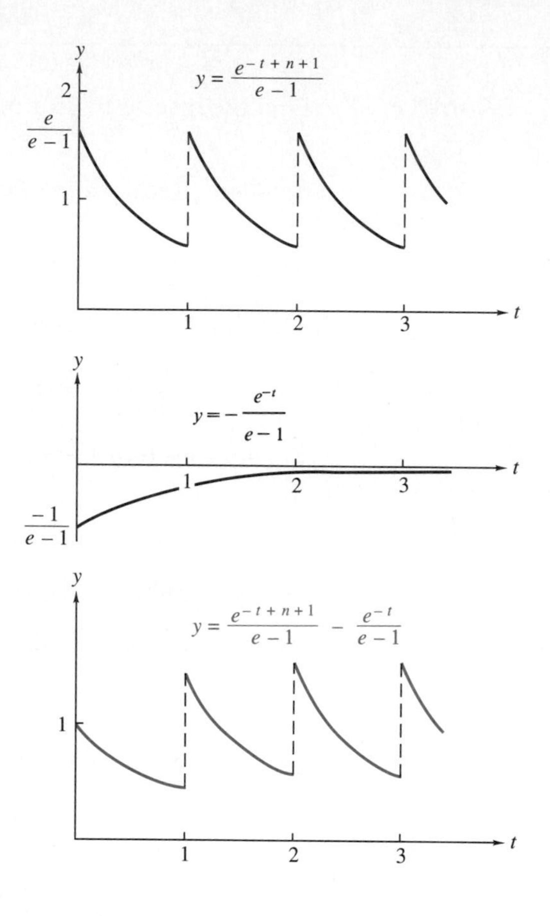

FIGURE 6.3
Solution to $\mathscr{L}^{-1}\{F(s)\}$ where $F(s) = 1/(s+1)(1-e^{-s})$. Note that the top and middle graphs plot the individual terms of the entire function shown in the bottom graph.

A graph of the inverse transform $f(t)$ is presented in Fig. 6.3.

A procedure similar to that used in Example 2 can be used to show that the graph of

$$f(t) = \mathscr{L}^{-1}\left\{\frac{1}{(s+a)(1-e^{-ks})}\right\}$$

is qualitatively the same as that presented in Fig. 6.3 (see Exercises 6.1, problem 20).

EXAMPLE 3 The differential equation for the current $i(t)$ in a series circuit containing a resistor and an inductor is

$$L\frac{di}{dt} + Ri = E(t) \tag{2}$$

where $E(t)$ is the impressed voltage. Assuming that $E(t)$ is the sawtooth function from Example 1 and that $i(0) = 0$, solve the equation for the current $i(t)$ if $L = 1$ henry and $R = 5$ ohms.

Solution. Using the Laplace transform technique we get

$$\mathcal{L}\left\{\frac{di}{dt}\right\} + 5\mathcal{L}\{i\} = \mathcal{L}\{E(t)\},$$

$$sI(s) - i(0) + 5I(s) = \frac{1-(s+1)e^{-s}}{s^2(1-e^{-s})}.$$

After solving for $I(s)$ and using the initial condition $i(0) = 0$, we have

$$\begin{aligned} I(s) &= \frac{1-(s+1)e^{-s}}{s^2(s+5)(1-e^{-s})} \\ &= \frac{1}{s^2(s+5)}\left(\frac{1}{1-e^{-s}}\right) - \frac{s+1}{s^2(s+5)}\left(\frac{e^{-s}}{1-e^{-s}}\right). \end{aligned}$$

From partial fraction decomposition (with the details left to you) we get

$$\frac{1}{s^2(s+5)} = \frac{1/25}{s+5} - \frac{1/25}{s} + \frac{1/5}{s^2}$$

and

$$\frac{s+1}{s^2(s+5)} = \frac{-4/25}{s+5} + \frac{4/25}{s} + \frac{1/5}{s^2}.$$

As in Example 2 the rule for the geometric series yields

$$\frac{1}{1-e^{-s}} = 1 + e^{-s} + e^{-2s} + \cdots, \qquad s > 0,$$

and

$$\frac{e^{-s}}{1-e^{-s}} = e^{-s} + e^{-2s} + e^{-3s} + \cdots, \qquad s > 0.$$

Substitution of these expressions into $I(s)$ yields

$$\begin{aligned} I(s) &= \frac{1}{25}\left(\frac{1}{s+5} - \frac{1}{s} + \frac{5}{s^2}\right)(1 + e^{-s} + e^{-2s} + \cdots) \\ &\quad + \frac{1}{25}\left(\frac{4}{s+5} - \frac{4}{s} - \frac{5}{s^2}\right)(e^{-s} + e^{-2s} + e^{-3s} + \cdots) \\ &= \frac{1}{25}\left(\frac{1}{s+5} - \frac{1}{s} + \frac{5}{s^2}\right) \\ &\quad + \frac{1}{25}\left(\frac{5}{s+5} - \frac{5}{s}\right)(e^{-s} + e^{-2s} + e^{-3s} + \cdots). \end{aligned}$$

From the second translation theorem, for each positive integer k we have

$$\mathscr{L}^{-1}\left\{\frac{e^{-ks}}{s+5}\right\} = U(t-k)e^{-5(t-k)}$$

and

$$\mathscr{L}^{-1}\left\{\frac{e^{-ks}}{s}\right\} = U(t-k).$$

Applying these results term by term to find the inverse of $I(s)$ leads to

$$\begin{aligned} i(t) &= \mathscr{L}^{-1}\{I(s)\} \\ &= \frac{1}{25}(e^{-5t} - 1 + 5t) + \frac{1}{5}\sum_{k=1}^{\infty}(e^{-5(t-k)} - 1)U(t-k). \end{aligned}$$

A procedure similar to that used in Example 2, although tedious, can be followed to determine the graph of $i(t)$.

EXERCISES 6.1

In problems 1–10 find the Laplace transform of the periodic functions.

1. $f(t) = \begin{cases} 1, & 0 \leq t < 1 \\ 0, & 1 \leq t < 2 \end{cases}$
$f(t+2) = f(t)$

2. $\begin{cases} 1, & 0 \leq t < 1 \\ -1, & 1 \leq t < 2 \end{cases}$
$f(t+2) = f(t)$

3. $f(t) = \begin{cases} t, & 0 \leq t < 1 \\ 0, & 1 \leq t < 2 \end{cases}$
$f(t+2) = f(t)$

4. $f(t) = \begin{cases} 1-t, & 0 \leq t < 1 \\ 0, & 1 \leq t < 2 \end{cases}$
$f(t+2) = f(t)$

5. $f(t) = \begin{cases} t, & 0 \leq t < 1 \\ 2-t, & 1 \leq t < 2 \end{cases}$
$f(t+2) = f(t)$

6. $f(t) = \begin{cases} 1-t, & 0 \leq t < 2 \\ t-3, & 2 \leq t < 4 \end{cases}$
$f(t+4) = f(t)$

7. $f(t) = \sin t,\ \ 0 \leq t < \pi$
$f(t+\pi) = f(t)$

8. $f(t) = \begin{cases} \sin t, & 0 \leq t < \pi \\ 0, & \pi \leq t < 2\pi \end{cases}$
$f(t+2\pi) = f(t)$

9. $f(t) = \cos t,\ \ 0 \leq t < 2\pi,$
$f(t+2\pi) = f(t)$

10. $f(t) = t^2,\ \ 0 \leq t < 1$
$f(t+1) = f(t)$

11. Solve Eq. (2) subject to the initial condition $i(0) = 0$ when $L = 1$ henry and $R = 10$ ohms for $f(t)$ as defined in problem 1.

12. Solve Eq. (2) subject to the initial condition $i(0) = 0$ when $L = 1$ henry and $R = 5$ ohms for $f(t)$ as defined in problem 5.

13. Solve Eq. (2) subject to the initial condition $i(0) = 0$ when $L = 1$ henry and $R = 10$ ohms for $f(t)$ as defined in problem 7.

14. Solve $y'' - 3y' + 2y = f(t)$, $y(0) = y'(0) = 0$ where $f(t)$ is given in problem 2.

15. Solve $y'' + 2y' + y = f(t)$, $y(0) = y'(0) = 0$ where $f(t)$ is given in problem 3.

16. Given

$$f(t) = \begin{cases} 1, & 0 < t < 1, \\ -1, & 1 < t < 2, \end{cases} \quad \text{and} \quad f(t+2) = f(t),$$

show that

$$\mathscr{L}\{f(t)\} = \frac{1}{s} \tanh \frac{s}{2}.$$

17. Use the periodic formula in Eq. (1) to find $\mathscr{L}\{\sin kt\}$.

18. The **half-wave rectification** of the sine wave is defined by

$$f(t) = \begin{cases} \sin t, & 0 < t < \pi, \\ 0, & \pi < t < 2\pi, \end{cases}$$

where $f(t + 2\pi) = f(t)$ (see Fig. 6.4b). Show that

$$\mathscr{L}\{f(t)\} = \frac{1}{(s^2 + 1)(1 - e^{-\pi s})}.$$

19. The **full-wave rectification** of the sine wave is the absolute value $|\sin t|$ (see Fig. 6.4a).

a) Show that the half-wave rectification of the sine wave in problem 18 satisfies the relation

$$f(t) = \frac{1}{2}(|\sin t| + \sin t).$$

b) Using the result in problem 18, establish that

$$\mathscr{L}\{|\sin t|\} = \frac{1}{s^2+1}\left(\frac{1 + e^{-\pi s}}{1 - e^{-\pi s}}\right) = \frac{1}{s^2+1} \coth \frac{\pi s}{2}.$$

20. Let $f(t) = \mathscr{L}^{-1}\left\{\dfrac{1}{(s + a)(1 - e^{-ks})}\right\}$

a) As in Example 2, show that

$$f(t) = e^{-at} + e^{-a(t-k)}U(t - k) + e^{-a(t-2k)}U(t - 2k) + \cdots = e^{-at}\sum_{n=0}^{\infty} e^{ank}U(t - nk).$$

b) Over the interval $nk < t < (n + 1)k$, show that

$$f(t) = \frac{e^{-a\tau}}{e^{ak} - 1} - \frac{e^{-at}}{e^{ak} - 1}$$

where $\tau = t - (n + 1)k$.

c) From part (b) show that the first term $e^{-a\tau}/(e^{ak} - 1)$ has a jump discontinuity equal to 1 at each of the points $t = k, 2k, 3k, \ldots$.

d) Deduce that the graph of $f(t)$ is qualitatively the same as that described in Fig. 6.3.

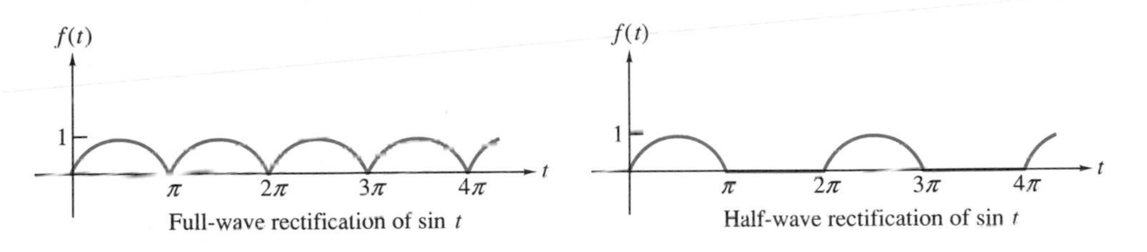

FIGURE 6.4
(a) Full-wave rectification of sin t; (b) Half-wave rectification of sin t.

6.2 IMPULSE FORCING FUNCTIONS

If a nail is struck by a hammer, or a golf ball is struck by a club, or the weighted end of an oscillating spring is struck by a sharp blow, the force applied is very large in magnitude but of short duration. A similar force might be experienced in driving an automobile over a New York pothole. Let us consider such a sharp blow to be a constant force acting over a short time interval $t_0 \leqslant t \leqslant t_1$. From Newton's second law we know that

$$F = \frac{d}{dt}(mv),$$

so integration with respect to t gives us

$$\int_{t_0}^{t_1} F\,dt = mv_1 - mv_0$$

or

$$F(t_1 - t_0) = mv(t_1) - mv(t_0).$$

The expression $F\,\Delta t$ on the lefthand side of this last equation is known as the **impulse.** Thus the impulse equals the change in momentum over the time interval Δt. If the time interval is arbitrarily small, we might think of this impulse as occurring from a force that changes the velocity *instantaneously* without altering mass or position. To this end it is useful to define formally a force "function" $\delta(t)$ satisfying the following conditions:

1. $\delta(t - t_0) = 0, \qquad t \neq t_0,$

2. $\displaystyle\int_{-\infty}^{\infty} \delta(t - t_0)\,dt = 1.$

Clearly $\delta(t)$ is no ordinary function since it is zero everywhere except at a single point yet is required to have a nonzero integral over the entire real number line. However, we might think of conditions 1 and 2 as suggestive of some kind of limiting process. A rigorous mathematical theory has been developed that establishes the validity of such a "generalized function," but the theory is beyond the scope of this text. Nevertheless, we can provide a description of the impulse function, and when working with impulse problems we operate formally on it as if it were an ordinary function. Physicists have found this approach to be very successful in modeling physical systems involving impulse forces.

Unit Impulse

To begin our description consider the following unit step function acting over a short time interval $-\epsilon < t < \epsilon$ and defined by

$$\delta_\epsilon(t) = \begin{cases} \dfrac{1}{2\epsilon}, & -\epsilon < t < \epsilon, \\ 0, & |t| \geqslant \epsilon. \end{cases} \tag{1}$$

A plot of $\delta_\epsilon(t)$ for three values of ϵ is presented in Fig. 6.5. Notice that the area

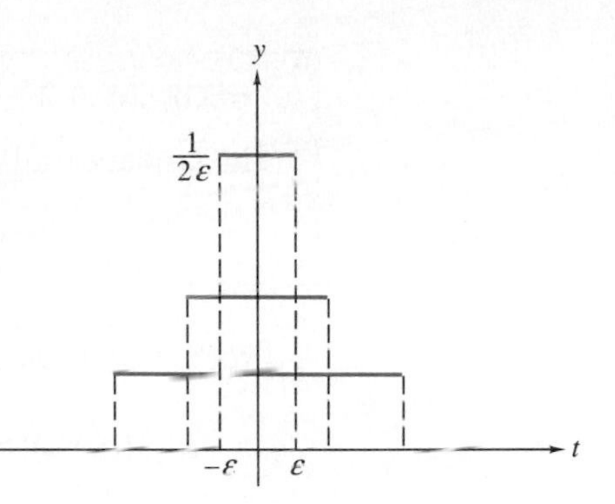

FIGURE 6.5
Function $\delta_\epsilon(t)$ for three different values of ϵ. The area under each curve is always 1.

under $\delta_\epsilon(t)$ for any value of ϵ is always 1:

$$\int_{-\infty}^{\infty} \delta_\epsilon(t)\, dt = \int_{-\epsilon}^{\epsilon} \frac{1}{2\epsilon}\, dt = \frac{1}{2\epsilon}(2\epsilon) = 1. \tag{2}$$

As ϵ becomes smaller and smaller, the function $\delta_\epsilon(t)$ acts over a shorter and shorter time period. Thus we formally define the following "generalized" function:

$$\delta(t) = \lim_{\epsilon \to 0} \delta_\epsilon(t) \tag{3}$$

with the formal property that

$$\int_{-\infty}^{\infty} \delta(t)\, dt = \lim_{\epsilon \to 0} \int_{-\infty}^{\infty} \delta_\epsilon(t)\, dt = 1. \tag{4}$$

The generalized function $\delta(t)$ is called the **Dirac delta** or **unit impulse function.*** The translation

$$\delta(t - a) \tag{5}$$

then gives a unit impulse at time $t = a$ where a is assumed to be positive.

Laplace Transform of the Unit Impulse

Let us treat $\delta(t - a)$ as if it were an ordinary function and formally compute its Laplace transform.

* This function was introduced by the physicist Paul A. M. Dirac in 1932. In 1933 he was a recipient of the Nobel Prize in physics for his work in quantum theory. Later Laurent Schwartz justified mathematically the use of the delta function.

THEOREM 6.2

The Laplace transform of the unit impulse function is given by

$$\mathscr{L}\{\delta(t-a)\} = e^{-sa}, \qquad a > 0, \quad s > 0. \tag{6}$$

Proof. From the definition of the Laplace transform we have

$$\begin{aligned}
\mathscr{L}\{\delta(t-a)\} &= \lim_{\epsilon\to 0} \int_0^\infty e^{-st}\delta_\epsilon(t-a)\,dt \\
&= \lim_{\epsilon\to 0} \int_{-a}^\infty e^{-s(u+a)}\delta_\epsilon(u)\,du, \qquad \text{where } u = t-a \\
&= \lim_{\epsilon\to 0} e^{-sa} \int_{-\epsilon}^{\epsilon} e^{-su}\cdot\frac{1}{2\epsilon}\,du \\
&= \lim_{\epsilon\to 0} e^{-sa}\left[-\frac{1}{2s\epsilon}e^{-su}\right]_{-\epsilon}^{\epsilon} \\
&= \lim_{\epsilon\to 0} e^{-sa}\left(\frac{e^{s\epsilon}-e^{-s\epsilon}}{2s\epsilon}\right) \\
&= e^{-sa}\lim_{\epsilon\to 0}\frac{\sinh s\epsilon}{s\epsilon}.
\end{aligned}$$

After applying l'Hôpital's rule to the righthand side (differentiating with respect to ϵ), we find

$$\begin{aligned}
\mathscr{L}\{\delta(t-a)\} &= e^{-sa}\lim_{\epsilon\to 0}\frac{s\cosh s\epsilon}{s} \\
&= e^{-sa}.
\end{aligned}$$

Theorem 6.2 can be extended to the case when $a = 0$ by formally defining

$$\mathscr{L}\{\delta(t)\} = \lim_{a\to 0}\mathscr{L}\{\delta(t-a)\} = 1. \tag{7}$$

Impulse Forcing Functions

The Laplace transform method can be used to solve constant-coefficient linear differential equations when the forcing function is a unit impulse.

EXAMPLE 1 Solve $y' + 2y = \delta(t-1)$ subject to the initial condition $y(0) = 0$.

Solution. Taking the Laplace transform of each side, we get

$$\begin{aligned}
\mathscr{L}\{y'\} + 2\mathscr{L}\{y\} &= \mathscr{L}\{\delta(t-1)\}, \\
sY(s) - y(0) + 2Y(s) &= e^{-s}.
\end{aligned}$$

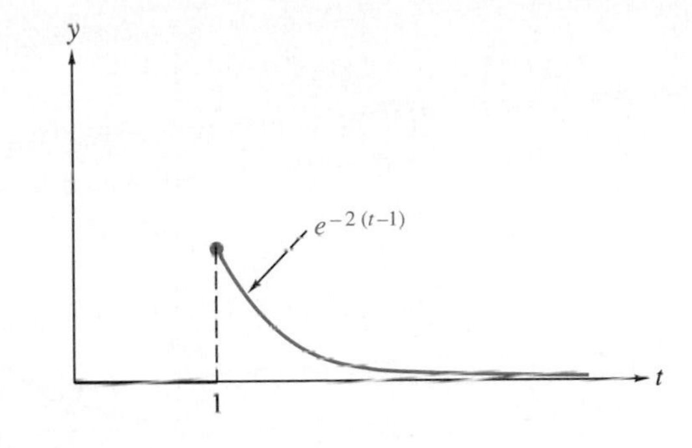

FIGURE 6.6
$y(t) = e^{-2(t-1)}U(t-1)$.

Solving for $Y(s)$ gives us

$$Y(s) = \frac{e^{-s}}{s+2}.$$

Applying the inverse form of the second translation theorem results in

$$\begin{aligned} y(t) &= \mathscr{L}^{-1}\left\{\frac{e^{-s}}{s+2}\right\} \\ &= e^{-2t}\big|_{t\to t-1}\, U(t-1) \\ &= e^{-2(t-1)}U(t-1). \end{aligned}$$

The graph of $y(t)$ is displayed in Fig. 6.6. In this example note that the impulse forcing function resulted in a discontinuous solution with distinct left and right derivatives at the point of discontinuity $t = 1$. The left derivative has a value of 0, and the right derivative has a value of -2.

EXAMPLE 2 Solve $y'' - 2y' + y = 3\delta(t-2)$ subject to the initial conditions $y(0) = 0$ and $y'(0) = 1$.

Solution. Taking the Laplace transform of each side gives us

$$\begin{aligned} \mathscr{L}\{y''\} - 2\mathscr{L}\{y'\} + \mathscr{L}\{y\} &= 3\mathscr{L}\{\delta(t-2)\}, \\ s^2Y(s) - sy(0) - y'(0) - 2sY(s) + 2y(0) + Y(s) &= 3e^{-2s}, \\ (s^2 - 2s + 1)Y(s) - 1 &= 3e^{-2s}, \end{aligned}$$

or

$$Y(s) = \frac{3e^{-2s}}{(s-1)^2} + \frac{1}{(s-1)^2}.$$

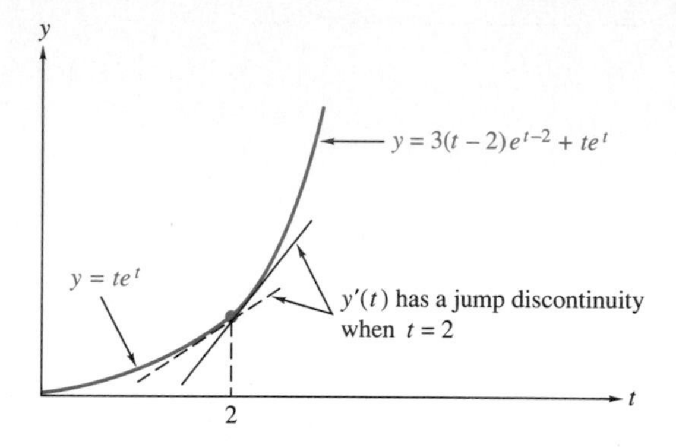

FIGURE 6.7
$y(t) = 3(t-2)e^{t-2}U(t-2) + te^t$.

Taking inverse transforms, we get

$$\begin{aligned}
y(t) &= \mathcal{L}^{-1}\left\{\frac{3e^{-2s}}{(s-1)^2}\right\} + \mathcal{L}^{-1}\left\{\frac{1}{(s-1)^2}\right\} \\
&= 3\mathcal{L}^{-1}\left\{\frac{1}{(s-1)^2}\right\}\Bigg|_{t\to t-2} U(t-2) + \mathcal{L}^{-1}\left\{\frac{1}{(s-1)^2}\right\} \\
&= 3e^t\mathcal{L}^{-1}\left\{\frac{1}{s^2}\right\}\Bigg|_{t\to t-2} U(t-2) + e^t\mathcal{L}^{-1}\left\{\frac{1}{s^2}\right\} \\
&= 3te^t|_{t\to t-2}\,U(t-2) + te^t \\
&= 3(t-2)e^{t-2}U(t-2) + te^t.
\end{aligned}$$

A plot of $y(t)$ is shown in Fig. 6.7. Note in this example that $y(t)$ is continuous at $t = 2$ and has one-sided derivatives there. However, the left and right derivatives differ in value, so $y(t)$ is not differentiable at $t = 2$.

Unit step functions, such as $\delta_\epsilon(t)$ in Eq. (1), and unit impulse functions $\delta(t)$ appear in practice only as approximations. When a physical quantity changes by a finite amount over a time interval so short in duration that a fraction of the interval has no practical interest, and when the new value of the quantity is assumed to be retained over the short interval, the quantity may be specified as a step function. For instance, the closing of a switch connecting a power source like a battery to a circuit can be analyzed by assuming that a step function of voltage is applied to the circuit. If the impulse voltage takes place over shorter and shorter time intervals, so for all practical purposes the length of the interval ϵ tends toward zero in Eq. (1), the corresponding sequence of solutions tends toward a function that is the solution to the problem with a unit impulse function. In that situation it is convenient to work directly with the impulse function $\delta(t-a)$ and its properties. This idea is illustrated in the next example.

EXAMPLE 3 In Section 3.2 we modeled a simple RC circuit (that is, having no inductance) by the equation

$$Rq' + \frac{1}{C}q = E(t) \tag{8}$$

where R is the resistance, C is the capacitance, q is the charge, and $E(t)$ is the applied voltage (see Fig. 6.8a). Differentiation of both sides of Eq. (8) leads to

$$Rq'' + \frac{1}{C}q' = E'(t).$$

Since the current satisfies $i = dq/dt$, this last result can be expressed as a first-order equation in i:

$$Ri' + \frac{1}{C}i = E'(t). \tag{9}$$

Suppose initially that $i(0) = 0$; also suppose that by switching on the circuit the voltage $E(t)$ is brought from zero to the constant voltage v_0 almost instantaneously and then held at that value for all subsequent time. Although we do not know exactly how $E(t)$ behaves over the short time interval, let us assume for definiteness that $E(t)$ is a linear function over $0 \leq t \leq \epsilon$. The situation is presented graphically in Fig. 6.8b.

Let us approximate $E'(t)$ by an impulse function:

$$E'(t) = v_0\delta(t), \qquad t > 0.$$

Substitution into Eq. (9) yields the equation

$$Ri' + \frac{1}{C}i = v_0\delta(t). \tag{10}$$

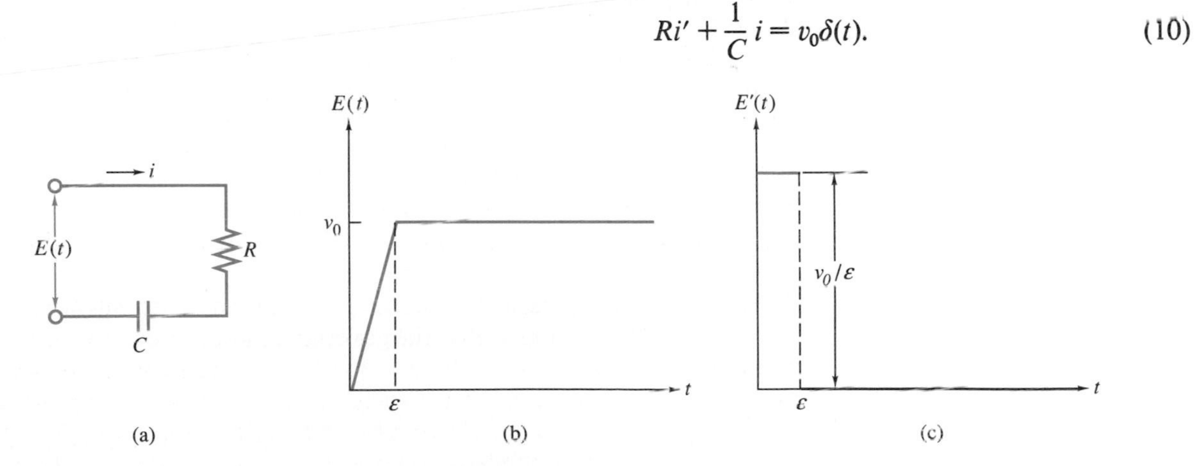

FIGURE 6.8
(a) An RC circuit that when switched on has an applied voltage $E(t)$ brought almost instantaneously from $v = 0$ to the constant value $v = v_0$. (b) For definiteness we assume the graph of $E(t)$ to be linear when $0 \leq t \leq \epsilon$. (c) The voltage derivative $E'(t)$.

After taking the Laplace transform of each side of Eq. (10) we find

$$RsI(s) - Ri(0) + \frac{1}{C} I(s) = v_0.$$

Since $i(0) = 0$, this last equation simplifies to

$$\left(Rs + \frac{1}{C}\right)I(s) = v_0.$$

Solving for the transform of the current we obtain

$$I(s) = \frac{v_0}{R}\left(\frac{1}{s + \frac{1}{RC}}\right). \tag{11}$$

Then taking the inverse transform we get the current

$$i(t) = \frac{v_0}{R} e^{-t/RC}. \tag{12}$$

It may seem disturbing that the initial condition $i(0) = 0$ is *not* satisfied in solution (12). The reason for this is that the derivative $E'(t)$ on the right side of Eq. (9) is approximated by the delta function $v_0\delta(t)$. This impulse as input at $t = 0$ causes a jump discontinuity at $t = 0$ from $i(0) = 0$ to the value $i(0^+) = v_0/R$ obtained from Eq. (12). In Exercises 6.2, problem 17 we will outline an alternative method for solving Eq. (9) by using step functions for $E'(t)$ rather than the delta function.

EXAMPLE 4 Consider a mass–spring system with no resisting force that is oscillating vertically according to simple harmonic motion. At time $t = T$ the weighted end of the spring is struck a hard and sudden vertical blow. Assuming that $y(0) = y_0$ and $y'(0) = v_0$, describe the motion of the spring for all time $t > 0$.

Solution. From our discussion of the automobile suspension system in Sections 3.4 and 4.5, we know that the governing initial value problem is given by

$$y'' + \omega^2 y = C\delta(t - T)$$

subject to the initial conditions $y(0) = y_0$ and $y'(0) = v_0$, where $\omega^2 = k/m$ is the design ratio associated with the spring and $C > 0$ multiplies the unit impulse force (quantifying the magnitude of the blow). The impulse force could model an automobile running over a curb. Then

$$\mathscr{L}\{y''\} + \omega^2\mathscr{L}\{y\} = C\mathscr{L}\{\delta(t-T)\},$$
$$s^2Y(s) - sy(0) - y'(0) + \omega^2Y(s) = Ce^{-sT},$$
$$(s^2+\omega^2)Y(s) - sy_0 - v_0 = Ce^{-sT},$$
$$Y(s) = \left(\frac{s}{s^2+\omega^2}\right)y_0 + \left(\frac{1}{s^2+\omega^2}\right)v_0 + \frac{Ce^{-sT}}{s^2+\omega^2}$$

Taking inverse transforms we get

$$y(t) = \left(y_0\cos\omega t + \frac{v_0}{\omega}\sin\omega t\right) + \frac{C}{\omega}\sin\omega t|_{t\to t-T}\,U(t-T)$$

or

$$y(t) = y_0\cos\omega t + \frac{v_0}{\omega}\sin\omega t + \left(\frac{C}{\omega}\right)U(t-T)\sin\omega(t-T).$$

Because $\sin\omega(t-T)=0$ when $t=T$, the function $y(t)$ is continuous for all values of $t>0$. However, the derivative $y'(t)$ has a jump discontinuity at $t=T$ due to the sudden impulse force at that time.

Products of Functions with the Delta Function

From calculus you are accustomed to thinking about a function in terms of its *value* at each point of the domain (often specified by a formula). Another way to think about a function, which is particularly revealing for generalized functions, is to consider how it behaves in association with other functions. The following result presents the Dirac delta function from this point of view.

THEOREM 6.3

If $f(t)$ is a continuous function for $t \geqslant 0$, then for each $a>0$

$$\int_0^\infty f(t)\delta(t-a)\,dt = f(a). \tag{13}$$

Proof. Applying the definition of the delta function

$$\int_0^\infty f(t)\delta(t-a)\,dt = \lim_{\epsilon\to 0}\int_0^\infty f(t)\delta_\epsilon(t-a)\,dt$$
$$= \lim_{\epsilon\to 0}\int_{a-\epsilon}^{a+\epsilon} f(t)\cdot\frac{1}{2\epsilon}\,dt.$$

The Mean Value Theorem for integral calculus tells us there exists a point t^* such that $a - \epsilon < t^* < a + \epsilon$ and

$$\frac{1}{2\epsilon}\int_{a-\epsilon}^{a+\epsilon} f(t)\,dt = \frac{1}{2\epsilon} f(t^*)[(a+\epsilon) - (a-\epsilon)] = f(t^*).$$

As $\epsilon \to 0$, $t^* \to a$ so the continuity of f yields

$$\int_0^\infty f(t)\delta(t-a)\,dt = \lim_{\epsilon\to 0} f(t^*) = f(a)$$

as claimed.

EXAMPLE 5 Solve $y'' + y = \delta(t-\pi)\cos t$ subject to the initial conditions $y(0) = 1$ and $y'(0) = 0$.

Solution. After taking the Laplace transform of each side we find that

$$\mathscr{L}\{y''\} + \mathscr{L}\{y\} = \mathscr{L}\{\delta(t-\pi)\cos t\},$$

$$s^2Y(s) - sy(0) - y'(0) + Y(s) = \int_0^\infty (e^{-st}\cos t)\delta(t-\pi)\,dt.$$

After combining the terms on the lefthand side and applying Theorem 6.3 to the integral on the right, we have

$$(s^2+1)Y(s) - s = e^{-s\pi}\cos\pi.$$

Solving for $Y(s)$ gives us

$$Y(s) = \frac{s}{s^2+1} - \frac{e^{-s\pi}}{s^2+1}.$$

Taking inverse transforms,

$$\begin{aligned} y(t) &= \cos t - [\sin t|_{t\to t-\pi} U(t-\pi)] \\ &= \cos t - \sin(t-\pi)U(t-\pi) \\ &= \cos t + U(t-\pi)\sin t. \end{aligned}$$

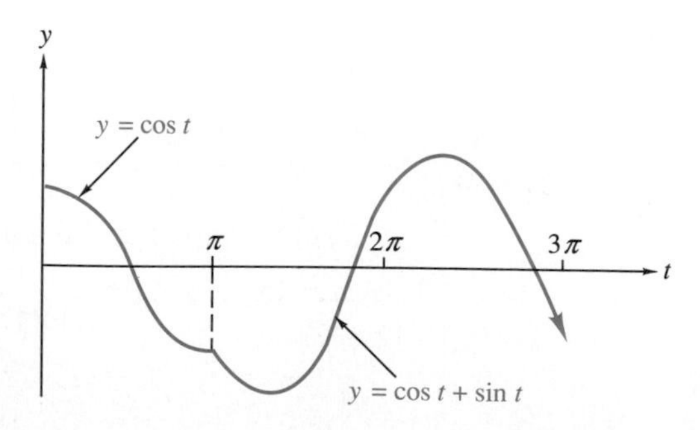

FIGURE 6.9
$y(t) = \cos t + U(t-\pi)\sin t$.

The graph of $y(t)$ is displayed in Fig. 6.9. Notice that $y(t)$ is everywhere continuous, but that $y'(t)$ has a jump discontinuity at $t = \pi$ (characterized by the sharp "corner" existing there on the graph).

EXERCISES 6.2

In problems 1–16 find the solution to the given initial value problem using the Laplace transform.

1. $y' - 2y = \delta(t-3), \quad y(0) = 0$
2. $y' + 3y = \delta(t-1), \quad y(0) = 1$
3. $y' - 3y = \delta(t-2), \quad y(0) = -1$
4. $y' + 4y = 3\delta\left(t - \frac{1}{2}\right), \quad y(0) = 2$
5. $y' + y = t\delta(t-2), \quad y(0) = 0$
6. $y'' - 5y' + 4y = \delta(t-1), \quad y(0) = 0, y'(0) = 0$
7. $y'' + 3y' - 4y = \delta(t-2), \quad y(0) = 0, y'(0) = 2$
8. $y'' + y' - 2y = 4\delta(t-1), \quad y(0) = 1, y'(0) = 1$
9. $y'' - 4y' - 5y = 2\delta(t-2), \quad y(0) = -1, y'(0) = 0$
10. $y'' + 16y = \delta(t - 2\pi), \quad y(0) = 0, y'(0) = 0$
11. $y'' + 4y' + 4y = \delta\left(t - \frac{1}{2}\right), y(0) = 0, y'(0) = -1$
12. $y'' - 6y' + 9y = \delta(t-2), \quad y(0) = -1, y'(0) = 1$
13. $y'' - 2y' = 3 + \delta(t-1), \quad y(0) = 0, y'(0) = 0$
14. $y'' + 4y = t\delta(t-\pi), \quad y(0) = 1, y'(0) = 0$
15. $y'' - 4y = t\delta(t-2), \quad y(0) = 0, y'(0) = 1$
16. $y'' + y = \delta(t - 2\pi)\sin t, \quad y(0) = 0, y'(0) = 1$
17. The derivative $E'(t)$ depicted in Fig. 6.8(c) can be written in terms of step functions as

$$E'(t) = \frac{v_0}{\epsilon} U(\epsilon - t).$$

a) Show that

$$\mathcal{L}\{E'(t)\} = \left(\frac{v_0}{\epsilon}\right)\left(\frac{1 - e^{-\epsilon s}}{s}\right)$$

(see Exercises 5.5, problem 45).

b) For a fixed value of $\epsilon > 0$, use Laplace transforms to derive the solution

$$i(t, \epsilon) = \frac{Cv_0}{\epsilon}\{1 - e^{-t/RC} - U(t-\epsilon)[1 - e^{-(t-\epsilon)/RC}]\}$$

to the equation

$$Ri' + \frac{1}{C}i = \frac{v_0}{\epsilon} U(\epsilon - t).$$

c) Verify that $i(0,\epsilon) = 0$ for every $\epsilon > 0$.

d) Use l'Hôpital's rule to establish the limit

$$\lim_{\epsilon \to 0} i(t,\epsilon) = \begin{cases} \frac{v_0}{R} e^{-t/RC}, & t > 0, \\ 0, & t = 0. \end{cases}$$

Thus, if $i(t) = \lim_{\epsilon \to 0} i(t,\epsilon)$, then $i(t)$ *does* include the jump discontinuity from 0 to v_0/R at $t = 0$.

6.3 CONVOLUTION INTEGRAL

We have seen that the inverse Laplace transform is *additive:* that is, the inverse transform of a sum of transforms is the sum of the inverse transforms:

$$\mathcal{L}^{-1}\{F(s) + G(s)\} = \mathcal{L}^{-1}\{F(s)\} + \mathcal{L}^{-1}\{G(s)\}.$$

But what about the inverse transform of a product? Quite often we encounter functions of s that are clearly products of two easy Laplace transforms.

However, we know the inverse transform of a product is *not* the product of the inverse transforms from the simple observation that

$$t = \mathscr{L}^{-1}\left\{\frac{1}{s^2}\right\} = \mathscr{L}^{-1}\left\{\frac{1}{s}\cdot\frac{1}{s}\right\} \neq \mathscr{L}^{-1}\left\{\frac{1}{s}\right\}\mathscr{L}^{-1}\left\{\frac{1}{s}\right\} = 1\cdot 1 = 1.$$

That is,

$$\mathscr{L}^{-1}\{F(s)G(s)\} \neq \mathscr{L}^{-1}\{F(s)\}\mathscr{L}^{-1}\{G(s)\}$$

in general. The following result, called the **convolution theorem** for reasons that will become apparent shortly, gives the inverse transform of a product.

THEOREM 6.4

Let $f(t)$ and $g(t)$ be piecewise continuous functions for $t \geqslant 0$ and of exponential order. Let $F(s) = \mathscr{L}\{f(t)\}$ and $G(s) = \mathscr{L}\{g(t)\}$. Then

$$\mathscr{L}^{-1}\{F(s)G(s)\} = \int_0^t f(z)g(t-z)\,dz. \tag{1}$$

We are going to postpone the proof of Theorem 6.4 until the end of this section; for now let us interpret Eq. (1). Note that the integral on the right-hand side of Eq. (1) is a function of the independent variable t: the integration of $f(z)g(t-z)$ is taking place with respect to the dummy variable z for each (fixed) value of the variable t as z varies from $z = 0$ to $z = t$. Thus for each fixed value of t the integral on the right produces a unique real number. The theorem states that the resulting function of t is precisely the inverse transform of the product $F(s)G(s)$. The integral on the righthand side of Eq. (1) is known as the **convolution of f and g,** denoted by $f * g$. Symbolically,

$$(f * g)(t) = \int_0^t f(z)g(t-z)\,dz \tag{2}$$

Basic Properties of the Convolution

The convolution of two functions produces another function $f * g$. Hence Eq. (2) defines a new operation on functions. Theorem 6.4 gives the Laplace transform of this new function:

$$\mathscr{L}\left\{\int_0^t f(z)g(t-z)\,dz\right\} = F(s)G(s)$$

or

$$\mathscr{L}\{(f * g)(t)\} = \mathscr{L}\{f(t)\}\mathscr{L}\{g(t)\}. \tag{3}$$

The convolution also satisfies the following properties.

Commutative law:	$f * g = g * f,$
Associative law:	$f * (g * h) = (f * g) * h,$
Distributive law:	$f * (g + h) = (f * g) + (f * h).$

Examples 1–3 illustrate how the convolution integral can be used to calculate inverse transforms of products.

EXAMPLE 1 Find $\mathscr{L}^{-1}\left\{\frac{1}{s^2(s-1)}\right\}$.

Solution. Here we have

$$F(s) = \frac{1}{s^2} = \mathscr{L}\{t\} \qquad \text{and} \qquad g(s) = \frac{1}{s-1} = \mathscr{L}\{e^t\}.$$

Thus, from Eq. (1), we get

$$\begin{aligned}
\mathscr{L}^{-1}\left\{\frac{1}{s^2(s-1)}\right\} &= \int_0^t ze^{t-z}\,dz \\
&= e^t \int_0^t ze^{-z}\,dz \\
&= e^t[-ze^{-z} - e^{-z}]_0^t \\
&= e^t(-te^{-t} - e^{-t} + 1) \\
&= e^t - (t + 1).
\end{aligned}$$

Note that we could also use partial fractions to find the inverse transform:

$$\frac{1}{s^2(s-1)} = \frac{1}{s-1} - \frac{1}{s^2} - \frac{1}{s}.$$

However, our purpose here was to demonstrate the use of convolution.

EXAMPLE 2 Find $\mathscr{L}^{-1}\left\{\frac{s}{(s^2+4)^2}\right\}$.

Solution. We first observe that

$$F(s) = \frac{s}{s^2+4} = \mathscr{L}\{\cos 2t\}$$

and

$$G(s) = \frac{1}{s^2+4} = \frac{1}{2}\mathcal{L}\{\sin 2t\}.$$

Thus

$$\mathcal{L}^{-1}\left\{\frac{s}{(s^2+4)^2}\right\} = \frac{1}{2}\int_0^t \cos 2z \sin 2(t-z)\, dz.$$

Recall from trigonometry that

$$\sin(A+B) = \sin A \cos B + \cos A \sin B$$

and

$$\sin(A-B) = \sin A \cos B - \cos A \sin B.$$

Subtracting the second equation from the first gives the identity

$$\cos A \sin B = \frac{1}{2}[\sin(A+B) - \sin(A-B)].$$

Substituting $A = 2z$ and $B = 2(t-z)$ in the above integral gives us

$$\begin{aligned}
\mathcal{L}^{-1}\left\{\frac{s}{(s^2+4)^2}\right\} &= \frac{1}{4}\int_0^t [\sin 2t - \sin(4z-2t)]\, dz \\
&= \left[\frac{z}{4}\sin 2t + \frac{1}{16}\cos(4z-2t)\right]_{z=0}^{z=t} \\
&= \frac{t}{4}\sin 2t + \frac{1}{16}\cos 2t - \frac{1}{16}\cos(-2t) \\
&= \frac{t}{4}\sin 2t.
\end{aligned}$$

Equation (3) can be used to calculate the Laplace transform of a convolution.

EXAMPLE 3 Evaluate $\mathcal{L}\{e^{-t} * e^t \cos t\}$.

Solution. From Eq. (3) we have

$$\begin{aligned}
\mathcal{L}\{e^{-t} * e^t\cos t\} &= \mathcal{L}\{e^{-t}\}\,\mathcal{L}\{e^t\cos t\} \\
&= \frac{1}{s+1}\left(\frac{s-1}{(s-1)^2+1}\right) \\
&= \frac{s-1}{(s+1)(s^2-2s+2)}.
\end{aligned}$$

Transform of an Integral

In the special case that $g(t) \equiv 1$, Eq. (3) gives the Laplace transform of an integral.

THEOREM 6.5

If $f(t)$ is piecewise continuous for $t \geq 0$ and of exponential order, then

$$\mathscr{L}\left\{\int_0^t f(z)\, dz\right\} = \frac{1}{s} F(s) \tag{4}$$

where $F(s) = \mathscr{L}\{f(t)\}$.

Proof. The function $g(t) \equiv 1$ is continuous and of exponential order with $\mathscr{L}\{1\} = 1/s = G(s)$. Thus from Eq. (3) we know that

$$\begin{aligned}\mathscr{L}\left\{\int_0^t f(z)\, dz\right\} &= \mathscr{L}\left\{\int_0^t f(z)g(t-z)\, dz\right\}\\ &= \mathscr{L}\{f(t)\}\, \mathscr{L}\{g(t)\}\\ &= \frac{F(s)}{s}.\end{aligned}$$

EXAMPLE 4 Evaluate $\mathscr{L}\left\{\int_0^t ze^z \sin 2z\, dz\right\}$.

Solution. Applying Eq. (4) we have

$$\begin{aligned}\mathscr{L}\left\{\int_0^t ze^z \sin 2z\, dz\right\} &= \frac{1}{s}\mathscr{L}\{te^t \sin 2t\}\\ &= \frac{1}{s}\left[\mathscr{L}\{t \sin 2t\}\right]_{s\to s-1}\\ &= \frac{1}{s}\left[\frac{4s}{(s^2+4)^2}\right]_{s\to s-1} \qquad \text{(Example 2)}\\ &= \frac{4(s-1)}{s[(s-1)^2+4]^2}.\end{aligned}$$

EXAMPLE 5 RESPONSE OF AN *RC* CIRCUIT TO A SQUARE WAVE VOLTAGE

Suppose an electrical circuit containing a resistor and capacitor has an induced voltage

$$E(t) = \begin{cases} v_0, & a < t < b,\\ 0, & \text{otherwise.}\end{cases}$$

The graph of $E(t)$ is a simple square wave with height v_0, as shown in Fig. 6.10.

The induced voltage $E(t)$ can be written as a difference of unit step functions times the constant voltage v_0:

$$E(t) = v_0[U(t-a) - U(t-b)].$$

Then, assuming that the initial magnitude of the current is $i(0) = 0$, the differential equation modeling the magnitude of the current is

$$Ri + \frac{1}{C}q = v_0[U(t-a) - U(t-b)]$$

(see Section 3.2). Since $i = dq/dt$, this last equation can be written as

$$Ri + \frac{1}{C}\int_0^t i\,dt = v_0[U(t-a) - U(t-b)].$$

After taking the Laplace transform of both sides of this equation and using Theorem 6.5 to give us the transform of an integral, we find that

$$RI(s) + \frac{1}{sC}I(s) = v_0\left[\frac{e^{-as}}{s} - \frac{e^{-bs}}{s}\right].$$

Solving algebraically for the transform yields

$$I(s) = \frac{v_0}{R\left(s + \dfrac{1}{RC}\right)}(e^{-as} - e^{-bs}).$$

After inverting the transform we get

$$\begin{aligned} i(t) &= \frac{v_0}{R}e^{-t/RC}\Big|_{t\to t-a}U(t-a) - \left[\frac{v_0}{R}e^{-t/RC}\Big|_{t\to t-b}U(t-b)\right] \\ &= \frac{v_0}{R}[e^{-(t-a)/RC}U(t-a) - e^{-(t-b)/RC}U(t-b)]. \end{aligned}$$

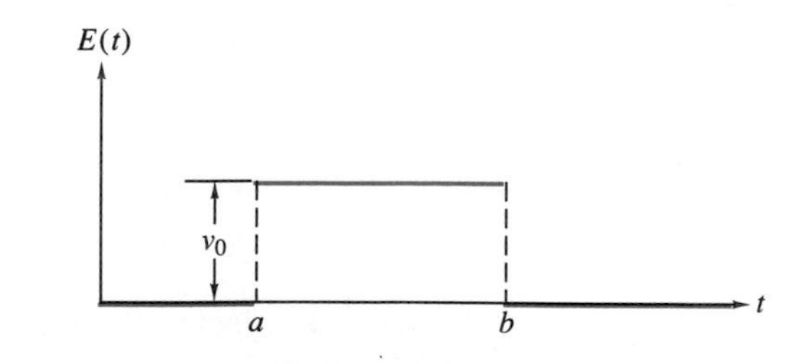

FIGURE 6.10
A square wave voltage $E(t)$.

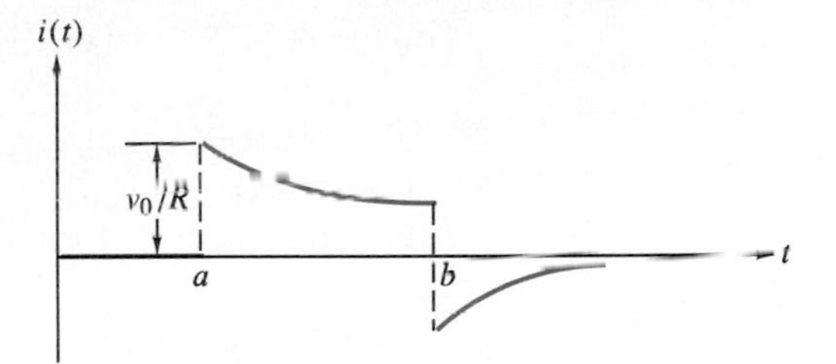

FIGURE 6.11
Plot of the current $i(t)$ flowing through an RC circuit with induced voltage given by a square wave of height v_0 over the interval $a < t < b$.

The current $i(t)$ is plotted in Fig. 6.11. Note the discontinuities in the solution at $t = a$ and $t = b$.

Some Integral Equations

Because of the convolution theorem, the Laplace transform is useful in solving certain **integral equations** of the form

$$\int_0^t y(z)g(t-z)\,dz = f(t) \tag{5}$$

or

$$y(t) + \int_0^t y(z)g(t-z)\,dz = f(t). \tag{6}$$

In Eqs. (5) and (6), f and g are known functions and $y(t)$ is the unknown function to be found that satisfies the given integral equation. The following examples illustrate this solution method.

EXAMPLE 6 Find the function $y(t)$ satisfying the integral equation

$$y(t) + \int_0^t y(z)(t-z)\,dz = t.$$

Solution. From the convolution theorem we have

$$\mathscr{L}\{y(t)\} + \mathscr{L}\left\{\int_0^t y(z)(t-z)dz\right\} = \mathscr{L}\{t\},$$

$$\mathscr{L}\{y(t)\} + \mathscr{L}\{y(t)\}\mathscr{L}\{t\} = \mathscr{L}\{t\},$$

$$Y(s) + Y(s)\frac{1}{s^2} = \frac{1}{s^2},$$

$$\frac{1+s^2}{s^2}Y(s) = \frac{1}{s^2}.$$

Thus

$$Y(s) = \frac{1}{1 + s^2}.$$

Taking inverse transforms of both sides gives us

$$y(t) = \sin t.$$

EXAMPLE 7 Find the function $y(t)$ satisfying

$$y(t) - \int_0^t ty(t - z)\, dz = te^t.$$

Solution. Applying the Laplace transform to both sides yields

$$\mathscr{L}\{y(t)\} - \mathscr{L}\{t\}\mathscr{L}\{y(t)\} = \mathscr{L}\{te^t\},$$

$$Y(s) - \frac{1}{s^2}\, Y(s) = \frac{1}{(s-1)^2}.$$

Thus

$$Y(s) = \frac{s^2}{(s^2 - 1)(s - 1)^2} = \frac{s^2}{(s + 1)(s - 1)^3}.$$

Using partial fraction decomposition, we get

$$\frac{s^2}{(s + 1)(s - 1)^3} = \frac{A}{s + 1} + \frac{B}{s - 1} + \frac{C}{(s - 1)^2} + \frac{D}{(s - 1)^3},$$

$$s^2 = A(s - 1)^3 + B(s - 1)^2(s + 1) + C(s - 1)(s + 1) + D(s + 1).$$

If $s = 1$, then $1 = 2D$, or $D = \frac{1}{2}$.
If $s = -1$, then $1 = -8A$, or $A = -\frac{1}{8}$.
If $s = 0$, then $0 = -A + B - C + D$, or $B - C = -\frac{5}{8}$.
If $s = 2$, then $4 = A + 3B + 3C + 3D$, or $B + C = \frac{7}{8}$.

Then $B = 1/8$ and $C = 3/4$ yielding

$$Y(s) = -\frac{1/8}{s + 1} + \frac{1/8}{s - 1} + \frac{3/4}{(s - 1)^2} + \frac{1/2}{(s - 1)^3}.$$

Taking the inverse transform of each term gives us

$$y(t) = -\frac{e^{-t}}{8} + \left(\frac{1}{8} + \frac{3t}{4} + \frac{t^2}{4}\right) e^t.$$

Nonhomogeneous Initial Value Problems with Initial Conditions Equal to Zero

The Laplace transform method, in conjunction with convolution, can be useful in solving certain constant-coefficient nonhomogeneous second-order initial value problems where the method of variation of parameters may be more difficult to apply. This application derives from the following result.

THEOREM 6.6

Suppose $y = f(t)$ is the solution of the constant-coefficient homogeneous initial value problem

$$y'' + Py' + Qy = 0, \qquad y(0) = 0, \quad y'(0) = 1.$$

Then the solution of the nonhomogeneous initial value problem

$$y'' + Py' + Qy = g(t), \qquad y(0) = 0, \quad y'(0) = 0,$$

is the convolution of f and g:

$$y(t) = (f * g)(t) = \int_0^t f(z)g(t - z)\, dz.$$

Proof. The Laplace transform of the homogeneous initial value problem yields

$$s^2Y(s) - sy(0) - y'(0) + P[sY(s) - y(0)] + QY(s) = 0$$

or

$$(s^2 + Ps + Q)Y(s) = 1.$$

Since $f(t)$ solves this initial value problem, we have

$$f(t) = \mathcal{L}^{-1}\{Y(s)\} = \mathcal{L}^{-1}\left\{\frac{1}{s^2 + Ps + Q}\right\}.$$

Now consider the nonhomogeneous case. Again taking the Laplace transform we get

$$s^2Y(s) - sy(0) - y'(0) + P[sY(s) - y(0)] + QY(s) = G(s)$$

or

$$(s^2 + Ps + Q)Y(s) = G(s).$$

In this case

$$Y(s) = \frac{G(s)}{s^2 + Ps + Q}.$$

The convolution theorem tells us the solution is given by

$$\begin{aligned}\mathscr{L}^{-1}\{Y(s)\} &= \mathscr{L}^{-1}\left\{\frac{1}{s^2+Ps+Q}\,G(s)\right\}\\ &= \int_0^t f(z)g(t-z)\,dz\\ &= f*g.\end{aligned}$$

EXAMPLE 8 Solve the nonhomogeneous initial value problem

$$y''-y=\frac{1}{e^t+e^{-t}}, \qquad y(0)=y'(0)=0.$$

Solution. The solution of the homogeneous initial value problem

$$y''-y=0, \qquad y(0)=0, \quad y'(0)=1$$

is

$$y=f(t)=\sinh t.$$

Thus according to Theorem 6.6 the solution to the given nonhomogeneous initial value problem is the convolution

$$\begin{aligned}(f*g)(t) &= \int_0^t \frac{\sinh(t-z)}{e^z+e^{-z}}\,dz\\ &= \int_0^t \frac{\sinh(t-z)}{2\cosh z}\,dz\\ &= \frac{1}{2}\int_0^t \frac{\sinh t\cosh z-\cosh t\sinh z}{\cosh z}\,dz\\ &= \frac{1}{2}\int_0^t (\sinh t-\cosh t\tanh z)\,dz\\ &= \frac{1}{2}\,[z\sinh t-(\cosh t)\ln(\cosh z)]_{z=0}^{z=t}\\ &= \frac{1}{2}\left[t\sinh t-(\cosh t)\ln(\cosh t)\right].\end{aligned}$$

Proof of the Convolution Theorem

By definition of the Laplace transform the product of $F(s)$ and $G(s)$ is given by

$$
\begin{aligned}
F(s)G(s) &= \int_0^\infty e^{-sz}f(z)\,dz \int_0^\infty e^{-s\tau}g(\tau)\,d\tau \\
&= \int_0^\infty f(z)\,dz \int_0^\infty e^{-s(z+\tau)}g(\tau)\,d\tau \\
&= \int_0^\infty f(z)\,dz \int_z^\infty e^{-st}g(t-z)\,dt, \qquad \text{where } \tau = t - z.
\end{aligned}
$$

The region of integration of the iterated integral is shown in Fig. 6.12.

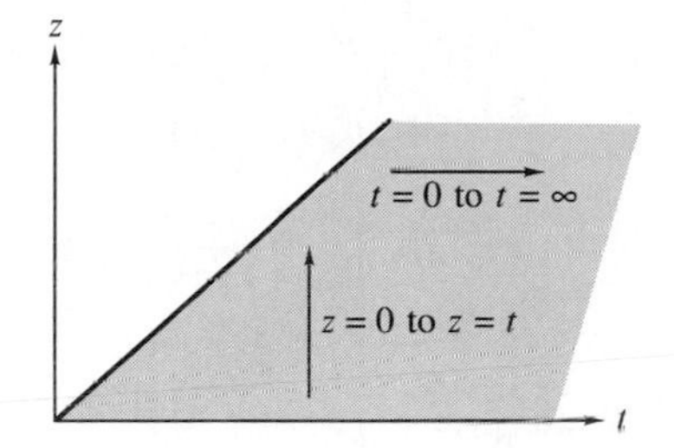

FIGURE 6.12
Region of integration for $F(s)G(s)$.

Interchanging the order of integration results in

$$
\begin{aligned}
F(s)G(s) &= \int_0^\infty \int_z^\infty e^{-st}f(z)g(t-z)\,dt\,dz \\
&= \int_0^\infty \int_0^t e^{-st}f(z)g(t-z)\,dz\,dt \\
&= \int_0^\infty e^{-st}\left[\int_0^t f(z)g(t-z)\,dz\right]dt \\
&= \int_0^\infty e^{-st}\,(f*g)(t)\,dt \\
&= \mathscr{L}\{(f*g)(t)\}.
\end{aligned}
$$

EXERCISES 6.3

In problems 1–12 determine the indicated Laplace transform.

1. $\mathscr{L}\left\{\int_0^t e^z \sin(t-z)\,dz\right\}$

2. $\mathscr{L}\left\{\int_0^t e^{-2z}\cos(t-z)\,dz\right\}$

3. $\mathscr{L}\left\{\int_0^t ze^{2(t-z)}\,dz\right\}$

4. $\mathscr{L}\left\{\int_0^t e^{-z}(t-z)^3\,dz\right\}$

5. $\mathscr{L}\left\{\int_0^t (t-z)^2\sin z\,dz\right\}$

6. $\mathscr{L}\left\{\int_0^t \sin z\cos(t-z)\,dz\right\}$

7. $\mathscr{L}\{1 * t^2\}$

8. $\mathscr{L}\{t^3 * t^2\}$

9. $\mathscr{L}\{e^{-t} * t^3\}$

10. $\mathscr{L}\{t^2 * te^t\}$

11. $\mathscr{L}\{e^{2t} * \cos t\}$

12. $\mathscr{L}\{e^{-2t} * e^t\sin t\}$

In problems 13–22 find the indicated inverse transform using convolution.

13. $\mathscr{L}^{-1}\left\{\dfrac{1}{(s+1)(s-2)}\right\}$

14. $\mathscr{L}^{-1}\left\{\dfrac{1}{s^3(s+1)}\right\}$

15. $\mathscr{L}^{-1}\left\{\dfrac{1}{(s-1)(s^2+1)}\right\}$

16. $\mathscr{L}^{-1}\left\{\dfrac{s}{(s-1)(s^2+9)}\right\}$

17. $\mathscr{L}^{-1}\left\{\dfrac{s}{(s+2)(s^2+2)}\right\}$

18. $\mathscr{L}^{-1}\left\{\dfrac{2}{s^2(s^2-4)}\right\}$

19. $\mathscr{L}^{-1}\left\{\dfrac{1}{(s+1)(s^2-2s+2)}\right\}$

20. $\mathscr{L}^{-1}\left\{\dfrac{s+1}{(s^2+2s+5)(s-1)}\right\}$

21. $\mathscr{L}^{-1}\left\{\dfrac{s-2}{s(s^2-4s+13)}\right\}$

22. $\mathscr{L}^{-1}\left\{\dfrac{1}{(s^2-4s+5)^2}\right\}$

In problems 23–30 find the indicated Laplace transform.

23. $\mathscr{L}\left\{\int_0^t ze^{2z}\,dz\right\}$

24. $\mathscr{L}\left\{\int_0^t e^{-z}\sin z\,dz\right\}$

25. $\mathscr{L}\left\{\int_0^t e^{z}\cos 2z\,dz\right\}$

26. $\mathscr{L}\left\{\int_0^t \delta(z-1)\,dz\right\}$

27. $\mathscr{L}\left\{\int_0^t e^{-z}\delta(z-1)\,dz\right\}$

28. $\mathscr{L}\left\{t\int_0^t ze^{-z}\,dz\right\}$

29. $\mathscr{L}\left\{t\int_0^t \cos z\,dz\right\}$

30. $\mathscr{L}\left\{\int_0^t e^{z}\delta(z-2)\,dz\right\}$

In problems 31–44 solve the given integral equation.

31. $\int_0^t zy(t-z)\,dz = t^2$

32. $\int_0^t y(t-z)\cos z\,dz = \sin t$

33. $\int_0^t e^{t-z}y(z)\,dz = t$

34. $\int_0^t y(z)\sin(t-z)\,dz = t^2$

35. $y(t) = 4t - 3\int_0^t y(z)\sin(t-z)\,dz$

36. $y(t) = t^2 + \int_0^t y(z)\sin(t-z)\,dz$

37. $y(t) = te^t + \int_0^t zy(t-z)\,dz$

38. $y(t) = \sin t + \int_0^t e^{t-z}y(z)\,dz$

39. $y(t) = t^2 - 2\int_0^t y(t-z)\sinh 2z\,dz$

40. $y(t) + \int_0^t y(z)(t-z)\,dz = 1$

41. $y(t) + \int_0^t y(z)\,dz = t$

42. $y(t) + 2\int_0^t y(t-z)\cos z\,dz = 1$

43. $y(t) + \int_0^t e^{t-z}y(z)\,dz = 1$

44. $y(t) - \int_0^t e^{-t}y(t-z)\,dz = e^t$

45. Using the Laplace transform technique, solve the following **integrodifferential equation**

$$y'(t) = 1 + \int_0^t y(z)\,dz, \quad y(0) = 0.$$

In problems 46–49 solve the nonhomogeneous initial value problem using convolution and Theorem 6.6.

46. $y'' + y = e^t, \quad y(0) = 0, y'(0) = 0$

47. $y'' - 3y' + 2y = \cos e^{-t}, \quad y(0) = 0, y'(0) = 0$

48. $y'' + y' = \sin t, \quad y(0) = 0, y'(0) = 0$

49. $y'' + y = \cos t, \quad y(0) = 0, y'(0) = 0$

6.4

AN APPLICATION TO CONTROL SYSTEMS (Optional)

Consider the nonhomogeneous second-order initial value problem

$$ay'' + by' + cy = u(t) \tag{1}$$

subject to the initial conditions

$$y(0) = y_0 \quad \text{and} \quad y'(0) = y_0'.$$

We assume that a, b, and c are constants and that $u(t)$ is a function possessing a Laplace transform. For instance, $u(t)$ could be the unit impulse function $\delta(t)$. For each such function $u(t)$ a solution $y(t)$ to initial value problem (1) exists. Thus we can think of Eq. (1) as a black-box machine: *input* $u(t)$ results in *output* $y(t)$. We call $y(t)$ the **response function** of the input $u(t)$.

Response Function

Let us examine more closely the nature of the output response. We know from our previous work in Chapter 4 that the solution $y(t)$ to Eq. (1) has the form

$$y = y_c + y_p \tag{2}$$

where y_c is the complementary function solving the associated homogeneous equation

$$ay'' + by' + cy = 0, \tag{3}$$

and y_p is any particular solution solving nonhomogeneous Eq. (1). In order for y to satisfy the initial conditions in Eq. (1), we choose y_p in such a way that $y_p(0) = y_p'(0) = 0$. Then $y_c(0) = y_0$ and $y_c'(0) = y_0'$.

Taking the Laplace transform of Eq. (3) yields the following equation for the transform $Y_c(s)$ of the complementary function:

$$as^2Y_c(s) - asy_0 - ay_0' + bsY_c(s) - by_0 + cY_c(s) = 0$$

or

$$Y_c(s) = \frac{1}{as^2 + bs + c}[a(sy_0 + y_0') + by_0]. \tag{4}$$

Likewise, taking the Laplace transform of Eq. (1) coupled with the initial conditions for y_p yields the following equation for $Y_p(s)$:

$$Y_p(s) = \frac{1}{as^2 + bs + c} U(s), \tag{5}$$

where

$$U(s) = \mathcal{L}\{u(t)\}.$$

Transfer Function

Equations (4) and (5) both contain the factor

$$H(s) = \frac{1}{as^2 + bs + c}. \tag{6}$$

The function $H(s)$ is called the **transfer function** of Eq. (1). The transfer function and the initial conditions $y(0) = y_0$ and $y'(0) = y_0'$ together produce the complementary solution $y_c(t) = \mathcal{L}^{-1}\{Y_c(s)\}$. On the other hand, the transfer function and the input function $u(t)$ produce the particular solution $y_p(t) = \mathcal{L}^{-1}\{Y_p(s)\}$. Therefore $u(t)$ is said to control the response $y_p(t)$ and is thus called the **control function.** This **control system** is depicted in Fig. 6.13. From the figure we can think of the black box for Eq. (1) as the transfer function H operating on the initial conditions and the forcing function separately to produce the output response. If we think of H as "fixed," then $u(t)$ controls the output $y_p(t)$.

Now what exactly is the transfer function? It is the Laplace transform of some function $h(t)$; that is, $H(s) = \mathcal{L}\{h(t)\}$. In fact, if the forcing function in Eq. (1) is set equal to the unit impulse function $\delta(t)$, then the Laplace transform of

$$ay'' + by' + cy = \delta(t) \tag{7}$$

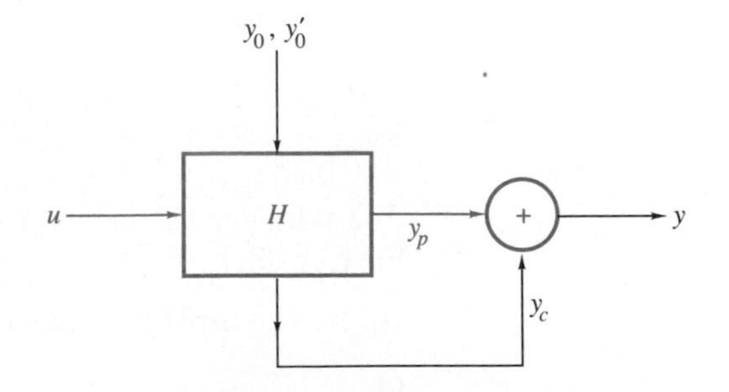

FIGURE 6.13

A control system with transfer function $H(s)$. The input function $u(t)$ controls the response $y_p(t)$, and the initial conditions determine $y_c(t)$.

with initial conditions

$$y(0)=0 \quad \text{and} \quad y'(0)=0$$

leads to

$$as^2Y(s)+bsY(s)+cY(s)=1$$

or

$$Y(s)=\frac{1}{as^2+bs+c}.$$

Thus $Y(s)$ is precisely the transfer function $H(s)$. This means that $H(s)$ is the Laplace transform of the response function $h(t)$ that solves Eq. (7). That is, $h(t)$ is a particular solution to Eq. (1) when the forcing function is the unit impulse function $\delta(t)$ applied at $t=0$. Thus it is natural to call $h(t)$ the **impulse response** of the control system. Since $H(s)=\mathscr{L}\{h(t)\}$, *the transfer function is therefore the Laplace transform of the impulse response.*

Response Function for Positive Coefficients

Suppose the constants a, b, and c in Eq. (1) are all positive. This is the situation for damped forced motion, for instance, as investigated in Section 4.5 and modeled by

$$my''+\delta y'+ky=f(t).$$

It is also the situation for an LRC circuit modeled by

$$Lq''+Rq'+\frac{1}{C}q=E(t).$$

The roots to the auxiliary polynomial for

$$ay''+by'+cy=0$$

are given by

$$m=\frac{-b\pm\sqrt{b^2-4ac}}{2a}. \tag{8}$$

Recall the three cases:

1. $\boldsymbol{b^2-4ac>0}$. Roots m_1 and m_2 are both real and negative, where

$$m_1=\frac{-b+\sqrt{b^2-4ac}}{2a} \quad \text{and} \quad m_2=\frac{-b-\sqrt{b^2-4ac}}{2a}.$$

Thus

$$y_c=c_1e^{m_1t}+c_2e^{m_2t}. \tag{9}$$

2. **$b^2 - 4ac = 0$.** The roots are repeated and negative such that

$$m_1 = m_2 = -\frac{b}{2a}.$$

Thus

$$y_c = c_1 e^{m_1 t} + c_2 t e^{m_1 t}. \tag{10}$$

3. **$b^2 - 4ac < 0$.** The roots are complex with a negative real part and are given by

$$m_1 = \alpha + \beta i \text{ and } m_2 = \alpha - \beta i,$$

where $\alpha = -b/2a$ and $\beta = \sqrt{4ac - b^2}/2a$. Thus

$$y_c = e^{\alpha t}(c_1 \cos \beta t + c_2 \sin \beta t). \tag{11}$$

In each of Eqs. (9–11) the solution $y_c(t) \to 0$ as $t \to +\infty$. Therefore $y_c(t)$ and the initial conditions y_0 and y_0' become increasingly less important to response function (2) as t gets larger and larger. The response tends toward the **steady-state** solution y_p. Moreover, from Eqs. (5) and (6),

$$y_p = \mathcal{L}^{-1}\{H(s)U(s)\} = h(t) * u(t). \tag{12}$$

Interpreting our results we see that *for positive coefficients, the response tends to the steady state given by the convolution of the impulse response with the control function.* Thus we can see what is meant by saying that the input u controls the (eventual) output response y_p. Initially, y_p and y_p', have the value 0.

EXAMPLE 1 Find the transfer function $H(s)$ and the impulse response for undamped forced motion

$$y'' + \omega^2 y = \frac{F_0}{m} \cos \gamma t, \qquad \frac{F_0}{m} = \text{constant}, \quad \gamma \neq \omega$$

such that $y(0) = y'(0) = 0$. Also determine the steady-state response.

Solution. The transfer function is

$$H(s) = \frac{1}{s^2 + \omega^2},$$

and the impulse response is

$$h(t) = \mathcal{L}^{-1}\left\{\frac{1}{s^2 + \omega^2}\right\} = \frac{1}{\omega} \sin \omega t.$$

Then the steady-state response is the convolution

$$h(t) * u(t) = \int_0^t \frac{1}{\omega} \sin \omega\tau \cdot \left(\frac{F_0}{m}\right) \cos \gamma(t-\tau)\, d\tau$$
$$= \frac{F_0}{m\omega} \int_0^t \sin \omega\tau \cos \gamma(t-\tau)\, d\tau.$$

If we take the trigonometric identity

$$\sin A \cos B = \frac{1}{2}[\sin(B+A) - \sin(B-A)],$$

and set $A = \omega\tau$ and $B = \gamma(t-\tau)$, then the convolution is

$$h(t) * u(t) = \frac{F_0}{2m\omega} \int_0^t \{\sin[(\omega-\gamma)\tau + \gamma t] - \sin[\gamma t - (\omega+\gamma)\tau]\}\, d\tau$$
$$= \frac{F_0}{2m\omega}\left[-\frac{\cos[(\omega-\gamma)\tau+\gamma t]}{\omega-\gamma} - \frac{\cos[\gamma t - (\omega+\gamma)\tau]}{\omega+\gamma}\right]_0^t$$
$$= \frac{F_0}{2m\omega}\left(-\frac{\cos \omega t}{\omega-\gamma} - \frac{\cos(-\omega t)}{\omega+\gamma} + \frac{\cos \gamma t}{\omega-\gamma} + \frac{\cos \gamma t}{\omega+\gamma}\right)$$
$$= \frac{F_0}{2m\omega}\left(\frac{-(\omega+\gamma)\cos \omega t - (\omega-\gamma)\cos \omega t + (\omega+\gamma)\cos \gamma t + (\omega-\gamma)\cos \gamma t}{\omega^2-\gamma^2}\right)$$
$$= \frac{F_0}{2m\omega(\omega^2-\gamma^2)}(-2\omega \cos \omega t + 2\omega \cos \gamma t)$$
$$= \frac{F_0}{m(\omega^2-\gamma^2)}(\cos \gamma t - \cos \omega t), \quad \omega \neq \gamma.$$

This result agrees with our findings for undamped forced motion presented in Section 4.5, Eq. (9).

EXERCISES 6.4

In problems 1–10 find the transfer function and the impulse response for the specified control system. Then find the steady-state response.

1. $y'' + 5y' + 6y = 2e^t, \quad y(0) = 1, y'(0) = 0$
2. $y'' + 3y' + 2y = 3e^{3t}, \quad y(0) = 0, y'(0) = 1$
3. $y'' + 4y' + 3y = te^{-t}, \quad y(0) = 1, y'(0) = 1$
4. $y'' + 7y' + 10y = \sin t, \quad y(0) = 0, y'(0) = 0$
5. $2y'' + 7y' + 3y = 2\cos t, \quad y(0) = 1, y'(0) = 0$
6. $y'' + 4y = t, \quad y(0) = 0, y'(0) = -1$
7. $y'' + 4y = 5\sin t, \quad y(0) = 1, y'(0) = -1$
8. $y'' + 9y = 6\sin 2t, \quad y(0) = 0, y'(0) = 1$
9. $y'' + 16y = 4\cos 3t, \quad y(0) = 0, y'(0) = -1$
10. $y'' + 2y' + y = 3t, \quad y(0) = 0, y'(0) = 0$
11. Find the transfer function and the impulse response for the undamped forced motion

$$y'' + \omega^2 y = \frac{F_0}{m}\cos \omega t, \quad y(0) = y'(0) = 0.$$

Determine the steady-state response.

12. Find the transfer function and the impulse response for the LR circuit

$$L\frac{di}{dt} + Ri = E(t), \quad i(0) = i_0$$

where L and R are positive constants. What is the steady-state response?

CHAPTER 6 REVIEW EXERCISES

1. Solve the initial value problem

$$y' + 4y + 3\int_0^t y\,dt = \begin{cases} 1, & 0 \le t < 2, \\ -1, & 2 \le t < 4, \\ 0, & 4 \le t \end{cases} \quad y(0) = 0.$$

2. We wish to determine information about a cannon system (see Chapter 3 Review Exercises, problem 20). The gun tube–breech block assembly has a mass of 1500 kg. The recoil spring has a spring constant of 19,500 N/m. The damping mechanism exerts a force (in newtons) numerically equal to 9000 times the instantaneous velocity of the gun tube–breech block assembly (in meters per second). When the cannon is fired, the gas pressure from the firing of the round exerts an impulse force of 40,500 N on the gun tube–breech block assembly. Except for the instant of firing, the force is otherwise zero.

a) Express the gas pressure force in terms of the unit impulse function.

b) Model this system as an initial value problem and find the particular solution that gives the displacement of the gun tube–breech block assembly with respect to time.

3. A shock absorber–coil spring system for an imported automobile is designed to support 350 kg. The spring has a constant of 140,000 kg/cm. The shock absorber exerts a damping force (in kilogram centimeters per second squared) numerically equal to 3500 times the instantaneous vertical velocity of the system (in centimeters per second). The system is in equilibrium at rest when it hits a pothole that exerts an impulse force of 5250 N upward on the system.

a) Express the force exerted by the pothole in terms of the unit impulse function.

b) Model this system with an initial value problem and solve it.

c) Graph the resulting equation for the first 3 sec of motion.

d) What is the maximum displacement from equilibrium that the system experiences?

4. You know that the differential equation that describes an LRC circuit is

$$E(t) = Lq'' + Rq' + \frac{1}{C}q,$$

where $q(t)$ is the charge on the capacitor at any time t. The charge on the capacitor is related to the current in the circuit by $i(t) = q'(t)$, where $i(t)$ represents the current in the circuit at any time t. We can then derive an integrodifferential equation describing the current in a circuit from the differential equation for charge on the capacitor:

$$E(t) = Li' + Ri + \frac{1}{C}\int_0^t i(\tau)\,d\tau.$$

A simple LRC electrical circuit consists of a capacitor of capacitance 1/60 F, a resistor of 50-Ω resistance, and an inductor with an inductance of 10 H. The circuit is connected to an alternating current source given (in volts) by $104\cos 2t$. Assuming no initial current in the circuit, use the integrodifferential equation for current shown above to find an equation for current in the circuit at any time. What is the current after 1 sec?

5. In problem 49 of the Chapter 5 Review Exercises, we investigated the shear stress on a structural girder of a tall building being designed to withstand the effects of earthquakes. A simple model that approximates the crosswise forces acting on a building girder is the familiar equation $mx'' + bx' + kx = f(t)$, where $x(t)$ is the crosswise displacement of the girder; $f(t)$ is the net applied shear stress; and m, b, and k are positive constants dependent on the characteristics of the girder. The constants of the girder that we investigated were $m = 1$, $b = 5$, and $k = 6$. Assume that this girder is in equilibrium at rest when an earthquake strikes, exerting an exponentially decaying periodic shear stress of $10\,e^{-t}$, which repeats every 2 secs.

a) Find the equation for the crosswise displacement of the girder over time.

b) What is the crosswise displacement of the girder after 1 sec? After 3 sec?

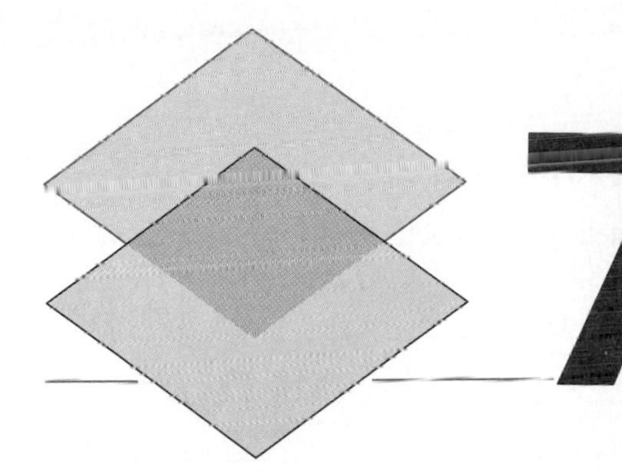

7

Systems of First-Order Differential Equations and Applications

In the model refinement process we may find it necessary to consider more than one dependent variable in order to obtain an adequate description of the behavior being studied. For instance, several lakes may be interconnected by canals allowing some of the lakes to flow into the others. We may wish to study the effects that polluting one lake has on the other lakes. Examples of other behaviors having more than one dependent variable are (1) electrical networks with more than one loop (giving rise to a current in each of the loops); (2) several masses connected together by springs with associated displacements for each mass; and (3) interacting populations of different species such as wolves and rabbits. Models of such behaviors are given by several differential equations involving the derivatives of the various dependent variables. A set of equations involving the ordinary derivatives of more than one dependent variable forms a *system* of ordinary differential equations. In Chapters 7 and 8 we will study solutions of *first-order systems*. We begin in Section 7.1 by investigating a variety of behaviors modeled by systems of differential equations.

MODELING WITH MORE THAN ONE DEPENDENT VARIABLE

In this section we will develop models of many behaviors from the real world that can be modeled by systems of differential equations.

EXAMPLE 1 SUPPLY AND DEMAND

Suppose we are interested in the variation of the price of a single good. It is observed that a high price for the good in the market attracts more suppliers. However, increasing the quantity of the good supplied tends to drive the price down. Over time there is an interaction between price and supply. Let us construct a simple model of this behavior. We define the following variables:

$$P(t) = \text{the price of the good at any time } t,$$
$$Q(t) = \text{the quantity supplied at time } t.$$

We assume that during a time interval Δt the change in price is proportional to the quantity supplied. We also assume that the change in the quantity supplied is proportional to the price. These assumptions result in the following difference equations:

$$P(t + \Delta t) - P(t) = -aQ(t)\Delta t,$$
$$Q(t + \Delta t) - Q(t) = bP(t)\Delta t.$$

The positive constants a and b are in units per time Δt. The minus sign occurs with a because price *decreases* as quantity increases. Although in reality the behavior occurs in discrete time intervals, we approximate it with a system of ordinary differential equations by dividing the difference equations by Δt and passing to the limit as Δt approaches zero. Thus

$$\begin{aligned} \frac{dP}{dt} &= -aQ, \\ \frac{dQ}{dt} &= bP. \end{aligned} \tag{1}$$

EXAMPLE 2 POLLUTION OF A SYSTEM OF LAKES

Suppose we are given a system of three lakes interconnected by channels flowing between them as shown in Fig. 7.1. We can imagine each lake as a large tank or compartment and the interconnecting channels as pipes between the tanks. The direction of flow in the channels or pipes is indicated by the arrows in the figure.

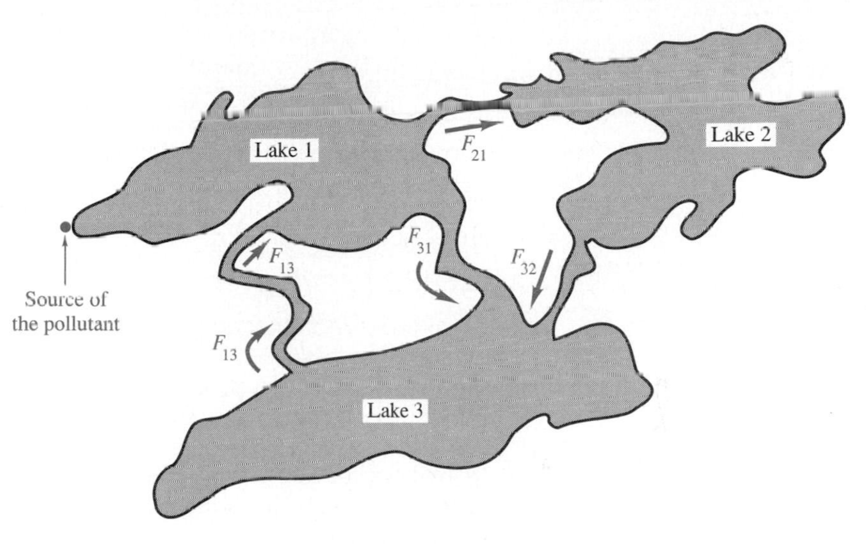

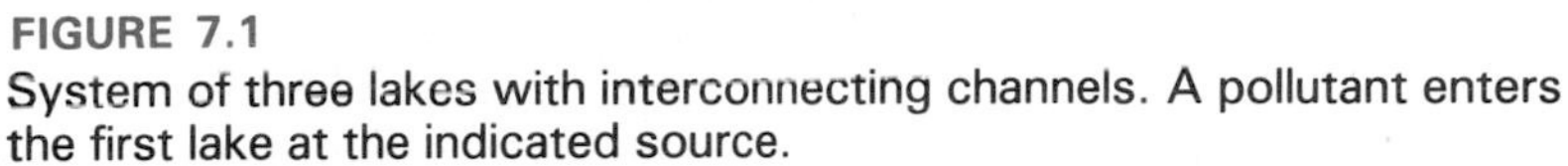
FIGURE 7.1
System of three lakes with interconnecting channels. A pollutant enters the first lake at the indicated source.

A pollutant is introduced into the first lake where $p(t)$ denotes the rate at which the pollutant enters the lake per unit time. The function $p(t)$ may be constant or may vary with time. We are interested in knowing the levels of pollution in each lake at any time.

We let $x_i(t)$ denote the amount of the pollutant in lake i at any time $t \geq 0$, where $i = 1, 2, 3$. We assume the pollutant in each lake is uniformly distributed throughout the lake by some mixing process (a rather unrealistic assumption). We assume that the volume of water V_i in lake i remains constant for each of the lakes. Then the concentration of the pollutant in lake i at any time is given by

$$c_i(t) = \frac{x_i(t)}{V_i}. \tag{2}$$

We assume that initially each lake is free of any contamination, so $x_i(0) = 0$ for each $i = 1, 2, 3$.

To model the dynamic behavior of the system of lakes, we let the constant F_{ji} denote the *flow rate from lake i to lake j*. These flow rates, which could be measured in gallons per minute, cubic feet per hour, or any other convenient units, are indicated in Fig. 7.1. Notice that $F_{12} = 0$, for instance, since there is no channel allowing any flow from lake 2 to lake 1. The *flux* of pollutant flowing from lake i into lake j at any time t, denoted by $r_{ji}(t)$, is defined by

$$\begin{aligned} r_{ji}(t) &= F_{ji} \quad \text{(gal/min)} \quad \cdot \quad c_i(t) \quad \text{(lb/gal)} \\ &= \frac{F_{ji}}{V_i} x_i(t) \quad \text{(lb/min)}. \end{aligned} \tag{3}$$

Thus $r_{ji}(t)$ measures the rate at which the concentration of pollutant in lake i flows into lake j at time t. In Section 2.2 we observed that

$$\text{rate of change of pollutant} = \text{input rate} - \text{output rate}.$$

Applying this principle to each lake results in the following system of first-order equations modeling the dynamic behavior of the lake system:

$$\begin{aligned}
\frac{dx_1}{dt} &= \frac{F_{13}}{V_3}\,x_3(t) + p(t) - \frac{F_{31}}{V_1}\,x_1(t) - \frac{F_{21}}{V_1}\,x_1(t),\\
\frac{dx_2}{dt} &= \frac{F_{21}}{V_1}\,x_1(t) - \frac{F_{32}}{V_2}\,x_2(t),\\
\frac{dx_3}{dt} &= \frac{F_{31}}{V_1}\,x_1(t) + \frac{F_{32}}{V_2}\,x_2(t) - \frac{F_{13}}{V_3}\,x_3(t).
\end{aligned}\tag{4}$$

In order for the volume of each lake to remain constant, the flow rate into each lake must balance the flow out of the lake. Thus we assume the following conditions:

$$\begin{aligned}
\text{lake 1:}&\quad F_{13} = F_{21} + F_{31},\\
\text{lake 2:}&\quad F_{21} = F_{32},\\
\text{lake 3:}&\quad F_{31} + F_{32} = F_{13}.
\end{aligned}\tag{5}$$

This example illustrates a certain type of model known as a **compartment model,** sometimes called a bathtub model. Compartment models are very useful in a variety of applications and studies including the following: the distribution, metabolism, absorption, and elimination of drugs in and from the body; energy flow through an aquatic ecosystem; the transfer of nutrients in food chains; the distribution of an insecticide through the food chain when crops are sprayed; and the return of certain types of vegetation to an area of land that has been destroyed by fire.

EXAMPLE 3 HEATING A SINGLE-STORY HOUSE

A single-story house is being heated with a forced-air central heating system. Imagine for simplicity that the house is composed of two main compartments: the lower living area and the upper attic area, as shown in Fig. 7.2. Only the living area is heated directly by the furnace, which generates 75,000 Btu/hr. Heat transfer takes place between the living and attic areas of the house as indicated by the vertical arrows in the figure. There is also heat loss through the walls of the house to the outside, as well as through the roof over the attic. We assume that initially the temperature inside the house and the attic is the same as that of the outside: a cold 35°F. At time $t = 0$ the furnace is turned on and begins to heat the house. We are interested in knowing when the temperature in the living area reaches a comfortable 68°F, assuming the outside temperature remains at a constant 35°F. Let us construct a model of the heat-transfer behavior.

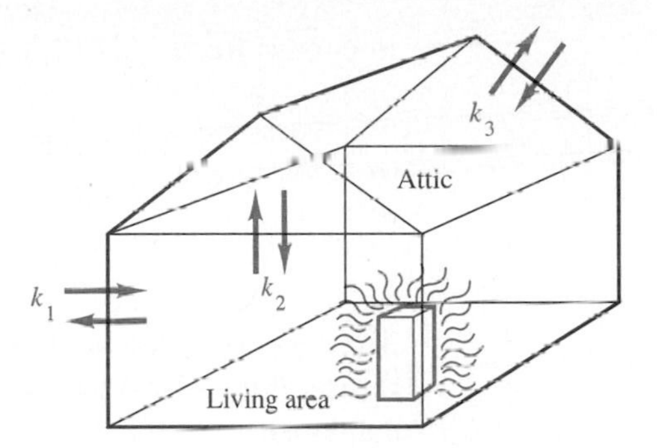

FIGURE 7.2
Heated single-story house with attic; the furnace is shown in the lower right corner.

We define $x(t)$ and $y(t)$ as the temperatures of the living and attic areas, respectively, at any time t. The change in temperature in the living area depends on the addition of heat from the furnace and the loss of heat to the outside and to the attic. The rate at which the furnace affects the temperature is the number of Btu per hour times the *heat capacity* of the living area. The heat capacity itself is a function of such variables as the size of the living area and the thermal characteristics of the objects inside the living area. Let us assume that the heat capacity of the living area is 0.2°F per thousand Btu. Then the furnace can provide 75(0.2) = 15°F each hour to the living area.

According to Newton's law of cooling (see Section 1.3), the rate of change of temperature of a region is proportional to the difference between the temperature of the region and the temperature of an adjacent region. For the living area the heat loss through the outside walls accounts for a change of $k_1(35 - x)$, and the heat loss to the attic is $k_2(y - x)$. The proportionality constants k_1 and k_2 are assumed to be positive and depend on the insulation and materials of the walls and ceiling. Thus for the living area the rate of change in temperature is given by $dx/dt = 15 + k_1(35 - x) + k_2(y - x)$. In a similar way we can derive the rate of change in temperature for the attic area. Together these two rates form the system

$$\begin{aligned} \frac{dx}{dt} &= 15 + k_1(35 - x) + k_2(y - x), \qquad \text{(living area)} \\ \frac{dy}{dt} &= k_2(x - y) + k_3(35 - y). \qquad \text{(attic)} \end{aligned} \tag{6}$$

The constant k_3 in Eq. (6) depends on the roofing materials.

In specific situations the proportionality constants k_1, k_2, and k_3 are often specified as the reciprocals of the **time constant** for the heat transfer to take place between the two adjacent regions involved. For instance, if the time constant for heat transfer between the living area and outside is 4 hr, then $k_1 = 1/4$. The time constant $1/k_1$ between the living area and the outside, for example, represents the time it takes for the temperature difference $35 - x(t)$

to change from

$$35 - x(0)$$

to

$$\frac{35 - x(0)}{e} \approx 0.368[35 - x(0)].$$

A typical value for the time constant of a building is 2–4 hr, but it can be shorter if there are open windows or doors and longer if the building is well insulated.

EXAMPLE 4 VOTING TENDENCY

Let us assume a political system consists of two parties, say Republicans and Democrats. Suppose pollsters have observed that about 30% of the Democrats reregister as Republicans before each next election. Likewise, Republicans migrate to the Democratic party at the rate of 25% between elections. As time advances, how many voters belong to each party?

We define the variables

$$D(t) = \text{number of Democrats at time } t,$$
$$R(t) = \text{number of Republicans at time } t.$$

If Δt denotes the time between elections, then the above assumptions translate to

$$D(t + \Delta t) = D(t) - 0.30D\,\Delta t + 0.25R\,\Delta t,$$
$$R(t + \Delta t) = R(t) + 0.30D\,\Delta t - 0.25R\,\Delta t.$$

Approximating these difference equations with differential equations yields the system

$$\begin{aligned} \frac{dD}{dt} &= -0.30D + 0.25R, \\ \frac{dR}{dt} &= 0.30D - 0.25R. \end{aligned} \tag{7}$$

EXAMPLE 5 CHEMICAL REACTIONS

Consider a chemical system involving three substances whose concentrations at any time t are $x(t)$, $y(t)$, and $z(t)$, respectively. We will assume that each substance is converted to either of the other two by simple reactions involving only the substance being converted. Also we will assume further that the rate of change per unit time from one substance a to substance b is proportional to the amount present of substance a with proportionality constant k_{ab}. Figure 7.3 shows the situation for the three substances.

To model the amount $x(t + \Delta t)$, we add to the amount present at time t those amounts converted to x over Δt but subtract the amount of x converted to y and z during the same time interval Δt. Thus

$$x(t + \Delta t) = x(t) + [k_{yx}y + k_{zx}z - (k_{xy} + k_{xz})x]\,\Delta t.$$

For practical purposes we assume the reactions are taking place continuously, which leads to the following system:

$$\begin{aligned}
\frac{dx}{dt} &= -(k_{xy} + k_{xz})x + k_{yx}y + k_{zx}z,\\
\frac{dy}{dt} &= k_{xy}x - (k_{yx} + k_{yz})y + k_{zy}z, \qquad (8)\\
\frac{dz}{dt} &= k_{xz}x + k_{yz}y - (k_{zx} + k_{zy})z.
\end{aligned}$$

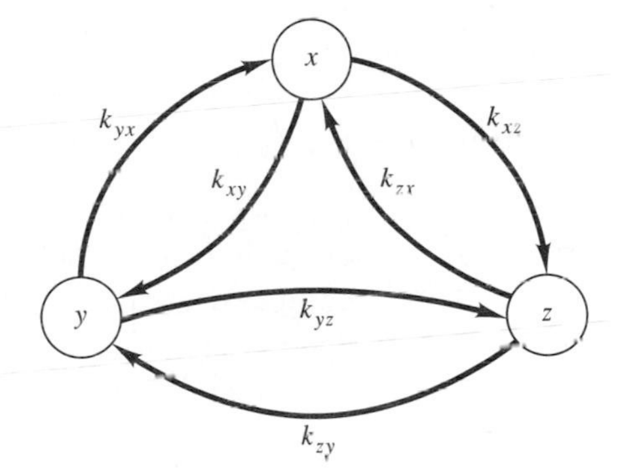

FIGURE 7.3
The constant k_{ab} represents the proportion of substance a converted to substance b during time interval Δt.

EXAMPLE 6 AN ELECTRICAL NETWORK

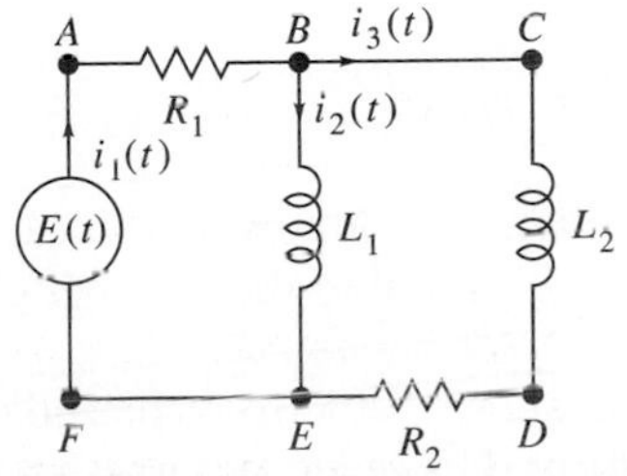

Electrical networks containing more than one loop also give rise to systems of differential equations. For instance, in the electrical network displayed in Fig. 7.4 there are two resistors and two inductors. At branch point B in the network, the current $i_1(t)$ splits in two directions. Thus

$$i_1(t) = i_2(t) + i_3(t). \qquad (9)$$

Kirchhoff's laws apply to each loop in the network (see Section 3.1). For the loop $ABEF$ we find that

$$E(t) = i_1 R_1 + L_1 \frac{di_2}{dt}. \qquad (10)$$

FIGURE 7.4
An electrical network.

The sum of the voltage drops across the loop *ABCDEF* is

$$E(t) = i_1 R_1 + L_2 \frac{di_3}{dt} + i_3 R_2. \tag{11}$$

Substituting the expression for i_1 from Eq. (9) into Eqs. (10) and (11) and isolating the derivatives results in the system

$$\begin{aligned} \frac{di_2}{dt} &= -\frac{R_1}{L_1} i_2 - \frac{R_1}{L_1} i_3 + \frac{E(t)}{L_1}, \\ \frac{di_3}{dt} &= -\frac{R_1}{L_2} i_2 - \frac{R_1 + R_2}{L_2} i_3 + \frac{E(t)}{L_2}. \end{aligned} \tag{12}$$

We assume that the currents satisfy the initial condition $i_2(0) = i_3(0) = 0$.

It is important to note that in any electrical network the algebraic sum of the currents flowing to a branch point must equal zero. In our example this fact resulted in Eq. (9).

EXAMPLE 7 A NUCLEAR ARMS RACE

Suppose two countries X and Y possess nuclear arms and each increases its weapons as the other side increases its weapons. We will let $x(t)$ denote the number of weapons possessed by country X and $y(t)$ be the number possessed by country Y. If b is a positive proportionality constant measuring "perceived threat," it might be reasonable to assume that

$$x(t + \Delta t) = x(t) + by(t)\,\Delta t.$$

That is, during the time interval Δt, country X increases its force by b missiles for every missile possessed by country Y.

Suppose that political and economic forces tend to dampen the arms race. For instance, a percentage of the weapons force may become obsolete over time, requiring expenditures for upgrading the systems. If we let the nonnegative constant a be a measure of this damping tendency, then we can assume that

$$x(t + \Delta t) = x(t) - ax(t)\,\Delta t + by(t)\,\Delta t.$$

Thus during the time period Δt, political and economic forces tend to decrease the number of arms at a constant rate proportional to the number of weapons possessed by X.

The arms race probably develops in discrete stages because of periodic program assessments and budgeting considerations. However, let us assume the continuous case and use differential equations. We also assume a symmetrical situation for country Y, yielding the system

$$\begin{aligned} \frac{dx}{dt} &= -ax + by, \\ \frac{dy}{dt} &= mx - ny, \end{aligned} \tag{13}$$

for nonnegative constants a, b, m, and n.

EXAMPLE 8 A SPRING–MASS SYSTEM

Two masses m_1 and m_2 are attached to each other and to outside walls by three springs with spring constants k_1, k_2, and k_3 in the straight-line horizontal fashion shown in Fig. 7.5. The spring–mass system is free to slide along a horizontal frictionless surface such that the ends of the outside springs are fixed to the rigid walls. Assume each spring has zero mass and that masses m_1 and m_2 are each concentrated at a point. The system is set in motion by holding mass m_1 at its equilibrium position, pulling mass m_2 to the right a short distance s, and then releasing both masses simultaneously. We are interested in knowing the displacements $x_1(t)$ and $x_2(t)$ of each mass from their respective equilibrium positions $x_1 = 0$ and $x_2 = 0$, as indicated in Fig. 7.5.

Let us apply Newton's second law to each mass separately. Since we are neglecting friction, only two forces act on mass m_1. The first force is due to the left spring and, according to Hooke's law (see Section 3.1), has a value of $-k_1x_1$. The middle spring also exerts a force on mass m_1, and it has a value of $+k_2(x_2 - x_1)$. Therefore from Newton's second law we have

$$m_1 \frac{d^2x_1}{dt^2} = -k_1x_1 + k_2(x_2 - x_1). \tag{14}$$

For mass m_2 we see that the middle spring exerts a force equal to $-k_2(x_2 - x_1)$, and the right spring exerts a force of $-k_3x_2$. Thus

$$m_2 \frac{d^2x_2}{dt^2} = -k_2(x_2 - x_1) - k_3x_2. \tag{15}$$

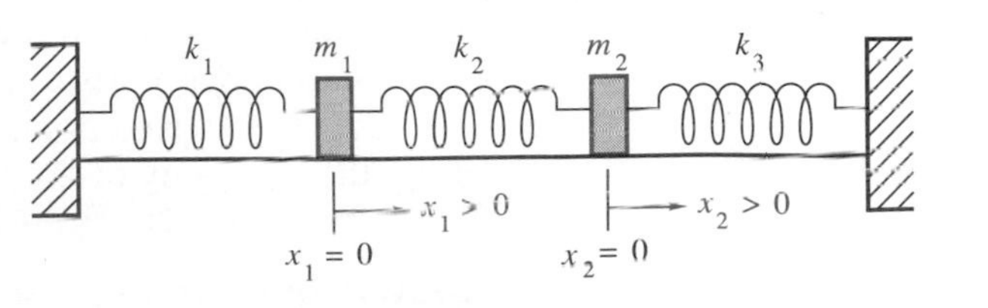

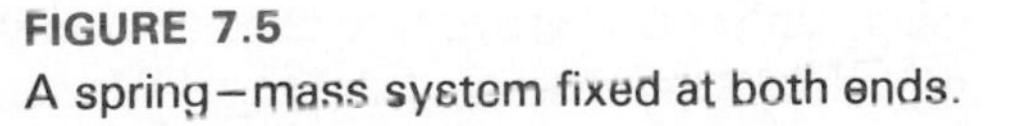

FIGURE 7.5
A spring–mass system fixed at both ends.

The initial conditions are $x_1(0) = 0$, $x_2(0) = s$, $x_1'(0) = 0$, and $x_2'(0) = 0$.

The second-order system comprised of Eqs. (14) and (15) can be converted into a first-order system by introducing new variables. We define $y_1 = x_1$, $y_2 = x_1'$, $y_3 = x_2$, and $y_4 = x_2'$. Then $m_1 y_2' = m_1 x_1''$ and $m_2 y_4' = m_2 x_2''$. Substituting these results into Eqs. (14) and (15) yields the equations

$$m_1 y_2' = -k_1 y_1 + k_2(y_3 - y_1), \tag{16}$$

$$m_2 y_4' = -k_2(y_3 - y_1) - k_3 y_3. \tag{17}$$

After combining Eqs. (16) and (17) with the derivatives for y_1 and y_2 (obtained from their respective definitions), we have the first-order system

$$\begin{aligned} y_1' &= y_2, \\ y_2' &= -\frac{k_1 + k_2}{m_1} y_1 + \frac{k_2}{m_1} y_3, \\ y_3' &= y_4, \\ y_4' &= \frac{k_2}{m_2} y_1 - \frac{k_2 + k_3}{m_2} y_3. \end{aligned} \tag{18}$$

The initial conditions translate to $y_1(0) = y_2(0) = y_4(0) = 0$ and $y_3(0) = s$.

In Chapter 1 we developed several models for the growth of a single species interacting with its environment only (Malthusian and limited growth models). In the present chapter we continue to refine our model by considering how a species not only interacts with its environment but also with other species in an ecological system. We next consider two types of interaction: competition and predator–prey.

EXAMPLE 9 COMPETITION BETWEEN SPECIES

Imagine a small pond that is mature enough to support wildlife. We desire to stock the pond with game fish, say trout and bass. Let $x(t)$ denote the population of the trout at any time t, and let $y(t)$ denote the bass population. We wish to answer questions such as, Is coexistence of the two species in the pond possible? If so, how sensitive are the population levels predicted by the model to the initial stockage levels and external influences on the environment such as floods, disasters, and epidemics?

The level of the trout population depends on many variables: the initial level x_0, the capability of the environment to support trout, the amount of competition for scarce resources, the existence of predators, and so forth.

Growth in Isolation For growth in isolation, we can assume Malthusian (unlimited) growth, logistic (limited) growth, or any of the submodels developed in Exercises 1.4. Let $x(t)$ denote the number of trout and $y(t)$ the number of bass at any time t. We assume the environment can initially support an unlimited number of trout, so that in isolation:

$$\frac{dx}{dt} = ax, \qquad a > 0.$$

Interaction with the Bass Now we model the effect of the interaction with the bass. Here we do not assume either species preys on the other. Rather, they compete with one another for living space and for a common food supply. We assume that the intensity of the competition is roughly proportional to the number of possible interactions between the two species. If either species is scarce, there are few interactions. There are many ways to model this interaction, but one simple submodel assumes that the decrease is proportional to the product xy. This leads to the model

$$\frac{dx}{dt} = ax - bxy, \qquad a, b > 0.$$

The situation for the bass population may be analyzed in the same manner. Thus we obtain the following system of two first-order differential equations:

$$\begin{aligned} \frac{dx}{dt} &= ax - bxy = (a - by)x, \\ \frac{dy}{dt} &= my - nxy = (m - nx)y. \end{aligned} \tag{19}$$

where $x(0) = x_0$, $y(0) = y_0$, and a, b, m, and n are all positive constants. The factor $a - by$ is called the **intrinsic growth rate** and decreases as the level of the bass population increases. The constants a and b indicate the degree of **self-regulation** of the trout population and its **competition** with the bass population, respectively. (In the problems you will be asked to consider a model that assumes the growth of each species in isolation is limited rather than Malthusian.)

Note that the product xy in Eq. (19) causes the system to be nonlinear. In general, nonlinear systems of equations are difficult to solve analytically. However, a graphical analysis of model (19) will suffice to answer the qualitative questions posed in the opening discussion. A graphical analysis procedure is presented in Section 7.6.

EXAMPLE 10 PREDATOR–PREY RELATIONSHIPS

We now develop a model of population growth for two species in which one is the primary food source for the other. An example of such a situation occurs in the Southern Ocean, where baleen whales eat Antarctic krill (a shrimplike crustacean) as their principal food source. Another example is wolves and rabbits in a closed forest; the wolves eat the rabbits for their principal food source, and the rabbits eat vegetation in the forest. Still other examples include sea otters as predators with abalone as prey, and ladybird beetles as predators with cottony cushion insects as prey.

Let us take a closer look at the situation of the baleen whales and the Antarctic krill. The whales eat the krill, and the krill live on plankton in the sea. If the whales eat too many krill, such that the krill cease to be abundant, the food supply of the whales is greatly reduced. The whales will either starve or leave the area in search of a new supply of krill. As the population of baleen whales dwindles, the krill population makes a comeback since not so many of them are being eaten. As the krill population increases, the food supply for the whales grows and consequently so also does the baleen whale population; now more baleen whales are eating increasingly more krill again. We want to know: in the pristine environment does this cycle continue indefinitely, or does one of the species eventually die out? What effect does exploitation of the whales have on the balance between the whale and krill populations?

Let $x(t)$ denote the Antarctic krill population at any time t, and let $y(t)$ denote the population of baleen whales in the Southern Ocean.

Krill Population The level of the krill population depends on many factors including the ability of the ocean to support them, the existence of competitors for the plankton they ingest, and the presence and levels of predators. For growth in isolation, let us assume Malthusian growth for the krill:

$$\frac{dx}{dt} = ax, \qquad a > 0.$$

Next we assume that the krill are eaten primarily by the baleen whales (so we neglect other predators). Then the growth rate of the krill is diminished in a way that is proportional to the number of interactions between them and the baleen whales. If either population is scarce, there will be few interactions. A simple submodel is to assume that the decrease in the krill growth rate is proportional to the product xy:

$$\frac{dx}{dt} = ax - bxy, \qquad a, b > 0.$$

Whale Population Now we consider the baleen whale population $y(t)$. In the absence of krill the whales have no food, so we assume their population declines at a rate proportional to their numbers. This assumption produces the exponential decay equation

$$\frac{dy}{dt} = -my, \qquad m > 0.$$

The interaction with the krill has a positive effect on the growth rate of the baleen whale. We assume an increase in the growth rate proportional to the number of interactions:

$$\frac{dy}{dt} = -my + nxy, \qquad m, n > 0.$$

Predator–Prey Model From our discussion so far, the predator–prey model becomes

$$\begin{aligned}\frac{dx}{dt} &= ax - bxy,\\ \frac{dy}{dt} &= -my + nxy.\end{aligned} \tag{20}$$

where $x(0) = x_0$, $y(0) = y_0$, and a, b, m, and n are all positive constants. The system (20), first proposed by Alfred Lotka (1925) and Vito Volterra (1931) as a simple model of predator–prey interaction, is called the **Lotka–Volterra model.**

Mutualism Note that the structural difference between competitive hunter model (19) and predator–prey model (20) is simply the signs of the coefficients in the second equation in each pair. Still a third type of interaction is **mutualism,** where neither species can survive without the presence of the other. An example is the bee's use of a plant's nectar as food while simultaneously pollinating the plant. For a system with only two species, a simple model incorporating mutualism is given by the system

$$\begin{aligned}\frac{dx}{dt} &= -ax + bxy,\\ \frac{dy}{dt} &= -my + nxy,\end{aligned} \tag{21}$$

where a, b, m, and n are positive constants.

EXERCISES 7.1

1. The gross national product (GNP) represents the sum of consumption purchases of goods and services, government purchases of goods and services, and gross private investment (which is the increase in inventories plus buildings constructed and equipment acquired). Assume that the GNP is increasing at the rate of 3% per year and that the national debt is increasing at a rate proportional to the GNP. Construct a system of two ordinary differential equations modeling the interaction between the GNP and the national debt.

2. Two large interconnected tanks each hold 100 gal of water, as shown in Fig. 7.6. Water containing 2 lb/gal of dye

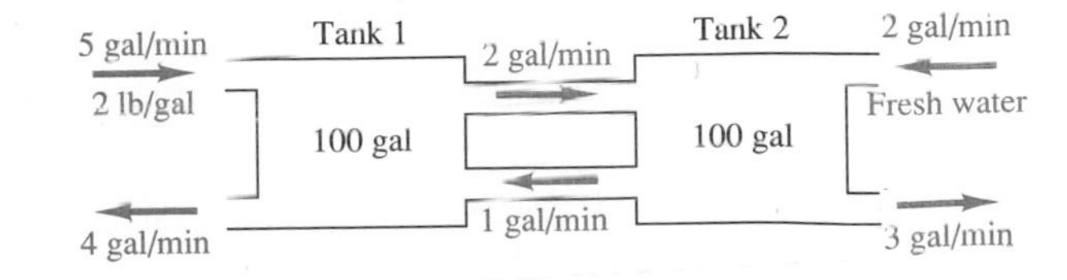

FIGURE 7.6
System of tanks for problem 2.

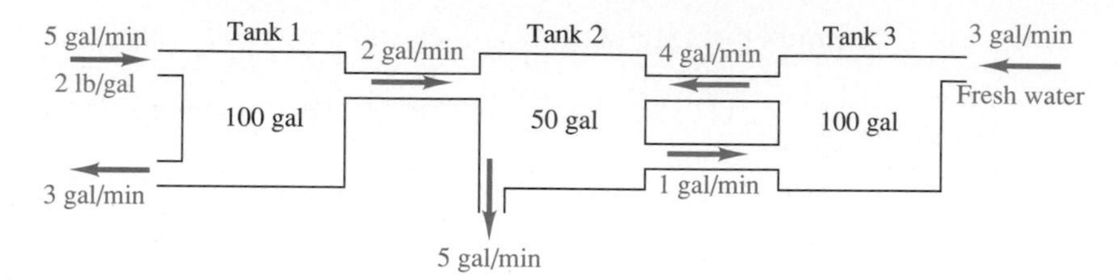

FIGURE 7.7
System of tanks for problem 3.

enters tank 1 at a rate of 5 gal/min. The mixture inside each tank is kept uniform by constant stirring. The liquid in tank 1 flows into tank 2 at a rate of 2 gal/min and also flows out a drain at the rate of 4 gal/min. At the same time, fresh water flows into tank 2 at the rate of 2 gal/min, and the mixture in tank 2 flows into tank 1 at a rate of 1 gal/min and out of tank 2 through a second drain at the rate of 3 gal/min. If initially both tanks contain only water (and no dye), find a system of differential equations for the amount of dye in each tank at any time.

3. Using the information displayed in Fig. 7.7, derive a system of differential equations specifying the amount of dye in each of tanks 1, 2, and 3 at any time (refer to problem 2). Assume initially that the tanks all contain fresh water and that the mixture in each tank is kept uniform by stirring.

4. The two-compartment model shown in Fig. 7.8 occurs frequently in drug kinetics. Suppose a concentration containing D mg of a drug is injected into the blood stream of a patient at time $t = 0$. If the flow rates are as indicated in Fig. 7.8, find a system of differential equations giving the amount of drug present in the blood and in the tissue at any time. Let V_b and V_t denote the volumes of the blood and tissue, respectively.

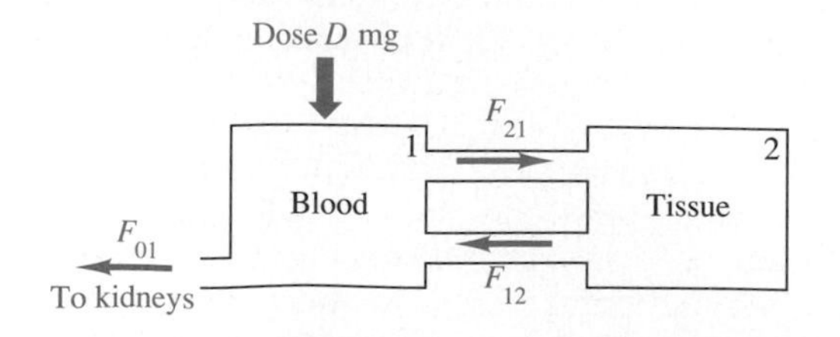

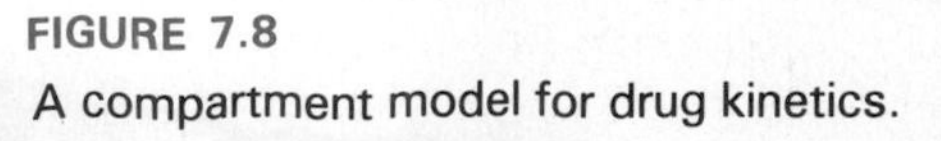
FIGURE 7.8
A compartment model for drug kinetics.

In problems 5–9 derive a system of differential equations to model the given electrical network.

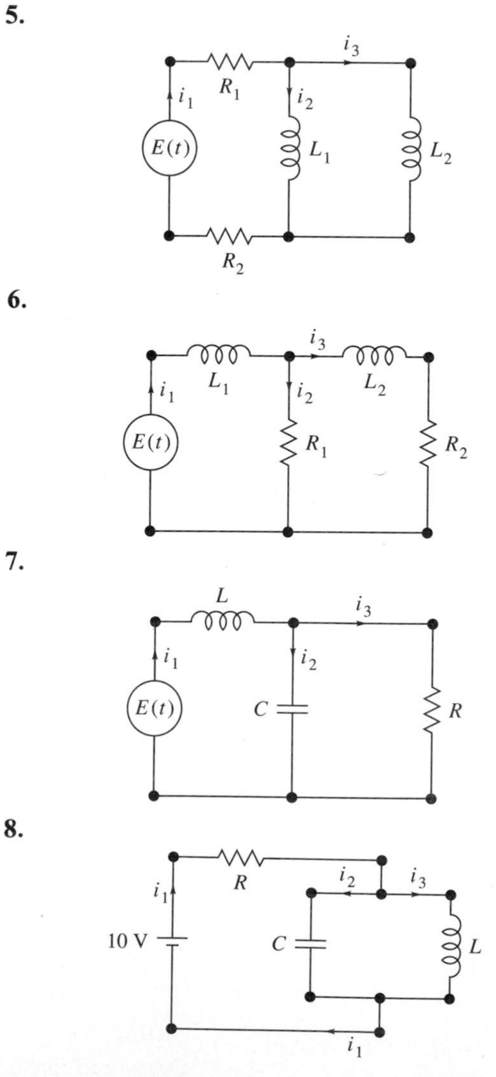

9.

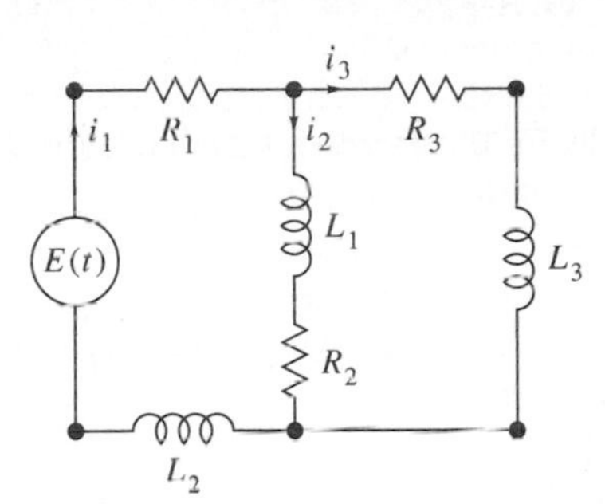

10. Two masses m_1 and m_2 are attached to two suspended springs with spring constants k_1 and k_2, as shown in Fig. 7.9. The upper spring is attached to a rigid support. Let $x_1 = 0$ and $x_2 = 0$ denote the vertical positions of the two masses, respectively, when the system is hanging freely in equilibrium. Assume that the springs themselves are massless. The system is set in motion by pulling vertically downward a short distance α on mass m_1 and then releasing the system from rest.

a) Derive a system of second-order differential equations, with appropriate initial conditions, describing the vertical position of each mass at any time. (*Note:* you do not need to consider the force of gravity acting on the masses because it is independent of the displacements. Therefore it does not contribute to the restoring forces of the springs. The force of gravity only shifts the equilibrium positions of the masses, and such shifts do not need to be determined.)

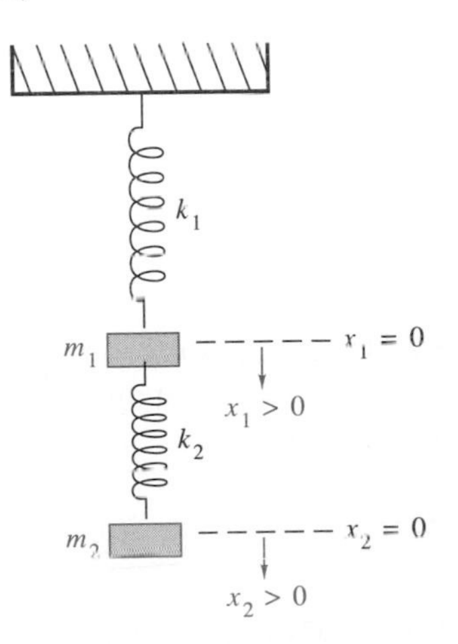

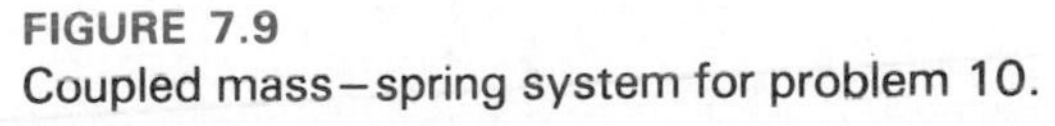

FIGURE 7.9
Coupled mass–spring system for problem 10.

b) By introducing new variables, as in Example 8 in the text, convert your system in part (a) to a first-order system. Be sure to translate your initial conditions.

11. Develop a model for the growth of trout and bass, assuming that in isolation trout demonstrate exponential decay and the bass population grows logistically with a population limit of M.

12. Consider the competitive hunter model defined by

$$\frac{dx}{dt} = a\left(1 - \frac{x}{k_1}\right)x - bxy,$$

$$\frac{dy}{dt} = m\left(1 - \frac{y}{k_2}\right)y - nxy,$$

where x represents the trout population and y the bass population.

a) What assumptions are implicitly being made about the growth of trout and bass in the absence of competition?

b) Interpret the constants a, b, m, n, k_1, and k_2 in terms of the physical problem.

13. Consider the following economic model. Let P be the price of a single item on the market. Let Q be the quantity of the item available on the market. Both P and Q are functions of time. If one considers price and quantity as two interacting species, the following model might be proposed:

$$\frac{dP}{dt} = aP\left(\frac{b}{Q} - P\right),$$

$$\frac{dQ}{dt} = cQ(fP - Q),$$

where a, b, c, and f are positive constants. Justify and discuss the adequacy of this model.

14. In 1868 the accidental introduction into the United States of the cottony cushion insect (*Icerya purchasi*), a species of scale insect from Australia, threatened to destroy the U.S. citrus industry. To counteract this situation, one of its natural Australian predators, a species of ladybird beetle, was imported. The beetles kept the scale insects down to a relatively low level. When DDT was discovered to kill scale insects, farmers applied it in the hopes of further reducing the scale insect population. However, DDT turned out to be fatal to the beetles as well, and the overall effect of using the insecticide was to *increase* the numbers of scale insects.

Modify the Lotka–Volterra model given by Eq. (20) to reflect a predator–prey system of two insect species where farmers on a continuing basis apply an insecticide that destroys both the insect predator and the insect prey at a common rate proportional to the numbers present.

15. In a 1969 study, E. R. Leigh concluded that the fluctuations in the numbers of Canadian lynx and its primary food source, the hare, trapped by the Hudson's Bay Company between 1847 and 1903 were periodic. The actual population levels of both species differed greatly from the predicted population levels obtained from the Lotka–Volterra predator–prey model (20).

Modify the Lotka–Volterra model to arrive at a more realistic model for the growth rates of both species. Answer the following questions:

a) How have you modified the basic assumptions of the predator–prey model?

b) Why are your modifications an improvement to the basic model?

7.2 SOLUTIONS TO SYSTEMS OF DIFFERENTIAL EQUATIONS

A general form of a system of two ordinary first-order differential equations in the dependent variables x and y with independent variable t is given by

$$\begin{aligned} \frac{dx}{dt} &= f(t, x, y), \\ \frac{dy}{dt} &= g(t, x, y). \end{aligned} \tag{1}$$

The two pairs of equations,

$$\begin{aligned} \frac{dx}{dt} &= 5x + y - 2e^t, \\ \frac{dy}{dt} &= -8x - 4y + 3e^t, \end{aligned} \tag{2}$$

and

$$\begin{aligned} \frac{dx}{dt} &= 6x - y, \\ \frac{dy}{dt} &= 5x + 2y, \end{aligned} \tag{3}$$

are examples of **linear systems** of differential equations because the terms involving the dependent variables x and y are linear on the righthand side of each equation. On the other hand, the system

$$\begin{aligned} \frac{dx}{dt} &= ax - bxy, \\ \frac{dy}{dt} &= my - nxy, \end{aligned} \tag{4}$$

which we developed in Section 7.1 to model the populations of two competing species, is an example of a **nonlinear system** because of the nonlinear term xy appearing on the righthand side.

If the variable t does not appear explicitly in the functions f and g in Eq. (1), so that f and g are functions only of x and y, then the system is said to be an **autonomous system**; otherwise it is a **nonautonomous system.** Systems (3) and (4) are autonomous, and system (2) is nonautonomous. If we think of the independent variable t as representing time, then autonomous systems are *time-independent* in the sense that the derivative relationships defined by the system do not change over time (although the solutions do vary with time). We will return to investigating autonomous systems after discussing solutions to general systems of form (1).

Solutions to Systems

We are going to study primarily linear systems in this text, especially when finding analytical solutions, but we will also discuss graphical solutions to certain autonomous systems that may be nonlinear, like the competitive hunter model. First we need to know more precisely what we mean by a *solution* to system (1).

DEFINITION 7.1

A **solution** to the system

$$\frac{dx}{dt} = f(t, x, y),$$

$$\frac{dy}{dt} = g(t, x, y),$$

is a pair of parametric equations

$$x = x(t) \qquad \text{and} \qquad y = y(t)$$

that satisfy the system over some open interval I.

Observe that, since $x = x(t)$ and $y = y(t)$ must be differentiable functions of t if they satisfy system (1), the solution curves to a system are necessarily continuous functions of t. Let us consider several examples illustrating the solution concept.

EXAMPLE 1 For any nonzero constant a, the functions

$$x = a\cos t,$$
$$y = a\sin t,$$

form a solution to the autonomous linear system

$$\frac{dx}{dt} = -y,$$
$$\frac{dy}{dt} = x,$$

over $-\infty < t < \infty$. This is easily verified by simply differentiating the two trigonometric functions.

EXAMPLE 2 For any nonzero constant a, the pair of functions

$$x = a\sec at,$$
$$y = a\tan at,$$

form a solution to the autonomous nonlinear system

$$\frac{dx}{dt} = xy,$$
$$\frac{dy}{dt} = x^2,$$

over the interval $-\pi/2a < t < \pi/2a$. This result is immediate from the derivative formulas

$$\frac{dx}{dt} = a^2 \sec at \tan at$$

and

$$\frac{dy}{dt} = a^2 \sec^2 at.$$

EXAMPLE 3 The two functions

$$x = (1 + t)e^t,$$
$$y = -te^t,$$

form a solution to the linear system

$$\frac{dx}{dt} = 2x + y,$$
$$\frac{dy}{dt} = -x,$$

for all real values of t. To see this we will show that each differential equation in the system is satisfied. Differentiating the function x we obtain

$$\frac{dx}{dt} = 2e^t + te^t.$$

Next note that

$$\begin{aligned} 2x + y &= 2(1 + t)e^t - te^t \\ &= 2e^t + te^t. \end{aligned}$$

Thus $dx/dt = 2x + y$, which verifies that the first equation is satisfied. Likewise, after differentiating y we obtain

$$\frac{dy}{dt} = -e^t - te^t = -x,$$

so the second equation is also satisfied.

We may wish to display a solution to a system of differential equations by plotting the functions $x(t)$ and $y(t)$ versus t individually. Thus in Example 1, for instance, we could display the cosine and sine functions. However, it is often more revealing to plot the path of points $(x(t), y(t))$ in the xy-plane as the independent variable t varies over the solution interval I. In the case of $x = a \cos t$ and $y = a \sin t$ we see that

$$x^2 + y^2 = a^2,$$

which yields the family of circles displayed in Fig. 7.10(a). Similarly, in

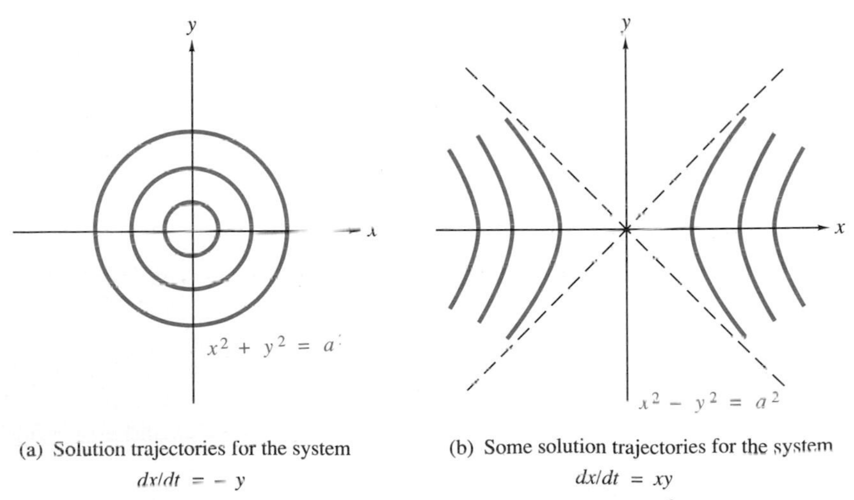

(a) Solution trajectories for the system
$dx/dt = -y$
$dy/dt = x$.

(b) Some solution trajectories for the system
$dx/dt = xy$
$dy/dt = x^2$.

FIGURE 7.10
Several trajectories associated with the solution curves presented in (a) Example 1 and (b) Example 2.

Example 2, where $x = a \sec at$ and $y = a \tan at$ over the interval $-\pi/2a < t < \pi/2a$, we have

$$x^2 - y^2 = a^2.$$

This last equation represents the family of hyperbolas displayed in Figure 7.10(b).

Some terminology will be helpful in studying systems of differential equations. A solution curve whose coordinates are $(x(t), y(t))$ as t varies over the solution interval I is called a **trajectory, path,** or **orbit** of the system. The xy-plane containing the trajectory is called the **phase plane** of system (1). It is convenient to think of a trajectory as the path of a particle moving in the xy-plane, and we will make use of this idea in our graphical analyses. By examining the differential equations themselves, we can study graphically the qualitative behavior of solutions to systems of differential equations. This technique is especially useful for investigating nonlinear autonomous systems where it may be very difficult or perhaps even impossible to find analytical solutions. (Of course, numerical approximation methods can also be used. We will give a brief introduction to numerical methods for systems in Section 7.8.) The qualitative studies of solutions of nonlinear differential equations was initiated by Henri Poincaré (1854–1912) in his search for periodic solutions to the equations governing planetary motion and the stability of orbits. We next present a brief introduction to the qualitative study of autonomous systems.*

Autonomous Systems

As defined earlier, an autonomous system takes the special form

$$\begin{aligned} \frac{dx}{dt} &= f(x, y), \\ \frac{dy}{dt} &= g(x, y). \end{aligned} \tag{5}$$

We assume that the functions f and g in Eq. (5), together with their first partial derivatives $\partial f/\partial x$, $\partial f/\partial y$, $\partial g/\partial x$, and $\partial g/\partial y$ are all continuous over a suitable region of the xy-plane. For an autonomous system, as t increases and the particle moves from a point (x, y) along a trajectory through the phase plane, the direction in which it moves depends only on the coordinates (x, y) and not on the time of its arrival at that point. The reason for this behavior is that the derivatives dx/dt and dy/dt depend only on the point (x, y) and not

* For a more thorough and complete presentation of qualitative studies of systems of differential equations, see the excellent text *Ordinary Differential Equations,* 2d ed., by Garrett Birkhoff and Gian-Carlo Rota (Waltham, Mass: Blaisdell Publishing Company, 1969), chap. 5.

on the independent parameter t. Hence, assuming that dx/dt is not zero at the point in question, the slope

$$\frac{dy}{dx} = \frac{\frac{dy}{dt}}{\frac{dx}{dt}}$$

is uniquely determined. (In this situation we can consider the trajectory through the point (x, y) to define y as a *function of* x near the point.)

If (x, y) is a point in the phase plane for which $f(x, y) = 0$ and $g(x, y) = 0$ simultaneously, then both dx/dt and dy/dt are zero. Hence there is no motion in either the x or the y direction, and the particle is stationary. Such a point is called a **rest point, critical point,** or **equilibrium point** of the system. Notice that whenever (x_0, y_0) is a rest point of system (5), the equations $x = x_0$ and $y = y_0$ give a solution to the system. In fact this **steady-state solution** is the only one passing through the point (x_0, y_0) in the phase plane. The trajectory associated with this solution is simply the rest point (x_0, y_0) itself. Hence the particle is "at rest" there. A trajectory $x = x(t)$, $y = y(t)$, is said to approach the rest point (x_0, y_0) if $x(t) \to x_0$ and $y(t) \to y_0$ as $t \to \infty$. In applications it is of interest to see what happens to a trajectory when it comes near a rest point.

The idea of stability is central to any discussion of the behavior of trajectories near a rest point. The rest point (x_0, y_0) is said to be **stable** if any trajectory starting "close" to the point stays close to it for all future time.* It is **asymptotically stable** if it is stable and if any trajectory that starts close to (x_0, y_0) approaches that point as t tends toward infinity. If it is not stable, the rest point is said to be **unstable.** These concepts are illustrated in Examples 4–6.

EXAMPLE 4 For the autonomous system

$$\frac{dx}{dt} = -y,$$

$$\frac{dy}{dt} = x,$$

presented in Example 1, the only rest point is the origin (0, 0). Notice that if a particle starts at any point (x_0, y_0) in the phase plane other than the origin, the trajectory through that point is a circle centered at the origin that is traversed counterclockwise for increasing values of t. Thus the origin is a

* More precisely, if a circle C_1 of radius $\epsilon > 0$ with center (x_0, y_0) is specified, then there is a circle C_2 of radius $\delta > 0$ with center (x_0, y_0) such that every trajectory that initially starts inside C_2 remains inside C_1 for all future time.

stable rest point but not asymptotically stable because the trajectory does not approach the origin as t approaches infinity.

EXAMPLE 5 The pair of functions

$$x = e^{-t}\sin t,$$
$$y = e^{-t}\cos t,$$

solve the linear autonomous system

$$\frac{dx}{dt} = -x + y,$$
$$\frac{dy}{dt} = -x - y.$$

We will present a method for finding solutions to such linear systems in Sections 7.3–7.5, but at this point we easily verify by differentiation that x and y do indeed satisfy the system of differential equations.

If simultaneously $dx/dt = 0$ and $dy/dt = 0$, then $x = y = 0$. Thus the origin (0, 0) is the only rest point of the system. Since

$$x^2 + y^2 = e^{-2t}\sin^2 t + e^{-2t}\cos^2 t = e^{-2t},$$

each trajectory is a circular spiral of decreasing radius around and approaching the origin as t approaches plus infinity. Therefore (0, 0) is an asymptotically stable rest point. A typical trajectory to the system, starting from the initial position $x(0) = x_0$ and $y(0) = y_0$ in the phase plane, is shown in Fig. 7.11.

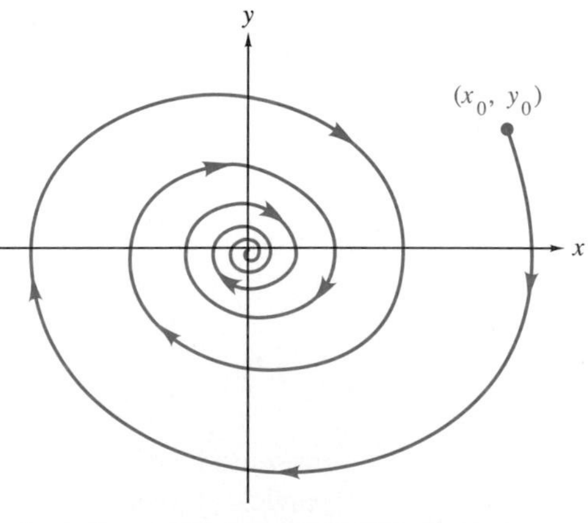

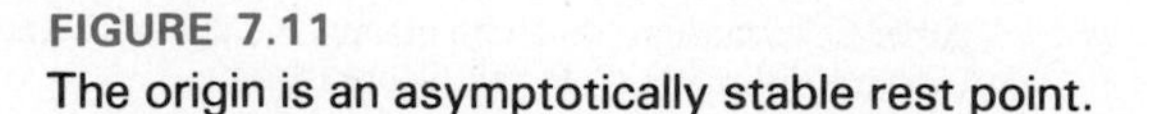
A spiral around the rest point (0, 0) for the system

$dx/dt = -x + y$

$dy/dt = -x - y.$

FIGURE 7.11
The origin is an asymptotically stable rest point.

EXAMPLE 6 Consider again the nonlinear autonomous system

$$\begin{aligned} dx/dt &= xy, \\ dy/dt &= x^2. \end{aligned} \tag{6}$$

In Example 2 we observed that the function pair $x = a \sec at$, $y = a \tan at$, $-\pi/2a < t < \pi/2a$, is a solution to the system for any real constant a. If $a > 0$, we obtain the family of rightward-opening hyperbolas shown in Fig. 7.12. If $a < 0$, we obtain a similar family of leftward-opening hyperbolas. The arrows indicate the direction of motion for a particle moving along a trajectory with increasing time.

Now for $-\infty < t < \infty$, the two function pairs

$$\begin{aligned} x &= -a \operatorname{csch} at, \\ y &= -a \coth at, \end{aligned} \tag{7}$$

and

$$\begin{aligned} x &= a \operatorname{csch} at, \\ y &= -a \coth at, \end{aligned} \tag{8}$$

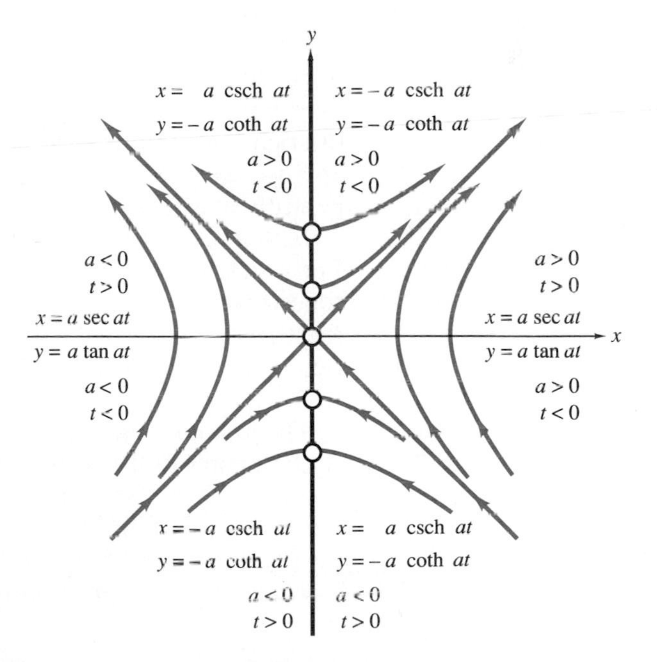

FIGURE 7.12

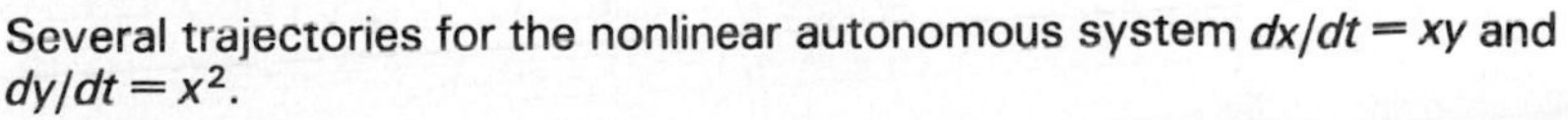

Several trajectories for the nonlinear autonomous system $dx/dt = xy$ and $dy/dt = x^2$.

also satisfy the system. These results are easily verified from the differentiation formulas

$$\frac{d}{dt}(a \operatorname{csch} at) = -a^2 \operatorname{csch} at \coth at,$$

$$\frac{d}{dt}(-a \coth at) = a^2 \operatorname{csch}^2 at.$$

(You may wish to consult an elementary calculus text to review the graphs and derivatives of the hyperbolic cosecant and cotangent functions, although it is not necessary to do so for this discussion.) Each of function pairs (7) and (8) satisfy the equation

$$y^2 - x^2 = a^2$$

because of the property that

$$\coth^2 u - \operatorname{csch}^2 u = 1.$$

Thus function pairs (7) and (8) represent upper and lower branches of the hyperbolas displayed in Fig. 7.12. The straight half-lines $y = x$ and $y = -x$, $x \neq 0$, are also solutions to system (6).

Let us find the rest points for system (6). Since $dx/dt = 0$ and $dy/dt = 0$ hold simultaneously whenever $x = 0$, *all* points along the y-axis are rest points for the system. We now classify these rest points.

Consider first the origin. If we select an initial point $(a, 0)$ close to the origin such that $a > 0$, for instance, then as t increases, the branch of the hyperbola trajectory passing through that point is traversed upward and to the right, as indicated by the arrows in Fig. 7.12. In fact, note that as t approaches $\pi/2a$, both x and y approach plus infinity. Therefore the rest point $(0, 0)$ is unstable.

If $(0, a)$ is any point along the y-axis for $a > 0$, the point $(0, a)$ is unstable. For instance, whenever a particle starts along a trajectory near $(0, a)$, the trajectory is traversed upward and away from the point $(0, a)$, as indicated by the trajectories shown in Fig. 7.12; whether the trajectory moves to the left or right depends on whether the second or first quadrant is selected, respectively, for the starting point.

The rest points $(0, b)$, where $b < 0$ are of an entirely different nature. If a starting point in the phase plane is near $(0, b)$ the trajectory through that point approaches some point $(0, a)$ on the negative y-axis as t approaches plus infinity. For instance, a particle starting very close to and to the left of the point $(0, b)$ will move along some trajectory given by

$$x = -a \operatorname{csch} at,$$
$$y = -a \coth at,$$

where $a < 0$ and $t > 0$ (see Fig. 7.12). As $t \to +\infty$, we have $at \to -\infty$. Then $(\operatorname{csch} at) \to 0^-$, so $x \to 0^-$. Also, $(\coth at) \to -1$ so $y \to a$. As time advances, the moving particle gets increasingly close to but never reaches the rest point $(0, a)$. If $(0, a)$ is close to $(0, b)$, then a particle that starts near $(0, b)$ stays near it for all future time [since the particle will also be near the point $(0, a)$]. Thus the rest points along the negative y-axis are stable. They are not asymptotically stable, however, because *not every* trajectory that starts near $(0, b)$ approaches $(0, b)$. Only two trajectories have that property; namely, the left and right branches of the lower hyperbola that approach that rest point, as indicated in Fig. 7.12. It is important to observe that no trajectory can ever cross the y-axis because every point $(0, b)$ along the y-axis is a rest point and hence a solution to the autonomous system.

Trajectories for Autonomous Systems The following results are useful when investigating solutions to autonomous system (5). We offer these results without proof.

1. There is at most one trajectory through any point in the phase plane.
2. A trajectory that starts at a point other than a rest point cannot reach a rest point in a finite amount of time.
3. No trajectory can cross itself unless it is a closed curve. If it is a closed curve, it is a periodic solution.

The implications of these three properties are that from a starting point that is not a rest point, the resulting motion

a) Will move along the same trajectory regardless of the starting time;
b) Cannot return to the starting point unless the motion is periodic;
c) Can never cross another trajectory;
d) Can only approach (never reach) a rest point.

Therefore the resulting motion of a particle along a trajectory behaves in one of three possible ways: (i) the particle approaches a rest point; (ii) the particle moves along or approaches asymptotically a closed path; or (iii) at least one of the trajectory components, $x(t)$ or $y(t)$, becomes arbitrarily large in absolute value as t tends toward infinity.

Existence and Uniqueness of Solutions

Just as was the case for a single differential equation, whenever we have a first-order system

$$\begin{aligned} \frac{dx}{dt} &= f(t, x, y), \\ \frac{dy}{dt} &= g(t, x, y), \end{aligned} \tag{9}$$

together with an initial condition $x(t_0) = x_0$ and $y(t_0) = y_0$, we must concern ourselves with the questions of existence and uniqueness of solutions:

1. **Existence:** Does a solution $x = x(t)$ and $y = y(t)$ to system (9) actually exist that satisfies the initial condition?

2. **Uniqueness:** Is there more than one pair of functions satisfying system (9) and the initial condition?

The following result, which we state without proof, answers those questions under fairly mild requirements for the functions defining the system.

EXISTENCE AND UNIQUENESS THEOREM

For the initial value problem

$$\frac{dx}{dt} = f(t, x, y),$$

$$\frac{dy}{dt} = g(t, x, y),$$

where $x(t_0) = x_0$ and $y(t_0) = y_0$, assume that the functions f, g, $\partial f/\partial x$, $\partial f/\partial y$, $\partial g/\partial x$, and $\partial g/\partial y$ are continuous throughout an open region of txy-space containing the point (t_0, x_0, y_0). Then there exists an open t interval I containing the time t_0 for which there is a unique solution $x = x(t)$ and $y = y(t)$ to the initial value problem.

For all of the systems in the remainder of this text, the existence and uniqueness theorem will apply.

Summary

The previous discussion of solutions to systems was intended to provide you with some geometric insight concerning solutions to systems of differential equations, especially autonomous systems. In Sections 7.3–7.5 we will concentrate our efforts on finding analytical solutions to *linear* autonomous systems. After presenting those solution methods, we will return our attention to graphical solutions. Graphical solution techniques can be very useful in studying some nonlinear autonomous systems, like the competitive hunter and predator–prey models developed in Section 7.1.

EXERCISES 7.2

In problems 1–10 verify that the given function pair is a solution to the first-order system. If an initial condition is specified, verify that it too is satisfied and that the conditions of the existence and uniqueness theorem are satisfied.

1. $x = -e^t$, $y = e^t$,

$$\frac{dx}{dt} = -y, \quad \frac{dy}{dt} = -x$$

2. $x = e^t \cos t, \quad y = -e^t \sin t$,

$$\frac{dx}{dt} = x + y, \quad \frac{dy}{dt} = -x + y$$

3. $x = -\frac{1}{2} + \frac{e^{2t}}{2}, \quad y = -\frac{3}{4} + \frac{3e^{2t}}{8} + \frac{3e^{-2t}}{8}$,

$$\frac{dx}{dt} = 2x + 1, \quad \frac{dy}{dt} = 3x - 2y$$

4. $x = 1 + \frac{e^{-t}}{3} - e^{-2t} + \frac{2e^{2t}}{3}$,

$y = -\frac{2e^{-t}}{3} + e^{-2t} + \frac{2e^{2t}}{3}$,

$$\frac{dx}{dt} = 2y + e^{-t}, \quad \frac{dy}{dt} = 2x - 2$$

5. $x = \cosh t + \sin t, \quad y = \sinh t$,

$$\frac{dx}{dt} = y + \cos t, \quad \frac{dy}{dt} = x - \sin t, \quad x(0) = 1,\ y(0) = 0$$

6. $x = -4 + 5t - 2t^2 + \frac{t^3}{3} + 5e^{-t}$,

$y = 5 - 5t + 2t^2 - 5e^{-t}$,

$$2\frac{dx}{dt} + \frac{dy}{dt} = y + t, \quad \frac{dx}{dt} + \frac{dy}{dt} = t^2$$

7. $x = (26t - 1)e^{4t}, \quad y = (13t + 6)e^{4t}$,

$$\frac{dx}{dt} = 2x + 4y, \quad \frac{dy}{dt} = -x + 6y,$$

$x(0) = -1,\ y(0) = 6$

8. $x = t + 3e^{-t} - 2e^{-3t}, \quad y = 1 - t + 2e^{-3t}$,

$$2\frac{dx}{dt} + \frac{dy}{dt} = -2x + y + 3t,$$

$$\frac{dx}{dt} + \frac{dy}{dt} = -x - y + 1, \quad x(0) = 1,\ y(0) = 3$$

9. $x = e^{2t}, \quad y = e^t$,

$$\frac{dx}{dt} = 2y^2, \quad \frac{dy}{dt} = y$$

10. $x = b \tanh bt, \quad y = b \operatorname{sech} bt, \quad b =$ any real number,

$$\frac{dx}{dt} = y^2, \quad \frac{dy}{dt} = xy$$

In problems 11–14 find and classify the rest points of the given autonomous system.

11. $\frac{dx}{dt} = 2y, \quad \frac{dy}{dt} = -3x$

12. $\frac{dx}{dt} = -(y - 1), \quad \frac{dy}{dt} = x - 2$

13. $\frac{dx}{dt} = -y(y - 1), \quad \frac{dy}{dt} = (x - 1)(y - 1)$

14. $\frac{dx}{dt} = \frac{1}{y}, \quad \frac{dy}{dt} = \frac{1}{x}$

15. Write the second-order equation

$$\frac{d^2y}{dt^2} + P(t)\frac{dy}{dt} + Q(t)y = 0$$

as a first-order system by introducing the new variables $x_1 = y$ and $x_2 = dy/dt$.

16. Show that if $x_1(t)$, $y_1(t)$ and $x_2(t)$, $y_2(t)$ are two pairs of solutions to the linear system

$$\frac{dx}{dt} = -x + 2y,$$

$$\frac{dy}{dt} = x - y,$$

then the pair

$$x(t) = c_1x_1(t) + c_2x_2(t),$$

$$y(t) = c_1y_1(t) + c_2y_2(t)$$

is also a solution, where c_1 and c_2 are arbitrary constants.

17. The simple undamped unforced pendulum is modeled by

$$m\ell\theta'' + mg \sin \theta = 0$$

where m is the mass of the pendulum, ℓ is the length, g is the gravitational constant, and θ is the angular displacement from the vertical equilibrium position (see Section 3.1).

a) Write the second-order equation as a first-order system by introducing the variables $x_1 = \theta$ and $x_2 = \theta'$.

b) Find the rest points of the system found in part (*a*).

c) Multiply each side of the original second-order equation by θ' and observe that

$$m\ell\theta'\theta'' + mg\theta' \sin\theta = \left[\frac{1}{2} m\ell(\theta')^2 - mg\cos\theta\right]'.$$

Show that

$$\tfrac{1}{2} m\ell(\theta')^2 - mg\cos\theta = c$$

where c is an arbitrary constant.

18. Consider the system

$$\frac{dx}{dt} = x^2, \quad \frac{dy}{dt} = 0, \quad x(0) = y(0) = 1.$$

a) Solve the system by integrating each equation.

b) Sketch the trajectory through the point $(1, 1)$ in the phase plane for $-\infty < t < 1$, indicating the direction of motion for increasing time. What happens near $t = 1$? Does a trajectory through $(1, 1)$ exist for all future times $t > 1$? Describe what occurs for $t > 1$.

7.3

HOMOGENEOUS 2 × 2 LINEAR SYSTEMS WITH CONSTANT COEFFICIENTS

In Section 7.2 we investigated the geometric aspects of solutions to first-order systems of differential equations. We now focus on the problem of finding analytical representations of solutions to a special class of first-order systems, namely, linear systems with constant coefficients. We are going to restrict our attention to **2 × 2 systems** in this chapter in order to avoid the need to invoke more advanced concepts from linear algebra. That is, we are going to consider systems of the form

$$\begin{aligned} \frac{dx}{dt} &= ax + by + f(t), \\ \frac{dy}{dt} &= cx + dy + g(t), \end{aligned} \tag{1}$$

where a, b, c, and d are constants. First we investigate *homogeneous* systems where $f(t)$ and $g(t)$ are identically zero over the interval $\alpha < t < \beta$ in which we seek solutions. In Section 7.9 we introduce the Laplace transform method for solving nonhomogeneous systems, and then study the latter systems in more depth in Chapter 8. We will see that there is a direct parallel in analytical solutions between those for 2×2 constant-coefficient linear systems and those for constant-coefficient linear second-order differential equations discussed in Sections 3.2 and 3.3.

Before considering the necessary preliminary theory, let us investigate an elementary example to gain some insight into the nature of analytical solutions to a linear system with constant coefficients. Suppose we are given the system

$$\frac{dx}{dt} = -2y,$$

$$\frac{dy}{dt} = -8x,$$

such that $x(0) = 100$ and $y(0) = 200$. These initial conditions specify the starting value of each dependent variable. Invoking the chain rule from calculus, we obtain the first-order differential equation

$$\frac{dy}{dx} = \frac{\frac{dy}{dt}}{\frac{dx}{dt}} = \frac{4x}{y}.$$

Separating the variables and integrating this last equation leads to the family of solutions

$$y^2 - 4x^2 = C,$$

where C is an arbitrary constant (see Fig. 7.13). If $C = 0$, the solutions represent the two lines $y = \pm 2x$. If $C > 0$, the solutions are hyperbolas with intercepts on the y-axis. If $C < 0$, the hyperbolas intersect the x-axis. We also want to know how the functions x and y vary with t. If $x = x(t)$ and $y = y(t)$, what form does each function take? We consider these questions next.

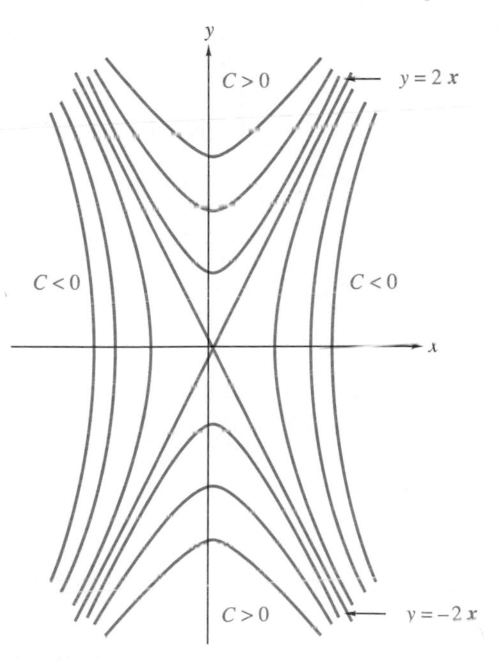

FIGURE 7.13
Family of solution curves $y^2 - 4x^2 = C$.

Form of the Solutions

Consider the following 2×2 constant-coefficient homogeneous linear system

$$\begin{aligned} \frac{dx}{dt} &= ax + by, \\ \frac{dy}{dt} &= cx + dy, \end{aligned} \tag{2}$$

where constants a, b, c, and d are assumed to be real numbers. It is notationally convenient to express system (2) in matrix form:

$$\begin{pmatrix} x' \\ y' \end{pmatrix} = \begin{pmatrix} a & b \\ c & d \end{pmatrix} \begin{pmatrix} x \\ y \end{pmatrix} \tag{3}$$

where $x' = dx/dt$ and $y' = dy/dt$. Introducing the vectors $\mathbf{X}'$, $\mathbf{X}$, and the matrix $\mathbf{A}$, system (3) can be written more compactly as the single matrix equation

$$\mathbf{X}' = \mathbf{A}\mathbf{X}. \tag{4}$$

The form of Eq. (4) is highly suggestive of the first-order separable equation

$$z' = \lambda z$$

where λ is a constant (see Chapter 2). We know the solution to this last equation is $z = ke^{\lambda t}$ for any constant k. It would therefore seem reasonable that the form of the solution to Eq. (4) should have a similar appearance. Let us assume the solution to Eq. (4) has the form

$$\mathbf{X} = \mathbf{K}e^{\lambda t}$$

where λ is a constant, and $\mathbf{X}$ and $\mathbf{K}$ are vectors. In terms of components, if we let

$$\mathbf{X} = \begin{pmatrix} x \\ y \end{pmatrix} \quad \text{and} \quad \mathbf{K} = \begin{pmatrix} k_1 \\ k_2 \end{pmatrix},$$

then the trial solution to system (2) takes the form

$$\begin{aligned} x &= k_1 e^{\lambda t}, \\ y &= k_2 e^{\lambda t}. \end{aligned} \tag{5}$$

In trial solution (5) we do not know the values of the constants λ, k_1, and k_2. Following the lead from the method of undetermined coefficients introduced in Chapter 4, we could expect to determine these constants by requiring that the functions in form (5) satisfy the original system (2). Let us try this procedure on the elementary example introduced in our opening discussion and see whether it works.

EXAMPLE 1 For the system

$$\frac{dx}{dt} = 2y,$$
$$\frac{dy}{dt} = -8x,$$

we assume solutions take the form of system (5). Once the value of λ is known, the ratio $y/x = k_2/k_1$ actually characterizes the solution. For simplicity in this example, we assume that $k_1 = 1$. Then from form (5) we have

$$x = e^{\lambda t} \qquad \text{with} \qquad \frac{dx}{dt} = \lambda e^{\lambda t},$$

$$y = k_2 e^{\lambda t} \qquad \text{with} \qquad \frac{dy}{dt} = \lambda k_2 e^{\lambda t}.$$

Substitution of these expressions into the system then yields the equations

$$\lambda e^{\lambda t} = -2k_2 e^{\lambda t},$$
$$\lambda k_2 e^{\lambda t} = -8e^{\lambda t}.$$

Since we know λ is not zero from the last equation above and $e^{\lambda t}$ is never zero, these last equations reduce to

$$\lambda = -2k_2,$$
$$k_2 = -\frac{8}{\lambda}.$$

Substitution of the second equation into the first and manipulating the result algebraically leads to $\lambda^2 = 16$. Thus simultaneous solution to the algebraic equations produces the two pairs of values

$$\lambda = 4, \quad k_2 = -2,$$

and

$$\lambda = -4, \quad k_2 = 2.$$

These values correspond to two pairs of solution functions:

$$\text{For } \lambda = 4: \quad x_1(t) = e^{4t}, \quad y_1(t) = -2e^{4t}. \qquad \text{For } \lambda = -4: \quad x_2(t) = e^{-4t}, \quad y_2(t) = 2e^{-4t}.$$

We can readily verify that both pairs satisfy the system by substituting each pair of solution functions and their derivatives into the original system. In fact, much more is true. It is straightforward to show that the pairs

$$x_1 = c_1 e^{4t}, \quad y_1 = c_1(-2e^{4t}), \qquad \text{and} \qquad x_2 = c_2 e^{-4t}, \quad y_2 = c_2(2e^{-4t}),$$

where c_1 and c_2 are arbitrary constants, also satisfy the system. Notice that, even though each c_i is an *arbitrary* constant, the ratio $y_i/x_i = k_2/k_1$ is constant for each solution pair. Finally, we can easily verify that the linear combinations

$$\begin{aligned} x &= c_1 e^{4t} + c_2 e^{-4t}, \\ y &= c_1(-2e^{4t}) + c_2(2e^{-4t}), \end{aligned} \tag{6}$$

also satisfy the system. We will see later on that solution (6) actually represents the general solution in this example, if c_1 and c_2 are interpreted to be arbitrary constants. Moreover, the discoveries revealed by this example hold in the more general case represented by system (2).

To complete this example, we can determine values for the constants c_1 and c_2 so that the initial conditions are satisfied. We substitute $x(0) = 100$ and $y(0) = 200$ into solution (6) to obtain

$$\begin{aligned} 100 &= \quad c_1 + \ c_2 \\ 200 &= -2c_1 + 2c_2. \end{aligned}$$

These equations imply that $c_1 = 0$ and $c_2 = 100$ to yield

$$\begin{aligned} x &= 100e^{-4t}, \\ y &= 200e^{-4t}. \end{aligned} \tag{7}$$

Thus solution (7) represents the particular solution to the system subject to the given initial conditions.

Finally, let us interpret solution (7) geometrically. In the xy-plane (that is, the phase plane) the initial conditions determine the particular solution curve given by solution (7). In this case the solution trajectory is simply a straight line with a slope of 2, that starts at the point (100, 200) and passes through the origin. This trajectory is displayed in Fig. 7.14(c) for positive time t. Graphs of x versus t and y versus t are also shown in Fig. 7.14.

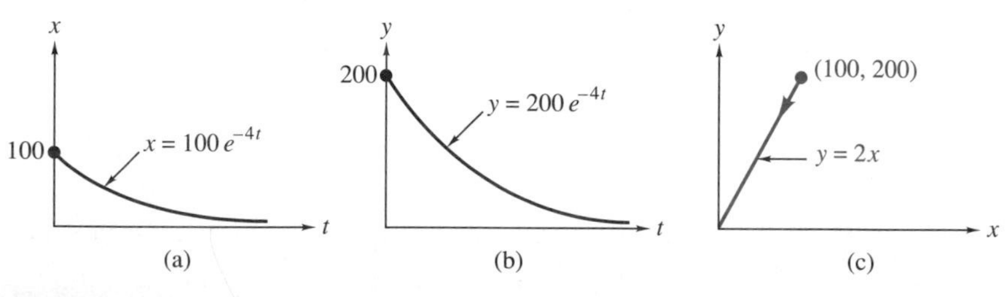

FIGURE 7.14

The functions x and y each decay exponentially. In this solution the ratio y/x is always 2.

To recapitulate our discussion so far, a solution to a 2×2 system of differential equations (2) is a pair of functions $x(t)$ and $y(t)$. In interpreting a solution, we are often interested in knowing the relationships between x and y, x and t, and y and t, as illustrated in Fig. 7.14. For instance, if the system models the population levels of two competing species, Fig. 7.14(c) depicts the relationship of one species to the other as they each approach a zero population level, while Figs. 7.14(a) and 7.14(b) depict the individual decay rate of each species. We next introduce some ideas and terminology basic to the study of linear systems.

Constant Coefficient Linear Systems

If we add a nonzero vector function

$$\mathbf{F}(t) = \begin{pmatrix} f(t) \\ g(t) \end{pmatrix}$$

to the righthand side of system (2), the system becomes *nonhomogeneous*. Thus the general form of a nonhomogeneous 2 × 2 linear system with constant coefficients is

$$\begin{aligned} \frac{dx}{dt} &= ax + by + f(t), \\ \frac{dy}{dt} &= cx + dy + g(t), \end{aligned} \tag{8}$$

or

$$\mathbf{X}' = \mathbf{AX} + \mathbf{F}(t).$$

The term *linear* here refers to the fact that x and y do not appear in (2) and (8) in such forms as e^x, $\sin y$, $\ln x$, xy, and $x^3y^{1/2}$. That is, each equation is linear in the dependent variables x and y. In this section and in Sections 7.4 and 7.5 we will be concerned with solving homogeneous 2 × 2 linear systems.

Superposition

From Definition 7.1 we know that a solution to the homogeneous system (2) is a pair of continuous functions $x(t)$ and $y(t)$ that identically satisfy the system over an interval I. If initial conditions are specified, say for instance

$$x(0) = x_0 \quad \text{and} \quad y(0) = y_0, \tag{9}$$

then $x(t)$ and $y(t)$ are required to satisfy them also. We will write

$$\mathbf{X}(0) = \mathbf{X}_0$$

for the initial condition when using the matrix notation for a system.

Suppose two solutions to (2) are given:

$$\mathbf{X}_1(t) = \begin{pmatrix} x_1(t) \\ y_1(t) \end{pmatrix},$$

and

$$\mathbf{X}_2(t) = \begin{pmatrix} x_2(t) \\ y_2(t) \end{pmatrix}.$$

Then any **linear combination**

$$c_1\mathbf{X}_1(t) + c_2\mathbf{X}_2(t) = \begin{pmatrix} c_1x_1(t) + c_2x_2(t) \\ c_1y_1(t) + c_2y_2(t) \end{pmatrix}$$

is also a solution to (2) for constants c_1 and c_2. This result is known as the **superposition principle** and is familiar to us from our study of homogeneous linear equations in Chapter 3. The principle follows by showing directly that the components satisfy system (2):

$$\begin{aligned} \frac{d}{dt}(c_1x_1 + c_2x_2) &= c_1\frac{dx_1}{dt} + c_2\frac{dx_2}{dt} \\ &= c_1(ax_1 + by_1) + c_2(ax_2 + by_2) \\ &= a(c_1x_1 + c_2x_2) + b(c_1y_1 + c_2y_2), \end{aligned}$$

and

$$\frac{d}{dt}(c_1y_1 + c_2y_2) = c(c_1x_1 + c_2x_2) + d(c_1y_1 + c_2y_2).$$

> In summary, whenever $\mathbf{X}_1(t)$ and $\mathbf{X}_2(t)$ are solutions to homogeneous system (2), then so is any linear combination
>
> $$\mathbf{X}(t) = c_1\mathbf{X}_1(t) + c_2\mathbf{X}_2(t).$$

Linear Dependence and Independence

Next we consider another key concept introduced in our discussion of single differential equations that can also be applied to systems of differential equations.

DEFINITION 7.2

> Two solution vectors $\mathbf{X}_1(t)$ and $\mathbf{X}_2(t)$ are said to be **linearly dependent** over the interval I if there exist constants c_1 and c_2, not both zero, such that for all t in I,
>
> $$c_1\mathbf{X}_1(t) + c_2\mathbf{X}_2(t) = \mathbf{0}.$$

Equivalently, two solutions are linearly dependent if one solution is a constant multiple of the other over the entire interval. Symbolically, this condition means, for instance, that

$$\mathbf{X}_2(t) = c\mathbf{X}_1(t)$$

or, in component form, that

$$x_2(t) = cx_1(t) \qquad \text{and} \qquad y_2(t) = cy_1(t) \tag{10}$$

for some constant c and all t in the interval I. If one solution is the zero solution, the vectors are automatically linearly dependent. This result follows immediately from the definition. Note that the zero vector is *always* a solution to homogeneous system (2), and is called the **trivial solution.** We are interested primarily in nontrivial solutions to systems.

By definition, the solutions $\mathbf{X}_1(t)$ and $\mathbf{X}_2(t)$ are **linearly independent** if and only if they are *not* linearly dependent. The following example illustrates these ideas.

EXAMPLE 2 It is easy to verify by direct substitution that

$$\mathbf{X}_1(t) = \begin{pmatrix} -e^{-t} \\ e^{-t} \end{pmatrix} \qquad \text{and} \qquad \mathbf{X}_2(t) = \begin{pmatrix} e^{5t} \\ e^{5t} \end{pmatrix}$$

are solutions to the homogeneous system

$$\frac{dx}{dt} = 2x + 3y,$$

$$\frac{dy}{dt} = 3x + 2y.$$

For instance, the components of $\mathbf{X}_1(t)$ satisfy the system because

$$\frac{d}{dt}(-e^{-t}) = e^{-t} = 2(-e^{-t}) + 3(e^{-t}),$$

$$\frac{d}{dt}(e^{-t}) = -e^{-t} = 3(-e^{-t}) + 2(e^{-t}).$$

Furthermore, the two solutions $\mathbf{X}_1(t)$ and $\mathbf{X}_2(t)$ are linearly independent for all t. To see this result, set an arbitrary linear combination of $\mathbf{X}_1$ and $\mathbf{X}_2$ equal to the zero vector:

$$c_1\begin{pmatrix} -e^{-t} \\ e^{-t} \end{pmatrix} + c_2\begin{pmatrix} e^{5t} \\ e^{5t} \end{pmatrix} = \begin{pmatrix} 0 \\ 0 \end{pmatrix}.$$

Then the vector components must satisfy identically the two equations

$$-c_1e^{-t} + c_2e^{5t} = 0,$$

$$c_1e^{-t} + c_2e^{5t} = 0.$$

Summing these equations gives us

$$2c_2e^{5t} = 0.$$

Therefore $c_2 = 0$ because the exponential e^{5t} is never zero. It then follows from the second equation that

$$c_1 e^{-t} = 0.$$

Thus $c_1 = 0$ also. Since the only linear combination of $\mathbf{X}_1(t)$ and $\mathbf{X}_2(t)$ that produces the zero vector is the case where the constants in the linear combination are both zero, we conclude that the solutions $\mathbf{X}_1(t)$ and $\mathbf{X}_2(t)$ are indeed linearly independent.

Fundamental Sets and the General Solution

In studying constant-coefficient homogeneous linear differential equations in Chapter 3, we were interested in finding the *general solution,* that is, the solution form that produced *all* solutions to the equation. We are just as interested in finding the general solution for homogeneous linear systems of differential equations. The concepts of a fundamental set of solutions and the general solution for single equations are easily extended to systems.

DEFINITION 7.3

Two solutions $\mathbf{X}_1(t)$ and $\mathbf{X}_2(t)$ of the homogeneous system

$$\begin{aligned} \frac{dx}{dt} &= ax + by, \\ \frac{dy}{dt} &= cx + dy, \end{aligned} \tag{11}$$

for $-\infty < t < \infty$ are said to form a **fundamental set** of solutions if every solution to (11) can be expressed as a linear combination of $\mathbf{X}_1(t)$ and $\mathbf{X}_2(t)$.

The following theorem is entirely analogous to that stated in Chapter 3 for homogeneous linear second-order equations. The result guarantees that every constant-coefficient homogeneous linear system always has a fundamental set of solutions. We state the theorem without proof.

THEOREM 7.1

The constant-coefficient homogeneous linear system

$$\mathbf{X}' = \mathbf{A}\mathbf{X} \tag{12}$$

has two linearly independent solutions $\mathbf{X}_1(t)$ and $\mathbf{X}_2(t)$ for $-\infty < t < \infty$.

Moreover, if $\mathbf{X}_1(t)$ and $\mathbf{X}_2(t)$ are *any* two linearly independent solutions and if $\mathbf{X}(t)$ is a solution to system (12), then there exist unique constants c_1 and c_2 such that, for $-\infty < t < \infty$,

$$\mathbf{X}(t) = c_1\mathbf{X}_1(t) + c_2\mathbf{X}_2(t). \tag{13}$$

From the statement of the theorem, every solution to the system (12) comes from solution form (13) for appropriate values of the constants c_1 and c_2. Thus, the **general solution** to constant-coefficient homogenous linear system (12) is expressed by

$$\mathbf{X}(t) = c_1\mathbf{X}_1(t) + c_2\mathbf{X}_2(t) \tag{14}$$

where c_1 and c_2 are arbitrary constants and $\mathbf{X}_1(t)$ and $\mathbf{X}_2(t)$ are two linearly independent solutions. Note also that, as a consequence of Theorem 7.1, $\{\mathbf{X}_1(t), \mathbf{X}_2(t)\}$ is a fundamental set of solutions to system (12) if and only if it is a linearly independent set. The situation is the same as that for solutions to homogeneous linear second-order differential equations, as presented in Section 3.2.

General solution (14) contains *all possible solutions* to homogeneous system (12). In terms of components, if

$$\mathbf{X}_1(t) = \begin{pmatrix} x_1(t) \\ y_1(t) \end{pmatrix} \quad \text{and} \quad \mathbf{X}_2(t) = \begin{pmatrix} x_2(t) \\ y_2(t) \end{pmatrix}$$

is a fundamental set, then

$$\begin{aligned} x(t) &= c_1x_1(t) + c_2x_2(t), \\ y(t) &= c_1y_1(t) + c_2y_2(t), \end{aligned} \tag{15}$$

forms the general solution to system (12). The next example illustrates these ideas.

EXAMPLE 3 From Example 2 and Definition 7.3 we know that

$$\mathbf{X}_1 = \begin{pmatrix} -e^{-t} \\ e^{-t} \end{pmatrix} \quad \text{and} \quad \mathbf{X}_2 = \begin{pmatrix} e^{5t} \\ e^{5t} \end{pmatrix}$$

comprise a fundamental set of solutions to the system

$$\begin{aligned} \frac{dx}{dt} &= 2x + 3y, \\ \frac{dy}{dt} &= 3x + 2y. \end{aligned}$$

Thus in component form the general solution is represented by

$$x = -c_1 e^{-t} + c_2 e^{5t},$$
$$y = c_1 e^{-t} + c_2 e^{5t},$$

for arbitrary constants c_1 and c_2.

It is customary to simplify the notation and write simply **X** instead of $\mathbf{X}(t)$ when referring to a solution function, as in Example 3. We will hereafter follow this custom except in cases where confusion might arise.

EXAMPLE 4 Determine the solution to the constant-coefficient homogeneous system in Example 3 that satisfies the initial conditions $x(0) = 1$ and $y(0) = 0$.

Solution. Substituting the initial conditions into the general solution found in Example 3 gives us

$$x(0) = -c_1 + c_2 = 1$$
$$y(0) = \quad c_1 + c_2 = 0.$$

Thus $c_2 = ½$ and $c_1 = -½$. The solution we seek is given by

$$x = \frac{1}{2} e^{-t} + \frac{1}{2} e^{5t}$$

$$y = -\frac{1}{2} e^{-t} + \frac{1}{2} e^{5t}.$$

Finding the General Solution: A Preview

We now know from Theorem 7.1 that *every* homogeneous constant-coefficient linear system (12) has a fundamental set of solutions. But the question still remains as to how to find a fundamental set. We do have a clue as to how we might proceed based on our opening remarks for this section. Because of the form of the linear system,

$$\mathbf{X}' = \mathbf{A}\mathbf{X}, \tag{16}$$

we assume as a trial solution the *form*

$$\mathbf{X} = \mathbf{K}e^{\lambda t}. \tag{17}$$

In terms of components, vector equation (17) translates to

$$\begin{aligned} x &= k_1 e^{\lambda t}, \\ y &= k_2 e^{\lambda t}. \end{aligned} \tag{18}$$

We are interested only in nontrivial solutions since we seek a fundamental set of solutions. Thus the set cannot contain the zero vector, and k_1 and k_2 cannot both be zero. In order to determine the unknown constants λ, k_1 and k_2, we require that solution (18) satisfy system (16). In terms of the components, differentiation and substitution yields

$$\frac{dx}{dt} = \lambda k_1 e^{\lambda t} = ak_1 e^{\lambda t} + bk_2 e^{\lambda t},$$

$$\frac{dy}{dt} = \lambda k_2 e^{\lambda t} = ck_1 e^{\lambda t} + dk_2 e^{\lambda t}.$$

Since the exponential term $e^{\lambda t}$ is never zero, it can be canceled from both sides of the last two equations. Then rearrangement of terms leads to the algebraic system

$$\begin{aligned} (a-\lambda)k_1 + \qquad bk_2 &= 0, \\ ck_1 + (d-\lambda)k_2 &= 0. \end{aligned} \tag{19}$$

Since we are seeking nonzero solutions, k_1 and k_2 cannot both be zero. Therefore the determinant of the coefficients in system (19) must be zero:

$$\det\begin{pmatrix} a-\lambda & b \\ c & d-\lambda \end{pmatrix} = 0;$$

that is,

$$(a-\lambda)(d-\lambda) - bc = 0$$

or

$$\lambda^2 - (a+d)\lambda + (ad-bc) = 0. \tag{20}$$

Equation (20) is called the **characteristic equation** of system (16). Its roots may be classified as

1. Real and distinct,
2. Complex,
3. Real and repeated.

Once a value of λ satisfying the characteristic equation is found, algebraic system (19) can be solved for suitable values of k_1 and k_2. These values are the components of the vector **K** in Eq. (17) associated with that particular value of λ. A value of λ satisfying characteristic equation (20) is called an **eigenvalue** of system (16), and an associated nonzero vector **K** whose components satisfy system (19) is called an **eigenvector.** Note that **0** is *never* an eigenvector since we have consistently assumed that k_1 and k_2 are not both zero (because we seek nontrivial solutions).

Summary

The previous discussion was intended only as a preview to the procedure we are going to use to find the general solution to a constant-coefficient homogeneous linear system. Essentially we will write algebraic equations (19) directly from linear system (16). Setting the determinant of that algebraic system equal to zero gives us characteristic equation (20). The roots of the characteristic equation are the eigenvalues of the system. In Sections 7.4 and 7.5 we will show how the eigenvalues and eigenvectors produce fundamental sets of solutions to constant-coefficient homogeneous linear system (16) for each of the three classes of roots for the characteristic equation.

We conclude this section by comparing the ideas associated with solving a 2×2 constant-coefficient homogeneous linear system with the analogous ideas presented in Chapter 3 for solving a constant-coefficient homogeneous linear second-order differential equation. Table 7.1 contrasts the preliminary theory and solution concepts underlying systems with the theory for second-order equations.

TABLE 7.1

Comparison Second-Order Differential Equations with 2×2 First-Order Systems, Where Both Are Constant-Coefficient, Homogeneous, and Linear

Second-Order Equations	2×2 First-Order Systems
$y'' + ay' + by = 0. \qquad (21)$	$\frac{dx}{dt} = ax + by,$ $\quad \frac{dy}{dt} = cx + dy. \qquad (22)$
Solution	
Any function $y(x)$ that identically satisfies (21), and any initial conditions if given, over an interval I is a solution.	Any pair of functions expressed in vector form as $\mathbf{X}(t) = \begin{pmatrix} x(t) \\ y(t) \end{pmatrix}$ whose components satisfy (22), and any initial conditions if given, over an interval I is a solution.
Linear Independence	
Two solutions $y_1(x)$ and $y_2(x)$ are linearly independent over I if $c_1 y_1(x) + c_2 y_2(x) = 0$ for all x in I implies that $c_1 = c_2 = 0$.	Two vector solutions $\mathbf{X}_1(t)$ and $\mathbf{X}_2(t)$ are linearly independent over I if $c_1\mathbf{X}_1(t) + c_2\mathbf{X}_2(t) = \mathbf{0}$ for all t in I implies that $c_1 = c_2 = 0$.

TABLE 7.1 *(Cont.)*

Superposition Principle	
If $y_1(x)$ and $y_2(x)$ are two solutions to (21), then so is $$y = c_1 y_1 + c_2 y_2$$ for any constants c_1 and c_2.	If $\mathbf{X}_1(t)$ and $\mathbf{X}_2(t)$ are two solution vectors to (22), then so is $$\mathbf{X} = c_1\mathbf{X}_1 + c_2\mathbf{X}_2$$ for any constants c_1 and c_2. In terms of components, $$x = c_1 x_1 + c_2 x_2, \quad y = c_1 y_1 + c_2 y_2,$$ solves (22).
General Solution	
If $\{y_1(x), y_2(x)\}$ is a fundamental set of solutions to (21), then the general solution is $$y = c_1 y_1 + c_2 y_2$$ for arbitrary constants c_1, c_2.	If $\{\mathbf{X}_1(t), \mathbf{X}_2(t)\}$ is a fundamental set of solutions to (22), then the general solution is $$x = c_1 x_1 + c_2 x_2, \quad y = c_1 y_1 + c_2 y_2,$$ for arbitrary constants c_1 and c_2.
Trial Solution Form	
$y = e^{rt}$	$x = k_1 e^{\lambda t}, \quad y = k_2 e^{\lambda t}.$
Auxiliary Equation	*Characteristic Equation*
$r^2 + ar + b = 0$	$\lambda^2 - (a + d)\lambda + (ad - bc) = 0$
Roots: Real and distinct, complex, real and repeated.	*Roots:* Real and distinct, complex, real and repeated.

EXERCISES 7.3

In problems 1–5 write the given system in matrix form.

1. $\dfrac{dx}{dt} = 3x - y$, $\quad \dfrac{dy}{dt} = x + 3y$

2. $\dfrac{dx}{dt} = -6x + y$, $\quad \dfrac{dy}{dt} = -4x - y$

3. $\dfrac{dx}{dt} = 3x - \dfrac{1}{2}y$, $\quad \dfrac{dy}{dt} = 8x - y$

4. $\dfrac{dx}{dt} = y$, $\quad \dfrac{dy}{dt} = -9x + 6y$

5. $\dfrac{dx}{dt} = 2x - y$, $\quad \dfrac{dy}{dt} = 4x - 3y$

In problems 6–10 write the given system without using matrices.

6. $\mathbf{X}' = \begin{pmatrix} 5 & 3 \\ -3 & 1 \end{pmatrix}\mathbf{X}$

7. $\mathbf{X}' = \begin{pmatrix} 5 & -1 \\ 4 & -1 \end{pmatrix}\mathbf{X}$

8. $\mathbf{X}' = \begin{pmatrix} 3 & 0 \\ -1 & 5 \end{pmatrix}\mathbf{X}$ **9.** $\mathbf{X}' = \begin{pmatrix} 2 & 3 \\ 0 & 4 \end{pmatrix}\mathbf{X}$

10. $\mathbf{X}' = \begin{pmatrix} -4 & 1 \\ -1 & -2 \end{pmatrix}\mathbf{X}$

In problems 11–15 verify by direct substitution that the given vector is a solution to the specified system.

11. $\mathbf{X} = \begin{pmatrix} -e^{-2t} \\ e^{-2t} \end{pmatrix}$, $\mathbf{X}' = \begin{pmatrix} 1 & 3 \\ 1 & -1 \end{pmatrix}\mathbf{X}$

12. $\mathbf{X} = \begin{pmatrix} -e^{t} \\ e^{t} \end{pmatrix}$, $\mathbf{X}' = \begin{pmatrix} 0 & -1 \\ -1 & 0 \end{pmatrix}\mathbf{X}$

13. $\mathbf{X} = \begin{pmatrix} e^{t}\cos t \\ -e^{t}\sin t \end{pmatrix}$, $\mathbf{X}' = \begin{pmatrix} 1 & 1 \\ -1 & 1 \end{pmatrix}\mathbf{X}$

14. $\mathbf{X} = \begin{pmatrix} e^{-t} + 2te^{-t} \\ e^{-t} + te^{-t} \end{pmatrix}$, $\mathbf{X}' = \begin{pmatrix} -3 & 4 \\ -1 & 1 \end{pmatrix}\mathbf{X}$

15. $\mathbf{X} = \begin{pmatrix} -e^{3t} \\ te^{3t} \end{pmatrix}$, $\mathbf{X}' = \begin{pmatrix} 3 & 0 \\ -1 & 3 \end{pmatrix}\mathbf{X}$

16. Show that any constant multiple $c\mathbf{X}$ also satisfies the system in problem 13.

17. Show that any constant multiple $c\mathbf{X}$ also satisfies the system in problem 14.

18. Show that any linear combination of

$$\mathbf{X}_1 = \begin{pmatrix} e^{-t} \\ e^{-t} \end{pmatrix} \quad \text{and} \quad \mathbf{X}_2 = \begin{pmatrix} -e^{t} \\ e^{t} \end{pmatrix}$$

satisfies the system

$$\mathbf{X}' = \begin{pmatrix} 0 & -1 \\ -1 & 0 \end{pmatrix}\mathbf{X}.$$

19. Show that any linear combination of

$$\mathbf{X}_1 = \begin{pmatrix} \cos 3t \\ -\sin 3t \end{pmatrix} e^{2t} \quad \text{and} \quad \mathbf{X}_2 = \begin{pmatrix} \sin 3t \\ \cos 3t \end{pmatrix} e^{2t}$$

also satisfies the system

$$\mathbf{X}' = \begin{pmatrix} 2 & 3 \\ -3 & 2 \end{pmatrix}\mathbf{X}.$$

20. Show that any linear combination of

$$\mathbf{X}_1 = \begin{pmatrix} 1 \\ 3 \end{pmatrix} \quad \text{and} \quad \mathbf{X}_2 = \begin{pmatrix} t \\ 3t - 1 \end{pmatrix}$$

also satisfies the system

$$\mathbf{X}' = \begin{pmatrix} 3 & -1 \\ 9 & -3 \end{pmatrix}\mathbf{X}.$$

In problems 21–25 assume the given pairs of vectors solve a linear system $\mathbf{X}' = \mathbf{A}\mathbf{X}$. Determine if the vectors are linearly independent.

21. $\mathbf{X}_1 = \begin{pmatrix} 1 \\ 1 \end{pmatrix} e^{2t}$, $\mathbf{X}_2 = \begin{pmatrix} 1 \\ t + \frac{1}{3} \end{pmatrix} e^{2t}$

22. $\mathbf{X}_1 = \begin{pmatrix} 1 \\ -2 \end{pmatrix} e^{-3t}$, $\mathbf{X}_2 = \begin{pmatrix} -1 \\ 1 \end{pmatrix} \mathbf{e}^{3t}$

23. $\mathbf{X}_1 = \begin{pmatrix} 3 \\ 5 \end{pmatrix} e^{6t}$, $\mathbf{X}_2 = \begin{pmatrix} 1 \\ -1 \end{pmatrix} e^{-2t}$

24. $\mathbf{X}_1 = \begin{pmatrix} 1 \\ -1 \end{pmatrix} e^{-t}$, $\mathbf{X}_2 = \begin{pmatrix} -2e^{-t} \\ 2e^{-t} \end{pmatrix}$

25. $\mathbf{X}_1 = \begin{pmatrix} \cos t \\ -\cos t - \sin t \end{pmatrix} e^{4t}$, $\mathbf{X}_2 = \begin{pmatrix} -\sin t \\ \sin t - \cos t \end{pmatrix} e^{4t}$

In problems 26–33 find the characteristic equation and eigenvalues for the given system.

26. $\dfrac{dx}{dt} = x + 4y$

$\dfrac{dy}{dt} = x + y$

27. $\dfrac{dx}{dt} = 2x$

$\dfrac{dy}{dt} = 3x - 2y$

28. $\mathbf{X}' = \begin{pmatrix} 5 & -2 \\ 2 & 3 \end{pmatrix}\mathbf{X}$ **29.** $\mathbf{X}' = \begin{pmatrix} 1 & 1 \\ -1 & 1 \end{pmatrix}\mathbf{X}$

30. $\mathbf{X}' = \begin{pmatrix} 4 & -8 \\ \frac{1}{2} & -1 \end{pmatrix}\mathbf{X}$ **31.** $\mathbf{X}' = \begin{pmatrix} 3 & -1 \\ 9 & -3 \end{pmatrix}\mathbf{X}$

32. $\mathbf{X}' = \begin{pmatrix} -1 & -1 \\ 1 & -3 \end{pmatrix}\mathbf{X}$

33. $\dfrac{dx}{dt} = 5x + y$

$\dfrac{dy}{dt} = -x + 4y$

7.4

DISTINCT EIGENVALUES

In this section we commence our study of finding the general solution to 2×2 homogeneous linear systems with constant coefficients. Here we consider the two cases when the eigenvalues (that is the roots to the characteristic equation) are real and distinct, or complex. Since complex roots always occur in conjugate pairs, and the characteristic equation is a quadratic with real coefficients, the complex eigenvalues of a 2×2 system are necessarily distinct.

Let us begin with a specific example. In Example 1 which follows, we assume the trial solution form that was suggested in Section 7.3. Substituting the trial solution into the system of differential equations leads to the characteristic equation. Then we find the eigenvalues or roots to the characteristic equation together with the associated eigenvectors. The eigenvalues and eigenvectors are then used to produce a fundamental set of solutions. In the example to follow the eigenvalues are real and distinct numbers.

EXAMPLE 1 Find the general solution of the system of equations

$$\frac{dx}{dt} = 2x + 3y, \qquad \frac{dy}{dt} = 2x + y. \tag{1}$$

Solution. As suggested in Section 7.3, we assume a trial solution of the form

$$x(t) = k_1 e^{\lambda t}, \qquad y(t) = k_2 e^{\lambda t}, \tag{2}$$

where λ, k_1, and k_2 are constants to be determined. Our goal is to find two values for λ and a vector $\mathbf{K}$ for each value. The numbers k_1 and k_2 are the components of the vector $\mathbf{K}$. To find these constants we substitute the trial solution into system (1) to obtain

$$\lambda k_1 e^{\lambda t} = 2k_1 e^{\lambda t} + 3k_2 e^{\lambda t},$$
$$\lambda k_2 e^{\lambda t} = 2k_1 e^{\lambda t} + k_2 e^{\lambda t}.$$

Since $e^{\lambda t}$ does not equal zero, we can cancel it from both sides and then rearrange terms to yield the algebraic system

$$(2-\lambda)k_1 + 3k_2 = 0, \qquad 2k_1 + (1-\lambda)k_2 = 0. \tag{3}$$

Notice that system (3) consists of two equations but three unknowns λ, k_1, and k_2. We will determine two quantities: λ and the ratio k_1/k_2.

System (3) always has the trivial solution $k_1 = k_2 = 0$, which corresponds to $x(t) \equiv 0$, $y(t) \equiv 0$. However, we seek nontrivial solutions to system (3). Solutions to (3) other than the trivial solution exist if and only if the determinant of the coefficients k_1 and k_2 is zero. Thus

$$\det\begin{pmatrix} 2-\lambda & 3 \\ 2 & 1-\lambda \end{pmatrix} = 0, \tag{4}$$

or

$$(2-\lambda)(1-\lambda)-6=0.$$

The last equation is simply the quadratic

$$\lambda^2-3\lambda-4=0, \tag{5}$$

which can be factored as

$$(\lambda-4)(\lambda+1)=0.$$

Equation (5) is the characteristic equation of system (1). The roots to the characteristic equation are the eigenvalues $\lambda_1=4$ and $\lambda_2=-1$. These eigenvalues are real and distinct.

To complete the solution procedure, we must find eigenvector components k_1 and k_2 for each eigenvalue λ. Then each of the eigenvalues λ_1 and λ_2 produces a solution vector:

$$\mathbf{X}_1=\mathbf{K}_1e^{\lambda_1 t} \qquad \text{and} \qquad \mathbf{X}_2=\mathbf{K}_2e^{\lambda_2 t}.$$

The set $\{\mathbf{X}_1, \mathbf{X}_2\}$ is a fundamental set of solutions.

To find the eigenvector $\mathbf{K}_1$, we substitute $\lambda_1=4$ into system (3) to yield the equations

$$\begin{aligned} -2k_1+3k_2&=0,\\ 2k_1-3k_2&=0. \end{aligned} \tag{6}$$

Note that the equations in (6) are dependent: each is simply the other multiplied by the constant -1. Thus we effectively have only one equation. Solution of the second equation leads to $2k_1=3k_2$ or

$$k_1=\frac{3}{2}k_2. \tag{7}$$

There are *infinitely many solutions* to system (6): for each value of k_2 the number k_1 is completely determined by Eq. (7). However, note that the *ratio* k_1/k_2 *is always a constant* (in this case $3/2$). Since we seek *any* fundamental set of solutions, any eigenvector $\mathbf{K}$ associated with the eigenvalue $\lambda_1=4$ will do. Thus we may choose any convenient nonzero value for k_2 in Eq. (7) to determine k_1 and the resulting vector

$$\mathbf{K}=\begin{pmatrix}k_1\\k_2\end{pmatrix}.$$

Let us set $k_2=2$ to avoid fractions, thereby obtaining $k_1=3$. Substitution of these results into trial solution (2) leads to the vector solution $\mathbf{X}_1(t)$ whose components are

$$\begin{aligned} x_1(t)&=3e^{4t},\\ y_1(t)&=2e^{4t}. \end{aligned}$$

In vector form we have

$$\mathbf{X}_1 = \begin{pmatrix} 3 \\ 2 \end{pmatrix} e^{4t} \tag{8}$$

Verifying that Eq. (8) does indeed satisfy system (1) is a straightforward matter. (Also the principle of superposition ensures that $c\mathbf{X}_1$ for any constant $c \neq 0$ also satisfies system (1). In scalar form the solution would be expressed

$$x = c(3e^{4t}),$$
$$y = c(2e^{4t}).$$

This fact emphasizes that *any* values chosen for k_1 and k_2 suffice as long as the ratio k_1/k_2 equals 3⁄2.)

To obtain a second, independent solution we substitute $\lambda_2 = -1$ into system (3) to obtain

$$3k_1 + 3k_2 = 0,$$
$$2k_1 + 2k_2 = 0.$$

Either equation implies that

$$k_1 = -k_2. \tag{9}$$

As before, we are free to choose the value of k_2 arbitrarily. Let us choose $k_2 = -1$, yielding $k_1 = 1$. Substitution of the values $\lambda_2 = -1$, $k_1 = 1$, and $k_2 = -1$ into trial solution (2) leads to our second vector solution:

$$x_2(t) = e^{-t},$$
$$y_2(t) = -e^{-t}.$$

In vector form this is expressed as

$$\mathbf{X}_2 = \begin{pmatrix} 1 \\ -1 \end{pmatrix} e^{-t}. \tag{10}$$

Observe that the solutions $\mathbf{X}_1$ and $\mathbf{X}_2$ are linearly independent. To see this, consider the first component of each solution vector. It is impossible to find a constant c so that $\mathbf{X}_1 = c\mathbf{X}_2$ since $3e^{4t} \neq ce^{-t}$ over any interval. The general solution to system (1) is therefore given by the vector equation

$$\mathbf{X} = c_1\mathbf{X}_1 + c_2\mathbf{X}_2.$$

In component form the general solution is therefore

$$\begin{aligned} x(t) &= 3c_1e^{4t} + c_2e^{-t}, \\ y(t) &= 2c_1e^{4t} - c_2e^{-t}. \end{aligned} \tag{11}$$

Since initial conditions are not given, the problem is complete. It is a straightforward process (which we leave up to you) to verify that the functions in solution (11) do indeed solve original system (1).

Case 1: General Solution for Distinct Real Eigenvalues

Let us review the solution procedure used in Example 1. We first substitute the trial solution form

$$x = k_1 e^{\lambda t},$$
$$y = k_2 e^{\lambda t}, \tag{12}$$

into the homogeneous system. This substitution leads to an algebraic 2×2 system of equations in the unknowns λ, k_1, and k_2. Since we seek nontrivial solutions to the system, the determinant of the coefficients of this resulting system must be zero, yielding the characteristic equation. The roots to the characteristic equation in our example produced *distinct real eigenvalues* λ_1 and λ_2.

After substituting each value of λ one at a time into system (3), we determined a unique relationship between k_1 and k_2 giving a constant ratio k_1/k_2. In Example 1 these relationships were given by Eqs. (7) and (9) for the two eigenvalues. In each instance of the example we chose a convenient nonzero value for k_2 in order to determine an associated value for k_1. Such a choice is permitted because *we seek only one nonzero solution* for each eigenvalue. Had k_2 turned out to be zero, any nonzero value of k_1 could be chosen and the ratio k_2/k_1 would always be zero.

Thus each eigenvalue λ produces an associated eigenvector $\mathbf{K}$ (whose components are k_1 and k_2), leading to a vector solution

$$\mathbf{X} = \mathbf{K}e^{\lambda t}. \tag{13}$$

The component form of Eq. (13) is the pair of equations given by trial solution (12). In this way we obtain the solution set $\{\mathbf{X}_1, \mathbf{X}_2\}$, where

$$\mathbf{X}_1 = \mathbf{K}_1 e^{\lambda_1 t},$$
$$\mathbf{X}_2 = \mathbf{K}_2 e^{\lambda_2 t}.$$

Since λ_1 and λ_2 are distinct values, it is not possible to find a constant c for which

$$\mathbf{X}_1 = c\mathbf{X}_2.$$

(The existence of such a constant c would imply that $e^{(\lambda_1 - \lambda_2)t}$ is constant, which is clearly impossible.) Thus $\{\mathbf{X}_1, \mathbf{X}_2\}$ is a *fundamental set* leading to the general solution

$$\mathbf{X} = c_1\mathbf{X}_1 + c_2\mathbf{X}_2 \tag{14}$$

where c_1 and c_2 are arbitrary constants.

If initial conditions

$$x(0) = x_0 \quad \text{and} \quad y(0) = y_0$$

are given, the constants c_1 and c_2 can be determined by solving the algebraic 2×2 system resulting from substituting the conditions into general solution (14). Following is an example of a system with initial conditions.

EXAMPLE 2 Solve the system

$$\frac{dx}{dt} = x + 3y,$$
$$\frac{dy}{dt} = 5x + 3y, \tag{15}$$

subject to the initial conditions $x(0) = 100$ and $y(0) = 300$.

Solution. We assume solutions of the form

$$x = k_1 e^{\lambda t},$$
$$y = k_2 e^{\lambda t}. \tag{16}$$

Computing derivatives and substituting into system (15) gives us

$$\lambda k_1 e^{\lambda t} = k_1 e^{\lambda t} + 3k_2 e^{\lambda t},$$
$$\lambda k_2 e^{\lambda t} = 5k_1 e^{\lambda t} + 3k_2 e^{\lambda t}.$$

Canceling $e^{\lambda t} \neq 0$ and simplifying yields the algebraic system

$$(1 - \lambda)k_1 + 3k_2 = 0,$$
$$5k_1 + (3 - \lambda)k_2 = 0. \tag{17}$$

Since we seek values of λ that permit solutions other than $k_1 = k_2 = 0$, the determinant of the coefficients must be zero:

$$\det\begin{pmatrix} 1 - \lambda & 3 \\ 5 & 3 - \lambda \end{pmatrix} = 0,$$

giving the characteristic equation

$$\lambda^2 - 4\lambda - 12 = 0. \tag{18}$$

The real distinct eigenvalues are $\lambda_1 = 6$ and $\lambda_2 = -2$. We seek a corresponding eigenvector for each λ.

Substituting $\lambda_1 = 6$ into system (17) results in the dependent system

$$-5k_1 + 3k_2 = 0,$$
$$5k_1 - 3k_2 = 0.$$

Either equation gives us $k_1 = 3/5\, k_2$, where k_2 is arbitrary. Setting $k_2 = 5$ yields $k_1 = 3$. Substituting λ_1, k_1, and k_2 into trial solution (16) gives us the vector solution $\mathbf{X}_1$, whose components are

$$x_1(t) = 3e^{6t},$$
$$y_1(t) = 5e^{6t}. \tag{19}$$

Substituting $\lambda_2 = -2$ into system (17) results in

$$3k_1 + 3k_2 = 0,$$
$$5k_1 + 5k_2 = 0,$$

which requires that $k_1 = -k_2$, where k_2 is arbitrary. Setting $k_2 = 1$, we find that $k_1 = -1$. Thus the combination of values $\lambda_2 = -2$, $k_1 = -1$, and $k_2 = 1$ yields a second vector solution $\mathbf{X}_2$ having components

$$\begin{aligned} x_2(t) &= -e^{-2t}, \\ y_2(t) &= e^{-2t}. \end{aligned} \tag{20}$$

The solutions $\mathbf{X}_1$ and $\mathbf{X}_2$ are linearly independent. The general solution is $\mathbf{X} = c_1\mathbf{X}_1 + c_2\mathbf{X}_2$, which in component form is

$$\begin{aligned} x(t) &= 3c_1e^{6t} - c_2e^{-2t}, \\ y(t) &= 5c_1e^{6t} + c_2e^{-2t}. \end{aligned} \tag{21}$$

Substituting the initial conditions into general solution (21) gives us

$$\begin{aligned} x(0) &= 100 = 3c_1 - c_2, \\ y(0) &= 300 = 5c_1 + c_2. \end{aligned}$$

The constants are $c_1 = 50$ and $c_2 = 50$. Thus the particular solution to the original system subject to the initial conditions is

$$\begin{aligned} x(t) &= 150e^{6t} - 50e^{-2t}, \\ y(t) &= 250e^{6t} + 50e^{-2t}. \end{aligned} \tag{22}$$

We check our solution by computing $dx/dt = 900e^{6t} + 100e^{-2t}$ and showing that it does equal $x + 3y$. Likewise, $dy/dt = 1500e^{6t} - 100e^{-2t}$ equals $5x + 3y$. Finally, we substitute $t = 0$ into particular solution (22) to check that the initial conditions are satisfied:

$$\begin{aligned} x(0) &= 150 - 50 = 100, \\ y(0) &= 250 + 50 = 300. \end{aligned}$$

We next turn our attention to the case in which the eigenvalues are complex numbers.

Case 2: General Solution for Complex Eigenvalues

For the constant-coefficient homogeneous linear system

$$\begin{aligned} \frac{dx}{dt} &= ax + by, \\ \frac{dy}{dt} &= cx + dy, \end{aligned} \tag{23}$$

substitution of the trial solution $x = k_1e^{\lambda t}$ and $y = k_2e^{\lambda t}$ leads to the characteristic equation

$$\lambda^2 - (a + d)\lambda + (ad - bc) = 0. \tag{24}$$

If the discriminant of the characteristic equation is negative, the values of λ are complex conjugates. Let us start with an example.

EXAMPLE 3 Solve the system

$$\begin{aligned}\frac{dx}{dt} &= 3x - 3y,\\ \frac{dy}{dt} &= 3x - y.\end{aligned} \tag{25}$$

Solution. Substituting the trial solution form

$$\begin{aligned}x &= k_1 e^{\lambda t},\\ y &= k_2 e^{\lambda t},\end{aligned} \tag{26}$$

into system (25) and simplifying gives us the algebraic system

$$\begin{aligned}(3-\lambda)k_1 - \qquad\quad 3k_2 &= 0,\\ 3k_1 + (-1-\lambda)k_2 &= 0.\end{aligned} \tag{27}$$

System (27) has nontrivial (nonzero) solutions if and only if the determinant of the coefficients is zero:

$$(3-\lambda)(-1-\lambda) + 9 = 0$$

or

$$\lambda^2 - 2\lambda + 6 = 0. \tag{28}$$

The roots of characteristic equation (28) are $\lambda_1 = 1 + \sqrt{5}i$ and $\lambda_2 = 1 - \sqrt{5}i$. We next determine the associated eigenvectors.

Substituting $\lambda_1 = 1 + \sqrt{5}i$ into system (27) gives us the system

$$\begin{aligned}(2-\sqrt{5}i)k_1 - \qquad\quad 3k_2 &= 0,\\ 3k_1 + (-2-\sqrt{5}i)k_2 &= 0.\end{aligned} \tag{29}$$

System (29) is dependent. This fact may not be clear at first glance, but a quick calculation shows that the determinant of the coefficients in system (29) is 0. Thus we need only solve the first equation in system (29) to find that

$$k_1 = \frac{3}{2-\sqrt{5}i}\,k_2. \tag{30}$$

Again notice that the ratio k_1/k_2 is constant, as in the case when the eigenvalues are real. Also, k_2 is arbitrary in Eq. (30), so we conveniently set $k_2 = 2 - \sqrt{5}i$ to yield $k_1 = 3$. Effectively, we have matched the numerators and denominators in the ratio k_1/k_2 to its constant value $3/(2-\sqrt{5}i)$. Thus one vector solution to system (25) is $\mathbf{X}_1$, whose components are

$$\begin{aligned} x_1(t) &= 3e^{(1+\sqrt{5}i)t}, \\ y_1(t) &= (2-\sqrt{5}i)e^{(1+\sqrt{5}i)t}. \end{aligned} \tag{31}$$

The second eigenvalue leads to a second solution vector. Substituting $\lambda_2 = 1 - \sqrt{5}i$ into algebraic system (27) produces the ratio

$$k_1/k_2 = \frac{3}{2+\sqrt{5}i}.$$

Again matching numerators and denominators, we have $k_1 = 3$ and $k_2 = 2 + \sqrt{5}i$. Thus a second solution vector to system (25) is $\mathbf{X}_2$, whose components are

$$\begin{aligned} x_2(t) &= 3e^{(1-\sqrt{5}i)t}, \\ y_2(t) &= (2+\sqrt{5}i)e^{(1-\sqrt{5}i)t}. \end{aligned} \tag{32}$$

Solutions (31) and (32) are linearly independent since $\lambda_1 \neq \lambda_2$ implies that $\mathbf{X}_1$ can never be a constant multiple of $\mathbf{X}_2$. In component form, the general solution to system (25) is therefore

$$\begin{aligned} x(t) &= 3c_1 e^{(1+\sqrt{5}i)t} + 3c_2 e^{(1-\sqrt{5}i)t}, \\ y(t) &= c_1(2-\sqrt{5}i)e^{(1+\sqrt{5}i)t} + c_2(2+\sqrt{5}i)e^{(1-\sqrt{5}i)t}. \end{aligned} \tag{33}$$

However, the functions in solution (33) are *complex valued,* whereas we seek *real* solutions to system (25). Just as we did when we found complex roots to the auxiliary equations for solving constant-coefficient linear second-order equations, we will use Euler's identity to obtain independent real solutions.

As illustrated in Example 3, the case of complex eigenvalues presents no difficulties in finding a general solution, since the distinct complex eigenvalues produce two complex linearly independent solutions. However, solution (33) would be difficult to interpret for a physical problem. For instance, does amplification, damping, and oscillation occur in the solution? Let us obtain a general form for two linearly independent real solutions when the eigenvalues are complex.

Finding Linearly Independent Real Solutions

To find real solutions for the complex solutions, we are going to follow a procedure similar to that used in Chapter 3 when we dealt with complex roots to the auxiliary equation. First, we make several very important observations from Example 3. Not only are the eigenvalues complex conjugates of each other, but so are the corresponding components k_1 and k_2 of the associated eigenvectors. In Example 3 we found the following pairs of eigenvalues and corresponding eigenvectors:

Eigenvalue	Eigenvector Components
$\lambda_1 = 1 + \sqrt{5}i$	$k_1 = 3,\ k_2 = 2 - \sqrt{5}i$
$\lambda_1^* = 1 - \sqrt{5}i$	$k_1^* = 3,\ k_2^* = 2 + \sqrt{5}i$

This situation generally holds in the case of complex eigenvalues. We state the result formally in the following theorem.

THEOREM 7.2

Let $\lambda = \alpha + i\beta$ be a complex eigenvalue associated with the homogeneous 2×2 system

$$\mathbf{X}' = \mathbf{A}\mathbf{X} \tag{34}$$

with corresponding eigenvector $\mathbf{K}$. Then

$$\begin{aligned} \mathbf{X}_1 &= \mathbf{K}e^{\lambda t}, \\ \mathbf{X}_2 &= \mathbf{K}^*e^{\lambda^* t} \end{aligned} \tag{35}$$

are linearly independent vector solutions to system (34).

The vector $\mathbf{K}^*$ in solutions (35) denotes the vector whose components are the complex conjugates of the vector components of $\mathbf{K}$.

The key to finding two real linearly independent solutions from complex solutions (35) is Euler's identity:

$$e^{i\theta} = \cos\theta + i\sin\theta. \tag{36}$$

For then $\mathbf{X}_1$ and $\mathbf{X}_2$ in solutions (35) can be rewritten as follows:

$$\begin{aligned} \mathbf{K}e^{\lambda t} &= \mathbf{K}e^{(\alpha+i\beta)t} = \mathbf{K}e^{\alpha t}(\cos\beta t + i\sin\beta t), \\ \mathbf{K}^*e^{\lambda^* t} &= \mathbf{K}^*e^{(\alpha-i\beta)t} = \mathbf{K}^*e^{\alpha t}(\cos\beta t - i\sin\beta t). \end{aligned} \tag{37}$$

Then, from system (37) we can form the vectors

$$\begin{aligned} \mathbf{K}e^{\lambda t} + \mathbf{K}^*e^{\lambda^* t} &= (\mathbf{K} + \mathbf{K}^*)e^{\alpha t}\cos\beta t + i(\mathbf{K} - \mathbf{K}^*)e^{\alpha t}\sin\beta t, \\ i(\mathbf{K}e^{\lambda t} - \mathbf{K}^*\mathbf{e}^{\lambda^* t}) &= i(\mathbf{K} - \mathbf{K}^*)e^{\alpha t}\cos\beta t - (\mathbf{K} + \mathbf{K}^*)e^{\alpha t}\sin\beta t. \end{aligned}$$

We now make an observation. For any complex number z with complex conjugate z^*, the sum $z + z^*$ equals twice the real part of z. Likewise, the number $i(z - z^*)$ equals twice the imaginary part of z. Therefore the vectors

$$\mathbf{B}_1 = \frac{1}{2}(\mathbf{K} + \mathbf{K}^*) \quad \text{and} \quad \mathbf{B}_2 = \frac{i}{2}(\mathbf{K} - \mathbf{K}^*) \tag{38}$$

have real components. These latter two vectors yield a fundamental set of real solutions, according to the following theorem.

THEOREM 7.3

Let $\lambda = \alpha + i\beta$ be a complex eigenvalue of homogeneous 2×2 system (34). Let $\mathbf{B}_1$ and $\mathbf{B}_2$ denote the column vectors with real components given by Eqs. (38), where $\mathbf{K}$ is the complex eigenvector associated with λ. Then

$$\mathbf{X}_1 = e^{\alpha t}(\mathbf{B}_1 \cos \beta t + \mathbf{B}_2 \sin \beta t),$$
$$\mathbf{X}_2 = e^{\alpha t}(\mathbf{B}_2 \cos \beta t - \mathbf{B}_1 \sin \beta t)$$

is a fundamental set of real solutions to system (34) for $-\infty < t < \infty$.

Remarks In stating the theorem we refer to the solution vectors in the fundamental set as $\mathbf{X}_1$ and $\mathbf{X}_2$; this is in keeping with the notation for naming the vectors in the fundamental sets used throughout the presentation. These vectors have real components, as desired, so they are not the same vectors as those in Eqs. (35), which have complex components.

It is not necessary to write both of the vectors $\mathbf{K}$ and $\mathbf{K}^*$ and then compute the real vectors $\mathbf{B}_1$ and $\mathbf{B}_2$ according to Eqs. (38). There is a simple procedure for obtaining the real vectors $\mathbf{B}_1$ and $\mathbf{B}_2$ from the single complex vector $\mathbf{K}$. If

$$\mathbf{K} = \begin{pmatrix} u_1 + iv_1 \\ u_2 + iv_2 \end{pmatrix},$$

then

$$\mathbf{B}_1 = \begin{pmatrix} u_1 \\ u_2 \end{pmatrix} \quad \text{and} \quad \mathbf{B}_2 = -\begin{pmatrix} v_1 \\ v_2 \end{pmatrix}. \tag{39}$$

An example will illustrate this procedure.

EXAMPLE 4 Find a fundamental set of real solutions for homogeneous system (25).

Solution. From our work in Example 3, we found the complex eigenvalue $\lambda = 1 + \sqrt{5}i$ with associated complex eigenvector

$$\mathbf{K} = \begin{pmatrix} 3 \\ 2 - \sqrt{5}i \end{pmatrix}.$$

Thus, from Eqs. (39) we have

$$\mathbf{B}_1 = \begin{pmatrix} 3 \\ 2 \end{pmatrix} \quad \text{and} \quad \mathbf{B}_2 = \begin{pmatrix} 0 \\ \sqrt{5} \end{pmatrix}.$$

From Theorem 7.3 we obtain the fundamental set

$$\mathbf{X}_1 = e^t\left[\begin{pmatrix}3\\2\end{pmatrix}\cos\sqrt{5}t + \begin{pmatrix}0\\\sqrt{5}\end{pmatrix}\sin\sqrt{5}t\right],$$

$$\mathbf{X}_2 = e^t\left[\begin{pmatrix}0\\\sqrt{5}\end{pmatrix}\cos\sqrt{5}t - \begin{pmatrix}3\\2\end{pmatrix}\sin\sqrt{5}t\right].$$

In component form we then have the two linearly independent real solutions

$$x_1(t) = 3e^t\cos\sqrt{5}t,$$
$$y_1(t) = e^t(2\cos\sqrt{5}t + \sqrt{5}\sin\sqrt{5}t),$$

and

$$x_2(t) = -3e^t\sin\sqrt{5}t,$$
$$y_2(t) = e^t(\sqrt{5}\cos\sqrt{5}t - 2\sin\sqrt{5}t).$$

Procedure for Complex Eigenvalues We conclude this section by summarizing the method for complex eigenvalues, after which we give an example of finding the particular solution to an initial value problem.

SUMMARY OF METHOD FOR COMPLEX ROOTS (EIGENVALUES) TO CHARACTERISTIC EQUATION

Step 1. Find the complex root $\lambda = \alpha + i\beta$ to the characteristic equation

$$\lambda^2 - (a+d)\lambda + (ad - bc) = 0.$$

Use the substitution method to find the characteristic equation.

Step 2. Find the complex eigenvector

$$\mathbf{K} = \begin{pmatrix}u_1 + iv_1\\u_2 + iv_2\end{pmatrix}$$

associated with the eigenvalue λ.

Step 3. Form the real vectors

$$\mathbf{B}_1 = \begin{pmatrix}u_1\\u_2\end{pmatrix} \quad\text{and}\quad \mathbf{B}_2 = -\begin{pmatrix}v_1\\v_2\end{pmatrix}.$$

Step 4. Form the fundamental set of real solutions

$$\mathbf{X}_1 = e^{\alpha t}(\mathbf{B}_1\cos\beta t + \mathbf{B}_2\sin\beta t),$$
$$\mathbf{X}_2 = e^{\alpha t}(\mathbf{B}_2\cos\beta t - \mathbf{B}_1\sin\beta t).$$

EXAMPLE 5 Solve the homogeneous 2×2 linear system

$$\begin{aligned} \frac{dx}{dt} &= x + 5y, \\ \frac{dy}{dt} &= -x - 3y, \end{aligned} \tag{40}$$

subject to the initial conditions $x(0) = 5$ and $y(0) = 4$.

Solution. Substitution of $x = k_1 e^{\lambda t}$ and $y = k_2 e^{\lambda t}$ into system (40) and simplification gives us

$$\begin{aligned} (1 - \lambda)k_1 + \qquad 5k_2 &= 0, \\ -k_1 + (-3 - \lambda)k_2 &= 0. \end{aligned} \tag{41}$$

The characteristic equation is then found by setting the determinant of the coefficients equal to zero:

$$\lambda^2 + 2\lambda + 2 = 0.$$

Therefore the eigenvalues are

$$\lambda = -1 \pm i.$$

For $\lambda_1 = -1 + i$, substitution into algebraic system (41) gives us

$$[1 - (-1 + i)]k_1 + 5k_2 = 0$$

or

$$\frac{k_1}{k_2} = \frac{5}{-2 + i}.$$

Then we may choose the components of the eigenvector $\mathbf{K}$ as

$$\begin{aligned} k_1 &= 5, \\ k_2 &= -2 + i. \end{aligned}$$

Thus, from Eqs. (39) we have

$$\mathbf{B}_1 = \begin{pmatrix} 5 \\ -2 \end{pmatrix} \quad \text{and} \quad \mathbf{B}_2 = -\begin{pmatrix} 0 \\ 1 \end{pmatrix}.$$

These vectors yield the fundamental set of solutions represented by

$$\mathbf{X}_1 = e^{-t}\left[\begin{pmatrix} 5 \\ -2 \end{pmatrix} \cos t + \begin{pmatrix} 0 \\ -1 \end{pmatrix} \sin t\right]$$

and

$$\mathbf{X}_2 = e^{-t}\left[\begin{pmatrix} 0 \\ -1 \end{pmatrix} \cos t - \begin{pmatrix} 5 \\ -2 \end{pmatrix} \sin t\right].$$

The general solution to system (40) is therefore

$$\mathbf{X} = c_1\mathbf{X}_1 + c_2\mathbf{X}_2.$$

In component form the general solution can be written as

$$x = e^{-t}[5c_1 \cos t - 5c_2 \sin t],$$
$$y = e^{-t}[(-2c_1 - c_2)\cos t + (-c_1 + 2c_2)\sin t].$$

To find the particular solution, we substitute the initial conditions into the general solution:

$$x(0) = 5 = 5c_1,$$
$$y(0) = 4 = -2c_1 - c_2.$$

Solution of these equations yields $c_1 = 1$ and $c_2 = -6$, so the particular solution (in component form) is

$$x = 5e^{-t}\cos t + 30e^{-t}\sin t,$$
$$y = 4e^{-t}\cos t - 13e^{-t}\sin t.$$

You should check to ensure that system (40) and the initial conditions are indeed satisfied by our results.

Summary

In each of Examples 1–5 presented in this section, we used the following solution procedure. First we substitute the trial solution

$$x = k_1 e^{\lambda t},$$
$$y = k_2 e^{\lambda t},$$

into the constant-coefficient linear system. This substitution results in a 2×2 system of algebraic equations in the unknowns λ, k_1, and k_2. Because we seek nontrivial solutions to the system, the determinant of the coefficients of k_1 and k_2 in that algebraic system must be zero, leading to the characteristic equation. If the eigenvalues are real and distinct, then each of them produces a solution vector, resulting in a fundamental set of solutions. If the eigenvalues are complex, one of them can be used to find an associated complex eigenvector. A fundamental set of real solution vectors can then be formed according to the summary steps outlined in this section for complex eigenvalues. In Section 7.5 we will take up the third case, when the characteristic equation has repeated (real) roots.

EXERCISES 7.4

In problems 1–15 find the general solution of the given system.

1. $\frac{dx}{dt} = 3x$, $\frac{dy}{dt} = 2x - y$

2. $\frac{dx}{dt} = 2x - 5y$, $\frac{dy}{dt} = -4y$

3. $\frac{dx}{dt} = 3x - 2y$, $\frac{dy}{dt} = 2x - 2y$

4. $\frac{dx}{dt} = 2x - y$, $\frac{dy}{dt} = 4x - 3y$

5. $\frac{dx}{dt} = 7x + y$, $\frac{dy}{dt} = -9x - 3y$

6. $\frac{dx}{dt} = -9x + 3y$, $\frac{dy}{dt} = -3x + y$

7. $\frac{dx}{dt} = 5x + 4y$, $\frac{dy}{dt} = -4x - 5y$

8. $\frac{dx}{dt} = \frac{3}{2}x - 2y$, $\frac{dy}{dt} = -x + 2y$

9. $\frac{dx}{dt} = x - y$, $\frac{dy}{dt} = 2x + 4y$

10. $\frac{dx}{dt} = 3y$, $\frac{dy}{dt} = 2x + y$

11. $\frac{dx}{dt} = -\frac{1}{2}x + 3y$, $\frac{dy}{dt} = 5x + 3y$

12. $\frac{dx}{dt} = -6x + y$, $\frac{dy}{dt} = -4x - y$

13. $\frac{dx}{dt} = 6x + 3y$, $\frac{dy}{dt} = 2x + y$

14. $\frac{dx}{dt} = -3y$, $\frac{dy}{dt} = -x - y$

15. $\frac{dx}{dt} = x + 4y$, $\frac{dy}{dt} = x + y$

In problems 16–30 find the general solution of the given system.

16. $\frac{dx}{dt} = 3x - y$, $\frac{dy}{dt} = x + 3y$

17. $\frac{dx}{dt} = -x + y$, $\frac{dy}{dt} = -x - y$

18. $\frac{dx}{dt} = 2x - y$, $\frac{dy}{dt} = 4x + 2y$

19. $\frac{dx}{dt} = -x - \frac{7}{4}y$, $\frac{dy}{dt} = 4x + 3y$

20. $\frac{dx}{dt} = 5x + 3y$, $\frac{dy}{dt} = -3x + y$

21. $\frac{dx}{dt} = -2x + 9y$, $\frac{dy}{dt} = -x + 2y$

22. $\frac{dx}{dt} = 2x + y$, $\frac{dy}{dt} = -2x + 4y$

23. $\frac{dx}{dt} = 3x + 2y$, $\frac{dy}{dt} = -2x + y$

24. $\frac{dx}{dt} = -5x + y$, $\frac{dy}{dt} = -4x - 3y$

25. $\frac{dx}{dt} = 4x + y$, $\frac{dy}{dt} = -x + 4y$

26. $\frac{dx}{dt} = -x - 3y$, $\frac{dy}{dt} = 3x - y$

27. $\frac{dx}{dt} = -5x - 8y$, $\frac{dy}{dt} = 2x - y$

28. $\frac{dx}{dt} = -3x - y$, $\frac{dy}{dt} = 9x - 5y$

29. $\frac{dx}{dt} = x + \frac{1}{6}y$, $\frac{dy}{dt} = -3x + 2y$

30. $\frac{dx}{dt} = -x - y$, $\frac{dy}{dt} = x - 2y$

In problems 31–35 solve the given system subject to the initial conditions.

31. $\dfrac{dx}{dt} = 5x - 2y$

$\dfrac{dy}{dt} = 4x - y$

$x(0) = 0,\ y(0) = 1$

32. $\dfrac{dx}{dt} = -4x + 3y$

$\dfrac{dy}{dt} = 2x + y$

$x(0) = -1,\ y(0) = 1$

35. $\dfrac{dx}{dt} = 2x - y$

$\dfrac{dy}{dt} = x + 3y$

$x(0) = 0,\ y(0) = -1$

33. $\dfrac{dx}{dt} = 3x + 2y$

$\dfrac{dy}{dt} = -x + 5y$

$x(0) = 1,\ y(0) = 0$

34. $\dfrac{dx}{dt} = x - 2y$

$\dfrac{dy}{dt} = 3x + 3y$

$x(0) = -1,\ y(0) = -1$

REPEATED EIGENVALUES

In this section we complete our study of solving homogeneous 2×2 linear systems with constant coefficients. Here we develop the general solution in the case of repeated *real* roots to the characteristic equation. (The roots cannot be complex since the characteristic equation is a quadratic for 2×2 systems.)

Case 3: General Solution for Repeated Real Roots

For the constant-coefficient homogeneous linear system

$$\begin{aligned}\frac{dx}{dt} &= ax + by,\\ \frac{dy}{dt} &= cx + dy,\end{aligned} \tag{1}$$

substituting the trial solution $x = k_1 e^{\lambda t}$ and $y = k_2 e^{\lambda t}$ leads, as before, to the characteristic equation

$$\lambda^2 + (a + d)\lambda + (ad - bc) = 0. \tag{2}$$

If the discriminant of quadratic Eq. (2) is zero, the values for λ are real and repeated. That is, $\lambda = -(a + d)/2$ is a double root. To gain some insight into this case, let us examine a specific example.

EXAMPLE 1 Find the general solution to the system

$$\begin{aligned} \frac{dx}{dt} &= x - y, \\ \frac{dy}{dt} &= x + 3y. \end{aligned} \tag{3}$$

Solution. We assume the solution form of $x = k_1 e^{\lambda t}$ and $y = k_2 e^{\lambda t}$ and substitute these equations into system (3) to get the following algebraic system:

$$\begin{aligned} (1-\lambda)k_1 - \quad\quad k_2 &= 0, \\ k_1 + (3-\lambda)k_2 &= 0. \end{aligned} \tag{4}$$

For a nontrivial solution the determinant of the coefficients in system (4) is zero, which leads to the characteristic equation

$$\lambda^2 - 4\lambda + 4 = 0$$

or

$$(\lambda - 2)^2 = 0.$$

The characteristic equation has the double root $\lambda = 2, 2$.

To find the eigenvectors for $\lambda_1 = \lambda_2 = 2$, we substitute this value into Eqs. (4) to obtain

$$\begin{aligned} -k_1 - k_2 &= 0, \\ k_1 + k_2 &= 0, \end{aligned} \tag{5}$$

or

$$k_1 = -k_2.$$

Since k_2 is arbitrary, we choose $k_2 = -1$, yielding $k_1 = 1$. Hence we find the vector solution $\mathbf{X}_1$ whose components are

$$\begin{aligned} x_1 &= e^{2t}, \\ y_1 &= -e^{2t}. \end{aligned}$$

To find the general solution to system (3) we must find a second, linearly independent solution. As we did with the situation for second-order equations, it would seem reasonable for us to try a solution of the form

$$\mathbf{X}_2 = t\mathbf{Q}e^{2t}.$$

In component form this would be

$$\begin{aligned} x &= q_1 t e^{2t}, \\ y &= q_2 t e^{2t}, \end{aligned} \tag{6}$$

where q_1 and q_2 are unknowns to be determined. Differentiating Eqs. (6), substituting the results in system (3), and simplifying gives us

$$(q_1 + q_2)t + q_1 = 0, \tag{7}$$

$$(q_1 + q_2)t - q_2 = 0. \tag{8}$$

Equations (7) and (8) are both linear polynomials in the variable t. Moreover, Eq. (7) requires that $q_1 = 0$, while Eq. (8) requires that $q_2 = 0$. Trial solution (6) yields the trivial solution $x(t) = 0$ and $y(t) = 0$. Thus *assumed trial solution (6) does not lead to a second linearly independent solution.* As it turns out, additional terms, $p_1 e^{2t}$ and $p_2 e^{2t}$, need to be added to x and y in trial solution (6) so that the q terms are not isolated. We will continue this example after a brief investigation into the more general case.

Finding a Second Linearly Independent Solution

Suppose homogeneous system (1) has a repeated eigenvalue λ. Two possibilities can occur. The eigenvalue may give rise to two linearly independent eigenvectors $\mathbf{K}_1$ and $\mathbf{K}_2$. In this case, just as for the case of distinct real roots,

$$\mathbf{X}_1 = \mathbf{K}_1 e^{\lambda t} \quad \text{and} \quad \mathbf{X}_2 = \mathbf{K}_2 e^{\lambda t} \tag{9}$$

form a fundamental set of solutions. We will illustrate this case in Example 3 later in the section. In the second case, λ produces only a single independent eigenvector $\mathbf{K}$, as occurred in Example 1. The following result, which we state without proof, provides the form of a second linearly independent solution.

THEOREM 7.4

Suppose λ is a repeated eigenvalue for homogeneous system (1) with only one independent eigenvector $\mathbf{K}$. Then a second linearly independent solution can be found that has the form

$$\mathbf{X}_2 = \mathbf{K}te^{\lambda t} + \mathbf{P}e^{\lambda t}. \tag{10}$$

By requiring that a trial solution in the form of Eq. (10) satisfy the given linear system, we will determine algebraic equations in the unknown components p_1 and p_2 for the vector $\mathbf{P}$. Solving the algebraic system for those components then leads immediately to the solution $\mathbf{X}_2$ expressed by Eq. (10). We now return to our first example to apply this procedure.

EXAMPLE 1 (CONTINUED) We have previously determined the repeated eigenvalue $\lambda = 2$ and single eigenvector

$$\mathbf{K} = \begin{pmatrix} 1 \\ -1 \end{pmatrix}$$

giving the solution

$$\mathbf{X}_1 = \begin{pmatrix} 1 \\ -1 \end{pmatrix} e^{2t}$$

to linear system (3). In component form we have

$$x_1 = e^{2t},$$
$$y_1 = -e^{2t}.$$

Now we assume a second trial solution in the form of Eq. (10):

$$\mathbf{X}_2 = \begin{pmatrix} 1 \\ -1 \end{pmatrix} te^{2t} + \begin{pmatrix} p_1 \\ p_2 \end{pmatrix} e^{2t} \tag{11}$$

where the components p_1 and p_2 are unknowns to be determined. In component form Eq. (11) translates to

$$x_2 = (t + p_1)e^{2t},$$
$$y_2 = (-t + p_2)e^{2t}.$$

The derivatives are

$$\frac{dx_2}{dt} = (2t + 2p_1 + 1)e^{2t},$$

$$\frac{dy_2}{dt} = (-2t + 2p_2 - 1)e^{2t}.$$

Requiring that x_2 and y_2 and their derivatives satisfy system (3) yields the equations

$$(2t + 2p_1 + 1)e^{2t} = (t + p_1)e^{2t} - (-t + p_2)e^{2t},$$
$$(-2t + 2p_2 - 1)e^{2t} = (t + p_1)e^{2t} + 3(-t + p_2)e^{2t}.$$

Cancellation of the nonzero term e^{2t} and algebraic simplification of the last two equations yields the system

$$\begin{aligned} -p_1 - p_2 &= 1, \\ p_1 + p_2 &= -1. \end{aligned} \tag{12}$$

If you compare algebraic system (12) with system (5) where we found the eigenvector components k_1 and k_2, you will notice that (12) has exactly the same lefthand-side coefficients as (5), except that the *right side consists of the eigenvector components k_1 and k_2*, rather than zeros. This situation always occurs when determining the components of the vector $\mathbf{P}$ in Eq. (10).

System (12) is dependent and has infinitely many solutions. Since we seek only one solution, we conveniently let $p_2 = 0$ in system (12) to obtain $p_1 = -1$. Then, substituting these values into Eq. (11) gives a second linearly independent solution to our problem:

$$\mathbf{X}_2 = \begin{pmatrix} 1 \\ -1 \end{pmatrix} te^{2t} + \begin{pmatrix} -1 \\ 0 \end{pmatrix} e^{2t}.$$

In component form we have

$$x_2 = (t-1)e^{2t},$$
$$y_2 = -te^{2t}.$$

Therefore the general solution in component form is

$$x = c_1 x_1 + c_2 x_2$$
$$y = c_1 y_1 + c_2 y_2,$$

or

$$x = c_1 e^{2t} + c_2(t-1)e^{2t},$$
$$y = -c_1 e^{2t} - c_2 te^{2t}.$$

The procedure illustrated in Example 1 generalizes to any system with a repeated eigenvalue but only one independent eigenvector. This procedure is justified, as we shall explain in the following discussion.

Finding the Vector P

We return now to the general problem for real repeated eigenvalues. To determine the components of the vector $\mathbf{P}$ in Eq. (10), we require that the trial solution form for $\mathbf{X}_2$ solve the system. From Eq. (10) the components of $\mathbf{X}_2$ are

$$x = k_1 te^{\lambda t} + p_1 e^{\lambda t},$$
$$y = k_2 te^{\lambda t} + p_2 e^{\lambda t}.$$

Differentiation gives us

$$\frac{dx}{dt} = k_1 e^{\lambda t} + \lambda k_1 te^{\lambda t} + \lambda p_1 e^{\lambda t},$$

$$\frac{dy}{dt} = k_2 e^{\lambda t} + \lambda k_2 te^{\lambda t} + \lambda p_2 e^{\lambda t}.$$

Substitution of these components and derivatives into homogeneous system (1) and canceling the $e^{\lambda t}$ terms throughout yields the following algebraic equations:

$$k_1 + \lambda k_1 t + \lambda p_1 = ak_1 t + ap_1 + bk_2 t + bp_2,$$
$$k_2 + \lambda k_2 t + \lambda p_2 = ck_1 t + cp_1 + dk_2 t + dp_2.$$

Collecting terms in like powers of t in these equations and rearranging terms results in the following system:

$$\begin{aligned}[(a-\lambda)k_1 + bk_2]t + (a-\lambda)p_1 + bp_2 &= k_1,\\ [ck_1 + (d-\lambda)k_2]t + cp_1 + (d-\lambda)p_2 &= k_2.\end{aligned} \tag{13}$$

Each equation in system (13) is a polynomial in t. Equating like powers of t on both sides in each equation produces the following two systems:

$$\begin{aligned}(a-\lambda)k_1 + \quad bk_2 &= 0,\\ ck_1 + (d-\lambda)k_2 &= 0,\end{aligned} \tag{14}$$

and

$$\begin{aligned}(a-\lambda)p_1 + \quad bp_2 &= k_1,\\ cp_1 + (d-\lambda)p_2 &= k_2.\end{aligned} \tag{15}$$

System (14) is the same one we obtained earlier for determining **K** as an eigenvector associated with the eigenvalue λ. System (15) gives the required conditions on the components p_1 and p_2 of the vector **P** so that $\mathbf{X}_2$ in Eq. (10) is a solution. The vector **P** is *not* itself an eigenvector; it simply allows Eq. (10) to be a solution to the system.

Notice that the determinant of the coefficients in system (15) is the characteristic polynomial $(a-\lambda)(d-\lambda) - bc$, which must equal zero since λ is an eigenvalue of system (1) (and therefore a root of the characteristic equation). Thus algebraic system (15) is dependent and has infinitely many solutions for p_1 and p_2. Since we seek only one solution, we can conveniently choose $p_2 = 0$. The following steps summarize the solution procedure for repeated real roots.

SUMMARY OF METHOD FOR REPEATED REAL ROOTS (EIGENVALUES) TO CHARACTERISTIC EQUATION

Step 1. Use the substitution method to find the characteristic equation

$$\lambda^2 - (a+d)\lambda + (ad - bc) = 0.$$

Then find the repeated eigenvalue λ to the characteristic equation.

Step 2. If λ yields two independent eigenvectors $\mathbf{K}_1$ and $\mathbf{K}_2$, a fundamental set of vector solutions is then

$$\mathbf{X}_1 = \mathbf{K}_1 e^{\lambda t} \qquad \text{and} \qquad \mathbf{X}_2 = \mathbf{K}_2 e^{\lambda t}.$$

Proceed to Step 4.

Step 3. If λ produces only one independent eigenvector $\mathbf{K}$, then one solution is

$$\mathbf{X}_1 = \mathbf{K}e^{\lambda t}.$$

A second linearly independent solution is given by

$$\mathbf{X}_2 = \mathbf{K}te^{\lambda t} + \mathbf{P}e^{\lambda t}$$

where the components of $\mathbf{P}$ must satisfy the algebraic system

$$\begin{aligned}(a-\lambda)p_1 + \quad bp_2 &= k_1,\\ cp_1 + (d-\lambda)p_2 &= k_2.\end{aligned}$$

Step 4. Form the general solution

$$\mathbf{X} = c_1\mathbf{X}_1 + c_2\mathbf{X}_2$$

from the fundamental set $\{\mathbf{X}_1, \mathbf{X}_2\}$ found in Steps 2 or 3, whichever is appropriate.

We now present several examples of the method.

EXAMPLE 2 Solve the homogeneous system

$$\begin{aligned}\frac{dx}{dt} &= 2x - y,\\ \frac{dy}{dt} &= 2y.\end{aligned} \tag{16}$$

Solution. We assume a solution takes the form of $x = k_1e^{\lambda t}$ and $y = k_2e^{\lambda t}$, substitute these into system (16), and simplify to get the equations

$$\begin{aligned}(2-\lambda)k_1 - k_2 &= 0,\\ (2-\lambda)k_2 &= 0.\end{aligned} \tag{17}$$

Since we seek nontrivial solutions, the determinant of the coefficients must be zero, giving us the characteristic equation

$$(2-\lambda)^2 = 0.$$

The eigenvalues are $\lambda = 2, 2$.

Next we find an eigenvector $\mathbf{K}$. Substitution of $\lambda = 2$ into Eqs. (17) leads to

$$\begin{aligned}0k_1 - k_2 &= 0,\\ 0k_2 &= 0.\end{aligned}$$

Thus $k_2 = 0$ and k_1 is arbitrary. Choose $k_1 = 1$ to obtain the solution

$$\mathbf{X}_1 = \begin{pmatrix} 1 \\ 0 \end{pmatrix} e^{2t}. \tag{18}$$

To find a second solution $\mathbf{X}_2$ we first find the vector $\mathbf{P}$ by solving the system

$$\begin{aligned} 0p_1 - p_2 &= 1, \\ 0p_2 &= 0, \end{aligned}$$

where the components of the eigenvector $\mathbf{K}$ yield the righthand side. Thus $p_2 = -1$ and p_1 is arbitrary. We choose $p_1 = 0$ and obtain the second solution,

$$\mathbf{X}_2 = \begin{pmatrix} 1 \\ 0 \end{pmatrix} te^{2t} + \begin{pmatrix} 0 \\ -1 \end{pmatrix} e^{2t}. \tag{19}$$

From Eqs. (18) and (19) the general solution to system (16) in component form is

$$\begin{aligned} x &= c_1 e^{2t} + c_2 te^{2t}, \\ y &= -c_2 e^{2t}. \end{aligned}$$

You should verify that system (16) is indeed satisfied by these functions.

EXAMPLE 3 Solve the homogeneous system

$$\begin{aligned} \frac{dx}{dt} &= -x, \\ \frac{dy}{dt} &= -y, \end{aligned} \tag{20}$$

by assuming a solution vector of the form $\mathbf{X} = \mathbf{K}e^{\lambda t}$.

Solution. If we assume solutions take the form of $x = k_1 e^{\lambda t}$, $y = k_2 e^{\lambda t}$, then substituting these into system (20) and simplifying leads to

$$\begin{aligned} (-1 - \lambda)k_1 + \qquad 0k_2 &= 0, \\ 0k_1 + (-1 - \lambda)k_2 &= 0. \end{aligned} \tag{21}$$

The determinant of the coefficients must be zero, thus yielding the characteristic equation

$$(-1 - \lambda)^2 = 0.$$

The eigenvalues are repeated: $\lambda = -1, -1$.

We now find an associated eigenvector. Substitution of $\lambda = -1$ into algebraic system (21) yields the system

$$\begin{aligned} 0k_1 + 0k_2 &= 0, \\ 0k_1 + 0k_2 &= 0. \end{aligned}$$

This system is satisfied for *all* values of k_1 and k_2. Thus we may choose *any* two linearly independent vectors as eigenvectors. Arbitrarily we choose

$$\mathbf{K}_1 = \begin{pmatrix} 1 \\ -1 \end{pmatrix} \quad \text{and} \quad \mathbf{K}_2 = \begin{pmatrix} 1 \\ 1 \end{pmatrix}.$$

The general solution to system (20) is then given by

$$\mathbf{X} = c_1 \begin{pmatrix} 1 \\ -1 \end{pmatrix} e^{-t} + c_2 \begin{pmatrix} 1 \\ 1 \end{pmatrix} e^{-t}. \tag{22}$$

System (20) can also be solved directly by simply integrating each equation. Thus

$$x = Ae^{-t} \quad \text{and} \quad y = Be^{-t}$$

where A and B are arbitrary constants. Indeed, from Eq. (22) you see that

$$x = (c_1 + c_2)e^{-t},$$
$$y = (-c_1 + c_2)e^{-t},$$

so that $A = c_1 + c_2$ and $B = -c_1 + c_2$ are arbitrary.

Example 3 does illustrate the form of the system when there is a repeated eigenvalue giving rise to two linearly independent eigenvectors. The system will have the **diagonal form**

$$\frac{dx}{dt} = ax,$$
$$\frac{dy}{dt} = \qquad dy,$$

where $a = d$. Such a system can be solved by direct integration of each "uncoupled" equation rather than by the systems method we have been discussing. However, we caution you that for higher first-order systems (say 3×3), a repeated eigenvalue producing two linearly independent eigenvectors can occur for "coupled" (that is, nondiagonal) systems. We pursue this idea in Chapter 8, when we present 3×3 systems.

Simplification of the Procedure

Consider the 2×2 constant-coefficient homogeneous linear system

$$\frac{dx}{dt} = ax + by,$$
$$\frac{dy}{dt} = cx + dy.$$

Substitution of $x = k_1 e^{\lambda t}$ and $y = k_2 e^{\lambda t}$ always leads to the algebraic system

$$(a - \lambda)k_1 + \qquad bk_2 = 0,$$
$$ck_1 + (d - \lambda)k_2 = 0.$$

These equations can be written down directly without going through the actual differentiation and substitution process. The characteristic equation is obtained by equating to zero the determinant of the coefficients in this system:

$$(a-\lambda)(d-\lambda)-bc=0.$$

We conclude this section by finding the particular solution to an initial value problem.

EXAMPLE 4 Solve the system

$$\begin{aligned}\frac{dx}{dt}&=3x-4y,\\ \frac{dy}{dt}&=x-y,\end{aligned}\tag{23}$$

subject to the initial conditions $x(0)=40$, and $y(0)=10$.

Solution. Assume $x=k_1e^{\lambda t}$ and $y=k_2e^{\lambda t}$ yielding

$$\begin{aligned}(3-\lambda)k_1-\qquad\quad 4k_2&=0,\\ k_1+(-1-\lambda)k_2&=0.\end{aligned}\tag{24}$$

The characteristic equation corresponding to system (24) is

$$\lambda^2-2\lambda+1=0$$

with the repeated eigenvalue $\lambda=1, 1$. Substituting $\lambda_1=1$ into system (24) results in $k_1=2k_2$. Choosing $k_2=1$ yields $k_1=2$ and the vector solution $\mathbf{X}_1=\mathbf{K}e^{\lambda t}$ with components

$$\begin{aligned}x_1&=2e^t,\\ y_1&=e^t.\end{aligned}$$

From Eq. (10) the form of the second solution is

$$\mathbf{X}_2=\begin{pmatrix}2\\1\end{pmatrix}te^t+\begin{pmatrix}p_1\\p_2\end{pmatrix}e^t.$$

The components p_1 and p_2 must satisfy the algebraic system

$$\begin{aligned}(3-1)p_1-\qquad\quad 4p_2&=2,\\ p_1+(-1-1)p_2&=1,\end{aligned}$$

or

$$\begin{aligned}2p_1-4p_2&=2,\\ p_1-2p_2&=1.\end{aligned}\tag{25}$$

System (25) is dependent. Choosing $p_2 = 0$, the second equation gives us $p_1 = 1$. Thus a second solution is

$$\mathbf{X}_2 = \begin{pmatrix} 2 \\ 1 \end{pmatrix} te^t + \begin{pmatrix} 1 \\ 0 \end{pmatrix} e^t.$$

Therefore the general solution to system (23) is

$$\begin{aligned} x &= 2c_1 e^t + c_2(2t + 1)e^t, \\ y &= c_1 e^t + c_2 te^t. \end{aligned} \tag{26}$$

Substituting the initial conditions into general solution (26) yields

$$\begin{aligned} 40 &= 2c_1 + c_2, \\ 10 &= c_1. \end{aligned}$$

These equations have the solution $c_1 = 10$ and $c_2 = 20$. Thus the particular solution to system (23) is

$$\begin{aligned} x &= 40e^t + 40te^t, \\ y &= 10e^t + 20te^t. \end{aligned} \tag{27}$$

We leave it to you to check that the functions in (27) do indeed satisfy the system (23) and the initial conditions.

EXERCISES 7.5

In problems 1–20 find the general solution to the given system.

1. $\dfrac{dx}{dt} = 4x - y$, $\dfrac{dy}{dt} = x + 2y$

2. $\dfrac{dx}{dt} = 3x - y$, $\dfrac{dy}{dt} = x + 5y$

3. $\dfrac{dx}{dt} = 2x - y$, $\dfrac{dy}{dt} = 4x - 2y$

4. $\dfrac{dx}{dt} = 4x - y$, $\dfrac{dy}{dt} = x + 6y$

5. $\dfrac{dx}{dt} = 4x - 3y$, $\dfrac{dy}{dt} = 3x - 2y$

6. $\dfrac{dx}{dt} = -2x + 2y$, $\dfrac{dy}{dt} = -2x + 2y$

7. $\dfrac{dx}{dt} = -x + 3y$, $\dfrac{dy}{dt} = -3x + 5y$

8. $\dfrac{dx}{dt} = 3x + y$, $\dfrac{dy}{dt} = -x + y$

9. $\dfrac{dx}{dt} = -4x - y$, $\dfrac{dy}{dt} = x - 2y$

10. $\dfrac{dx}{dt} = -x - 4y$, $\dfrac{dy}{dt} = 4x + 7y$

11. $\dfrac{dx}{dt} = x + 9y$, $\dfrac{dy}{dt} = -x - 5y$

12. $\dfrac{dx}{dt} = -5x - y$, $\dfrac{dy}{dt} = 4x - y$

13. $\dfrac{dx}{dt} = 3x - \dfrac{1}{2}y$, $\dfrac{dy}{dt} = 8x - y$

14. $\dfrac{dx}{dt} = 5x - y$, $\dfrac{dy}{dt} = x + 3y$

15. $\dfrac{dx}{dt} = 7x - 2y$, $\dfrac{dy}{dt} = 2x + 3y$

16. $\dfrac{dx}{dt} = 2x + y$, $\dfrac{dy}{dt} = -x + 4y$

17. $\dfrac{dx}{dt} = x - 3y$, $\dfrac{dy}{dt} = 3x + 7y$

18. $\dfrac{dx}{dt} = -5x - 4y$, $\dfrac{dy}{dt} = 4x + 3y$

19. $\dfrac{dx}{dt} = -x - 8y$, $\dfrac{dy}{dt} = \dfrac{1}{2}x + 3y$

20. $\dfrac{dx}{dt} = y$, $\dfrac{dy}{dt} = -9x + 6y$

In problems 21–25 solve the given system subject to the initial conditions.

21. $\dfrac{dx}{dt} = -x - y$, $\dfrac{dy}{dt} = x - 3y$, $x(0) = 1,\ y(0) = -2$

22. $\dfrac{dx}{dt} = -4x + y$, $\dfrac{dy}{dt} = -x - 2y$, $x(0) = -1,\ y(0) = 1$

23. $\dfrac{dx}{dt} = -2x - 8y$, $\dfrac{dy}{dt} = 2x + 6y$, $x(0) = 0,\ y(0) = -1$

24. $\dfrac{dx}{dt} = 3x + 8y$, $\dfrac{dy}{dt} = -2x - 5y$, $x(0) = 1,\ y(0) = 0$

25. $\dfrac{dx}{dt} = 5x - 2y$, $\dfrac{dy}{dt} = 2x + y$, $x(0) = -2,\ y(0) = 3$

A NONLINEAR SYSTEM: A COMPETITIVE HUNTER MODEL AND PHASE PLANE ANALYSIS (Optional)*

In Section 7.1, Example 9, we developed the following model for the growth of trout and bass in a pond:

$$\begin{aligned}\frac{dx}{dt} &= (a - by)x,\\ \frac{dy}{dt} &= (m - nx)y,\end{aligned} \tag{1}$$

where $x(t)$ represents the trout population and $y(t)$ the bass population. The constants $x(0) = x_0$, $y(0) = y_0$, and a, b, m, and n are all positive. This model is useful in studying the growth patterns of two species exhibiting competitive behavior, like the trout and bass. In this section we use a graphical analysis to study the solution trajectories of system (1).

* Adapted from *A First Course in Mathematical Modeling* by the authors (Pacific Grove, Calif.: Brooks/Cole Publishing Company, 1985), Chapter 10. Adaptation presented by permission of Brooks/Cole Publishing Company.

One of our concerns is whether the trout and bass populations reach equilibrium levels. If so, then we will know whether coexistence of the two species in the pond is possible. The only way such a state can be achieved is if both populations stop growing; that is, $dx/dt = 0$ and $dy/dt = 0$. Thus we seek the rest points or equilibrium points of system (1). We analyze the steady-state or equilibrium outcomes predicted by the model using a graphical analysis. (You may want to review briefly the graphical analysis presented in Sections 1.4 and 7.2 before continuing.)

After setting the right-hand sides of the system (1) equations equal to zero and solving for x and y simultaneously, we find the rest points $(x, y) = (0, 0)$ and $(x, y) = (m/n, a/b)$ in the phase plane. Along the vertical line $x = m/n$ and the x-axis in the phase plane, the growth dy/dt in the bass population is zero; along the horizontal line $y = a/b$ and the y-axis, the growth dx/dt in the trout population is zero. If the initial stockage is at these population levels, there would be no growth in either population. These features are depicted in Fig. 7.15.

Considering the approximations necessary in any model, it is inconceivable that we would estimate precisely the values for the constants a, b, m, and n in system (1). So the pertinent behavior we need to investigate is what happens to the solution trajectories in the vicinity of the rest points $(0, 0)$ and $(m/n, a/b)$. Specifically, are these points stable or unstable?

To investigate this question graphically, let us analyze the directions of dx/dt and dy/dt in the phase plane. Although $x(t)$ and $y(t)$ represent the trout and bass populations, respectively, it is helpful to think of the trajectories as paths of a moving particle, in accordance with our previous discussion. When dx/dt is positive, the horizontal component $x(t)$ of the trajectory is increasing and the particle is moving towards the right; when dx/dt is negative, the particle is moving to the left. Likewise, when dy/dt is positive, the component $y(t)$ is increasing and the particle is moving upward; when dy/dt is negative, the particle is moving downward. In system (1) the vertical line $x = m/n$ divides the phase plane into two half-planes. In the left half-plane

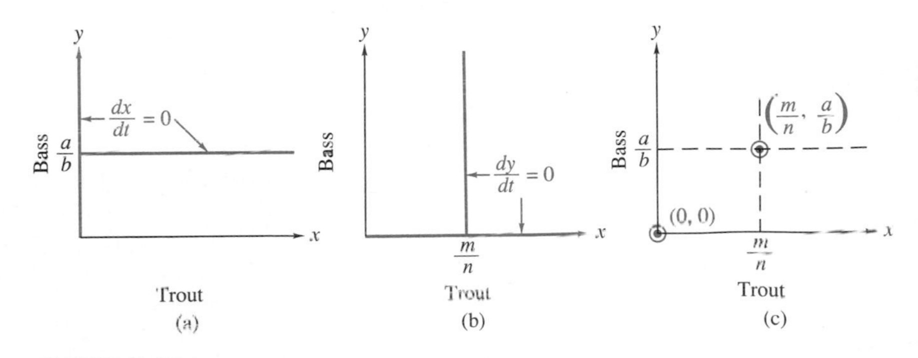

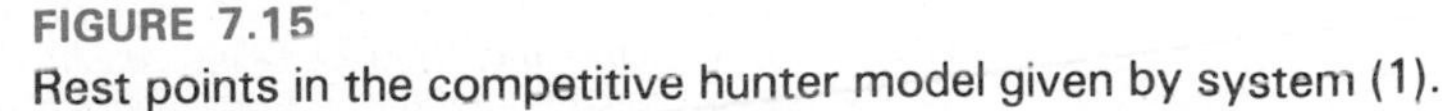

FIGURE 7.15
Rest points in the competitive hunter model given by system (1).

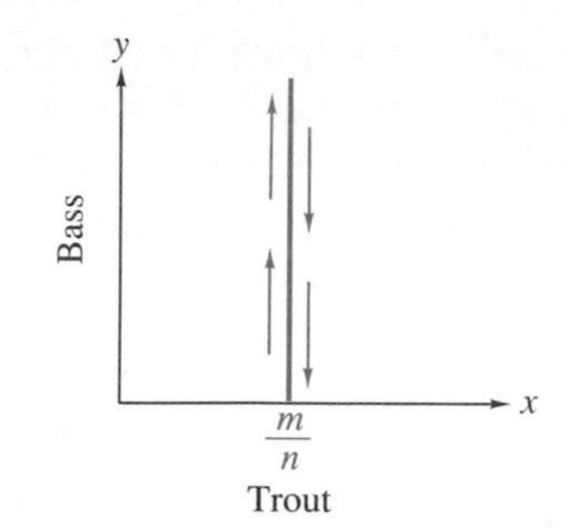

FIGURE 7.16
To the left of the line $x = m/n$ the trajectories move upward, and to the right they move downward.

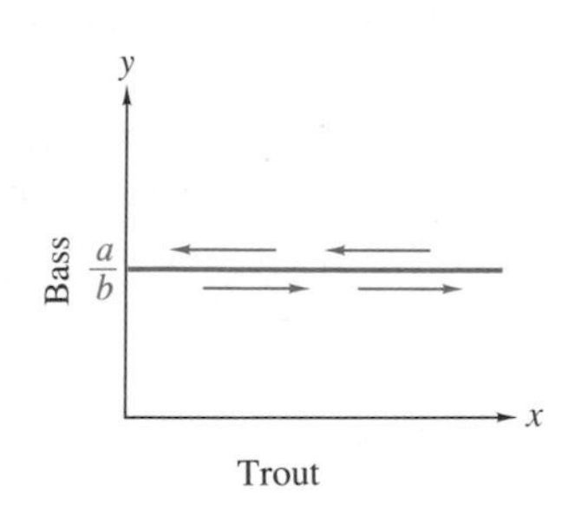

FIGURE 7.17
Above the line $y = a/b$ the trajectories move to the left, and below it they move to the right.

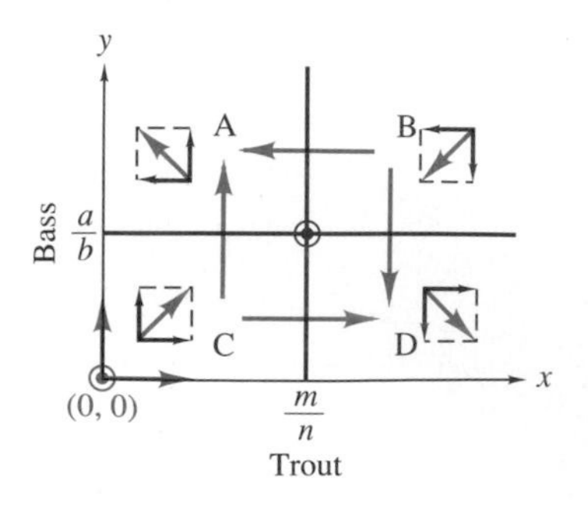

FIGURE 7.18
Composite graphical analysis of the trajectory directions in the four regions determined by $x = m/n$ and $y = a/b$.

dy/dt is positive, and in the right half-plane it is negative. The directions of the associated trajectories are indicated in Fig. 7.16. Likewise, the horizontal line $y = a/b$ determines the half-planes where dx/dt is positive or negative. The directions of the associated trajectories are indicated in Fig. 7.17. Along the line $y = a/b$ itself, $dx/dt = 0$. Therefore any trajectory crossing this line will do so vertically. Similarly, along the line $x = m/n$, $dy/dt = 0$, so the line will be crossed horizontally. Finally, along the y-axis motion is vertically upward because dy/dt is positive and $dx/dt = 0$. Similarly, along the x-axis motion is horizontal and to the right because dx/dt is positive and $dy/dt = 0$. Combining all this information together into a single graph gives the four distinct regions A, B, C, and D with their respective trajectory directions as depicted in Fig. 7.18.

Now let us analyze the motion in the vicinity of the rest points. For (0, 0) we can see that all motion is away from it—upward and toward the right. In the vicinity of the rest point $(m/n, a/b)$ the behavior depends on the region in which the trajectory begins. If the trajectory starts in region B, for instance, then it will move downward and leftward toward the rest point. However, as it gets nearer to the rest point, the derivatives dx/dt and dy/dt approach zero. Depending on where the trajectory begins and the relative sizes of the constants a, b, m, and n, either the trajectory will continue moving downward and into region D as it swings past the rest point, or it will move leftward into region A. Once it enters either one of these latter two regions, it will move away from the rest point. Thus both rest points are unstable. These features are suggested in Fig. 7.19.

Model Interpretation

Now let us consider the half-planes $y < a/b$ and $y > a/b$. In each half-plane there is exactly one trajectory approaching the rest point $(m/n, a/b)$. A proof of this fact is outlined in problem 7 of Exercises 7.6. Above these two trajectories the bass population increases, and below them the bass population decreases. The trajectory for $y < a/b$ is shown as a line joining (0, 0) to $(m/n, a/b)$ in Fig. 7.20 for simplicity, but it is not likely to be a line.

The graphical analysis conducted so far leads us to the preliminary conclusion that, under the assumptions of our model, it is highly unlikely for both species to reach equilibrium levels. Furthermore, the initial stockage levels turn out to be important in determining which of the two species might survive. Perturbations of the system may also affect the outcome of the competition. Thus mutual coexistence of the species is highly improbable. This phenomenon is known as the **principle of competitive exclusion,** or Gause's principle.* Moreover, the initial conditions completely determine

* Named after G. F. Gause, 1910– , who furthered the work of Joseph Grinnel, Alfred Lotka, and Vita Volterra in population ecology with his book *The Struggle for Existence,* (Baltimore: Williams and Wilkens, 1934). Actually it was Joseph Grinnel who first expressed the exclusion principle in 1904. For an interesting historical account, see the article by G. Hardin, "The Competitive Exclusion Principle," *Science* 131:1291–1297 (1960).

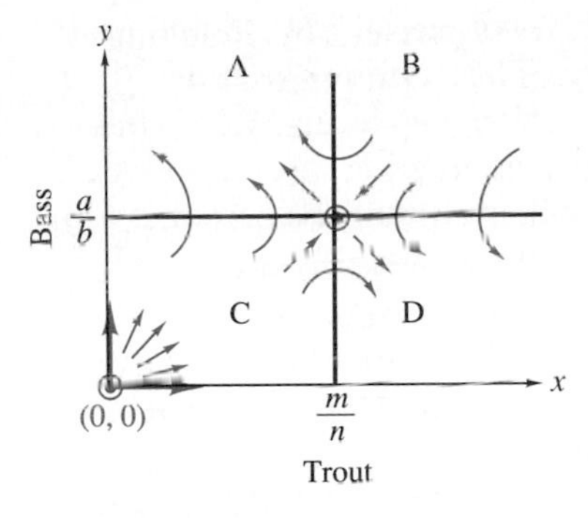

FIGURE 7.19
Motion along the trajectories near the rest points (0, 0) and $(m/n, a/b)$.

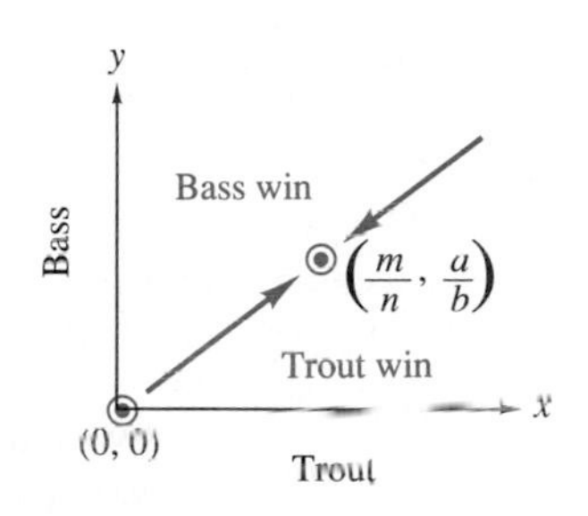

FIGURE 7.20
Qualitative results of analyzing the competitive hunter model. There are exactly two trajectories approaching the point $(m/n, a/b)$.

FIGURE 7.21
Trajectory direction near the rest point (0, 0).

the outcome as depicted in Figure 7.20. You can see from that graph that any perturbation causing a switch from one region (say below the two trajectories approaching the rest point $(m/n, a/b)$) to the other region (above the two trajectories) would change the outcome. One of the limitations of our graphical analysis is that we have not determined those separating trajectories precisely. If we are satisfied with our model we may very well want to determine that separating boundary for the two regions.

Limitations of a Graphical Analysis

It is not always possible to determine the nature of the motion near a rest point using only graphical analysis. To understand this limitation, let us consider the rest point (0, 0) and the direction of motion of the trajectories shown in Fig. 7.21. The information given in the figure is insufficient to distinguish between the three possible motions shown in Fig. 7.22. Moreover, even if we have determined by some other means that Fig. 7.22(c) correctly portrays the motion near the rest point (0, 0), we might be tempted to deduce that the motion will grow without bound in both the x and y directions. However, consider the system given by

$$\begin{aligned} \frac{dx}{dt} &= y + x - x(x^2 + y^2), \\ \frac{dy}{dt} &= -x + y - y(x^2 + y^2). \end{aligned} \tag{2}$$

It can be shown that (0, 0) is the only rest point for system (2). Yet any trajectory starting on the unit circle $x^2 + y^2 = 1$ will traverse the unit circle in a periodic solution because in that case $dy/dx = -x/y$ (see Exercises 7.6, problem 2). Moreover, if a trajectory starts inside the circle (provided it does not start at the origin), it will spiral outward asymptotically, getting closer and closer to the circular path as t tends toward infinity. Likewise, if the

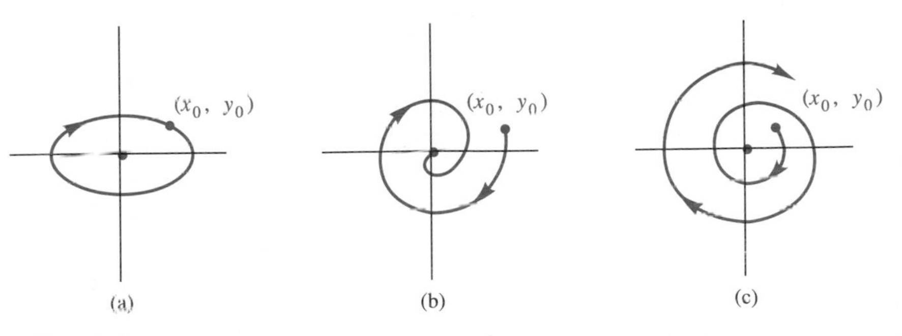

FIGURE 7.22
Three possible trajectory motions: (a) periodic motion, (b) motion toward an asymptotically stable rest point, and (c) motion near an unstable rest point.

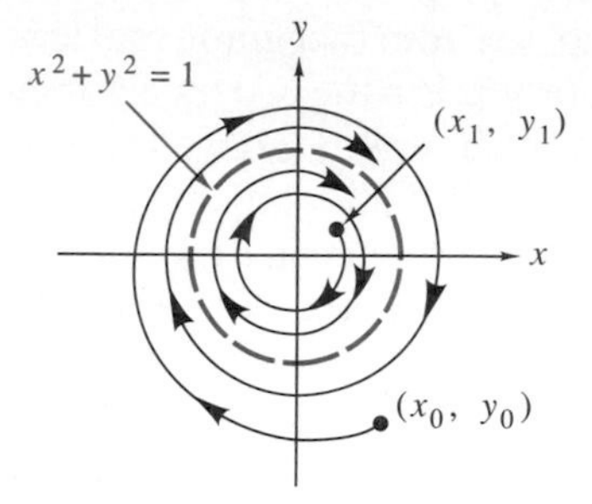

FIGURE 7.23
The solution $x^2 + y^2 = 1$ is a limit cycle.

trajectory starts outside the circular region, it will spiral inward and again approach the circular path asymptotically. The solution $x^2 + y^2 = 1$ is called a **limit cycle.** The trajectory behavior is sketched in Fig. 7.23. Thus, if system (2) models population behavior for two competing species, we would have to conclude that the population levels will eventually be periodic. This example illustrates that the results of a graphical analysis are useful for determining the motion *in the immediate vicinity of an equilibrium point only.* (Here we have assumed that negative values for x and y have a physical meaning in Fig. 7.23 or that the point (0, 0) represents a translation of a rest point from the first quadrant to the origin.)

EXERCISES 7.6

1. List three important considerations that are ignored in the competitive hunter model as presented so far in the text.

2. For system (2), show that any trajectory starting on the unit circle $x^2 + y^2 = 1$ will traverse the unit circle in a periodic solution. First introduce polar coordinates and rewrite the system as $dr/dt = r(1 - r^2)$ and $d\theta/dt = 1$.

3. Develop a model for the growth of trout and bass assuming that in isolation trout demonstrate exponential decay [so that $a < 0$ in Eq. (1)] and that the bass population grows logistically with a population limit M. Analyze graphically the motion in the vicinity of the rest points in your model. Is coexistence possible?

4. How might competitive hunter model (1) be validated? Include a discussion of how the various constants a, b, m, and n might be estimated. How could state conservation authorities use the model to ensure the survival of both species?

5. Consider the competitive hunter model defined by

$$\frac{dx}{dt} = a\left(1 - \frac{x}{k_1}\right)x - bxy,$$

$$\frac{dy}{dt} = m\left(1 - \frac{y}{k_2}\right)y - nxy,$$

where x and y represent the trout and bass populations, respectively.

a) What assumptions are implicitly being made about the growth of trout and bass in the absence of competition?

b) Interpret the constants a, b, m, n, k_1, and k_2 in terms of the physical problem.

c) Perform a graphical analysis:

i) Find the possible equilibrium levels.

ii) Determine whether coexistence is possible.

iii) Pick several typical starting points and sketch typical trajectories in the phase plane.

iv) Interpret the outcomes predicted by your graphical analysis in terms of the constants a, b, m, n, k_1, and k_2.

Note: When you get to part (iii), you should realize that five cases exist. You will need to analyze all five cases.

6. Consider the following economic model. Let P be the price of a single item on the market. Let Q be the quantity of the item available on the market. Both P and Q are functions of time. If one considers price and quantity as two interacting species, the following model might be proposed:

$$\frac{dP}{dt} = aP\left(\frac{b}{Q} - P\right),$$

$$\frac{dQ}{dt} = cQ(fP - Q),$$

where a, b, c, and f are positive constants. Justify and discuss the adequacy of the model.

a) If $a = 1$, $b = 20{,}000$, $c = 1$, and $f = 30$, find the equilibrium points of this system. If possible, classify each equilibrium point with respect to its stability. If a point cannot be readily classified, give some explanation.

b) Perform a graphical stability analysis to determine what will happen to the levels of P and Q as time increases.

c) Give an economic interpretation of the curves that determine the equilibrium points.

7. Show that the two trajectories leading to $(m/n, a/b)$ shown in Fig. 7.20 are unique.

a) From system (1) derive the following equation:

$$\frac{dy}{dx} = \frac{(m - nx)y}{(a - by)x}.$$

b) Separate variables, integrate, and exponentiate to obtain

$$y^a e^{-by} = K x^m e^{-nx}$$

where K is a constant of integration.

c) Let $f(y) = y^a/e^{by}$ and $g(x) = x^m/e^{nx}$. Show that $f(y)$ has a unique maximum of $M_y = (a/eb)^a$ when $y = a/b$ as shown in Fig. 7.24. Similarly, show that $g(x)$ has a unique maximum $M_x = (m/en)^m$ when $x = m/n$, also shown in Fig. 7.24.

d) Consider what happens as (x, y) approaches $(m/n, a/b)$. Take limits in part (b) as $x \to m/n$ and $y \to a/b$ to show that

$$\lim_{\substack{x \to m/n \\ y \to a/b}} \left[\left(\frac{y^a}{e^{by}}\right)\left(\frac{e^{nx}}{x^m}\right)\right] = K$$

or $M_y/M_x = K$. Thus any solution trajectory that approaches $(m/n, a/b)$ must satisfy

$$\frac{y^a}{e^{by}} = \left(\frac{M_y}{M_x}\right)\left(\frac{x^m}{e^{nx}}\right).$$

e) Show that only one trajectory can approach $(m/n, a/b)$ from below the line $y = a/b$. Pick $y_0 < a/b$. From Fig. 7.24 you can see that $f(y_0) < M_y$, which implies that

$$\frac{M_y}{M_x}\left(\frac{x^m}{e^{nx}}\right) = y_0^a/e^{by_0} < M_y.$$

This in turn implies that

$$\frac{x^m}{e^{nx}} < M_x.$$

Figure 7.24 tells you that for $g(x)$ there is a unique value $x_0 < m/n$ satisfying this last inequality. That is, for each $y < a/b$ there is a unique value of x satisfying the equation in part (d). Thus there can exist only one trajectory solution approaching $(m/n, a/b)$ from below, as shown in Fig. 7.25.

f) Use a similar argument to show that the solution trajectory leading to $(m/n, a/b)$ is unique if $y_0 > a/b$.

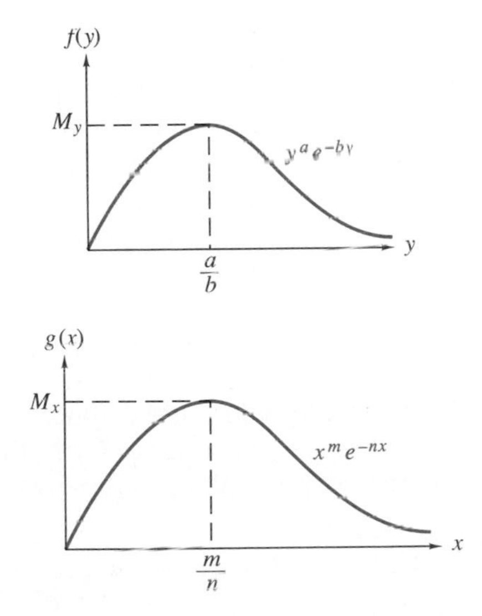

FIGURE 7.24
Graphs of the functions $f(y) = y^a/e^{by}$ and $g(x) = x^m/e^{nx}$.

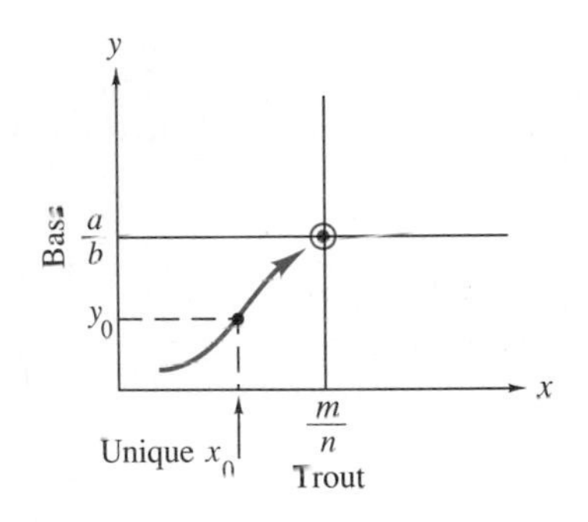

FIGURE 7.25
For any $y < a/b$ only one solution trajectory leads to the rest point $(m/n, a/b)$.

A PREDATOR–PREY MODEL (Optional)*

In this section we will study a model of population growth of two species in which one species is the primary food source for the other. One example is the situation we introduced in Section 7.1, in which baleen whales in the Southern Ocean eat Antarctic krill. The model we developed there was the Lotka–Volterra model:

$$\frac{dx}{dt} = (a - by)x, \qquad \frac{dy}{dt} = (-m + nx)y, \tag{1}$$

where $x(t)$ represents the krill population (the prey) and $y(t)$ represents the baleen whale population (the predator). The constants $x(0) = x_0$, $y(0) = y_0$, a, b, m, and n are all positive. We are interested in knowing if the predator–prey interaction is stable, or if one of the species will eventually die out. We are also interested in knowing what effect the harvesting of some of the krill might have on the two population levels. In this section we use a graphical analysis to study the behavior and answer these questions.

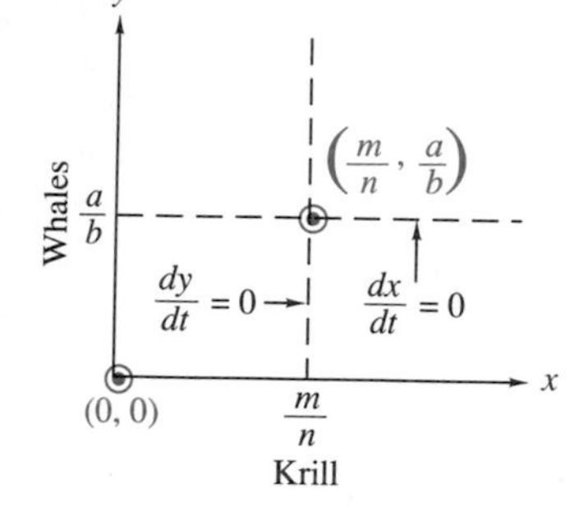

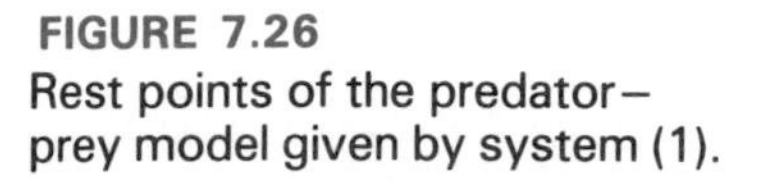

FIGURE 7.26
Rest points of the predator–prey model given by system (1).

Let us determine whether the krill and whale populations reach equilibrium levels. The rest points (equilibrium levels) occur when $dx/dt = dy/dt = 0$. Setting the righthand sides of Eqs. (1) equal to zero and solving for x and y simultaneously gives the rest points $(x, y) = (0, 0)$ and $(x, y) = (m/n, a/b)$. Along the vertical line $x = m/n$ in the phase plane, the growth dy/dt in the baleen whale population is zero; along the horizontal line $y = a/b$, the growth dx/dt in the krill population is zero. These features are depicted in Fig. 7.26.

Because the values for the constants a, b, m, and n in system (1) will only be estimates, we need to investigate the behavior of the solution trajectories near the two rest points $(0, 0)$ and $(m/n, a/b)$. To do this we will analyze the directions of dx/dt and dy/dt in the phase plane. In system (1) the vertical line $x = m/n$ divides the phase plane into two half-planes. In the left half-plane dy/dt is negative, and in the right half-plane it is positive. In a similar way, the horizontal line $y = a/b$ determines two half-planes. In the upper half-plane dx/dt is negative, and in the lower half-plane it is positive. Along the y-axis motion must be vertical and toward the rest point $(0, 0)$, and along the x-axis

* This section is adapted from *A First Course in Mathematical Modeling* by the authors (Pacific Grove, Calif.: Brooks/Cole Publishing Company, 1985), Chapter 10. Adaptation presented by permission of Brooks/Cole Publishing Company.

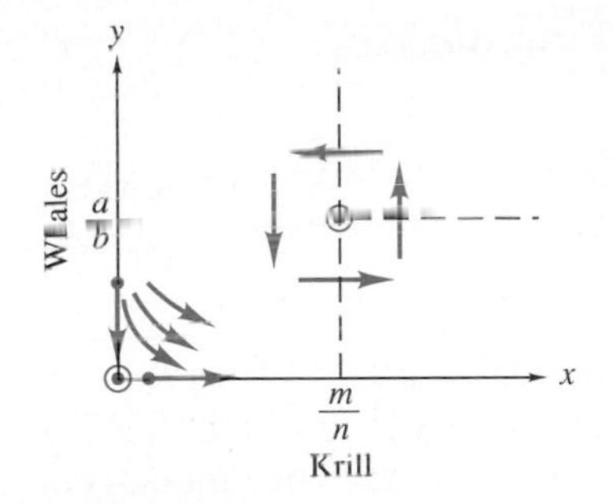

FIGURE 7.27
Trajectory directions in the predator–prey model.

motion must be horizontal and away from the rest point (0, 0). The directions of the associated trajectories are indicated in Fig. 7.27.

From Fig. 7.27 we see that the rest point (0, 0) is unstable. Entrance to the rest point is along the vertical line $x = 0$, where there are no krill. Thus the whale population declines to zero in the absence of its primary food supply. All other trajectories recede from the rest point. The rest point $(m/n, a/b)$ is more complicated to analyze. The information given by Fig. 7.27 is insufficient to determine which of the three possible kinds of motion shown in Fig. 7.22 is occurring: periodic, asymptotically stable, or unstable. Thus we must perform a further analysis.

Analytical Solution of the Model

Since the number of baleen whales will depend on the number of Antarctic krill available for food, we assume that y is a function of x. Then from the chain rule in calculus we know that

$$\frac{dy}{dx} = \frac{\dfrac{dy}{dt}}{\dfrac{dx}{dt}}$$

or,

$$\frac{dy}{dx} = \frac{(-m + nx)y}{(a - by)x}. \tag{2}$$

Equation (2) is a separable first-order differential equation and may be rewritten as

$$\left(\frac{a}{y} - b\right) dy = \left(n - \frac{m}{x}\right) dx. \tag{3}$$

Integration of each side of Eq. (3) yields

$$a \ln y - by = nx - m \ln x + k_1$$

or

$$a \ln y + m \ln x - by - nx = k_1,$$

where k_1 is a constant.

Using properties of the natural logarithm and exponential functions, this last equation can be rewritten as

$$\frac{y^a x^m}{e^{by+nx}} = K \tag{4}$$

where K is a constant. Equation (4) defines the solution trajectories in the phase plane. We will now show that these trajectories are closed and represent periodic motion.

Periodic Nature of Predator–Prey Trajectories

Equation (4) can be rewritten as

$$\frac{y^a}{e^{by}} = K\left(\frac{e^{nx}}{x^m}\right). \tag{5}$$

Let us determine the behavior of the function $f(y) = y^a/e^{by}$. Using the first derivative test (see problem 1 in Exercises 7.7), it is easy to show that $f(y)$ has a relative maximum at $y = a/b$ and no other critical points. For simplicity of notation, call this maximum value M_y. Moreover, $f(0) = 0$ and from l'Hôpital's rule in calculus, we know that $f(y)$ approaches zero as y tends toward infinity. Similar arguments apply to the function $g(x) = x^m/e^{nx}$, which achieves its maximum value M_x at $x = m/n$. Functions f and g are plotted in Fig. 7.28.

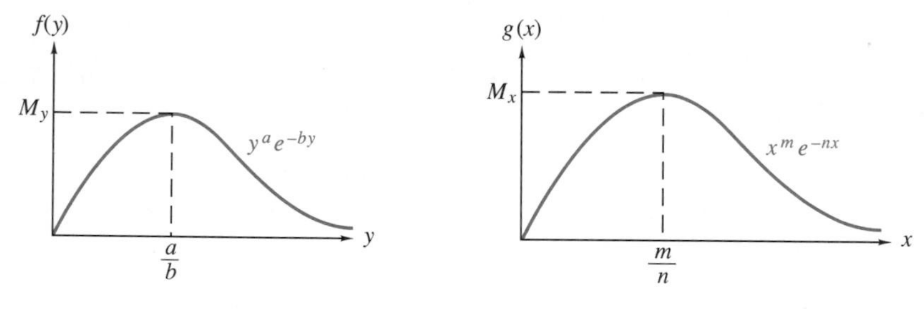

FIGURE 7.28
Functions $f(y) = y^a/e^{by}$ and $g(x) = x^m/e^{nx}$.

From Fig. 7.28, the largest value for $y^a e^{-by} x^m e^{-nx}$ is $M_y M_x$. That is, Eq. (5) has no solutions if $K > M_y M_x$ and exactly one solution, $x = m/n$ and $y = a/b$, when $K = M_y M_x$. Let us consider what happens when $K < M_y M_x$.

Suppose $K = sM_y$ where $s < M_x$ is a constant. Then the equation

$$x^m e^{-nx} = s$$

has exactly two solutions: $x_m < m/n$ and $x_M > m/n$ (see Fig. 7.29). Now if

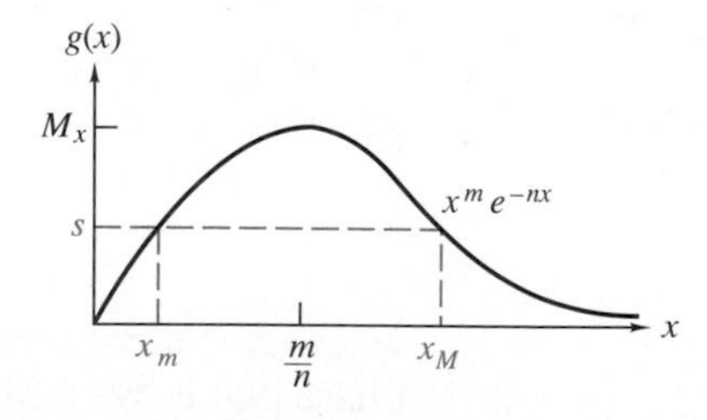

FIGURE 7.29
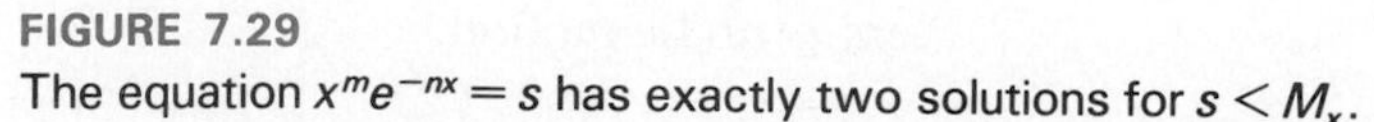
The equation $x^m e^{-nx} = s$ has exactly two solutions for $s < M_x$.

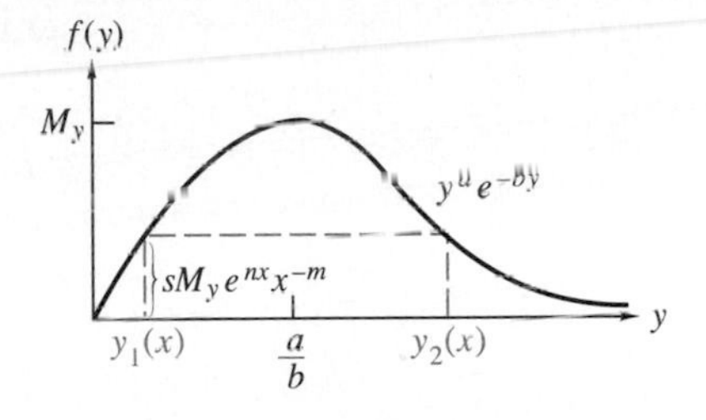

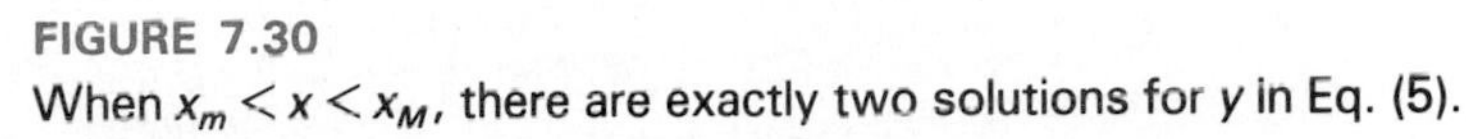
FIGURE 7.30
When $x_m < x < x_M$, there are exactly two solutions for y in Eq. (5).

$x < x_m$, then $x^m e^{-nx} < s$ so that $se^{nx}x^{-m} > 1$ and

$$f(y) = y^a e^{-by} = Ke^{nx}x^{-m} = sM_y e^{nx}x^{-m} > M_y.$$

Therefore there is no solution for y in Eq. (5) when $x < x_m$. Likewise, there is no solution when $x > x_M$. If $x = x_m$ or $x = x_M$, then Eq. (5) has exactly the one solution $y = a/b$. Finally, if x lies between x_m and x_M, Eq. (5) has exactly two solutions. The smaller solution $y_1(x)$ is less than a/b, and the larger solution $y_2(x)$ is greater than a/b. This situation is depicted in Fig. 7.30. Moreover, as x approaches either x_m or x_M, $f(y)$ approaches M_y so that both $y_1(x)$ and $y_2(x)$ approach a/b. It follows that the trajectories defined by Eq. (5) are periodic and have the form depicted qualitatively in Fig. 7.31.

Model Interpretation

What conclusions can we draw from the trajectories in Fig. 7.31? First, because the trajectories are *closed curves*, they predict that, under the assumptions of model (1), neither the baleen whales nor the Antarctic krill will

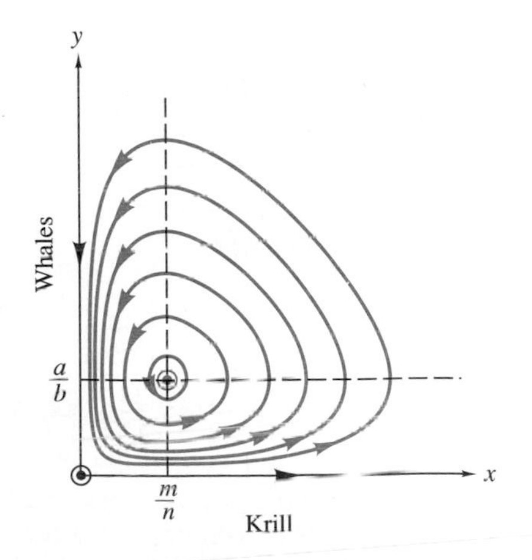

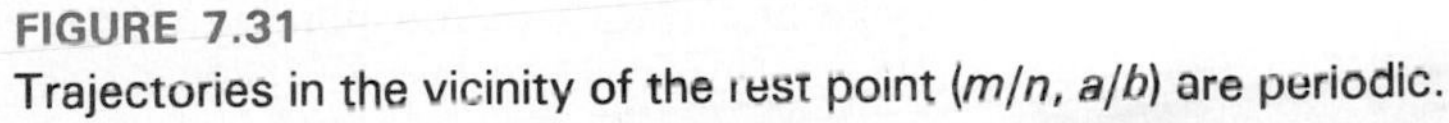
FIGURE 7.31
Trajectories in the vicinity of the rest point $(m/n, a/b)$ are periodic.

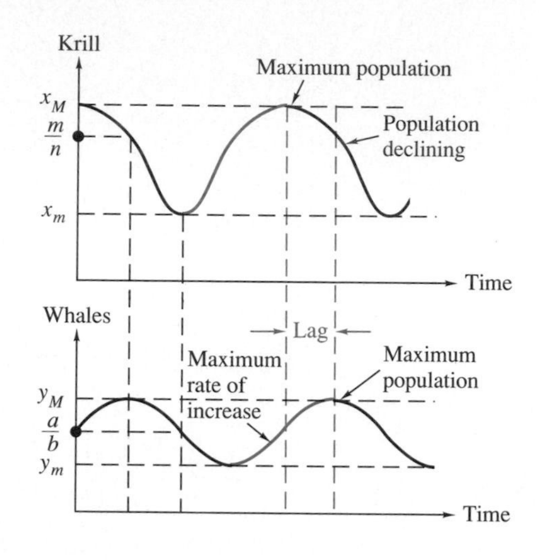

FIGURE 7.32
The whale population lags behind the krill population as both populations fluctuate cyclically between their maximum and minimum values.

become extinct. (Remember, the model assumes the ideal situation, with no other factors affecting it). The second observation is that along a single (periodic) trajectory the two populations fluctuate between their maximum and minimum values. That is, starting with populations in the region where $x > m/n$ and $y > a/b$, the krill population will decline and the whale population increase until the krill population reaches the level $x = m/n$, after which point the whale population also begins to decline. Both populations continue to decline until the whale population reaches the level $y = a/b$, after which the krill population begins to increase. This behavior continues as we "travel" around the trajectory. Recall from our discussion in Section 7.2 that the trajectories never cross each other. A sketch of the population curves is shown in Fig. 7.32. Notice that the krill population fluctuates between its maximum and minimum values over one complete cycle. When the krill are plentiful, the whale population has its maximum rate of increase, and the whale population reaches its maximum value after the krill population is on the decline. Thus the predator *lags behind* the prey in a cyclic fashion.

Effects of Harvesting

For given initial population levels $x(0) = x_0$ and $y(0) = y_0$, the whale and krill populations will fluctuate with time around one of the closed trajectories depicted in Fig. 7.31. Let T denote the time it takes to complete one full cycle and return to the starting point. The *average* levels of the krill and baleen whale populations over the time cycle are defined by the integrals

$$\bar{x} = \frac{1}{T}\int_0^T x(t)\,dt \qquad \text{and} \qquad \bar{y} = \frac{1}{T}\int_0^T y(t)\,dt,$$

respectively. From the differential equation for the krill population, we know that

$$\frac{1}{x}\left(\frac{dx}{dt}\right) = a - by.$$

Then integration of both sides from $t = 0$ to $t = T$ leads to

$$\int_0^T \frac{1}{x}\left(\frac{dx}{dt}\right) dt = \int_0^T (a - by)\,dt$$

or

$$\ln x(T) - \ln x(0) = aT - b\int_0^T y(t)\,dt.$$

Because of the periodicity of the trajectory, $x(T) = x(0)$, which gives us the average value

$$\bar{y} = \frac{a}{b}.$$

In an analogous manner, it can be shown that

$$\bar{x} = \frac{m}{n}$$

(see Exercises 7.7, problem 2). Therefore, the average levels of the predator and prey populations are also their equilibrium levels. Let us see what this means in terms of harvesting krill.

We will assume that the effect of fishing for krill is to decrease its population level at a rate $rx(t)$. The constant r indicates the intensity of fishing, which is comprised of such factors as the number of fishing vessels at sea and the number of people casting nets for krill. Since less food is now available for the baleen whales, we assume the whale population also decreases at a rate $ry(t)$. We incorporate these fishing assumptions to obtain the following refined model:

$$\begin{aligned}\frac{dx}{dt} &= (a - by)x - rx = [(a - r) - by]x,\\ \frac{dy}{dt} &= (-m + nx)y - ry = [-(m + r) + nx]y.\end{aligned} \tag{6}$$

Autonomous system (6) takes the same form as system (1) (provided that $a - r > 0$) except that a is replaced by $a - r$ and m is replaced by $m + r$. Thus the new average population levels will be

$$\bar{x} = \frac{m+r}{n} \quad \text{and} \quad \bar{y} = \frac{a-r}{b}.$$

Consequently, harvesting krill only a moderate amount (so that $r < a$) actually *increases* the average level of krill and *decreases* the average baleen whale population (under the assumptions for the model). The fact that some fishing increases the number of krill is known as **Volterra's principle.** The increase in krill population is beneficial to other species in the Southern Ocean that depend on the krill for their main food source, such as seals, seabirds, penguins, and fish.

The Lotka–Volterra model (1) can be modified to reflect the situation in which both predator and prey are diminished by some kind of depleting force, such as when applying insecticide treatments that destroy both the insect predator and its insect prey (see problem 3 in Exercises 7.7).

A number of biologists and ecologists argue that the Lotka–Volterra model (1) is unrealistic because the system is not asymptotically stable, whereas most observable natural predator–prey systems tend toward equilibrium levels as time evolves. Nevertheless, regular population cycles, as suggested by the trajectories in Fig. 7.31, do occur in nature. Other models have been proposed that do exhibit asymptotically stable oscillations (so that the trajectories approach equilibrium solutions). One such model is given by

$$\begin{aligned} \frac{dx}{dt} &= ax + bxy - rx^2, \\ \frac{dy}{dt} &= -my + nxy - sy^2. \end{aligned} \tag{7}$$

In autonomous system (7) the term rx^2 indicates the degree of internal competition of the prey for their limited resource (such as food and space), and the term sy^2 indicates the degree of competition among the predators for the finite amount of available prey. An analysis of this model is more difficult than that presented for the Lotka–Volterra model, but it can be shown that the trajectories of the model are not periodic and tend toward equilibrium levels. The constants r and s are positive and would be determined by experimentation or historical data.

EXERCISES 7.7

1. Apply the first and second derivative tests to the function $f(y) = y^a/e^{by}$ to show that $y = a/b$ is a unique critical point that yields the relative maximum $f(a/b)$. Show also that $f(y)$ approaches zero as y tends toward infinity.

2. Derive the result that the average value $\bar{x}$ of the prey population modeled by Lotka–Volterra system (1) is given by the population level m/n.

3. a) Do problem 14 in Exercises 7.1 using a graphical analysis.

b) What conclusions do you reach concerning the effects of the application of the insecticide?
c) What is the effect of using the insecticide once on an irregular basis?

4. In a 1969 study of the fluctuations in the numbers of Canadian lynx and its primary food source, the hare, trapped by the Hudson's Bay Company between 1847 and 1903, E. R. Leigh concluded that the fluctuations were periodic. The actual population levels of both species differed greatly from the predicted population levels obtained from the Lotka–Volterra predator–prey model.

Use the entire model-building process to modify the Lotka–Volterra model to arrive at a more realistic model for the growth rates of both species. Answer the following questions in the appropriate sections of the model-building process:

a) How have you modified the basic assumptions of the predator–prey model?
b) Why are your modifications an improvement to the basic model?
c) What are the equilibrium points for your model?
d) Is each equilibrium point stable, unstable, or not classifiable?
e) Based on your equilibrium analysis, what values will the population levels of lynx and hare approach as t tends toward infinity?
f) How would you use your revised model to suggest hunting policies for Canadian lynx and hare? *Hint:* you are introducing a second predator (the human) into the system.

5. Consider two species each of whose survival depends on their mutual cooperation. An example would be a species of bee that feeds primarily on the nectar of one plant species and simultaneously pollinates that plant. One simple model of this mutualism is given by the autonomous system

$$\frac{dx}{dt} = -ax + bxy,$$

$$\frac{dy}{dt} = -my + nxy.$$

a) What assumptions are implicitly being made about the growth of each species in the absence of cooperation?
b) Interpret the constants a, b, m, and n in terms of the physical problem.
c) What are the equilibrium levels?
d) Perform a graphical analysis and indicate the trajectory directions in the phase plane.
e) Find an analytical solution and sketch typical trajectories in the phase plane.
f) Interpret the outcomes predicted by your graphical analysis. Do you believe the model is realistic?

FURTHER READING

Clark, Colin W. *Mathematical Bioeconomics: The Optimal Management of Renewable Resources.* New York: John Wiley & Sons, 1976.

May, R. M. *Stability and Complexity in Model Ecosystems,* Monographs in Population Biology VI. Princeton, N.J.: Princeton University Press, 1973.

May, R. M., J. R. Beddington, C. W. Clark, S. J. Holt, and R. M. Lewis. "Management of Multispecies Fisheries," *Science 205* (July 1979): 256–277.

May, R. M., Ed. *Theoretical Ecology: Principles and Applications.* Philadelphia: W. B. Saunders, 1976.

NUMERICAL METHODS FOR SYSTEMS OF EQUATIONS (Optional)

In Chapter 2 three numerical methods for approximating the solution to a first-order initial value problem were presented:

- Euler's method,
- Improved Euler's method,
- Fourth-order Runge–Kutta method.

In Section 4.6 we studied the numerical approximation of the second-order equation

$$y'' = g(x, y, y') \tag{1}$$

with the initial conditions

$$y(x_0) = y_0,$$
$$y'(x_0) = v_0.$$

Using the substitutions $v = y'(x)$ and $v' = y''$ we converted second-order Eq. (1) to a system of first-order equations:

$$\begin{aligned} y' &= v, \\ v' &= g(x, y, v) \end{aligned} \tag{2}$$

subject to the initial conditions

$$y(x_0) = y_0,$$
$$v(x_0) = v_0.$$

Algorithms for approximating solutions to system (2) were then presented using Euler's method and the Runge–Kutta method. We presented several examples and noted that the method applied even when the function $g(x, y, v)$ included variable coefficients or was a nonlinear function. Thus you have already studied numerical approximations for a system of first-order equations having the special form of Eqs. (2). In this section we present a numerical method for approximating the solution to a more general system of first-order equations:

$$\begin{aligned} \frac{dx}{dt} &= f(t, x, y), \\ \frac{dy}{dt} &= g(t, x, y), \end{aligned} \tag{3}$$

subject to the initial conditions

$$x(t_0) = x_0,$$
$$y(t_0) = y_0$$

Note that system (3) may be both nonlinear and nonautonomous. We assume the functions f and g are continuous over a suitable region of txy-space containing the point (t_0, x_0, y_0). Any of the three methods listed above can easily be adapted to solve system (3). For brevity we present only the Runge–Kutta method since it is the most widely used of the three.

Recall from Section 2.6 that the Runge–Kutta methods obtained a weighted average of four estimates to the derivative over the step interval. If we call the four estimates $M_1, M_2, M_3, M_4, L_1, L_2, L_3$, and L_4 for each of the derivatives in system (3), respectively, then the method estimates the approximations x_1 and y_1 using x_0 and y_0 in the following manner:

$$x_1 = x_0 + \frac{h}{6}(M_1 + 2M_2 + 2M_3 + M_4),$$

$$y_1 = y_0 + \frac{h}{6}(L_1 + 2L_2 + 2L_3 + L_4).$$

Successive approximations x_{n+1} and y_{n+1} are made in a similar manner. The following algorithm summarizes the method.

FOURTH-ORDER RUNGE–KUTTA METHOD

For solving $\frac{dx}{dt} = f(t, x, y)$, $\frac{dy}{dt} = g(t, x, y)$

Subject to $x(t_0) = x_0$, $y(t_0) = y_0$ over an Interval I.

Step 1. First divide the interval $t_0 \leq t \leq b$ into n subintervals using the equally spaced points

$$t_1 = t_0 + h, \quad t_2 = t_1 + h, \quad \ldots, \quad t_n = t_{n-1} + h = b.$$

Step 2. Next, assume that the kth estimates x_k and y_k have been computed and calculate the following slope estimates in the order listed:

$$M_1 = f(t_k, x_k, y_k),$$
$$L_1 = g(t_k, x_k, y_k),$$
$$M_2 = f\left(t_k + \frac{h}{2}, x_k + \frac{hM_1}{2}, y_k + \frac{hL_1}{2}\right),$$
$$L_2 = g\left(t_k + \frac{h}{2}, x_k + \frac{hM_1}{2}, y_k + \frac{hL_1}{2}\right),$$

$$M_3 = f\left(t_k + \frac{h}{2}, x_k + \frac{hM_2}{2}, y_k + \frac{hL_2}{2}\right),$$

$$L_3 = g\left(t_k + \frac{h}{2}, x_k + \frac{hM_2}{2}, y_k + \frac{hL_2}{2}\right),$$

$$M_4 = f(t_k + h, x_k + hM_3, y_k + hL_3),$$

$$L_4 = g(t_k + h, x_k + hM_3, y_k + hL_3).$$

Step 3. Calculate the next estimates to the solution:

$$x_{k+1} = x_k + \frac{h}{6}(M_1 + 2M_2 + 2M_3 + M_4),$$

$$y_{k+1} = y_k + \frac{h}{6}(L_1 + 2L_2 + 2L_3 + L_4).$$

Step 4. Repeat Steps 2 and 3 for $k = 0, 1, 2, \ldots, n-1$.

EXAMPLE 1 Solve the system

$$\begin{aligned} \frac{dx}{dt} &= 3x - xy, \\ \frac{dy}{dt} &= xy - 2y, \end{aligned} \tag{4}$$

subject to the initial conditions $x(0) = 1$ and $y(0) = 2$ over the interval $0 \leq t \leq 1$. Use $h = 1/4$.

Solution. We apply the fourth-order Runge–Kutta method where $f(t, x, y) = 3x - xy$ and $g(t, x, y) = xy - 2y$. The slope estimates for the first step are as follows:

$$M_1 = 3x_0 - x_0 y_0 = 3 - 2 = 1,$$

$$L_1 = x_0 y_0 - 2y_0 = 2 - 4 = -2,$$

$$M_2 = 3\left(x_0 + \frac{M_1}{8}\right) - \left(x_0 + \frac{M_1}{8}\right)\left(y_0 + \frac{L_1}{8}\right) = 3\left(\frac{9}{8}\right) - \frac{9}{8}\left(\frac{7}{4}\right)$$

$$\approx 1.40625,$$

$$L_2 = \left(x_0 + \frac{M_1}{8}\right)\left(y_0 + \frac{L_1}{8}\right) - 2\left(y_0 + \frac{L_1}{8}\right) = \frac{9}{8}\left(\frac{7}{4}\right) - 2\left(\frac{7}{4}\right)$$

$$\approx -1.53125,$$

$$M_3 = 3\left(x_0 + \frac{M_2}{8}\right) - \left(x_0 + \frac{M_2}{8}\right)\left(y_0 + \frac{L_2}{8}\right) = 3\left(\frac{301}{256}\right) - \frac{301}{256}\left(\frac{463}{256}\right)$$

$$\approx 1.40083,$$

$$L_3 = \left(x_0 + \frac{M_2}{8}\right)\left(y_0 + \frac{L_2}{8}\right) - 2\left(y_0 + \frac{L_2}{8}\right) = \frac{301}{256}\left(\frac{463}{256}\right) - 2\left(\frac{463}{256}\right)$$
$$\approx -1.49068,$$
$$M_4 = 3\left(x_0 + \frac{M_3}{4}\right) - \left(x_0 + \frac{M_3}{4}\right)\left(y_0 + \frac{L_3}{4}\right) \approx 1.85339,$$
$$L_4 = \left(x_0 + \frac{M_3}{4}\right)\left(y_0 + \frac{L_3}{4}\right) - 2\left(y_0 + \frac{L_3}{4}\right) \approx -1.05743.$$

Thus

$$x_1 = x_0 + \frac{h}{6}(M_1 + 2M_2 + 2M_3 + M_4) \approx 1.35281,$$

$$y_1 = y_0 + \frac{h}{6}(L_1 + 2L_2 + 2L_3 + L_4) \approx 1.62078.$$

The slope estimates for the next step are calculated as follows:

$$M_1 = 3x_1 - x_1 y_1 \approx 1.86583,$$
$$L_1 = x_1 y_1 - 2y_1 \approx -1.04894,$$
$$M_2 = 3\left(x_1 + \frac{M_1}{8}\right) - \left(x_1 + \frac{M_1}{8}\right)\left(y_1 + \frac{L_1}{8}\right) \approx 2.39546,$$
$$L_2 = \left(x_1 + \frac{M_1}{8}\right)\left(y_1 + \frac{L_1}{8}\right) - 2\left(y_1 + \frac{L_1}{8}\right) \approx -0.61666,$$
$$M_3 = 3\left(x_1 + \frac{M_2}{8}\right) - \left(x_1 + \frac{M_2}{8}\right)\left(y_1 + \frac{L_2}{8}\right) \approx 2.40617,$$
$$L_3 = \left(x_1 + \frac{M_2}{8}\right)\left(y_1 + \frac{L_2}{8}\right) - 2\left(y_1 + \frac{L_2}{8}\right) \approx -0.53682,$$
$$M_4 = 3\left(x_1 + \frac{M_3}{4}\right) - \left(x_1 + \frac{M_3}{4}\right)\left(y_1 + \frac{L_3}{4}\right) \approx 2.95778,$$
$$L_4 = \left(x_1 + \frac{M_3}{4}\right)\left(y_1 + \frac{L_3}{4}\right) - 2\left(y_1 + \frac{L_3}{4}\right) \approx -0.06785.$$

Thus

$$x_2 = x_1 + \frac{h}{6}(M_1 + 2M_2 + 2M_3 + M_4) \approx 1.95393,$$

$$y_2 = y_1 + \frac{h}{6}(L_1 + 2L_2 + 2L_3 + L_4) \approx 1.47812.$$

The slope estimates for the third step are calculated as follows:

$$M_1 = 3x_2 - x_2 y_2 \approx 2.97365,$$
$$L_1 = x_2 y_2 - 2y_2 \approx -0.06809,$$

$$M_2 = 3\left(x_2 + \frac{M_1}{8}\right) - \left(x_2 + \frac{M_1}{8}\right)\left(y_2 + \frac{L_1}{8}\right) \approx 3.55913,$$
$$L_2 = \left(x_2 + \frac{M_1}{8}\right)\left(y_2 + \frac{L_1}{8}\right) - 2\left(y_2 + \frac{L_1}{8}\right) \approx 0.47856,$$
$$M_3 = 3\left(x_2 + \frac{M_2}{8}\right) - \left(x_2 + \frac{M_2}{8}\right)\left(y_2 + \frac{L_2}{8}\right) \approx 3.50722,$$
$$L_3 = \left(x_2 + \frac{M_2}{8}\right)\left(y_2 + \frac{L_2}{8}\right) - 2\left(y_2 + \frac{L_2}{8}\right) \approx 0.61337,$$
$$M_4 = 3\left(x_2 + \frac{M_3}{4}\right) - \left(x_2 + \frac{M_3}{4}\right)\left(y_2 + \frac{L_3}{4}\right) \approx 3.87396,$$
$$L_4 = \left(x_2 + \frac{M_3}{4}\right)\left(y_2 + \frac{L_3}{4}\right) - 2\left(y_2 + \frac{L_3}{4}\right) \approx 1.35532.$$

Thus

$$x_3 = x_2 + \frac{h}{6}(M_1 + 2M_2 + 2M_3 + M_4) \approx 2.82811,$$
$$y_3 = y_2 + \frac{h}{6}(L_1 + 2L_2 + 2L_3 + L_4) \approx 1.62275.$$

The slope estimates for the final step are calculated as follows:

$$M_1 = 3x_3 - x_3 y_3 \approx 3.89501,$$
$$L_1 = x_3 y_3 - 2y_3 \approx 1.34382,$$
$$M_2 = 3\left(x_3 + \frac{M_1}{8}\right) - \left(x_3 + \frac{M_1}{8}\right)\left(y_3 + \frac{L_1}{8}\right) \approx 4.00872,$$
$$L_2 = \left(x_3 + \frac{M_1}{8}\right)\left(y_3 + \frac{L_1}{8}\right) - 2\left(y_3 + \frac{L_1}{8}\right) \approx 2.35479,$$
$$M_3 = 3\left(x_3 + \frac{M_2}{8}\right) - \left(x_3 + \frac{M_2}{8}\right)\left(y_3 + \frac{L_2}{8}\right) \approx 3.60519,$$
$$L_3 = \left(x_3 + \frac{M_2}{8}\right)\left(y_3 + \frac{L_2}{8}\right) - 2\left(y_3 + \frac{L_2}{8}\right) \approx 2.54822,$$
$$M_4 = 3\left(x_3 + \frac{M_3}{4}\right) - \left(x_3 + \frac{M_3}{4}\right)\left(y_3 + \frac{L_3}{4}\right) \approx 2.76048,$$
$$L_4 = \left(x_3 + \frac{M_3}{4}\right)\left(y_3 + \frac{L_3}{4}\right) - 2\left(y_3 + \frac{L_3}{4}\right) \approx 3.90814.$$

Thus

$$x_4 = x_3 + \frac{h}{6}(M_1 + 2M_2 + 2M_3 + M_4) \approx 3.73992,$$
$$y_4 = y_3 + \frac{h}{6}(L_1 + 2L_2 + 2L_3 + L_4) \approx 2.25017.$$

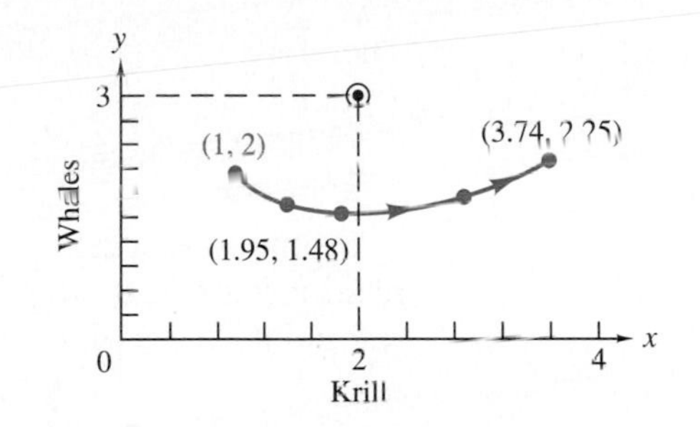

FIGURE 7.33

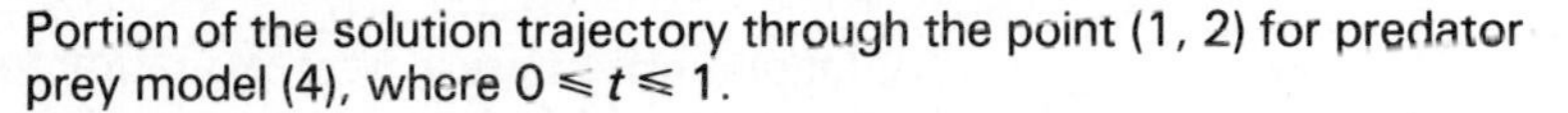

Portion of the solution trajectory through the point (1, 2) for predator prey model (4), where $0 \leq t \leq 1$.

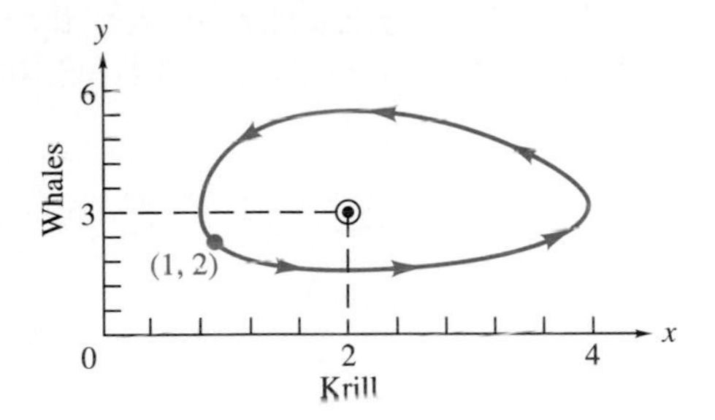

FIGURE 7.34

Complete cycle of the solution trajectory through the point (1, 2) for predator–prey model (4).

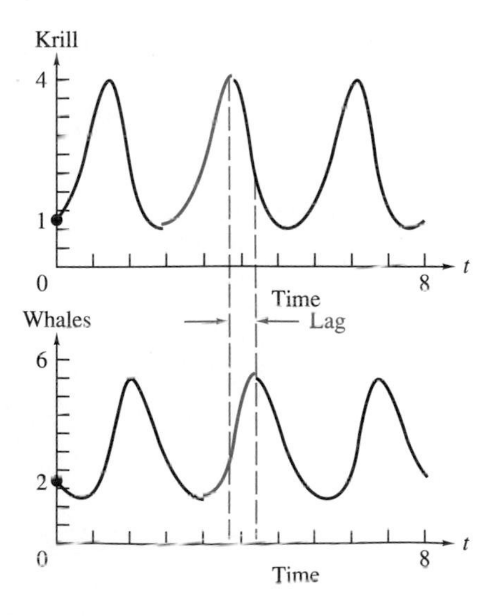

FIGURE 7.35

Plots revealing the fluctuations of the individual population levels over time. Notice that the whale (predator) population lags behind the krill (prey) population.

Therefore

$$x(1) \approx x_4 \approx 3.73992,$$
$$y(1) \approx y_4 \approx 2.25017.$$

From our discussion in Section 7.1, system (4) models a predator–prey relationship (see Section 7.7). The solution trajectories are periodic about the rest point (2, 3), which is obtained from the equations

$$3x - xy = 0,$$
$$xy - 2y = 0.$$

A plot of our numerical approximation to the system for $0 \leq t \leq 1$ is shown in Fig. 7.33, which shows a segment of the periodic trajectory moving counterclockwise about the rest point. Figure 7.34 plots one complete cycle of the trajectory satisfying the initial conditions $x(0) = 1$ and $y(0) = 2$. In both

TABLE 7.2

Numerical Computations for

$\frac{dx}{dt} = 3x - xy, \quad \frac{dy}{dt} = -2y + xy$ where $x(0) = 1$, $y(0) = 2$.

Note the Variable Step Size $h = \Delta t$ from Point to Point.

t	$x(t)$	$y(t)$
0.000000	1.00000	2.00000
0.126491E-03	1.00013	1.99975
0.685900E-02	1.00700	1.98647
0.135915E-01	1.01401	1.97337
0.203240E-01	1.02115	1.96045
0.558871E-01	1.06090	1.89468
0.914501E-01	1.10469	1.83383
0.127013	1.15268	1.77783
0.162576	1.20506	1.72662
0.224314	1.30688	1.64899
0.286052	1.42343	1.58554
0.347790	1.55577	1.53629
0.409528	1.70482	1.50156
0.471266	1.87126	1.48199
0.541295	2.08150	1.47941
0.611324	2.31397	1.49986
0.681353	2.56639	1.54665
0.751382	2.83402	1.62430
0.821411	3.10867	1.73864
0.891440	3.37773	1.89693
0.965430	3.63606	2.12107
1.00000	3.74016	2.24861

figures the x-axis represents the krill (prey) population and the y-axis represents the whale (predator) population, in consonance with our discussion in Section 7.7.

In Fig. 7.33 we show each of the two population levels as they fluctuate over time. Note the lag of the maximum whale population behind the maximum level of the krill population, as discussed in Section 7.7. The graphs in Figs. 7.34 and 7.35 were computer generated, with varying step size h for each computed point (to satisfy a specified tolerance for the truncation error). Table 7.2 is a table of values computed for the trajectory but only for the time interval $0 \leq t \leq 1$. Notice the improvement in the accuracy of $x(1) \approx 3.74$ and $y(1) \approx 2.25$ over the estimates in our hand calculation when the step size had the (relatively) large value of $h = \Delta t = 0.25$.

EXERCISES 7.8

In problems 1–7 use the fourth-order Runge–Kutta method to solve the linear first-order system subject to the given initial conditions. Use the specified step size h, and calculate the first three approximations x_1, y_1, x_2, y_2, and x_3, y_3. Then repeat your calculations for $h/2$. Finally, compare your numerical results with the values of the given analytical solution.

1. $\dfrac{dx}{dt} = 2x + 3y$

$\dfrac{dy}{dt} = 3x + 2y$

$x(0) = 1,\ y(0) = 0,\ h = \frac{1}{4}$,
$x(t) = \frac{1}{2}e^{-t} + \frac{1}{2}e^{5t},\quad y(t) = -\frac{1}{2}e^{-t} + \frac{1}{2}e^{5t}$

2. $\dfrac{dx}{dt} = 2x + 3y$

$\dfrac{dy}{dt} = 2x + y$

$x(0) = 5,\ y(0) = 0,\ h = \frac{1}{4}$,
$x(t) = 3e^{4t} + 2e^{-t},\quad y(t) = 2e^{4t} - 2e^{-t}$

3. $\dfrac{dx}{dt} = 2x - y$

$\dfrac{dy}{dt} = 2y$

$x(0) = 1,\ y(0) = 1,\ h = \frac{1}{4}$,
$x(t) = e^{2t} - te^{2t},\quad y(t) = e^{2t}$

4. $\dfrac{dx}{dt} = x + 5y$

$\dfrac{dy}{dt} = -x - 3y$

$x(0) = 5,\ y(0) = 4,\ h = \frac{1}{4}$,
$x(t) = 5e^{-t}(\cos t + 6 \sin t),\quad y(t) = e^{-t}(4 \cos t - 13 \sin t)$

5. $\dfrac{dx}{dt} = x + 3y$

$\dfrac{dy}{dt} = x - y + 2e^{t}$

$x(0) = 0,\ y(0) = 2,\ h = \frac{1}{4}$,
$x(t) = -e^{-2t} + 3e^{2t} - 2e^{t},\quad y(t) = e^{-2t} + e^{2t}$

6. $\dfrac{dx}{dt} = 3x - y + e^{-t}$

$\dfrac{dy}{dt} = 9x - 3y + \sin t$

$x(0) = 0,\ y(0) = 0,\ h = \frac{1}{4}$,
$x(t) = -2 + 2t + 2e^{-t} + \sin t$,
$y(t) = -8 + 6t + 9e^{-t} + 3 \sin t - \cos t$

7. $\dfrac{dx}{dt} = 3x + e^{2t}$

$\dfrac{dy}{dt} = -x + 3y + te^{2t}$

$x(0) = 2,\ y(0) = -1,\ h = \frac{1}{4}$,
$x(t) = 3e^{3t} - e^{2t},\quad y(t) = e^{3t} - 3te^{2t} - 2e^{2t} - te^{2t}$

In problems 8–11 use the fourth-order Runge–Kutta method to solve the nonlinear first-order system subject to the given initial conditions. Use the specified step size h, and calculate the first three approximations x_1, y_1, x_2, y_2, and x_3, y_3. Then repeat your calculations for $h/2$.

8. $\dfrac{dx}{dt} = 2x - xy$

$\dfrac{dy}{dt} = xy - y$

$x(0) = 1,\ y(0) = 2,\ h = \frac{1}{4}$

9. $\dfrac{dx}{dt} = x - 2xy$

$\dfrac{dy}{dt} = 2y - xy$

$x(0) = 2,\ y(0) = 1,\ h = \frac{1}{4}$

10. $\dfrac{dx}{dt} = y + x - x(x^2 + y^2)$

$\dfrac{dy}{dt} = -x + y - y(x^2 + y^2)$

$x(0) = 1,\ y(0) = 0,\ h = \frac{1}{4}$

11. $\dfrac{dx}{dt} = (1 - x)x - xy$

$\dfrac{dy}{dt} = (1 - 2y)y - xy$

$x(0) = 2,\ y(0) = 1,\ h = \frac{1}{4}$

12. Write a computer program in a language of your choice to implement the fourth-order Runge–Kutta method for solving a 2×2 first-order system. Test your program using the examples in the text.

7.9

NONHOMOGENEOUS 2×2 SYSTEMS: THE LAPLACE TRANSFORM METHOD (Optional)

In Chapter 5 we saw that with the Laplace transform a linear differential equation with constant coefficients and given initial values can be converted into an algebraic equation. The Laplace transform also converts a *system* of linear differential equations with constant coefficients and initial conditions into an algebraic system of equations.

In our work in Chapters 5 and 6 we found that the method of Laplace transforms works as well for nonhomogeneous as for homogeneous equations, assuming the Laplace transforms and their inverses can be determined. Nothing new is needed beyond the ideas presented in those chapters in order to use Laplace transforms to solve linear systems of differential equations. The Laplace transform simply operates on each equation in the system, converting it into an algebraic equation as before. The resulting system of algebraic equations is then solved for the transform of each dependent variable in the original system. Finally, an inverse transform method is used to find each dependent variable as a function of the independent variable. Our purpose in this section is to provide one method for solving nonhomogeneous 2×2 constant-coefficient linear systems. In Chapter 8 we will address more generally the question of solving nonhomogeneous linear systems. However, the methods discussed there require additional techniques from matrix algebra.

We now present several examples illustrating the use of Laplace transforms to solve nonhomogeneous 2×2 linear systems.

EXAMPLE 1 Consider the nonhomogeneous system

$$\frac{dx}{dt} = y,$$
$$\frac{dy}{dt} = 5y - 6x + 2e^{-t}, \tag{1}$$

with initial values $x(0) = 0$ and $y(0) = 1$. After taking the Laplace transform of each equation we have

$$sX(s) - x(0) = Y(s),$$
$$sY(s) - y(0) = 5Y(s) - 6X(s) + \frac{2}{s+1},$$

where $\mathscr{L}\{y\} = Y(s)$ and $\mathscr{L}\{x\} = X(s)$. Substituting the initial conditions and rearranging the terms in these algebraic equations gives us

$$sX(s) = Y(s) \tag{2}$$

and

$$(s-5)Y(s) + 6X(s) = 1 + \frac{2}{s+1}. \tag{3}$$

Substituting for $Y(s)$ from (2) into (3) eliminates $Y(s)$:

$$(s^2 - 5s + 6)X(s) = 1 + \frac{2}{s+1}$$

or

$$(s-3)(s-2)X(s) = \frac{s+3}{s+1}.$$

We then solve this last equation for $X(s)$:

$$X(s) = \frac{s+3}{(s-3)(s-2)(s+1)}. \tag{4}$$

We use the partial fraction decomposition

$$\frac{s+3}{(s-3)(s-2)(s+1)} = \frac{A}{s-3} + \frac{B}{s-2} + \frac{C}{s+1}$$

to find that $A = 3/2$, $B = -5/3$, and $C = 1/6$. Thus,

$$X(s) = \frac{3/2}{s-3} - \frac{5/3}{s-2} + \frac{1/6}{s+1}.$$

Taking the inverse transform of $X(s)$ then yields the first of the desired component functions:

$$x(t) = \frac{3}{2}e^{3t} - \frac{5}{3}e^{2t} + \frac{1}{6}e^{-t}.$$

We now seek the second component function $y(t)$. We can substitute for $X(s)$ from Eq. (2) into Eq. (4) to obtain the transform

$$Y(s) = \frac{s^2 + 3s}{(s-3)(s-2)(s+1)}.$$

Partial fraction decomposition of the expression for $Y(s)$ results in the following:

$$Y(s) = \frac{9/2}{s-3} - \frac{10/3}{s-2} - \frac{1/6}{s+1}$$

(we leave the details to you). Finally the inverse transform of $Y(s)$ yields the second component function,

$$y(t) = \frac{9}{2}e^{3t} - \frac{10}{3}e^{2t} - \frac{1}{6}e^{-t}.$$

Thus the solution to nonhomogeneous linear system (1) subject to the initial conditions is

$$x(t) = \frac{3}{2}e^{3t} - \frac{5}{3}e^{2t} + \frac{1}{6}e^{-t},$$

$$y(t) = \frac{9}{2}e^{3t} - \frac{10}{3}e^{2t} - \frac{1}{6}e^{-t}.$$

EXAMPLE 2 Solve the nonhomogeneous system

$$\begin{aligned} \frac{dx}{dt} &= 3x - 4y + 1, \\ \frac{dy}{dt} &= 2x - 3y + t, \end{aligned} \tag{5}$$

subject to the initial conditions $x(0) = 1$ and $y(0) = -1$.

Solution. Taking the Laplace transform of each equation results in the algebraic system,

$$sX(s) - x(0) = 3X(s) - 4Y(s) + \frac{1}{s},$$

$$sY(s) - y(0) = 2X(s) - 3Y(s) + \frac{1}{s^2}.$$

Rearrangement of the terms in the previous set of equations so the unknown transforms $X(s)$ and $Y(s)$ are isolated on the lefthand side yields

$$\begin{aligned} (s-3)X(s) + 4Y(s) &= 1 + \frac{1}{s}, \\ -2X(s) + (s+3)Y(s) &= -1 + \frac{1}{s^2}. \end{aligned} \tag{6}$$

We solve algebraic system (6) by Cramer's rule (the method of determinants):

$$X(s) = \frac{\begin{vmatrix} 1+s^{-1} & 4 \\ -1+s^{-2} & s+3 \end{vmatrix}}{s^2-1} = \frac{s^3+8s^2+3s-4}{s^2(s-1)(s+1)},$$

and

$$Y(s) = \frac{\begin{vmatrix} s-3 & 1+s^{-1} \\ -2 & -1+s^{-2} \end{vmatrix}}{s^2-1} = \frac{-s^3+5s^2+3s-3}{s^2(s-1)(s+1)}.$$

From partial fraction decomposition we have

$$\frac{s^3+8s^2+3s-4}{s^2(s-1)(s+1)} = \frac{4}{s-1} + \frac{-3s+4}{s^2}$$

and

$$\frac{-s^3+5s^2+3s-3}{s^2(s-1)(s+1)} = \frac{2}{s-1} + \frac{-3s+3}{s^2}.$$

(You should work out the details of arriving at these last two equations.) Therefore we find the transforms

$$X(s) = \frac{4}{s-1} - \frac{3}{s} + \frac{4}{s^2},$$

$$Y(s) = \frac{2}{s-1} - \frac{3}{s} + \frac{3}{s^2}.$$

Taking inverse transforms of each equation then yields the unique solution

$$x(t) = 4e^t - 3 + 4t,$$

$$y(t) = 2e^t - 3 + 3t.$$

EXAMPLE 3 Solve the system

$$\begin{aligned} 2\frac{dx}{dt} + \frac{dy}{dt} &= 2x + 1, \\ \frac{dx}{dt} + \frac{dy}{dt} &= 3x + 3y + 2, \end{aligned} \tag{7}$$

subject to the initial conditions $x(0) = 0$ and $y(0) = 0$.

Solution. Using the Laplace transform, we convert the system to

$$2[sX(s) - x(0)] + sY(s) - y(0) = 2X(s) + \frac{1}{s},$$

$$sX(s) - x(0) + sY(s) - y(0) = 3X(s) + 3Y(s) + \frac{2}{s}.$$

Rearranging terms as before, we find the algebraic system

$$\begin{aligned} 2(s-1)X(s) + \qquad sY(s) &= \frac{1}{s}, \\ (s-3)X(s) + (s-3)Y(s) &= \frac{2}{s}. \end{aligned} \tag{8}$$

Solving system (8) by the method of determinants results in

$$X(s) = -\frac{s+3}{s(s-3)(s-2)},$$

$$Y(s) = \frac{3s-1}{s(s-3)(s-2)}.$$

(Again we leave it to you to work out the details.) The partial fraction decomposition of each of these last expressions is straightforward and gives us the following:

$$X(s) = -\frac{1/2}{s} + \frac{-2}{s-3} + \frac{5/2}{s-2},$$

$$Y(s) = -\frac{1/6}{s} + \frac{8/3}{s-3} + \frac{-5/2}{s-2}.$$

Finally, we take inverse transforms to obtain the solution:

$$x(t) = -\frac{1}{2} - 2e^{3t} + \frac{5}{2}e^{2t},$$

$$y(t) = -\frac{1}{6} + \frac{8}{3}e^{3t} - \frac{5}{2}e^{2t}.$$

From Examples 1–3 we see that the method of Laplace transforms is easy to use for solving linear systems of differential equations with constant coefficients. Essentially the Laplace transform converts the *analytical* problem of a system of differential equations into an *algebraic* problem of a system of linear algebraic equations. It is especially easy to solve 2×2 linear algebraic systems, so the only real difficulty that may occur in implementing the method is in inverting the transforms to obtain the two individual component functions at the end of the process. Of course, that is no more a problem than when we used the Laplace transform to solve nonhomogeneous second-order initial value problems (except that there are two inverse transforms to find instead of one).

We complete this section by solving a real-world behavior modeled by a 2×2 constant-coefficient nonhomogeneous linear system.

EXAMPLE 4 Consider the electrical network depicted in Fig. 7.36. It consists of two circuits between which no current flows; however, applied voltages are introduced to each circuit through a mutual inductance M between the inductors and through a linear amplifier. The linear amplifier is an electrical device that is capable of "looking back" to the voltage across resistor R_1 in the circuit on the left and amplifying it to the voltage $\mu i_1 R_1$, where μ is the constant amplifying factor. The amplified voltage is applied to the feedback circuit pictured on the right in Fig. 7.36. Notice that the current flows clockwise in the circuit on the left and counterclockwise in the circuit on the right.

Assume the following values for the network:

$$\begin{aligned} L_1 &= 2 \text{ henrys} \qquad & R_1 &= 10 \text{ ohms} \\ L_2 &= 3 \text{ henrys} \qquad & R_2 &= 75 \text{ ohms} \\ M &= 2 \text{ henrys} \qquad & \mu &= 5. \end{aligned}$$

The induced voltage is $E(t) = 2e^{-t}$ (in volts). Assuming the initial currents $i_1(0) = i_2(0) = 0$, we wish to know the currents $i_1(t)$ and $i_2(t)$ in the two circuits for $t > 0$.

Applying Kirchhoff's second law to the circuit on the left results in the relation

$$E(t) - L_1 \frac{di_1}{dt} - M \frac{di_2}{dt} - R_1 i_1 = 0.$$

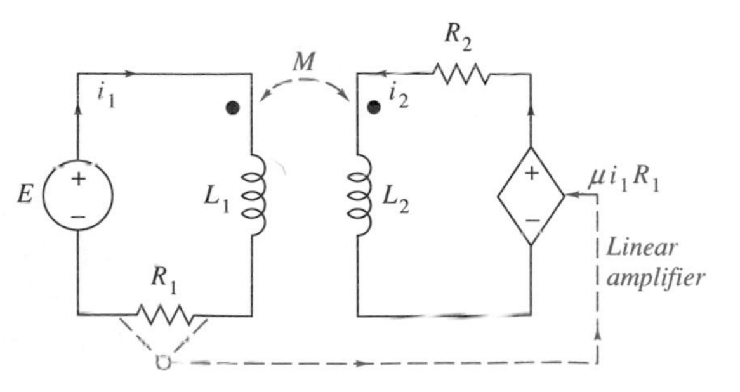

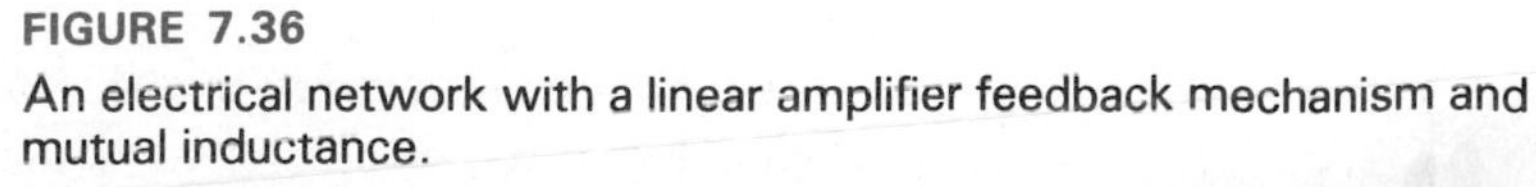
FIGURE 7.36
An electrical network with a linear amplifier feedback mechanism and mutual inductance.

Substituting the given values into this equation and rearranging terms gives us

$$2\frac{di_1}{dt} + 2\frac{di_2}{dt} = -10i_1 + 2e^{-t}. \tag{9}$$

Similarly, for the feedback loop on the right, Kirchhoff's law results in the equation

$$-\mu i_1 R_1 - R_2 i_2 - L_2\frac{di_2}{dt} - M\frac{di_1}{dt} = 0.$$

Substituting the given values into this last equation and rearranging terms gives us

$$2\frac{di_1}{dt} + 3\frac{di_2}{dt} = -50i_1 - 75i_2. \tag{10}$$

Using the method of Laplace transforms we now solve the nonhomogeneous system expressed by Eqs. (9) and (10) subject to the initial conditions $i_1(0) = i_2(0) = 0$.

Applying the Laplace transform to each equation results in the algebraic system

$$\begin{aligned} 2sI_1 + 2sI_2 &= -10I_1 + \frac{2}{s+1}, \\ 2sI_1 + 3sI_2 &= -50I_1 - 75I_2, \end{aligned}$$

where $I_1 = I_1(s) = \mathscr{L}\{i_1\}$ and $I_2 = I_2(s) = \mathscr{L}\{i_2\}$. This algebraic system is easily rewritten as

$$\begin{aligned} (s+5)I_1 + sI_2 \qquad &= \frac{1}{s+1}, \\ 2(s+25)I_1 + 3(s+25)I_2 &= 0. \end{aligned} \tag{11}$$

We now solve system (11) by Cramer's rule:

$$I_1 = \frac{\begin{vmatrix} \dfrac{1}{s+1} & s \\ 0 & 3 \end{vmatrix}}{\begin{vmatrix} s+5 & s \\ 2 & 3 \end{vmatrix}} = \frac{3}{(s+1)(3s+15-2s)}$$

$$= \frac{3}{(s+1)(s+15)}$$

and

$$I_2 = \frac{\begin{vmatrix} s+5 & \frac{1}{s+1} \\ 2 & 0 \end{vmatrix}}{\begin{vmatrix} s+5 & s \\ 2 & 3 \end{vmatrix}} = \frac{-2}{(s+1)(s+15)}.$$

Partial fraction decomposition yields

$$I_1 = \frac{3/14}{s+1} + \frac{-3/14}{s+15}$$

and

$$I_2 = \frac{-1/7}{s+1} + \frac{1/7}{s+15}.$$

After applying the inverse transform we find:

$$i_1 = \frac{3}{14}e^{-t} - \frac{3}{14}e^{-15t},$$

$$i_2 = \frac{-1}{7}e^{-t} + \frac{1}{7}e^{-15t}.$$

EXERCISES 7.9

In problems 1–13 use the Laplace transform method to solve the given first-order system.

1. $\frac{dx}{dt} = x + 3y$
   $\frac{dy}{dt} = x - y + 2e^t$
   $x(0) = 0,\ y(0) = 0$

2. $\frac{dx}{dt} = 2x + 1$
   $\frac{dy}{dt} = 3x - 2y$
   $x(0) = 0,\ y(0) = 0$

3. $\frac{dx}{dt} + y = 0$
   $\frac{dy}{dt} + x = 0$
   $x(0) = 1,\ y(0) = -1$

4. $\frac{dx}{dt} = 2y + e^{-t}$
   $\frac{dy}{dt} = 2x - 2$
   $x(0) = 1,\ y(0) = 1$

5. $\frac{dx}{dt} = x + 2y + e^{2t}$
   $\frac{dy}{dt} = 4x - y - e^{2t}$
   $x(0) = -1,\ y(0) = 1$

6. $\frac{dx}{dt} = x - 1$
   $\frac{dy}{dt} = 2x + 3y + t$
   $x(0) = 1,\ y(0) = 0$

7. $\frac{dx}{dt} + y = 2e^{3t}$
   $\frac{dy}{dt} + x = e^{3t} - e^{-3t}$
   $x(0) = 0,\ y(0) = 0$

8. $\frac{dx}{dt} = y + \cos t$
   $\frac{dy}{dt} = x - \sin t$
   $x(0) = 1,\ y(0) = 0$

9. $\frac{dx}{dt} + 4x - y = t$

$\frac{dy}{dt} + 2x + y = 1$

$x(0) = 0,\ y(0) = 1$

10. $\frac{dx}{dt} = 3x + 2y + 1$

$\frac{dy}{dt} = -2x - y - 1$

$x(0) = 1,\ y(0) = -1$

11. $\frac{dx}{dt} = -y + e^t$

$\frac{dy}{dt} = x - e^{-t}$

$x(0) = 0,\ y(0) = 1$

12. $\frac{dx}{dt} + 3x + 2y = \sin t$

$\frac{dy}{dt} + 4x + y = 0$

$x(0) = 0,\ y(0) = 0$

13. $3\frac{dx}{dt} + 3x + 2y = e^t$

$3\frac{dy}{dt} - 4x - 3y = -3t$

$x(0) = 1,\ y(0) = -1$

In problems 14–18 use the Laplace transform method to solve the more general linear first-order system.

14. $\frac{dx}{dt} + \frac{dy}{dt} = 1,\quad y - x = t,\quad x(0) = 1,\ y(0) = 1$

15. $2\frac{dx}{dt} + \frac{dy}{dt} = y - x,\quad \frac{dx}{dt} - \frac{dy}{dt} = -x - y,$
$x(0) = 1,\quad y(0) = 1$

16. $2\frac{dx}{dt} + \frac{dy}{dt} = 2x + 1,\quad \frac{dx}{dt} + \frac{dy}{dt} = 3x + 3y,$
$x(0) = 0,\ y(0) = 0$

17. $2\frac{dx}{dt} + \frac{dy}{dt} = -2x + y + 3t,$
$\frac{dx}{dt} + \frac{dy}{dt} = -x - y + 1,\quad x(0) = 1,\ y(0) = 3$

18. $\frac{dx}{dt} + \frac{dy}{dt} = 1 - 4x,\quad \frac{dx}{dt} = 2x - y + t^2,\quad x(0) = 2,$
$y(0) = -1$

CHAPTER 7 REVIEW EXERCISES

1. Consider the system of differential equations

$$\frac{d}{dt}\begin{pmatrix} x \\ y \end{pmatrix} = \begin{pmatrix} 2 & -3 \\ 1 & -2 \end{pmatrix}\begin{pmatrix} x \\ y \end{pmatrix}.$$

a) Verify that

$$\mathbf{X}_1 = \begin{pmatrix} 3 \\ 1 \end{pmatrix} e^t \text{ and } \mathbf{X}_2 = \begin{pmatrix} 1 \\ 1 \end{pmatrix} e^{-t}$$

are solutions to the given system of differential equations.

b) Are $\mathbf{X}_1$ and $\mathbf{X}_2$, as defined in (a) above, linearly independent? Justify your answer.

c) Find the general solution to the system of differential equations.

In problems 2–21, find the general solution to the given system $\mathbf{X}' = \mathbf{AX}$.

2. $x' = -2x - 8y$
$y' = 2x + 6y$

3. $u' = -8u - 5v$
$v' = -5u + 2v$

4. $x' = 4x - 9y$
$y' = 4x - 8y$

5. $x_1' = 3x_1 + 2x_2$
$x_2' = -x_1 + x_2$

6. $x_1' = 3x_1 - 2x_2$
$x_2' = 2x_1 + 3x_2$

7. $x'(t) = 2x - 2y$
$y'(t) = 2x + 2y$

8. $x_1' = x_1 + x_2$
$x_2' = -x_1 + 3x_2$

9. $\frac{dx}{dt} = -2x + 3y$

$\frac{dy}{dt} = 3x - 2y$

10. $\frac{dx}{dt} = 3x - 2y$

$\frac{dy}{dt} = 2x - y$

11. $\frac{dx}{dt} = 2x - y$

$\frac{dy}{dt} = 9x + 2y$

12. $\frac{dx}{dt} = 3x + 4y$

$\frac{dy}{dt} = -x - y$

13. $\frac{dx}{dt} = \frac{3}{2}x - \frac{5}{2}y$

$\frac{dy}{dt} = -\frac{5}{2}x - \frac{3}{2}y$

14. $\mathbf{A} = \begin{pmatrix} 1 & -1 \\ 1 & 3 \end{pmatrix}$

15. $\mathbf{A} = \begin{pmatrix} 1 & -4 \\ 4 & -7 \end{pmatrix}$

16. $\mathbf{A} = \begin{pmatrix} 3 & -4 \\ 1 & -1 \end{pmatrix}$

17. $\mathbf{A} = \begin{pmatrix} 4 & -2 \\ 8 & -4 \end{pmatrix}$

18. $\mathbf{A} = \begin{pmatrix} 1 & -2 \\ 13/2 & -5 \end{pmatrix}$ **19.** $\mathbf{A} = \begin{pmatrix} 2 & -1 \\ 1 & 2 \end{pmatrix}$

20. $\mathbf{A} = \begin{pmatrix} 3 & -2 \\ 4 & -1 \end{pmatrix}$ **21.** $\mathbf{A} = \begin{pmatrix} -1 & -2 \\ 1/2 & -1 \end{pmatrix}$

In problems 22–24, find the particular solution $\mathbf{X}_p$ of a nonhomogeneous system $\mathbf{X}' = \mathbf{AX} + \mathbf{F}(t)$ where

$$\mathbf{X} = \begin{pmatrix} x_1(t) \\ x_2(t) \end{pmatrix}$$

and $\mathbf{A}$ and $\mathbf{F}(t)$ are as given below, $x_1(0) = x_2(0) = 0$.

22. $\mathbf{A} = \begin{pmatrix} 1 & -3 \\ 2 & -4 \end{pmatrix}$, $\mathbf{F}(t) = \begin{pmatrix} e^t \\ e^t \end{pmatrix}$

23. $\mathbf{A} = \begin{pmatrix} 3 & 5 \\ 6 & 2 \end{pmatrix}$, $\mathbf{F}(t) = \begin{pmatrix} e^{7t} \\ e^{7t} \end{pmatrix}$

24. $\mathbf{A} = \begin{pmatrix} 1 & -2 \\ 3 & -4 \end{pmatrix}$, $\mathbf{F}(t) = \begin{pmatrix} e^t \\ e^t \end{pmatrix}$

In problems 25–29, solve the following initial value problems $\mathbf{X}' = \mathbf{AX}$ where

$$\mathbf{X} = \begin{pmatrix} x(t) \\ y(t) \end{pmatrix}$$

and $\mathbf{A}$ and the initial conditions are given below.

25. $\mathbf{A} = \begin{pmatrix} 4 & 13 \\ -1 & -2 \end{pmatrix}$, $\begin{pmatrix} x(0) \\ y(0) \end{pmatrix} = \begin{pmatrix} 13 \\ 9 \end{pmatrix}$

26. $\mathbf{A} = \begin{pmatrix} -5 & 10 \\ -2 & 3 \end{pmatrix}$, $\begin{pmatrix} x(0) \\ y(0) \end{pmatrix} = \begin{pmatrix} 10 \\ 8 \end{pmatrix}$

27. $\mathbf{A} = \begin{pmatrix} 4 & -17 \\ 2 & -6 \end{pmatrix}$, $\begin{pmatrix} x(0) \\ y(0) \end{pmatrix} = \begin{pmatrix} 17 \\ 20 \end{pmatrix}$

28. $\begin{pmatrix} x \\ y \end{pmatrix}' = \begin{pmatrix} 4 & 13 \\ -2 & 2 \end{pmatrix}\begin{pmatrix} x \\ y \end{pmatrix}$, $\begin{pmatrix} x(0) \\ y(0) \end{pmatrix} = \begin{pmatrix} -16 \\ 12 \end{pmatrix}$

29. $\dfrac{dx}{dt} = 7x + 6y$

$\dfrac{dy}{dt} = 2x + 6y$

$x(0) = 8$, $y(0) = -3$

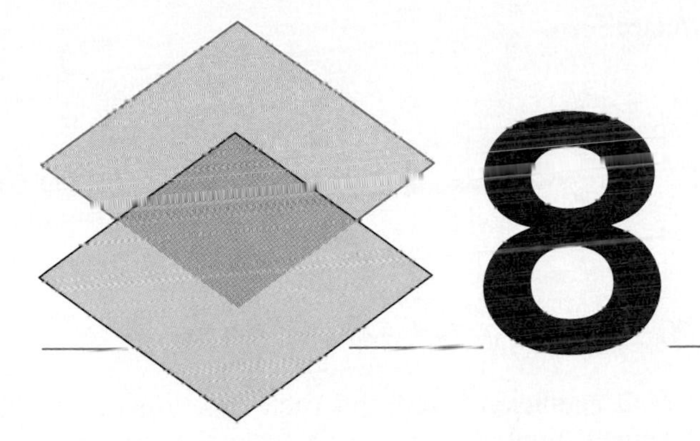

8

Matrices and First-Order Linear Systems

In this chapter we continue our investigations of constant-coefficient linear first-order systems. In Chapter 7 the focus was on solving 2×2 homogeneous systems. We will now extend the ideas discussed there to solving nonhomogeneous and higher-dimensional $n \times n$ linear first-order systems. For computational purposes, 2×2 and 3×3 systems will serve as most of the examples and exercises here, in both the homogeneous and nonhomogeneous cases. Nevertheless, we will present the theoretical underpinnings for the general $n \times n$ system. We will introduce matrix notation and use elementary matrix algebra to solve nonhomogeneous linear first-order systems. We assume you are familiar with the matrix operations of addition, multiplication, and scalar multiplication and that you can find the determinant of a matrix, invert a matrix, and form the transpose of a matrix. A brief review of these elementary matrix operations is presented in Appendix D.

8.1 BASIC THEORY

In this section we present the basic theoretical ideas underlying the *general solution* concept for $n \times n$ linear first-order systems. The $n \times n$ linear first-order system has the standard form

$$\begin{aligned}
\frac{dx_1}{dt} &= a_{11}(t)x_1 + a_{12}(t)x_2 + \cdots + a_{1n}(t)x_n + f_1(t),\\
\frac{dx_2}{dt} &= a_{21}(t)x_1 + a_{22}(t)x_2 + \cdots + a_{2n}(t)x_n + f_2(t),\\
&\vdots\\
\frac{dx_n}{dt} &= a_{n1}(t)x_1 + a_{n2}(t)x_2 + \cdots + a_{nn}(t)x_n + f_n(t).
\end{aligned} \tag{1}$$

Here it is assumed that each coefficient $a_{ij}(t)$ and each function $f_i(t)$ is continuous over some specified interval I. In matrix form the system is expressible as

$$\mathbf{X}'(t) = \mathbf{A}(t)\mathbf{X}(t) + \mathbf{F}(t) \tag{2}$$

where

$$\mathbf{X}(t) = \begin{pmatrix} x_1(t) \\ x_2(t) \\ \vdots \\ x_n(t) \end{pmatrix}, \qquad \mathbf{X}'(t) = \begin{pmatrix} x_1'(t) \\ x_2'(t) \\ \vdots \\ x_n'(t) \end{pmatrix}, \qquad \mathbf{F}(t) = \begin{pmatrix} f_1(t) \\ f_2(t) \\ \vdots \\ f_n(t) \end{pmatrix}$$

and

$$\mathbf{A}(t) = \begin{pmatrix} a_{11}(t) & a_{12}(t) & \cdots & a_{1n}(t) \\ a_{21}(t) & a_{22}(t) & \cdots & a_{2n}(t) \\ \vdots & \vdots & \vdots & \vdots \\ a_{n1}(t) & a_{n2}(t) & \cdots & a_{nn}(t) \end{pmatrix}.$$

Furthermore, if initial conditions $x_1(t_0) = x_{10}$, $x_2(t_0) = x_{20}$, . . ., $x_n(t_0) = x_{n0}$ are imposed, the resulting initial value problem can be expressed as

$$\mathbf{X}' = \mathbf{A}\mathbf{X} + \mathbf{F} \tag{3}$$

subject to the initial condition

$$\mathbf{X}(t_0) = \mathbf{X}_0$$

where

$$\mathbf{X}_0 = \begin{pmatrix} x_{10} \\ x_{20} \\ \vdots \\ x_{n0} \end{pmatrix}.$$

If all the components $f_1(t), f_2(t), \ldots, f_n(t)$ of $\mathbf{F}(t)$ are identically zero over the interval I, then system (2) is said to be **homogeneous;** otherwise it is **nonhomogeneous.**

In modeling a spring-mass system in Section 7.1, we converted a system of two second-order equations into a first-order 4×4 system by making an appropriate substitution. The same idea was used in converting a second-order *equation* into a first-order 2×2 *system* when studying numerical approximations to second-order equations in Section 4.6. In general, it is always possible to express a single linear nth-order equation with constant coefficients as a system of linear first-order equations. The original nth-order equation may be either homogeneous or nonhomogeneous. We next present this conversion process.

Conversion of an *nth*-Order Equation to a First-Order System

Suppose the linear nth-order equation with constant coefficients is written in the form

$$\frac{d^n y}{dt^n} = -\frac{a_0}{a_n}y - \frac{a_1}{a_n}y' - \cdots - \frac{a_{n-1}}{a_n}y^{(n-1)} + g(t). \tag{4}$$

We introduce the new variables $x_1, x_2, \ldots, x_n$ as follows:

$$\begin{aligned} x_1 &= y, \\ x_2 &= y', \\ x_3 &= y'', \\ &\vdots \\ x_n &= y^{(n-1)}. \end{aligned} \tag{5}$$

Then $x_1' = y' = x_2$, $x_2' = y'' = x_3$, and so forth. Also, $x_n' = y^{(n)}$ is given by Eq. (4). Substitution of these results into system (5) leads to the linear first-order system

$$\begin{aligned} \frac{dx_1}{dt} &= x_2, \\ \frac{dx_2}{dt} &= x_3, \\ \frac{dx_3}{dt} &= x_4, \\ &\vdots \\ \frac{dx_{n-1}}{dt} &= x_n, \\ \frac{dx_n}{dt} &= -\frac{a_0}{a_n}x_1 - \frac{a_1}{a_n}x_2 - \cdots - \frac{a_{n-1}}{a_n}x_n + g(t). \end{aligned} \tag{6}$$

EXAMPLE 1 Write the following fourth-order differential equation as a system:

$$2y^{(4)} - 3y''' - y'' + 2y' - y = 2e^{-t}.$$

Solution. First we write the equation in the standard form:

$$y^{(4)} = \frac{1}{2}y - y' + \frac{1}{2}y'' + \frac{3}{2}y''' + e^{-t}.$$

Introducing the variables defined in system (5) gives us $x_1 = y$, $x_2 = y'$, $x_3 = y''$, and $x_4 = y'''$. Thus substitution leads to the following linear first-order system:

$$\begin{aligned} x_1' &= y' = x_2, \\ x_2' &= y'' = x_3, \\ x_3' &= y''' = x_4, \\ x_4' &= y^{(4)} = \frac{1}{2}x_1 - x_2 + \frac{1}{2}x_3 + \frac{3}{2}x_4 + e^{-t}. \end{aligned}$$

Solutions

A **solution** to $n \times n$ linear system (1) or (2) is a vector function

$$\mathbf{X}(t) = \begin{pmatrix} x_1(t) \\ x_2(t) \\ \vdots \\ x_n(t) \end{pmatrix}$$

whose components are differentiable functions satisfying each of the equations in system (1), and any initial condition(s) that may be specified.

EXAMPLE 2 The vector function

$$\mathbf{X} = \begin{pmatrix} e^{3t} \\ e^{t} - e^{3t} \\ -e^{3t} \end{pmatrix}$$

is a solution to the 3×3 system

$$\mathbf{X}' = \begin{pmatrix} 4 & 0 & 1 \\ -2 & 1 & 0 \\ -2 & 0 & 1 \end{pmatrix} \mathbf{X}.$$

To check this result we differentiate each component of **X**:

$$\frac{dx_1}{dt} = \frac{d}{dt}(e^{3t}) \qquad = 3e^{3t},$$

$$\frac{dx_2}{dt} = \frac{d}{dt}(e^t - e^{3t}) = e^t - 3e^{3t},$$

$$\frac{dx_3}{dt} = \frac{d}{dt}(-e^{3t}) \qquad = -3e^{3t}.$$

Next we find the components of the product **AX**:

$$\begin{pmatrix} 4 & 0 & 1 \\ -2 & 1 & 0 \\ -2 & 0 & 1 \end{pmatrix} \begin{pmatrix} e^{3t} \\ e^t - e^{3t} \\ -e^{3t} \end{pmatrix} = \begin{pmatrix} 4e^{3t} - e^{3t} \\ -2e^{3t} + e^t - e^{3t} \\ -2e^{3t} - e^{3t} \end{pmatrix}$$

$$= \begin{pmatrix} 3e^{3t} \\ e^t - 3e^{3t} \\ -3e^{3t} \end{pmatrix}.$$

Since the components of **AX** agree with the components of **X**′, the given vector **X** solves the system, as claimed.

Existence and Uniqueness of Solutions

The following result gives sufficient conditions for the existence and uniqueness of solutions to the initial value problem for linear first-order systems.

> **THEOREM 8.1**
>
> If the components of the matrix $\mathbf{A}(t)$ and column vector $\mathbf{F}(t)$ are continuous functions on a common interval $I: a \leq t \leq b$ containing t_0, then there exists a unique solution $\mathbf{X}(t)$ to the initial value problem
>
> $$\mathbf{X}' = \mathbf{AX} + \mathbf{F},$$
>
> with
>
> $$\mathbf{X}(t_0) = \mathbf{X}_0$$
>
> that is valid over the entire interval I.

While Theorem 8.1 applies equally well to both homogeneous and nonhomogeneous systems, we delay further discussion of nonhomogeneous systems until the end of this section. The solution procedure discussed in this chapter requires solving the associated homogeneous system $\mathbf{X}' = \mathbf{AX}$ before solving nonhomogeneous system (2), which is analogous to the process of solving the linear nth-order equation we studied in Chapter 4.

Homogeneous Linear Systems

Our goal is to find all possible solutions to the constant-coefficient $n \times n$ system

$$\mathbf{X}' = \mathbf{AX}. \tag{7}$$

As we saw in Chapter 7, the 2×2 homogeneous system has two linearly independent solutions in any fundamental set. Any fundamental set can produce the general solution to the system. The concepts discussed in Section 7.3 easily extend to general $n \times n$ linear systems.

As was the case for homogeneous linear equations or systems studied previously, any linear combination of solutions is also a solution to the homogeneous $n \times n$ linear system. More precisely, we let $\{\mathbf{X}_1, \mathbf{X}_2, \ldots, \mathbf{X}_k\}$ be any set of k solution vectors to homogeneous system (7) on the interval I. Then any linear combination

$$\mathbf{X} = c_1\mathbf{X}_1 + c_2\mathbf{X}_2 + \cdots + c_k\mathbf{X}_k \tag{8}$$

where the c_i are constants is also a solution to system (7). Thus any scalar multiple of a solution is also a solution, and the sum of any set of solutions is again a solution. As before, this result is known as the *superposition principle.* The concepts of *linear dependence* and *linear independence* generalize as well.

DEFINITION 8.1

Let $\{\mathbf{X}_1(t), \mathbf{X}_2(t), \ldots, \mathbf{X}_k(t)\}$ be a set of vector functions on the interval I. The set is **linearly dependent** over I if there exist constants c_1, $c_2, \ldots, c_k$, not all zero, such that

$$c_1\mathbf{X}_1(t) + c_2\mathbf{X}_2(t) + \cdots + c_k\mathbf{X}_k(t) = \mathbf{0} \tag{9}$$

for every t in I. If the set is not linearly dependent, it is **linearly independent.**

In Chapter 7 we stated that we could produce all solutions to a 2×2 linear system by forming linear combinations of any two linearly independent solutions. The next theorem generalizes that result to $n \times n$ systems.

THEOREM 8.2

For the constant-coefficient homogeneous system

$$\mathbf{X}' = \mathbf{AX}$$

there exist n linearly independent solution vectors $\{\mathbf{X}_1(t), \mathbf{X}_2(t), \ldots, \mathbf{X}_n(t)\}$. Moreover, given any solution $\mathbf{X}(t)$ to this system over the interval I, there exist unique constants $c_1, c_2, \ldots, c_n$ such that

$$\mathbf{X}(t) = c_1\mathbf{X}_1(t) + c_2\mathbf{X}_2(t) + \cdots + c_n\mathbf{X}_n(t). \tag{10}$$

Theorem 8.2 tells us that every solution to an $n \times n$ linear system comes from solution form (10) for appropriate values of the constants $c_1, c_2, \ldots, c_n$. Thus the **general solution** to $n \times n$ constant-coefficient homogeneous linear system (7) is expressed by

$$\mathbf{X}(t) = c_1\mathbf{X}_1(t) + c_2\mathbf{X}_2(t) + \cdots + c_n\mathbf{X}_n(t) \tag{11}$$

where $c_1, c_2, \ldots, c_n$ are arbitrary constants, and $\mathbf{X}_1(t), \mathbf{X}_2(t), \ldots, \mathbf{X}_n(t)$ form a set of n linearly independent solutions. *The general solution contains all possible solutions* to homogeneous system (7). In consonance with the notion of a fundamental set of solutions for 2×2 systems, we then have the following definition for $n \times n$ systems.

DEFINITION 8.2

Any set of n linearly independent solutions to the $n \times n$ homogeneous linear system

$$\mathbf{X}' = \mathbf{AX}$$

is called a **fundamental set** of solutions.

Theorem 8.2 guarantees the existence of at least one set of n linearly independent solutions to system (7). That set is then a fundamental set of solutions. Thus to find the general solution to an $n \times n$ homogeneous linear system you find any fundamental set of solution vectors (that is, any linearly independent set of n solution vectors) satisfying the system $\mathbf{X}' = \mathbf{AX}$, and then form sum (11). If n initial conditions are given, the arbitrary constants c_i in the sum can then be evaluated.

Wronskian Test for Linear Independence

A set of vectors $\{\mathbf{X}_1, \mathbf{X}_2, \ldots, \mathbf{X}_k\}$ is linearly dependent over the interval I if and only if one of the vectors is a linear combination of the remaining ones. (You will be asked to prove this in Exercises 8.1, problem 31). However, in the special case that the vectors are solutions to a homogeneous system, there is a simple test for linear dependence or independence. We first introduce the idea of the Wronskian for systems.

DEFINITION 8.3

Suppose the n vector functions

$$\mathbf{X}_1(t) = \begin{pmatrix} x_{11}(t) \\ x_{21}(t) \\ \vdots \\ x_{n1}(t) \end{pmatrix}, \quad \mathbf{X}_2(t) = \begin{pmatrix} x_{12}(t) \\ x_{22}(t) \\ \vdots \\ x_{n2}(t) \end{pmatrix}, \ldots, \quad \mathbf{X}_n(t) = \begin{pmatrix} x_{1n}(t) \\ x_{2n}(t) \\ \vdots \\ x_{nn}(t) \end{pmatrix}$$

are solutions to the system $\mathbf{X}' = \mathbf{AX}$. The **Wronskian** of the solutions is the determinant of the matrix whose columns are the vectors $\mathbf{X}_i$:

$$W[\mathbf{X}_1, \mathbf{X}_2, \ldots, \mathbf{X}_n] = \det \begin{pmatrix} x_{11}(t) & x_{12}(t) & \cdots & x_{1n}(t) \\ x_{21}(t) & x_{22}(t) & \cdots & x_{2n}(t) \\ \vdots & \vdots & \vdots & \vdots \\ x_{n1}(t) & x_{n2}(t) & \cdots & x_{nn}(t) \end{pmatrix}.$$

Note that unlike the Wronskian defined in Chapter 3, Definition 8.3 does not require any differentiation. However, one can demonstrate that the Wronskian for an nth-order equation is the same, up to a constant multiple, as that defined in Definition 8.3 when the equation is written as a first-order system in accordance with our previous discussion (see problem 32). The next result provides a convenient test for linear independence of a set of solution vectors.

THEOREM 8.3

A necessary and sufficient condition that a set $\{\mathbf{X}_1(t), \mathbf{X}_2(t), \ldots, \mathbf{X}_n(t)\}$ of n solution vectors to the system $\mathbf{X}' = \mathbf{AX}$ be linearly independent over the interval I is that the Wronskian $W[\mathbf{X}_1, \mathbf{X}_2, \ldots, \mathbf{X}_n]$ not equal zero everywhere on I.

As is the case for nth-order equations, it can be shown that whenever the vector functions $\mathbf{X}_i$ are solutions to system (7), the Wronskian is *either always zero* on I or it is *never zero* on I. Hence, for linear independence of a set of solutions to system (7) it is sufficient to show that the Wronskian is not equal to zero for a single value t_0 in I. Let us consider an example.

EXAMPLE 3 Each of the vector functions

$$\mathbf{X}_1 = \begin{pmatrix} 0 \\ e^t \\ 0 \end{pmatrix}, \qquad \mathbf{X}_2 = \begin{pmatrix} e^{2t} \\ -2e^{2t} \\ -2e^{2t} \end{pmatrix}, \qquad \mathbf{X}_3 = \begin{pmatrix} e^{3t} \\ -e^{3t} \\ -e^{3t} \end{pmatrix}$$

solves the system

$$\mathbf{X}' = \begin{pmatrix} 4 & 0 & 1 \\ -2 & 1 & 0 \\ -2 & 0 & 1 \end{pmatrix} \mathbf{X}. \tag{12}$$

For instance, to verify that $\mathbf{X}_2$ is a solution we first list the component functions:

$$\begin{aligned} x_1 &= e^{2t}, \\ x_2 &= -2e^{2t}, \\ x_3 &= -2e^{2t}. \end{aligned}$$

The derivatives then satisfy the following relations:

$$\begin{aligned} x_1' &= 2e^{2t} &&= 4x_1 + x_3, \\ x_2' &= -4e^{2t} &&= -2x_1 + x_2, \\ x_3' &= -4e^{2t} &&= -2x_1 + x_3. \end{aligned}$$

Thus, $\mathbf{X}_2' = \mathbf{A}\mathbf{X}_2$ for the matrix $\mathbf{A}$ in system (12). Moreover, the Wronskian is given by

$$\begin{aligned} W[\mathbf{X}_1, \mathbf{X}_2, \mathbf{X}_3] &= \det \begin{pmatrix} 0 & e^{2t} & e^{3t} \\ e^t & -2e^{2t} & -e^{3t} \\ 0 & -2e^{2t} & -e^{3t} \end{pmatrix} \\ &= -e^t(-e^{5t} + 2e^{5t}) \\ &= -e^{6t} \neq 0. \end{aligned}$$

Thus $\{\mathbf{X}_1, \mathbf{X}_2, \mathbf{X}_3\}$ is a fundamental set of solutions. From Theorem 8.2 we know that the general solution to system (12) is given by

$$\begin{aligned} \mathbf{X} &= c_1\mathbf{X}_1 + c_2\mathbf{X}_2 + c_3\mathbf{X}_3 \\ &= c_1 \begin{pmatrix} 0 \\ 1 \\ 0 \end{pmatrix} e^t + c_2 \begin{pmatrix} 1 \\ -2 \\ -2 \end{pmatrix} e^{2t} + c_3 \begin{pmatrix} 1 \\ -1 \\ -1 \end{pmatrix} e^{3t}. \end{aligned}$$

A Fundamental Matrix

Suppose the set of vectors $\{\mathbf{X}_1, \mathbf{X}_2, \ldots, \mathbf{X}_n\}$ is a fundamental set of solutions to homogeneous system (7) on the interval I. The matrix Φ whose *columns* are formed from the vectors $\mathbf{X}_i$ is called a **fundamental matrix** of system (7):

$$\Phi = \begin{pmatrix} x_{11}(t) & x_{12}(t) & \cdots & x_{1n}(t) \\ x_{21}(t) & x_{22}(t) & \cdots & x_{2n}(t) \\ \vdots & \vdots & \vdots & \vdots \\ x_{n1}(t) & x_{n2}(t) & \cdots & x_{nn}(t) \end{pmatrix}. \tag{13}$$

General solution (11) to the homogeneous system can be expressed in terms of the fundamental matrix Φ. To see this, we write the general solution as follows:

$$\begin{aligned}\mathbf{X} &= c_1\mathbf{X}_1 + c_2\mathbf{X}_2 + \cdots + c_n\mathbf{X}_n \\ &= c_1\begin{pmatrix} x_{11} \\ x_{21} \\ \vdots \\ x_{n1} \end{pmatrix} + c_2\begin{pmatrix} x_{12} \\ x_{22} \\ \vdots \\ x_{n2} \end{pmatrix} + \cdots + c_n\begin{pmatrix} x_{1n} \\ x_{2n} \\ \vdots \\ x_{nn} \end{pmatrix}.\end{aligned}$$

Therefore

$$\mathbf{X} = \Phi\mathbf{C} \tag{14}$$

where

$$\mathbf{C} = \begin{pmatrix} c_1 \\ c_2 \\ \vdots \\ c_n \end{pmatrix} \tag{15}$$

and Φ is the fundamental matrix (13).

Properties of Fundamental Matrices

We first list the properties of any fundamental matrix and then examine them in an example.

1. The columns of Φ are linearly independent solution vectors.

2. The determinant of Φ is the Wronskian of its columns:

$$\det \Phi = W[\mathbf{X}_1, \mathbf{X}_2, \ldots, \mathbf{X}_n]. \tag{16}$$

3. The general solution $\mathbf{X}$ satisfies the relation

$$\mathbf{X} = \Phi\mathbf{C}. \tag{17}$$

4. If initial conditions are given, then

$$\mathbf{X}_0 = \Phi(t_0)\mathbf{C}.$$

Therefore

$$\mathbf{C} = \Phi^{-1}(t_0)\mathbf{X}_0. \tag{18}$$

5. From $\mathbf{X}' = \mathbf{A}\mathbf{X}$ and $\mathbf{X}' = \Phi'\mathbf{C}$ (Property 3), we have

$$\mathbf{A}\mathbf{X} = \Phi'\mathbf{C}.$$

Substitution of $\mathbf{X} = \Phi\mathbf{C}$ on the lefthand side yields

$$\mathbf{A}\Phi\mathbf{C} = \Phi'\mathbf{C}.$$

Since $\mathbf{C}$ is arbitrary,

$$\Phi' = \mathbf{A}\Phi. \tag{19}$$

EXAMPLE 4 In Example 3 we found that the vectors

$$\mathbf{X}_1 = \begin{pmatrix} 0 \\ e^t \\ 0 \end{pmatrix}, \quad \mathbf{X}_2 = \begin{pmatrix} e^{2t} \\ -2e^{2t} \\ -2e^{2t} \end{pmatrix}, \quad \mathbf{X}_3 = \begin{pmatrix} e^{3t} \\ -e^{3t} \\ -e^{3t} \end{pmatrix}$$

are linearly independent solutions to the system

$$\mathbf{X}' = \begin{pmatrix} 4 & 0 & 1 \\ -2 & 1 & 0 \\ -2 & 0 & 1 \end{pmatrix} \mathbf{X}. \tag{20}$$

Thus a fundamental matrix is

$$\Phi(t) = \begin{pmatrix} 0 & e^{2t} & e^{3t} \\ e^t & -2e^{2t} & -e^{3t} \\ 0 & -2e^{2t} & -e^{3t} \end{pmatrix}.$$

As we saw before, $\det \Phi = -e^{6t} \neq 0$. The inverse of Φ turns out to be

$$\Phi^{-1}(t) = \begin{pmatrix} 0 & e^{-t} & -e^{-t} \\ -e^{-2t} & 0 & -e^{-2t} \\ 2e^{-3t} & 0 & e^{-3t} \end{pmatrix}.$$

You should verify for yourself that

$$\Phi\Phi^{-1} = \Phi^{-1}\Phi = \begin{pmatrix} 1 & 0 & 0 \\ 0 & 1 & 0 \\ 0 & 0 & 1 \end{pmatrix}.$$

Given the initial condition

$$\mathbf{X}(0) = \begin{pmatrix} 1 \\ 0 \\ -1 \end{pmatrix},$$

the unique solution solving system (20) and satisfying the initial condition is

$$\mathbf{X} = \mathbf{\Phi C}$$

where

$$\begin{aligned} \mathbf{C} &= \mathbf{\Phi}^{-1}(0)\mathbf{X}_0 \\ &= \begin{pmatrix} 0 & 1 & -1 \\ -1 & 0 & -1 \\ 2 & 0 & 1 \end{pmatrix} \begin{pmatrix} 1 \\ 0 \\ -1 \end{pmatrix} \\ &= \begin{pmatrix} 1 \\ 0 \\ 1 \end{pmatrix}. \end{aligned}$$

Thus

$$\mathbf{X} = 1\begin{pmatrix} 0 \\ e^t \\ 0 \end{pmatrix} + 0\begin{pmatrix} e^{2t} \\ -2e^{2t} \\ -2e^{2t} \end{pmatrix} + 1\begin{pmatrix} e^{3t} \\ -e^{3t} \\ -e^{3t} \end{pmatrix}.$$

In component form we have the unique solution

$$\begin{aligned} x_1 &= e^{3t}, \\ x_2 &= e^t - e^{3t}, \\ x_3 &= -e^{3t}. \end{aligned}$$

Nonhomogeneous Systems

We now turn our attention to finding the general solution of a nonhomogeneous system. For the nonhomogeneous linear system

$$\mathbf{X}' = \mathbf{AX} + \mathbf{F}(t), \tag{21}$$

a **particular solution** is any solution vector over the interval I that satisfies system (21). In Section 8.2 we will investigate one method for finding a particular solution vector for Eq. (21).

We let $\mathbf{X}_c$ denote the general solution to the associated homogeneous system

$$\mathbf{X}' = \mathbf{AX}. \tag{22}$$

We let $\mathbf{X}_p$ be any particular solution to Eq. (21). Then for the sum

$$\mathbf{X} = \mathbf{X}_c + \mathbf{X}_p, \tag{23}$$

differentiation yields the relations

$$\begin{aligned} \mathbf{X}' &= (\mathbf{X}_c + \mathbf{X}_p)' \\ &= \mathbf{X}_c' + \mathbf{X}_p' \\ &= \mathbf{AX}_c + [\mathbf{AX}_p + \mathbf{F}(t)] \\ &= \mathbf{A}(\mathbf{X}_c + \mathbf{X}_p) + \mathbf{F}(t) \\ &= \mathbf{AX} + \mathbf{F}(t). \end{aligned}$$

Therefore vector sum (23) is a solution to nonhomogeneous linear system (21). This result is analogous to the solution we found for nonhomogeneous nth-order differential equations in Chapter 4.

DEFINITION 8.4

Let $\mathbf{X}_p$ denote any particular solution to nonhomogeneous system $\mathbf{X}' = \mathbf{A}\mathbf{X} + \mathbf{F}$; and let $\mathbf{X}_c$ denote the general solution on the same interval of the corresponding homogeneous system $\mathbf{X}' = \mathbf{A}\mathbf{X}$. The **general solution** of the nonhomogeneous system on the interval is defined to be

$$\mathbf{X} = \mathbf{X}_c + \mathbf{X}_p.$$

The general solution $\mathbf{X}_c$ of the homogeneous system is called the **complementary solution** of the nonhomogeneous system $\mathbf{X}' = \mathbf{A}\mathbf{X} + \mathbf{F}$.

In Section 8.2 we will apply these ideas to find the general solution to 2×2 nonhomogeneous systems.

EXERCISES 8.1

In problems 1–5 write the given system in matrix form.

1. $\dfrac{dx}{dt} = 3x - 4y + 1$

   $\dfrac{dy}{dt} = 2x - 3y + t$

2. $\dfrac{dx}{dt} = x + 3y$

   $\dfrac{dy}{dt} = x - y + 2e^t$

3. $\dfrac{dx}{dt} = -y + 2e^{3t}$

   $\dfrac{dy}{dt} = -x + e^{3t} - e^{-3t}$

4. $\dfrac{dx}{dt} = -x + 3y + 3te^t - 2e^t$

   $\dfrac{dy}{dt} = -3x + 5y + 3te^t - 2e^t$

5. $\dfrac{dx}{dt} = -x + y + e^{-t}\cos t$

   $\dfrac{dy}{dt} = -y + te^{-t}$

In problems 6–10 write the system without the use of matrices.

6. $\mathbf{X}' = \begin{pmatrix} 0 & -1 \\ -1 & 0 \end{pmatrix}\mathbf{X} + \begin{pmatrix} 1 \\ e^t \end{pmatrix}$

7. $\mathbf{X}' = \begin{pmatrix} 2 & 3 \\ -3 & 2 \end{pmatrix}\mathbf{X} + \begin{pmatrix} te^t \\ e^t \end{pmatrix}$

8. $\mathbf{X}' = \begin{pmatrix} 3 & 1 \\ -1 & 1 \end{pmatrix}\mathbf{X} + \begin{pmatrix} 3t \\ -2 \end{pmatrix}e^{2t}$

9. $\mathbf{X}' = \begin{pmatrix} -1 & 0 & 0 \\ 0 & -1 & -4 \\ 0 & 1 & -1 \end{pmatrix}\mathbf{X}$

10. $\mathbf{X}' = \begin{pmatrix} 1 & -1 & 1 \\ -1 & 0 & 0 \\ 2 & -1 & 0 \end{pmatrix}\mathbf{X} + \begin{pmatrix} 0 \\ 3te^t \\ e^t \end{pmatrix}$

In problems 11–15 verify that vector $\mathbf{X}$ is a solution to system $\mathbf{X}' = \mathbf{A}\mathbf{X} + \mathbf{F}$.

11. $\mathbf{X} = \begin{pmatrix} 4e^t + 4t - 3 \\ 2e^t + 3t - 3 \end{pmatrix}, \quad \mathbf{X}' = \begin{pmatrix} 3 & -4 \\ 2 & -3 \end{pmatrix}\mathbf{X} + \begin{pmatrix} 1 \\ t \end{pmatrix}$

12. $\mathbf{X} = \begin{pmatrix} e^{-t} \\ -e^{-t} \end{pmatrix}, \quad \mathbf{X}' = \begin{pmatrix} 2 & 3 \\ 2 & 1 \end{pmatrix}\mathbf{X}$

13. $\mathbf{X} = \begin{pmatrix} 3e^t + 4te^t \\ 3e^t + 2te^t \end{pmatrix}, \quad \mathbf{X}' = \begin{pmatrix} 0 & 2 \\ -1 & 3 \end{pmatrix}\mathbf{X} + \begin{pmatrix} 1 \\ -1 \end{pmatrix} e^t$

14. $\mathbf{X} = \begin{pmatrix} -\frac{1}{5} \\ -\frac{3}{5} \end{pmatrix} e^{2t}, \quad \mathbf{X}' = \begin{pmatrix} 1 & 2 \\ 4 & -1 \end{pmatrix}\mathbf{X} + \begin{pmatrix} e^{2t} \\ -e^{2t} \end{pmatrix}$

15. $\mathbf{X} = \begin{pmatrix} \frac{1}{2}\sin t \\ \frac{1}{2}\cos t \end{pmatrix}, \quad \mathbf{X}' = \begin{pmatrix} 0 & -1 \\ 1 & 0 \end{pmatrix}\mathbf{X} + \begin{pmatrix} \cos t \\ -\sin t \end{pmatrix}$

In problems 16–20 write the given nth-order differential equation as a linear first-order system.

16. $y'' + 3y' + 2y = -x + 4x^2$

17. $y'' - 3y' = 5e^{-2x} + \sin x$

18. $y'' - 2y' + 2y = \dfrac{e^x}{1 + x^2}$

19. $y'' + y = \sec x$

20. $y''' - 5y'' + 6y' = 3\cos x - e^x$

In problems 21–25 the given vectors are solutions to a 3×3 system $\mathbf{X}' = \mathbf{AX}$. Determine whether they form a fundamental set.

21. $\begin{pmatrix} 0 \\ e^t \\ 0 \end{pmatrix}, \begin{pmatrix} e^{2t} \\ -2e^{2t} \\ -2e^{2t} \end{pmatrix}, \begin{pmatrix} e^{3t} \\ -e^{3t} \\ -e^{3t} \end{pmatrix}$

22. $\begin{pmatrix} 0 \\ e^{-2t} \\ 0 \end{pmatrix}, \begin{pmatrix} -3e^{-2t} \\ 0 \\ e^{-2t} \end{pmatrix}, \begin{pmatrix} -1 \\ -1 \\ 1 \end{pmatrix}$

23. $\begin{pmatrix} 7e^{-t} \\ -2e^{-t} \\ 13e^{-t} \end{pmatrix}, \begin{pmatrix} e^t \\ 0 \\ e^t \end{pmatrix}, \begin{pmatrix} e^{2t} \\ e^{2t} \\ e^{2t} \end{pmatrix}$

24. $\begin{pmatrix} e^t\cos t \\ -e^t\sin t \\ 0 \end{pmatrix}, \begin{pmatrix} -e^t\sin t \\ -e^t\cos t \\ 0 \end{pmatrix}, \begin{pmatrix} 0 \\ 0 \\ e^{-t} \end{pmatrix}$

25. $\begin{pmatrix} e^{-2t} \\ 0 \\ 0 \end{pmatrix}, \begin{pmatrix} te^{-2t} \\ e^{-2t} \\ 0 \end{pmatrix}, \begin{pmatrix} \frac{t^2}{2}e^{-2t} \\ (t-3)e^{-2t} \\ e^{-2t} \end{pmatrix}$

In problems 26–30, given the system and corresponding fundamental set, find a particular solution satisfying the specified initial conditions. Use formula (18) and write your answer in scalar form.

26. $\mathbf{X}' = \begin{pmatrix} 2 & 0 \\ 3 & -2 \end{pmatrix}\mathbf{X}, \quad \left\{\begin{pmatrix} 0 \\ e^{-2t} \end{pmatrix}, \begin{pmatrix} 4e^{2t} \\ 3e^{2t} \end{pmatrix}\right\},$
$x(0) = 4, \quad y(0) = -2$

27. $\mathbf{X}' = \begin{pmatrix} 0 & -1 \\ -1 & 0 \end{pmatrix}\mathbf{X}, \quad \left\{\begin{pmatrix} e^{-t} \\ e^{-t} \end{pmatrix}, \begin{pmatrix} -e^t \\ e^t \end{pmatrix}\right\},$
$x(0) = 1, \quad y(0) = -1$

28. $\mathbf{X}' = \begin{pmatrix} 2 & 3 \\ -3 & 2 \end{pmatrix}\mathbf{X}, \quad \left\{\begin{pmatrix} -e^{2t}\sin 3t \\ e^{2t}\cos 3t \end{pmatrix}, \begin{pmatrix} -e^{2t}\cos 3t \\ -e^{2t}\sin 3t \end{pmatrix}\right\},$
$x(0) = -1, y(0) = 0$

29. $\mathbf{X}' = \begin{pmatrix} -1 & 3 \\ -3 & 5 \end{pmatrix}\mathbf{X}, \quad \left\{\begin{pmatrix} e^{2t} \\ e^{2t} \end{pmatrix}, \begin{pmatrix} te^{2t} \\ (t + \frac{1}{3})e^{2t} \end{pmatrix}\right\},$
$x(0) = 1, y(0) = 3$

30. $\mathbf{X}' = \begin{pmatrix} -3 & -2 \\ -4 & -1 \end{pmatrix}\mathbf{X}, \quad \left\{\begin{pmatrix} e^{-5t} \\ e^{-5t} \end{pmatrix}, \begin{pmatrix} e^t \\ -2e^t \end{pmatrix}\right\},$
$x(0) = -2, y(0) = 1$

31. Prove that a set of vectors is linearly dependent if and only if one of the vectors is a linear combination of the remaining ones.

32. Suppose that $\{y_1, y_2\}$ is a fundamental set of solutions to a linear second-order homogeneous equation. Let $\{\mathbf{X}_1, \mathbf{X}_2\}$ be a fundamental set of solutions to the corresponding linear first-order system. Show that the Wronskians satisfy the relation

$$W[y_1, y_2] = cW[\mathbf{X}_1, \mathbf{X}_2]$$

for some constant c.

33. Prove that the Wronskian of the system $\mathbf{X}' = \mathbf{AX}$ satisfies the differential equation

$$\frac{dW}{dt} = (a_{11} + a_{22} + \cdots + a_{nn})W.$$

Show that W is an exponential function.

34. Prove that the Wronskians of two fundamental sets of solutions of the system $\mathbf{X}' = \mathbf{AX}$ differ at most by a multiplicative constant. (*Hint:* see problem 33.)

35. Let

$$\mathbf{X}_1(t) = \begin{pmatrix} t^2 \\ 2t \end{pmatrix} \quad \text{and} \quad \mathbf{X}_2(t) = \begin{pmatrix} t|t| \\ 2|t| \end{pmatrix}.$$

Verify that $W[\mathbf{X}_1, \mathbf{X}_2] \equiv 0$ on the interval $(-\infty, \infty)$, but that $\{\mathbf{X}_1, \mathbf{X}_2\}$ is a linearly independent set on $(-\infty, \infty)$.

36. Using problem 33, prove that the Wronskian of n solutions to the system $\mathbf{X}' = \mathbf{AX}$ on an interval I is either identically zero or never zero on I.

SOLUTIONS TO NONHOMOGENEOUS 2 × 2 SYSTEMS: VARIATION OF PARAMETERS

In this section we discuss a procedure for obtaining the general solution to the 2 × 2 constant-coefficient nonhomogeneous linear system. Using matrix notation this system can be written in the form

$$\mathbf{X}' = \mathbf{A}\mathbf{X} + \mathbf{F}(t) \tag{1}$$

where

$$\mathbf{X}(t) = \begin{pmatrix} x(t) \\ y(t) \end{pmatrix}, \qquad \mathbf{F}(t) = \begin{pmatrix} f(t) \\ g(t) \end{pmatrix}, \qquad \mathbf{A} = \begin{pmatrix} a & b \\ c & d \end{pmatrix}.$$

The component functions $f(t)$ and $g(t)$ in **F** are assumed to be continuous throughout some interval I, and a, b, c, and d are real constants.

In Section 8.1 we established that the general solution to system (1) has the form

$$\mathbf{X} = \mathbf{X}_c + \mathbf{X}_p \tag{2}$$

where $\mathbf{X}_c$ is the general solution to the associated homogeneous system $\mathbf{X}' = \mathbf{A}\mathbf{X}$ and $\mathbf{X}_p$ is *any* solution vector satisfying nonhomogeneous system (1). We learned how to obtain $\mathbf{X}_c$ for 2 × 2 systems in Sections 7.4 and 7.5. Now we are going to describe the method of variation of parameters for finding a suitable particular solution $\mathbf{X}_p$.

Variation of Parameters

In finding the complementary solution $\mathbf{X}_c$ in Sections 7.4 and 7.5 we found two linearly independent solution vectors to the homogeneous system

$$\mathbf{X}' = \mathbf{A}\mathbf{X}.$$

Let us denote these two linearly independent solution vectors by

$$\mathbf{X}_1 = \begin{pmatrix} x_1(t) \\ y_1(t) \end{pmatrix} \quad \text{and} \quad \mathbf{X}_2 = \begin{pmatrix} x_2(t) \\ y_2(t) \end{pmatrix}.$$

We then form the fundamental matrix

$$\Phi(t) = \begin{pmatrix} x_1(t) & x_2(t) \\ y_1(t) & y_2(t) \end{pmatrix}.$$

Recall that the columns of Φ are the linearly independent solution vectors $\mathbf{X}_1$ and $\mathbf{X}_2$. In terms of Φ, the general solution to the homogeneous system can be written as follows:

$$\mathbf{X}_c = c_1\mathbf{X}_1 + c_2\mathbf{X}_2 = \begin{pmatrix} x_1(t) & x_2(t) \\ y_1(t) & y_2(t) \end{pmatrix}\begin{pmatrix} c_1 \\ c_2 \end{pmatrix}$$

or

$$\mathbf{X}_c = \mathbf{\Phi C} \tag{3}$$

where

$$\mathbf{C} = \begin{pmatrix} c_1 \\ c_2 \end{pmatrix}$$

is the vector of arbitrary constants.

Emulating the procedure for variation of parameters in Section 4.4, we *vary the parameters* c_1 and c_2 and replace them with *functions* $u_1(t)$ and $u_2(t)$. Our goal is to find the unknown functions u_1 and u_2. Thus we demand that the solution form

$$\mathbf{X}_p = u_1(t)\mathbf{X}_1 + u_2(t)\mathbf{X}_2, \tag{4}$$

which is equivalent to

$$\mathbf{X}_p = \mathbf{\Phi U}, \tag{5}$$

solve nonhomogeneous system (1). Thus we must determine the unknown functions u_1 and u_2 that are the components of **U**.

Using the product rule to differentiate Eq. (4) gives us

$$\mathbf{X}_p' = u_1\mathbf{X}_1' + u_1'\mathbf{X}_1 + u_2\mathbf{X}_2' + u_2'\mathbf{X}_2 = \begin{pmatrix} x_1' & x_2' \\ y_1' & y_2' \end{pmatrix}\begin{pmatrix} u_1 \\ u_2 \end{pmatrix} + \begin{pmatrix} x_1 & x_2 \\ y_1 & y_2 \end{pmatrix}\begin{pmatrix} u_1' \\ u_2' \end{pmatrix}$$

or

$$\mathbf{X}_p' = \mathbf{\Phi}'\mathbf{U} + \mathbf{\Phi U}'. \tag{6}$$

Then substitution of $\mathbf{X}_p$ and $\mathbf{X}_p'$ from Eqs. (5) and (6) into nonhomogeneous system (1) results in

$$\mathbf{\Phi}'\mathbf{U} + \mathbf{\Phi U}' = \mathbf{A\Phi U} + \mathbf{F}. \tag{7}$$

Property 5 for fundamental matrices (see Section 8.1) states that

$$\mathbf{\Phi}' = \mathbf{A\Phi}. \tag{8}$$

Substituting Eq. (8) into the lefthand side of Eq. (7) gives us

$$\mathbf{A\Phi U} + \mathbf{\Phi U}' = \mathbf{A\Phi U} + \mathbf{F},$$

or

$$\mathbf{\Phi U}' = \mathbf{F}. \tag{9}$$

The fundamental matrix Φ is invertible since its columns are linearly independent. Thus we can solve Eq. (9) for the derivative vector:

$$\mathbf{U}' = \Phi^{-1}\mathbf{F} \tag{10}$$

By integrating each of the components u_1' and u_2' of $\mathbf{U}'$ calculated from the righthand side of Eq. (10), we then determine the unknown functions $u_1(t)$ and $u_2(t)$, so that

$$\mathbf{U} = \int \Phi^{-1}(t)\mathbf{F}(t)\, dt. \tag{11}$$

Finally, we substitute the functions u_1 and u_2 into solution form (4) to obtain the solution vector $\mathbf{X}_p$. Let us summarize this method for finding a particular solution to the nonhomogeneous system, after which we will present several examples.

SUMMARY OF METHOD OF VARIATION OF PARAMETERS FOR SOLVING X′ = AX + F

Step 1. Find the complementary solution $\mathbf{X}_c$. Solve the associated 2 × 2 homogeneous system

$$\mathbf{X}' = \mathbf{AX} \tag{12}$$

and form the fundamental matrix Φ whose columns are the linearly independent solutions $\mathbf{X}_1$ and $\mathbf{X}_2$.

Step 2. Vary the parameters. Write the form of the particular solution $\mathbf{X}_p$ as

$$\mathbf{X}_p = u_1(t)\mathbf{X}_1(t) + u_2(t)\mathbf{X}_2(t). \tag{13}$$

Step 3. Invert the fundamental matrix Φ:

$$\Phi^{-1} = \frac{1}{\det \Phi}\begin{pmatrix} y_2 & -x_2 \\ -y_1 & x_1 \end{pmatrix}. \tag{14}$$

Step 4. Determine the parameters u_1 and u_2:

$$\begin{pmatrix} u_1 \\ u_2 \end{pmatrix} = \int \Phi^{-1}(t)\mathbf{F}(t)\, dt. \tag{15}$$

Step 5. Calculate the particular solution $\mathbf{X}_p$:

$$\mathbf{X}_p = \Phi\mathbf{U} \tag{16}$$

where

$$\mathbf{U} = \begin{pmatrix} u_1 \\ u_2 \end{pmatrix}.$$

Step 6. Form the general solution:

$$\mathbf{X} = \mathbf{X}_c + \mathbf{X}_p. \tag{17}$$

EXAMPLE 1 Find the general solution to the nonhomogeneous system

$$\begin{aligned} \frac{dx}{dt} &= 3x - 4y + 1, \\ \frac{dy}{dt} &= 2x - 3y + t. \end{aligned} \tag{18}$$

Solution. We follow the steps in the method of variation of parameters.

Step 1. To solve the associated homogeneous system substitute $x = k_1 e^{\lambda t}$ and $y = k_2 e^{\lambda t}$ into the homogeneous system to obtain the algebraic equations

$$\begin{aligned} (3 - \lambda)k_1 - \quad\quad 4k_2 &= 0, \\ 2k_1 + (-3 - \lambda)k_2 &= 0. \end{aligned} \tag{19}$$

The determinant of the coefficients must equal zero, giving us the characteristic equation

$$(3 - \lambda)(-3 - \lambda) + 8 = 0$$

or

$$\lambda^2 - 1 = 0.$$

The eigenvalues are $\lambda = -1, 1$.

Substitution of $\lambda = -1$ into system (19) leads to the algebraic system

$$\begin{aligned} 4k_1 - 4k_2 &= 0, \\ 2k_1 - 2k_2 &= 0. \end{aligned}$$

Thus $k_1 = k_2$ where k_2 is arbitrary. We choose $k_2 = 1$ to obtain the vector solution

$$\mathbf{X}_1 = \begin{pmatrix} 1 \\ 1 \end{pmatrix} e^{-t}.$$

Next we substitute $\lambda = 1$ into system (19) to obtain the algebraic system

$$\begin{aligned} 2k_1 - 4k_2 &= 0, \\ 2k_1 - 4k_2 &= 0. \end{aligned}$$

Thus $k_1 = 2k_2$ where k_2 is arbitrary. We choose $k_2 = 1$ to obtain the vector solution

$$\mathbf{X}_2 = \begin{pmatrix} 2 \\ 1 \end{pmatrix} e^{t}.$$

A fundamental matrix is thus

$$\Phi = \begin{pmatrix} e^{-t} & 2e^{t} \\ e^{-t} & e^{t} \end{pmatrix}.$$

Step 2. Write the form of the particular solution:

$$\begin{aligned}\mathbf{X}_p &= \Phi\mathbf{U}\\ &= \begin{pmatrix} e^{-t} & 2e^{t} \\ e^{-t} & e^{t}\end{pmatrix}\begin{pmatrix} u_1 \\ u_2 \end{pmatrix}.\end{aligned}$$

Step 3. Invert the fundamental matrix Φ:

$$\det \Phi = e^{-t}e^{t} - 2e^{-t}e^{t} = -1 \neq 0.$$

Thus, according to Eq. (14),

$$\Phi^{-1} = \frac{1}{(-1)}\begin{pmatrix} e^{t} & -2e^{t} \\ -e^{-t} & e^{-t}\end{pmatrix}.$$

Step 4. We next determine the functions u_1 and u_2. First we find their derivatives:

$$\begin{aligned}\mathbf{U}' = \Phi^{-1}\mathbf{F} &= -\begin{pmatrix} e^{t} & -2e^{t} \\ -e^{-t} & e^{-t}\end{pmatrix}\begin{pmatrix} 1 \\ t \end{pmatrix}\\ &= \begin{pmatrix} -e^{t} + 2te^{t} \\ e^{-t} - te^{-t}\end{pmatrix}.\end{aligned}$$

Integration of each component of $\mathbf{U}'$ then leads to

$$\begin{aligned}\mathbf{U} = \begin{pmatrix} u_1 \\ u_2 \end{pmatrix} &= \begin{pmatrix} -e^{t} + 2(t-1)e^{t} \\ -e^{-t} + (t+1)e^{-t}\end{pmatrix}\\ &= \begin{pmatrix} 2te^{t} - 3e^{t} \\ te^{-t}\end{pmatrix}.\end{aligned}$$

Note that the constants of integration are omitted in determining u_1 and u_2 since they would be absorbed into the arbitrary constants in the complementary solution $\mathbf{X}_c$. Moreover we seek only one particular solution rather than a family of solutions.

Step 5. The particular solution is

$$\begin{aligned}\mathbf{X}_p &= \Phi\mathbf{U}\\ &= \begin{pmatrix} e^{-t} & 2e^{t} \\ e^{-t} & e^{t}\end{pmatrix}\begin{pmatrix} 2te^{t} - 3e^{t} \\ te^{-t}\end{pmatrix}\\ &= \begin{pmatrix} 2t - 3 + 2t \\ 2t - 3 + t\end{pmatrix}\\ &= \begin{pmatrix} 4t - 3 \\ 3t - 3\end{pmatrix}.\end{aligned}$$

It is straightforward to check that $x_p = 4t - 3$ and $y_p = 3t - 3$ do indeed satisfy nonhomogeneous system (18).

Step 6. The general solution to system (18) is thus

$$\mathbf{X} = \mathbf{X}_c + \mathbf{X}_p$$
$$= c_1 \begin{pmatrix} 1 \\ 1 \end{pmatrix} e^{-t} + c_2 \begin{pmatrix} 2 \\ 1 \end{pmatrix} e^{t} + \begin{pmatrix} 4t - 3 \\ 3t - 3 \end{pmatrix}$$

or, in component form,

$$x = c_1 e^{-t} + 2c_2 e^{t} + 4t - 3,$$
$$y = c_1 e^{-t} + c_2 e^{t} + 3t - 3.$$

EXAMPLE 2 Solve the system

$$\mathbf{X}' = \begin{pmatrix} 0 & -1 \\ 1 & 0 \end{pmatrix} \mathbf{X} + \begin{pmatrix} \sec t \\ 0 \end{pmatrix} \tag{20}$$

for $-\pi/2 < t < \pi/2$.

Solution. Following the steps for variation of parameters:

Step 1. We first solve the associated homogeneous equation:

$$\begin{aligned} (0 - \lambda)k_1 - \qquad k_2 &= 0, \\ k_1 + (0 - \lambda)k_2 &= 0, \end{aligned} \tag{21}$$

yields the characteristic equation

$$\lambda^2 + 1 = 0$$

with complex eigenvalues $\lambda = \pm i$. When $\lambda = i$, system (21) becomes

$$-ik_1 - k_2 = 0,$$
$$k_1 - ik_2 = 0.$$

Thus $k_1 = ik_2$ where k_2 is arbitrary. Setting $k_2 = 1$, we have the eigenvector

$$\mathbf{K} = \begin{pmatrix} i \\ 1 \end{pmatrix}.$$

Then

$$\mathbf{B}_1 = \begin{pmatrix} 0 \\ 1 \end{pmatrix} \quad \text{and} \quad \mathbf{B}_2 = \begin{pmatrix} -1 \\ 0 \end{pmatrix}.$$

The general solution to the associated homogeneous equation is

$$\mathbf{X}_c = c_1\mathbf{X}_1 + c_2\mathbf{X}_2$$

where

$$\mathbf{X}_1 = \begin{pmatrix} 0 \\ 1 \end{pmatrix} \cos t + \begin{pmatrix} -1 \\ 0 \end{pmatrix} \sin t,$$
$$\mathbf{X}_2 = \begin{pmatrix} -1 \\ 0 \end{pmatrix} \cos t - \begin{pmatrix} 0 \\ 1 \end{pmatrix} \sin t.$$

Thus a fundamental matrix is

$$\Phi = \begin{pmatrix} -\sin t & -\cos t \\ \cos t & -\sin t \end{pmatrix}.$$

Step 2. The form of the particular solution is

$$\begin{aligned} \mathbf{X}_p &= u_1\mathbf{X}_1 + u_2\mathbf{X}_2 \\ &= \begin{pmatrix} -\sin t & -\cos t \\ \cos t & -\sin t \end{pmatrix}\begin{pmatrix} u_1 \\ u_2 \end{pmatrix}. \end{aligned}$$

Step 3. We invert the fundamental matrix Φ using Eq. (14),

$$\det \Phi = \sin^2 t + \cos^2 t = 1 \neq 0,$$

to give us the inverse

$$\Phi^{-1} = \begin{pmatrix} -\sin t & \cos t \\ -\cos t & -\sin t \end{pmatrix}.$$

Step 4. The derivatives of u_1 and u_2 are given in matrix form by the following:

$$\begin{aligned} \mathbf{U}' = \Phi^{-1}\mathbf{F} &= \begin{pmatrix} -\sin t & \cos t \\ -\cos t & -\sin t \end{pmatrix}\begin{pmatrix} \sec t \\ 0 \end{pmatrix} \\ &= \begin{pmatrix} -\tan t \\ -1 \end{pmatrix}. \end{aligned}$$

Integration of each component of $\mathbf{U}'$ results in

$$\mathbf{U} = \begin{pmatrix} u_1 \\ u_2 \end{pmatrix} = \begin{pmatrix} \ln|\cos t| \\ -t \end{pmatrix}.$$

Since $-\pi/2 < t < \pi/2$, $\cos t > 0$ and $u_1 = \ln(\cos t)$.

Step 5. The particular solution is as follows:

$$\begin{aligned} \mathbf{X}_p &= \Phi\mathbf{U} \\ &= \begin{pmatrix} -\sin t & -\cos t \\ \cos t & -\sin t \end{pmatrix}\begin{pmatrix} \ln(\cos t) \\ -t \end{pmatrix} \\ &= \begin{pmatrix} t\cos t - \sin t \ln(\cos t) \\ t \sin t + \cos t \ln(\cos t) \end{pmatrix}. \end{aligned}$$

It is a straightforward matter to verify that $\mathbf{X}_p$ does indeed satisfy system (20).

Step 6. The general solution to system (20) is

$$\mathbf{X} = \mathbf{X}_c + \mathbf{X}_p$$

or, in component form,

$$\begin{aligned} x &= -c_1 \sin t - c_2 \cos t + t\cos t - \sin t \ln(\cos t) \\ y &= c_1 \cos t - c_2 \sin t + t \sin t + \cos t \ln(\cos t). \end{aligned}$$

EXAMPLE 3 Solve the system

$$\frac{dx}{dt} = -3x + 2y + 2t, \qquad \frac{dy}{dt} = -3y + t + 1. \tag{22}$$

Solution. In this example we omit designation of the steps, but be certain you are able to identify each one as we move through the variation of parameters solution procedure.

To solve the associated homogeneous system we substitute the forms $x = k_1 e^{\lambda t}$ and $y = k_2 e^{\lambda t}$ to obtain

$$(-3 - \lambda)k_1 + 2k_2 = 0, \qquad (-3 - \lambda)k_2 = 0. \tag{23}$$

The characteristic equation is thus

$$(-3 - \lambda)^2 = 0,$$

which has the repeated eigenvalue $\lambda = -3, -3$. Substitution of $\lambda = -3$ into system (23) yields

$$0k_1 + 2k_2 = 0, \qquad 0k_2 = 0.$$

Thus $k_2 = 0$ and k_1 is arbitrary. Setting $k_1 = 1$ yields the eigenvector

$$\mathbf{K} = \begin{pmatrix} 1 \\ 0 \end{pmatrix}$$

and the vector solution

$$\mathbf{X}_1 = \begin{pmatrix} 1 \\ 0 \end{pmatrix} e^{-3t}.$$

To obtain a second independent solution to the homogeneous system we solve

$$0p_1 + 2p_2 = 1, \qquad 0p_2 = 0.$$

Then $p_2 = ½$ and p_1 is arbitrary. We choose $p_1 = 0$ giving the vector

$$\mathbf{P} = \begin{pmatrix} 0 \\ \frac{1}{2} \end{pmatrix}.$$

Then

$$\mathbf{X}_2 = (\mathbf{K}t + \mathbf{P})e^{-3t} = \begin{pmatrix} t \\ \frac{1}{2} \end{pmatrix} e^{-3t}.$$

A fundamental matrix is

$$\Phi = \begin{pmatrix} e^{-3t} & te^{-3t} \\ 0 & \dfrac{e^{-3t}}{2} \end{pmatrix}.$$

Next we invert Φ. The determinant of Φ is

$$\det \Phi = \frac{e^{-6t}}{2} \neq 0.$$

Hence

$$\Phi^{-1} = 2e^{6t} \begin{pmatrix} \dfrac{e^{-3t}}{2} & -te^{-3t} \\ 0 & e^{-3t} \end{pmatrix} = \begin{pmatrix} e^{3t} & -2te^{3t} \\ 0 & 2e^{3t} \end{pmatrix}.$$

Then

$$\mathbf{U}' = \Phi^{-1}\mathbf{F} = \begin{pmatrix} e^{3t} & -2te^{3t} \\ 0 & 2e^{3t} \end{pmatrix} \begin{pmatrix} 2t \\ t+1 \end{pmatrix} = \begin{pmatrix} -2t^2e^{3t} \\ 2(t+1)e^{3t} \end{pmatrix}.$$

We integrate each component of $\mathbf{U}'$ (leaving the details to you):

$$u_1 = -2\int t^2 e^{3t}\,dt = \left(-\frac{2t^2}{3} + \frac{4t}{9} - \frac{4}{27}\right)e^{3t},$$

$$u_2 = 2\int (t+1)e^{3t}\,dt = \left(\frac{2t}{3} + \frac{4}{9}\right)e^{3t}.$$

A particular solution to the nonhomogeneous system is thus

$$\mathbf{X}_p = \Phi\mathbf{U} = \begin{pmatrix} e^{-3t} & te^{-3t} \\ 0 & \dfrac{e^{-3t}}{2} \end{pmatrix} \begin{pmatrix} -\dfrac{2t^2}{3} + \dfrac{4t}{9} - \dfrac{4}{27} \\ \dfrac{2t}{3} + \dfrac{4}{9} \end{pmatrix} e^{3t} = \begin{pmatrix} \dfrac{8t}{9} - \dfrac{4}{27} \\ \dfrac{t}{3} + \dfrac{2}{9} \end{pmatrix}.$$

Therefore the general solution to system (22) is

$$\begin{aligned}\mathbf{X} &= \mathbf{X}_c + \mathbf{X}_p \\ &= c_1\mathbf{X}_1 + c_2\mathbf{X}_2 + \mathbf{X}_p \\ &= c_1\begin{pmatrix}1\\0\end{pmatrix}e^{-3t} + c_2\begin{pmatrix}t\\ \frac{1}{2}\end{pmatrix}e^{-3t} + \begin{pmatrix}\frac{8t}{9}-\frac{4}{27}\\ \frac{t}{3}+\frac{2}{9}\end{pmatrix}\end{aligned}$$

or, in component form,

$$x = c_1e^{-3t} + c_2te^{-3t} + \frac{8t}{9} - \frac{4}{27},$$

$$y = \frac{1}{2}c_2e^{-3t} + \frac{t}{3} + \frac{2}{9}.$$

Initial Value Problems

The general solution to nonhomogeneous system (1) is given by

$$\mathbf{X} = \mathbf{X}_c + \mathbf{X}_p.$$

Substituting for $\mathbf{X}_c$ and $\mathbf{X}_p$ from Eqs. (3), (15), and (16) results in the equation

$$\mathbf{X} = \Phi\mathbf{C} + \Phi\int \Phi^{-1}(t)\mathbf{F}(t)\,dt. \tag{24}$$

The general solution on an interval I: $t_0 < t < b$ can be expressed as

$$\mathbf{X}(t) = \Phi(t)\mathbf{C} + \Phi(t)\int_{t_0}^{t} \Phi^{-1}(s)\mathbf{F}(s)\,ds. \tag{25}$$

(Here s is a dummy variable of integration.) Given the initial condition

$$\mathbf{X}(t_0) = \mathbf{X}_0,$$

evaluation of Eq. (25) at $t = t_0$ leads to the relation

$$\mathbf{X}_0 = \Phi(t_0)\mathbf{C}.$$

Solving this last equation for $\mathbf{C}$ yields the vector of constants

$$\mathbf{C} = \Phi^{-1}(t_0)\mathbf{X}_0. \tag{26}$$

Equation (26) is often convenient to use when calculating $\mathbf{C}$ because Φ^{-1} is already determined in the process of finding $\mathbf{X}_p$ using Eqs. (15) and (16).

Caution If you choose to evaluate $\mathbf{C}$ using Eq. (26), then you *must* determine the vector of unknown parameters $\mathbf{U}$ by integrating from t_0 to t:

$$\mathbf{U} = \int_{t_0}^{t} \Phi^{-1}(s)\mathbf{F}(s)\,ds. \tag{27}$$

EXAMPLE 4 Solve the initial value problem

$$\begin{aligned}\frac{dx}{dt} &= 3x - 4y + 1,\\ \frac{dy}{dt} &= 2x - 3y + t,\end{aligned} \tag{28}$$

subject to the initial conditions $x(0) = 1$ and $y(0) = -1$.

Solution. This system is the same as that presented in Example 1. There we determined that

$$\Phi^{-1}(t) = \begin{pmatrix} -e^t & 2e^t \\ e^{-t} & -e^{-t} \end{pmatrix}.$$

To take advantage of Eq. (26) we need to find $\mathbf{U}$ by integrating from 0 to t:

$$\begin{aligned}\mathbf{U}(t) &= \int_0^t \Phi^{-1}(s)\mathbf{F}(s)\,ds = \int_0^t \begin{pmatrix} -e^s + 2se^s \\ e^{-s} - se^{-s} \end{pmatrix} ds\\ &= \begin{pmatrix} -e^s + 2(s-1)e^s|_0^t \\ -e^{-s} + (s+1)e^{-s}|_0^t \end{pmatrix}\\ &= \begin{pmatrix} -e^t + 2(t-1)e^t + 3 \\ -e^{-t} + (t+1)e^{-t} \end{pmatrix}.\end{aligned}$$

Then

$$\begin{aligned}\mathbf{X}_p = \Phi(t)\mathbf{U} &= \begin{pmatrix} e^{-t} & 2e^t \\ e^{-t} & e^t \end{pmatrix}\begin{pmatrix} -e^t + 2(t-1)e^t + 3 \\ -e^{-t} + (t+1)e^{-t} \end{pmatrix}\\ &= \begin{pmatrix} 3e^{-t} + 4t - 3 \\ 3e^{-t} + 3t - 3 \end{pmatrix}.\end{aligned}$$

Notice that this particular solution vector is *not the same* as the one found in Example 1. This is because in integrating $\mathbf{U}'$ from 0 to t we are finding appropriate constants of integration in the components of $\mathbf{U}$ consistent with Eq. (25).

Thus, according to Eq. (25), the general solution to system (28) is

$$\mathbf{X} = c_1\begin{pmatrix} e^{-t} \\ e^{-t} \end{pmatrix} + c_2\begin{pmatrix} 2e^t \\ e^t \end{pmatrix} + \begin{pmatrix} 3e^{-t} + 4t - 3 \\ 3e^{-t} + 3t - 3 \end{pmatrix}.$$

To evaluate the constants c_1 and c_2 so that the initial conditions $x(0) = 1$ and $y(0) = -1$ are satisfied, we can now use Eq. (26):

$$\begin{aligned}\mathbf{C} = \Phi^{-1}(0)\mathbf{X}(0) &= \begin{pmatrix} -1 & 2 \\ 1 & -1 \end{pmatrix}\begin{pmatrix} 1 \\ -1 \end{pmatrix}\\ &= \begin{pmatrix} -3 \\ 2 \end{pmatrix}.\end{aligned}$$

In component form the solution to the initial value problem is

$$x = -3e^{-t} + 4e^{t} + 3e^{-t} + 4t - 3,$$
$$y = -3e^{-t} + 2e^{t} + 3e^{-t} + 3t - 3,$$

or

$$x = 4e^{t} + 4t - 3,$$
$$y = 2e^{t} + 3t - 3.$$

You should verify that x and y do indeed satisfy system (28) subject to the initial conditions.

Remark The arbitrary constants c_1 and c_2 can be computed directly from the general solution to the system by substituting the initial condition values into the solution. However, Eq. (26) for $\mathbf{C}$ produces the solution formula

$$\mathbf{X}(t) = \Phi(t)\Phi^{-1}(t_0)\mathbf{X}_0 + \Phi(t)\int_{t_0}^{t} \Phi^{-1}(s)\mathbf{F}(s)\,ds. \tag{29}$$

The solution form on the righthand side of Eq. (29) is useful for theoretical purposes as well.

A Simplification for Finding the Eigenvalues

In determining the complementary solution $\mathbf{X}_c$ to solve the associated homogeneous system

$$\mathbf{X}' = \mathbf{A}\mathbf{X}, \tag{30}$$

we have substituted $x = k_1e^{\lambda t}$ and $y = k_2e^{\lambda t}$ into the system. This substitution *always* produces the following system of algebraic equations:

$$(\mathbf{A} - \lambda\mathbf{I})\mathbf{K} = \mathbf{0} \tag{31}$$

or

$$\begin{pmatrix} a-\lambda & b \\ c & d-\lambda \end{pmatrix}\begin{pmatrix} k_1 \\ k_2 \end{pmatrix} = \begin{pmatrix} 0 \\ 0 \end{pmatrix}. \tag{32}$$

Then a nontrivial solution for k_1 and k_2 exists if and only if

$$\begin{vmatrix} a-\lambda & b \\ c & d-\lambda \end{vmatrix} = 0, \tag{33}$$

or

$$\det(\mathbf{A} - \lambda\mathbf{I}) = 0. \tag{34}$$

The symbol **I** denotes the 2 × 2 identity matrix

$$\mathbf{I} = \begin{pmatrix} 1 & 0 \\ 0 & 1 \end{pmatrix}.$$

Formulas (33) and (34) are two expressions for the characteristic equation whose solution provides the eigenvalues of the system. Once an eigenvalue is determined, Eq. (32) can be used to solve for the components k_1 and k_2 of the associated eigenvector. This procedure is the one we have used in previous sections. However, we can write Eqs. (32) and (33) directly from original system (30) without first having to make the substitutions $x = k_1e^{\lambda t}$ and $y = k_2e^{\lambda t}$ in order to derive those equations.

EXERCISES 8.2

In problems 1–20 use the variation-of-parameters method to solve the given nonhomogeneous system.

1. $\dfrac{dx}{dt} = x + 3y$, $\quad \dfrac{dy}{dt} = x - y + 2e^t$

2. $\dfrac{dx}{dt} = 2x + 1$, $\quad \dfrac{dy}{dt} = 3x - 2y$

3. $\mathbf{X}' = \begin{pmatrix} 0 & -1 \\ 1 & 0 \end{pmatrix}\mathbf{X} + \begin{pmatrix} 1 \\ e^t \end{pmatrix}$

4. $\mathbf{X}' = \begin{pmatrix} 0 & 2 \\ 2 & 0 \end{pmatrix}\mathbf{X} + \begin{pmatrix} e^{-t} \\ -2 \end{pmatrix}$

5. $\mathbf{X}' = \begin{pmatrix} 1 & 2 \\ 4 & -1 \end{pmatrix}\mathbf{X} + \begin{pmatrix} e^{2t} \\ -e^{2t} \end{pmatrix}$

6. $\mathbf{X}' = \begin{pmatrix} 1 & 0 \\ 2 & 3 \end{pmatrix}\mathbf{X} + \begin{pmatrix} -1 \\ t \end{pmatrix}$

7. $\mathbf{X}' = \begin{pmatrix} 0 & -1 \\ -1 & 0 \end{pmatrix}\mathbf{X} + \begin{pmatrix} 2e^{3t} \\ e^{3t} - e^{-3t} \end{pmatrix}$

8. $\mathbf{X}' = \begin{pmatrix} 0 & -1 \\ 1 & 0 \end{pmatrix}\mathbf{X} + \begin{pmatrix} e^t \\ -e^{-t} \end{pmatrix}$

9. $\mathbf{X}' = \begin{pmatrix} 0 & -1 \\ 1 & 0 \end{pmatrix}\mathbf{X} + \begin{pmatrix} \cos t \\ -\sin t \end{pmatrix}$

10. $\mathbf{X}' = \begin{pmatrix} -4 & 1 \\ -2 & -1 \end{pmatrix}\mathbf{X} + \begin{pmatrix} t \\ 1 \end{pmatrix}$

11. $\mathbf{X}' = \begin{pmatrix} 2 & 3 \\ -3 & 2 \end{pmatrix}\mathbf{X} + \begin{pmatrix} -1 \\ t \end{pmatrix} e^{2t}$

12. $\mathbf{X}' = \begin{pmatrix} 1 & -1 \\ 1 & 1 \end{pmatrix}\mathbf{X} + \begin{pmatrix} te^t \\ e^t \end{pmatrix}$

13. $\mathbf{X}' = \begin{pmatrix} 1 & 1 \\ -1 & 1 \end{pmatrix}\mathbf{X} + \begin{pmatrix} t \\ \cos t \end{pmatrix} e^t$

14. $\mathbf{X}' = \begin{pmatrix} 3 & -1 \\ 9 & -3 \end{pmatrix}\mathbf{X} + \begin{pmatrix} e^{-t} \\ \sin t \end{pmatrix}$

15. $\mathbf{X}' = \begin{pmatrix} -1 & 3 \\ -3 & 5 \end{pmatrix}\mathbf{X} + \begin{pmatrix} 3t - 2 \\ 3t - 2 \end{pmatrix} e^t$

16. $\mathbf{X}' = \begin{pmatrix} 3 & 0 \\ -1 & 3 \end{pmatrix}\mathbf{X} + \begin{pmatrix} 1 \\ t \end{pmatrix} e^{2t}$

17. $\mathbf{X}' = \begin{pmatrix} -1 & 1 \\ 0 & -1 \end{pmatrix}\mathbf{X} + \begin{pmatrix} \cos t \\ t \end{pmatrix} e^{-t}$

18. $\mathbf{X}' = \begin{pmatrix} 3 & -18 \\ 2 & -9 \end{pmatrix}\mathbf{X} + \begin{pmatrix} 2e^{-2t} \\ e^{-2t} \end{pmatrix}$

19. $\mathbf{X}' = \begin{pmatrix} 3 & 1 \\ -1 & 1 \end{pmatrix}\mathbf{X} + \begin{pmatrix} 3t \\ -2 \end{pmatrix} e^{2t}$

20. $\mathbf{X}' = \begin{pmatrix} -3 & -2 \\ -4 & -1 \end{pmatrix}\mathbf{X} + \begin{pmatrix} \sin t \\ 0 \end{pmatrix}$

21. Suppose a single-story home with attic and living area is being heated, as described in Section 7.1, Example 3. Assume the temperatures are the same as those given in that example, and that the time constants are $1/k_1 = 5$, $1/k_2 = 10$, and $1/k_3 = 5$. Determine the time when the temperature $x(t)$ in the living area reaches 68°F. Assume initially that $x(0) = y(0) = 35$°F.

22. For the electrical network displayed in Fig. 8.1, assume that $R_1 = 1$ ohm, $R_2 = 4$ ohms, $L_1 = 0.01$ henry, and $L_2 = 0.0125$ henry. The induced voltage is $E(t) = 50$ volts. Determine the currents i_1, i_2, and i_3 at any time $t > 0$ assuming they are initially all zero.

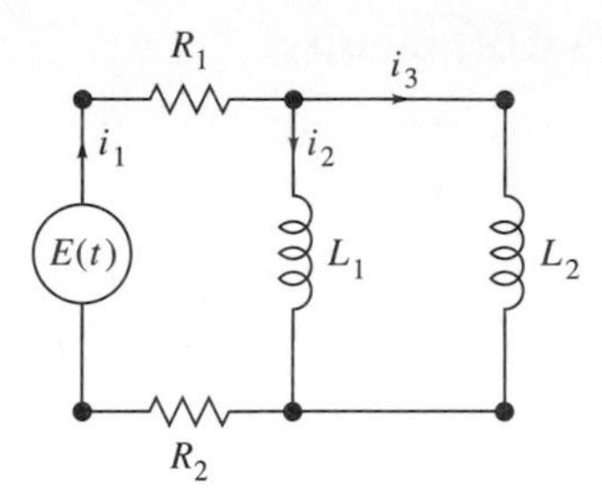

FIGURE 8.1

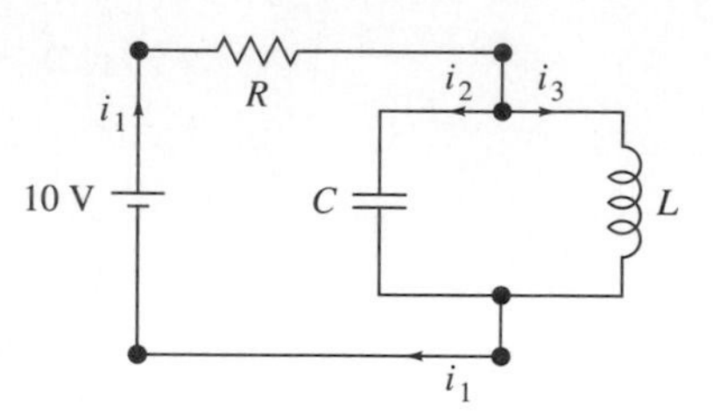

FIGURE 8.2

23. For the electrical network displayed in Fig. 8.2, assume that $R = 10$ ohms, $C = 1.6 \times 10^{-4}$ farads, $L = 0.1$ henry, and $E(t) = 30$ volts is the induced voltage. If initially the currents are zero, find the currents i_1, i_2, and i_3 at any time $t > 0$.

24. Solve problem 23 assuming the same values and initial conditions but with an induced voltage of $E(t) = e^{-10t}$ (in volts).

8.3 HIGHER-DIMENSIONAL LINEAR SYSTEMS

The theory for higher-dimensional linear systems is the same as that for 2×2 systems. The $n \times n$ constant-coefficient linear first-order system has the form

$$\begin{aligned} \frac{dx_1}{dt} &= a_{11}x_1 + a_{12}x_2 + \cdots + a_{1n}x_n + f_1(t), \\ \frac{dx_2}{dt} &= a_{21}x_1 + a_{22}x_2 + \cdots + a_{2n}x_n + f_2(t), \\ &\vdots \\ \frac{dx_n}{dt} &= a_{n1}x_1 + a_{n2}x_2 + \cdots + a_{nn}x_n + f_n(t). \end{aligned} \tag{1}$$

Here it is assumed that all the a_{ij} coefficients are constants and that each function $f_i(t)$ is continuous over some interval I. In matrix form the system is expressible as

$$\mathbf{X}' = \mathbf{AX} + \mathbf{F}(t). \tag{2}$$

If the components $f_i(t)$ of $\mathbf{F}(t)$ are identically zero over I, then system (2) is *homogeneous;* otherwise it is *nonhomogeneous.*

As was the case for 2×2 systems, the general solution to nonhomogeneous system (2) takes the form

$$\mathbf{X} = \mathbf{X}_c + \mathbf{X}_p, \tag{3}$$

where $\mathbf{X}_c$ is the complementary solution solving the associated homogeneous system

$$\mathbf{X}' = \mathbf{A}\mathbf{X}, \tag{4}$$

and $\mathbf{X}_p$ is any vector solution to nonhomogeneous system (2). We need first to focus attention on finding the general solution to Eq. (4).

Homogeneous Systems

Based on Eq. (4), we assume as before that a solution to Eq. (4) takes the form

$$\mathbf{X} = \mathbf{K}e^{\lambda t} \tag{5}$$

where λ is a constant and

$$\mathbf{K} = \begin{pmatrix} k_1 \\ k_2 \\ \vdots \\ k_n \end{pmatrix}.$$

We require that Eq. (5) solve system (4). Thus differentiation of Eq. (5) gives us

$$\mathbf{X}' = \lambda\mathbf{K}e^{\lambda t}. \tag{6}$$

Then substitution of Eqs. (5) and (6) into system (4) results in the relation

$$\lambda\mathbf{K}e^{\lambda t} = \mathbf{A}\mathbf{K}e^{\lambda t}.$$

Since $e^{\lambda t} \neq 0$, this last equation can be simplified to

$$(\mathbf{A} - \lambda\mathbf{I})\mathbf{K} = \mathbf{0} \tag{7}$$

where $\mathbf{I}$ is the $n \times n$ identity matrix. We are interested only in nontrivial (that is, nonzero) solutions to system (4). Thus a nonzero vector $\mathbf{K}$ satisfies algebraic system (7) if and only if the determinant of the coefficients is zero:

$$\det(\mathbf{A} - \lambda\mathbf{I}) = 0. \tag{8}$$

Equation (8) is the characteristic equation of system (4). In component form the characteristic equation is

$$\det\begin{pmatrix} a_{11} - \lambda & a_{12} & \cdots & a_{1n} \\ a_{21} & a_{22} - \lambda & \cdots & a_{2n} \\ \vdots & \vdots & \vdots & \vdots \\ a_{n1} & a_{n2} & \cdots & a_{nn} - \lambda \end{pmatrix} = 0. \tag{9}$$

The characteristic equation is a polynomial of degree n in the unknown λ. This polynomial has exactly n complex roots (some of which may be real). Once an eigenvalue or root to the characteristic equation is known, it can be substituted into algebraic system (7) to find the associated eigenvectors **K**. In scalar form this algebraic system is

$$\begin{aligned}(a_{11}-\lambda)k_1 + \quad a_{12}k_2 + \cdots + \quad a_{1n}k_n &= 0,\\ a_{21}k_1 + (a_{22}-\lambda)k_2 + \cdots + \quad a_{2n}k_n &= 0,\\ &\vdots\\ a_{n1}k_1 + \quad a_{n2}k_2 + \cdots + (a_{nn}-\lambda)k_n &= 0.\end{aligned} \tag{10}$$

As was the case for 2×2 linear systems, the roots of the characteristic polynomial (9) can be categorized as follows:

1. Real and distinct.
2. Complex (occurring in conjugate pairs).
3. Real and repeated.

While the theory does hold for arbitrary $n \times n$ systems, we will confine our attention in the examples to solving 3×3 systems. The nature of the solutions depends on the roots.

Distinct Real Eigenvalues

If the eigenvalues λ_i are real and distinct, then for each λ_i we find a corresponding eigenvector $\mathbf{K}_i$. This procedure yields a fundamental set of solutions

$$\{\mathbf{K}_1 e^{\lambda_1 t}, \mathbf{K}_2 e^{\lambda_2 t}, \ldots, \mathbf{K}_n e^{\lambda_n t}\}.$$

Thus the general solution to system (4) on the interval $-\infty < t < \infty$ is

$$\mathbf{X} = c_1 \mathbf{K}_1 e^{\lambda_1 t} + c_2 \mathbf{K}_2 e^{\lambda_2 t} + \cdots + c_n \mathbf{K}_n e^{\lambda_n t}. \tag{11}$$

Following is an example illustrating this case.

EXAMPLE 1 Solve the system

$$\mathbf{X}' = \begin{pmatrix} 4 & 0 & 1 \\ -2 & 1 & 0 \\ -2 & 0 & 1 \end{pmatrix} \mathbf{X}.$$

Solution. The characteristic equation is given by

$$\det \begin{pmatrix} 4-\lambda & 0 & 1 \\ -2 & 1-\lambda & 0 \\ -2 & 0 & 1-\lambda \end{pmatrix} = 0$$

or

$$(4-\lambda)(1-\lambda)^2 + 2(1-\lambda) = 0.$$

Thus

$$(1-\lambda)(\lambda^2-5\lambda+6)=0,$$
$$(1-\lambda)(\lambda-2)(\lambda-3)=0.$$

The eigenvalues are $\lambda = 1, 2, 3$.

For each value of λ we find an associated eigenvector $\mathbf{K}$ by solving the system

$$\begin{aligned}(4-\lambda)k_1 \quad\quad\quad + \quad\quad k_3 &= 0,\\ -2k_1+(1-\lambda)k_2 \quad\quad\quad\quad &= 0,\\ -2k_1 \quad\quad\quad +(1-\lambda)k_3 &= 0.\end{aligned}$$

For $\lambda = 1$ the system $(\mathbf{A}-\lambda\mathbf{I})\mathbf{K}=\mathbf{0}$ becomes

$$\begin{aligned}3k_1 \quad\quad + \; k_3 &= 0,\\ -2k_1+0k_2 \quad\quad &= 0,\\ -2k_1 \quad\quad +0k_3 &= 0.\end{aligned}$$

Thus $k_1 = k_3 = 0$ and k_2 is arbitrary. We choose $k_2 = 1$ and obtain the eigenvector

$$\mathbf{K}_1 = \begin{pmatrix}0\\1\\0\end{pmatrix},$$

which yields the solution

$$\mathbf{X}_1 = \begin{pmatrix}0\\1\\0\end{pmatrix} e^t.$$

For $\lambda = 2$ the system $(\mathbf{A}-\lambda\mathbf{I})\mathbf{K}=\mathbf{0}$ becomes

$$\begin{aligned}2k_1 \quad\quad +k_3 &= 0,\\ -2k_1-k_2 \quad\quad &= 0,\\ -2k_1 \quad\quad -k_3 &= 0.\end{aligned}$$

Then $k_3 = k_2 = -2k_1$ and k_1 is arbitrary. We choose $k_1 = 1$ to obtain the eigenvector

$$\mathbf{K}_3 = \begin{pmatrix}1\\-2\\-2\end{pmatrix},$$

which yields the solution

$$\mathbf{X}_2 = \begin{pmatrix}1\\-2\\-2\end{pmatrix} e^{2t}.$$

For $\lambda = 3$ the system $(\mathbf{A} - \lambda\mathbf{I})\mathbf{K} = \mathbf{0}$ becomes

$$\begin{aligned} k_1 \qquad\quad + k_3 &= 0, \\ -2k_1 - 2k_2 \qquad &= 0, \\ -2k_1 \qquad - 2k_3 &= 0. \end{aligned}$$

Then $k_3 = k_2 = -k_1$ and k_1 is arbitrary. We choose $k_1 = 1$ to yield the eigenvector

$$\mathbf{K}_3 = \begin{pmatrix} 1 \\ -1 \\ -1 \end{pmatrix},$$

which yields the solution

$$\mathbf{X}_3 = \begin{pmatrix} 1 \\ -1 \\ -1 \end{pmatrix} e^{3t}.$$

The general solution is thus

$$\mathbf{X} = c_1 \begin{pmatrix} 0 \\ 1 \\ 0 \end{pmatrix} e^{t} + c_2 \begin{pmatrix} 1 \\ -2 \\ -2 \end{pmatrix} e^{2t} + c_3 \begin{pmatrix} 1 \\ -1 \\ -1 \end{pmatrix} e^{3t},$$

which in component form is expressed as

$$\begin{aligned} x_1 &= c_2 e^{2t} + c_3 e^{3t}, \\ x_2 &= c_1 e^{t} - 2c_2 e^{2t} - c_3 e^{3t}, \\ x_3 &= -2c_2 e^{2t} - c_3 e^{3t}. \end{aligned}$$

Complex Eigenvalues

Solutions for complex eigenvalues are written as real solutions using the ideas presented in Section 7.4 for complex roots. In the case of a 3×3 system there will be exactly two complex roots: $\lambda_1 = \alpha + \beta i$ and $\lambda_2 = \alpha - \beta i$. Since complex roots always occur in conjugate pairs, the third root λ_3 must be real. Therefore the general solution has the form

$$\mathbf{X} = e^{\alpha t}(c_1 \mathbf{B}_1 \cos \beta t + c_2 \mathbf{B}_2 \sin \beta t) + c_3 \mathbf{K}_3\, e^{\lambda_3 t}. \tag{12}$$

EXAMPLE 2 Solve the system

$$\mathbf{X}' = \begin{pmatrix} 5 & 1 & -1 \\ -2 & 3 & 0 \\ 0 & 0 & 3 \end{pmatrix} \mathbf{X}.$$

Solution. The characteristic equation gives the eigenvalues:

$$\det(\mathbf{A} - \lambda\mathbf{I}) = \det\begin{pmatrix} 5-\lambda & 1 & -1 \\ -2 & 3-\lambda & 0 \\ 0 & 0 & 3-\lambda \end{pmatrix} = 0$$

or

$$\begin{aligned} (3-\lambda)[(5-\lambda)(3-\lambda)+2] &= 0, \\ (3-\lambda)(\lambda^2 - 8\lambda + 17) &= 0. \end{aligned}$$

The eigenvalues are $\lambda = 3$ and $\lambda = 4 \pm i$. For each value of λ we find an associated eigenvector by solving the algebraic system

$$\begin{aligned} (5-\lambda)k_1 + k_2 - k_3 &= 0, \\ -2k_1 + (3-\lambda)k_2 &= 0, \\ (3-\lambda)k_3 &= 0. \end{aligned}$$

For $\lambda = 3$ the system $(\mathbf{A} - \lambda\mathbf{I})\mathbf{K} = \mathbf{0}$ becomes

$$\begin{aligned} 2k_1 + k_2 - k_3 &= 0, \\ -2k_1 + 0k_2 &= 0, \\ 0k_3 &= 0. \end{aligned}$$

Then $k_1 = 0$ and $k_2 = k_3$ where k_3 is arbitrary. We choose $k_3 = 1$ to obtain the eigenvector

$$\mathbf{K}_1 = \begin{pmatrix} 0 \\ 1 \\ 1 \end{pmatrix},$$

which results in the vector solution

$$\mathbf{X}_1 = \begin{pmatrix} 0 \\ 1 \\ 1 \end{pmatrix} e^{3t}.$$

For $\lambda = 4 + i$ the system $(\mathbf{A} - \lambda\mathbf{I})\mathbf{K} = \mathbf{0}$ becomes

$$\begin{aligned} (1-i)k_1 + k_2 - k_3 &= 0, \\ -2k_1 + (-1-i)k_2 &= 0, \\ (-1-i)k_3 &= 0. \end{aligned}$$

Thus $k_3 = 0$ and $k_2 = (-1 + i)k_1$ where k_1 is arbitrary. We choose $k_1 = 1$ to obtain the eigenvector

$$\mathbf{K} = \begin{pmatrix} 1 \\ -1+i \\ 0 \end{pmatrix}.$$

Using the procedure for complex eigenvectors developed in Section 7.4, we find

$$\mathbf{B}_1 = \begin{pmatrix} 1 \\ -1 \\ 0 \end{pmatrix} \quad \text{and} \quad \mathbf{B}_2 = \begin{pmatrix} 0 \\ -1 \\ 0 \end{pmatrix}.$$

From our discussion concerning complex eigenvalues in Section 7.4, the two real vectors $\mathbf{B}_1$ and $\mathbf{B}_2$ account for both the eigenvalue $\lambda = 4 + i$ and its complex conjugate $\lambda^* = 4 - i$. Thus the real solution vectors are

$$\mathbf{X}_2 = e^{4t}\left[\begin{pmatrix} 1 \\ -1 \\ 0 \end{pmatrix} \cos t + \begin{pmatrix} 0 \\ -1 \\ 0 \end{pmatrix} \sin t\right],$$

$$\mathbf{X}_3 = e^{4t}\left[\begin{pmatrix} 0 \\ -1 \\ 0 \end{pmatrix} \cos t - \begin{pmatrix} 1 \\ -1 \\ 0 \end{pmatrix} \sin t\right].$$

The general solution is

$$\mathbf{X} = c_1\mathbf{X}_1 + c_2\mathbf{X}_2 + c_3\mathbf{X}_3,$$

which in component form is

$$x_1 = (c_2 \cos t - c_3 \sin t)e^{4t},$$
$$x_2 = c_1 e^{3t} + [(-c_2 - c_3)\cos t + (-c_2 + c_3)\sin t]e^{4t},$$
$$x_3 = c_1 e^{3t}.$$

Repeated Eigenvalues

Finding the general solution for repeated eigenvalues for higher-dimensional systems is more complicated than for 2×2 systems. The form of the solution depends on the number of independent eigenvectors that can be determined for the repeated eigenvalue. We will restrict our attention to 3×3 systems and treat only those cases that do not require advanced topics in linear algebra. First we will discuss the cases, and then we will present several examples illustrating them.

Case 1: An Eigenvalue Repeated Twice

For a 3×3 homogeneous system, assume the three eigenvalues are λ_1, λ_1, and λ_2, where $\lambda_1 \neq \lambda_2$ and λ_1 is repeated twice. There are two possibilities that can occur for the repeated eigenvalue.

a) Two Linearly Independent Eigenvectors For the repeated eigenvalue λ_1 assume there are two independent eigenvectors $\mathbf{K}_1$ and $\mathbf{K}_2$ solving the system

$$(\mathbf{A} - \lambda_1 \mathbf{I})\mathbf{K} = \mathbf{0}.$$

Then there are two linearly independent solutions:

$$\mathbf{X}_1 = \mathbf{K}_1 e^{\lambda_1 t}, \tag{13}$$

$$\mathbf{X}_2 = \mathbf{K}_2 e^{\lambda_1 t}. \tag{14}$$

The third independent solution is obtained from the nonrepeated eigenvalue and associated eigenvector:

$$\mathbf{X}_3 = \mathbf{K}_3 e^{\lambda_2 t}. \tag{15}$$

b) One Independent Eigenvector For the repeated eigenvalue, assume $\mathbf{K}_1$ denotes the single independent eigenvector. One solution is

$$\mathbf{X}_1 = \mathbf{K}_1 e^{\lambda_1 t}. \tag{16}$$

Since $\mathbf{K}_2$ does not exist, find a second vector $\mathbf{P}$, as in the 2×2 case, by solving the system

$$(\mathbf{A} - \lambda_1 \mathbf{I})\mathbf{P} = \mathbf{K}_1. \tag{17}$$

A second independent solution is then

$$\mathbf{X}_2 = (\mathbf{K}_1 t + \mathbf{P}) e^{\lambda_1 t}. \tag{18}$$

The vector $\mathbf{P}$ in Eq. (18) is *not* an eigenvector but is sometimes called a **generalized eigenvector.** The third independent solution comes from the nonrepeated eigenvalue λ_2 and associated eigenvector $\mathbf{K}_3$:

$$\mathbf{X}_3 = \mathbf{K}_3 e^{\lambda_2 t}. \tag{19}$$

Case 2: An Eigenvalue Repeated Thrice

Assume that λ represents an eigenvalue repeated three times as a root to the characteristic polynomial. The nature of the three linearly independent solutions to the 3×3 homogeneous system depends on the number of linearly independent eigenvectors that can be obtained for λ. We discuss the three possibilities.

a) Three Linearly Independent Eigenvectors Suppose there are three linearly independent eigenvectors $\mathbf{K}_1$, $\mathbf{K}_2$, and $\mathbf{K}_3$ solving the equation

$$(\mathbf{A} - \lambda \mathbf{I})\mathbf{K} = \mathbf{0}.$$

These eigenvectors give three independent solutions:

$$\mathbf{X}_1 = \mathbf{K}_1 e^{\lambda t}, \tag{20}$$

$$\mathbf{X}_2 = \mathbf{K}_2 e^{\lambda t}, \tag{21}$$

$$\mathbf{X}_3 = \mathbf{K}_3 e^{\lambda t}. \tag{22}$$

In this case the matrix $\mathbf{A}$ is diagonal, with λ appearing along the main diagonal and zeros elsewhere.

b) Two Linearly Independent Eigenvectors Adequate treatment of this case is not possible without a knowledge of advanced topics in linear algebra that are beyond the scope of this text, so we omit its presentation here.

c) One Independent Eigenvector Suppose there is only one independent eigenvector $\mathbf{K}_1$. One solution is

$$\mathbf{X}_1 = \mathbf{K}_1 e^{\lambda t}. \tag{23}$$

Since $\mathbf{K}_2$ and $\mathbf{K}_3$ do not exist, find two additional vectors $\mathbf{P}$ and $\mathbf{Q}$ by solving the following algebraic systems in order:

$$(\mathbf{A} - \lambda\mathbf{I})\mathbf{P} = \mathbf{K}_1, \tag{24}$$

$$(\mathbf{A} - \lambda\mathbf{I})\mathbf{Q} = \mathbf{P}. \tag{25}$$

These vectors yield two linearly independent solutions:

$$\mathbf{X}_2 = (\mathbf{K}_1 t + \mathbf{P})e^{\lambda t}, \tag{26}$$

$$\mathbf{X}_3 = \left(\frac{1}{2}\mathbf{K}_1 t^2 + \mathbf{P}t + \mathbf{Q}\right)e^{\lambda t}. \tag{27}$$

Again, the vectors $\mathbf{P}$ and $\mathbf{Q}$ are not true eigenvectors but are often called generalized eigenvectors.

We now present several examples to illustrate the various cases.

EXAMPLE 3 Solve the system

$$\mathbf{X}' = \begin{pmatrix} -3 & 0 & -3 \\ 1 & -2 & 3 \\ 1 & 0 & 1 \end{pmatrix} \mathbf{X}.$$

Solution. The characteristic equation is

$$\det \begin{pmatrix} -3-\lambda & 0 & -3 \\ 1 & -2-\lambda & 3 \\ 1 & 0 & 1-\lambda \end{pmatrix} = 0$$

or

$$\lambda^3 + 4\lambda^2 + 4\lambda = \lambda(\lambda + 2)^2 = 0.$$

The eigenvalues are $\lambda = -2, -2, 0$. From these we find the associated eigenvectors.

For $\lambda = -2$ the system $(\mathbf{A} - \lambda\mathbf{I})\mathbf{K} = \mathbf{0}$ becomes

$$\begin{aligned} -k_1 + 0k_2 - 3k_3 &= 0, \\ k_1 + 0k_2 + 3k_3 &= 0, \\ k_1 + 0k_2 + 3k_3 &= 0. \end{aligned}$$

Thus $k_1 = -3k_3$ where k_2 and k_3 are both arbitrary. Choosing $k_2 = 1$ and $k_3 = 0$ yields one eigenvector,

$$\mathbf{K}_1 = \begin{pmatrix} 0 \\ 1 \\ 0 \end{pmatrix}.$$

The choice $k_2 = 0$ and $k_3 = 1$ yields a second eigenvector,

$$\mathbf{K}_2 = \begin{pmatrix} -3 \\ 0 \\ 1 \end{pmatrix}.$$

Thus

$$\mathbf{X}_1 = \begin{pmatrix} 0 \\ 1 \\ 0 \end{pmatrix} e^{-2t} \quad \text{and} \quad \mathbf{X}_2 = \begin{pmatrix} -3 \\ 0 \\ 1 \end{pmatrix} e^{-2t}$$

are linearly independent solutions.

For $\lambda = 0$ the system $(\mathbf{A} - \lambda\mathbf{I})\mathbf{K} = \mathbf{0}$ becomes

$$\begin{aligned} -3k_1 \qquad - 3k_3 &= 0, \\ k_1 - 2k_2 + 3k_3 &= 0, \\ k_1 \qquad + \ k_3 &= 0. \end{aligned}$$

Thus $k_1 = -k_3$ and $k_2 = k_3$ where k_3 is arbitrary. Choosing $k_3 = 1$ yields the eigenvector

$$\mathbf{K}_3 = \begin{pmatrix} -1 \\ 1 \\ 1 \end{pmatrix}.$$

Thus

$$\mathbf{X}_3 = \begin{pmatrix} -1 \\ 1 \\ 1 \end{pmatrix} e^{0t}$$

is a third linearly independent solution. The general solution is

$$\mathbf{X} = c_1 \begin{pmatrix} 0 \\ 1 \\ 0 \end{pmatrix} e^{-2t} + c_2 \begin{pmatrix} -3 \\ 0 \\ 1 \end{pmatrix} e^{-2t} + c_3 \begin{pmatrix} -1 \\ 1 \\ 1 \end{pmatrix},$$

which in component form is

$$x_1 = -3c_2e^{-2t} - c_3,$$
$$x_2 = c_1e^{-2t} + c_3,$$
$$x_3 = c_2e^{-2t} + c_3.$$

EXAMPLE 4 Solve the system

$$\mathbf{X}' = \begin{pmatrix} 1 & 1 & 1 \\ 2 & 1 & -1 \\ 0 & -1 & 1 \end{pmatrix} \mathbf{X}.$$

Solution. The characteristic equation is

$$\det\begin{pmatrix} 1-\lambda & 1 & 1 \\ 2 & 1-\lambda & -1 \\ 0 & -1 & 1-\lambda \end{pmatrix} = 0$$

or

$$\lambda^3 - 3\lambda^2 + 4 = (\lambda - 2)^2(\lambda + 1) = 0.$$

The eigenvalues are $\lambda = 2, 2, -1$.

For $\lambda = 2$ the system $(\mathbf{A} - \lambda\mathbf{I})\mathbf{K} = \mathbf{0}$ becomes

$$-k_1 + k_2 + k_3 = 0,$$
$$2k_1 - k_2 - k_3 = 0,$$
$$-k_2 - k_3 = 0.$$

The solution to this algebraic system is $k_1 = 0$ and $k_2 = -k_3$ where k_3 is arbitrary. Choosing $k_3 = 1$ yields the eigenvector

$$\mathbf{K}_1 = \begin{pmatrix} 0 \\ -1 \\ 1 \end{pmatrix}.$$

Since there is no second independent eigenvector, we determine the generalized eigenvector $\mathbf{P}$:

$$-p_1 + p_2 + p_3 = 0,$$
$$2p_1 - p_2 - p_3 = -1,$$
$$-p_2 - p_3 = 1.$$

The solution is $p_1 = -1$ and $p_2 = -1 - p_3$ where p_3 is arbitrary. Choosing $p_3 = 0$ yields the vector

$$\mathbf{P} = \begin{pmatrix} -1 \\ -1 \\ 0 \end{pmatrix}.$$

The two linearly independent solutions are

$$\mathbf{X}_1 = \begin{pmatrix} 0 \\ -1 \\ 1 \end{pmatrix} e^{2t} \quad \text{and} \quad \mathbf{X}_2 = \left[\begin{pmatrix} 0 \\ 1 \\ 1 \end{pmatrix} t + \begin{pmatrix} -1 \\ -1 \\ 0 \end{pmatrix}\right] e^{2t}.$$

For $\lambda = -1$ the system $(\mathbf{A} - \lambda\mathbf{I})\mathbf{K} = \mathbf{0}$ becomes

$$\begin{aligned} 2k_1 + \quad k_2 + \quad k_3 &= 0, \\ 2k_1 + 2k_2 - \quad k_3 &= 0, \\ -k_2 + 2k_3 &= 0. \end{aligned}$$

The solution to this algebraic system is $k_1 = -\tfrac{3}{2}k_3$ and $k_2 = 2k_3$ where k_3 is arbitrary. Choosing $k_3 = 2$ to avoid fractions yields the eigenvector

$$\mathbf{K}_3 = \begin{pmatrix} -3 \\ 4 \\ 2 \end{pmatrix}.$$

A third linearly independent solution is

$$\mathbf{X}_3 = \begin{pmatrix} -3 \\ 4 \\ 2 \end{pmatrix} e^{-t}.$$

The general solution is

$$\mathbf{X} = c_1 \begin{pmatrix} 0 \\ -1 \\ 1 \end{pmatrix} e^{2t} + c_2 \left[\begin{pmatrix} 0 \\ -1 \\ 1 \end{pmatrix} t + \begin{pmatrix} -1 \\ -1 \\ 0 \end{pmatrix}\right] e^{2t} + c_3 \begin{pmatrix} -3 \\ 4 \\ 2 \end{pmatrix} e^{-t}.$$

EXAMPLE 5 Solve the system

$$\mathbf{X}' = \begin{pmatrix} 1 & 0 & 0 \\ -1 & 2 & -1 \\ 0 & 1 & 0 \end{pmatrix} \mathbf{X}.$$

Solution. The characteristic equation is

$$\det \begin{pmatrix} 1-\lambda & 0 & 0 \\ -1 & 2-\lambda & -1 \\ 0 & 1 & -\lambda \end{pmatrix} = 0$$

or

$$(\lambda - 1)^3 = 0.$$

The eigenvalues are $\lambda = 1, 1, 1$.

For $\lambda = 1$ the system $(\mathbf{A} - \lambda\mathbf{I})\mathbf{K} = \mathbf{0}$ becomes

$$\begin{aligned} 0k_1 + 0k_2 + 0k_3 &= 0, \\ -k_1 + k_2 - k_3 &= 0, \\ k_2 - k_3 &= 0. \end{aligned}$$

The solution is $k_1 = 0$ and $k_2 = k_3$ where k_3 is arbitrary. Choosing $k_3 = 1$ yields the eigenvector

$$\mathbf{K}_1 = \begin{pmatrix} 0 \\ 1 \\ 1 \end{pmatrix}.$$

One solution vector is

$$\mathbf{X}_1 = \begin{pmatrix} 0 \\ 1 \\ 1 \end{pmatrix} e^t.$$

Since no other linearly independent eigenvectors exist, we determine the generalized eigenvectors $\mathbf{P}$ and $\mathbf{Q}$ in order:

$$\begin{aligned} 0p_1 + 0p_2 + 0p_3 &= 0, \\ -p_1 + p_2 - p_3 &= 1, \\ p_2 - p_3 &= 1. \end{aligned}$$

The solution is $p_1 = 0$ and $p_2 = 1 + p_3$ where p_3 is arbitrary. Choosing $p_3 = 0$ gives us the vector

$$\mathbf{P} = \begin{pmatrix} 0 \\ 1 \\ 0 \end{pmatrix}.$$

Next we find $\mathbf{Q}$:

$$\begin{aligned} 0q_1 + 0q_2 + 0q_3 &= 0, \\ -q_1 + q_2 - q_3 &= 1, \\ q_2 - q_3 &= 0. \end{aligned}$$

The solution is $q_1 = -1$ and $q_2 = q_3$ where q_3 is arbitrary. We choose $q_3 = 0$ for the vector

$$\mathbf{Q} = \begin{pmatrix} -1 \\ 0 \\ 0 \end{pmatrix}.$$

According to Eqs. (26) and (27), these generalized eigenvectors yield two additional linearly independent solutions:

$$\mathbf{X}_2 = \left[\begin{pmatrix} 0 \\ 1 \\ 1 \end{pmatrix} t + \begin{pmatrix} 0 \\ 1 \\ 0 \end{pmatrix} \right] e^t$$

and

$$\mathbf{X}_3 = \left[\begin{pmatrix} 0 \\ 1 \\ 1 \end{pmatrix} \frac{t^2}{2} + \begin{pmatrix} 0 \\ 1 \\ 0 \end{pmatrix} t + \begin{pmatrix} -1 \\ 0 \\ 0 \end{pmatrix}\right] e^t.$$

The general solution is

$$\mathbf{X} = c_1\mathbf{X}_1 + c_2\mathbf{X}_2 + c_3\mathbf{X}_3,$$

which in component form is

$$\begin{aligned} x_1 &= -c_3 e^t, \\ x_2 &= \left[c_1 + c_2(t+1) + c_3\left(\frac{1}{2}t^2 + t\right)\right]e^t, \\ x_3 &= \left(c_1 + c_2 t + \frac{1}{2}c_3 t^2\right)e^t. \end{aligned}$$

EXAMPLE 6 Solve the system

$$\mathbf{X}' = \begin{pmatrix} -2 & 0 & 0 \\ 0 & -2 & 0 \\ 0 & 0 & -2 \end{pmatrix} \mathbf{X}.$$

Solution. The matrix $\mathbf{A}$ is diagonal, and the eigenvalues are the diagonal elements: $\lambda = -2, -2, -2$. The system $(\mathbf{A} - \lambda\mathbf{I})\mathbf{K} = \mathbf{0}$ is

$$\begin{aligned} 0k_1 + 0k_2 + 0k_3 &= 0, \\ 0k_1 + 0k_2 + 0k_3 &= 0, \\ 0k_1 + 0k_2 + 0k_3 &= 0. \end{aligned}$$

Thus k_1, k_2, and k_3 are all arbitrary so *every* nonzero vector $\mathbf{K}$ is an eigenvector. We choose the three linearly independent eigenvectors

$$\mathbf{K}_1 = \begin{pmatrix} 1 \\ 0 \\ 0 \end{pmatrix}, \qquad \mathbf{K}_2 = \begin{pmatrix} 0 \\ 1 \\ 0 \end{pmatrix}, \qquad \mathbf{K}_3 = \begin{pmatrix} 0 \\ 0 \\ 1 \end{pmatrix}.$$

The general solution is therefore

$$\mathbf{X} = c_1 \begin{pmatrix} 1 \\ 0 \\ 0 \end{pmatrix} e^{-2t} + c_2 \begin{pmatrix} 0 \\ 1 \\ 0 \end{pmatrix} e^{-2t} + c_3 \begin{pmatrix} 0 \\ 0 \\ 1 \end{pmatrix} e^{-2t}.$$

Nonhomogeneous Higher-dimensional Systems

The method of variation of parameters studied in Section 8.2 extends to higher-dimensional systems of the form

$$\mathbf{X}' = \mathbf{A}\mathbf{X} + \mathbf{F}(t). \tag{28}$$

If $\mathbf{\Phi}$ denotes a fundamental matrix of solutions to the associated homogeneous system, then a particular solution to system (28) is given by

$$\mathbf{X}_p = \mathbf{\Phi} \int \mathbf{\Phi}^{-1}\mathbf{F}(t)\, dt, \tag{29}$$

as in the 2×2 case. The general solution to system (28) is

$$\mathbf{X} = \mathbf{X}_c + \mathbf{X}_p. \tag{30}$$

EXAMPLE 7 Solve the nonhomogeneous system

$$\mathbf{X}' = \begin{pmatrix} 4 & 0 & 1 \\ -2 & 1 & 0 \\ -2 & 0 & 1 \end{pmatrix} \mathbf{X} + \begin{pmatrix} -1 \\ t \\ e^{2t} \end{pmatrix}.$$

Solution. From Example 1 we know that a fundamental matrix for the associated homogeneous system is

$$\mathbf{\Phi} = \begin{pmatrix} 0 & e^{2t} & e^{3t} \\ e^t & -2e^{2t} & -e^{3t} \\ 0 & -2e^{2t} & -e^{3t} \end{pmatrix}.$$

Inverting $\mathbf{\Phi}$ gives us

$$\mathbf{\Phi}^{-1} = \begin{pmatrix} 0 & e^{-t} & -e^{-t} \\ -e^{-2t} & 0 & -e^{-2t} \\ 2e^{-3t} & 0 & e^{-3t} \end{pmatrix}.$$

Then integration of $\mathbf{\Phi}^{-1}\mathbf{F}$ yields the following:

$$\begin{aligned} \int \mathbf{\Phi}^{-1}\mathbf{F}\, dt &= \int \begin{pmatrix} 0 & e^{-t} & -e^{-t} \\ -e^{-2t} & 0 & -e^{-2t} \\ 2e^{-3t} & 0 & e^{-3t} \end{pmatrix} \begin{pmatrix} -1 \\ t \\ e^{2t} \end{pmatrix} dt \\ &= \int \begin{pmatrix} te^{-t} - e^t \\ e^{-2t} - 1 \\ -2e^{-3t} + e^{-t} \end{pmatrix} dt \\ &= \begin{pmatrix} -te^{-t} - e^{-t} - e^t \\ -\frac{1}{2}e^{-2t} - t \\ \frac{2}{3}e^{-3t} - e^{-t} \end{pmatrix}. \end{aligned}$$

(We leave the details of the calculation to you.) Hence, from Eq. (29) we know that a particular solution vector is given by

$$\mathbf{X}_p = \begin{pmatrix} 0 & e^{2t} & e^{3t} \\ e^{t} & -2e^{2t} & -e^{3t} \\ 0 & -2e^{2t} & -e^{3t} \end{pmatrix} \begin{pmatrix} -te^{-t} - e^{-t} - e^{t} \\ -\frac{1}{2}e^{-2t} - t \\ \frac{2}{3}e^{-3t} - e^{-t} \end{pmatrix}$$

$$= \begin{pmatrix} \frac{1}{6} - te^{2t} - e^{2t} \\ -\frac{2}{3} + 2te^{2t} - t \\ \frac{1}{3} + 2te^{2t} + e^{2t} \end{pmatrix}.$$

Then, from Eq. (30), the general solution in component form is:

$$x_1 = c_2 e^{2t} + c_3 e^{3t} + \frac{1}{6} - te^{2t} - e^{2t},$$

$$x_2 = c_1 e^{t} - 2c_2 e^{2t} - c_3 e^{3t} - \frac{2}{3} + 2te^{2t} - t,$$

$$x_3 = -2c_2 e^{2t} - c_3 e^{3t} + \frac{1}{3} + 2te^{2t} + e^{2t}.$$

EXERCISES 8.3

In problems 1–20 solve the homogeneous system $\mathbf{X}' = \mathbf{A}\mathbf{X}$ for the given matrix $\mathbf{A}$.

1. $\begin{pmatrix} 1 & -1 & 4 \\ 3 & 2 & -1 \\ 2 & 1 & -1 \end{pmatrix}$

2. $\begin{pmatrix} 1 & 1 & 2 \\ 1 & 0 & 1 \\ 2 & 1 & 3 \end{pmatrix}$

3. $\begin{pmatrix} -1 & 0 & 0 \\ 0 & 1 & 2 \\ 0 & 3 & 1 \end{pmatrix}$

4. $\begin{pmatrix} 4 & -2 & 1 \\ -2 & 1 & 2 \\ 1 & 2 & 4 \end{pmatrix}$

5. $\begin{pmatrix} 3 & 1 & -2 \\ -1 & 2 & 1 \\ 4 & 1 & -3 \end{pmatrix}$

6. $\begin{pmatrix} -1 & 4 & -2 \\ -3 & 4 & 0 \\ -3 & 1 & 3 \end{pmatrix}$

7. $\begin{pmatrix} -1 & 0 & 0 \\ 0 & -1 & -4 \\ 0 & 1 & -1 \end{pmatrix}$

8. $\begin{pmatrix} 2 & -5 & 0 \\ 1 & -2 & 0 \\ 0 & 0 & 1 \end{pmatrix}$

9. $\begin{pmatrix} 1 & -3 & 2 \\ 0 & -2 & 2 \\ 1 & -5 & 2 \end{pmatrix}$

10. $\begin{pmatrix} 1 & -1 & 2 \\ -1 & 2 & -1 \\ 2 & -1 & 1 \end{pmatrix}$

11. $\begin{pmatrix} 1 & -1 & 2 \\ -1 & 1 & 0 \\ -1 & 0 & 1 \end{pmatrix}$

12. $\begin{pmatrix} 5 & -5 & -5 \\ -1 & 4 & 2 \\ 3 & -5 & -3 \end{pmatrix}$

13. $\begin{pmatrix} 3 & 0 & 0 \\ 0 & -1 & 3 \\ 0 & -3 & 5 \end{pmatrix}$

14. $\begin{pmatrix} 5 & -4 & 0 \\ 1 & 0 & 2 \\ 0 & 2 & 5 \end{pmatrix}$

15. $\begin{pmatrix} 1 & 1 & 1 \\ 2 & 1 & -1 \\ -3 & 2 & 4 \end{pmatrix}$

16. $\begin{pmatrix} 2 & 1 & 0 \\ 0 & 2 & 1 \\ 0 & 0 & 2 \end{pmatrix}$

17. $\begin{pmatrix} 2 & 1 & 0 \\ 0 & 2 & 0 \\ 0 & 0 & 1 \end{pmatrix}$

18. $\begin{pmatrix} -1 & 1 & 0 \\ 0 & -1 & 1 \\ 0 & 0 & -1 \end{pmatrix}$

19. $\begin{pmatrix} -3 & 0 & -3 \\ 3 & -2 & 1 \\ 1 & 0 & 1 \end{pmatrix}$

20. $\begin{pmatrix} 3 & -1 & 0 \\ -4 & 1 & 2 \\ 8 & -3 & 0 \end{pmatrix}$

In problems 21–28 solve the nonhomogeneous system.

21. $\mathbf{X}' = \begin{pmatrix} 1 & -2 & 0 \\ -1 & 0 & 0 \\ 0 & 0 & 2 \end{pmatrix}\mathbf{X} + \begin{pmatrix} 0 \\ 3te^{t} \\ e^{t} \end{pmatrix}$

22. $\mathbf{X}' = \begin{pmatrix} 1 & 1 & 2 \\ 1 & 2 & 1 \\ 2 & 1 & 1 \end{pmatrix}\mathbf{X} + \begin{pmatrix} 1 \\ t \\ e^{-t} \end{pmatrix}$

23. $\mathbf{X}' = \begin{pmatrix} 1 & -1 & 0 \\ -1 & 2 & -1 \\ 0 & 1 & 1 \end{pmatrix} \mathbf{X} + \begin{pmatrix} 1 \\ e^t \\ 0 \end{pmatrix}$

24. $\mathbf{X}' = \begin{pmatrix} 4 & 0 & 0 \\ 0 & 1 & 1 \\ 0 & 1 & 1 \end{pmatrix} \mathbf{X} + \begin{pmatrix} e^t \\ e^{-t} \\ 2e^{3t} \end{pmatrix}$

25. $\mathbf{X}' = \begin{pmatrix} 0 & 1 & 0 \\ 1 & 0 & 0 \\ 0 & 0 & -1 \end{pmatrix} \mathbf{X} + \begin{pmatrix} 1 \\ t \\ e^{-t} \end{pmatrix}$

26. $\mathbf{X}' = \begin{pmatrix} 2 & 0 & 0 \\ 0 & 0 & 1 \\ 0 & 1 & 0 \end{pmatrix} \mathbf{X} + \begin{pmatrix} 1 \\ 1 \\ 0 \end{pmatrix}$

27. $\mathbf{X}' = \begin{pmatrix} 1 & -1 & 0 \\ 1 & 1 & 0 \\ 0 & 0 & -1 \end{pmatrix} \mathbf{X} + \begin{pmatrix} -e^t \\ 2e^t \\ 3t^2e^{-t} \end{pmatrix}$

28. $\mathbf{X}' = \begin{pmatrix} 0 & -1 & 0 \\ 1 & 0 & 0 \\ 0 & 0 & 1 \end{pmatrix} \mathbf{X} + \begin{pmatrix} \sec t \\ 0 \\ 1 \end{pmatrix}$

In problems 29–32 solve the given system subject to the indicated initial condition.

29. $\mathbf{X}' = \begin{pmatrix} -2 & 1 & 3 \\ 0 & -2 & 1 \\ 0 & 0 & -2 \end{pmatrix} \mathbf{X}, \quad \mathbf{X}(0) = \begin{pmatrix} 1 \\ 0 \\ 1 \end{pmatrix}$

30. $\mathbf{X}' = \begin{pmatrix} 2 & 0 & -1 \\ 0 & 2 & -1 \\ 0 & 0 & 1 \end{pmatrix} \mathbf{X}, \quad \mathbf{X}(0) = \begin{pmatrix} 1 \\ -1 \\ 1 \end{pmatrix}$

31. $\mathbf{X}' = \begin{pmatrix} 1 & -1 & 0 \\ -1 & 1 & 0 \\ 0 & 0 & 3 \end{pmatrix} \mathbf{X} + \begin{pmatrix} 0 \\ t \\ e^{4t} \end{pmatrix}, \quad \mathbf{X}(0) = \begin{pmatrix} 1 \\ 0 \\ 0 \end{pmatrix}$

32. $\mathbf{X}' = \begin{pmatrix} 3 & 0 & 0 \\ 0 & 2 & 0 \\ 0 & 0 & 1 \end{pmatrix} \mathbf{X} + \begin{pmatrix} t \\ t^2 \\ e^t \end{pmatrix}, \quad \mathbf{X}(0) = \begin{pmatrix} 0 \\ 1 \\ -1 \end{pmatrix}$

33. Consider the model for polluting the system of lakes given in Section 7.1, Example 2. Solve the model assuming the volumes of the lakes are $V_1 = 340$ mi^3, $V_2 = 680$ mi^3, and $V_3 = 204$ mi^3. Assume the flow rates $F_{21} = 68$ and $F_{31} = 34$ mi^3 per year, and that the pollutant is entering lake 1 according to the rule $p(t) = 5e^{0.07t}$. Approximate the eigenvalues of the system. This illustrates a system you would want to solve numerically.

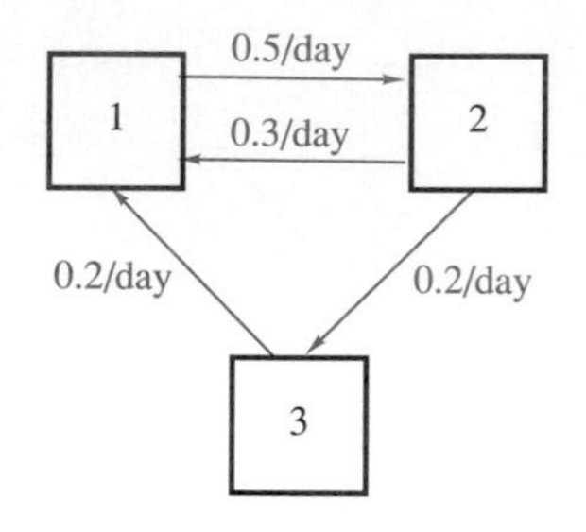

FIGURE 8.3

34. For the compartment model shown in Fig. 8.3 (see also Section 7.1, Example 2), each tank initially contains 100 gal of fresh water. At time $t = 0$ assume that 20 lb of salt are mixed in tank 1. Determine the concentration of salt in tank 3 when $t = 4$ days. All tanks are kept uniformly mixed by constant stirring.

35. For the spring–mass system shown in Fig. 7.5, assume that $k_1 = k_2 = k_3 = k$ and $m_1 = m_2 = m$. Solve the system if the initial conditions are $x_1(0) = a$, $x_1'(0) = 0$, $x_2(0) = 0$, and $x_2'(0) = 0$.

36. In problem 34, instead of the 20 lbs of salt added at only time $t = 0$, assume that a quantity of salt given by $f_1(t) = 20e^{0.07t}$ (also in pounds) is added to tank 1 over time. Determine the concentration of salt in tank 2 when $t = 5$ days.

37. Find a fundamental set of solutions to the compartment model shown in Fig. 8.4, where r_i is the transfer rate between compartments. Show that each solution has a limit as t tends toward infinity.

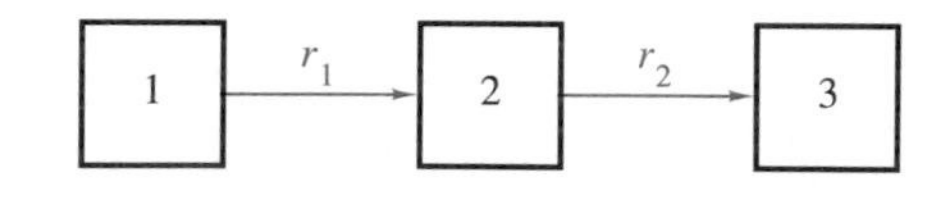

FIGURE 8.4

38. For the spring–mass system depicted in Fig. 8.5, find a system of three second-order equations in x_1, x_2, and x_3 to describe the motion of the masses. Then write an equivalent first-order system in terms of the variables for position and velocity of the masses.

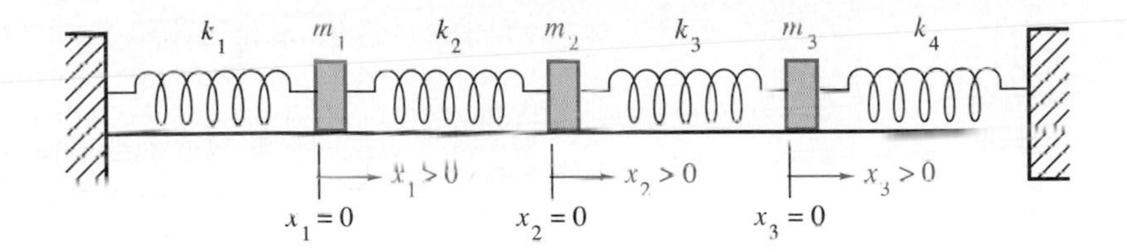

FIGURE 8.5

CHAPTER 8 REVIEW EXERCISES

In problems 1–3, write the third-order differential equation as a system of first-order equations.

1. $4y''' - 8y'' + 16y' + 4y = 16 \sin 2t + 8e^{-t}$

2. $3y''' - 6y'' + y' - 2y = t + 3e^{-2t}$

3. $2y''' + y'' - 12y' + 2y = 2 \cos t + \sin t$

In problems 4–7, verify that the given vector solves the system of differential equations.

4. $\dfrac{dx}{dt} = x + 2y + 3z - \sin 3t$

$\dfrac{dy}{dt} = -4x + 2y + 4 \sin 3t$

$\dfrac{dz}{dt} = -6x + y + 3 \sin 3t$

where

$$\mathbf{X}_p = \begin{pmatrix} \sin 3t \\ 0 \\ 0 \end{pmatrix} + \begin{pmatrix} 0 \\ 0 \\ \cos 3t \end{pmatrix}$$

5. $\dfrac{dx}{dt} = -x - y + t^2 + 4t - 1,$

$\dfrac{dy}{dt} = -x + y + t^2 - 6t + 5$

where

$$\mathbf{X}_p = \begin{pmatrix} 1 \\ 0 \end{pmatrix} t^2 + \begin{pmatrix} -2 \\ 4 \end{pmatrix} t + \begin{pmatrix} 1 \\ 0 \end{pmatrix}$$

6. $\mathbf{X}' = \begin{pmatrix} 1 & 4 \\ 2 & 3 \end{pmatrix} \mathbf{X} + \begin{pmatrix} 2e^t \\ -e^t \end{pmatrix},$

$$\mathbf{X}_p = \begin{pmatrix} 1 \\ 1 \end{pmatrix} e^{5t} + \begin{pmatrix} 1 \\ -1/2 \end{pmatrix} e^t$$

7. $\mathbf{X}' = \begin{pmatrix} -4 & 2 \\ 2 & -1 \end{pmatrix} \mathbf{X} + \begin{pmatrix} \frac{1}{t} \\ 4 + \frac{2}{t} \end{pmatrix},$

$$\mathbf{X}_p = \begin{pmatrix} \frac{8}{5} t + \ln t - \frac{8}{25} \\ \frac{16}{5} t + 2 \ln t + \frac{4}{25} \end{pmatrix}$$

In problems 8–11, find the general solution to the given nonhomogeneous system of differential equations.

8. $x_1' = 4x_1 - 3x_2 + e^{-t}$
$x_2' = 2x_1 - 3x_2 - e^{-t}$

9. $x_1' = x_1 + 5x_2 + 2$
$x_2' = -2x_1 - x_2 - 1$

10. $\mathbf{X}' = \begin{pmatrix} 3 & 5 \\ 6 & 2 \end{pmatrix} \mathbf{X} + \begin{pmatrix} e^{7t} \\ e^{7t} \end{pmatrix}$

11. $\mathbf{X}' = \begin{pmatrix} 4 & 3 \\ 1 & 2 \end{pmatrix} \mathbf{X} + \begin{pmatrix} 4e^{4t} \\ 8 \end{pmatrix}$

In problems 12–18, solve the nonhomogeneous system subject to the specified initial conditions.

12. $\dfrac{dx}{dt} = x - 2y + e^t,$

$\dfrac{dy}{dt} = 3x - 4y + e^t,$

$x(0) = 0, \quad y(0) = 2$

13. $\mathbf{X}' = \begin{pmatrix} 0 & 1 \\ -6 & 5 \end{pmatrix} \mathbf{X} + \begin{pmatrix} 0 \\ e^{3t} \end{pmatrix}, \quad \mathbf{X}(0) = \begin{pmatrix} 0 \\ 1 \end{pmatrix}$

14. $\mathbf{X}' = \begin{pmatrix} 4 & -5 \\ 5 & -4 \end{pmatrix} \mathbf{X} + \begin{pmatrix} 0 \\ 7 \end{pmatrix}, \quad \mathbf{X}(0) = \begin{pmatrix} 1 \\ 7 \end{pmatrix}$

15. $\mathbf{X}' = \begin{pmatrix} 1 & 9 \\ -1 & 1 \end{pmatrix} \mathbf{X} + \begin{pmatrix} e^t \\ 0 \end{pmatrix}, \quad \mathbf{X}(0) = \begin{pmatrix} 2 \\ -3 \end{pmatrix}$

16. $\mathbf{X}' = \begin{pmatrix} -2 & 1 \\ -4 & 3 \end{pmatrix} \mathbf{X} + \begin{pmatrix} t \\ e^{2t} \end{pmatrix}, \quad \mathbf{X}(0) = \begin{pmatrix} 6 \\ -4 \end{pmatrix}$

17. $\mathbf{X}' = \begin{pmatrix} 3 & -4 \\ 1 & -1 \end{pmatrix} \mathbf{X} + \begin{pmatrix} e^t \\ 2e^t \end{pmatrix}, \quad \mathbf{X}(0) = \begin{pmatrix} 4 \\ 5 \end{pmatrix}$

18. $\mathbf{X}' = \begin{pmatrix} -7 & 3 \\ 1 & -5 \end{pmatrix} \mathbf{X} + \begin{pmatrix} 12 \sin 4t \\ 92 \sin 4t \end{pmatrix}, \quad \mathbf{X}(0) = \begin{pmatrix} -14 \\ -10 \end{pmatrix}$

In problems 19–25, solve the given system of differential equations.

19. $\mathbf{X}' = \begin{pmatrix} 1 & 0 & 0 \\ 3 & 1 & -2 \\ 2 & 2 & 1 \end{pmatrix} \mathbf{X}$

20. $\mathbf{X}' = \begin{pmatrix} 3 & 1 & -1 \\ 1 & 3 & -1 \\ 3 & 3 & -1 \end{pmatrix} \mathbf{X}$

21. $\mathbf{X}' = \begin{pmatrix} 1 & -1 & 2 \\ -1 & 1 & 0 \\ -1 & 0 & 1 \end{pmatrix} \mathbf{X}$

22. $\mathbf{X}' = \begin{pmatrix} 1 & 0 & 0 \\ 3 & 1 & -2 \\ 2 & 2 & 1 \end{pmatrix} \mathbf{X}$

23. $\mathbf{X}' = \begin{pmatrix} 0 & 1 & 0 \\ 0 & 0 & 1 \\ 1 & -3 & 3 \end{pmatrix} \mathbf{X}$

24. $\mathbf{X}' = \begin{pmatrix} 0 & 0 & -1 \\ 1 & -1 & 1 \\ 0 & 2 & 1 \end{pmatrix} \mathbf{X}$

25. $\mathbf{X}' = \begin{pmatrix} 2 & 4 & 4 \\ -1 & -2 & 0 \\ -1 & 0 & -2 \end{pmatrix} \mathbf{X}$

26. The interaction of several animal species that inhabit the same environs can be modeled using a system of differential equations. Consider the following situation, where $x(t)$ and $y(t)$ denote the populations of two animal species at any given time:

$$\frac{dx}{dt} = 3x - 3y - 3t, \quad x(0) = 11,$$

$$\frac{dy}{dt} = 2x - 2y, \quad y(0) = 7.$$

Solve this nonhomogeneous system of differential equations.

27. The interaction of several animal species that inhabit the same environs can be modeled using a system of differential equations. Consider the following situation, where $x(t)$ and $y(t)$ denote the populations of two animal species at any given time:

$$\frac{dx}{dt} = 4x - 2y - 8e^{3t}, \quad x(0) = 10,$$

$$\frac{dy}{dt} = \frac{5}{2}x - 2y, \quad y(0) = 2.$$

Solve this nonhomogeneous system of differential equations.

28. Pricing Policy

A general pricing policy for a manufacturer is described by the following differential equation for the forecast price $p(t)$:

$$\frac{dp}{dt} = \delta[L(t) - L_0].$$

In this equation, δ is a constant, $L(t)$ is the inventory level at time t, and L_0 is the desired inventory level. The inventory changes according to

$$\frac{dL}{dt} = Q(t) - S(t),$$

where $Q(t)$ and $S(t)$ represent production and sales, respectively, and are modeled by

$$Q(t) = a - bp - c\frac{dp}{dt},$$

$$S(t) = \alpha - \beta p - v\frac{dp}{dt},$$

where a, b, c, α, β, and v are positive constants. Using the above equations, one can show that the forecast price is determined by the second-order differential equation

$$\frac{d^2p}{dt^2} + \delta(v - c)\frac{dp}{dt} + \delta(\beta - b)p = \delta(\alpha - a).$$

We desire to forecast the price of a commodity at certain time intervals for the upcoming year. Our marketing research office has provided the values for the constants in the last equation:

$$\delta = -1, \quad a = 2, \quad b = \tfrac{1}{4}, \quad c = 7,$$
$$\alpha = 14, \quad \beta = 3, \quad v = 2.$$

a) Rewrite the second-order equation as a system of first-order differential equations. Solve for the forecast price p using the initial conditions $p(0) = 100$ and $p'(0) = -526$. Use the method of variation of parameters for a nonhomogeneous linear system.

b) Using their defining equations, calculate Q and S.

Use these values to calculate L, the inventory level. The inventory at $t = 0$ is 474 units.

c) Plot the graph of $p(t_k)$ for $t_k = k\,\Delta t$, $k = 0, 1, 2, \ldots, 20$, where $\Delta t = 0.05$ yr. Using the same time intervals, plot the graph of L. What are the values of p and L at time $t = 0.5$ yr?

29. **a)** Show that the electrical network in Fig. 8.6 can be modeled by system

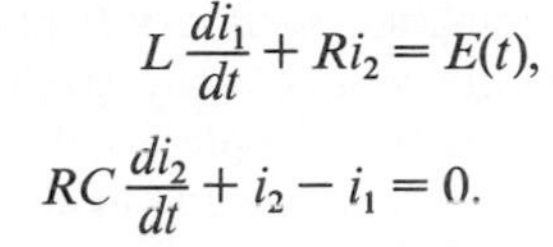

$$L\frac{di_1}{dt} + Ri_2 = E(t),$$

$$RC\frac{di_2}{dt} + i_2 - i_1 = 0.$$

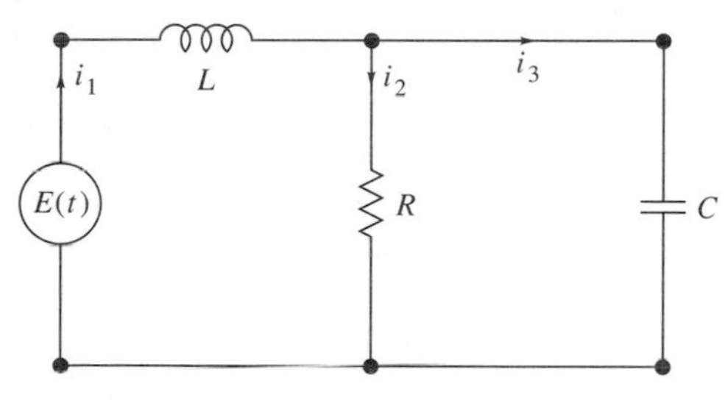

FIGURE 8.6

b) Let $E = 60$ V, $L = \frac{1}{2}$ H, $R = 50\ \Omega$, $C = 0.0001$ Fa, $i_1(0) = 0$, and $i_2(0) = 0$. Solve the system of equations subject to the given conditions.

c) Using Kirchhoff's first law we can determine that $i_3 = i_1 - i_2$. Describe the long-term (i.e., steady-state) behavior of the currents i_1, i_2, and i_3.

30. **Defense Budget Growth**

Each year the Department of Defense must prepare its annual budget, an incredibly complex estimate of proposed expenditures.

Even if all of the Defense Department's eventual commitments could be predicted at the time of the budget's preparation, the budget—which represents the cost of carrying out these commitments—would remain an inexact estimate. The probabilistic nature of the calculations and the uncertainty in the interrelationships among cost categories are two major factors that defy exact description. Consider two examples that illustrate these factors.

(i) The effects of inflation are always considered in budget development by estimating a percentage factor, because no one can predict with certainty the rise in the cost of goods and services during the next fiscal year. (ii) When defense managers consider increasing the maintenance interval for military equipment, they must attempt to estimate the impact on a unit's equipment breakdown rate and repair costs. In the aggregate, many of these budget uncertainties average out to relatively small errors that can be accommodated in the short term by reallocation of available funds and resources at the expense of scheduled expenditures of a lesser priority.

Over the long term, however, the impact of decisions and errors in estimation made each year can cause astronomical growth of expenditure requirements in future budgets. This in turn must result in enormously increased budgets, severely curtailed capabilities or services, or a combination of the two. To illustrate this point, consider the fact that one effect of military pay raises is an increase in military retirement costs so drastic that the current retirement system will probably have to be changed eventually. The cost growth problems of major weapon system programs frequently result from decisions to delay the program or extend the system's development time in order to reduce annual expenditures on the program. Such decisions, which often are made in order to accommodate fiscal constraints, can easily double the cost of a weapon system.

With a budget as complex as that of the Defense Department, predicting the cost growth of future budgets is extremely difficult. However, some sense of the interrelationships among costs can be determined by solving a simple system of linear differential equations.

As a simple model, consider the defense budget to be the sum of three expenditure categories. Let x_1 be the total of procurement and R & D costs, let x_2 be the total of personnel costs, and let x_3 be the total of all operations and maintenance costs. If $\mathbf{I}(t)$ represents an inflation factor, the time rate of change of the three major categories of defense expenditure can be represented by the following system of differential equations:

$$\frac{dx_1}{dt} = a_{11}x_1 + a_{12}x_2 + a_{13}x_3 + I_1(t),$$

$$\frac{dx_2}{dt} = a_{21}x_1 + a_{22}x_2 + a_{23}x_3 + I_2(t),$$

$$\frac{dx_3}{dt} = a_{31}x_1 + a_{32}x_2 + a_{33}x_3 + I_3(t).$$

If specific values for $\mathbf{I}(t)$ and the coefficients of the x's are provided, this system of differential equations can be solved for x_1, x_2, and x_3. For this model, let

$$\mathbf{A} = \begin{pmatrix} 3 & 1 & 0 \\ 1 & 2 & 1 \\ 2 & 1 & 1 \end{pmatrix}, \quad \mathbf{I}(t) = \begin{pmatrix} 2 \\ 2 \\ 2 \end{pmatrix}$$

a) Find the general solution $\mathbf{X}(t)$ of the nonhomogeneous system of differential equations, subject to the specific conditions for **A** and **I**(t).

b) Find the particular solution of the system, if $x_1(0) = 2$, $x_2(0) = 5$, and $x_3(0) = 2$.

c) Write the expression $S(t)$ that represents the total defense budget.

d) What is the value of the total defense budget at time $t = 0$?

e) What fraction of the total budget is devoted to procurement and R & D? To personnel costs? To operations and maintenance?

f) Calculate the ratio of the budget at time $t = 1$ to the budget at time $t = 0$. This represents the multiplication of the budget after one time unit (1 yr). Does this growth ratio make sense?

31. Double Pendulum

A double pendulum consists of a pendulum attached to a pendulum. As shown in Figure 8.7, a pendulum of mass m_2 and length l_2 is attached to a pendulum of mass m_1 and length l_1. Under the influence of gravity, this double pendulum oscillates in a vertical plane. It can be shown that for small displacements $\theta_1(t)$ and $\theta_2(t)$, as shown, the differential equations of motion are

$$(m_1 + m_2)l_1^2\theta_1'' + m_2 l_1 l_2 \theta_2'' + (m_1 + m_2)l_1 g\theta_1 = 0,$$
$$m_2 l_1 l_2 \theta_1'' + m_2 l_2^2 \theta_2'' + m_2 l_2 g\theta_2 = 0,$$

where g is the acceleration due to gravity. Since the solutions $\theta_1(t)$ and $\theta_2(t)$ each contain two arbitrary constants, four initial conditions must be specified: $\theta_1(0)$, $\theta_1'(0)$, $\theta_2(0)$, and $\theta_2'(0)$. By using the method of Laplace transforms, this system can be solved quite easily. (Refer to Section 7.9.)

A CH-53 Chinook helicopter is delivering an emergency resupply of ammunition to a unit in the field. The ammunition is being carried on two pallets slung underneath the helicopter, approximating a double pendulum. The upper pallet mass $m_1 = 937.5$ slugs. The lower pallet mass $m_2 = 312.5$ slugs. The lengths of the cables, l_1 and l_2, are 16 ft long. The hook holding the lower pallet cable is bent open, and if the lower pallet oscillates through an arc of 60° or more toward the open end of the hook, the cable will come off the hook and the lower pallet will be lost. As a result of a sudden evasive maneuver, the pallets begin moving subject to the following initial conditions:

$$\theta_1(0) = -\tfrac{1}{2} \text{ rad}, \quad \theta_1'(0) = -1 \text{ rad/sec},$$
$$\theta_2(0) = 1 \text{ rad}, \quad \theta_2'(0) = 2 \text{ rad/sec}.$$

a) Solve the system of equations for $\theta_1(t)$ and $\theta_2(t)$ using Laplace transforms.

b) Graph $\theta_1(t)$ and $\theta_2(t)$ for $0 \le t < 2\pi$ using $t = 0, 0.1, 0.2, \ldots 6.1, 6.2$.

c) Will the lower pallet be lost and, if so, at approximately what time?

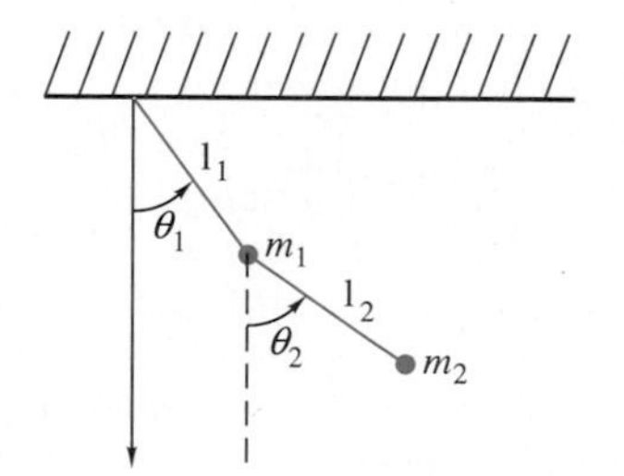

FIGURE 8.7

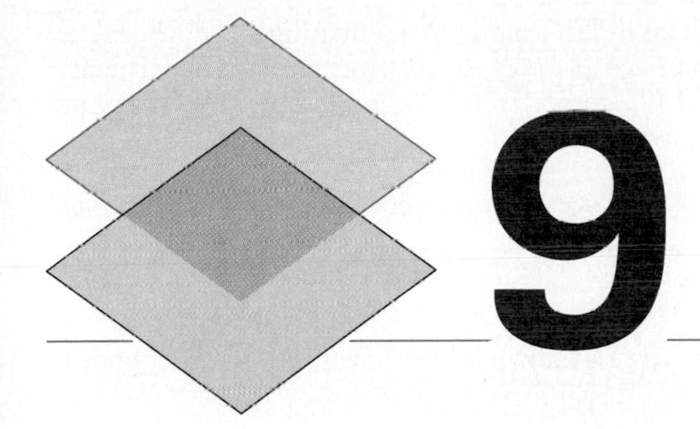

9

Linear Equations with Variable Coefficients

In this chapter we will investigate the solutions to differential equations with variable coefficients. Two important classes of equations of this type are Bessel's equation and Legendre's equation. *Bessel's equation* takes the form

$$x^2y'' + xy' + (x^2 - p^2)y = 0 \tag{1}$$

where p is a constant. The equation occurs in the studies of fluid motion, wave propagation, potential theory, and boundary-value problems. *Legendre's equation* has the form

$$(1 - x^2)y'' - 2xy' + n(n+1)y = 0 \tag{2}$$

where n is a nonnegative integer. The equation first occurred in the study of gravitation but today has applications in numerical analysis.

The general form of the second-order equation we shall study is the homogeneous equation

$$a_2(x)y'' + a_1(x)y' + a_0(x)y = 0. \tag{3}$$

Our main concern will be with equations in which the coefficients $a_i(x)$ in Eq. (3) are polynomials. A special case is the *Euler equation,* which has an elementary analytical solution form. The Euler equation is discussed in Section 9.2. However, most variable-coefficient equations fail to have elementary analytical solutions of the sort encountered in previous chapters. Rather, the analytical solutions often take the form of an infinite series. (If initial conditions are given, it may also be possible to obtain a numerical approximation to the solution, as discussed in Chapter 4.) We introduce these series solution methods in this chapter. After treating the more general cases, *Bessel functions* and *Legendre polynomials* are presented as particular solutions to Eqs. (1) and (2), respectively. Some of their properties and applications are also discussed.

9.1 REDUCTION OF ORDER

Sometimes a solution $y = y_1(x)$ is known for the linear second-order homogeneous equation

$$a_2(x)y'' + a_1(x)y' + a_0(x)y = 0. \tag{1}$$

For example, it is easy to verify that $y_1 = x^2$ is a solution to the homogeneous equation

$$x^2y'' - 2y = 0. \tag{2}$$

In this situation it is possible to construct a second linearly independent solution $y_2(x)$. The procedure, known as **reduction of order,** is like the one used in the variation-of-parameters method. We set $y_2 = u(x)y_1$ for some function $u(x)$, which is then determined from the requirement that y_2 be a solution to Eq. (1). Let us start with an example to illustrate the method before presenting the general case.

EXAMPLE 1 Given that $y_1 = x^2$ is a solution to Eq. (2), find a second linearly independent solution $y_2(x)$ for $0 < x < \infty$.

Solution. We define

$$y_2 = u(x)x^2.$$

Then

$$y_2' = u'x^2 + 2ux,$$
$$y_2'' = u''x^2 + 4u'x + 2u.$$

Substitution of y_2 and y_2'' into Eq. (2) yields

$$x^4u'' + 4x^3u' + 2x^2u - 2x^2u = 0.$$

Now we make the change of variables $w = u'$ and $w' = u''$. Then this last equation becomes a first-order equation in the new variables w and w':

$$x^4w' + 4x^3w = 0$$

or

$$w' + \frac{4}{x}w = 0.$$

Separation of variables and integration produces the result

$$\ln w = -4 \ln x$$

or

$$w = x^{-4}.$$

Since we seek only one solution for y_2, we can omit any constants of integration here. A second integration then gives the original variable

$$\begin{aligned} u &= \int w\,dx = \int x^{-4}\,dx \\ &= -\frac{1}{3}x^{-3}, \end{aligned}$$

where again we have omitted the constant of integration. Finally, substitution of u into the form for y_2 yields

$$y_2 = ux^2 = -\frac{1}{3}x^{-1}.$$

The general solution to Eq. (2) is

$$\begin{aligned} y &= c_1 y_1 + c_2 y_2 \\ &= c_1 x^2 + c_2 x^{-1} \end{aligned}$$

(where the constant multiplier $-1/3$ has been absorbed into the arbitrary constant c_2).

General Case

Assume the coefficient $a_2(x)$ is never zero over the interval I. Divide Eq. (1) by $a_2(x)$ to obtain the *standard form*

$$y'' + p(x)y' + q(x)y = 0. \tag{3}$$

Assume $y_1(x)$ is a known solution to Eq. (3) over I and that $y_1(x) \neq 0$ for every x in I. Define

$$y_2(x) = u(x)y_1(x) \tag{4}$$

where $u(x)$ is a function to be determined from the requirement that y_2 satisfy Eq. (3). Differentiating Eq. (4) twice yields

$$\begin{aligned} y_2' &= u'y_1 + uy_1', \\ y_2'' &= u''y_1 + 2u'y_1' + uy_1''. \end{aligned}$$

Substitution of y_2 and its derivatives into Eq. (3) then gives us

$$u''y_1 + 2u'y_1' + uy_1'' + p(x)(u'y_1 + uy_1') + q(x)uy_1 = 0.$$

Collecting the terms in u and its derivatives we have

$$y_1 u'' + (2y_1' + py_1)u' + \underbrace{(y_1'' + py_1' + qy_1)}_{=0}u = 0.$$

Since the last term on the left equals zero (because y_1 is a solution to the homogeneous equation), this last equation reduces to

$$y_1 u'' + (2y_1' + py_1)u' = 0. \tag{5}$$

After substituting $w = u'$ and $w' = u''$, Eq. (5) becomes the first-order separable (and also linear) equation

$$y_1 w' + (2y_1' + py_1)w = 0$$

or

$$\frac{dw}{w} = -\left(\frac{2y_1'}{y_1} + p\right) dx. \tag{6}$$

Integration of Eq. (6) and algebraic manipulation produces the function $w = u'$; integration again yields the unknown function u. Finally, substitution of u into Eq. (4) yields the second solution y_2.

Note that $y_1(x)$ and $y_2(x)$ are linearly independent over I. Calculation of the Wronskian yields

$$\begin{aligned} W[y_1(x), y_2(x)] &= \begin{vmatrix} y_1 & uy_1 \\ y_1' & u'y_1 + uy_1' \end{vmatrix} \\ &= u'y_1^2. \end{aligned}$$

Since $y_1 \neq 0$ on I and u is not a constant, the Wronskian is never zero on I. Thus the solutions are linearly independent by Theorem 3.10.

Let us summarize the reduction-of-order procedure.

CONSTRUCTING A SECOND SOLUTION FROM A KNOWN SOLUTION

Step 1. Given that $y_1(x)$ is a known solution over I to

$$a_2(x)y'' + a_1(x)y' + a_0(x)y = 0 \tag{7}$$

and that $y_1(x) \neq 0$ and $a_2(x) \neq 0$ over I, write the solution form

$$y_2(x) = u(x)y_1(x). \tag{8}$$

Step 2. Substitute y_2 and its derivatives y_2', y_2'' into second-order Eq. (7). The terms involving $u(x)$ alone will cancel.

Step 3. Substitute $w = u'$ and $w' = u''$ into the equation obtained in Step 2. Solve the resulting separable (or linear) first-order equation for w.

Step 4. Integrate the function w obtained in the preceding step:

$$u(x) = \int w(x)\, dx. \tag{9}$$

Step 5. Substitute the function $u(x)$ from Eq. (9) into form (8) and write the general solution to Eq. (7):

$$y(x) = c_1 y_1(x) + c_2 y_2(x). \tag{10}$$

EXAMPLE 2 The function $y_1 = e^x$ is a solution of

$$xy'' - (x+1)y' + y = 0$$

on the interval $0 < x < \infty$. Find the general solution.

Solution. We let $y_2 = ue^x$. Then

$$y_2' = (u' + u)e^x,$$
$$y_2'' = (u'' + 2u' + u)e^x.$$

Substitution into the second-order equation yields

$$x(u'' + 2u' + u)e^x - (x+1)(u' + u)e^x + ue^x = 0$$

or, after algebraic manipulation,

$$xe^x u'' + (x-1)e^x u' = 0.$$

After we set $w = u'$ and $w' = u''$, this equation becomes

$$w' + \left(1 - \frac{1}{x}\right) w = 0,$$

since $e^x \neq 0$ and $x > 0$. Solution of this first-order equation gives us

$$\ln w = \ln x - x.$$

Applying the exponential to both sides results in the relation

$$w = xe^{-x}.$$

Integration yields

$$u = \int w\, dx$$
$$= \int xe^{-x}\, dx$$
$$= -(x+1)e^{-x}.$$

Thus

$$y_2 = uy_1 = -(x+1).$$

You should check that y_2 does indeed solve the second-order equation. The general solution is

$$y = c_1 e^x + c_2(x+1).$$

EXERCISES 9.1

In problems 1–22 find the general solution of each second-order equation using reduction of order. Assume an appropriate interval of validity. Check to ensure that y_1 is a solution to the given equation.

1. $y'' - 2y' - 3y = 0, \quad y_1 = e^{-x}$

2. $y'' + y' - 6y = 0, \quad y_1 = e^{2x}$

3. $y'' - 4y' + 4y = 0, \quad y_1 = e^{2x}$

4. $y'' + 2y' + y = 0, \quad y_1 = e^{-x}$

5. $y'' + 4y = 0, \quad y_1 = \sin 2x$

6. $y'' + 2y' + 2y = 0, \quad y_1 = e^{-x}\cos x$

7. $(x+1)y'' + xy' - y = 0, \quad y_1 = e^{-x}$

8. $2x^2y'' - xy' - 2y = 0, \quad y_1 = x^2$

9. $x^2y'' - 6y = 0, \quad y_1 = x^3$

10. $x^2y'' - 12y = 0, \quad y_1 = x^4$

11. $xy'' - y' = 0, \quad y_1 = 1$

12. $xy'' + y' = 0, \quad y_1 = 1$

13. $(1+x^2)y'' - 2xy' = 0, \quad y_1 = 1$

14. $xy'' + y' - x^{-1}y = 0, \quad y_1 = x^{-1}$

15. $x^2y'' - xy' + y = 0, \quad y_1 = x$

16. $x^2y'' + 5xy' + 4y = 0, \quad y_1 = x^{-2}$

17. $2x^2y'' + 3xy' - y = 0, \quad y_1 = \sqrt{x}, x > 0$

18. $2xy'' + (1-4x)y' + (2x-1)y = 0, \quad y_1 = e^x$

19. $x^2y'' - 5xy' + 9y = 0, \quad y_1 = x^3$

20. $x^2y'' - 2xy' + (x^2+2)y = 0, \quad y_1 = x\sin x$

21. $x^2y'' + xy' + y = 0, \quad y_1 = \cos(\ln x)$

22. $x^2y'' - xy' + 2y = 0, \quad y_1 = x\sin(\ln x)$

9.2 EULER EQUATION

The simplest type of linear second-order equation with variable coefficients is the **Euler equation** (also called the **Euler–Cauchy** or **equidimensional equation**). This equation has the special form

$$x^2y'' + axy' + by = 0, \qquad x \neq 0 \tag{1}$$

where a and b are constants. Notice that the power of x in each term on the lefthand side of Eq. (1) matches exactly the order of the derivative appearing in that term: x^2 with the second derivative, x with the first derivative, and x^0 with y itself. First we will seek solutions to the Euler equation over the interval $0 < x < \infty$, and then we will consider solutions when $x < 0$.

Solutions for Positive x

Equation (1) states that some function y and its derivatives, when multiplied by certain powers of x, produce zero. Recall that the derivative of a power x^r is another power rx^{r-1}. Thus it might seem reasonable to assume that the solution function is just a power of x. Therefore we let

$$y = x^r \tag{2}$$

where r is some power to be determined from the requirement that Eq. (2) satisfy Eq. (1). Differentiating Eq. (2) twice yields

$$y' = rx^{r-1},$$
$$y'' = r(r-1)x^{r-2}.$$

Substitution of y and its derivatives into Eq. (1) then gives us

$$x^2 r(r-1)x^{r-2} + axrx^{r-1} + bx^r = 0,$$
$$r(r-1)x^r + arx^r + bx^r = 0$$

or, since $x^r \neq 0$,

$$r(r-1) + ar + b = 0. \tag{3}$$

Thus the power function $y = x^r$ solves the Euler equation (1) if and only if r is a root to quadratic Eq. (3). As we did in solving constant-coefficient linear second-order equations, to solve the Euler equation we will consider three cases for the roots of quadratic Eq. (3).

Case 1: Roots Are Real and Distinct

In this case r_1 and r_2 $(r_1 \neq r_2)$ are the roots to Eq. (3) and yield the two solutions

$$y_1 = x^{r_1} \qquad \text{and} \qquad y_2 = x^{r_2}.$$

Moreover, the functions y_1 and y_2 are linearly independent from the Wronskian test (Theorem 3.10, Section 3.5):

$$W[y_1, y_2] = \begin{vmatrix} x^{r_1} & x^{r_2} \\ r_1 x^{r_1-1} & r_2 x^{r_2-1} \end{vmatrix}$$
$$= (r_2 - r_1)x^{r_1+r_2-1} \neq 0.$$

Thus the general solution to the Euler equation in this case is

$$y = c_1 x^{r_1} + c_2 x^{r_2}. \tag{4}$$

EXAMPLE 1 Find the general solution to

$$x^2 y'' - xy' - 3y = 0, \qquad x > 0.$$

Solution. Substituting $y = x^r$ into the differential equation leads to

$$r(r-1) - r - 3 = 0$$

or

$$r^2 - 2r - 3 = 0.$$

The roots are $r_1 = -1$ and $r_2 = 3$. Thus the general solution is

$$y = c_1 x^{-1} + c_2 x^3.$$

Case 2: Roots Are Complex

The roots occur in conjugate pairs $r_1 = \alpha + \beta i$ and $r_2 = \alpha - \beta i$. The corresponding solutions are

$$x^{\alpha+\beta i} = x^\alpha x^{\beta i} \qquad \text{and} \qquad x^{\alpha-\beta i} = x^\alpha x^{-\beta i}. \tag{5}$$

Since $x = e^{\ln x}$, and since $e^{i\theta} = \cos\theta + i\sin\theta$ (from Euler's identity), by substitution into Eqs. (5) we have

$$x^{\beta i} = e^{i(\beta \ln x)} = \cos(\beta \ln x) + i \sin(\beta \ln x).$$

In the same way we treated complex roots for the constant-coefficient linear second-order differential equation, we thus obtain from Eqs. (5) the two linearly independent solutions

$$y_1 = x^\alpha \cos(\beta \ln x) \qquad \text{and} \qquad y_2 = x^\alpha \sin(\beta \ln x).$$

It is an elementary exercise to verify that the Wronskian $W[y_1, y_2]$ is never zero for $x > 0$. Thus the general solution to the Euler equation (1) in this case is

$$y = x^\alpha[c_1 \cos(\beta \ln x) + c_2 \sin(\beta \ln x)]. \tag{6}$$

EXAMPLE 2 Find the general solution to

$$x^2 y'' - 3xy' + 5y = 0, \qquad x > 0.$$

Solution. Substituting $y = x^r$ into the differential equation leads to

$$r(r-1) - 3r + 5 = 0$$

or

$$r^2 - 4r + 5 = 0.$$

The roots are $r = 2 \pm i$. The general solution is thus

$$y = x^2[c_1 \cos(\ln x) + c_2 \sin(\ln x)].$$

Case 3: Repeated Roots

By assumption, r is a double root to quadratic Eq. (3). Since Eq. (3) has the equivalent form

$$r^2 + (a-1)r + b = 0,$$

it must be the case that

$$r = -\frac{a-1}{2}$$

or

$$2r + a = 1. \tag{7}$$

In this case the repeated root r leads to only one solution

$$y_1 = x^r.$$

Using the method of reduction of order we now obtain a second linearly independent solution.

Let

$$y_2 = u(x)x^r. \tag{8}$$

Then

$$\begin{aligned} y_2' &= u'x^r + rux^{r-1}, \\ y_2'' &= u''x^r + 2ru'x^{r-1} + r(r-1)ux^{r-2}. \end{aligned}$$

Substitution of y_2 and its derivatives into the Euler equation (1) followed by algebraic simplification leads to

$$x^{r+2}u'' + (2r+a)x^{r+1}u' + \underbrace{[r(r-1) + ar + b]}_{=0}x^r u = 0$$

The last term on the lefthand side is equal to zero since r is a root of quadratic Eq. (3). Since $x^r \neq 0$, the equation further reduces to

$$xu'' + (2r+a)u' = 0. \tag{9}$$

Substitution of (7) into (9) gives us

$$xu'' + u' = 0. \tag{10}$$

Setting $w = u'$ and $w' = u''$ in Eq. (10) we obtain the first-order equation

$$\frac{w'}{w} = -\frac{1}{x}$$

with a solution

$$\ln w = -\ln x$$

or

$$w = x^{-1}.$$

Then

$$u = \int w\,dx = \ln x. \tag{11}$$

Finally, after substituting Eq. (11) into form (8) we obtain the second solution

$$y_2 = ux^r = x^r \ln x.$$

Thus the general solution to the Euler equation (1) in this case is

$$y = x^r(c_1 + c_2 \ln x) \tag{12}$$

EXAMPLE 3 Find the general solution to

$$4x^2y'' + 16xy' + 9y = 0, \qquad x > 0.$$

Solution. Substituting $y = x^r$ into the differential equation leads to

$$4r(r-1) + 16r + 9 = 0$$

or

$$4r^2 + 12r + 9 = 0.$$

This last equation factors into

$$(2r+3)^2 = 0,$$

which yields the repeated roots $r = -3/2, -3/2$. The general solution is thus

$$y = x^{-3/2}(c_1 + c_2 \ln x).$$

Solution for Negative x

We now consider the Euler equation (1) when $x < 0$. By making the change of variable $z = -x$, we find $z > 0$ and

$$\frac{dy}{dz} = \frac{dy}{dx}\frac{dx}{dz} = -\frac{dy}{dx},$$

$$\frac{d^2y}{dz^2} = \frac{d}{dz}\left(-\frac{dy}{dx}\right) = \left[\frac{d}{dx}\left(-\frac{dy}{dx}\right)\right]\frac{dx}{dz}$$

$$= -\frac{d^2y}{dx^2}(-1) = \frac{d^2y}{dx^2}.$$

Thus, in terms of the variable z the Euler equation (1) becomes

$$(-z)^2 \frac{d^2y}{dz^2} + a(-z)\left(-\frac{dy}{dz}\right) + by = 0$$

or

$$z^2 \frac{d^2y}{dz^2} + az\frac{dy}{dz} + by = 0. \tag{13}$$

Since $z > 0$, we solve Eq. (13) according to the methods just discussed. Thus, setting $y = z^r = |x|^r$ we obtain solutions in the form of Eqs. (4), (6), or (12) depending on the nature of the roots to quadratic Eq. (3). The only difference is that the variable x is replaced by $|x|$ in each case. Since $x = |x|$ when $x > 0$, *using those solution forms with x replaced by $|x|$ works in every situation for which $x \neq 0$.*

Let us summarize our results.

Summary of Solutions to Second-Order Euler Equation

To solve the equation

$$x^2y'' + axy' + by = 0,$$

substitute $y = x^r$ to obtain the quadratic equation

$$r(r-1) + ar + b = 0.$$

If the roots are real and distinct, then

$$y = c_1|x|^{r_1} + c_2|x|^{r_2}. \tag{14}$$

If the roots are complex, where $r = \alpha \pm \beta i$, then

$$y = |x|^\alpha[c_1 \cos(\beta \ln|x|) + c_2 \sin(\beta \ln|x|)]. \tag{15}$$

If the roots are repeated (equal), then

$$y = |x|^r (c_1 + c_2 \ln|x|). \tag{16}$$

EXERCISES 9.2

In problems 1–24 find the general solution to the given Euler equation.

1. $x^2y'' + 2xy' - 2y = 0$
2. $x^2y'' + xy' - 4y = 0$
3. $x^2y'' - 6y = 0$
4. $x^2y'' + xy' - y = 0$
5. $x^2y'' - 5xy' + 8y = 0$
6. $2x^2y'' + 7xy' + 2y = 0$
7. $3x^2y'' + 4xy' = 0$
8. $x^2y'' + 6xy' + 4y = 0$
9. $x^2y'' - xy' + y = 0$
10. $x^2y'' - xy' + 2y = 0$
11. $x^2y'' - xy' + 5y = 0$
12. $x^2y'' + 7xy' + 13y = 0$
13. $x^2y'' + 3xy' + 10y = 0$
14. $x^2y'' - 5xy' + 10y = 0$

15. $4x^2y'' + 8xy' + 5y = 0$

16. $4x^2y'' - 4xy' + 5y = 0$

17. $x^2y'' + 3xy' + y = 0$

18. $x^2y'' - 3xy' + 9y = 0$

19. $x^2y'' + xy' = 0$

20. $4x^2y'' + y = 0$

21. $9x^2y'' + 15xy' + y = 0$

22. $16x^2y'' - 8xy' + 9y = 0$

23. $16x^2y'' + 56xy' + 25y = 0$

24. $4x^2y'' - 16xy' + 25y = 0$

In problems 25–30 solve the given initial value problem.

25. $x^2y'' + 3xy' - 3y = 0, \quad y(1) = 1, y'(1) = -1$

26. $6x^2y'' + 7xy' - 2y = 0, \quad y(1) = 0, y'(1) = 1$

27. $x^2y'' - xy' + y = 0, \quad y(1) = 1, y'(1) = 1$

28. $x^2y'' + 7xy' + 9y = 0, \quad y(1) = 1, y'(1) = 0$

29. $x^2y'' - xy' + 2y = 0, \quad y(1) = -1, y'(1) = 1$

30. $x^2y'' + 3xy' + 5y = 0, \quad y(1) = 1, y'(1) = 0$

In problems 31–35 find the general solution to the given nonhomogeneous Euler equation. Use variation of parameters to find a particular solution. Assume that $x > 0$.

31. $x^2y'' + xy' - y = x$

32. $x^2y'' + 2xy' - 2y = x^2$

33. $x^2y'' - 3xy' + 4y = x^3$

34. $x^2y'' + xy' + y = 1$

35. $x^2y'' + xy' + y = x^{-1}$

36. Verify that $y_1 = x^\alpha \cos(\beta \ln x)$ and $y_2 = x^\alpha \sin(\beta \ln x)$ are linearly independent.

37. Verify that $y_1 = x^r$ and $y_2 = x^r \ln x$ are linearly independent.

9.3

POWER-SERIES SOLUTIONS

The second-order Euler equation we studied in Section 9.2 is a special case of the more general variable-coefficient equation

$$a_2(x)y'' + a_1(x)y' + a_0(x)y = 0. \tag{1}$$

Let us assume that $a_2(x)$ is never zero over the interval I in which we seek solutions to Eq. (1). We assume further that the coefficient functions $a_2(x)$, $a_1(x)$, and $a_0(x)$ are all continuous and have Taylor series representations over I. In this situation we say that each function $a_i(x)$ is **analytic** over I. For purposes of convenience we also assume that zero belongs to I and use *Maclaurin series expansions* (Taylor series expansions about $x = 0$). This assumption allows for complete generality because the change of variable $z = x - x_0$ for any x_0 in I gives rise to a differential equation equivalent to Eq. (1), where $z = 0$ does belong to I.

The coefficients $a_i(x)$ in the Euler equation studied in Section 9.2 were of such a special form that it seemed reasonable to assume a solution as being a *single* power x^r. (Of course, two powers might occur in the general solution because of the two values for r that might be obtained in the solution procedure.) Perhaps what we need in the more general setting of Eq. (1) is *all possible integral powers of* x. That is, we assume that Eq. (1) has a power-series solution of the form

$$y = \sum_{n=0}^{\infty} c_n x^n \tag{2}$$

where the coefficients c_n in the power series are to be determined from the requirement that Eq. (2) must solve Eq. (1). The class of functions represented by a power series in the form of Eq. (2) is much larger than the class of continuous functions studied in elementary calculus, which include the algebraic, trigonometric, exponential, and logarithmic functions together with various combinations of these types. Thus, seeking solutions within this larger class of continuous functions may be helpful to us.

Power-Series Method of Solution

Before presenting the power-series method of solution, let us gain some perspective as to its usefulness, at least from a theoretical point of view. (We will discuss the issue of practicality further on.) You have seen in calculus that many functions important to science and engineering can be represented in power-series form; that is, the functions are analytic. Moreover, analytic functions, such as the exponential and trigonometric functions, have proven very useful in our quest for solutions to ordinary differential equations. But any power series, due to its convergence and infinite differentiability, also represents an analytic function. Thus the use of power series makes available a vast source of analytic functions as potential candidates for solutions to given differential equations. Many of these functions you have not encountered before in your mathematical studies. An example is the family of **Legendre polynomials,** which are solutions to second-order equations of the form

$$(1 - x^2)y'' - 2xy' + n(n + 1)y = 0$$

where $n = 0, 1, 2, 3, \ldots$. For each nonnegative integer n there is one Legendre polynomial solution. We will investigate solutions to this equation in Section 9.6. Thus power series provide us with a whole new arsenal of functions to use for tackling real-world problems of interest to science and engineering.

The **power-series method** for solving a differential equation consists of substituting form (2) into the differential equation and solving for the unknown coefficients $c_0, c_1, c_2, \ldots$. In a sense the method is very similar to the method of undetermined coefficients except that there are *infinitely many* coefficients to solve for. For some differential equations it is possible to find a **closed-form solution:** a formula or function for the coefficients from which the nth coefficient c_n can be calculated directly from the integer n. For other differential equations it is not possible to find a convenient closed-form solution. In that case we choose a finite value of n, calculate the first n coefficients, and truncate the power series after the nth term. Of course, this latter procedure provides us only with a **polynomial approximation** to the actual solution since we do not actually have an analytical expression for the entire series solution itself. Nevertheless, such approximations can be very useful in qualitative (and even quantitative) analyses of the solution.

It should be remarked that closed-form solutions, even when they do exist, may be difficult to obtain algebraically. Moreover, even if a closed-form solution does exist, it would usually not be recognizable as a known function studied in calculus. Most of the difficulties students encounter with power-series solution methods stem from the algebraic manipulation and recognition of patterns for the coefficients in finding a closed-form solution. In order to work with solution form (2), it will be helpful to now review some results from elementary calculus concerning power series.

Review of Power Series

A series of the form

$$\sum_{n=0}^{\infty} c_n x^n = c_0 + c_1 x + c_2 x^2 + \cdots \tag{3}$$

is called a **power series in x.** A series of the form

$$\sum_{n=0}^{\infty} c_n (x - x_0)^n = c_0 + c_1(x - x_0) + c_2(x - x_0)^2 + \cdots$$

is called a **power series in $(x - x_0)$.** By setting $z = x - x_0$, a power series in $x - x_0$ is simply a power series in z. Thus it suffices to study power series in x only.

Now we fix the value of x, and power series (3) becomes simply a series of constants. If that series of constants converges (so that it equals a *finite* real number), then we say that the power series **converges at x.** If it does not converge, we say it **diverges at x.** If a power series converges at every point on the interval I, then we say simply that the series **converges over I.** If the series of absolute values converges, we say that the power series **converges absolutely over I.**

> Thus a convergent power series is a function over any interval for which it converges.

Notice that power series (3) always converges when $x = 0$; in fact, it has the value c_0 there. One of the key results concerning the convergence of a power series is that it always converges over an interval if it converges for any value of x other than zero.

THEOREM 9.1

Given the power series $\Sigma c_n x^n$, there is a nonnegative R such that if R is finite and nonzero, then

1. $\Sigma c_n x^n$ converges absolutely if $|x| < R$, and
2. $\Sigma c_n x^n$ diverges if $|x| > R$.
3. If $R = +\infty$, then $\Sigma c_n x^n$ converges for all x.
4. If $R = 0$, then $\Sigma c_n x^n$ converges nowhere except $x = 0$.

The R in Theorem 9.1 is called the **radius of convergence** of the power series. It can usually be calculated from the limit

$$R = \lim_{n\to\infty} \left| \frac{c_n}{c_{n+1}} \right| \tag{4}$$

whenever the limit exists. If the limit in Eq. (4) yields $R = 0$, then the power series *converges only for* $x = 0$. If the limit gives $R = +\infty$, then the power series converges *for all real values of* x.

Theorem 9.1 does not say if the power series converges or diverges when $|x| = R$. For a specific power series we must test for convergence or divergence at the endpoints $x = -R, R$. Some power series will converge at one endpoint but diverge at the other. The entire interval over which a power series converges is called its **interval of convergence.**

EXAMPLE 1 For the power series $\sum_{n=1}^{\infty} \frac{x^n}{n}$ we calculate the radius of convergence:

$$R = \lim_{n\to\infty} \frac{n+1}{n} = \lim_{n\to\infty} \left(1 + \frac{1}{n}\right) = 1.$$

Thus the series converges whenever $-1 < x < 1$. What about convergence at the endpoints?

If $x = -1$ the series is $\sum_{n=1}^{\infty} (-1)^n/n$, which converges by the alternating-series test. However, for $x = 1$ the series is the divergent harmonic series $\sum_{n=1}^{\infty} 1/n$. Therefore the interval of convergence is $-1 \leqslant x < 1$.

Two power series can be added or subtracted term by term:

$$\sum_{n=0}^{\infty} b_n x^n \pm \sum_{n=0}^{\infty} c_n x^n = \sum_{n=0}^{\infty} (b_n \pm c_n) x^n. \tag{5}$$

The radius of convergence of the sum is the smallest radius associated with the two series: $R = \min\{R_1, R_2\}$.

Within its interval of convergence, every power series can be differentiated and integrated term by term:

$$\begin{aligned} \frac{d}{dx} \sum_{n=0}^{\infty} c_n x^n &= c_1 + 2c_2 x + 3c_3 x^2 + \cdots \\ &= \sum_{n=1}^{\infty} n c_n x^{n-1} \end{aligned} \tag{6}$$

and

$$\begin{aligned} \int_a^b \sum_{n=0}^{\infty} c_n x^n \, dx &= \sum_{n=0}^{\infty} \int_a^b c_n x^n \, dx \\ &= \sum_{n=0}^{\infty} c_n \left[\frac{x^{n+1}}{n+1} \right]_{x=a}^{x=b} \end{aligned} \tag{7}$$

Repeated application of rule (6) illustrates that a power series represents an infinitely differentiable, hence continuous, function on its interval of convergence. Moreover, all the derivatives are computed by term by term differentiation. The next result establishes that the coefficients themselves can be expressed in terms of these derivatives.

THEOREM 9.2

Let $\Sigma c_n x^n$ have a nonzero radius of convergence R and let $f(x)$ denote the function that is its sum. Then

$$c_n = \frac{f^{(n)}(0)}{n!}. \tag{8}$$

Because the coefficients of a convergent power series are uniquely computable according to Eq. (8), two power series are equal if and only if they have exactly the same coefficients:

$$\sum_{n=0}^{\infty} b_n x^n = \sum_{n=0}^{\infty} c_n x^n \quad \text{if and only if} \quad b_n = c_n \text{ for all } n.$$

TABLE 9.1

Maclaurin Series Expansions of Elementary Functions

$$e^x = \sum_{n=0}^{\infty} \frac{1}{n!} x^n$$

$$= 1 + x + \frac{x^2}{2} + \frac{x^3}{3!} + \cdots, \qquad -\infty < x < \infty. \tag{9}$$

$$\cos x = \sum_{n=0}^{\infty} \frac{(-1)^n}{(2n)!} x^{2n}$$

$$= 1 - \frac{x^2}{2} + \frac{x^4}{4!} - \cdots, \qquad -\infty < x < \infty. \tag{10}$$

$$\sin x = \sum_{n=0}^{\infty} \frac{(-1)^n}{(2n+1)!} x^{2n+1}$$

$$= x - \frac{x^3}{3!} + \frac{x^5}{5!} - \cdots, \qquad -\infty < x < \infty. \tag{11}$$

$$\ln(1+x) = \sum_{n=1}^{\infty} \frac{(-1)^{n+1}}{n} x^n$$

$$= x - \frac{x^2}{2} + \frac{x^3}{3} - \cdots, \qquad -1 < x < 1. \tag{12}$$

$$(1-x)^{-1} = \sum_{n=0}^{\infty} x^n = 1 + x + x^2 + \cdots, \qquad -1 < x < 1. \tag{13}$$

You learned in calculus that many familiar functions can be represented in power-series form. That is, if $y = f(x)$ is a known function, it may be possible to represent $f(x)$ by a power series whose coefficients are computed according to formula (8). This representation is known as the **Maclaurin series expansion** of the function. Well-known Maclaurin series expansions you studied in calculus are summarized in Table 9.1.

Not every familiar function has a Maclaurin series expansion. For instance, the functions $y = |x|$ and $y = \sqrt{x}$ do not possess Maclaurin series expansions. The difficulty arises because these particular functions fail to have derivatives at $x = 0$.

Applying the Solution Method

The next example illustrates how a power-series solution can be applied for a familiar first-order differential equation. The example is instructive for investigating the nature of the solution.

EXAMPLE 2 Solve the equation $y' - y = 0$ using the power-series method.

Solution. We substitute

$$y = \sum_{n=0}^{\infty} c_n x^n$$

and its derivative

$$y' = \sum_{n=1}^{\infty} n c_n x^{n-1}$$

into the differential equation to obtain

$$\sum_{n=1}^{\infty} n c_n x^{n-1} - \sum_{n=0}^{\infty} c_n x^n = 0.$$

Since all the coefficients are zero on the righthand side of this last equation, the coefficients of each power of x on the lefthand side must also equal zero. The following table summarizes these coefficients.

Power of x	Coefficient Equation		
x^0	$1c_1 - c_0 = 0$	or	$c_1 = c_0$
x^1	$2c_2 - c_1 = 0$	or	$c_2 = \frac{1}{2}c_1$
x^2	$3c_3 - c_2 = 0$	or	$c_3 = \frac{1}{3}c_2$
x^3	$4c_4 - c_3 = 0$	or	$c_4 = \frac{1}{4}c_3$
$\vdots$	$\vdots$		$\vdots$
x^{n-1}	$nc_n - c_{n-1} = 0$	or	$c_n = \frac{1}{n} c_{n-1}$

The formula

$$c_n = \frac{1}{n} c_{n-1}$$

obtained in the table is known as a **recursion formula** or **recursive relation.** It expresses the nth coefficient in terms of its predecessor c_{n-1}. (In some problems several predecessors may be involved in the expression.) The formula holds for all values of n, so we can use it next to calculate c_{n-1}:

$$c_{n-1} = \frac{1}{n-1} c_{n-2}.$$

Then substitution for c_{n-1} into our expression for c_n gives us

$$c_n = \frac{1}{n} \cdot \frac{1}{n-1} c_{n-2}.$$

Then we use the original recursion formula again to calculate the next preceding coefficient c_{n-2}:

$$c_{n-2} = \frac{1}{n-2} c_{n-3}.$$

Substitution into our form for c_n now yields

$$c_n = \frac{1}{n} \cdot \frac{1}{n-1} \cdot \frac{1}{n-2} c_{n-3}.$$

Continuing to work backwards in this fashion we eventually arrive at the expression

$$c_n = \frac{1}{n} \cdot \frac{1}{n-1} \cdot \frac{1}{n-2} \cdots \frac{1}{2} c_1.$$

Finally we substitute for $c_1 : c_1 = c_0$ to obtain

$$c_n = \frac{1}{n} \cdot \frac{1}{n-1} \cdot \frac{1}{n-2} \cdots \frac{1}{2} c_0$$

or

$$c_n = \frac{1}{n!} c_0.$$

We now have a formula for c_n *in terms of the index n*. Substituting c_n into the solution form for y yields the power-series solution

$$y = c_0 \sum_{n=0}^{\infty} \frac{1}{n!} x^n.$$

Here c_0 is an arbitrary constant or parameter that could be determined from an initial condition $y(x_0) = y_0$.

Comparing our power-series solution with Maclaurin series expansion (9) for e^x, we see that

$$y = c_0 e^x.$$

Our results are consistent with the solution found in Chapter 1 when we solved the equation $y' - y = 0$ by separating the variables. We will not, however, always be able to recognize our series solution as some known elementary function.

EXAMPLE 3 Solve the equation $y'' + y = 0$ by the power-series method.

Solution. We assume the series solution takes the form of

$$y = \sum_{n=0}^{\infty} c_n x^n$$

and calculate the derivatives:

$$y' = \sum_{n=1}^{\infty} nc_n x^{n-1} \quad \text{and} \quad y'' = \sum_{n=2}^{\infty} n(n-1)c_n x^{n-2}.$$

Substitution of these forms into the second-order equation gives us

$$\sum_{n=2}^{\infty} n(n-1)c_n x^{n-2} + \sum_{n=0}^{\infty} c_n x^n = 0.$$

Again we equate the coefficients of each power of x to zero, as summarized in the following table.

Power of x	Coefficient Equation			
x^0	$2(1)c_2 + c_0$	$= 0$	or	$c_2 = -\frac{1}{2}c_0$
x^1	$3(2)c_3 + c_1$	$= 0$	or	$c_3 = -\frac{1}{3 \cdot 2}c_1$
x^2	$4(3)c_4 + c_2$	$= 0$	or	$c_4 = -\frac{1}{4 \cdot 3}c_2$
x^3	$5(4)c_5 + c_3$	$= 0$	or	$c_5 = -\frac{1}{5 \cdot 4}c_3$
x^4	$6(5)c_6 + c_4$	$= 0$	or	$c_6 = -\frac{1}{6 \cdot 5}c_4$
$\vdots$		$\vdots$		$\vdots$
x^{n-2}	$n(n-1)c_n + c_{n-2}$	$= 0$	or	$c_n = -\frac{1}{n(n-1)}c_{n-2}$

From the table we notice that the coefficients with even indices ($n = 2k$, $k = 1, 2, 3, \ldots$) are related to each other, and the coefficients with odd indices ($n = 2k + 1$) are also interrelated. We treat each group in turn.

Even indices: Here $n = 2k$, so the power is x^{2k-2}. From the last line of the table we have

$$2k(2k-1)c_{2k} + c_{2k-2} = 0$$

or

$$c_{2k} = -\frac{1}{2k(2k-1)} c_{2k-2}.$$

From this recursive relation we find

$$\begin{aligned} c_{2k} &= \left[-\frac{1}{2k(2k-1)}\right]\left[-\frac{1}{(2k-2)(2k-3)}\right] \cdots \left[-\frac{1}{4(3)}\right]\left[-\frac{1}{2}\right] c_0 \\ &= \frac{(-1)^k}{(2k)!} c_0. \end{aligned}$$

Odd indices: Here $n = 2k + 1$, so the power is x^{2k-1}. Substituting this into the last line of the table yields

$$(2k+1)(2k)c_{2k+1} + c_{2k-1} = 0$$

or

$$c_{2k+1} = -\frac{1}{(2k+1)(2k)} c_{2k-1}.$$

Thus

$$\begin{aligned} c_{2k+1} &= \left[-\frac{1}{(2k+1)(2k)}\right]\left[-\frac{1}{(2k-1)(2k-2)}\right] \cdots \left[-\frac{1}{5(4)}\right]\left[-\frac{1}{3(2)}\right] c_1 \\ &= \frac{(-1)^k}{(2k+1)!} c_1. \end{aligned}$$

Writing the power series by grouping its even and odd powers together and substituting for the coefficients yields

$$\begin{aligned} y &= \sum_{n=0}^{\infty} c_n x^n \\ &= \sum_{k=0}^{\infty} c_{2k} x^{2k} + \sum_{k=0}^{\infty} c_{2k+1} x^{2k+1} \\ &= c_0 \sum_{k=0}^{\infty} \frac{(-1)^k}{(2k)!} x^{2k} + c_1 \sum_{k=0}^{\infty} \frac{(-1)^k}{(2k+1)!} x^{2k+1}. \end{aligned}$$

Comparing our series solution with Maclaurin series expansions (10) and (11), we see that

$$y = c_0 \cos x + c_1 \sin x$$

where c_0 and c_1 are arbitrary constants. This agrees with our solution of the homogeneous equation $y'' + y = 0$ found in Chapter 3.

Our purpose in presenting Examples 2 and 3 was to demonstrate that the power-series method leads to the same general solution obtained by earlier methods. This gives you some understanding of the form of the solution when you can recognize it as a familiar Maclaurin series. The examples were intended to help you be more comfortable with a solution expressed in series form. Because of the amount of algebra involved, especially in working with a recursive relation to obtain a closed-form solution, we acknowledge that the power-series method is not easy to employ. Let us consider now a variable-coefficient example not solvable by other methods.

EXAMPLE 4 Find the general solution to $y'' + xy' + y = 0$.

Solution. We assume the series solution form

$$y = \sum_{n=0}^{\infty} c_n x^n$$

and calculate the derivatives

$$y' = \sum_{n=1}^{\infty} nc_n x^{n-1} \qquad \text{and} \qquad y'' = \sum_{n=2}^{\infty} n(n-1)c_n x^{n-2}.$$

Substitution of these forms into the second-order equation yields

$$\sum_{n=2}^{\infty} n(n-1)c_n x^{n-2} + \sum_{n=1}^{\infty} nc_n x^n + \sum_{n=0}^{\infty} c_n x^n = 0.$$

We equate the coefficients of each power of x to zero, as summarized in the following table.

Power of x	Coefficient Equation		
x^0	$2(1)c_2 + c_0 = 0$	or	$c_2 = -\frac{1}{2}c_0$
x^1	$3(2)c_3 + c_1 + c_1 = 0$	or	$c_3 = -\frac{1}{3}c_1$
x^2	$4(3)c_4 + 2c_2 + c_2 = 0$	or	$c_4 = -\frac{1}{4}c_2$
x^3	$5(4)c_5 + 3c_3 + c_3 = 0$	or	$c_5 = -\frac{1}{5}c_3$
x^4	$6(5)c_6 + 4c_4 + c_4 = 0$	or	$c_6 = -\frac{1}{6}c_4$
$\vdots$	$\vdots$		$\vdots$
x^n	$(n+2)(n+1)c_{n+2} + (n+1)c_n = 0$	or	$c_{n+2} = -\dfrac{1}{n+2}c_n$

From the table notice that the coefficients with even indices are interrelated, and the coefficients with odd indices are also interrelated.

Even indices: Here $n = 2k - 2$, so the power is x^{2k-2}. From the last line in the table, we have

$$c_{2k} = -\frac{1}{2k}c_{2k-2}.$$

From this recurrence relation we obtain

$$c_{2k} = \left(-\frac{1}{2k}\right)\left(-\frac{1}{2k-2}\right) \cdots \left(-\frac{1}{6}\right)\left(-\frac{1}{4}\right)\left(-\frac{1}{2}\right) c_0$$
$$= \frac{(-1)^k}{(2)(4)(6) \ldots (2k)} c_0.$$

Odd indices: Here $n = 2k - 1$, so the power is x^{2k-1}. From the last line in the table, we have

$$c_{2k+1} = -\frac{1}{2k+1} c_{2k-1}.$$

From this recurrence relation we obtain

$$c_{2k+1} = \left(-\frac{1}{2k+1}\right)\left(-\frac{1}{2k-1}\right) \cdots \left(-\frac{1}{5}\right)\left(-\frac{1}{3}\right) c_1$$
$$= \frac{(-1)^k}{(3)(5) \cdots (2k+1)} c_1.$$

Writing the power series by grouping its even and odd powers and substituting for the coefficients yields

$$y = \sum_{k=0}^{\infty} c_{2k} x^{2k} + \sum_{k=0}^{\infty} c_{2k+1} x^{2k+1}$$
$$= c_0 \sum_{k=0}^{\infty} \frac{(-1)^k}{(2)(4) \cdots (2k)} x^{2k} + c_1 \sum_{k=0}^{\infty} \frac{(-1)^k}{(3)(5) \cdots (2k+1)} x^{2k+1}.$$

EXAMPLE 5 Find the general solution to

$$(1 - x^2)y'' - 6xy' - 4y = 0.$$

Solution. Notice that the leading coefficient is zero when $x = \pm 1$. Thus we assume the solution interval I: $-1 < x < 1$. Substitution of the series form

$$y = \sum_{n=0}^{\infty} c_n x^n$$

and its derivatives gives us

$$(1 - x^2) \sum_{n=2}^{\infty} n(n-1)c_n x^{n-2} - 6 \sum_{n=1}^{\infty} nc_n x^n - 4 \sum_{n=0}^{\infty} c_n x^n = 0,$$

$$\sum_{n=2}^{\infty} n(n-1)c_n x^{n-2} - \sum_{n=2}^{\infty} n(n-1)c_n x^n - 6 \sum_{n=1}^{\infty} nc_n x^n - 4 \sum_{n=0}^{\infty} c_n x^n = 0.$$

Next we equate the coefficients of each power of x to zero.

Power of x	Coefficient Equation		
x^0	$2(1)c_2 - 4c_0 = 0$	or	$c_2 = \frac{4}{2}c_0$
x^1	$3(2)c_3 - 6(1)c_1 - 4c_1 = 0$	or	$c_3 = \frac{5}{3}c_1$
x^2	$4(3)c_4 - 2(1)c_2 - 6(2)c_2 - 4c_2 = 0$	or	$c_4 = \frac{6}{4}c_2$
x^3	$5(4)c_5 - 3(2)c_3 - 6(3)c_3 - 4c_3 = 0$	or	$c_5 = \frac{7}{5}c_3$
$\vdots$	$\vdots$		$\vdots$
x^n	$(n+2)(n+1)c_{n+2} - [n(n-1) + 6n + 4]c_n = 0$		
	$(n+2)(n+1)c_{n+2} - (n+4)(n+1)c_n = 0$	or	$c_{n+2} = \frac{n+4}{n+2} c_n$

Again we notice that the coefficients with even indices are interrelated, and those with odd indices are interrelated.

Even indices: Here $n = 2k - 2$, so the power is x^{2k}. From the righthand column and last line of the table we get

$$
\begin{aligned}
c_{2k} &= \frac{2k+2}{2k} c_{2k-2} \\
&= \left(\frac{2k+2}{2k}\right)\left(\frac{2k}{2k-2}\right)\left(\frac{2k-2}{2k-4}\right) \cdots \frac{6}{4}\left(\frac{4}{2}\right) c_0 \\
&= (k+1)c_0.
\end{aligned}
$$

Odd indices: $n = 2k - 1$, so the power is x^{2k+1}. The righthand column and last line of the table gives us

$$
\begin{aligned}
c_{2k+1} &= \frac{2k+3}{2k+1} c_{2k-1} \\
&= \left(\frac{2k+3}{2k+1}\right)\left(\frac{2k+1}{2k-1}\right)\left(\frac{2k-1}{2k-3}\right) \cdots \frac{7}{5}\left(\frac{5}{3}\right) c_1 \\
&= \frac{2k+3}{3} c_1.
\end{aligned}
$$

The general solution is thus

$$
\begin{aligned}
y &= \sum_{n=0}^{\infty} c_n x^n \\
&= \sum_{k=0}^{\infty} c_{2k} x^{2k} + \sum_{k=0}^{\infty} c_{2k+1} x^{2k+1} \\
&= c_0 \sum_{k=0}^{\infty} (k+1) x^{2k} + c_1 \sum_{k=0}^{\infty} \frac{2k+3}{3} x^{2k+1}.
\end{aligned}
$$

Example 5 illustrates that a power-series solution may exist only over a finite interval of values of the independent variable x. This is because of some theoretical considerations concerning power series, which we now discuss briefly.

Solutions around Ordinary Points

For purposes of discussion it is useful to place the second-order differential equation

$$a_2(x)y'' + a_1(x)y' + a_0(x)y = 0 \tag{14}$$

in the standard form

$$y'' + p(x)y' + q(x)y = 0 \tag{15}$$

where $p(x) = a_1(x)/a_2(x)$ and $q(x) = a_0(x)/a_2(x)$.

DEFINITION 9.1

A point $x = x_0$ is an **ordinary point** of Eq. (15) if both $p(x)$ and $q(x)$ are **analytic** at x_0; that is, if both $p(x)$ and $q(x)$ have Taylor series representations about $x = x_0$. A point that is not an ordinary point is called a **singular point** of the equation.

If in Eq. (14) the coefficients $a_i(x)$ are all *polynomials with no common factor,* then $x = x_0$ is an ordinary point if $a_2(x_0) \neq 0$. If $a_2(x_0) = 0$, then $x = x_0$ is a singular point.

EXAMPLE 6 The differential equation

$$(1 - x^2)y'' - 6xy' - 4y = 0$$

has an ordinary point at $x = 0$. The points $x = -1$ and $x = 1$ are singular points of the equation.

EXAMPLE 7 Singular points need not be real numbers. The equation

$$(x^2 + 4)y'' + 2xy' - 12y = 0$$

has singular points at $x = \pm 2i$. The point $x = 0$ is an ordinary point.

EXAMPLE 8 The Euler equation

$$x^2y'' + axy' + by = 0$$

has a singular point at $x = 0$. All other values of x, real or complex, are ordinary points.

The following is the fundamental result concerning power-series solutions. We state the result without proof.

THEOREM 9.3

Assume that $x = x_0$ is an ordinary point of

$$a_2(x)y'' + a_1(x)y' + a_0(x)y = 0.$$

Then there are two different power-series solutions of the form

$$y = \sum_{n=0}^{\infty} c_n(x - x_0)^n.$$

A series solution converges at least for all values of x satisfying the relation $|x - x_0| < R$, where R is the distance in the complex plane from x_0 to the nearest singular point.

Applying Theorem 9.3 to the differential equation given in Example 5, we see that the two solutions we found there,

$$y_1 = \sum_{k=0}^{\infty} (k+1)x^{2k} \qquad \text{and} \qquad y_2 = \sum_{k=0}^{\infty} \frac{2k+3}{3} x^{2k+1}$$

converge for all values of x satisfying $|x| < 1$ (because $x = \pm 1$ are singular points).

EXAMPLE 9 Find the general solution to $y'' - 2xy' + y = 0$.

Solution. We note that $x = 0$ is an ordinary point of the differential equation and that no singular points exist in the complex plane. Thus any series solution will converge for all x. Assuming that

$$y = \sum_{n=0}^{\infty} c_n x^n,$$

substitution into the differential equation gives us

$$\sum_{n=2}^{\infty} n(n-1)c_n x^{n-2} - 2\sum_{n=1}^{\infty} nc_n x^n + \sum_{n=0}^{\infty} c_n x^n = 0.$$

We next determine the coefficients, listing them in the following table.

Power of x	Coefficient Equation		
x^0	$2(1)c_2 + c_0 = 0$	or	$c_2 = -\frac{1}{2} c_0$
x^1	$3(2)c_3 - 2c_1 + c_1 = 0$	or	$c_3 = \frac{1}{3 \cdot 2} c_1$
x^2	$4(3)c_4 - 4c_2 + c_2 = 0$	or	$c_4 = \frac{3}{4 \cdot 3} c_2$
x^3	$5(4)c_5 - 6c_3 + c_3 = 0$	or	$c_5 = \frac{5}{5 \cdot 4} c_3$
x^4	$6(5)c_6 - 8c_4 + c_4 = 0$	or	$c_6 = \frac{7}{6 \cdot 5} c_4$
⋮	⋮		⋮
x^n	$(n+2)(n+1)c_{n+2} - (2n-1)c_n = 0$	or	$c_{n+2} = \frac{2n-1}{(n+2)(n+1)} c_n$

In this problem there is no simple formula for the even and odd indexed coefficients resulting from the recursive relation

$$c_{n+2} = \frac{2n-1}{(n+2)(n+1)} c_n.$$

Thus we simply write out the first few terms of each series for the general solution

$$y = c_0\left(1 - \frac{1}{2}x^2 - \frac{3}{4!}x^4 - \frac{21}{6!}x^6 - \cdots\right) + c_1\left(x + \frac{1}{3!}x^3 + \frac{5}{5!}x^5 + \frac{45}{7!}x^7 + \cdots\right).$$

We now comment on the practicality of power-series solutions. Although Theorem 9.3 does guarantee the existence of power-series solutions about an ordinary point, our examples show that a considerable amount of computational algebra is required to obtain the coefficients in the series. Moreover, it may be exceedingly difficult, even impossible, to obtain a simple formula for c_n in terms of n from the recursive relation obtained in the solution procedure. In such situations we are going to obtain polynomial approximations only, as we did in Example 9. Therefore, if initial conditions are known, it might be far better to proceed with a numerical approximation to the solution, as discussed in Chapter 4. In practice that is the procedure most likely to be followed in engineering or applied science. In general, power-series solutions are a last resort for approximating solutions.

EXERCISES 9.3

In problems 1–6 find a power-series solution of the given differential equation. Use Eq. (4) to find the radius of convergence of the resulting series.

1. $y' = 2y$

2. $y' + y = 0$

3. $y' - xy = 0$

4. $(1 - x)y' = y$

5. $(1 + x^2)y' + 2xy = 0$

6. $y' = x^2 y$

In problems 7–30 find two linearly independent solutions in powers of x for the given differential equation. Begin by finding only the first four terms in the series solution, thereby obtaining a polynomial function *approximating* the analytical solution. Then find a closed-form solution, if possible.

7. $y'' + 2y' = 0$

8. $y'' + 2y' + y = 0$

9. $y'' + 4y = 0$

10. $y'' - 3y' + 2y = 0$

11. $x^2 y'' - 2xy' + 2y = 0$

12. $y'' - xy' + y = 0$

13. $(1 + x)y'' - y = 0$

14. $(1 - x^2)y'' - 4xy' + 6y = 0$

15. $(x^2 - 1)y'' + 2xy' - 2y = 0$

16. $y'' + y' - x^2 y = 0$

17. $y'' + xy' = 0$

18. $x(x + 1)y'' + (x - 1)y' - y = 0$

19. $(x^2 - 1)y'' - 6y = 0$

20. $xy'' - (x + 2)y' + 2y = 0$

21. $(x^2 - 1)y'' + 4xy' + 2y = 0$

22. $y'' - 2xy' + 4y = 0$

23. $y'' - 2xy' + 3y = 0$

24. $(1 - x^2)y'' - xy' + 4y = 0$

25. $y'' - xy' + 3y = 0$

26. $x^2 y'' - 4xy' + 6y = 0$

27. $y'' + y = \sin x$ (*Hint:* expand $\sin x$ about $x = 0$)

28. $(x^2 + 4)y'' + 6xy' + 4y = 0$

29. $x^2 y'' - 2xy' + (2 - x^2)y = 0$

30. $(1 - x^2)y'' - 2xy' + 2y = 0$ (This is Legendre's equation for $n = 1$.)

In problems 31–33 use the power-series method to solve the given differential equation subject to the indicated initial conditions.

31. $y'' + xy' - x^2 y = 0$, $y(0) = 1$, $y'(0) = -1$

32. $x(3 - 2x)y'' - 6(1 - x)y' - 6y = 0$, $y(1) = 1$, $y'(1) = 0$

33. $(x^2 + 1)y'' + 6xy' + 6y = 0$, $y(0) = 0$, $y'(0) = 1$

34. **a)** Show that the power-series method gives only a particular solution of the differential equation

$$4xy'' + 2y' + y = 0.$$

b) Write the particular solution found in part (a) in terms of the cosine function.

c) Using the method of reduction of order, find the general solution to the equation.

35. **Hermite's equation** is

$$y'' - 2xy' + 2\alpha y = 0$$

where α is a constant.

a) Find two linearly independent power-series solutions.

b) If $\alpha = n$ is a nonnegative integer, Hermite's equation has a polynomial solution. Find polynomial solutions for $\alpha = 0, 1, 2, 3$.

9.4

FROBENIUS SERIES

Suppose we attempt a power-series solution of the form

$$y = \sum_{n=0}^{\infty} c_n x^n$$

for the Euler–Cauchy equation

$$x^2y'' - xy' - 3y = 0. \tag{1}$$

Substitution of the power-series form and its derivatives into Eq. (1) leads to

$$\sum_{n=2}^{\infty} n(n-1)c_n x^n - \sum_{n=1}^{\infty} nc_n x^n - 3\sum_{n=0}^{\infty} c_n x^n = 0.$$

We next determine the coefficients.

Power of x	Coefficient Equation		
x^0	$-3c_0 = 0$	or	$c_0 = 0$
x^1	$-c_1 - 3c_1 = 0$	or	$c_1 = 0$
x^2	$2c_2 - 2c_2 - 3c_2 = 0$	or	$c_2 = 0$
x^3	$6c_3 - 3c_3 - 3c_3 = 0$	or	c_3 is arbitrary
x^4	$12c_4 - 4c_4 - 3c_4 = 0$	or	$c_4 = 0$
x^5	$20c_5 - 5c_5 - 3c_5 = 0$	or	$c_5 = 0$
$\vdots$	$\vdots$		$\vdots$
x^n	$n(n-1)c_n - nc_n - 3c_n = 0$	or	$(n-3)(n+1)c_n = 0$

From the last line of the table we see that $c_n = 0$ unless $n = 3$. Thus we obtain only one independent solution:

$$y_1 = x^3.$$

However, in solving the equation given in Example 1 of Section 9.2 we found a second solution, $y_2 = x^{-1}$. The fact that the power-series method did not produce y_2 comes as no surprise when you realize that $y_2 = x^{-1}$ does not possess a Maclaurin series expansion.

The power-series method studied in Section 9.3 applies in the situation when $x = x_0$ is an ordinary point of the differential equation. In certain cases where $x = x_0$ is a singular point, a power series multiplied by an appropriate power of $x - x_0$, say $(x - x_0)^r$, will yield a solution. We will pursue this idea in this section.

DEFINITION 9.2

A singular point $x = x_0$ of the equation

$$y'' + p(x)y' + q(x)y = 0 \qquad (2)$$

is said to be a **regular singular point** if both terms $(x - x_0)p(x)$ and $(x - x_0)^2 q(x)$ are analytic at x_0. Otherwise $x = x_0$ is an **irregular singular point.**

Equations with Polynomial Coefficients

If $p(x)$ and $q(x)$ are rational functions (that is, quotients of polynomials) reduced to lowest terms, then a singular point $x = x_0$ is regular provided that the factor $x - x_0$ appears *at most* to the first power in the denominator of $p(x)$ and *at most* to the second power in the denominator of $q(x)$.

EXAMPLE 1

The point $x = 0$ is a singular point of the Euler–Cauchy equation

$$x^2y'' - xy' - 3y = 0.$$

Since the factor $x - 0$ occurs to only the first power in the denominator of $p(x) = -x/x^2 = -1/x$ and to only the second power in $q(x) = -3/x^2$, we conclude that $x = 0$ is a regular singular point.

EXAMPLE 2

The points $x = -3$ and $x = 3$ are singular points of the equation

$$(x^2 - 9)^2y'' + (x - 3)y' + 2y = 0.$$

Since

$$p(x) = \frac{1}{(x + 3)^2(x - 3)} \qquad \text{and} \qquad q(x) = \frac{2}{(x + 3)^2(x - 3)^2},$$

the point $x = 3$ is a regular singular point. But $x = -3$ is an irregular singular point because the factor $x + 3$ occurs to the *second* power in the denominator of $p(x)$.

EXAMPLE 3

The points $x = -i$ and $x = i$ are singular points of the equation

$$(x^2 + 1)y'' + (x + 1)y' + 5y = 0.$$

Moreover, since

$$p(x) = \frac{x + 1}{(x + i)(x - i)} \qquad \text{and} \qquad q(x) = \frac{5}{(x + i)(x - i)},$$

both points are regular singular points.

The following result, due to Georg Frobenius (1884–1917), is useful for solving second-order differential equations about a regular singular point.

THEOREM 9.4

Assume that $x = x_0$ is a regular singular point of the differential equation

$$a_2(x)y'' + a_1(x)y' + a_0(x)y = 0. \tag{3}$$

Then there exists at least one series solution of the form

$$\begin{aligned} y &= (x - x_0)^r \sum_{n=0}^{\infty} c_n(x - x_0)^n \\ &= \sum_{n=0}^{\infty} c_n(x - x_0)^{n+r} \end{aligned} \tag{4}$$

where the number r is some real constant. The series will converge on some interval $0 < |x - x_0| < R$.

The series in Eq. (4) is known as a **Frobenius series**. Since the number r in Eq. (4) need not be an integer, the powers of $(x - x_0)$ appearing in a Frobenius series are not necessarily integral powers. Moreover, if r happens to be a negative integer, series (4) will contain negative powers of $(x - x_0)$. *As was the case for power series, we henceforth assume that $x_0 = 0$ without any loss of generality.* We also assume that $c_0 \neq 0$, so that the first power appearing in the Frobenius series is x^r.

EXAMPLE 4 Let us solve Euler–Cauchy Eq. (1) by assuming a Frobenius series solution. Thus

$$\begin{aligned} y &= \sum_{n=0}^{\infty} c_n x^{n+r}, \\ y' &= \sum_{n=0}^{\infty} (n + r)c_n x^{n+r-1}, \\ y'' &= \sum_{n=0}^{\infty} (n + r)(n + r - 1)c_n x^{n+r-2}. \end{aligned}$$

Note that the lower index of summation is still $n = 0$ after the differentiation process for a Frobenius series because of the presence of the x^r factor. Substitution of y and its derivatives into Eq. (1) gives us

$$\sum_{n=0}^{\infty} (n + r)(n + r - 1)c_n x^{n+r} - \sum_{n=0}^{\infty} (n + r)c_n x^{n+r} - 3 \sum_{n=0}^{\infty} c_n x^{n+r} = 0.$$

As in the case of power-series solutions, we equate the coefficient of each power of x to zero.

Power of x	Coefficient Equation
x^r	$r(r-1)c_0 - rc_0 - 3c_0 = 0,$
	or, since $c_0 \neq 0$,
	$r(r-1) - r - 3 = (r-3)(r+1) = 0.$
	Thus $r = -1$ or $r = 3$.
x^{n+r}	$[(n+r)(n+r-1) - (n+r) - 3]c_n = 0,$
	$[(n+r)^2 - 2(n+r) - 3]c_n = 0,$
	$(n+r-3)(n+r+1)c_n = 0.$

If we substitute $r = -1$ into the last equation in the table above, we obtain

$$(n-4)nc_n = 0.$$

Thus $c_n = 0$ if $n \neq 0$ or $n \neq 4$, where c_0 and c_4 are arbitrary. Substitution into the Frobenius series form yields

$$y_1 = \sum_{n=0}^{\infty} c_n x^{n+r} = \sum_{n=0}^{\infty} c_n x^{n-1}$$

or

$$y_1 = c_0 x^{-1} + c_4 x^3. \tag{5}$$

If $r = 3$, the coefficient equation for x^{n+r} becomes

$$n(n+4)c_n = 0,$$

which gives us $c_n = 0$ if $n \neq 0$. Substitution into the Frobenius series form yields

$$y_2 = \sum_{n=0}^{\infty} c_n x^{n+3} = c_0 x^3. \tag{6}$$

When the superposition principle is applied to Eqs. (5) and (6), we obtain the general solution

$$y = C_1 x^{-1} + C_2 x^3$$

where C_1 and C_2 are arbitrary constants. This solution agrees with that found in Example 1 of Section 9.2.

Although usually it will not be so easy to find the general solution using the Frobenius series method, Example 4 does illustrate the technique for finding the general solution about a regular singular point. We now discuss some important features of the method.

Indicial Equation

In Example 4 we calculated the **coefficient equation for the lowest power of x** occurring in the Frobenius series form of the differential equation (x^r in the example), and we obtained

$$(r^2 - 2r - 3)c_0 = 0.$$

The example illustrates what happens in the general case when $x = 0$ is a regular singular point of second-order Eq. (3). The coefficient equation for the lowest power of x has the form

$$P(r)c_0 = 0 \tag{7}$$

where $P(r)$ is a *quadratic polynomial in* r. Since $c_0 \neq 0$, we must have

$$P(r) = 0. \tag{8}$$

Equation (8) is called the **indicial equation** associated with the Frobenius series solution. Since $P(r)$ is a quadratic polynomial, three possible cases are distinguished according to the nature of the roots to the indicial equation. For purposes of discussion we assume that r_1 and r_2 are real roots of the indicial equation and that r_1 *denotes the largest root*. The three cases are as follows:

1. The roots r_1 and r_2 are unequal and *do not* differ by an integer.
2. The roots r_1 and r_2 are unequal but differ by a positive integer.
3. The roots r_1 and r_2 are equal.

The form of the solutions obtained by the Frobenius series method in each of the three cases is summarized in the following theorem. We omit the proof.

THEOREM 9.5

Assume that $x = 0$ is a regular singular point of second-order differential equation

$$a_2(x)y'' + a_1(x)y' + a_0(x)y = 0. \tag{9}$$

Suppose that $r_1 \geqslant r_2$ are two real roots to the indicial equation $P(r) = 0$.

1. If $r_1 \neq r_2$ and $r_1 - r_2$ is *not* an integer, then there exist two linearly independent solutions to Eq. (9) of the form

$$y_1 = \sum_{n=0}^{\infty} c_n x^{n+r_1}, \qquad c_0 \neq 0, \tag{10}$$

$$y_2 = \sum_{n=0}^{\infty} b_n x^{n+r_2}, \qquad b_0 \neq 0. \tag{11}$$

2. If $r_1 - r_2$ is a positive integer, then there exist two linearly independent solutions to Eq. (9) of the form

$$y_1 = \sum_{n=0}^{\infty} c_n x^{n+r_1}, \quad c_0 \neq 0, \tag{12}$$

$$y_2 = Cy_1(x) \ln x + \sum_{n=0}^{\infty} b_n x^{n+r_2}, \quad b_0 \neq 0 \tag{13}$$

where C is a constant that could be zero.

3. If $r_1 = r_2$, then there exist two linearly independent solutions to Eq. (9) of the form

$$y_1 = \sum_{n=0}^{\infty} c_n x^{n+r_1}, \qquad c_0 \neq 0, \tag{14}$$

$$y_2 = y_1(x) \ln x + \sum_{n=1}^{\infty} b_n x^{n+r_1}. \tag{15}$$

Examples 5–7 illustrate each case.

EXAMPLE 5 INDICIAL ROOTS DO NOT DIFFER BY AN INTEGER

Find the general solution to

$$2x^2 y'' + 3xy' - (1 + x)y = 0.$$

Solution. Substituting the Frobenius series form

$$y = \sum_{n=0}^{\infty} c_n x^{n+r}$$

and its derivatives into the differential equation gives us

$$2x^2 \sum_{n=0}^{\infty} (n+r)(n+r-1)c_n x^{n+r-2} + 3x \sum_{n=0}^{\infty} (n+r)c_n x^{n+r-1}$$
$$-(1+x) \sum_{n=0}^{\infty} c_n x^{n+r} = 0$$

or

$$\sum_{n=0}^{\infty} [2(n+r)(n+r-1)+3(n+r)-1]c_n x^{n+r} - \sum_{n=0}^{\infty} c_n x^{n+r+1} = 0.$$

The indicial equation comes from the coefficient of x^r:

$$[2r(r-1)+3r-1]c_0 = 0$$

or, since $c_0 \neq 0$,

$$2r^2 + r - 1 = 0,$$
$$(2r-1)(r+1) = 0.$$

Thus, $r_1 = \frac{1}{2}$ and $r_2 = -1$.

The coefficients of the remaining powers of x are summarized in the following table.

Power of x	Coefficient Equation
x^{r+1}	$[2(r+1)r+3(r+1)-1]c_1 - c_0 = 0$
	or $c_1 = \dfrac{1}{(2r+1)(r+2)} c_0.$
x^{r+2}	$[2(r+2)(r+1)+3(r+2)-1]c_2 - c_1 = 0$
	or $c_2 = \dfrac{1}{(2r+3)(r+3)} c_1$
$\vdots$	$\vdots$
x^{r+n}	$[2(r+n)(r+n-1)+3(r+n)-1]c_n - c_{n-1} = 0$
	or $c_n = \dfrac{1}{[2(r+n)-1][(r+n)+1]} c_{n-1}.$

We let $r = r_1 = \frac{1}{2}$, the larger indicial root. Then

$$c_n = \frac{1}{2n(n+\frac{3}{2})} c_{n-1} = \frac{1}{n(2n+3)} c_{n-1}, \qquad n \geqslant 1.$$

Hence

$$c_1 = \frac{1}{1 \cdot 5} c_0,$$

$$c_2 = \frac{1}{2 \cdot 7} c_1 = \frac{1}{(1 \cdot 2) \cdot 5 \cdot 7} c_0,$$

$$c_3 = \frac{1}{3 \cdot 9} c_2 = \frac{1}{(1 \cdot 2 \cdot 3) \cdot 5 \cdot 7 \cdot 9} c_0,$$

$$c_4 = \frac{1}{4 \cdot 11} c_3 = \frac{1}{(1 \cdot 2 \cdot 3 \cdot 4) \cdot 5 \cdot 7 \cdot 9 \cdot 11} c_0,$$

$$\vdots$$

$$c_n = \frac{1}{n! \cdot 5 \cdot 7 \cdot 9 \cdots (2n+3)} c_0.$$

Thus

$$y_1 = c_0 \sum_{n=0}^{\infty} \frac{1}{n! \cdot 5 \cdot 7 \cdots (2n+3)} x^{n+1/2}.$$

Next we let $r = r_2 = -1$. Then (using the notation b_n instead of c_n to avoid ambiguity with our previous results) from the recurrence relation for the coefficient of x^{r+n},

$$b_n = \frac{1}{n(2n-3)} b_{n-1}, \quad n \geqslant 1.$$

Hence

$$b_1 = \frac{1}{1(-1)} b_0,$$

$$b_2 = \frac{1}{2(1)} b_1 = \frac{1}{(1 \cdot 2)(-1)(1)} b_0,$$

$$b_3 = \frac{1}{3(3)} b_2 = \frac{1}{(1 \cdot 2 \cdot 3)(-1)(1 \cdot 3)} b_0,$$

$$b_4 = \frac{1}{4(5)} b_3 = \frac{1}{(1 \cdot 2 \cdot 3 \cdot 4)(-1)(1 \cdot 3 \cdot 5)} b_0,$$

$$\vdots$$

$$b_n = \frac{1}{n!(-1)[1 \cdot 3 \cdots (2n-3)]} b_0.$$

Thus

$$y_2 = b_0 \left[x^{-1} - 1 - \sum_{n=2}^{\infty} \frac{1}{n![1 \cdot 3 \cdots (2n-3)]} x^{n-1} \right].$$

The general solution is

$$y = C_1 y_1 + C_2 y_2$$

where C_1 and C_2 are arbitrary constants.

EXAMPLE 6 INDICIAL ROOTS DIFFER BY AN INTEGER

Solve the second-order equation

$$x^2 y'' + x(2+3x)y' - 2y = 0.$$

Solution. Substituting the series form

$$y = \sum_{n=0}^{\infty} c_n x^{n+r}$$

and its derivatives gives us

$$x^2 \sum_{n=0}^{\infty} (n+r-1)(n+r)c_n x^{n+r-2}$$
$$+ x(2+3x) \sum_{n=0}^{\infty} (n+r)c_n x^{n+r-1}$$
$$- 2 \sum_{n=0}^{\infty} c_n x^{n+r} = 0$$

or

$$\sum_{n=0}^{\infty} [(n+r-1)(n+r) + 2(n+r) - 2]c_n x^{n+r} + \sum_{n=0}^{\infty} 3(n+r)c_n x^{n+r+1} = 0.$$

The indicial equation comes from the coefficient of x^r:

$$[r(r-1) + 2r - 2]c_0 = 0$$

or, since $c_0 \neq 0$,

$$r^2 + r - 2 = 0,$$
$$(r+2)(r-1) = 0.$$

Thus $r_1 = 1$ and $r_2 = -2$.

Using the smaller indicial root $r_2 = -2$, the second-order equation becomes

$$\sum_{n=0}^{\infty} [(n-3)(n-2) + 2(n-2) - 2]c_n x^{n-2} + \sum_{n=0}^{\infty} 3(n-2)c_n x^{n-1} = 0.$$

The coefficients of the powers x^k for $k > -2$ are now summarized in the following table.

Power of x	Coefficient Equation		
x^{-1}	$[-2(-1) + 2(-1) - 2]c_1 + 3(-2)c_0 = 0$	or	$c_1 = -3c_0$
x^0	$[-1(0) + 2(0) - 2]c_2 + 3(-1)c_1 = 0$	or	$c_2 = -\frac{3}{2}c_1$
x^1	$[0(1) + 2(1) - 2]c_3 + 3(0)c_2 = 0$	or	c_3 is arbitrary
x^2	$[1(2) + 2(2) - 2]c_4 + 3(1)c_3 = 0$	or	$c_4 = -\frac{3}{4}c_3$
$\vdots$	$\vdots$		$\vdots$
x^{n-2}	$[(n-3)(n-2) + 2(n-2) - 2]c_n + 3(n-3)c_{n-1} = 0,$		
	$(n^2 - 5n + 6 + 2n - 6)c_n + 3(n-3)c_{n-1} = 0,$		
	$n(n-3)c_n + 3(n-3)c_{n-1} = 0$	or	$c_n = -\dfrac{3}{n}c_{n-1}$

From the table we have

$$c_1 = -3c_0,$$

$$c_2 = -\frac{3}{2}c_1 = \frac{9}{2}c_0,$$

c_3 is arbitrary,

$$c_4 = -\frac{3}{4}c_3,$$

$$c_5 = -\frac{3}{5}c_4 = \frac{(-3)(-3)}{(4)(5)}c_3 = \frac{(2)(3)(-3)^2}{5!}c_3,$$

$$\vdots$$

$$\begin{aligned} c_n &= -\frac{3}{n}c_{n-1} = \frac{(-3)(-3)}{n(n-1)}c_{n-2} = \cdots \\ &= \left[\frac{(-3)^{n-3}}{n(n-1)\cdots 4}\right]c_3 \\ &= \left[\frac{2(-1)^{n-3}3^{n-2}}{n!}\right]c_3, \qquad n \geqslant 4. \end{aligned}$$

Therefore the general solution is given by

$$\begin{aligned} y &= \sum_{n=0}^{\infty} c_n x^{n-2} \\ &= c_0\left(x^{-2} - 3x^{-1} + \frac{9}{2}\right) + c_3 \sum_{n=3}^{\infty}\left[\frac{2(-1)^{n-3}3^{n-2}}{n!}\right]x^{n-2}. \end{aligned}$$

In this example $r_2 - r_1 = 1 - (-2) = 3$ and c_3 is arbitrary. Thus no logarithm term appears. Comparison with Eqs. (12) and (13) reveals that

$$y_1 = c_3 \sum_{n=3}^{\infty}\left[\frac{2(-1)^{n-3}3^{n-2}}{n!}\right]x^{n-2}$$

and

$$y_2 = c_0\left(x^{-2} - 3x^{-1} + \frac{9}{2}\right)$$

(where the arbitrary constants c_3 and c_0 have different names).

EXAMPLE 7 INDICIAL ROOTS ARE EQUAL

Solve the second-order equation

$$x^2y'' + x(x-1)y' + (1-x)y = 0.$$

Solution. Using the series form

$$y = \sum_{n=0}^{\infty} c_n x^{n+r}$$

the equation becomes

$$x^2 \sum_{n=0}^{\infty} (n+r-1)(n+r)c_n x^{n+r-2} + x(x-1) \sum_{n=0}^{\infty} (n+r)c_n x^{n+r-1} + (1-x) \sum_{n=0}^{\infty} c_n x^{n+r} = 0$$

or

$$\sum_{n=0}^{\infty} [(n+r-1)(n+r) - (n+r) + 1]c_n x^{n+r} + \sum_{n=0}^{\infty} [(n+r) - 1]c_n x^{n+r+1} = 0.$$

To obtain the indicial equation we consider the coefficient of x^r:

$$[(r-1)r - r + 1]c_0 = 0$$

or, since $c_0 \neq 0$,

$$r^2 - 2r + 1 = 0.$$

Thus $r_1 = r_2 = 1$.

For $r_1 = 1$ the series form of the differential equation is

$$\sum_{n=0}^{\infty} [n(n+1) - (n+1) + 1]c_n x^{n+1} + \sum_{n=0}^{\infty} [(n+1) - 1]c_n x^{n+2} = 0$$

or

$$\sum_{n=0}^{\infty} n^2 c_n x^{n+1} + \sum_{n=0}^{\infty} n c_n x^{n+2} = 0.$$

The coefficient for x^{n+1} is

$$n^2 c_n + (n-1)c_{n-1} = 0.$$

Therefore

$$c_n = -\frac{n-1}{n^2} c_{n-1}, \qquad n \geqslant 1.$$

Thus c_0 is arbitrary and $c_n = 0$ whenever $n \geqslant 1$. These calculations lead to the solution

$$y_1 = c_0 x.$$

To find a second linearly independent solution we use reduction of order.

We let

$$\begin{aligned} y_2 &= ux, \\ y_2' &= u'x + u, \\ y_2'' &= u''x + 2u'. \end{aligned}$$

Then

$$\begin{aligned} x^2(u''x + 2u') + x(x-1)(u'x+u) + (1-x)ux &= 0, \\ x^3u'' + (2x^2 + x^3 - x^2)u' + [x(x-1) + x(1-x)]u &= 0, \end{aligned}$$

or

$$xu'' + (x+1)u' = 0.$$

We let $w = u'$ and $w' = u''$ to obtain

$$\frac{w'}{w} = -\left(1 + \frac{1}{x}\right).$$

Then

$$\ln w = -x - \ln x$$

or

$$\begin{aligned} w &= \frac{1}{x}e^{-x} \\ &= \frac{1}{x}\sum_{n=0}^{\infty}\frac{(-x)^n}{n!} \\ &= \frac{1}{x} + \sum_{n=1}^{\infty}\frac{(-1)^n x^{n-1}}{n!}. \end{aligned}$$

Integrating w term by term gives us

$$u = \int w\,dx = \ln x + \sum_{n=1}^{\infty}\frac{(-1)^n x^n}{n(n!)}.$$

Therefore

$$\begin{aligned} y_2 &= x\ln x + \sum_{n=1}^{\infty}\frac{(-1)^n x^{n+1}}{n(n!)} \\ &= y_1 \ln x + \sum_{n=1}^{\infty}\frac{(-1)^n x^{n+1}}{n(n!)}. \end{aligned}$$

Note that this second solution has the form indicated by Eq. (15).

In Example 7 we used reduction of order to obtain a second linearly independent solution. This procedure may be impractical if the first solution y_1 consists of an infinite Frobenius series instead of a few nonzero terms. A methodology does exist for finding a second solution in this situation. The method consists of assuming that the Frobenius series solution form is a function of *two* variables x and r:

$$y(x, r) = \sum_{n=0}^{\infty} c_n x^{n+r}.$$

When the equal indicial roots $r_1 = r_2$ are found, as in Example 7, the two solutions are

$$y_1 = y(x, r_1) = \sum_{n=0}^{\infty} c_n x^{n+r_1}, \qquad c_0 \neq 0,$$

and

$$y_2 = \frac{\partial}{\partial r} y(x, r)\bigg|_{r=r_1}.$$

We will not pursue this idea further because of the tediousness of the calculations.

Finally, observe that we considered only the case when the roots to the indicial equation are real. In the case of complex roots the solutions have the form indicated by Eqs. (10) and (11) where $r_1 = \alpha + \beta i$ and $r_2 = \alpha - \beta i$. However, both solutions will be complex valued, and appropriate linear combinations are needed to produce real solutions. (See, for instance, the case for complex roots when solving the Euler equation in Section 9.2).

EXERCISES 9.4

In problems 1–10 determine the singular points of each equation. Classify each singular point as regular or irregular.

1. $2x^3y'' - 4x^2y' + y = 0$
2. $x^3y'' + 4xy' - xy = 0$
3. $(x-1)y'' + y' + xy = 0$
4. $x(x-1)y'' + 2y' + x^2y = 0$
5. $(x^2+1)y'' + y' - 2y = 0$
6. $x^2(x+1)y'' + (x-1)y' + y = 0$
7. $x^3(x-1)y'' + x(x+1)y' - xy = 0$
8. $x^2(x+1)^2y'' + x(x-1)y' - xy = 0$
9. $(x^2 - x - 2)y'' + xy' + y = 0$
10. $(x^3 - 3x - 2)y'' + 2xy' + y = 0$

In problems 11–16 solve the given Euler–Cauchy equation by use of a Frobenius series.

11. $x^2y'' + 2xy' - 2y = 0$
12. $x^2y'' - 6y = 0$
13. $x^2y'' + 6xy' + 4y = 0$
14. $2x^2y'' + 7xy' + 2y = 0$
15. $3x^2y'' + 4xy' = 0$
16. $6x^2y'' + 7xy' - 2y = 0$

In problems 17–20 one solution can be found using a Frobenius series. Use reduction of order to find a second linearly independent solution.

17. $x^2y'' - xy' + y = 0$
18. $x^2y'' + xy' = 0$

19. $x^2y'' + 3xy' + y = 0$

20. $4x^2y'' + y = 0$

In problems 21–33 solve each equation by use of a Frobenius series. Find the first five terms in your series solutions. For extra credit, find the recurrence relation and general nth term. Express your solutions in finite form whenever possible.

21. $4xy'' + 2y' + y = 0$

22. $2x^2y'' + x(2x + 1)y' + 2xy = 0$

23. $xy'' + (3 + 2x)y' + 4y = 0$

24. $(2x^3 - x^2)y'' + (6x^2 - 4x)y' - 2y = 0$

25. $xy'' + 2y' - xy = 0$

26. $(x + 3x^3)y'' + 2y' - 6xy = 0$

27. $2x^2y'' + 3xy' - (1 + x)y = 0$

28. $x(1 - x)y'' + 3y' - 2y = 0$

29. $(2x^3 + 6x^2)y'' + (x^2 + 9x)y' - 3y = 0$

30. $4x(x - 1)y'' + 2(1 - 2x)y' + y = 0$

31. $x^2y'' - 2xy' + (2 + x^2)y = 0$

32. $xy'' + 2y' + xy = 0$

33. $4x^2y'' - 2x(x + 2)y' + (x + 3)y = 0$

In problems 34–35 one solution can be found using a Frobenius series. Find a second linearly independent solution using reduction of order.

34. $x^2y'' + x^2y' - xy = 0$

35. $4x^2y'' - 4x^2y' + (1 + 2x)y = 0$

9.5

BESSEL FUNCTIONS

In studying perturbations of planetary orbits Friedrich Wilhelm Bessel (1784–1846) obtained a differential equation of the form

$$x^2y'' + xy' + (x^2 - p^2)y = 0. \tag{1}$$

The number p may assume any real or complex value, but we restrict our attention to the case when p is a nonnegative integer: $p = 0, 1, 2, \ldots$. Eq. (1) is known as **Bessel's equation of order *p*,** and any particular solution to Eq. (1) is called a **Bessel function.** Today Bessel functions are used in the study of a variety of problems in applied science and engineering including the motion of fluids, wave propagation, elasticity, acoustics, and potential theory. They are also useful in the investigation of solutions to partial differential equations.

Notice that $x = 0$ is a regular singular point of Bessel's equation. Thus we can use a Frobenius series to find a particular solution to Eq. (1). Substitution of

$$y = \sum_{n=0}^{\infty} c_n x^{n+r} \tag{2}$$

and its derivatives

$$y' = \sum_{n=0}^{\infty} (n + r)\, c_n x^{n+r-1}$$

and

$$y'' = \sum_{n=0}^{\infty} (n+r)(n+r-1)c_n x^{n+r-2}$$

into Bessel's equation produces

$$\sum_{n=0}^{\infty} [(n+r)(n+r-1) + (n+r) - p^2]c_n x^{n+r} + \sum_{n=0}^{\infty} c_n x^{n+r+2} = 0.$$

The indicial equation is obtained by setting the coefficient of the lowest power of x (in this case x^r) equal to zero:

$$[r(r-1) + r - p^2]c_0 = 0,$$

or, since it is assumed that $c_0 \neq 0$,

$$r^2 - p^2 = 0. \tag{3}$$

Thus $r = p$ and $r = -p$ are the indicial roots.

Setting the coefficient of x^{r+1} to zero gives us

$$[(r+1)r + (r+1) - p^2]c_1 = 0,$$
$$[r^2 + 2r + (1 - p^2)]c_1 = 0,$$

or, when $r = p$,

$$(2p+1)c_1 = 0. \tag{4}$$

Since we are assuming p is a nonnegative integer, Eq. (4) implies that $c_1 = 0$.

More generally, the coefficient of x^{n+r} must be zero yielding

$$[(n+r)(n+r-1) + (n+r) - p^2]c_n + c_{n-2} = 0,$$
$$[(n+r)^2 - p^2]c_n + c_{n-2} = 0,$$

or

$$c_n = -\frac{1}{(n+r-p)(n+r+p)} c_{n-2}. \tag{5}$$

Setting $r = p$ in Eq. (5) then gives the recurrence relation

$$c_n = -\frac{1}{n(n+2p)} c_{n-2}, \qquad n \geqslant 2. \tag{6}$$

Calculation of the coefficients from Eq. (6) leads to the following:

$$c_2 = -\frac{1}{2(2+2p)} c_0,$$

$$c_3 = -\frac{1}{3(3+2p)} c_1 = 0,$$

$$\begin{aligned}
c_4 &= -\frac{1}{4(4+2p)}c_2,\\
&= \frac{(-1)^2}{2\cdot 4(2+2p)(4+2p)}c_0,\\
c_5 &= -\frac{1}{5(5+2p)}c_3 = 0,\\
&\vdots
\end{aligned}$$

$$c_{2k+1} = 0 \qquad \text{for all } k = 0, 1, 2, \ldots,$$

$$c_{2k} = \frac{(-1)^k}{(2\cdot 4\cdot 6\cdots 2k)(2+2p)(4+2p)(6+2p)\cdots(2k+2p)}c_0. \tag{7}$$

Now

$$2\cdot 4\cdot 6\cdots 2k = 2^k(1\cdot 2\cdot 3\cdots k) = 2^k k!$$

and

$$\begin{aligned}
&(2+2p)(4+2p)(6+2p)\cdots(2k+2p)\\
&= 2^k(1+p)(2+p)(3+p)\cdots(k+p).
\end{aligned}$$

Thus, substitution of these identities into Eq. (7) gives us

$$c_{2k} = \frac{(-1)^k}{2^{2k}k!(1+p)(2+p)\cdots(k+p)}c_0.$$

After multiplying both numerator and denominator by $2^p p!$, we have

$$c_{2k} = \frac{2^p p!(-1)^k}{2^{p+2k}k!p!(1+p)(2+p)\cdots(k+p)}c_0$$

or

$$c_{2k} = \frac{(-1)^k 2^p p!}{2^{p+2k}k!(p+k)!}c_0, \tag{8}$$

where $k = 0, 1, 2, 3, \ldots$. Therefore a particular solution to Bessel's equation of order p is

$$y = Cx^0\left[\frac{x^p}{2^p p!} - \frac{x^{p+2}}{2^{p+2}1!(p+1)!} + \frac{x^{p+4}}{2^{p+4}2!(p+2)!} - \cdots\right]$$

where $C = c_0 2^p p!$ is a constant. The solution function

$$J_p(x) = \sum_{n=0}^{\infty}\frac{(-1)^n x^{p+2n}}{2^{p+2n}n!(p+n)!} \tag{9}$$

is called the **Bessel function of nonnegative integral order p.**

Bessel Functions J_0 and J_1

The Bessel functions of orders zero and one are of special interest. For $J_0(x)$ we find from Eq. (9) that

$$J_0(x) = \sum_{n=0}^{\infty} \frac{(-1)^n x^{2n}}{2^{2n} n! n!}$$

or

$$J_0(x) = 1 - \frac{x^2}{2^2} + \frac{x^4}{2^2 \cdot 4^2} - \frac{x^6}{2^2 \cdot 4^2 \cdot 6^2} + \cdots \tag{10}$$

where the denominators are obtained from the observation that

$$\begin{aligned} 2^{2n} n! n! &= (2 \cdot 2)(2 \cdot 2) \cdots (2 \cdot 2)(1 \cdot 2 \cdot 3 \cdots n)(1 \cdot 2 \cdot 3 \cdots n) \\ &= (2 \cdot 1)^2 (2 \cdot 2)^2 (2 \cdot 3)^2 \cdots (2n)^2. \end{aligned}$$

Likewise,

$$J_1(x) = \frac{x}{2} - \frac{x^3}{2^2 \cdot 4} + \frac{x^5}{2^2 \cdot 4^2 \cdot 6} - \frac{x^7}{2^2 \cdot 4^2 \cdot 6^2 \cdot 8} + \cdots \tag{11}$$

The graphs of $J_0(x)$ and $J_1(x)$, as well as $J_2(x)$, are presented in Fig. 9.1. Notice from the graphs the oscillatory behavior of $J_0(x)$ and $J_1(x)$. These two Bessel functions appear to behave like damped cosine and sine functions, respectively. This may not be so surprising if we observe that Bessel's equation (1) can be written in the form

$$y'' + \left(\frac{1}{x}\right) y' + \left(1 - \frac{p^2}{x^2}\right) y = 0. \tag{12}$$

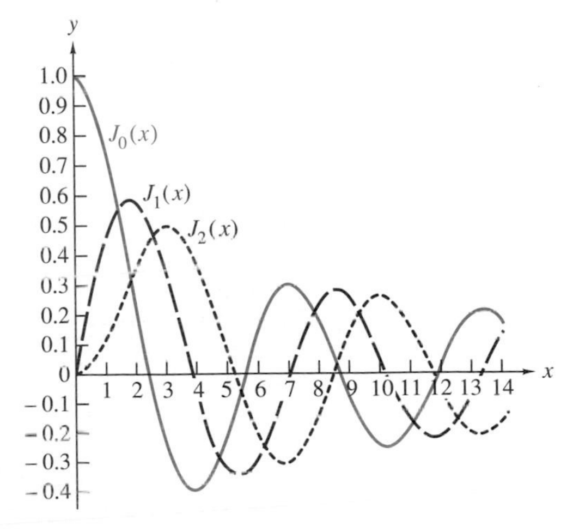

FIGURE 9.1
Bessel functions $J_0(x)$, $J_1(x)$, and $J_2(x)$.

Thus for very large values of x the equation is approximated by

$$y'' + y = 0. \tag{13}$$

Since the solutions to Eq. (13) are cosines and sines we might suspect that the solutions to Bessel's equation will resemble these trigonometric functions in some way for large x. Comparison of $J_0(x)$ and $J_1(x)$ with the Maclaurin series expansions for $\cos x$ and $\sin x$,

$$\cos x = 1 - \frac{x^2}{2!} + \frac{x^4}{4!} - \frac{x^6}{6!} + \cdots,$$

$$\sin x = x - \frac{x^3}{3!} + \frac{x^5}{5!} - \frac{x^7}{7!} + \cdots,$$

also reveals a similarity between $J_0(x)$ and $\cos x$ and between $J_1(x)$ and $\sin x$. In fact, as $x \to +\infty$,

$$J_0(x) \approx \left(\frac{2}{\pi x}\right)^{1/2} \cos\left(x - \frac{\pi}{4}\right) \tag{14}$$

and

$$J_1(x) \approx \left(\frac{2}{\pi x}\right)^{1/2} \sin\left(x - \frac{\pi}{4}\right). \tag{15}$$

Perhaps more striking are the following comparisons:

$$\begin{aligned}
J_0(0) &= 1, & \cos 0 &= 1, \\
J_0(-x) &= J_0(x), & \cos(-x) &= \cos x, \\
J_0'(0) &= 0, & \frac{d}{dx}\cos x\bigg|_{x=0} &= 0, \\
J_1(0) &= 0, & \sin 0 &= 0, \\
J_1(-x) &= -J_1(x), & \sin(-x) &= -\sin x, \\
J_0'(x) &= -J_1(x), & \frac{d}{dx}\cos x &= -\sin x.
\end{aligned}$$

Zeros of Bessel Functions

Notice in Fig. 9.1 how the zeros of the Bessel functions appear to *interlace*. That is, between successive zeros of J_0 there is a zero of J_1, and between successive zeros of J_1 there is a zero of J_0. This interlacing property holds for all the Bessel functions of nonnegative integral order.

THEOREM 9.6

Between successive zeros of the Bessel function $J_p(x)$ there is a zero of $J_{p+1}(x)$, and between successive zeros of $J_{p+1}(x)$ there is a zero of $J_p(x)$.

The zeros of $J_p(x)$ assume particular importance in the solution of boundary value problems that arise in engineering applications. Approximate values of the first few zeros of $J_0(x)$ and $J_1(x)$ are listed below.

	1st	2d	3d	4th	5th	6th
$J_0(x)$	2.405	5.520	8.654	11.792	14.931	18.071
$J_1(x)$	0	3.832	7.016	10.173	13.323	16.471

More generally, the mth root λ_m of the Bessel function $J_p(x)$ can be approximated by

$$\lambda_m \approx \left(m + \frac{p}{2} - \frac{1}{4}\right) \qquad \text{as } m \to \infty. \tag{16}$$

It is important to note, however, that the *zeros of $J_p(x)$ are not evenly spaced along the x-axis.*

Properties of Bessel Functions*

Two properties relating the Bessel functions of orders zero and one are

$$\frac{d}{dx} J_0(x) = -J_1(x), \tag{17}$$

$$\frac{d}{dx} J_1(x) = J_0(x) - \frac{1}{x} J_1(x). \tag{18}$$

More general properties of Bessel functions include the following:

$$\frac{d}{dx}[x^p J_p(x)] = x^p J_{p-1}(x), \qquad p \geqslant 1, \tag{19}$$

$$\frac{d}{dx}[x^{-p} J_p(x)] = -x^{-p} J_{p+1}(x), \qquad p \geqslant 0, \tag{20}$$

$$J_{p+1}(x) = \frac{2p}{x} J_p(x) - J_{p-1}(x), \qquad p \geqslant 1, \tag{21}$$

$$J_{p+1}(x) = -2J_p'(x) + J_{p-1}(x), \qquad p \geqslant 1. \tag{22}$$

* For a deeper investigation into Bessel functions and their properties, including Bessel functions of noninteger and negative orders, see *A Treatise on the Theory of Bessel Functions*, 2d ed., by G. N. Watson (Cambridge: The University Press, 1941). A very useful reference to the practicing engineer is the *Handbook of Mathematical Functions* by M. Abramovitz and I. A. Stegun (Washington, D.C.: National Bureau of Standards, 1964).

Note that we found only one particular solution $J_p(x)$ to Bessel's equation of order p. A second linearly independent solution may be found having one of the forms discussed in Section 9.4. This second solution is often called the **Bessel function of second kind of order *p*,** and its form depends on the nature of the number p. However, we will not pursue the study of these functions in this introductory text.

EXERCISES 9.5

1. Express the following functions using Σ notation:

a) $J_0(\sqrt{x})$

b) $\sqrt{x}\, J_1(\sqrt{x})$

2. Compute the following to four decimal places from the series representations of J_0 and J_1.

a) $J_0(2)$ **b)** $J_1(1)$

c) $J_0(3)$ **d)** $J_1(2)$

e) $J_0(\frac{1}{2})$ **f)** $J_1(\frac{1}{2})$

In problems 3–6 assume properties (17–22) together with the chain rule to establish that the specified function solves the given differential equation.

3. $xy'' - y' + xy = 0, \quad y = xJ_1(x)$

4. $xy'' + y' + 2y = 0, \quad y = J_0(2\sqrt{2x})$

5. $xy'' + 3y' + xy = 0, \quad y = x^{-1}J_1(x)$

6. $xy'' + y' + x^3y = 0, \quad y = J_0\left(\dfrac{x^2}{2}\right)$

7. If $u = u(x)$ is a differentiable function of x, show that

a) $\dfrac{d}{dx} J_0(u) = -J_1(u)\dfrac{du}{dx},$

b) $\dfrac{d}{dx} uJ_1(u) = uJ_0(u)\dfrac{du}{dx}.$

8. Find a particular solution to the differential equation

$$x^2y'' + xy' + (x^2 - 1)y = 0$$

subject to the initial condition $y(1) = 1$.

9. If $u = \sqrt{x}$, establish that the function $y = uJ_0(u)$ satisfies the differential equation

$$4x^2y'' + (x + 1)y = 0.$$

Use the results in problem 7 above. Then find a particular solution satisfying $y(4) = 1$.

10. Establish that the function $y = \dfrac{1}{x}J_1(x)$ satisfies the differential equation

$$xy'' + 3y' + xy = 0.$$

Find a particular solution satisfying $y(\frac{1}{2}) = 1$.

11. Using the substitution $u = \sqrt{x}$ convert the equation

$$4x^2y'' + 4xy' + (x - 1)y = 0$$

into a Bessel equation. Find a particular solution satisfying $y = 1$ when $x = 4$.

12. Using the substitution $u = x^{-1}$ convert the equation

$$x^4y'' + x^3y' + y = 0$$

into a Bessel equation. Find a particular solution satisfying $y(\frac{1}{2}) = 3$.

13. Show that the series for $J_0(x)$ converges absolutely for $0 < x < \infty$.

14. Show that the series for $J_1(x)$ converges absolutely for $0 < x < \infty$.

15. Using term-by-term differentiation prove that

$$\frac{d}{dx}[x^pJ_p(x)] = x^pJ_{p-1}(x), \quad p \geq 1.$$

16. Carry out the differentiation process in problem 15, and divide your result by x^p to establish that

$$J_p'(x) + px^{-1}J_p(x) = J_{p-1}(x), \quad p \geq 1.$$

17. Using term-by-term differentiation prove that

$$\frac{d}{dx}[x^{-p}J_p(x)] = -x^{-p}J_{p+1}(x), \quad p \geq 0.$$

18. Carry out the differentiation process in problem 17 and multiply your result by x^p to establish that

$$J_p'(x) - px^{-1}J_p(x) = -J_{p+1}(x), \quad p \geq 0.$$

19. From the results in problems 16 and 18 show that, for $p \geqslant 1$,

a) $J_{p-1}(x) - J_{p+1}(x) = 2J_p'(x)$

b) $J_{p-1}(x) + J_{p+1}(x) = \frac{2p}{x} J_p(x)$.

20. Use problem 19(a) to prove that

$$4J_p''(x) = J_{p-2}(x) - 2J_p(x) + J_{p+2}(x), \quad p \geqslant 2.$$

21. Verify that the differential equation

$$xy'' + (1 - 2p)y' + xy = 0, \quad x > 0$$

has the particular solution $y = x^p J_p(x)$.

22. Using the result of problem 21 find a particular solution to

$$xy'' - 3y' + xy = 0, \quad x > 0$$

satisfying the initial condition $y(2) = 1$.

23. Use problem 16 to establish that

$$\int_0^x tJ_0(t)\,dt = xJ_1(x).$$

24. Use problem 17 to establish that

$$\int_0^x t^{-1}J_2(t)\,dt = -x^{-1}J_1(x) + \frac{1}{2}.$$

25. Use problem 14 and *Rolle's theorem* from elementary calculus to prove that between successive positive zeros of $J_0(x)$ there is a zero of $J_1(x)$.

26. Undamped Aging Spring

a) Justify the model

$$my''(t) + ke^{-\alpha t}y(t) = 0, \quad \alpha > 0,$$

for an undamped aging spring. Refer to Section 3.4.

b) Make the change of variable $s = \gamma e^{\beta t}$ for constants γ, β. Show that the differential equation in part (a) can be rewritten as

$$\beta^2 s^2 \frac{d^2y}{ds^2} + \beta^2 s \frac{dy}{ds} + \frac{k}{m}\left(\frac{s}{\gamma}\right)^{-\alpha/\beta} y = 0.$$

c) Choose γ and β so that

$$-\frac{\alpha}{\beta} = 2 \quad \text{and} \quad \frac{k}{m\beta^2\gamma^{-\alpha/\beta}} = 1.$$

Show that the equation in part (b) can be written as the Bessel equation

$$s^2 \frac{d^2y}{ds^2} + s\frac{dy}{ds} + s^2 y = 0.$$

One solution is therefore

$$J_0(s) = J_0\left(\frac{2}{\alpha}\sqrt{\frac{k}{m}} \cdot e^{-\alpha t/2}\right).$$

A sketch of the graph of this solution is shown in Fig. 9.2. Interpret the graph.

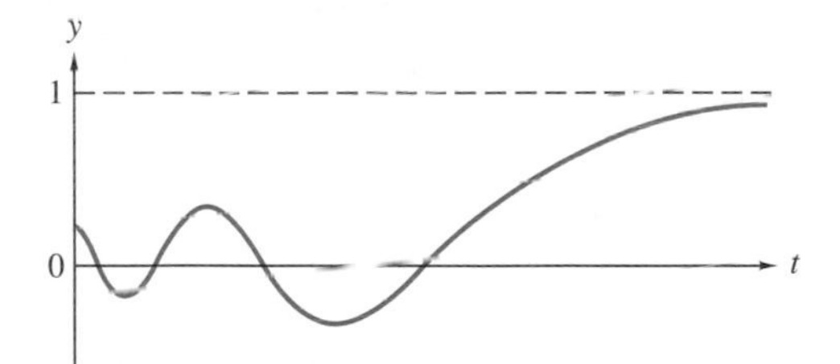

FIGURE 9.2
Undamped aging spring;
$y = J_0\left(\frac{2}{\alpha}\sqrt{\frac{k}{m}}\, e^{-\alpha t/2}\right)$.

9.6

LEGENDRE POLYNOMIALS

In studying gravitation Adrien Marie Legendre (1752–1833) encountered the differential equation

$$(1 - x^2)y'' - 2xy' + n(n+1)y = 0 \tag{1}$$

where n is a nonnegative integer, $n = 0, 1, 2, \ldots$. Equation (1) is known as **Legendre's equation.** The values $x = 1$ and $x = -1$ yield regular singular points of Eq. (1), and $x = 0$ is an ordinary point. When n is a fixed nonnegative integer, one solution to Legendre's equation turns out to be a polynomial. We assume a power-series solution valid near the origin to obtain such a polynomial for each n.

Since $x = 0$ is an ordinary point of Eq. (1), we substitute

$$y = \sum_{k=0}^{\infty} c_k x^k$$

and its derivatives

$$y' = \sum_{k=1}^{\infty} k c_k x^{k-1}$$

and

$$y'' = \sum_{k=2}^{\infty} k(k-1) c_k x^{k-2}$$

into Legendre's equation. The substitution leads to

$$\sum_{k=2}^{\infty} k(k-1)c_k x^{k-2} - \sum_{k=2}^{\infty} k(k-1)c_k x^k - 2\sum_{k=1}^{\infty} k c_k x^k + n(n+1)\sum_{k=0}^{\infty} c_k x^k = 0.$$

Equating the coefficients of each power of x to zero produces the following results:

x^0: $\quad 2(1)c_2 + n(n+1)c_0 = 0$, or

$$c_2 = -\frac{1}{2} n(n+1)c_0.$$

x^1:

$$3(2)c_3 - 2c_1 + n(n+1)c_1 = 0,$$
$$3(2)c_3 + [n(n+1) - 2]c_1 = 0,$$
$$3(2)c_3 + (n-1)(n+2)c_1 = 0, \text{ or}$$
$$c_3 = -\frac{1}{2 \cdot 3}(n-1)(n+2)c_1.$$

x^2:

$$4(3)c_4 - 2(1)c_2 - 2(2)c_2 + n(n+1)c_2 = 0,$$
$$4(3)c_4 + [n(n+1) - 6]c_2 = 0,$$
$$4(3)c_4 + (n-2)(n+3)c_2 = 0, \text{ or}$$
$$c_4 = -\frac{1}{3 \cdot 4}(n-2)(n+3)c_2.$$

x^3:

$$5(4)c_5 - 3(2)c_3 - 2(3)c_3 + n(n+1)c_3 = 0,$$
$$5(4)c_5 + [n(n+1) - 12]c_3 = 0,$$
$$5(4)c_5 + (n-3)(n+4)c_3 = 0, \text{ or}$$
$$c_5 = -\frac{1}{4 \cdot 5}(n-3)(n+4)c_3.$$

$$x^k: \quad (k+2)(k+1)c_{k+2} + [-k(k-1) - 2k + n(n+1)]c_k = 0,$$
$$(k+2)(k+1)c_{k+2} + [n^2 + n - k(k+1)]c_k = 0,$$
$$(k+2)(k+1)c_{k+2} + (n-k)(n+k+1)c_k = 0,$$

or

$$c_{k+2} = -\frac{(n-k)(n+k+1)}{(k+1)(k+2)} c_k, \qquad k \geqslant 0. \tag{2}$$

Notice that the coefficients with even indices are interrelated, and similarly, the coefficients with odd indices are interrelated. By using recursive relation (2) we find the following coefficients:

$$c_4 = -\frac{(n-2)(n+3)}{3 \cdot 4} c_2$$
$$= \frac{(-1)^2(n-2)n(n+1)(n+3)}{4!} c_0,$$

$$c_5 = -\frac{(n-3)(n+4)}{4 \cdot 5} c_3$$
$$= \frac{(-1)^2(n-3)(n-1)(n+2)(n+4)}{5!} c_1,$$

$$c_6 = -\frac{(n-4)(n+5)}{5 \cdot 6} c_4$$
$$= \frac{(-1)^3(n-4)(n-2)n(n+1)(n+3)(n+5)}{6!} c_0,$$

$$c_7 = -\frac{(n-5)(n+6)}{6 \cdot 7} c_5$$
$$= \frac{(-1)^3(n-5)(n-3)(n-1)(n+2)(n+4)(n+6)}{7!} c_1,$$

$$c_8 = -\frac{(n-6)(n+7)}{7 \cdot 8} c_6,$$
$$= \frac{(-1)^4(n-6)(n-4)(n-2)n(n+1)(n+3)(n+5)(n+7)}{8!} c_0,$$

$$c_9 = -\frac{(n-7)(n+8)}{8 \cdot 9} c_7,$$
$$= \frac{(-1)^4(n-7)(n-5)(n-3)(n-1)(n+2)(n+4)(n+6)(n+8)}{9!} c_1,$$

and so forth. Thus we obtain the following two linearly independent solutions valid for at least the interval $-1 < x < 1$:

$$y_1(x) = c_0\left[1 - \frac{n(n+1)}{2!}x^2 + \frac{(n-2)n(n+1)(n+3)}{4!}x^4 - \frac{(n-4)(n-2)n(n+1)(n+3)(n+5)}{6!}x^6 + \frac{(n-6)(n-4)(n-2)n(n+1)(n+3)(n+5)(n+7)}{8!}x^8 - \cdots\right]$$

$$y_2(x) = c_1\left[x - \frac{(n-1)(n+2)}{3!}x^3 + \frac{(n-3)(n-1)(n+2)(n+4)}{5!}x^5 - \frac{(n-5)(n-3)(n-1)(n+2)(n+4)(n+6)}{7!}x^7 + \frac{(n-7)(n-5)(n-3)(n-1)(n+2)(n+4)(n+6)(n+8)}{9!}x^9 - \cdots\right]$$

If n is even, the infinite series for y_1 terminates and becomes an nth-degree polynomial. If n is odd, the series for y_2 terminates and becomes an nth-degree polynomial. Since c_0 and c_1 are arbitrary constants, we choose their value *for each n* so that the solution obtained has the value 1 at $x = 1$. The solutions obtained in this way are called **Legendre polynomials.** Following is a list of the first four Legendre polynomials:

$n = 0$: $c_0 = 1$, and the solution y_1 yields

$$P_0(x) = 1. \tag{3}$$

$n = 1$: $c_1 = 1$, and the solution y_2 yields

$$P_1(x) = x. \tag{4}$$

$n = 2$: $c_0 = -\tfrac{1}{2}$, and the solution y_1 yields

$$P_2(x) = \frac{1}{2}(3x^2 - 1). \tag{5}$$

$n = 3$: $c_1 = -\tfrac{3}{2}$, and the solution y_2 yields

$$P_3(x) = \frac{1}{2}(5x^3 - 3x). \tag{6}$$

In general, for $n = 2, 4, 6, \ldots$, we choose

$$c_0 = (-1)^{n/2}\left[\frac{1 \cdot 3 \cdots (n-1)}{2 \cdot 4 \cdots n}\right], \tag{7}$$

and for $n = 1, 3, 5, \ldots$, we choose

$$c_1 = (-1)^{(n-1)/2}\left[\frac{1 \cdot 3 \cdots n}{2 \cdot 4 \cdots (n-1)}\right]. \tag{8}$$

When the values for c_0 and c_1 from Eqs. (7) and (8) are substituted into the formulas for $y_1(x)$ and $y_2(x)$, we obtain the various Legendre polynomials. For example, if $n = 4$,

$$c_0 = (-1)^2\left(\frac{1 \cdot 3}{2 \cdot 4}\right) = \frac{3}{8},$$

and the solution y_1 becomes

$$\begin{aligned} P_4(x) &= \frac{3}{8}\left(1 - 10x^2 + \frac{35}{3}x^4\right) \\ &= \frac{1}{8}(35x^4 - 30x^2 + 3). \end{aligned} \tag{9}$$

It is important to note that *each Legendre polynomial is a particular solution to a Legendre differential equation.* For example, $P_2(x)$ is a particular solution to

$$(1 - x^2)y'' - 2xy' + 6y = 0,$$

and $P_3(x)$ is a particular solution to

$$(1 - x^2)y'' - 2xy' + 12y = 0.$$

In general, $P_n(x)$ is a particular solution to

$$(1 - x^2)y'' - 2xy' + n(n+1)y = 0.$$

The first five Legendre polynomials for $-1 \leq x \leq 1$ are plotted in Fig. 9.3.

It is possible to write a closed-form summation formula for the nth Legendre polynomial:

$$P_n(x) = \frac{1}{2^n}\sum_{k=0}^{N}\left[\frac{(-1)^k(2n-2k)!}{k!(n-2k)!(n-k)!}\right]x^{n-2k} \tag{10}$$

where $N = n/2$ if n is even and $N = (n-1)/2$ if n is odd.

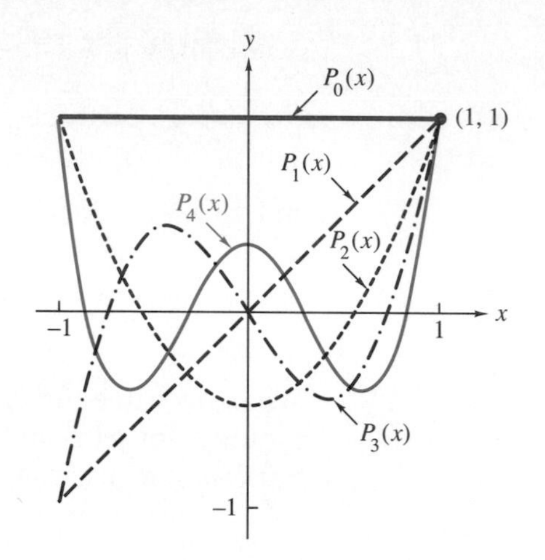

FIGURE 9.3
Legendre polynomials $P_0(x)$, $P_1(x)$, $P_2(x)$, $P_3(x)$, and $P_4(x)$.

EXAMPLE 1 Use formula (10) to find $P_3(x)$.

Solution. Here $N = (3 - 1)/2 = 1$ and

$$\begin{aligned} P_3(x) &= \frac{1}{8}\sum_{k=0}^{1}\left[\frac{(-1)^k(6-2k)!}{k!(3-2k)!(3-k)!}\right]x^{3-2k} \\ &= \frac{1}{8}\left[\left(\frac{6!}{0!3!3!}\right)x^3 - \left(\frac{4!}{1!1!2!}\right)x\right] \\ &= \frac{1}{8}(20x^3 - 12x) \\ &= \frac{1}{2}(5x^3 - 3x). \end{aligned}$$

Rodrigues' Formula

An important and useful formula for obtaining $P_n(x)$ is obtained in the following way. From the binomial theorem for any positive integer n,

$$(x^2 - 1)^n = \sum_{k=0}^{n}\left[\frac{(-1)^k n!}{k!(n-k)!}\right]x^{2n-2k}. \tag{11}$$

Differentiation of Eq. (11) n times yields

$$\begin{aligned}\frac{d^n}{dx^n}(x^2-1)^n &= \sum_{k=0}^{N}\left[(-1)^k n!\,\frac{(2n-2k)(2n-2k-1)\cdots(2n-2k-n+1)}{k!(n-k)!}\right]x^{n-2k} \\ &= \sum_{k=0}^{N}\left[(-1)^k n!\,\frac{(2n-2k)!}{k!(n-k)!(n-2k)!}\right]x^{n-2k}.\end{aligned} \tag{12}$$

where the last term in the series must be constant. But $n-2N=0$ implies that $N=n/2$, and $n-1-2N=0$ implies that $N=(n-1)/2$. Since N must be a nonnegative integer, it is defined as $n/2$ if n is even and $(n-1)/2$ if n is odd. If we multiply the righthand side of Eq. (10) by $2^n n!$ we obtain the righthand side of Eq. (12). Thus we have the relation

$$\frac{d^n}{dx^n}(x^2-1)^n = 2^n n!\; P_n(x)$$

or

$$P_n(x) = \frac{1}{2^n n!}\frac{d^n}{dx^n}(x^2-1)^n. \tag{13}$$

Equation (13) is known as **Rodrigues' formula,** after Olinde Rodrigues (1794–1851), a French economist and mathematician.

EXAMPLE 2 Use Rodrigues' formula to find $P_3(x)$.

Solution. Substituting $n=3$ into Eq. (13) results in

$$\begin{aligned}P_3(x) &= \frac{1}{2^3 3!}\frac{d^3}{dx^3}(x^2-1)^3 \\ &= \frac{1}{48}\frac{d^3}{dx^3}(x^6-3x^4+3x^2-1) \\ &= \frac{1}{48}\frac{d^2}{dx^2}(6x^5-12x^3+6x) \\ &= \frac{1}{48}\frac{d}{dx}(30x^4-36x^2+6) \\ &= \frac{1}{48}(120x^3-72x) \\ &= \frac{1}{2}(5x^3-3x).\end{aligned}$$

Properties of Legendre Polynomials

Some properties of Legendre polynomials are as follows:

$$(n+1)P_{n+1}(x) + nP_{n-1}(x) = (2n+1)xP_n(x), \tag{14}$$

$$\frac{d}{dx}P_{n+1}(x) - \frac{d}{dx}P_{n-1}(x) = (2n+1)P_n(x), \tag{15}$$

$$\frac{d}{dx}P_{n+1}(x) - x\frac{d}{dx}P_n(x) = (n+1)P_n(x), \tag{16}$$

where $n = 1, 2, 3 \ldots$.

Orthogonality of Legendre Polynomials

The following integral property of the Legendre polynomials is very useful in the study of boundary-value problems. It is known as the **orthogonality property** of the Legendre polynomials:

$$\int_{-1}^{1} P_n(x)P_m(x)\,dx = 0 \qquad \text{if} \qquad m \neq n. \tag{17}$$

To establish Eq. (17), we note first that $P_n(x)$ and $P_m(x)$ solve Legendre's equation (1). Thus

$$(1-x^2)P_n'' - 2xP_n' + n(n+1)P_n = 0 \tag{18}$$

and

$$(1-x^2)P_m'' - 2xP_m' + m(m+1)P_m = 0. \tag{19}$$

Multiplication of Eq. (18) by P_m and Eq. (19) by P_n and subtracting the results gives us

$$(1-x^2)(P_mP_n'' - P_nP_m'') - 2x(P_mP_n' - P_nP_m') + [n(n+1) - m(m+1)]P_nP_m = 0. \tag{20}$$

Now

$$(P_mP_n' - P_nP_m')' = P_mP_n'' - P_nP_m'',$$

so Eq. (20) reduces to

$$(1-x^2)\frac{d}{dx}(P_mP_n' - P_nP_m') - 2x(P_mP_n' - P_nP_m') + [n(n+1) - m(m+1)]P_nP_m = 0$$

or

$$\frac{d}{dx}[(1-x^2)(P_m P_n' - P_n P_m')] = [m(m+1) - n(n+1)]P_n P_m. \quad (21)$$

Integration of both sides of Eq. (21) for $-1 \leq x \leq 1$ yields

$$(1-x^2)(P_m P_n' - P_n P_m')\Big|_{-1}^{1} = [m(m+1) - n(n+1)] \int_{-1}^{1} P_n(x)P_m(x)\,dx.$$

Since the lefthand side evaluates to zero, this last equation gives us

$$\int_{-1}^{1} P_n(x)P_m(x)dx = 0 \qquad \text{if} \qquad m \neq n.$$

In the case where $m = n$ it can be shown that

$$\int_{-1}^{1} [P_n(x)]^2\,dx = \frac{2}{2n+1}. \quad (22)$$

We leave the proof of Eq. (22) as an exercise (see problem 19).

Representing a Function by Legendre Polynomial Series

Our opening discussion on power-series solutions reviewed the property that analytic functions can be represented by a power series

$$f(x) = \sum_{n=0}^{\infty} c_n x^n$$

where $c_n = f^{(n)}(0)/n!$.

If instead we represent f in the form

$$f(x) = \sum_{n=0}^{\infty} b_n P_n(x) \quad (23)$$

where $-1 \leq x \leq 1$ and $P_n(x)$ is the nth Legendre polynomial, then Eq. (23) is called the **Legendre polynomial series representing f.** To compute the coefficients b_n in Eq. (23), we multiply both sides of Eq. (23) by $P_k(x)$ and integrate:

$$\int_{-1}^{1} f(x)P_k(x)\,dx = \sum_{n=0}^{\infty}\left[b_n \int_{-1}^{1} P_n(x)P_k(x)\,dx\right]$$
$$= b_k\left(\frac{2}{2k+1}\right)$$

where the integral on the righthand side has been obtained from orthogonality property (17) and result (22). Thus we obtain the formula

$$b_k = \frac{2k+1}{2}\int_{-1}^{1} f(x)P_k(x)\,dx \quad (24)$$

for the kth coefficient required in Eq. (23). We present an example illustrating the expansion of a function in terms of Legendre polynomials.

EXAMPLE 3 Find the Legendre polynomial series representation of the function

$$f(x)=\begin{cases}0, & -1<x<0,\\ 1, & 0<x<1.\end{cases}$$

Solution. From Eq. (24) we calculate the Legendre coefficients,

$$b_0=\frac{1}{2}\int_0^1 (1)(1)\,dx=\frac{1}{2},$$

$$b_1=\frac{3}{2}\int_0^1 (1)(x)\,dx=\frac{3}{4},$$

$$b_2=\frac{5}{2}\int_0^1 (1)\left(\frac{1}{2}\right)(3x^2-1)\,dx=0,$$

$$b_3=\frac{7}{2}\int_0^1 (1)\left(\frac{1}{2}\right)(5x^3-3x)\,dx=-\frac{7}{16},$$

$$b_4=\frac{9}{2}\int_0^1 (1)\left(\frac{1}{8}\right)(35x^4-30x^2+3)\,dx=0,$$

$$b_5=\frac{11}{12}\int_0^1 (1)\left(\frac{1}{8}\right)(63x^5-70x^3+15x)\,dx=\frac{11}{32},$$

and so on. Therefore

$$f(x)=\frac{1}{2}P_0(x)+\frac{3}{4}P_1(x)-\frac{7}{16}P_3(x)+\frac{11}{32}P_5(x)-\cdots.$$

Zeros of Legendre Polynomials

The zeros of Legendre polynomials are particularly useful in a numerical method for approximating a definite integral known as **Gaussian quadrature.** It is beyond the scope of this text to develop the method, but the main idea is to approximate an integral $\int_b^a f(x)\,dx$ by a suitable sum

$$\sum_{i=1}^{n} a_i f(x_i). \tag{25}$$

The zeros of the Legendre polynomials can be used for the x_i and the a_i computed. Formula (25) becomes exact when $f(x)$ is a polynomial of degree less than or equal to $2n-1$. In order to apply the method we need to know that the zeros of $P_n(x)$ are real. This is guaranteed by the next result, which we state without proof.

TABLE 9.2

Values to Approximate a Definite Integral for Various Integers n Using Gaussian Quadrature

n	Roots	Coefficients
2	0.5773502692	1.0000000000
	−0.5773502692	1.0000000000
3	0.7745966692	0.5555555556
	0.0000000000	0.8888888889
	−0.7745966692	0.5555555556
4	0.8611363116	0.3478548451
	0.3399810436	0.6521451549
	−0.3399810436	0.6521451549
	−0.8611363116	0.3478548451
5	0.9061798459	0.2369268850
	0.5384693101	0.4786286705
	0.0000000000	0.5688888889
	−0.5384693101	0.4786286705
	−0.9061798459	0.2369268850

THEOREM 9.7

The Legendre polynomial $P_n(x)$ has exactly n distinct roots over the interval $-1 < x < 1$.

Table 9.2 lists the roots of the Legendre polynomials for various values of n together with the coefficients a_i in Eq. (25) that are associated with them in order to approximate the definite integral

$$\int_{-1}^{1} f(x)\, dx \approx \sum_{i=1}^{n} a_i f(x_i) \tag{26}$$

when the function f can be evaluated at each root x_i.

EXAMPLE 4 Let us illustrate the approximation of an integral by the method of quadratures for two different values of n.

$$n = 2: \quad \int_{-1}^{1} \frac{e^{-x^2/2}}{\sqrt{2\pi}}\, dx \approx \frac{1}{\sqrt{2\pi}} [e^{(0.5773502692)^2/2} + e^{(-0.5773502692)^2/2}]$$

$$= \frac{1}{\sqrt{2\pi}} (1.69296345) = 0.6753946993.$$

$$n=5: \quad \int_{-1}^{1} \frac{e^{-x^2/2}}{\sqrt{2\pi}}\,dx \approx \frac{1}{\sqrt{2\pi}}[(0.2369268850\, e^{(0.9061798459)^2/2}$$
$$+(0.4786286705)\, e^{(0.5384693101)^2/2}$$
$$+(0.5688888889)(1)$$
$$+(0.4786286705)\, e^{(-0.5384693101)^2/2}$$
$$+(0.2369268850)\, e^{(-0.9061798459)^2/2}]$$
$$=\frac{1}{\sqrt{2\pi}}(1.711249393)=0.6826897354.$$

The correct answer to 6 decimal places is 0.682689.

EXERCISES 9.6

1. From summation formula (10) write out the following Legendre polynomials.

a) $P_2(x)$ **b)** $P_4(x)$
c) $P_5(x)$ **d)** $P_6(x)$
e) $P_7(x)$ **f)** $P_8(x)$

2. Use Rodrigues' formula to find the following Legendre polynomials.

a) $P_2(x)$ **b)** $P_4(x)$
c) $P_5(x)$ **d)** $P_6(x)$
e) $P_7(x)$ **f)** $P_8(x)$

3. Using identity (14) and mathematical induction, verify that $P_n(1)=1$ and $P_n(-1)=(-1)^n$ for every Legendre polynomial.

4. Graph the Legendre polynomials $P_5(x)$, $P_6(x)$, and $P_7(x)$. Use the results in problems 1 or 2.

In problems 5–11 find the first four nonzero terms in the Legendre series representation of the given function over the interval $-1<x<1$.

5. $f(x)=\begin{cases}-1, & -1<x<0\\ 1, & 0<x<1\end{cases}$

6. $f(x)=x$

7. $f(x)=\begin{cases}0, & -1<x<1\\ x, & 0<x<1\end{cases}$

8. $f(x)=x^2$

9. $f(x)=|x|$

10. $f(x)=x^3$

11. $f(x)=x^4$

In problems 12–17 approximate the value of the definite integral using formula (26) and Table 9.2.

12. $\int_{-1}^{1}(2x-1)\,dx$

13. $\int_{-1}^{1}(x^3-x^2+2x+1)\,dx$

14. $\int_{-1}^{1}e^{-x^2}\,dx$

15. $\int_{-1}^{1}\sin \pi x\,dx$

16. $\int_{-1}^{1}xe^x\,dx$

17. $\int_{-1}^{1}e^x\cos x\,dx$

18. Verify that integral approximation (26) is exact when $n=2$ and $f(x)$ is a polynomial of degree less than or equal to $2n-1=3$. Note that the two roots of $P_2(x)$ are $\pm 1/\sqrt{3}$.

19. Prove integral result (22) by carrying out the following procedures.

a) Multiply identity (14) by $P_{n+1}(x)$ and integrate the result from -1 to 1. Use the orthogonality condition to obtain

$$(n+1)\int_{-1}^{1}[P_{n+1}(x)]^2\,dx = (2n+1)\int_{-1}^{1} xP_{n+1}(x)P_n(x)\,dx.$$

b) Multiply identity (14) by $P_{n-1}(x)$ and integrate to obtain

$$n\int_{-1}^{1}[P_{n-1}(x)]^2\,dx = (2n+1)\int_{-1}^{1} xP_{n-1}(x)P_n(x)\,dx.$$

c) Replace n by $n+1$ in your result from part (b).

d) Use parts (a) and (c) to establish the identity

$$(2n+1)\int_{-1}^{1}[P_n(x)]^2\,dx = (2n+3)\int_{-1}^{1}[P_{n+1}(x)]^2\,dx.$$

e) Argue from the result of part (d) that the value of the integral

$$(2n+1)\int_{-1}^{1}[P_n(x)]^2\,dx$$

is a constant, independent of n. Find that constant.

f) From part (e) obtain desired result (22).

20. Using summation formula (10), prove property (16).

21. Using summation formula (10), prove the identity

$$xP_n'(x) - P_{n-1}'(x) = nP_n(x), \quad n = 1, 2, 3, \ldots.$$

22. Using the properties (14–16), obtain the formula

$$\int_x^1 P_n(t)\,dt = \frac{1}{2n+1}[P_{n-1}(x) - P_{n+1}(x)],$$
$$n = 1, 2, 3, \ldots.$$

23. a) Use Rodrigues' formula to obtain

$$2^n n! P_{n+1}(x) = (2n+1)\frac{d^{n-1}}{dx^{n-1}}u^n + 2n\frac{d^{n-1}}{dx^{n-1}}u^{n-1}$$

where $u = x^2 - 1$.

b) In the relation in part (a) make the substitution

$$\frac{d^{n-1}}{dx^{n-1}}u^{n-1} = 2^{n-1}(n-1)!P_{n-1}(x)$$

to obtain

$$P_{n+1}(x) - P_{n-1}(x) = \left[\frac{2n+1}{2^n n!}\right]\frac{d^{n-1}}{dx^{n-1}}u^n.$$

24. The expression $(1 - 2xt + t^2)^{-1/2}$ is called a **generating function** for the Legendre polynomials. Use the binomial theorem to show formally that

$$(1 - 2xt + t^2)^{-1/2} = \sum_{n=0}^{\infty} P_n(x)t^n.$$

CHAPTER 9 REVIEW EXERCISES

In problems 1–5, classify the specified point x_0 as an ordinary or singular point of the given differential equation. If x_0 is a singular point, state whether it is regular or irregular. If a series solution exists, write the form of the series solution to the differential equation and find the interval for which convergence of the series is guaranteed.

1. $(x^2 - 2x + 2)y'' + y' + y = 0, \quad x_0 = 2$

2. $(x^3 - 2x^2 + x)y'' + \dfrac{x-1}{2}y' + y = 0, \quad x_0 = 1$

3. $(x^3 + x^2)^2y'' + (x+1)y' + y = 0, \quad x_0 = -1$

4. $x^2(x+2)^2y'' + (x^2 - 4)y' + 2y = 0, \quad x_0 = 0$

5. $(x-6)^2y'' - 4xy' + 3y = 0, \quad x_0 = 0$

In problems 6–10, solve the given Euler equation.

6. $x^2y'' + 3xy' + 3y = 0$

7. $x^2y'' - y = 0$

8. $x^2y'' + xy' + y = 0$

9. $x^2y'' + 5xy' + 4y = 0$

10. $x^2y'' - 3xy' + 13y = 0$

In problems 11–21 find one series solution to the given differential equation by assuming an appropriate series solution form. Find at least the first four nonzero terms of the series.

11. $x^2y'' - xy' + (1 - x)y = 0$

12. $x^2y'' - xy' + (x - 3)y = 0$

13. $(x^2 + 1)y'' + xy' + 3y = 0$

14. $y'' - 8xy' - 4y = 0$

15. $x^2y'' + 3xy' + (1 + x)y = 0$

16. $x^2y'' + x(3 - x^2)y' - 3y = 0$

17. $x^2y'' + x^2y' - \frac{3}{4}y = 0$

18. $xy'' - x(2x + 3)y' + 4y = 0$

19. $(x^2 - 1)y'' + xy' + 2y = 0$

20. $y'' - 2xy' + 8y = 0$

21. $(x^2 + 1)y'' - 3xy' + y = 0$

22. Neutron Flux in a Nuclear Reactor

In a nuclear reactor, energy is produced through the process of nuclear fission. The specific mechanism that results in the release of nuclear energy is the chain reaction, a regenerative procedure in which free neutrons collide with fissionable atoms (such as uranium or plutonium) and cause those atoms to split, thus generating energy and a number of fission products, some of which are more neutrons. These newly released neutrons collide with other fissionable atoms, and the cycle (or "chain") continues.

In order to control the chain reaction, the neutron population in the reactor must be controlled. Otherwise, the fission process can increase the population at an exponential rate. In practice, however, the number of free neutrons present in the reactor can be controlled quite routinely as long as their density and distribution within the reactor can be accurately and quickly predicted. This is done by calculating what is known as the **neutron flux,** the number of neutrons of each energy E (or equivalently, speed v) passing a unit area in each direction at any time t. In complete detail this calculation is quite complex and best solved on a computer. However, reasonable estimates of the neutron flux for some simple types of reactors can be made.

Consider a cylindrical reactor composed of a homogeneous mass of properly cooled fissionable material with radius $r = R_0$ and height H (Fig. 9.4). In the cylindrical coordinate system, if some simplifying assumptions are made concerning the behavior of the neutrons, the neutron flux can be represented by the functional relationship $\phi(r, z)$, where r is the radial distance from the center line of the reactor and z is the vertical distance from the reactor's horizontal center plane. In effect, we are assuming that the reaction process is independent of the angle θ in the cylindrical coordinate system, and therefore the neutron flux $\phi(r, z)$ is not a function of θ but only of r and z.

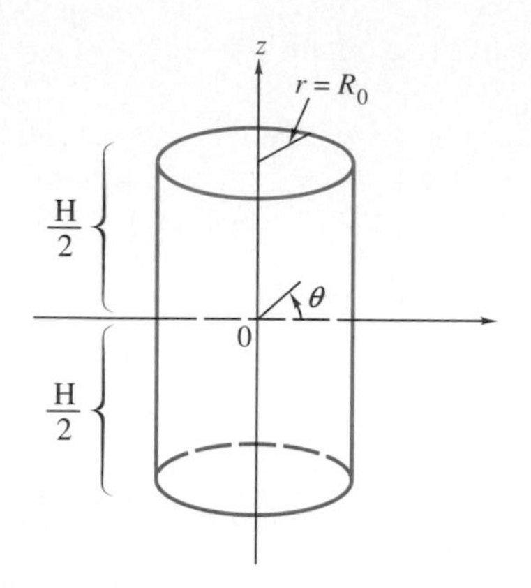

FIGURE 9.4

The differential equation used to describe the behavior of the neutron flux $\phi(r, z)$ in the reactor is a homogeneous partial differential equation of the form

$$\frac{\partial^2\phi}{\partial r^2} + \frac{1}{r}\frac{\partial\phi}{\partial r} + \frac{\partial^2\phi}{\partial z^2} + B^2\phi = 0,$$

where B^2 is a positive constant that depends on the reactor's physical properties. By assuming a solution of the form

$$\phi(r, z) = R(r)Z(z),$$

where $R(r)$ is a function of the radial displacement alone and $Z(z)$ is a function of the vertical displacement, and by letting $B^2 = \alpha^2 + \beta^2$, this partial differential equation can be reduced to the following pair of homogeneous ordinary differential equations:

$$\frac{d^2Z}{dz^2} + \alpha^2 Z = 0, \tag{1}$$

$$r^2\frac{d^2R}{dr^2} + r\frac{dR}{dr} + \beta^2r^2R = 0. \tag{2}$$

Equation (1) is solvable by the methods taught earlier in the course. Equation (2) is the parametric form of Bessel's equation of order 0.

A change of variables will remove the parameter β^2: Let $t = \beta r$, which means

$$\frac{dt}{dr} = \beta, \quad \frac{dR}{dr} = \beta\frac{dR}{dt}, \quad \text{and} \quad \frac{d^2R}{dt^2} = \beta^2\frac{d^2R}{dt^2}.$$

Then Eq. (2) reduces to Bessel's equation of order 0:

$$t^2\frac{d^2R}{dt^2} + t\frac{dR}{dt} + t^2R = 0. \tag{3}$$

a) Find the general solution $Z(z)$ by solving Eq. (1).

b) One solution of Eq. (3) is $J_0(t)$. State the general form for the second solution $R_2(t)$. This equation, when fully derived, is known as the *Bessel function of the second kind of order 0*, denoted $Y_0(t)$.
c) Make the substitution $\beta r = t$ and state the general solution $R(r)$ of Eq. (3).
d) State the solution $\phi(r, z)$ for the neutron flux in the finite cylindrical reactor using the solutions from (a–c) and the relation for $\phi(r, z)$.

23. Temperature Distribution in a Cannon Barrel

The barrel of a cannon, shown at Fig. 9.5, may be described as a long, hollow cylindrical tube of inner radius ρ and outer radius ρ_0, where ρ is the distance from the barrel's longitudinal axis. The tube must be made of material that is strong enough to withstand the shock it receives during firing and conductive enough that it rapidly dissipates the heat generated by firing. Obviously the ability to predict the temperature distribution as a function of time in the barrel is a fundamental requirement for cannon manufacturers. Let us investigate one aspect of that problem—prediction of the temperature distribution after the cannon has stopped firing.

If longitudinal and axial differences in temperature are considered to be insignificant, the differential equation that may be used to describe the behavior of the temperature $u(\rho, t)$ in the barrel is the partial differential equation

$$\frac{\partial u}{\partial t} = k\left(\frac{\partial^2 u}{\partial \rho^2} + \frac{1}{\rho}\frac{\partial u}{\partial \rho}\right),$$

where k is the diffusivity of the barrel's metal. This partial differential equation can be reduced to a pair of ordinary differential equations by assuming a solution of the form

$$u(\rho, t) = R(\rho)T(t),$$

where $R(\rho)$ is the radial dependence of the temperature in the barrel and $T(t)$ is the temperature difference between the barrel and the surrounding air, as a function of time. The resulting equations are

$$\frac{dT}{dt} + \lambda^2 kT = 0 \tag{1}$$

and

$$\rho^2\frac{d^2R}{d\rho^2} + \rho\frac{dR}{d\rho} + \lambda^2\rho^2 R = 0. \tag{2}$$

a) Find the general solution to Eq. (1).
b) Show that $J_0(\lambda\rho)$ solves Eq. (2).
c) State the general form of the second solution $R_2(\lambda\rho)$ to Eq. (2). When fully derived, this solution is the Bessel function of the second kind of order 0, denoted $Y_0(\lambda\rho)$.
d) Using the solutions from (a), (b), and (c), state the general solution $u(\rho, t)$ to the temperature distribution in the cannon barrel.
e) Why will the fact that $\ln \rho$ tends toward ∞ as ρ approaches 0 not create a physically unreasonable solution for the temperature in the barrel?
f) What will the steady-state solution of $u(\rho, t)$ be?
g) Use the method of reduction of order to find the second solution $R_2(\lambda\rho)$ to Eq. (2).

Note: We will discuss the heat equation

$$\frac{\partial u}{\partial t} = k\left(\frac{\partial^2 u}{\partial \rho^2} + \frac{1}{\rho}\frac{\partial u}{\partial \rho}\right)$$

in Chapter 10.

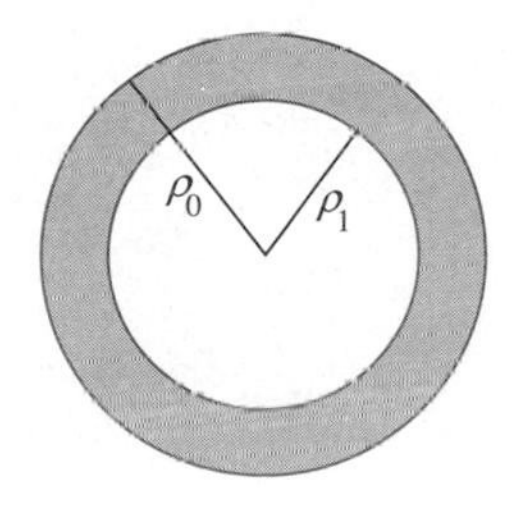

FIGURE 9.5

24. Displacement of a Surface

In general, the displacement of surfaces under tension is a difficult problem to model and solve mathematically. Consider the motion of waves in a tub of water or the vibration of the surface of a drum. Solutions to these problems might require the use of partial differential equations and perhaps a numerical solution technique. This problem deals with the *amplitude of vibrations* z of the surface of a drum with a tightly stretched circular drumhead. Under the assumption that the vibrations are symmetric about the origin, the amplitude of vibration z at any point $P(r, \theta)$ is modeled by the differential equation

$$\frac{d^2z}{dr^2} + \frac{1}{r}\frac{dz}{dr} + k^2 z = 0 \tag{1}$$

where r represents the distance from the center of the surface of the drum to any point $P(r, \theta)$, and θ is a positive clockwise angle measured from the x-axis (Fig. 9.6). In Eq. (1),

$$k^2 = \frac{\sigma\omega^2}{T}$$

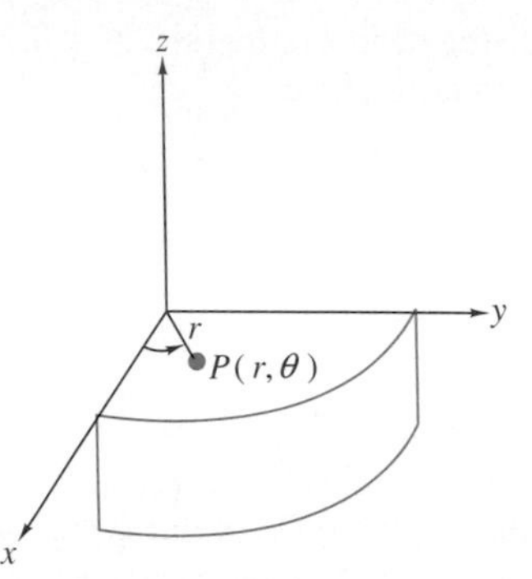

FIGURE 9.6

where T is the tension per unit length of the drumhead, σ is the mass per unit area, and ω is the angular frequency of vibration. Note that, since the vibrations are symmetric about the origin, the differential equation (and therefore the solution) does not depend on θ.

a) By substituting $x = kr$, show that the second-order differential equation has the form

$$x^2 \frac{d^2z}{dx^2} + x\frac{dz}{dx} + x^2 z = 0, \tag{2}$$

which is Bessel's equation of order 0.

b) Consider a circular drumhead whose radius is 5 in (so the drumhead is clamped at $r = 5$). Let $T = 4$ lb/in and $\sigma = 2 \times 10^{-6}$ slugs/in². The amplitude of vibration at $r = 0$ is $1/16$ in (i.e., $z(0) = 1/16$). Write a solution z to Eq. (2) in terms of r and ω using the given values of σ and T.

c) Write the form of a second solution z_2 to the differential Eq. (2) in part (a). Explain why it must be excluded from the general solution $z = C_1 z_1 + C_2 z_2$; that is, $C_2 = 0$.

d) The correct solution is $z = C_1 z_1$ where it can be shown that $C_1 = 1/16$. Suppose $\omega = 800$ cycles/sec. Then for $r = 0, 0.5, 1.0, 1.5, 2.0, 2.5, 3.0, 3.5, 4.0, 4.5$, and 5.0, complete the following:

(i) Graph the first term of your series solution for $z(r)$.

(ii) Using the first two terms of your series solution, graph $z(r)$.

(iii) Using the first three terms of your series solution, graph $z(r)$.

(iv) Which graph do you think best represents the solution $z(r)$ and why?

(v) Why should $C_1 = 1/16$, and why is $z = (1/16)z_1$ a solution even when $r = 0$? (*Hint:* Consider the limit as $r \to 0$.)

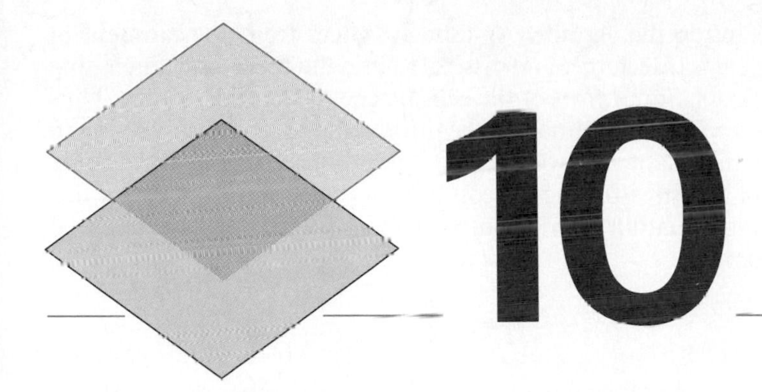

10 Partial Differential Equations: Modeling Position- and Time-Dependent Behaviors

Often when we are modeling a physical behavior using derivatives to express rates of change, we find that we need several independent variables to obtain a reasonable approximation of the behavior. In these situations the derivatives are necessarily partial derivatives and give rise to **partial differential equations** relating the dependent and independent variables. Although we will not construct the models here, we shall describe several real-world scenarios that give rise to partial differential equations modeling their behaviors.

EXAMPLE 1 HEAT TRANSFER IN A SOLID BODY

In Section 1.3 we considered the cooling of a space capsule. By treating the capsule as a single point we derived the approximation for the behavior known as Newton's law of cooling. However, we know that the space capsule does have dimension. As the space capsule reenters and travels through the earth's atmosphere, heat is generated by friction. At the instant the capsule splashes into the ocean, it has an initial temperature distribution varying with position inside the capsule. As time advances from the moment of splashdown, heat is transferred and dissipated into the ocean and air causing the capsule to cool. Some parts of the capsule cool faster than others. Thus the temperature varies with time and the three variables locating position in the capsule. In this chapter we will simplify this problem by considering only one dimension and investigate heat conduction in a long thin rod. For this one-dimensional situation we will construct and solve the partial differential equation model

$$\frac{\partial^2 u}{\partial x^2} = \frac{1}{k}\frac{\partial u}{\partial t}$$

where x is the position along the rod, t is time, u is the temperature along the rod, and k is a constant known as **thermal diffusivity.**

EXAMPLE 2 CONCENTRATION OF CHEMICALS AND POLLUTANTS

We will consider again the problem of the dissipation and transfer of a liquid pollutant in a large body of water such as one of the Great Lakes, as discussed in Section 2.2. We will no longer assume the mixture is kept uniform by constant stirring but will be concerned with the concentration of the pollutant over time at *different locations* in the water. There is a source discharging the pollutant into the lake that may be varying with time as well as location. The pollutant is being transported throughout the lake by two processes: (1) *diffusion* causes the molecules of the pollutant to combine chemically with neighboring water molecules, and (2) *convection* (caused by currents in the lake and the discharging apparatus) moves the polluting molecules from one region of the lake to another. We will restrict our attention to studying what happens at one depth only (say just beneath the lake's surface) and assume the liquid pollutant has a constant velocity and enters a region between two wide parallel barriers. With these assumptions the transport behavior can be modeled by a partial differential equation like the one-dimensional heat equation we will study later in this chapter:

$$\frac{\partial^2 C}{\partial Y^2} = \frac{\partial C}{\partial X}$$

where C is the concentration of the pollutant, and X and Y are scaled distances expressed in terms of the velocity, diffusivity, and distance between the parallel barriers.

EXAMPLE 3 VIBRATING STRINGS AND MEMBRANES

Another important class of partial differential equations arises from modeling the behavior of a vibrating string or membrane. In this situation the description of the movement of a point on the vibrating string or membrane in space depends on its location on the string or membrane as well as time. For instance, if we think of the road surface of the Golden Gate Bridge as a two-dimensional membrane, we would be interested in knowing the up-and-down vibrations of the road as traffic moves across the bridge. In the one-dimensional case we might be interested in the sideways movement of the bridge, thinking of the bridge as a long, thin string being "plucked" by the wind or some other energy source. We will construct and analyze a model for the one-dimensional wave equation in Sections 10.5 and 10.7. The model is described by the partial differential equation

$$\frac{\partial^2 y}{\partial t^2} = a^2 \frac{\partial^2 y}{\partial x^2}$$

where y is the displacement of the string away from its equilibrium position over time, and a is an important velocity (the meaning of which will become clear during the development in Section 10.5).

Other Examples

Physical behaviors that give rise to partial differential equations abound. For instance, when an electrical current is passed through a conductor, an electromagnetic field is formed causing changes in the current and electrical potential over time. Oscillation takes place in the conductor as a result of these changes, and the behavior can be approximated by a partial differential equation in which the independent variables are position x along the axis of the conductor and time t. Here the governing equation is

$$\frac{\partial i}{\partial x} + C\frac{\partial v}{\partial t} + Gv = 0$$

where i is the current, v is the potential, C is the capacitance, and G is the leakage of electricity through the imperfect insulation.

Another example involves the changes in voltage v along a transmission cable. The behavior can be approximated by the equation

$$\frac{\partial^2 v}{\partial x^2} = LC\frac{\partial^2 v}{\partial t^2} + (RC + GL)\frac{\partial v}{\partial t} + RGv$$

where x is the position along the cable, t is time, R is the resistance per unit length, L is the inductance per unit length, C is the capacitance per unit length, and G is the conductance to ground per unit length.

Yet another example occurs in the study of one-dimensional **Brownian motion,** the ceaseless, irregular, and apparently random motion of tiny particles immersed in a gas or liquid. A given particle undergoes one-dimensional motion caused by random collisions with other particles in the fluid or gas. There are many collisions over an interval of time, and they are assumed to be independent of one another. If $v(x, t)$ denotes the probability that the particle in question is at position x at time t, then the following equation approximates one-dimensional Brownian motion:

$$\frac{\partial v}{\partial t} = -c\frac{\partial v}{\partial x} + \frac{D}{2}\left(\frac{\partial^2 v}{\partial x^2}\right)$$

where c is the average displacement of the particle per unit time, and D is the variance of the observed displacement about the average.

From these examples you can see that partial differential equations provide powerful models for describing myriad phenomena in the physical universe. Because of their importance in modeling physical phenomena, much effort has been spent by mathematicians to solve partial differential equations. These efforts have been enormously fruitful and are directly responsible for the development of a great deal of modern mathematics. Moreover, the solutions themselves have provided insight into the underlying physical processes being modeled and have led to even further discoveries.

In this chapter we will introduce you to several classes of partial differential equations important to engineering: the *heat* or *diffusion equation* in one dimension, the one-dimensional *wave equation*, and *Laplace's equation*. We begin in Section 10.1 by constructing a model approximating the behavior of heat flowing in a long, thin rod. In Sections 10.2–10.4 we investigate both analytical and numerical methods for solving the model in order to predict the temperature at various locations in the rod over time. In Sections 10.5–10.7 some of the other partial differential equations described so far are then developed and studied as well.

10.1

INTRODUCTION: MODELING ONE-DIMENSIONAL HEAT FLOW

Consider the problem of predicting the temperature at any time along a thin long rod in which heat is flowing. Assume that the rod is of homogeneous material and has cross-sectional area A, where A is constant and "small" throughout the length of the rod. We assume the x-axis is aligned with the length of the rod, so x varies from $x = 0$ to $x = L$ as illustrated in Fig. 10.1. The rod is laterally insulated along its entire length so that we may assume no heat is conducted in the y or z directions. This assumption is also satisfied if the y and z dimensions of the rod are negligible with respect to the length of the rod. Thus we say that the rod is **one-dimensional** and is approximated by a long thin length of wire.

Let $u(x, t)$ denote the temperature of the rod, measured in degrees (the units of which are unimportant to our development), at position x along the rod and time t. We define **temperature** as the amount of heat energy per unit volume and assume the temperature is constant at each cross section (in accordance with our assumption that the rod is one-dimensional). The

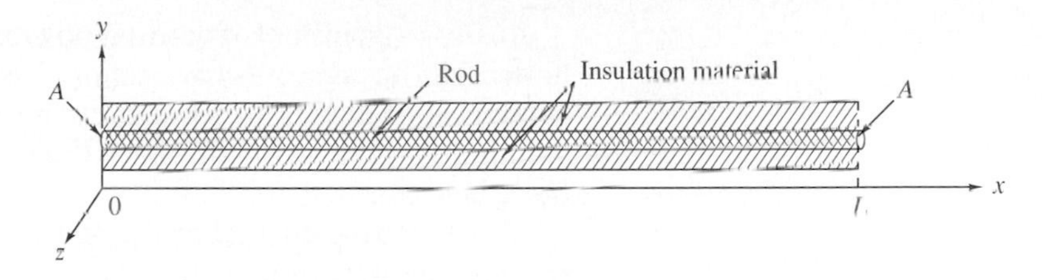

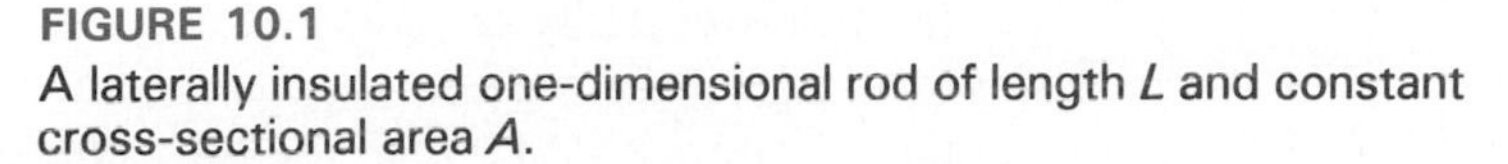

FIGURE 10.1
A laterally insulated one-dimensional rod of length L and constant cross-sectional area A.

amount of heat energy that must be supplied to a unit mass of the material from which the rod is made in order to raise its temperature by 1° is called the **specific heat** of the material. The **thermal conductivity** is a property of the material and is a measure of its ability to conduct heat. The values of these constants for various materials have been determined experimentally, and we assume they are constant for the rod.

Suppose at time $t = 0$ the rod has an initial temperature $u(x, 0)$ at each position x along its length. The ends of the rod $x = 0$ and $x = L$ may or may not be insulated. Suppose first that they are *not* insulated but are held at some constant temperature, say 0°. For instance, you can think of the ends of the rod as being immersed in a fluid like the ocean, acting as a vast heat sink having a constant temperature of 0°. We assume that no heat is generated internally, either chemically or electrically, and consider what happens to the temperature of the rod as time advances. At time $t = 0$ the rod has some initial temperature distribution $u(x, 0)$. The ends of the rod are held at the constant temperature $u(0, t) = u(L, t) = 0$ for all time. What is the temperature $u(x, t)$ interior to the rod at any time $t > 0$? This situation is depicted in Fig. 10.2.

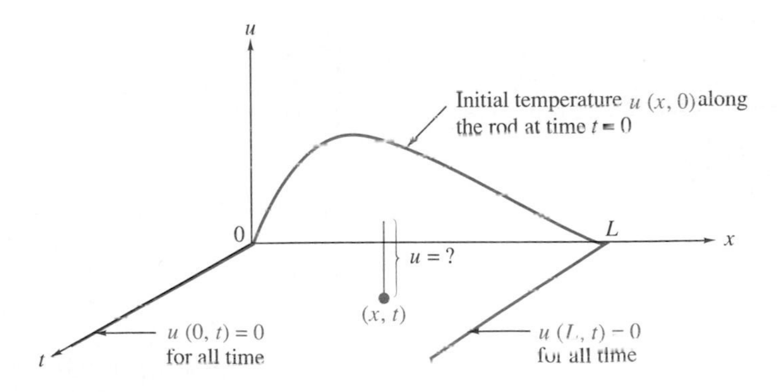

FIGURE 10.2
The initial temperature distribution as a function of position x along the rod shown in Fig. 10.1. The temperature is held constant at the two ends $x = 0$ and $x = L$ for all time. What is the temperature at any point interior to the rod at any time $t > 0$?

At any given instant of time it is possible that one cross section of the rod is hotter than an adjacent cross section. Temperature differences along the rod cause heat to flow from a hotter cross section (cooling it) to a colder cross section (which is warmed). Since the temperature at the ends of the rod is being held constant at $0°$, the heat will simply flow out of these two noninsulated ends. As the internal heat of the rod is transferred from hotter to colder regions and then to the ends of the rod (where it is lost), the entire length of the rod eventually cools to the constant temperature of $0°$. The material from which the rod is made, its specific heat, and its thermal conductivity govern the speed at which the rod cools down. Some materials are able to retain heat longer than others. Nevertheless, this cooling-down process is one you have experienced many times.

Suppose instead that the rod has an initial temperature at each position but that the ends of the rod *are* insulated. Then no heat escapes from the rod at all. In this case the heat is eventually transported from hotter to colder regions throughout the rod, diffusing the heat until the rod has one constant temperature. We want to mathematically model this heat flow for the one-dimensional rod.

Assumptions for the One-Dimensional Heat Equation

Before constructing the model, let us first explicitly state our assumptions:

1. The rod is one-dimensional, and its lateral surface is insulated. Thus no heat escapes or is input through the lateral surface (*not* including the ends of the rod).
2. No heat is generated internally.
3. The rod is uniform, so it must be made of one material having constant density throughout the entire length.
4. The rod's material has constant specific heat and constant thermal conductivity. This assumption is reasonable if the temperature range is not too large.

Constructing the Heat-Conduction Model

We define the temperature function

$$u(x, t) = \text{temperature at position } x, \text{ at time } t.$$

Assuming we know the initial temperature at each position along the rod $0 \leq x \leq L$ when $t = 0$, we want to predict the temperature $u(x, t)$ at each position x for future times $t > 0$. Various conditions will affect the temperature, for example, heating the rod at one end or insulating the rod at one or both ends. We will elaborate on these conditions later on in the discussion.

The initial temperature distribution along the rod also affects the behavior of the heat flow, and this initial temperature can be given by different distributions. We want to model mathematically the change in temperature along the rod. Since the temperature is a function of two independent variables, it will require a partial differential equation to approximate its behavior.

Temperature differences along the rod cause heat to flow from hotter to colder regions. Let us define this heat flow function, called the **heat flux,** by the following:

$q(x, t) =$ amount of heat energy per unit time flowing through the rod at position x, at time t.

Consider a small cross section of the rod of width Δx as depicted in Fig. 10.3. Suppose at time t the amount of heat energy per unit time flowing into the rod on the left of the cross section is $q(x, t)$ and that in the time increment Δt heat flows through the cross section and out the righthand side. Then we have

$$\text{energy in} = q(x, t)\, \Delta t,$$
$$\text{energy out} = q(x + \Delta x, t + \Delta t)\, \Delta t.$$

The fundamental heat-flow process is described by the relation

$$\text{energy in} = \text{energy out} + \text{energy absorbed}.$$

Thus

$$q(x, t)\, \Delta t = q(x + \Delta x, t + \Delta t)\, \Delta t + \text{energy absorbed}.$$

This principle is called **conservation of heat energy.** To complete our derivation we need an expression for the absorbed energy and we need to relate heat flux q to temperature u.

The energy absorbed is related to the temperature difference across the thin slice of rod over the time increment Δt. From our understanding of Newton's law of cooling discussed in Section 1.3 and the definition of temperature, it seems reasonable to assume that the energy absorbed is proportional to the change in temperature times the cross-sectional length. (After all, temperature has the dimension of energy per unit volume, and volume in

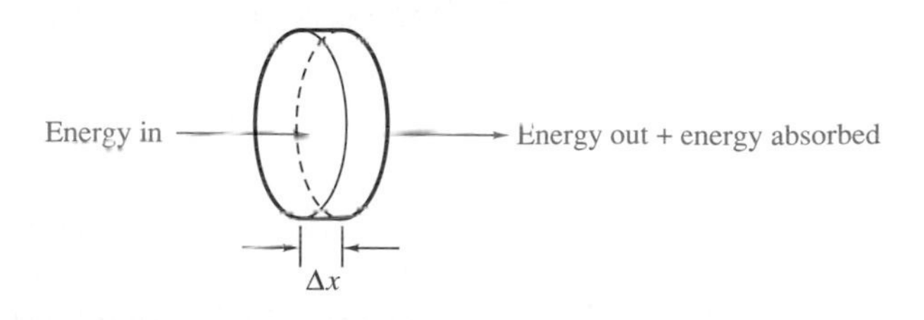

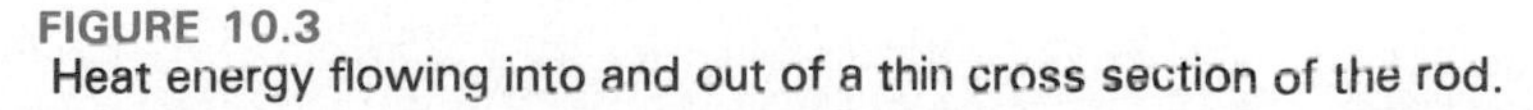
FIGURE 10.3
Heat energy flowing into and out of a thin cross section of the rod.

the one-dimensional case is length.) Symbolically, this proportionality assumption translates to

$$\text{energy absorbed} = k_1[u(x + \Delta x, t + \Delta t) - u(x, t)]\, \Delta x.$$

Thus we obtain the equation

$$q(x, t)\, \Delta t = q(x + \Delta x, t + \Delta t)\, \Delta t + k_1[u(x + \Delta x, t + \Delta t) - u(x, t)]\, \Delta x$$

or

$$\frac{q(x, t) - q(x + \Delta x, t + \Delta t)}{\Delta x} = \frac{k_1[u(x + \Delta x, t + \Delta t) - u(x, t)]}{\Delta t}.$$

Although more is involved mathematically than just the notion of a partial derivative, it can be shown that passage to the limit as Δx and Δt both tend toward zero independently gives us the equation

$$-\frac{\partial q}{\partial x} = k_1 \frac{\partial u}{\partial t}. \tag{1}$$

Since Eq. (1) involves two dependent variables (as well as the two independent variables x and t), we need to relate q and u. We use two principles of heat flow:

1. If there are temperature differences, heat flows from the hotter region to the colder region.
2. The flow of heat is proportional to the change in temperature per unit length.

Expressing these principles mathematically yields the equation

$$q = -\alpha \frac{\partial u}{\partial x}, \qquad \alpha > 0. \tag{2}$$

Equation (2) is known as **Fourier's law of heat conduction.** Substituting Eq. (2) into Eq. (1) then yields

$$-\frac{\partial}{\partial x}\left(-\alpha \frac{\partial u}{\partial x}\right) = k_1 \frac{\partial u}{\partial t}$$

or

$$\frac{\partial^2 u}{\partial x^2} = \frac{1}{k} \frac{\partial u}{\partial t}. \tag{3}$$

The constant $k\,(=\alpha/k_1)$ is positive and called the **thermal diffusivity.** Equation (3) is called the **one-dimensional heat equation** for the laterally insulated rod. (We emphasize that the ends of the rod may not be insulated.)

Notice that partial differential Eq. (3) is of the form

$$A(x, t)u_{xx} + B(x, t)u_{xt} + C(x, t)u_{tt} + D(x, t)u_x + E(x, t)u_t + F(x, t)u = G(x, t). \tag{4}$$

Here we use the subscripted notation to denote partial derivatives. Equation (4) is said to be a **linear second-order partial differential equation.** When $G(x, t) \equiv 0$, the equation is said to be **homogeneous;** otherwise it is **nonhomogeneous.** Thus heat equation (3) is an example of a homogeneous linear partial differential equation. As we saw with homogeneous linear *ordinary* differential equations in Section 3.2, any linear combination of solutions to the equation also gives a solution to the equation. This same principle holds for partial differential equations.

SUPERPOSITION PRINCIPLE

If $u_1, u_2, \ldots, u_k$ are solutions of a homogeneous linear partial differential equation, then for any constants $c_1, c_2, \ldots, c_k$ the linear combination

$$u(x, t) = c_1 u_1 + c_2 u_2 + \cdots + c_k u_k$$

is also a solution.

We assert without proof that whenever infinitely many solutions $u_1, u_2, u_3, \ldots$ to a homogeneous linear partial differential equation exist, then the infinite series

$$u(x, t) = \sum_{k=1}^{\infty} u_k$$

is also a solution.*

Initial Temperature Distribution

We return now to the heat conduction problem. In addition to partial differential Eq. (3), what else do we need to know to fully determine the temperature $u(x, t)$? We note that the equation describing the flow of heat energy has one time derivative u_t. When solving an ordinary differential equation with one time derivative, we need an initial condition to evaluate the arbitrary constant for the particular solution. Likewise, for the partial differential equation describing heat flow, we need an **initial condition** that gives the initial state of the rod's temperature distribution. That is, at time $t = 0$ (the initial time) we need to know the temperature function $u(x, 0) = f(x)$. Any function can serve for $u(x, 0)$ as long as it and its derivative are piecewise continuous over the length of the rod, where $0 < x < L$. We can depict the graph of the initial condition using a three-dimensional coordinate system for $u(x, t)$, as shown before in Fig. 10.2.

* Technical issues concerned with the idea of *uniform convergence* are needed to guarantee the validity of our assertion. If interested in these issues, consult any standard text on advanced calculus for the detailed information.

Boundary Conditions

The partial differential equation for heat flow also contains a second derivative with respect to position, u_{xx}. In solving a second-order ordinary differential equation two conditions are needed to evaluate the arbitrary constants for a particular solution, often given as the initial value of the dependent variable and the initial value of its derivative, $y(0)$ and $y'(0)$. Likewise, we cannot predict the future temperature $u(x, t)$ knowing only the initial temperature distribution $u(x, 0)$ and partial differential Eq. (3). We also need to know what happens to the temperature at the two boundaries of the rod, $x = 0$ and $x = L$, for all time $t > 0$. The conditions specifying the state of the temperature at the two ends or boundaries of the rod are called **boundary conditions.** They fall into two categories: fixed and insulated.

Fixed Boundary Conditions In a fixed boundary condition the temperature of the rod at the end in question is prescribed to be a certain fixed value for all time. For instance, the condition

$$u(0, t) = T,$$

where T is the constant temperature of a fluid or substance with which the end of the rod is in contact, fixes the temperature at the left end of the rod. If the temperature $T = 0$, the fixed boundary condition is said to be *homogeneous;* otherwise it is *nonhomogeneous.* Homogeneous boundary conditions are depicted in Fig. 10.2.

Physically we can think of the end of the rod as being in contact with a large volume of substance whose temperature is kept constant by some means (such as stirring) so that the temperature of the substance is not affected by contact with the rod. Any heat flowing out of the end of the rod, for instance, would instantaneously be absorbed uniformly into this substance, whose volume is so vast there is virtually no change in its temperature. Of course this assumption is an idealization of the real world in order to simplify our model. If the temperature of the substance is less than the temperature of the rod, heat will flow out of the rod and the substance acts as a heat sink. On the other hand, if the substance is warmer than the rod, heat flows into the rod and warms it, and the substance acts as a heat source.

Insulated Boundary Conditions In this situation the end of the rod in question is insulated so that no heat flow occurs across this boundary. How do we model this condition mathematically? If the heat flow is zero, then there is no change in temperature, so $\partial u/\partial x = 0$ at that end. For instance, the condition

$$u_x(0, t) = 0$$

for all time $t > 0$ says that the left end of the rod is insulated.

If one end of the rod has a fixed boundary condition and the other end is insulated, we say that the heat equation has **mixed boundary conditions.**

A Boundary-Value Problem

Let us summarize the heat-flow model for a rod with fixed boundary conditions. The following model is an example of a **boundary-value problem.**

ONE-DIMENSIONAL HEAT EQUATION WITH FIXED BOUNDARY CONDITIONS

PDE:	$u_{xx} = \frac{1}{k} u_t,$	$k > 0, \quad 0 < x < L, \quad t > 0;$	(5)
BC:	$u(0, t) = T_1,$ $u(L, t) = T_2,$	$t > 0;$	(6)
IC:	$u(x, 0) = f(x),$	$0 < x < L.$	(7)

Here, as in similar models presented throughout the chapter, PDE represents the partial differential equation, BC is the boundary condition(s), and IC is the initial condition.

The *solution* to this heat-flow model is a function $u(x, t)$ that satisfies the partial differential equation, the boundary conditions, and the initial condition. We next examine how a solution might be obtained, and how it could be depicted.

Representing the Solution

For illustrative purposes, let us consider the special problem where $k = L = 1$, $T_1 = T_2 = 0$ (homogeneous boundary conditions), and the initial condition is given by $f(x) = 1$ if $0 < x < 0.5$, but $f(x) = 0$ if $0.5 < x < 1$. Since the solution is a function of two independent variables, its graph is represented by a surface in *txu*-space as depicted in Fig. 10.4. However, surfaces in space are usually difficult to illustrate graphically without the aid of a computer.

Another method used in calculus to investigate functions of two variables shows *level curves* or *contours* in the *tx*-plane. Along a given contour the temperature has a constant value. Level curves for our three-dimensional surface are illustrated in Fig. 10.5. To use a contour plot, select some position value along the rod, say $x = 0.25$. As time advances, read the temperatures along the horizontal line associated with that position. For instance, at $t = 0.03$ the temperature is approximately $u \approx 0.57$; at $t = 0.09$, $u \approx 0.20$; at $t = 0.15$, $u \approx 0.10$; at $t = 0.21$, $u \approx 0.06$; and so forth.

Another option is to depict the curves of intersection of the surface with planes parallel to the *xu*-plane for various values of t. This procedure is illustrated in Fig. 10.6. Usually these curves of intersection are displayed on a single two-dimensional graph as "snapshots" of the solution function $u(x, t)$ for selected values of t. This representation is shown in Fig. 10.7 for our

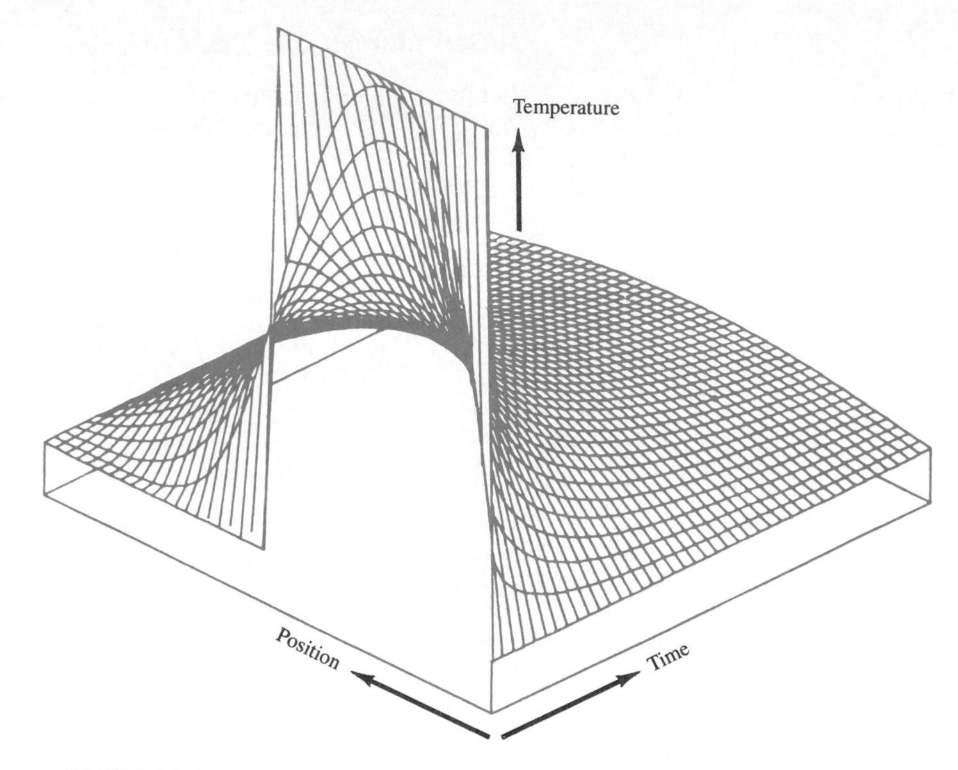

FIGURE 10.4
Temperature function $u(x, t)$ depicted as a surface lying above the tx-plane.

example. Notice in the graph that the temperature of the left half of the rod is initially $u = 1$, but the left end (boundary) is constantly held at the temperature $T_1 = 0$. Thus the extreme left half cools rapidly as heat flows out the left end, and our mathematical idealization assumes the left endpoint cools "instantaneously" to 0°. Notice too that heat flows from the warmer left half of the rod into the right half, warming it. However, the right end is being held at the constant temperature $T_2 = 0$ so that heat also flows out this end. When $t = 0.3$, the rod has cooled significantly throughout. Eventually the temperature will be the constant $u(x, t) \equiv 0$.

Now let us consider how we might obtain the solution function analytically. It will turn out that you do not yet have all the mathematical tools needed to obtain the analytical solution, but our preliminary work will motivate the mathematics you will need and study in Section 10.2.

Preliminary Solution to the Heat Equation with Fixed Homogeneous Boundary Conditions

Consider the model for one-dimensional heat flow in a rod given by Eqs. (5–7). For definiteness in our discussion here we assume homogeneous boundary temperatures $T_1 = T_2 = 0$. Let us work out a solution to the

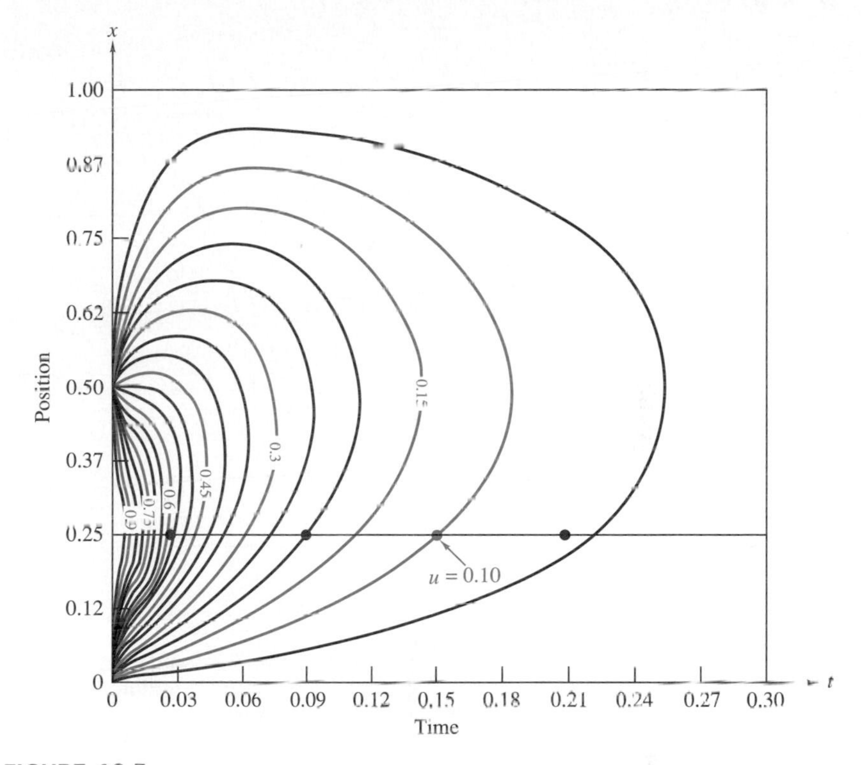

FIGURE 10.5
Level curves or contours associated with the surface in Fig. 10.4. The contour curves are plotted for differences in temperature values of 0.05°. When $t = 0.15$, the temperature at position $x = 0.25$ is approximately $u \approx 0.10°$.

model. Observe first that $u(x, t) = 0$ solves the partial differential equation and the homogeneous boundary conditions but *not* the initial condition. Hence we will not consider this *trivial solution* $u(x, t) = 0$ any further. We are interested in nontrivial solutions.

Suppose the solution function can be factored into a function only of x times a function only of t. We employ the notation $X(x)$ and $T(t)$ for these factors to remind us that the variables are independent and that X is a function of x alone whereas T is a function of t alone. Thus we suppose that

$$u(x, t) = X(x)T(t). \tag{8}$$

Then the partial derivatives in Eq. (5) become

$$u_t(x, t) = X(x)\frac{dT}{dt} = XT'$$

and

$$u_{xx}(x, t) = \frac{d^2X}{dx^2}T(t) = X''T.$$

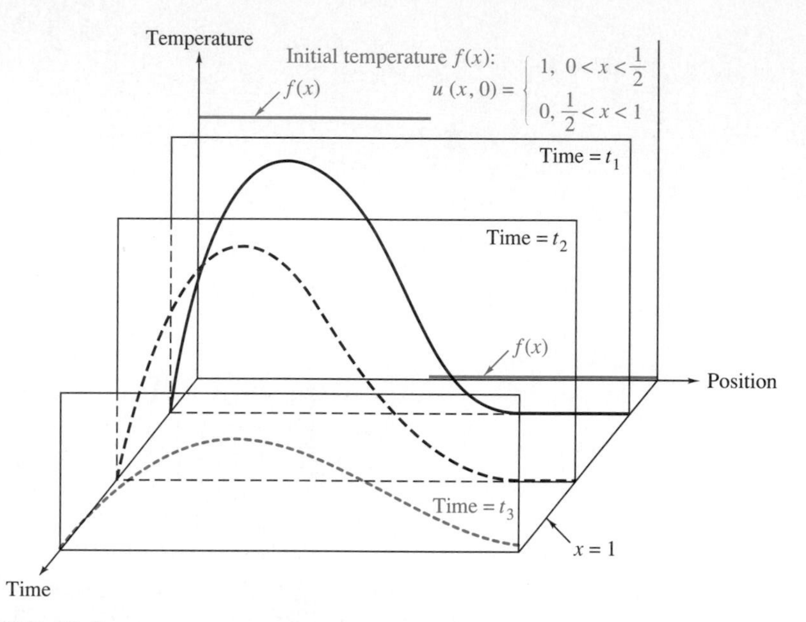

FIGURE 10.6

Temperature curves for the one-dimensional heat equation with fixed homogeneous boundary conditions $T_1 = T_2 = 0$, $k = L = 1$, and the initial condition $f(x)$ as shown.

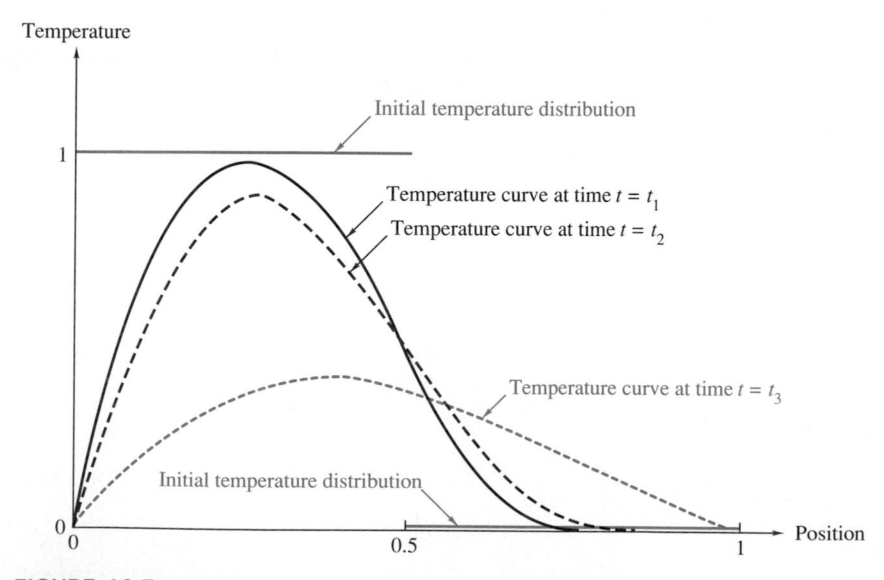

FIGURE 10.7

Solution curves for the one-dimensional heat equation shown in a two-dimensional graph.

Here T' and X'' are ordinary derivatives with respect to the appropriate independent variable. Then substitution into heat Eq. (5) yields,

$$X''T = \frac{1}{k} XT'$$

or,

$$\underbrace{\frac{1}{X} X''}_{\substack{\text{function}\\ \text{of } x\\ \text{alone}}} = \underbrace{\frac{1}{kT} T'}_{\substack{\text{function}\\ \text{of } t\\ \text{alone}}} \tag{9}$$

Note that in Eq. (9) the variables have been separated in the sense that the lefthand side is a function of x only and the righthand side is a function of t only. Now x and t are both independent variables. Equation (9) must hold for every x and every t. In particular, if we vary x while holding t fixed, the righthand side of Eq. (9) remains constant; if we vary t while holding x constant, the lefthand side of Eq. (9) remains constant. That is, the only way a function of the independent variable x alone can equal a function of the independent variable t alone is when both functions equal the same constant. Therefore Eq. (9) becomes the two *ordinary* differential equations,

$$\frac{X''}{X} = \text{constant}, \tag{10}$$

$$\frac{T'}{kT} = \text{same constant}. \tag{11}$$

There are three cases to consider for the constant in Eqs. (10) and (11): whether it is positive, zero, or negative. We investigate each case in turn.

Case 1: Positive Constant $= \lambda^2$

Here Eq. (10) becomes

$$X'' - \lambda^2 X = 0. \tag{12}$$

From our study of second-order ordinary differential equations, we know the general solution to Eq. (12) is

$$X = c_1 e^{\lambda x} + c_2 e^{-\lambda x}. \tag{13}$$

We use boundary conditions (6) to evaluate the constants c_1 and c_2:

$$u(0, t) = 0 \quad \text{translates to} \quad X(0)T(t) = 0,$$
$$u(L, t) = 0 \quad \text{translates to} \quad X(L)T(t) = 0.$$

Since $T(t) \not\equiv 0$ in order to obtain nontrivial solutions for $u = XT$, these last two conditions imply that $X(0) = 0$ and $X(L) = 0$. Substitution of these conditions into Eq. (13) gives us

$$\begin{aligned} 0 &= c_1 + c_2, \\ 0 &= c_1 e^{\lambda L} + c_2 e^{-\lambda L}. \end{aligned} \tag{14}$$

The first equation implies that $c_2 = -c_1$. If $c_1 \neq 0$, then the second equation implies that $e^{\lambda L} = e^{-\lambda L}$, which is impossible since $\lambda \neq 0$. However, $c_1 = c_2 = 0$ implies that $u(x, t)$ is the trivial solution. We conclude there are no nontrivial solutions in this case.

Case 2: Zero Constant

Now Eq. (10) becomes

$$X'' = 0,$$

so

$$X = c_3 x + c_4.$$

Application of boundary conditions (6) to evaluate the constants produces

$$\begin{aligned} 0 &= c_4, \\ 0 &= c_3 L + c_4. \end{aligned}$$

Again, these values yield $u(x, t) \equiv 0$, which is impossible as before. Thus no solutions exist in this case.

Case 3: Negative Constant $= -\lambda^2$

In this case Eq. (10) becomes

$$X'' + \lambda^2 X = 0,$$

yielding the second-order solution

$$X = c_5 \cos \lambda x + c_6 \sin \lambda x.$$

The boundary conditions are used to evaluate the arbitrary constants c_5 and c_6 as follows:

$$\begin{aligned} 0 &= c_5 \cdot 1 + c_6 \cdot 0, \\ 0 &= c_5 \cos \lambda L + c_6 \sin \lambda L. \end{aligned}$$

Thus $c_5 = 0$. For nontrivial solutions $u(x, t)$ we must have $c_6 \neq 0$ and $\sin \lambda L = 0$ implying that λL is an integer multiple of π:

$$\lambda L = n\pi, \qquad n = \pm 1, \pm 2, \pm 3, \ldots .$$

Note that n cannot be zero because $\lambda L \neq 0$ in this case. These values of λ yield an entire family of solutions for the function $X(x)$:

$$X(x) = c_6 \sin \frac{n\pi x}{L}.$$

Since c_6 is an arbitrary constant and $\sin(-n\pi x/L) = -\sin(n\pi x/L)$, we can choose n to range only over the positive integers. Thus

$$\lambda = \frac{n\pi}{L}$$

and

$$X(x) = c_6 \sin \frac{n\pi x}{L}, \qquad n = 1, 2, 3, \ldots \tag{15}$$

yield the X factor of each solution. The values of λ are called **eigenvalues** and the associated functions $\sin(n\pi x/L)$ are called **eigenfunctions** for Eq. (5) subject to boundary conditions (6).

Now we must determine the factor $T(t)$ associated with each function $X(x)$. From Eq. (11),

$$T' + k\lambda^2 T = 0.$$

The general solution to this first-order ordinary differential equation is

$$T(t) = c_7 e^{-k\lambda^2 t} = c_7 e^{-k(n\pi/L)^2 t}. \tag{16}$$

From Eqs. (15) and (16) each integer value of $n = 1, 2, 3, \ldots$ yields a solution of the form XT:

$$u_n(x, t) = B_n e^{-k(n\pi/L)^2 t} \sin \frac{n\pi x}{L}$$

where the constant $B_n = c_6 c_7$ depends on the positive integer n. From the superposition principle, the infinite series

$$u(x, t) = \sum_{n=1}^{\infty} B_n e^{-k(n\pi/L)^2 t} \sin \frac{n\pi x}{L} \tag{17}$$

is a solution of Eq. (5) subject to boundary conditions (6). We want to find the values of the constants B_n in order to determine the particular solution to the heat conduction problem; to do so we must employ initial condition (7). Thus the constants B_n must satisfy the equation

$$u(x, 0) = f(x), \qquad 0 < x < L.$$

That is

$$\sum_{n=1}^{\infty} B_n \sin \frac{n\pi x}{L} = f(x). \tag{18}$$

The initial condition function $f(x)$ can be any reasonable function representing the initial temperature distribution. For instance, any function depicted in Fig. 10.8 would be suitable. We will discuss the properties that must be satisfied by the initial-condition function in Section 10.2. We thus have an interesting mathematical problem: how to choose constants B_n in order to represent a known function $f(x)$ in terms of sines only, as expressed by Eq.

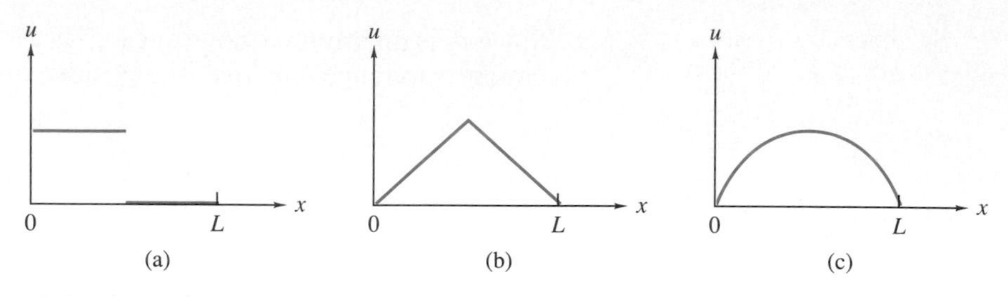

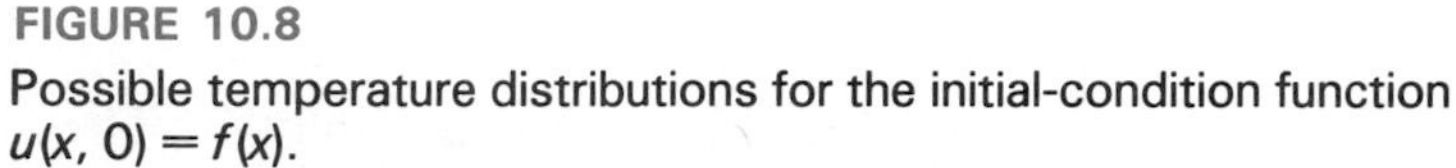
FIGURE 10.8
Possible temperature distributions for the initial-condition function $u(x, 0) = f(x)$.

(18). We will see in Section 10.2 that the constants B_n can be calculated and that the first several terms of the series in Eq. (18) sum to an approximation of $f(x)$ as depicted in Fig. 10.9 for the square wave. Moreover, the approximation can be made as accurate as we choose simply by summing together more terms. After completing our study of this representation problem in Section 10.2, we will return to investigating the solution of the heat conduction model with fixed homogeneous boundary conditions, and other boundary conditions as well.

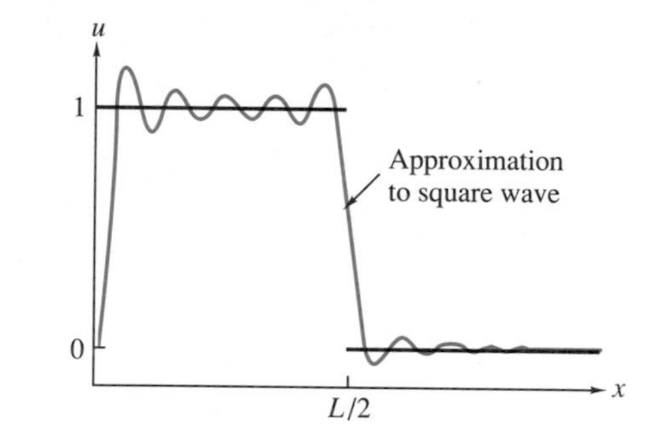

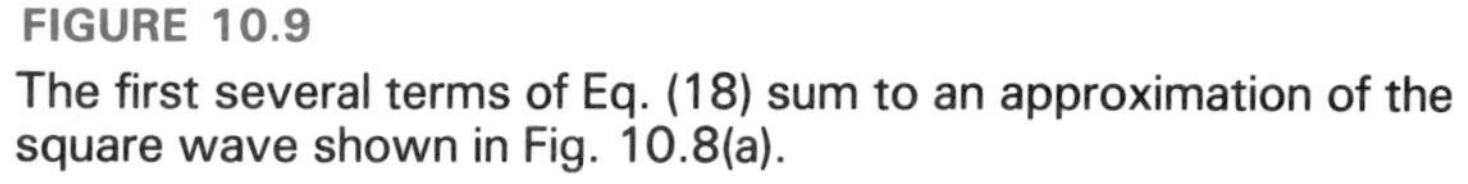
FIGURE 10.9
The first several terms of Eq. (18) sum to an approximation of the square wave shown in Fig. 10.8(a).

EXERCISES 10.1

1. A homogeneous iron rod 50 cm long is immersed in steam until its temperature is 100°C throughout. At time $t = 0$ its lateral surface is insulated and its two ends embedded in ice at 0°C. The thermal diffusivity of iron is 0.15 cm²/sec. Determine a boundary-value problem that models this situation.

2. Consider a one-dimensional rod whose lateral surface is not insulated. Assume that the heat flowing out of the lateral sides per unit length per unit time is proportional to the temperature difference between the rod, $u(x, t)$, and the constant outside temperature T. Assume that the remaining assumptions for our heat conduction model hold. Construct a partial differential equation model of this heat flow.

In problems 3–10 determine whether the method of separation of variables can be used to replace the given partial differential equation by a pair of ordinary differential equations. If so, find the equations

3. $\dfrac{\partial^2 u}{\partial x^2} = \dfrac{\partial^2 u}{\partial t^2}$

4. $c^2 \dfrac{\partial^2 u}{\partial x^2} + \dfrac{\partial^2 u}{\partial y^2} = 0$

5. $x \dfrac{\partial^2 u}{\partial x^2} + \dfrac{\partial u}{\partial t} = 0$

6. $t \dfrac{\partial^2 u}{\partial x^2} + x \dfrac{\partial u}{\partial t} = 0$

7. $\dfrac{\partial^2 u}{\partial x^2} = \dfrac{\partial u}{\partial x} + \dfrac{\partial u}{\partial t}$

8. $\dfrac{\partial^2 u}{\partial x^2} + \dfrac{\partial^2 u}{\partial t^2} = -\dfrac{\partial^2 u}{\partial x \partial t}$

9. $\dfrac{\partial^2 u}{\partial x^2} + (x+t)\dfrac{\partial^2 u}{\partial t^2} = 0$

10. $\dfrac{\partial^2 u}{\partial x^2} = u + \dfrac{\partial u}{\partial t}$

10.2 FOURIER SERIES

In Section 10.1 we realized the need to express the initial-condition function $u(x, 0)$ over $0 < x < L$ in the form of the infinite series

$$\sum_{n=1}^{\infty} B_n \sin \frac{n\pi x}{L}.$$

That is, knowing the function $u(x, 0) = f(x)$, we seek to determine the coefficients B_n of the series. In addressing this problem, let us consider a more general form of the series.

Coefficients in the Fourier Series Expansion

Suppose that f is a function defined over the *symmetric* interval $-L < x < L$. Assume that f is expressible as the following trigonometric series:

$$f(x) = \frac{a_0}{2} + \sum_{n=1}^{\infty} \left[a_n \cos \frac{n\pi x}{L} + b_n \sin \frac{n\pi x}{L} \right]. \tag{1}$$

Let us determine the coefficients $a_0, a_1, a_2, \ldots, b_1, b_2, \ldots$ in the series. In our derivation we will need to use the following results on integrals for sines and cosines (see Exercises 10.2, problems 33–35):

$$\int_{-L}^{L} \cos \frac{n\pi x}{L} \cos \frac{m\pi x}{L}\, dx = \begin{cases} 0, & m \neq n, \\ L, & m = n, \end{cases} \tag{2}$$

$$\int_{-L}^{L} \sin \frac{n\pi x}{L} \cos \frac{m\pi x}{L}\, dx = 0, \tag{3}$$

$$\int_{-L}^{L} \sin \frac{n\pi x}{L} \sin \frac{m\pi x}{L}\, dx = \begin{cases} 0, & m \neq n, \\ L, & m = n. \end{cases} \tag{4}$$

Calculation of a_0 We integrate both sides of Eq. (1) from $-L$ to L and assume that the operations of integration and summation can be interchanged to obtain

$$\int_{-L}^{L} f(x)\,dx = \frac{a_0}{2}\int_{-L}^{L} dx + \sum_{n=1}^{\infty} a_n \int_{-L}^{L} \cos\frac{n\pi x}{L}\,dx + \sum_{n=1}^{\infty} b_n \int_{-L}^{L} \sin\frac{n\pi x}{L}\,dx. \tag{5}$$

For every positive integer n the last two integrals on the righthand side of Eq. (5) are zero (see problems 31 and 32). Therefore

$$\int_{-L}^{L} f(x)\,dx = \frac{a_0}{2}\int_{-L}^{L} dx = \left.\frac{a_0 x}{2}\right|_{-L}^{L} = La_0.$$

Solving for a_0 yields

$$a_0 = \frac{1}{L}\int_{-L}^{L} f(x)\,dx. \tag{6}$$

Calculation of a_m We multiply both sides of Eq. (1) by $\cos m\pi x/L$, $m > 0$, and integrate the result from $-L$ to L:

$$\int_{-L}^{L} f(x)\cos\frac{m\pi x}{L}\,dx = \frac{a_0}{2}\int_{-L}^{L} \cos\frac{m\pi x}{L}\,dx + \sum_{n=1}^{\infty} a_n \int_{-L}^{L} \cos\frac{n\pi x}{L}\cos\frac{m\pi x}{L}\,dx + \sum_{n=1}^{\infty} b_n \int_{-L}^{L} \sin\frac{n\pi x}{L}\cos\frac{m\pi x}{L}\,dx. \tag{7}$$

The first integral on the righthand side of Eq. (7) is zero (see problem 31). Moreover, from the results (2–4) we know the equation further reduces to

$$\int_{-L}^{L} f(x)\cos\frac{m\pi x}{L}\,dx = a_m \int_{-L}^{L} \cos\frac{m\pi x}{L}\cos\frac{m\pi x}{L}\,dx = La_m.$$

Therefore

$$a_m = \frac{1}{L}\int_{-L}^{L} f(x)\cos\frac{m\pi x}{L}\,dx. \tag{8}$$

Calculation of b_m We multiply both sides of Eq. (1) by $\sin m\pi x/L$, $m > 0$, and integrate the result from $-L$ to L:

$$\int_{-L}^{L} f(x)\sin\frac{m\pi x}{L}\,dx = \frac{a_0}{2}\int_{-L}^{L} \sin\frac{m\pi x}{L}\,dx + \sum_{n=1}^{\infty} a_n \int_{-L}^{L} \cos\frac{n\pi x}{L}\sin\frac{m\pi x}{L}\,dx + \sum_{n=1}^{\infty} b_n \int_{-L}^{L} \sin\frac{n\pi x}{L}\sin\frac{m\pi x}{L}\,dx.$$

Using the results in problem 32, combined with Eqs. (3) and (4), we obtain

$$\int_{-L}^{L} f(x) \sin \frac{m\pi x}{L}\, dx = b_m \int_{-L}^{L} \sin \frac{m\pi x}{L} \sin \frac{m\pi x}{L}\, dx = Lb_m.$$

Therefore

$$b_m = \frac{1}{L} \int_{-L}^{L} f(x) \sin \frac{m\pi x}{L}\, dx. \tag{9}$$

Trigonometric series (1), whose coefficients a_0, a_n, b_n are determined by formulas (6), (8), and (9), respectively, is called the **Fourier series expansion** of the function f over the interval $-L < x < L$. The constants a_0, a_n, and b_n are called the **Fourier coefficients** of the function f. The French mathematician John Baptiste Joseph Fourier (1768–1830) first studied the problem of expressing a function as a sum of sine and cosine functions in his investigation of the heat-conduction problem. Notice that in order to have a Fourier series expansion, the interval over which f is defined *must* be symmetric about $x = 0$. Shortly we will discuss the situation when f is defined over a nonsymmetric interval.

In our formal derivation of the Fourier coefficients we assumed the function f could be represented as Eq. (1). However, at this point we do not know if the Fourier series given by the righthand side of Eq. (1) is even a function at all, let alone whether the sum equals $f(x)$. Below we state (without proof) the condition under which a Fourier series actually represents a function. The condition guarantees that for suitable functions f we can calculate the integrals given by Eqs. (6), (8), and (9) to obtain the Fourier coefficients. Hence we will write

$$f(x) \approx \frac{a_0}{2} + \sum_{n=1}^{\infty} \left(a_n \cos \frac{n\pi x}{L} + b_n \sin \frac{n\pi x}{L} \right)$$

to signify the Fourier expansion of f whenever we can calculate the Fourier coefficients a_0, a_n, and b_n.

Convergence of the Fourier Series

We now state without proof the result concerning the convergence of the Fourier series expansion for a wide class of functions commonly encountered in simplified models of several physical behaviors. (If interested, consult any standard advanced calculus text for a proof.) Recall from Definition 5.3 that a function f is *piecewise continuous* over an interval I if both limits

$$\lim_{x \to c^+} f(x) = f(c^+) \qquad \text{and} \qquad \lim_{x \to c^-} f(x) = f(c^-)$$

exist at every point c in I and, moreover, if f has at most finitely many discontinuities in I. If c is an endpoint of I, we require only the existence of the appropriate one-sided limit. Notice that a piecewise continuous function over a closed interval must be bounded (so it cannot tend toward infinity).

THEOREM 10.1

If the functions f and its derivative f' are piecewise continuous over the interval $-L < x < L$, then the function f equals its Fourier series at all points of continuity. At a point c where a jump discontinuity occurs in f, the Fourier series converges to the midpoint of the jump, $[f(c^+) + f(c^-)]/2$. The Fourier series of $f(x)$ over the symmetric interval $-L < x < L$ is given by

$$f(x) \approx \frac{a_0}{2} + \sum_{n=1}^{\infty} \left[a_n \cos \frac{n\pi x}{L} + b_n \sin \frac{n\pi x}{L} \right] \tag{10}$$

where

$$a_0 = \frac{1}{L} \int_{-L}^{L} f(x)\, dx, \tag{11}$$

$$a_n = \frac{1}{L} \int_{-L}^{L} f(x) \cos \frac{n\pi x}{L}\, dx, \tag{12}$$

$$b_n = \frac{1}{L} \int_{-L}^{L} f(x) \sin \frac{n\pi x}{L}\, dx. \tag{13}$$

EXAMPLE 1 Let us find the Fourier series of the function

$$f(x) = \begin{cases} 1, & -\pi < x < 0, \\ x, & 0 < x < \pi, \end{cases} \tag{14}$$

depicted in Fig. 10.10.

Solution. Notice from Fig. 10.10 that $L = \pi$. Thus from Eq. (11) we have

$$\begin{aligned} a_0 &= \frac{1}{\pi} \int_{-\pi}^{\pi} f(x)\, dx \\ &= \frac{1}{\pi} \int_{-\pi}^{0} dx + \frac{1}{\pi} \int_{0}^{\pi} x\, dx \\ &= 1 + \frac{\pi}{2}. \end{aligned}$$

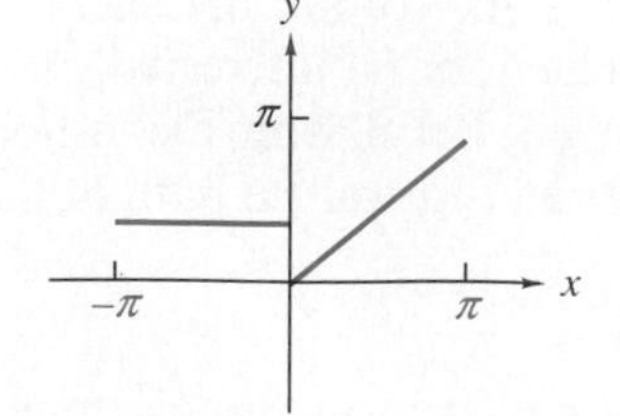

FIGURE 10.10
Piecewise continuous function (14).

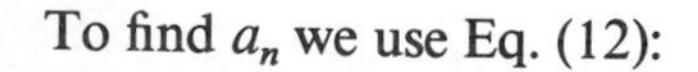

To find a_n we use Eq. (12):

$$\begin{aligned} a_n &= \frac{1}{\pi} \int_{-\pi}^{\pi} f(x) \cos nx\, dx \\ &= \frac{1}{\pi} \int_{-\pi}^{0} \cos nx\, dx + \frac{1}{\pi} \int_{0}^{\pi} x \cos nx\, dx \end{aligned}$$

$$= \frac{1}{n\pi} \sin nx \Big|_{-\pi}^{0} + \frac{1}{\pi}\left[\frac{x}{n}\sin nx\right]_0^{\pi} - \frac{1}{\pi n}\int_0^{\pi} \sin nx\, dx$$

$$= \frac{1}{\pi n^2} \cos nx \Big|_0^{\pi}$$

$$= \frac{1}{\pi n^2}(\cos n\pi - 1)$$

$$= \frac{(-1)^n - 1}{\pi n^2}.$$

In a similar manner, from Eq. (13) we have

$$b_n = \frac{1}{\pi}\int_{-\pi}^{\pi} f(x) \sin nx\, dx$$

$$= \frac{1}{\pi}\int_{-\pi}^{0} \sin nx\, dx + \frac{1}{\pi}\int_0^{\pi} x \sin nx\, dx$$

$$= \frac{(-1)^n(1-\pi) - 1}{n\pi}.$$

Therefore

$$f(x) = \frac{1}{2} + \frac{\pi}{4} + \sum_{n=1}^{\infty} \frac{(-1)^n - 1}{\pi n^2} \cos nx + \sum_{n=1}^{\infty} \frac{(-1)^n(1-\pi) - 1}{\pi n} \sin nx.$$

A graph depicting the Fourier series approximations as n varies up to one, five, and 20 terms is given in Fig. 10.11. Notice how the approximations get

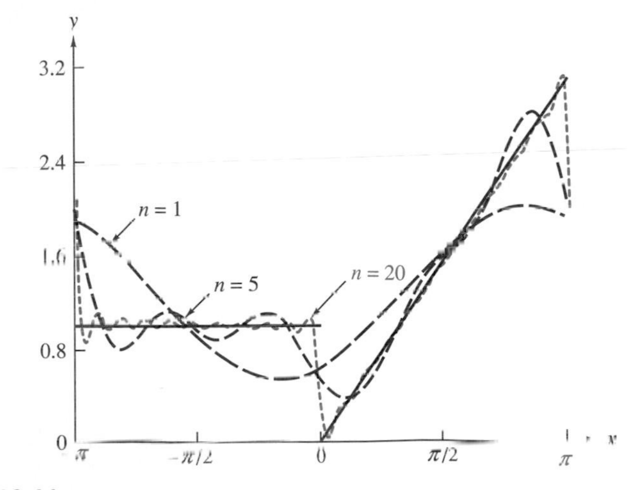

FIGURE 10.11
Fourier series approximations of function (14) as n varies up to one, five, and 20 terms in the infinite series. As n increases, the Fourier approximations approach the actual $f(x)$ values.

closer and closer to the graph of function (14) at all points of continuity as n increases. At the point $x = 0$, where f is discontinuous, the Fourier approximations approach the value 0.5 halfway between the jump. These results are in accord with Theorem 10.1.

Extensions of the Function *f*

Consider again the one-dimensional heat equation discussed in Section 10.1. In that problem the initial condition specifies $u(x, 0) = f(x)$, where $f(x)$ is defined over the *nonsymmetric* interval $0 < x < L$. We want to know how we can calculate a Fourier series expansion for f. To do so we extend the function so that it is defined over the symmetric interval $-L < x < L$. But how do we define the extension of f for $-L < x < 0$? The answer is that we can define the extension to be *any function* over $-L < x < 0$ we choose as long as the extension and its derivative are piecewise continuous (in order to satisfy the hypothesis of Theorem 10.1). No matter what piecewise continuous function we define as the extension over $-L < x < 0$, Fourier series (10) will equal the given function $f(x)$ at all points of continuity over the original domain $0 < x < L$ (in accordance with the conclusion of Theorem 10.1). Of course, the Fourier series also converges to whatever extension function we have chosen for $-L < x < 0$. Nevertheless there are two special extensions that are particularly useful and whose Fourier coefficients are especially easy to calculate; these are the even and odd extensions of f.

Even and Odd Functions

A function g is said to be an **even function** if

$$g(-x) = g(x).$$

On the other hand, if

$$g(-x) = -g(x),$$

then g is said to be an **odd function.**

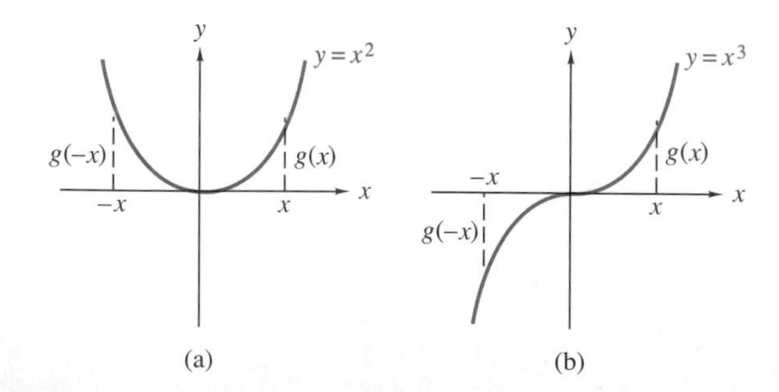

FIGURE 10.12
(a) An even function is symmetric about the y-axis. (b) An odd function is symmetric about the origin.

The graph of an even function is *symmetric about the y-axis*, whereas the graph of an odd function is *symmetric about the origin*. These results are depicted in Fig. 10.12.

Since $\cos(-x) = \cos x$, the cosine is an example of an even function. Likewise, $g(x) = x^2$ and $g(x) = |x|$ are also examples of even functions. On the other hand, $\sin(-x) = -\sin x$, so the sine is an odd function; $g(x) = x^3$ and $g(x) = x$ are also odd functions.

Integrals of even and odd functions over a symmetric interval are convenient to calculate. For instance, if we consider the "appropriately signed" portions of the graphs in Fig. 10.12, we obtain the following:

Odd function:

$$\int_{-L}^{L} g(x)\, dx = 0. \tag{15}$$

Even function:

$$\int_{-L}^{L} g(x)\, dx = 2\int_{0}^{L} g(x)\, dx. \tag{16}$$

It is because of rules (15) and (16) that even and odd extensions of a function are so convenient to use. The following results also hold for even and odd functions.

1. The product of two even functions is even.
2. The product of an even function with an odd function is odd.
3. The product of two odd functions is even.

Even Extension: Fourier Cosine Series

Suppose the function $y = f(x)$ is specified for the interval $0 < x < L$. We define the **even extension of f** by requiring that

$$f(-x) = f(x), \qquad -L < x < L.$$

Graphically we obtain the even extension by reflecting $y = f(x)$ about the y-axis. The even extension of a function is illustrated in Fig. 10.13. Therefore if we use the even extension for a function f, we obtain the Fourier coefficients

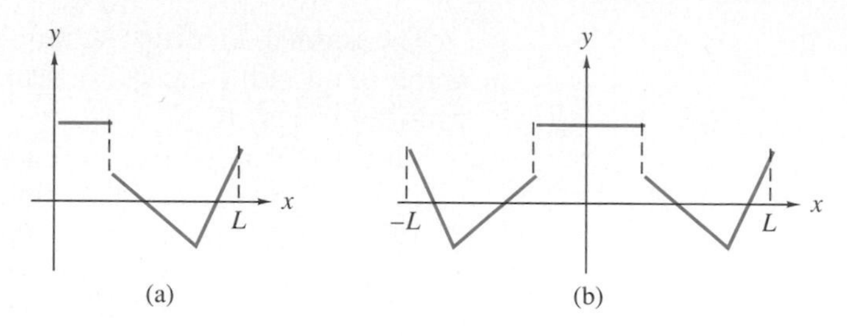

FIGURE 10.13
(a) Original piecewise continuous function f defined over $0 < x < L$. (b) The even extension of f over $-L < x < L$.

$$a_0 = \frac{1}{L}\int_{-L}^{L} f(x)\,dx = \frac{2}{L}\int_{0}^{L} f(x)\,dx,$$

$$a_n = \frac{1}{L}\underbrace{\int_{-L}^{L} f(x)\cos\frac{n\pi x}{L}}_{\text{even}}\,dx = \frac{2}{L}\int_{0}^{L} f(x)\cos\frac{n\pi x}{L}\,dx,$$

$$b_n = \frac{1}{L}\underbrace{\int_{-L}^{L} f(x)\sin\frac{n\pi x}{L}}_{\text{odd}}\,dx = 0.$$

The Fourier series of f is then given by

$$f(x) \approx \frac{a_0}{2} + \sum_{n=1}^{\infty} a_n \cos\frac{n\pi x}{L}.$$

Because the Fourier coefficients b_n are all zero, no sine terms appear in the Fourier series expansion, and the series is called the **Fourier cosine series** of the function f. It converges to the original function f over the interval $0 < x < L$ and to the even extension over the interval $-L < x < 0$ (assuming the piecewise continuity of f and f'). We summarize this result.

FOURIER COSINE SERIES

The Fourier series of an even function on the interval $-L < x < L$ is the **cosine series**

$$f(x) \approx \frac{a_0}{2} + \sum_{n=1}^{\infty} a_n \cos\frac{n\pi x}{L} \tag{17}$$

where

$$a_0 = \frac{2}{L}\int_{0}^{L} f(x)\,dx, \tag{18}$$

$$a_n = \frac{2}{L}\int_{0}^{L} f(x)\cos\frac{n\pi x}{L}\,dx. \tag{19}$$

Odd Extension: Fourier Sine Series

Consider again the function $y = f(x)$ specified for the interval $0 < x < L$. We define the **odd extension of f** by requiring that

$$f(-x) = -f(x), \qquad -L < x < L.$$

Graphically we obtain the odd extension by reflecting $y = f(x)$ about the origin. The odd extension of a function is illustrated in Fig. 10.14. Therefore if we use the odd extension for a function f we obtain the Fourier coefficients

$$a_0 = \frac{1}{L}\int_{-L}^{L} f(x)\,dx = 0,$$

$$a_n = \frac{1}{L}\int_{-L}^{L} \underbrace{f(x)\cos\frac{n\pi x}{L}}_{\text{odd}}\,dx = 0,$$

$$b_n = \frac{1}{L}\int_{-L}^{L} \underbrace{f(x)\sin\frac{n\pi x}{L}}_{\text{even}}\,dx = \frac{2}{L}\int_{0}^{L} f(x)\sin\frac{n\pi x}{L}\,dx.$$

The Fourier series of f is then given by

$$f(x) \approx \sum_{n=1}^{\infty} b_n \sin\frac{n\pi x}{L}.$$

Because the Fourier coefficients a_0 and a_n are all zero, no cosine terms appear in the Fourier series expansion, and the series is called the **Fourier sine series** of the function f. This series also converges to the original function f over the interval $0 < x < L$, but it converges to the *odd* extension over the interval $-L < x < 0$ (assuming the piecewise continuity of f and f'). We summarize this result.

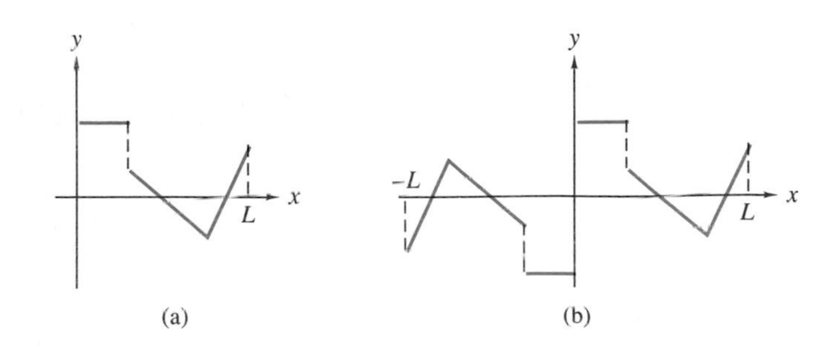

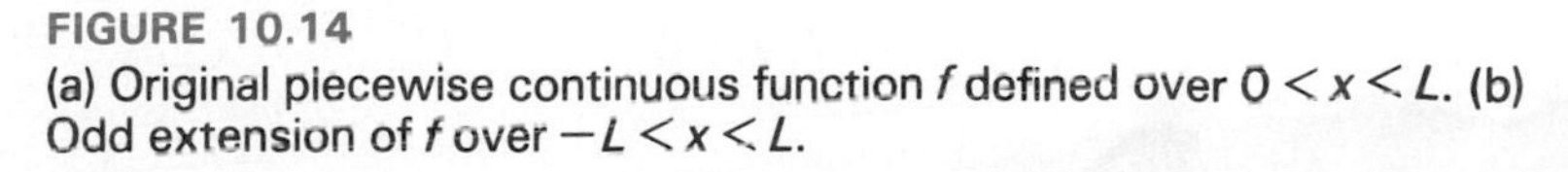

FIGURE 10.14
(a) Original piecewise continuous function f defined over $0 < x < L$. (b) Odd extension of f over $-L < x < L$.

FOURIER SINE SERIES

The Fourier series of an odd function on the interval $-L < x < L$ is the **sine series**

$$f(x) \approx \sum_{n=1}^{\infty} b_n \sin \frac{n\pi x}{L} \tag{20}$$

where

$$b_n = \frac{2}{L} \int_0^L f(x) \sin \frac{n\pi x}{L}\, dx. \tag{21}$$

EXAMPLE 2 Find the Fourier cosine series for the function

$$f(x) = \begin{cases} 1, & 0 < x < \dfrac{\pi}{2}, \\ 0, & \dfrac{\pi}{2} < x < \pi, \end{cases} \tag{22}$$

depicted in Fig. 10.15.

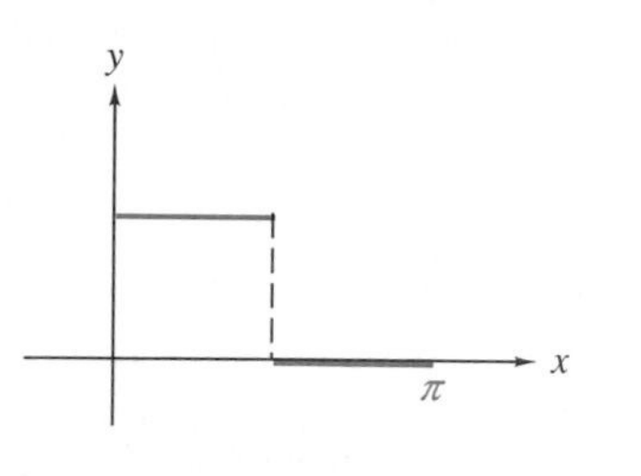

FIGURE 10.15
Function given in Eq. (22).

Solution. For the Fourier cosine series we select the even extension of the function f over $-\pi < x < \pi$ as shown in Fig. 10.16. Calculation of the Fourier coefficients gives us:

$$\begin{aligned} a_0 &= \frac{2}{\pi} \int_0^{\pi} f(x)\, dx \\ &= \frac{2}{\pi} \int_0^{\pi/2} dx \\ &= \frac{2x}{\pi} \bigg|_0^{\pi/2} = 1, \end{aligned}$$

$$\begin{aligned} a_n &= \frac{2}{\pi} \int_0^{\pi} f(x) \cos \frac{n\pi x}{\pi}\, dx \\ &= \frac{2}{\pi} \int_0^{\pi/2} \cos nx\, dx \\ &= \frac{2}{n\pi} \sin \frac{n\pi}{2}. \end{aligned}$$

Therefore

$$f(x) = \frac{1}{2} + \sum_{n=1}^{\infty} \frac{2}{n\pi} \sin \frac{n\pi}{2} \cos nx.$$

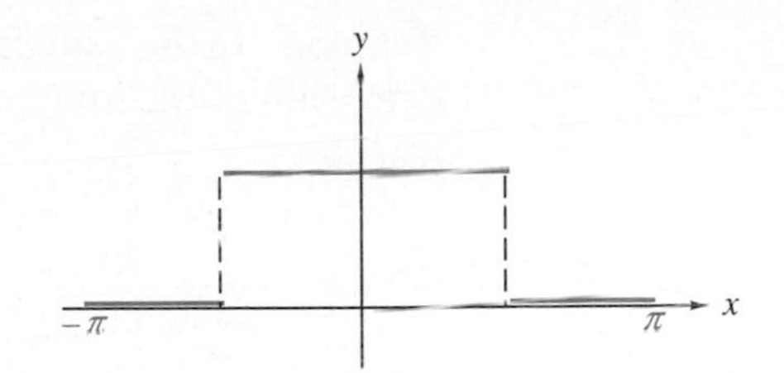

FIGURE 10.16
Even extension of f given by eq. (22).

The Fourier cosine series equals exactly the values of $f(x)$ given in Eq. (22); at the point $x = \pi/2$ the value of the Fourier cosine series is ½. Plots of Fourier cosine approximations for $f(x)$ as n varies up to one, five, and 20 terms is given in Fig. 10.17.

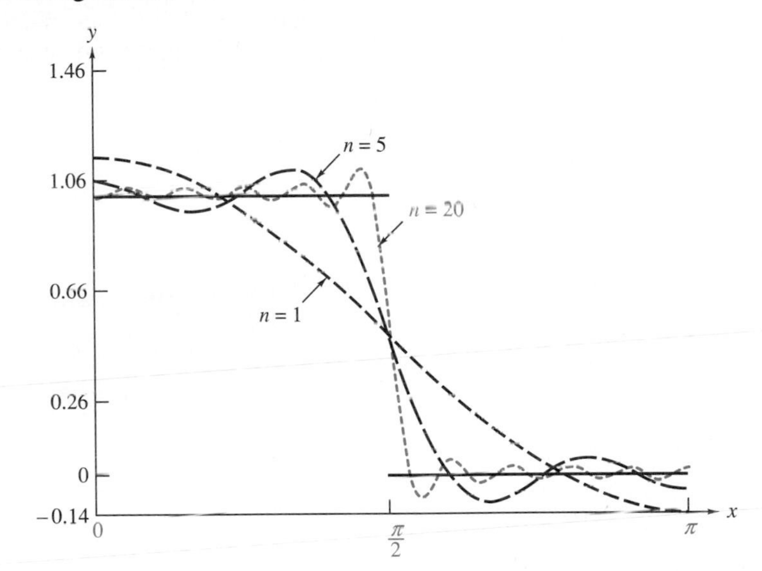

FIGURE 10.17
Fourier cosine series approximations of function (22) as n varies up to one, five, and 20 terms in the infinite series. As n increases, the Fourier cosine approximations approach the actual values of function $f(x)$. Note that each Fourier cosine approximation passes through the value $y = 0.5$, the midvalue of the jump, at the point of discontinuity $x = \pi/2$.

EXAMPLE 3 Represent function (22) of Example 2 in the form

$$f(x) = \sum_{n=1}^{\infty} B_n \sin \frac{n\pi x}{L}$$

where $L = \pi$.

Solution. In this case the Fourier sine series is desired. Thus we select the *odd* extension of the function $f(x)$. Calculation of the Fourier coefficients gives us:

$$\begin{aligned} B_n &= \frac{2}{\pi}\int_0^{\pi} f(x)\sin\frac{n\pi x}{\pi}\,dx \\ &= \frac{2}{\pi}\int_0^{\pi/2} \sin nx\,dx \\ &= -\frac{2}{n\pi}\cos nx\,\Big|_0^{\pi/2} = \frac{2}{n\pi}\left(1-\cos\frac{n\pi}{2}\right). \end{aligned}$$

Therefore

$$f(x) = \sum_{n=1}^{\infty} \frac{2}{n\pi}\left(1-\cos\frac{n\pi}{2}\right)\sin nx.$$

A graph depicting the Fourier sine approximations for $f(x)$ as n varies up to one, five, and 20 terms is shown in Fig. 10.18.

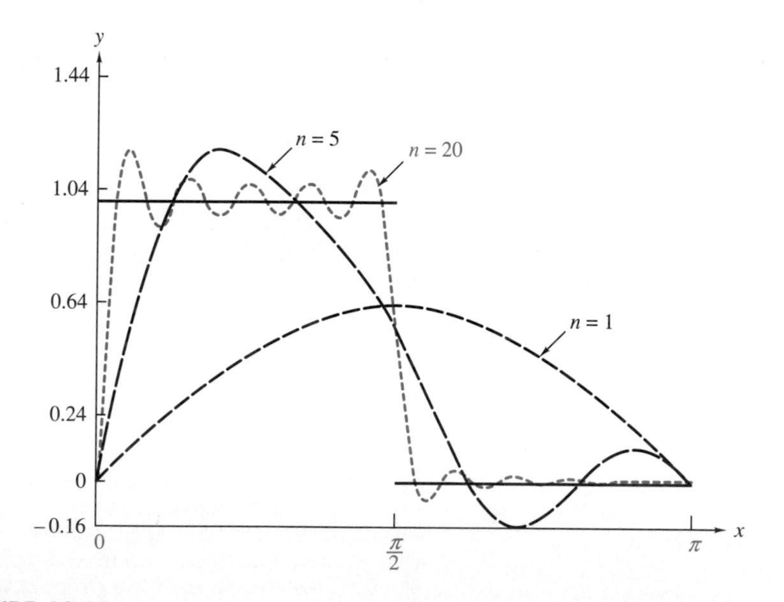

FIGURE 10.18
Fourier sine approximations of the function (22) as n varies up to one, five, and 20 terms in the infinite series. As n increases, the Fourier sine approximations approach the actual function values of $f(x)$. The Fourier sine approximations converge to the midpoint of the jump at the point of discontinuity.

Periodic Extension

Observe that the trigonometric terms $\sin n\pi x/L$ and $\cos n\pi x/L$ in the Fourier series are **periodic** with period $2L$:

$$\sin \frac{n\pi(x+2L)}{L} = \sin \frac{n\pi x}{L} \cos 2n\pi + \cos \frac{n\pi x}{L} \sin 2n\pi$$
$$= \sin \frac{n\pi x}{L},$$

and

$$\cos \frac{n\pi(x+2L)}{L} = \cos \frac{n\pi x}{L} \cos 2n\pi - \sin \frac{n\pi x}{L} \sin 2n\pi$$
$$= \cos \frac{n\pi x}{L}.$$

It follows that the Fourier series is also periodic with period $2L$. Thus the Fourier series not only represents the function f over the interval $-L < x < L$, but it also produces the **periodic extension** of f over the entire real-number line. From Theorem 10.1 we see that the series will converge to the midpoint value $[f(L^-) + f(-L^+)]/2$ at the endpoints of the interval, as well as to this value extended periodically to $\pm 3L, \pm 5L, \pm 7L$, and so forth. The situation is depicted in Fig. 10.19 for the function $f(x) = x$.

Fourier Series and Boundary-Value Problems

Returning now to the heat-conduction problem with fixed boundary conditions studied in Section 10.1, we see that the initial condition $u(x, 0) = f(x)$ over $0 < x < L$ requires that

$$f(x) = \sum_{n=1}^{\infty} B_n \sin \frac{n\pi x}{L}.$$

That is, the constants B_n are precisely the Fourier coefficients of the function f in its Fourier sine series expansion. In Section 10.3 we will complete our solution to the heat equation with fixed boundary conditions and investigate solutions for other boundary conditions as well.

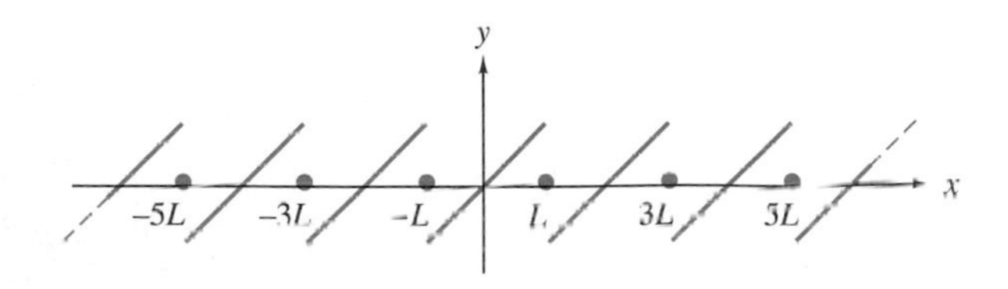

FIGURE 10.19
The Fourier series of $f(x) = x$ converges to f over the interval $-L < x < L$ and to its periodic extension along the real axis, in accordance with Theorem 10.1.

EXERCISES 10.2

In problems 1–14 find the Fourier series expansion for the functions over the specified intervals.

1. $f(x) = 1, \quad -\pi < x < \pi$

2. $f(x) = \begin{cases} -1, & -\pi < x < 0 \\ 1, & 0 < x < \pi \end{cases}$

3. $f(x) = x, \quad -\pi < x < \pi$

4. $f(x) = 1 - x, \quad -\pi < x < \pi$

5. $f(x) = \dfrac{x^2}{4}, \quad -\pi < x < \pi$

6. $f(x) = \begin{cases} 0, & -\pi < x < 0 \\ x^2, & 0 < x < \pi \end{cases}$

7. $f(x) = e^x, \quad -\pi < x < \pi$

8. $f(x) = \begin{cases} 0, & -\pi < x < 0 \\ e^x, & 0 < x < \pi \end{cases}$

9. $f(x) = \begin{cases} 0, & -\pi < x < 0 \\ \cos x, & 0 < x < \pi \end{cases}$

10. $f(x) = \begin{cases} -x, & -2 < x < 0 \\ 2, & 0 < x < 2 \end{cases}$

11. $f(x) = \begin{cases} 0, & -\pi < x < -\dfrac{\pi}{2} \\ 1, & -\dfrac{\pi}{2} < x < \dfrac{\pi}{2} \\ 0, & \dfrac{\pi}{2} < x < \pi \end{cases}$

12. $f(x) = |x|, \quad -1 < x < 1$

13. $f(x) = |2x - 1|, \quad -1 < x < 1$

14. $f(x) = x|x|, \quad -\pi < x < \pi$

In problems 15–22 find the Fourier cosine series expansion for the functions. Draw a graph showing your even extension.

15. $f(x) = x, \quad 0 < x < \pi$

16. $f(x) = \sin x, \quad 0 < x < \pi$

17. $f(x) = e^x, \quad 0 < x < 1$

18. $f(x) = \cos x, \quad 0 < x < \pi$

19. $f(x) = \begin{cases} 1, & 0 < x < 1 \\ -x, & 1 < x < 2 \end{cases}$

20. $f(x) = \begin{cases} -1, & 0 < x < 0.5 \\ 1, & 0.5 < x < 1 \end{cases}$

21. $f(x) = |2x - 1|, \quad 0 < x < 1$

22. $f(x) = |2x - \pi|, \quad 0 < x < \pi$

In problems 23–30 find the Fourier sine series expansion for the functions. Draw a graph showing your odd extension.

23. $f(x) = -x, \quad 0 < x < 1$

24. $f(x) = x^2, \quad 0 < x < \pi$

25. $f(x) = \cos x, \quad 0 < x < \pi$

26. $f(x) = e^x, \quad 0 < x < 1$

27. $f(x) = \sin x, \quad 0 < x < \pi$

28. $f(x) = \begin{cases} x, & 0 < x < 1 \\ 1, & 1 < x < 2 \end{cases}$

29. $f(x) = \begin{cases} 1 - x, & 0 < x < 1 \\ 0, & 1 < x < 2 \end{cases}$

30. $f(x) = |2x - \pi|, \quad 0 < x < \pi$

In problems 31–35 establish the results for the given sine and cosine functions.

31. $\displaystyle\int_{-L}^{L} \cos \frac{m\pi x}{L}\, dx = 0$ for all m.

32. $\displaystyle\int_{-L}^{L} \sin \frac{m\pi x}{L}\, dx = 0$ for all m.

33. $\displaystyle\int_{-L}^{L} \cos \frac{n\pi x}{L} \cos \frac{m\pi x}{L}\, dx = \begin{cases} 0, & m \neq n \\ L, & m = n \end{cases}$

34. $\displaystyle\int_{-L}^{L} \sin \frac{n\pi x}{L} \sin \frac{m\pi x}{L}\, dx = \begin{cases} 0, & m \neq n \\ L, & m = n \end{cases}$

35. $\displaystyle\int_{-L}^{L} \sin \frac{n\pi x}{L} \cos \frac{m\pi x}{L}\, dx = 0$ for all m and n.

36. Use problem 5 to show that

$$1 + \frac{1}{4} + \frac{1}{9} + \frac{1}{16} + \frac{1}{25} + \cdots = \frac{\pi^2}{6}.$$

SOLVING THE HEAT EQUATION

Let us review the solution procedure for the one-dimensional heat-conduction problem presented in Section 10.1. The specific model under investigation is summarized as follows:

ONE-DIMENSIONAL HEAT EQUATION WITH FIXED HOMOGENEOUS BOUNDARY CONDITIONS

PDE:	$u_{xx} = \frac{1}{k} u_t,$	$k > 0, \quad 0 < x < L, \quad t > 0;$	(1)
BC:	$u(0, t) = 0,$ $u(L, t) = 0,$	$t > 0;$	(2)
IC:	$u(x, 0) = f(x),$	$0 < x < L.$	(3)

We now discuss the steps in our solution procedure. Since $f(x) \not\equiv 0$, we are interested only in nontrivial solutions to Eq. (1).

Step 1: Separate the Variables

Substitute $u = XT$ into partial differential Eq. (1), where it is assumed that X is a function only of x, and T a function only of t, and interpret the boundary conditions in terms of this substitution. The substitution separates the variables in Eq. (1) and produces two ordinary differential equations:

$$\frac{X''}{X} = \text{constant}, \tag{4}$$

$$\frac{T'}{kT} = \text{same constant}. \tag{5}$$

Step 2: Solve the Equation $X''/X = $ Constant

You must find *all* possible solutions X to Eq. (4) when the indicated constant is positive, zero, or negative. You will need to interpret the boundary conditions for $u = XT$ and apply them in order to evaluate the two arbitrary constants resulting from your solution to this homogeneous second-order equation. The boundary conditions actually determine which constants are admissible in Eqs. (4) and (5) as well as the corresponding solution functions for X: these are the *eigenvalues* and corresponding *eigenfunctions* for the partial differential equation, respectively.

Step 3: Solve the Equation $T'/kT =$ Same Constant

For each eigenvalue (permissible constant) you found in Step 2, determine the solution to homogeneous first-order Eq. (5). The solution will include the eigenvalue, the variable t, and an arbitrary constant of integration. For each such solution T and its associated eigenfunction X (both coming from the *same* eigenvalue), form the product XT. This procedure yields an entire family of product solutions, with one solution for each eigenvalue.

Step 4: Apply the Superposition Principle

Form the (possibly infinite) linear combination of all the solutions in the family from Step 3. The result is a function $u(x, t)$ expressed as a sum with possibly infinitely many arbitrary constants—one constant for each term in the linear combination.

Step 5: Apply the Initial Condition to Determine $u(x, t)$

Interpret the initial condition for the linear combination you determined in Step 4. This interpretation produces a *single* equation that must be satisfied by all the arbitrary constants in the expression for $u(x, t)$. These constants are then determined by a suitable Fourier analysis, that is, by finding the appropriate Fourier series associated with the function $f(x)$ specified on the righthand side of initial condition (3).

Let us consider a specific example.

EXAMPLE 1 Solve the heat equation

$$\text{PDE:} \quad u_{xx} = u_t, \qquad 0 < x < 1, \quad t > 0; \tag{6}$$

$$\text{BC:} \quad u(0, t) = u(1, t) = 0; \tag{7}$$

$$\text{IC:} \quad u(x, 0) = \begin{cases} 1, & 0 < x < 0.5, \\ 0, & 0.5 < x < 1. \end{cases} \tag{8}$$

Solution. We follow the five steps of the solution procedure.

Step 1 Substitution of $u = XT$ into Eq. (6) yields

$$\frac{X''}{X} = \text{constant}, \tag{9}$$

$$\frac{T'}{T} = \text{same constant}. \tag{10}$$

Interpretation of boundary conditions (7) in terms of this substitution produces

$$u(0, t) = X(0)T(t) = 0 \qquad \text{and} \qquad u(1, t) = X(1)T(t) = 0.$$

Thus for nontrivial solutions, we have the relations

$$X(0) = 0 \qquad \text{and} \qquad X(1) = 0. \tag{11}$$

Step 2 There are three cases to consider in solving ordinary differential Eq. (9): the constant is positive, zero, or negative. As we found in Section 10.1, no solutions are produced if the constant is positive or zero. If the constant is negative and designated by $-\lambda^2$, the solution to Eq. (9) is given by

$$X = c_1 \cos \lambda x + c_2 \sin \lambda x. \tag{12}$$

By applying boundary conditions (11) we can evaluate the constant c_1 and determine the eigenvalues:

$$0 = c_1 \qquad \text{and} \qquad 0 = c_2 \sin \lambda.$$

In order for a nontrivial solution to exist for X and hence for $u(x, t)$, we must require that $c_2 \neq 0$ and $\sin \lambda = 0$. The latter equation means that λ must equal an integer multiple of π. However, as we discussed in Section 10.1, it suffices to choose the eigenvalues for only positive integer multiples of π:

$$\lambda = n\pi, \qquad n = 1, 2, 3, \ldots. \tag{13}$$

The eigenfunctions are now found by substituting for c_1 and λ in Eq. (12):

$$X(x) = c_2 \sin n\pi x. \tag{14}$$

Step 3 Now we solve first-order ordinary differential Eq. (10):

$$\frac{T'}{T} = -\lambda^2 = -n^2\pi^2.$$

Integration of this separable differential equation leads to

$$T(t) = c_3\, e^{-n^2\pi^2 t}.$$

Therefore, for each positive integer n we obtain the product solution

$$u_n(x, t) = XT = B_n\, e^{-n^2\pi^2 t} \sin n\pi x.$$

The constants B_n are to be determined.

Step 4 The superposition principle tells us that the infinite series

$$u(x, t) = \sum_{n=1}^{\infty} B_n\, e^{-n^2\pi^2 t} \sin n\pi x \tag{15}$$

solves Eq. (6) subject to boundary conditions (7).

Step 5 Next we determine the constants B_n. Application of initial condition (8) to (15) yields

$$\sum_{n=1}^{\infty} B_n \sin n\pi x = \begin{cases} 1, & 0 < x < 0.5, \\ 0, & 0.5 < x < 1. \end{cases}$$

Therefore the constants B_n are the coefficients of the Fourier sine series for the function $u(x, 0)$ specified in the righthand side of Eq. (8). That is,

$$\begin{aligned} B_n &= \frac{2}{1}\int_0^{0.5} 1 \cdot \sin n\pi x \, dx \\ &= -\frac{2}{n\pi}\cos n\pi x \Big|_0^{0.5} \\ &= \frac{2}{n\pi}\left(1 - \cos\frac{n\pi}{2}\right). \end{aligned}$$

The solution to the one-dimensional heat equation of this example is then given by

$$u(x, t) = \sum_{n=1}^{\infty} \frac{2}{n\pi}\left(1 - \cos\frac{n\pi}{2}\right) e^{-n^2\pi^2 t} \sin n\pi x.$$

The solution surface for $u(x, t)$ was illustrated in Fig. 10.4. We repeat the graph of the solution curves in Fig. 10.20 using the Fourier representations

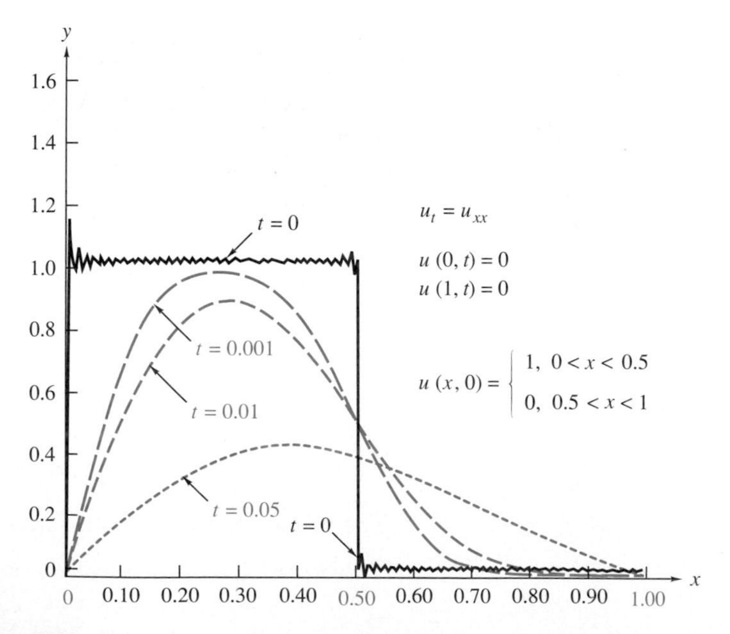

FIGURE 10.20
Solution curves to the heat equation with fixed homogeneous boundary conditions for $t = 0.0$, $t = 0.001$, $t = 0.01$, $t = 0.05$.

for $n = 250$. Note that $u(x, t) \to 0$ as $t \to \infty$, so the rod is cooling down at every location as heat escapes across the two noninsulated ends. This process is reasonable from our assumptions.

Insulated Boundary Conditions

Next let us investigate the heat equation when the two ends of the rod are insulated so that no heat flows out of or into either end.

ONE-DIMENSIONAL HEAT EQUATION WITH INSULATED BOUNDARY CONDITIONS

PDE:	$u_{xx} = \frac{1}{k} u_t,$	$k > 0, \quad 0 < x < L, \quad t > 0;$	(16)
BC:	$u_x(0, t) = 0,$ $u_x(L, t) = 0,$	$t > 0;$	(17)
IC:	$u(x, 0) = f(x),$	$0 < x < L.$	(18)

Solution to the Heat Equation with Insulated Boundary Conditions

The solution procedure here is the same as that followed for solving the heat equation with fixed boundary conditions. As was the case for fixed homogeneous boundary conditions, we are only interested in nontrivial solutions to the partial differential equation.

Step 1 We substitute $u = XT$ into partial differential Eq. (16) as before and obtain the ordinary differential equations

$$\frac{X''}{X} = \text{constant}, \tag{19}$$

$$\frac{T'}{kT} = \text{same constant}. \tag{20}$$

Now $u_x = \frac{\partial(XT)}{\partial x} = X'T$. Thus, interpretation of boundary conditions (17) in terms of the substitution yields

$$u_x(0, t) = X'(0)T(t) = 0 \quad \text{and} \quad u_x(L, t) = X'(L)T(t) = 0.$$

It follows that

$$X'(0) = 0 \quad \text{and} \quad X'(L) = 0. \tag{21}$$

Step 2 Again there are three cases to consider in solving Eq. (19): the constant is positive, zero, or negative.

Case 1: Positive Constant $= \lambda^2$. In this case Eq. (19) becomes

$$X'' - \lambda^2 X = 0,$$

giving rise to the solution

$$X = c_1 e^{\lambda x} + c_2 e^{-\lambda x}.$$

We apply interpreted boundary conditions (21) to evaluate the constants c_1 and c_2:

$$X' = c_1 \lambda e^{\lambda x} - c_2 \lambda e^{-\lambda x},$$

so

$$c_1\lambda - c_2\lambda = 0, \tag{23}$$

$$c_1\lambda e^{\lambda L} - c_2\lambda e^{-\lambda L} = 0. \tag{24}$$

Since $\lambda^2 \neq 0$, Eq. (23) implies that $c_1 = c_2$. In order to obtain a nontrivial solution $u(x, t)$, c_1 must not equal zero. Thus from Eq. (24), $e^{\lambda L} = e^{-\lambda L}$. But this is impossible since $\lambda \neq 0$ in this case. Therefore no nontrivial solutions exist in this case.

Case 2: Zero constant. Now Eq. (19) becomes

$$X'' = 0$$

resulting in the solution

$$X = c_3 x + c_4. \tag{25}$$

Since $X' = c_3$, application of Eq. (21) to evaluate c_3 and c_4 gives us $c_3 = 0$. Therefore

$$X = c_4 \tag{26}$$

yielding a constant solution to the partial differential equation subject to the insulated boundary conditions.

Case 3: Negative constant $= -\lambda^2$. In this case Eq. (19) becomes

$$X'' + \lambda^2 X = 0$$

yielding the solution

$$X = c_5 \cos \lambda x + c_6 \sin \lambda x. \tag{27}$$

Furthermore,

$$X' = -c_5\lambda \sin \lambda x + c_6\lambda \cos \lambda x,$$

so application of Eq. (21) to evaluate c_5 and c_6 gives us

$$c_6\lambda = 0 \tag{28}$$

and

$$-c_5\lambda \sin \lambda L + c_6\lambda \cos \lambda L = 0. \tag{29}$$

Since $\lambda^2 \neq 0$, Eq. (28) demands that $c_6 = 0$. In order to obtain a nontrivial solution $u(x, t)$, c_5 must not equal zero. Thus Eq. (29) implies that $\sin \lambda L = 0$. Again, as in Section 10.1, it suffices to choose the eigenvalues for only positive integer multiples of π:

$$\lambda = \frac{n\pi}{L}, \qquad n = 1, 2, 3, \ldots . \tag{30}$$

The corresponding eigenfunctions are found by substituting for $c_6 = 0$ and λ in Eq. (27):

$$X(x) = c_5 \cos \frac{n\pi x}{L}. \tag{31}$$

Step 3 Now we solve first-order ordinary differential Eq. (20) for each admissible value of the constant in Step 2.

Zero constant, $\lambda = 0$. The ordinary differential equation

$$\frac{T'}{kT} = 0$$

yields the constant solution

$$T = c_7.$$

Combined with result (26) this case produces the constant product

$$XT = \frac{A_0}{2}. \tag{32}$$

Here we name the constant $A_0/2$ to emphasize further on its relationship to a Fourier cosine series.

Negative constant, $-\lambda^2$. From Eq. (20) we obtain

$$T' + k\lambda^2 T = 0,$$

resulting in the solution

$$T = c_8 e^{-k\lambda^2 t}.$$

Substituting Eq. (30) into the above equation then yields

$$T = c_8 e^{-k(n\pi/L)^2 t}. \tag{33}$$

Therefore, after combining this result with Eq. (31), for each positive integer n we obtain the product solution

$$u_n(x, t) = XT = A_n e^{-k(n\pi/L)^2 t} \cos \frac{n\pi x}{L}. \tag{34}$$

The constants A_n are to be determined. (We have renamed the constant expression $c_5 c_8$ as A_n for convenience because it depends on n.)

Step 4 From the superposition principle the infinite series

$$u(x, t) = \frac{A_0}{2} + \sum_{n=1}^{\infty} A_n e^{-k(n\pi/L)^2 t} \cos \frac{n\pi x}{L} \tag{35}$$

solves Eq. (16) subject to insulated boundary conditions (17).

Step 5 Next we determine the constants $A_0/2$ and A_n. Application of initial condition (18) to Eq. (35) yields

$$f(x) = \frac{A_0}{2} + \sum_{n=1}^{\infty} A_n \cos \frac{n\pi x}{L}. \tag{36}$$

It follows that the constants A_0 and A_n are the coefficients of the Fourier cosine series for the initial condition function $f(x) = u(x, 0)$. Let us consider a specific example.

EXAMPLE 2 Find the solution to heat Eq. (16) when $k = L = 1$ and initial condition (18) is specified as the function

$$f(x) = \begin{cases} 1, & 0 < x < 0.5, \\ 0, & 0.5 < x < 1. \end{cases} \tag{37}$$

Solution. Notice that the initial condition is the same as that given for the heat equation with fixed (that is, noninsulated) boundary conditions in Example 1.

To complete solution (35), we apply the initial condition as in Step 5 to find the Fourier cosine series for $f(x)$:

$$A_0 = \frac{2}{1} \int_0^{0.5} dx = 1,$$

$$\begin{aligned} A_n &= \frac{2}{1} \int_0^{0.5} 1 \cdot \cos n\pi x \, dx \\ &= \frac{2}{n\pi} \sin n\pi x \Big|_0^{0.5} \\ &= \frac{2}{n\pi} \sin \frac{n\pi}{2}. \end{aligned}$$

The solution to the one-dimensional heat equation of this example is then given by

$$u(x, t) = \frac{1}{2} + \sum_{n=1}^{\infty} \left(\frac{2}{n\pi} \sin \frac{n\pi}{2} \right) e^{-n^2\pi^2 t} \cos n\pi x. \tag{38}$$

The surface represented by this solution is shown in Fig. 10.21, and the associated contours are given in Fig. 10.22. Finally, Fig. 10.23 shows the

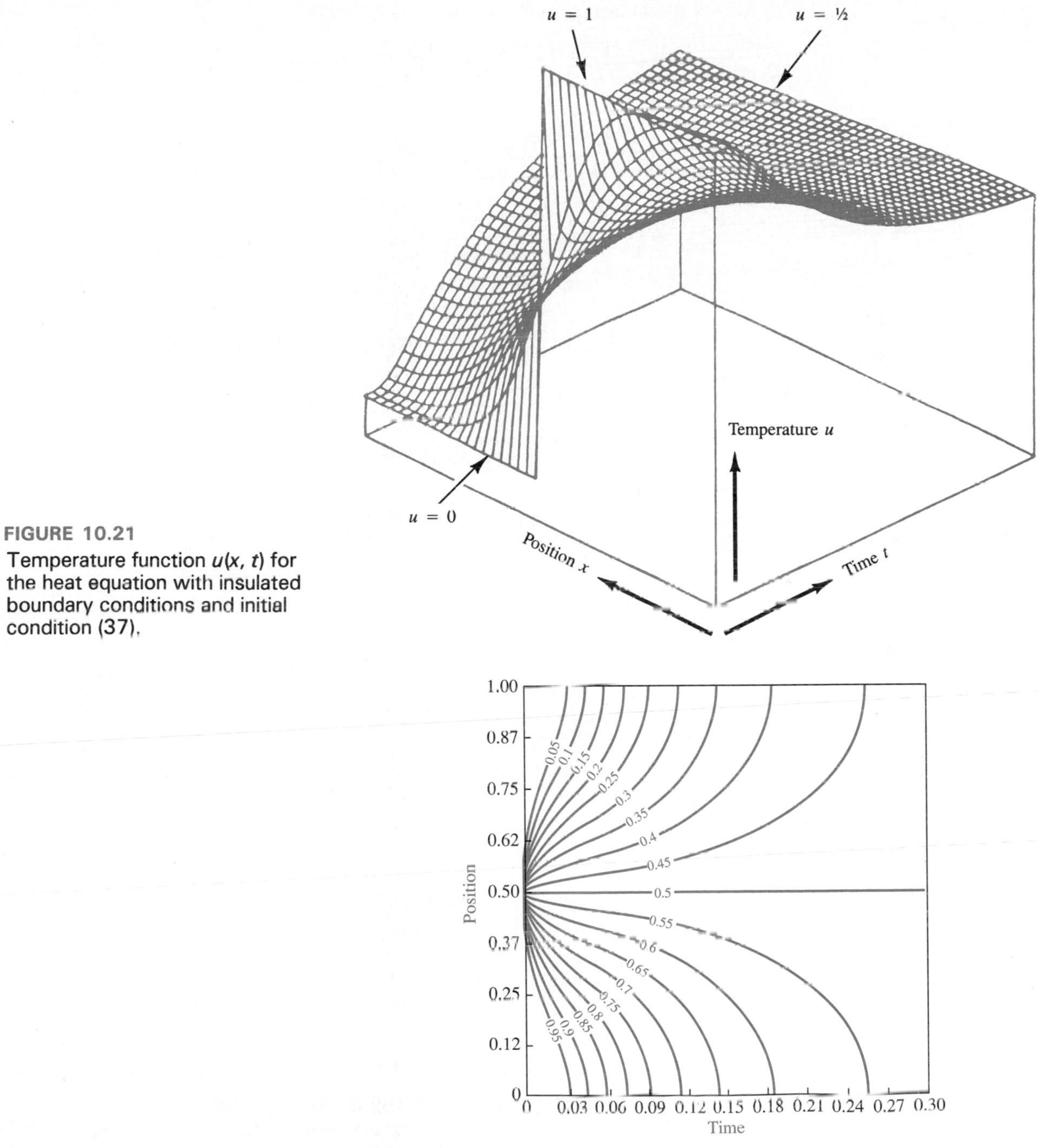

FIGURE 10.21
Temperature function $u(x, t)$ for the heat equation with insulated boundary conditions and initial condition (37).

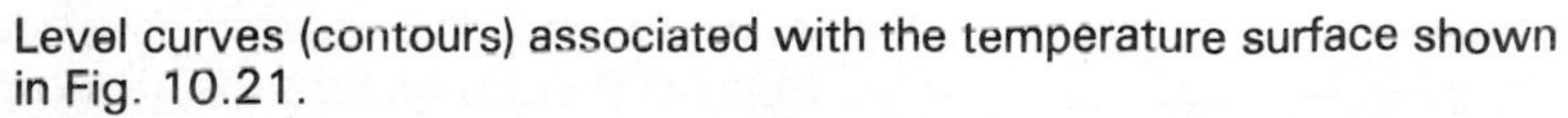

FIGURE 10.22
Level curves (contours) associated with the temperature surface shown in Fig. 10.21.

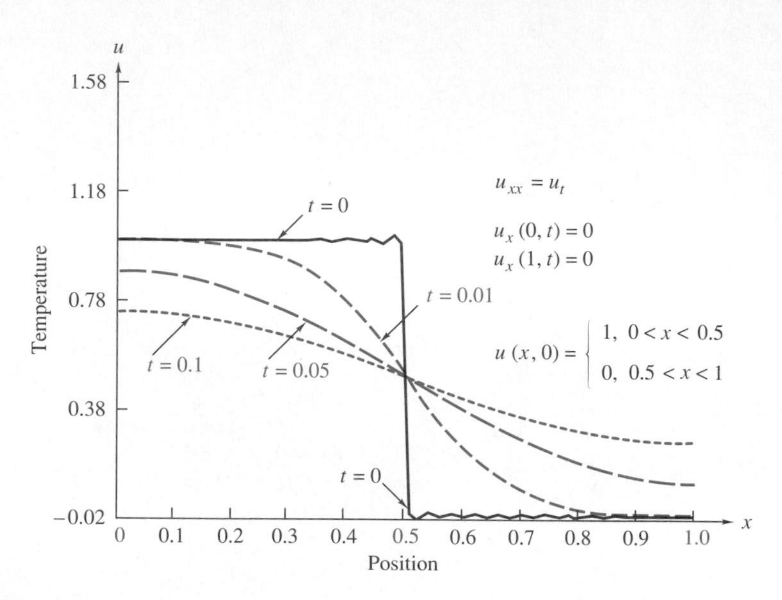

FIGURE 10.23
Curves depicting the solution to the heat equation with insulated boundary conditions for $t = 0.0$, $t = 0.01$, $t = 0.05$, $t = 0.1$.

solution curves in the xu-plane for selected values of t. Let us interpret our analytical solution.

Notice from solution (38) that as $t \to \infty$ the exponential factor $e^{-n^2\pi^2 t} \to 0$ and $u(x, t) \to 1/2$. The temperature $u = 1/2$ is called the **steady-state solution** to the heat-conduction problem with insulated boundary conditions. It is not too surprising: one would expect a rod of temperature $u = 1$ on the left half and $u = 0$ on the right half, for which no heat is escaping, to diffuse its heat uniformly throughout itself to the average temperature $u = 1/2$. The part of solution (38) containing the factor $e^{-n^2\pi^2 t}$ and expressed by the infinite series is called the **transient solution** of the heat equation. The transient solution is the part that "fades out" as t becomes arbitrarily large and the steady-state solution becomes dominant. The steady-state solution is independent of time. You can see in the graph in Fig. 10.23 how the solution curves are converging to this steady-state temperature.

Mixed Boundary Conditions

So far we have investigated solutions to the heat-conduction problem with homogeneous fixed boundary conditions and with insulated boundary conditions. What about the situation when the boundary conditions are mixed? One of the simplest formulations of this situation from a physical point of view is to consider heat conduction in a thin circular ring. Here we think of the ring as a thin rod, laterally insulated, satisfying the assumptions discussed

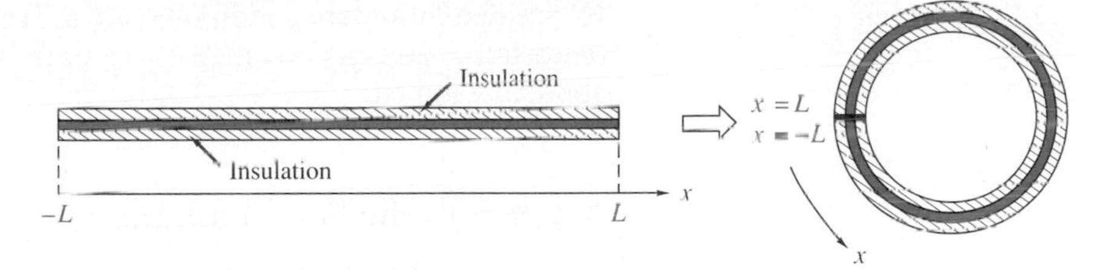

FIGURE 10.24
A one-dimensional rod of length $2L$ and constant cross-sectional area, laterally insulated and bent to form a thin circular ring.

in Section 10.1, and bent into the shape of a circular ring by tightly joining the two ends. We choose the length of the rod or circumference of the circular ring to be $2L$, labeled from $-L$ to $+L$. This convention simplifies the form of the Fourier series solution so that it appears similar to our previous solutions. The ring is depicted in Fig. 10.24.

Since we assume the same properties for the ring as for the rod, the partial differential equation modeling the heat flow continues to be reasonable:

$$u_{xx} = \frac{1}{k} u_t, \qquad k > 0, \quad -L < x < L, \quad t > 0.$$

However, the boundary conditions now change. Let us assume that the ends of the rod are perfectly joined together in forming the ring so that the temperature is continuous there:

$$u(-L, t) = u(L, t).$$

Moreover, if we assume the heat flow is perfect across the joined ends, then the derivative of the temperature function with respect to length is also continuous:

$$u_x(-L, t) = u_x(L, t).$$

Finally, as was the case for the straight rod, there must also be a condition specifying the initial temperature as a function of position around the ring:

$$u(x, 0) = f(x), \qquad -L < x < L.$$

We summarize the heat-flow model for a circular ring.

ONE-DIMENSIONAL HEAT EQUATION FOR A CIRCULAR RING

PDE: $u_{xx} = \frac{1}{k} u_t, \qquad k > 0, \quad -L < x < L, \quad t > 0;$ (39)

BC: $u(-L, t) = u(L, t),$
$u_x(-L, t) = u_x(L, t), \qquad t > 0;$ (40)

IC: $u(x, 0) = f(x), \qquad -L < x < L.$ (41)

Notice that boundary conditions (40) are of the *mixed type* involving both temperature and its derivative at the location where the two ends of the rod are tightly joined.

Solution to the Heat Equation for a Ring

Again the solution procedure is the same as that for solving the heat equation for a one-dimensional rod, and we are interested only in nontrivial solutions to the partial differential equation.

Step 1 Substitute $u = XT$ into Eq. (39) and obtain the following ordinary differential equations:

$$\frac{X''}{X} = \text{constant}, \tag{42}$$

$$\frac{T'}{kT} = \text{same constant}. \tag{43}$$

Interpretation of Eq. (40) in terms of the substitution yields

$$X(-L) = X(L), \tag{44}$$

$$X'(-L) = X'(L). \tag{45}$$

Step 2 There are three cases to consider in solving Eq. (42): whether the constant is positive, zero, or negative.

Case 1: Positive constant $= \lambda^2$. In this case Eq. (42) becomes

$$X'' - \lambda^2 X = 0,$$

giving rise to the solution

$$X = c_1 e^{\lambda x} + c_2 e^{-\lambda x}.$$

To apply the boundary conditions to evaluate c_1 and c_2 we also need the derivative

$$X' = c_1 \lambda e^{\lambda x} - c_2 \lambda e^{-\lambda x}.$$

Then boundary conditions (44) and (45) yield the equations

$$c_1 e^{-\lambda L} + c_2 e^{\lambda L} = c_1 e^{\lambda L} + c_2 e^{-\lambda L} \tag{46}$$

and

$$c_1 \lambda e^{-\lambda L} - c_2 \lambda e^{\lambda L} = c_1 \lambda e^{\lambda L} - c_2 \lambda e^{-\lambda L}. \tag{47}$$

Multiplying Eq. (46) by λ and adding the result to Eq. (47) gives us

$$2c_1 \lambda e^{-\lambda L} = 2c_1 \lambda e^{\lambda L}. \tag{48}$$

Since $\lambda^2 \neq 0$, Eq. (48) implies that $c_1 = 0$. Then from Eq. (46) we have that $c_2 e^{\lambda L} = c_2 e^{-\lambda L}$, which implies that $c_2 = 0$. This situation leads only to the trivial function $u(x, t) \equiv 0$, which fails to satisfy initial condition (41). Therefore no nontrivial solutions exist in this case.

Case 2: Zero constant. Equation (42) now becomes

$$X'' = 0,$$

yielding the solution

$$X = c_3 x + c_4. \tag{49}$$

Since $X' = c_3$, application of boundary conditions (44) and (45) produces

$$c_3(-L) + c_4 = c_3 L + c_4$$

and

$$c_3 = -c_3.$$

Therefore $c_3 = 0$, and we obtain the constant solution

$$X = c_4. \tag{50}$$

Case 3: Negative constant $= -\lambda^2$. Here Eq. (42) becomes

$$X'' + \lambda^2 X = 0,$$

giving us once again the solution

$$X = c_5 \cos \lambda x + c_6 \sin \lambda x. \tag{51}$$

Furthermore,

$$X' = -c_5 \lambda \sin \lambda x + c_6 \lambda \cos \lambda x,$$

so application of boundary conditions (44) and (45) produces

$$c_5 \cos(-\lambda L) + c_6 \sin(-\lambda L) = c_5 \cos \lambda L + c_6 \sin \lambda L$$

and

$$-c_5 \lambda \sin(-\lambda L) + c_6 \lambda \cos(-\lambda L) = -c_5 \lambda \sin \lambda L + c_6 \lambda \cos \lambda L.$$

Due to the evenness of the cosine and the oddness of the sine, these last two equations reduce to

$$-c_6 \sin \lambda L = c_6 \sin \lambda L \tag{52}$$

and

$$c_5 \lambda \sin \lambda L = -c_5 \lambda \sin \lambda L. \tag{53}$$

If $\sin \lambda L \neq 0$, then Eqs. (52) and (53) imply $c_5 = c_6 = 0$ leading only to the (unacceptable) trivial solution $u(x, t) \equiv 0$. Therefore we must have $\sin \lambda L = 0$, leading to the eigenvalues

$$\lambda = \frac{n\pi}{L}, \qquad n = 1, 2, 3, \ldots. \tag{54}$$

Since the constants c_5 and c_6 need not satisfy any constraints other than Eqs. (52) and (53), they are arbitrary. Therefore the eigenfunctions are found by substituting Eq. (54) into (51):

$$X(x) = c_5 \cos \frac{n\pi x}{L} + c_6 \sin \frac{n\pi x}{L}. \tag{55}$$

Step 3 We now solve ordinary differential equation (43) for each admissible value of the constant found in Step 2.

Zero constant, $\lambda = 0$: From Eq. (43) we obtain

$$\frac{T'}{kT} = 0,$$

which leads to

$$T = c_7.$$

After combining this relation with result (50) we obtain the constant solution

$$u = XT = \frac{A_0}{2}. \tag{56}$$

Negative constant, $-\lambda^2$: Eq. (43) now becomes

$$T' + k\lambda^2 T = 0,$$

resulting in

$$T = c_8 e^{-k(n\pi/L)^2 t}.$$

Combining this last result with Eq. (55), for each positive integer n we obtain the product solution XT given by

$$u_n(x, t) = \left(A_n \cos \frac{n\pi x}{L} + B_n \sin \frac{n\pi x}{L}\right) e^{-k(n\pi/L)^2 t} \tag{57}$$

We have renamed the constants $c_5 c_8$ and $c_6 c_8$ as A_n and B_n, respectively, because they depend on n (which determines the eigenvalue λ).

Step 4 From the superposition principle, the infinite series

$$u(x, t) = \frac{A_0}{2} + \sum_{n=1}^{\infty} \left[A_n \cos \frac{n\pi x}{L} + B_n \sin \frac{n\pi x}{L}\right] e^{-k(n\pi/L)^2 t} \tag{58}$$

solves partial differential equation (39) subject to boundary conditions (40). Next we determine the constants A_0, A_n, and B_n.

Step 5 Application of initial condition (41) to Eq. (58) yields

$$f(x) = \frac{A_0}{2} + \sum_{n=1}^{\infty} \left[A_n \cos \frac{n\pi x}{L} + B_n \sin \frac{n\pi x}{L}\right].$$

Therefore we recognize that the constants A_0, A_n, and B_n are precisely the *Fourier series coefficients* for the initial condition function $f(x) = u(x, 0)$.

In Exercises 10.3 you are asked to solve the heat-conduction problem for a circular ring when a particular initial-condition function $u(x, 0) = f(x)$ is specified.

There are mixed boundary conditions other than those specified by Eq. (40). For instance, we might specify the homogeneous conditions

$$u(0, t) = 0 \qquad \text{and} \qquad u_x(L, t) = 0.$$

In this situation the left end of the rod is kept at the constant temperature zero and the right end is insulated to prevent any heat flow. The solution of this heat problem involves eigenvalues that are only *odd* multiples of $\pi/2L$, with corresponding eigenfunctions. Thus it is a bit more complicated to determine the coefficients in the trigonometric series expansion for the initial-condition function in Step 5. A problem of this kind is presented in the exercises where we lead you through the solution procedure.

Fixed Nonhomogeneous Boundary Conditions

As a last example let us consider the situation of solving the heat equation with fixed but *nonhomogeneous* boundary conditions, that is, when the temperatures specified at the two ends of the rod are constant but nonzero. The solution method for this problem is slightly more complicated than previous methods. First let us summarize the model.

ONE-DIMENSIONAL HEAT EQUATION WITH FIXED NONHOMOGENEOUS BOUNDARY CONDITIONS

PDE: $u_{xx} = \frac{1}{k} u_t, \qquad k > 0, \quad 0 < x < L, \quad t > 0;$ (59)

BC: $u(0, t) = T_1,$
$u(L, t) = T_2, \qquad t > 0;$ (60)

IC: $u(x, 0) = f(x), \qquad 0 < x < L.$ (61)

Solution to the Heat Equation with Fixed Nonhomogeneous Boundary Conditions

The idea behind the solution procedure is to reduce the nonhomogeneous problem to solving a homogeneous one. Whenever we solve the heat equation we can ask, What is the *steady-state solution* as t approaches infinity? For instance, in the heat equation with insulated boundary conditions we observed that the solution $u(x, t)$ given by Eq. (35) approaches the steady-state solution $A_0/2$ as t tends toward infinity. The *transient solution*, given by that part of Eq. (38) expressed as an infinite series and containing the factor $e^{-n^2\pi^2 t}$, fades out to zero as t tends toward infinity; it is the part of the solution that changes over time but becomes increasingly negligible as time advances. Equipped with this information, suppose we write the solution to the heat equation as the sum of a steady-state solution and a transient solution:

$$u(x, t) = \underbrace{w(x, t)}_{\substack{\text{transient} \\ \text{solution}}} + \underbrace{v(x)}_{\substack{\text{steady-state} \\ \text{solution}}}. \tag{62}$$

Since $v(x)$ is the steady-state solution, $\partial v/\partial t = 0$. Therefore, since $v_{xx} = v_t/k$, $v(x)$ must satisfy the second-order boundary value problem

$$v'' = 0, \tag{63}$$

$$v(0) = T_1 \quad \text{and} \quad v(L) = T_2 \tag{64}$$

because these conditions hold for all time $t > 0$. But we also know from boundary conditions (60) that $u(0, t) = T_1$ and $u(L, t) = T_2$. It then follows from Eq. (62) that the transient solution $w(x, t)$ satisfies

$$w(0, t) = w(L, t) = 0. \tag{65}$$

Conditions (65) are *homogeneous* boundary conditions for $w(x, t)$, and we already know how to solve a partial differential equation with such boundary conditions. What partial differential equation does $w(x, t)$ satisfy? From Eqs. (62) and (63),

$$u_{xx} = w_{xx} + v'' = w_{xx}$$

and

$$u_t = w_t + 0.$$

Therefore, substitution into Eq. (59) gives us

$$w_{xx} = \frac{1}{k} w_t. \tag{66}$$

In other words, the transient solution satisfies exactly the same partial differential equation as does the original function $u(x, t)$ except that $w(x, t)$ satisfies *homogeneous* boundary conditions (65). All that remains to be done is to translate initial condition (61) for Eq. (62):

$$f(x) = u(x, 0) = w(x, 0) + v(x),$$

so

$$w(x, 0) = f(x) - v(x). \tag{67}$$

In other words, the *transient solution* $w(x, t)$ satisfies the following problem:

PDE: $w_{xx} = \frac{1}{k} w_t$, $\quad k > 0, \quad 0 < x < L, \quad t > 0;$ (68)

BC: $w(0, t) = 0,$
$w(L, t) = 0, \quad t > 0;$ (69)

IC: $w(x, 0) = f(x) - v(x), \quad 0 < x < L.$ (70)

We can solve this homogeneous boundary-value problem just as before provided we can first determine the *steady-state function* $v(x)$. The latter is an easy task. From Eq. (63) we obtain

$$v(x) = c_1 x + c_2.$$

Application of the conditions $v(0) = T_1$ and $v(L) = T_2$ from Eqs. (64) gives us $c_2 = T_1$ and $c_1 = (T_2 - T_1)/L$. Therefore

$$v(x) = T_1 + \left(\frac{T_2 - T_1}{L}\right) x. \tag{71}$$

In Exercises 10.3 you will be asked to solve the one-dimensional heat equation with fixed nonhomogeneous boundary conditions for particular temperatures T_1 and T_2, and specified initial-condition functions $f(x)$.

Conclusion

There are other combinations of mixed boundary conditions that we have not investigated here. These combinations can lead to solutions that are not

TABLE 10.1

Summary for $u_{xx} = \frac{1}{k} u_t$, $k > 0$, $0 < x < L$, $t > 0$

Boundary Conditions	Eigenvalues	Series Solution
Fixed, homogeneous: $u(0, t) = 0$, $u(L, t) = 0$	$\lambda = \frac{n\pi}{L}$, $n = 1, 2, 3, \ldots$	$u(x, t) = \sum_{n=1}^{\infty} B_n e^{-k\lambda^2 t} \sin \lambda x$ where B_n are the Fourier sine series coefficients of $f(x)$
Insulated: $u_x(0, t) = 0$, $u_x(L, t) = 0$	$\lambda = \frac{n\pi}{L}$, $n = 0, 1, 2, \ldots$	$u(x, t) = \frac{A_0}{2} + \sum_{n=1}^{\infty} A_n e^{-k\lambda^2 t} \cos \lambda x$ where A_0 and A_n are the Fourier cosine series coefficients of $f(x)$
Ring: $u(-L, t) = u(L, t)$, $u_x(-L, t) = u_x(L, t)$	$\lambda = \frac{n\pi}{L}$, $n = 0, 1, 2, \ldots$	$u(x, t) = \frac{A_0}{2} + \sum_{n=1}^{\infty} [A_n \cos \lambda x + B_n \sin \lambda x] e^{-k\lambda^2 t}$ where A_0, A_n, and B_n are the Fourier coefficients of $f(x)$
Fixed, nonhomogeneous: $u(0, t) = T_1$, $u(L, t) = T_2$	$\lambda = \frac{n\pi}{L}$, $n = 1, 2, 3, \ldots$	$u(x, t) = w(x, t) + v(x)$; $v(x) = T_1 + \left(\frac{T_2 - T_1}{L}\right) x$; $w(x, t) = \sum_{n=1}^{\infty} B_n e^{-k\lambda^2 t} \sin \lambda x$ where B_n are the Fourier sine series coefficients of $f(x) - v(x)$

given by an ordinary type of Fourier series. One example of this kind is given in Exercises 10.3, problem 24, where the solution is a trigonometric series consisting only of the terms $\sin(\pi x/2L)$, $\sin(3\pi x/2L)$, $\sin(5\pi x/2L)$, and so forth.

Table 10.1 summarizes the solutions to the heat-conduction model we obtained in this chapter. Do not memorize the chart. It is only intended to give you a picture of the whole sweep of the analytical results we obtained. In Section 10.4 we will consider numerical solutions to this model.

EXERCISES 10.3

In problems 1–22 solve the one-dimensional heat equation for the models. Use the five-step method developed in the text including an analysis for each case of the separation constant. What is the steady-state temperature at the midpoint of the rod or ring in each problem?

1. $u_{xx} = 2u_t,\quad u(0, t) = u(3, t) = 0,\ u(x, 0) = \sin \pi x,$ $0 < x < 3$

2. $u_{xx} = 3u_t,\quad u(0, t) = u(\pi, t) = 0,\ u(x, 0) = x,$ $0 < x < \pi$

3. $u_{xx} = u_t,\quad u(0, t) = u(\pi, t) = 0,\ u(x, 0) = \sin x,$ $0 < x < \pi$

4. $u_{xx} = u_t,\quad u(0, t) = u(\pi, t) = 0,$ $u(x, 0) = -1 + \cos 2x,\quad 0 < x < \pi$

5. $u_{xx} = u_t,\quad u_x(0, t) = u_x(\pi, t) = 0,\ u(x, 0) = \cos x,$ $0 < x < \pi$

6. $u_{xx} = 0.5u_t,\quad u_x(0, t) = u_x(1, t) = 0,\ u(x, 0) = x,$ $0 < x < 1$

7. $u_{xx} = 3u_t,\quad u_x(0, t) = u_x(2, t) = 0,$ $u(x, 0) = 2 + \cos \pi x,\quad 0 < x < 2$

8. $u_{xx} = 0.25u_t,\quad u(0, t) = u(2\pi, t) = 0,$ $u(x, 0) = e^{2x},\quad 0 < x < 2\pi$

9. $u_{xx} = 0.1u_t,\quad u(0, t) = u(100, t) = 0,$

$$u(x, 0) = \begin{cases} 0, & 0 < x < 25 \\ 50, & 25 \leq x \leq 75 \\ 0, & 75 < x < 100 \end{cases}$$

10. $u_{xx} = u_t,\quad u_x(0, t) = u_x(2, t) = 0,$

$$u(x, 0) = \begin{cases} x, & 0 < x < 1 \\ 0, & 1 < x < 2 \end{cases}$$

11. $u_{xx} = u_t,\quad u_x(0, t) = u_x(2, t) = 0,$

$$u(x, 0) = \begin{cases} x, & 0 < x < 1 \\ 2 - x, & 1 < x < 2 \end{cases}$$

12. $u_{xx} = \frac{1}{3} u_t,\quad u_x(0, t) = u_x(\pi, t) = 0,$

$$u(x, 0) = \begin{cases} 0, & 0 < x < \pi/2 \\ 1, & \pi/2 < x < \pi \end{cases}$$

13. $u_{xx} = 2u_t,\quad u(0, t) = u(\pi, t) = 100,$

$$u(x, 0) = \begin{cases} 0, & 0 < x < \pi/2 \\ 1, & \pi/2 < x < \pi \end{cases}$$

14. $u_{xx} = u_t,\quad u(0, t) = u(\pi, t) = 100,$ $u(x, 0) = \sin 2x,\quad 0 < x < \pi$

15. $u_{xx} = 2u_t,\quad u(0, t) = 0,\ u(3, t) = 30,$ $u(x, 0) = 30,\quad 0 < x < 3$

16. $u_{xx} = 0.5u_t,\quad u(0, t) = 100,\ u(1, t) = 50,$ $u(x, 0) = 100,\quad 0 < x < 1$

17. $u_{xx} = 0.5u_t,\quad u(0, t) = 100, u(1, t) = 50, u(x, 0) = 50,$ $0 < x < 1$

18. $u_{xx} = u_t,\quad u(0, t) = 100,\ u(2, t) = 50,$ $u(x, 0) = 100 - 13x,\quad 0 < x < 2$

19. $u_{xx} = 0.5u_t,\quad u(-1, t) = u(1, t),\quad u_x(-1, t) = u_x(1, t),$ $u(x, 0) = |x|,\quad -1 < x < 1$

20. $u_{xx} = u_t,\quad u(-\pi, t) = u(\pi, t),\quad u_x(-\pi, t) = u_x(\pi, t),$

$$u(x, 0) = \begin{cases} 0, & -\pi < x < -\pi/2 \\ 1, & -\pi/2 < x < \pi/2 \\ 0, & \pi/2 < x < \pi \end{cases}$$

21. $u_{xx} = \frac{1}{3} u_t,\quad u(-\pi, t) = u(\pi, t),\quad u_x(-\pi, t) = u_x(\pi, t),$ $u(x, 0) = x + \pi,\quad -\pi < x < \pi$

22. $u_{xx} = u_t$, $u(-2, t) = u(2, t)$, $u_x(-2, t) = u_x(2, t)$,

$$u(x, 0) = \begin{cases} -x, & -2 < x < 0 \\ 2, & 0 < x < 2 \end{cases}$$

23. The voltage $E(x, t)$ in a transmission line, grounded at $x = 0$ and $x = L$, with the constant initial-voltage distribution of 2400 volts, satisfies the diffusion equation

$$E_{xx} = \frac{1}{a} E_t, \quad \text{where } a = 1,$$

the boundary conditions

$$E(0, t) = E(L, t) = 0,$$

and the initial condition

$$E(x, 0) = 2400.$$

Find the solution $E(x, t)$ of the model.

24. a) Find the trigonometric series expansion

$$f(x) \approx \sum_{n=1}^{\infty} B_{2n-1} \sin \frac{(2n-1)\pi x}{L}, \quad 0 < x < L,$$

by completing the following three steps:

Step 1. Extend $f(x)$ to the interval $0 < x < 2L$ by making the extension $F(x)$ symmetric across the line $x = L$, as illustrated in Fig. 10.25. Then take the Fourier sine series of the extension $F(x)$:

$$F(x) \approx \sum_{k=1}^{\infty} B_k \sin \frac{k\pi x}{2L}$$

where

$$B_k = \frac{2}{2L} \int_0^{2L} F(x) \sin \frac{k\pi x}{2L}\, dx.$$

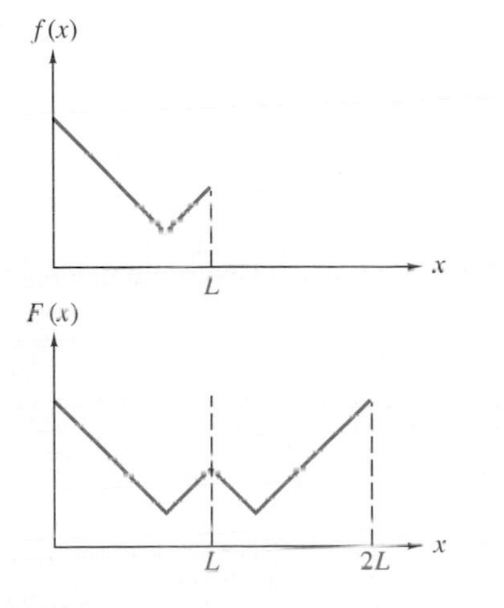

FIGURE 10.25

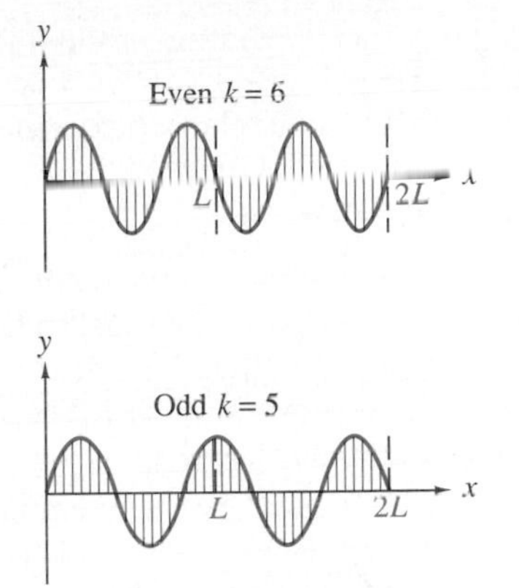

FIGURE 10.26

Step 2. Show that the function $\sin k\pi x/2L$ has period $T = 4L/k$. Graph this function for even and odd values of k and convince yourself that the patterns shown in Fig. 10.26 are obtained. Conclude from your graphs that

$$\int_0^{2L} \sin \frac{k\pi x}{2L}\, dx = 0$$

if $k = 2n$ is even, and that

$$\int_0^{2L} \sin \frac{k\pi x}{2L}\, dx = 2\int_0^{L} \sin \frac{k\pi x}{2L}\, dx$$

if $k = 2n - 1$ is odd.

Step 3. From your graphs in Step 2 and the definition of $F(x)$ in Step 1, convince yourself that

$$B_k = 0$$

if $k = 2n$ is even, and that

$$B_k = \frac{2}{L} \int_0^{L} f(x) \sin \frac{k\pi x}{L}\, dx$$

if $k = 2n - 1$ is odd.

b) Using part (a), solve the heat equation

PDE:	$u_{xx} = u_t$,	$0 < x < \pi$, $t > 0$;
BC:	$u(0, t) = 0$,	
	$u_x(\pi, t) = 0$,	$t > 0$;
IC:	$u(x, 0) = x$,	$0 < x < \pi$.

25. A homogeneous copper rod 100 cm long is heated to a uniform temperature of 100°C throughout. At time $t = 0$, its lateral surface is insulated and its two ends embedded in ice at 0°C. The thermal diffusivity of copper is 1.14 cm²/sec.

a) Determine the temperature at the midpoint of the rod when $t = 0.5$ sec.

b) How much time must elapse before the center of the rod cools to 20°C? Use only the first term in the series expansion for $u(x, t)$.

26. A homogeneous aluminum rod 50 cm long is heated according to the initial temperature distribution $f(x) = x$, $0 \leq x \leq 50$. At time $t = 0$, both its lateral surface and its two ends are insulated so no heat can escape. The thermal diffusivity of aluminum is 0.86 cm²/sec.

a) Determine the steady-state temperature of the rod.

b) How much time elapses before the left end of the rod at $x = 0$ reaches this steady-state temperature? Use only the first term in the series expansion for $u(x, t)$.

27. Suppose a homogeneous silver wire 100 cm long is heated so that its center section from $25 < x < 75$ has a uniform temperature of 100°C while the two quarter ends $0 < x < 25$ and $75 < x < 100$ are maintained at the temperature 0°C throughout. At time $t = 0$ the lateral surface of the wire is insulated and the two ends are perfectly joined to form a ring. The thermal diffusivity of silver is 1.71 cm²/sec.

a) What is the steady-state temperature of the wire?

b) How much time elapses before the point where the two ends are perfectly joined reaches the steady-state temperature? Use only the first term in the series expansion for $u(x, t)$.

28. Consider a homogeneous iron rod 100 cm long that is heated to the initial temperature distribution $f(x) = 100 - x$, $0 \leq x \leq 100$. At time $t = 0$ the lateral surface is insulated, and the left end of the rod is plunged into a reservoir of oil kept at 20°C while the right end remains at the constant temperature of 70°C. The thermal diffusivity of iron is 0.15 cm²/sec.

a) What is the temperature at the midpoint of the rod when $t = 1$ sec?

b) How much time elapses before the point $x = 95$ reaches the steady-state temperature? Use only the first term in the series expansion for $w(x, t)$.

29. Solve the heat equation

PDE: $u_{xx} = u_t + u;$

BC: $u(0, t) = u(1, t) = 0;$

IC: $u(x, 0) = \begin{cases} 1, & 0 < x < 0.5, \\ 0, & 0.5 < x < 1. \end{cases}$

Use the method of separation of variables. Describe the physical situation approximated by the partial differential equation.

30. Diffusion of a Dye

Consider the following laboratory experiment. As suggested in Fig. 10.27, a long thin glass tube is filled with water and sealed at both ends. A thin membrane (the baffle) is placed at the exact center of the tube, dividing the tube into two separate sections. Before the tube is sealed, a blue dye is injected into each section of the tube. Upon sealing the ends, the concentration of the dye is measured. The baffle then dissolves, and the dye is free to diffuse throughout the tube. Let $C(x, t)$ represent the concentration of the dye per unit volume. Let k denote the diffusivity of the dye. Then C obeys the following partial differential equation:

$$C_{xx}(x, t) = kC_t(x, t).$$

Moreover, since the tube is sealed at each end, no dye can pass through the tube walls there. In other words, the tube ends essentially "insulate" the dye. That is, the following boundary conditions are satisfied:

$$C_x(0, t) = C_x(L, t) = 0.$$

Finally, for this particular experiment, measurements indicate that at the instant the baffle dissolves, the dye concentration is zero at each end. A maximum value of 2 is observed at the exact center of the tube. This concentration can be approximated analytically by

$$C(x, 0) = 2 \sin^2 \frac{\pi x}{L}.$$

a) Briefly describe some assumptions that must hold in order for the mathematical model to be valid.

b) Using the method of separation of variables, solve the partial differential equation subject to the boundary conditions and initial conditions.

c) Explain why the Fourier series obtained in part (b) has only a finite number of nonzero terms.

d) Using the result found in part (b), determine where in the tube the dye concentration initially is chang-

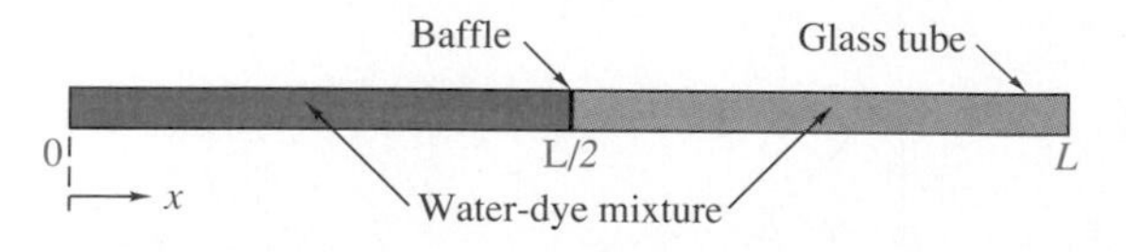

FIGURE 10.27
Water–dye mixture in a glass tube.

ing at its fastest rate. Provide a physical explanation of your answer.

e) What is the limiting value of the dye concentration for very large values of time? Compare this value to the first term of the Fourier series you obtained in part (b). Explain why they both have the same value.

10.4

NUMERICAL SOLUTION TO THE ONE-DIMENSIONAL HEAT EQUATION (Optional)

In this section we introduce you to one numerical method for solving the heat equation with both fixed and insulated boundary conditions. Like Euler's method for first-order ordinary differential equations presented in Chapter 2, the method we discuss here is the simplest although not the most accurate one, and has errors accumulating at each step of the approximation process. However, our goal here is not to develop the most accurate or computationally efficient numerical methods for solving the heat equation; such methods are best investigated in a course in numerical partial differential equations. Rather, we want you to see how to approach numerically the solving of partial differential equations; we also want you to be aware of the reality that, with today's computers, most partial differential equations *are* solved numerically by engineers and scientists.

To begin, recall from Sections 2.5 and 2.6 what we did in numerically solving the first-order initial value problem for ordinary differential equations defined by

$$y' = g(x, y), \qquad y(x_0) = y_0, \quad x_0 \leq x \leq b.$$

First we subdivided the interval $x_0 \leq x \leq b$ into n subintervals each of length $\Delta x = (b - x_0)/n$. Then we used an approximation procedure (the Euler or the Runge–Kutta methods) based on the idea of the derivative as the slope of the line tangent to the solution curve. At each stage we proceeded from one approximate value to the next, moving along a line with some appropriately defined slope. In this way we successively achieved approximate values y_1, $y_2, y_3, \ldots, y_n$ to the solution $y = y(x)$ at each subinterval point $x_0 < x_1 < x_2 < \cdots < x_n = b$. Therefore the numerical solution is a discrete *one-dimensional table of (x, y) values* instead of an analytical formula for $y(x)$:

x	x_0	x_1	x_2	$\ldots$	x_n
y	y_0	y_1	y_2	$\ldots$	y_n

(2)

How might these ideas be extended to numerically solving the heat equation?

Interpreting the Numerical Solution of a Partial Differential Equation

In the case of the heat-conduction problem we have the partial differential equation, the boundary conditions, and the initial condition. The equations in this model are all defined over a suitable strip of the x,t-plane as illustrated again in Fig. 10.28 for fixed homogeneous boundary conditions. What might we expect for a numerical solution to the model? For each selected time value it seems reasonable to expect a discrete approximation to the solution curve at predetermined values of x. This idea is depicted in Fig. 10.29, where we select three time values and graph an approximate solution curve by plotting the temperatures for certain position values x. In this way the numerical solution consists of a *two-dimensional table of values* of u for selected time steps and certain values of position x, as shown below.

Numerical Solution Table for a Partial Differential Equation

	Position					
Time	**0**	x_1	x_2	x_3	. . .	L
0	$u(0, 0)$		. . . initial condition . . .			$u(L, 0)$
t_1	0	•	•	•	• •	0
t_2	0	•	•	$u(x_3, t_2)$	• •	0
t_3	0	•	•	•	• •	0
⋮	⋮	⋮	⋮	⋮	⋮	⋮
	0	•	•	•	• •	0
	↑ BC $u(0, t)$					↑ BC $u(L, t)$

Notice that the first row of the numerical solution table contains the values of the initial-value function $u(x, 0) = f(x)$ for the chosen values of x. The next row contains the discrete approximation to the solution function at the first time step t_1, the following row contains the approximation values to the solution at the next time step t_2, and so on. Moreover, the left column of the table contains the left-end boundary values $u(0, t)$ and the right column the right-end boundary values $u(L, t)$. Therefore, since we know the initial condition and the boundary conditions from the given model, we can always calculate and enter all the values of the first row, and then enter the left-column and right-column values of the table. We still need to know how we can obtain reasonable interior entries in the table to approximate $u(x, t)$. We next describe one procedure to accomplish this task.

Setting Up the Numerical Solution Table

The first step in the numerical solution procedure is to subdivide the position interval $0 < x < L$ into n subdivisions of equal length $\Delta x = L/n$. As with Euler's algorithm, we label the equally spaced subdivision points:

$$x_0 = 0, x_1 = x_0 + \Delta x, x_2 = x_1 + \Delta x, \ldots, x_n = x_{n-1} + \Delta x = L. \quad (3)$$

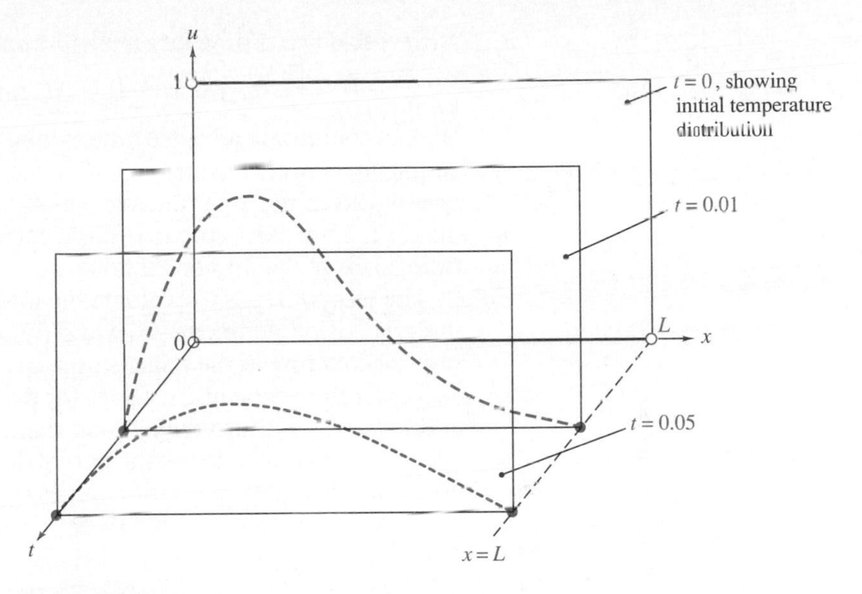

FIGURE 10.28
The inequalities $0 < x < L$ and $t > 0$ define the strip in the xt-plane over which the heat conduction model is defined. The solution $u(x, t)$ is defined over this strip by depicting temperature curves for various time steps.

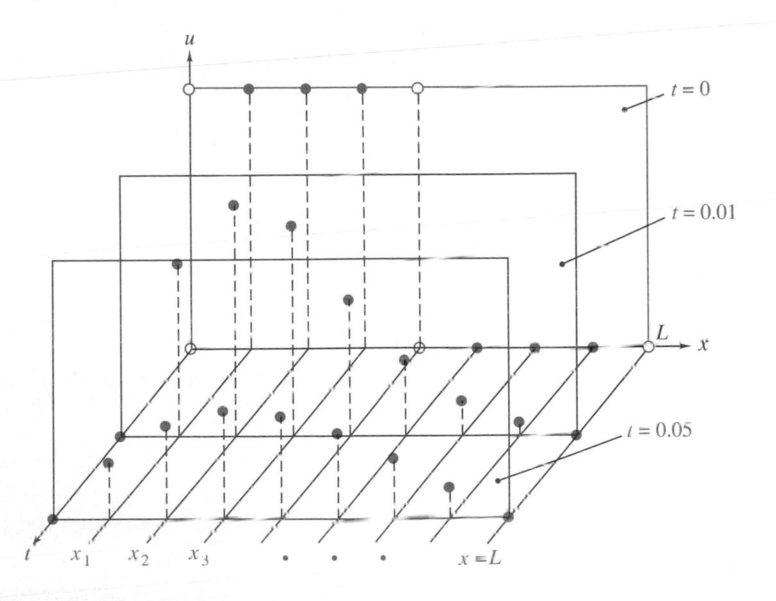

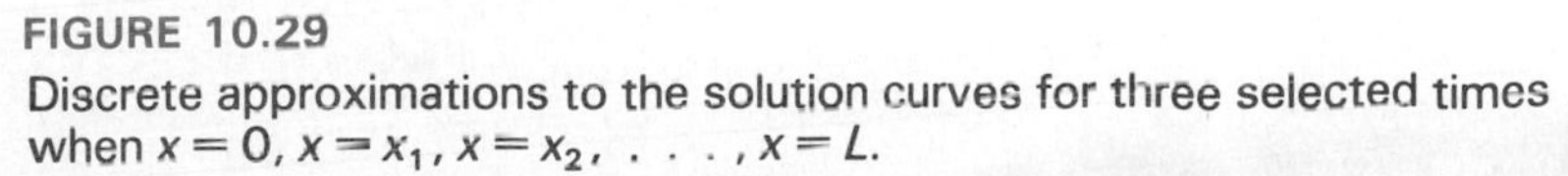

FIGURE 10.29
Discrete approximations to the solution curves for three selected times when $x = 0$, $x = x_1$, $x = x_2$, . . . , $x = L$.

Next, decide on a time increment Δt and label the time stages:

$$t_0 = 0,\ t_1 = t_0 + \Delta t,\ t_2 = t_1 + \Delta t,\ \ldots . \tag{4}$$

We can continue to advance time as long as we like. These choices for Δx and Δt are not entirely independent if our procedure is going to produce reasonable results in the long run. We will say more about this issue later on. For now it is enough to note that these choices do determine the position and time labels of the numerical table.

The second step is to evaluate the initial-value function $u(x, 0) = f(x)$ at the subdivision points $x_0, x_1, x_2, \ldots, x_n = L$. We then enter these values into the first row of the table. In the case of fixed boundary conditions, we also enter the temperature values for the boundary conditions, $u(0, t) = T_1$ and $u(L, t) = T_2$, into the left and right columns of the table, respectively. We will see next how to use the partial differential equation for the heat-flow model to calculate the interior values of the table, advancing one complete row at a time with each time step.

Numerical Formulation of the Partial Differential Equation

Consider the partial differential equation

$$u_{xx} = \frac{1}{k} u_t, \qquad k > 0. \tag{5}$$

It involves a first derivative with respect to time and a second derivative with respect to position. These derivatives can be approximated by finite divided differences similar to what was done for Euler's method. For instance, the first derivative u_t can be approximated by the divided difference

$$u_t(x, t) \approx \frac{u(x, t + \Delta t) - u(x, t)}{\Delta t}. \tag{6}$$

Since we will be evaluating the function $u(x, t)$ in Eq. (6) at **grid points** (x_i, t_j) for our table, it will simplify the notation to write $u_{i,j}$ in place of $u(x_i, t_j)$. Thus

$$u_{i,j} = u(x_i, t_j). \tag{7}$$

Using this notational convention, Eq. (6) can be written as

$$u_t(x_i, t_j) \approx \frac{u_{i,j+1} - u_{i,j}}{\Delta t}. \tag{8}$$

Likewise the second derivative, being the derivative of the first derivative, can be approximated by second divided differences. It is a straightforward algebraic calculation to produce the following formula (which we give as a problem in the exercises):

$$u_{xx}(x_i, t_j) \approx \frac{u_{i-1,j} - 2u_{i,j} + u_{i+1,j}}{(\Delta x)^2}. \tag{9}$$

Substitution of approximations (8) and (9) into Eq. (5) yields

$$\frac{u_{i-1,j} - 2u_{i,j} + u_{i+1,j}}{(\Delta x)^2} = \frac{u_{i,j+1} - u_{i,j}}{k\Delta t}. \tag{10}$$

If we define the number r by

$$r = k\frac{\Delta t}{(\Delta x)^2} \tag{11}$$

and rewrite Eq. (10), we obtain

$$r(u_{i-1,j} - 2u_{i,j} + u_{i+1,j}) = u_{i,j+1} - u_{i,j}.$$

Solving algebraically for $u_{i,j+1}$ gives us the **PDE finite difference formula for the heat equation:**

$$u_{i,j+1} = ru_{i-1,j} + (1-2r)u_{i,j} + ru_{i+1,j}. \tag{12}$$

How can we use numerical formulation (12) to determine the unknown interior entries of the solution table?

Using the PDE Finite Difference Formula

Let us interpret formula (12) in terms of our grid setup. Remember that the subscript i refers to the *position* or *column* number of the table and j refers to the *time* or *row* label. We have already stated that we are going to obtain the entries of the numerical solution table row by complete row, and the first row is completed when we enter the initial-condition function values. Suppose then we have done row j, which corresponds to having the numerical solution approximation at time $t = t_j$. We want to calculate the values for the next row $j+1$. The lefthand side of formula (12) represents those row values $u_{i,j+1}$ as i advances across the columns from 1 to $n-1$. To obtain the value $u_{i,j+1}$ for a particular column i, we calculate the righthand side of Eq. (12), which involves solution values only over the *previous row* j. The grid configuration is displayed in Fig. 10.30. In other words, *to calculate the value $u_{i,j+1}$, multiply the value $u_{i-1,j}$, in the immediately preceding row and column, by r; add in the value of $u_{i,j}$, occurring in the immediately preceding row and same column, multiplied by $(1-2r)$; and finally add the value of $u_{i+1,j}$, occurring in the immediately preceding row and next column, multiplied by r.*

		Column i ↓	
Known row j ⟶	$u_{i-1,j}$	$u_{i,j}$	$u_{i+1,j}$
Unknown row $j+1$ ⟶	•	$u_{i,j+1}$	

FIGURE 10.30
Grid entries involved in PDE finite difference formula (12).

To illustrate the procedure, suppose $r = 0.25$ and we have the following configuration of entries:

$$\begin{array}{llll} \text{time row } j & \longrightarrow 100 & 45.31 & 14.06 \\ \text{time row } j+1 \longrightarrow & & * & \end{array}$$

Then the value at the position marked * is computed as

$$0.25(100) + (1 - 0.5)45.31 + 0.25(14.06) = 51.17.$$

More completely, if the entire row j is

	0	x_1	x_2	x_3	x_4	L
Time t_j	100	45.31	14.06	14.06	45.31	100

then for $r = 0.25$ the next row in the table is

	0	x_1	x_2	x_3	x_4	L
Time t_{j+1}	100	51.17	21.87	21.87	51.17	100

Convergence of Numerical Solutions

The previously described method does not necessarily produce numerical solutions that converge. Appropriate choices have to be made for the increments Δx and Δt that determine the value of the constant multiplier r. We state the following result without proof.

THEOREM 10.2

If $0 < r \leq 0.5$, then the approximations $u_{i,j}$ converge to the solution $u(x, t)$.

Thus, if we choose Δx and Δt so that $r = k\,\Delta t/(\Delta x)^2$ is positive and less than ½, the numerical solution curves are guaranteed to converge over time.

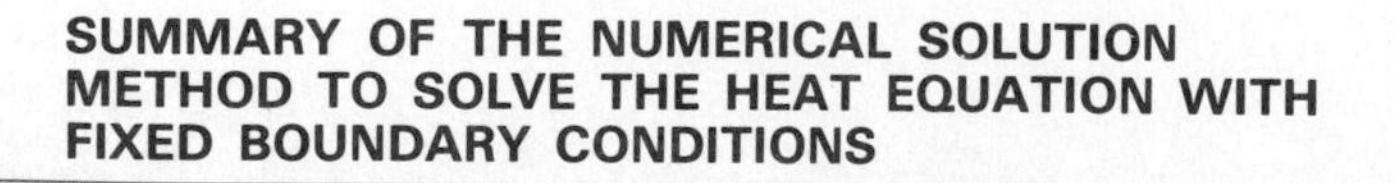

SUMMARY OF THE NUMERICAL SOLUTION METHOD TO SOLVE THE HEAT EQUATION WITH FIXED BOUNDARY CONDITIONS

Step 1. Subdivide the position interval $0 < x < L$ and choose the time increment so that $r = k\,\Delta t/(\Delta x)^2$ satisfies $0 < r \leq ½$. Then

$\Delta x = L/n$ and Δt yield the grid points

$$x_0 = 0,\ x_1 = x_0 + \Delta x,\ x_2 = x_1 + \Delta x,\ \ldots,$$
$$x_n = x_{n-1} + \Delta x = L,$$

$$t_0 = 0,\ t_1 = t_0 + \Delta t,\ t_2 = t_1 + \Delta t,\ \ldots,$$
$$t_j = t_{j-1} + \Delta t,\ \ldots .$$

Step 2. Evaluate the initial-condition function $u(x, 0) = f(x)$ at each subdivision point:

$$f(x_0), f(x_1), f(x_2), \ldots, f(x_{n-1}), f(x_n).$$

These values form the first row of the numerical solution table.

Step 3. Set the first and last columns of the solution table to the fixed boundary-condition values:

$$u(0, t_j) = T_1 \qquad \text{and} \qquad u(L, t_j) = T_2$$

for every time row t_j.

Step 4. Calculate and record in the table the interior numerical solution approximations, row by complete row in succession, beginning with row t_1, according to the formula

$$u_{i,j+1} = ru_{i-1,j} + (1 - 2r)u_{i,j} + ru_{i+1,j}.$$

EXAMPLE 1 Solve numerically the following heat-conduction model with fixed boundary conditions until time $t = 0.05$:

$$\text{PDE:} \quad u_{xx} = u_t, \qquad 0 < x < 1;$$
$$\text{BC:} \quad u(0, t) = u(1, t) = 0, \qquad t > 0;$$
$$\text{IC:} \quad u(x, 0) = \begin{cases} 1, & 0 < x < 0.5, \\ 0, & 0.5 < x < 1. \end{cases}$$

Solution. We determined the analytical solution to this model in Section 10.3. Let us calculate a numerical solution.

Step 1 We subdivide the interval $0 < x < 1$ into five subdivisions of equal length $\Delta x = 0.20$. We choose $\Delta t = 0.01$ for the time increment. Then $r = k\,\Delta t/(\Delta x)^2 = 0.25$ satisfies the convergence criterion.

Step 2 We evaluate the initial condition function to obtain the following first row of the solution table:

	Position Interval x					
Time Interval t	0.0	0.2	0.4	0.6	0.8	1.0
0	0	1	1	0	0	0

Step 3 We next set the boundary-condition values in the left and right columns of the table for each time increment:

	Position Interval x					
Time Interval t	0.0	0.2	0.4	0.6	0.8	1.0
0	0	1	1	0	0	0
0.01	0					0
0.02	0					0
0.03	0					0
0.04	0					0
0.05	0					0

Step 4 We can now verify the following tabulated values using finite difference formula (12) with $r = 0.25$:

$$u_{i,j+1} = \frac{u_{i-1,j} + 2u_{i,j} + u_{i+1,j}}{4}.$$

	Position Interval x					
Time Interval t	0.0	0.2	0.4	0.6	0.8	1.0
0	0	1	1	0	0	0
0.01	0	0.75	0.75	0.25	0	0
0.02	0	0.56	0.63	0.31	0.06	0
0.03	0	0.44	0.53	0.33	0.11	0
0.04	0	0.35	0.46	0.33	0.14	0
0.05	0	0.29	0.40	0.32	0.15	0

We have rounded the values in our table to two decimal places, and the numerical results compare favorably with the solution for $t = 0.05$ graphed in Fig. 10.20.

Numerically Solving the Heat Equation with Insulated Boundary Conditions

In this situation we must translate the insulated boundary conditions to finite difference form:

$$0 = u_x(0, t) \approx \frac{u(\Delta x, t) - u(0, t)}{\Delta x}$$

and

$$0 = u_x(L, t) \approx \frac{u(L, t) - u(L - \Delta x, t)}{\Delta x}.$$

In other words, for all $t > 0$,

$$u(0, t) = u(\Delta x, t), \tag{13}$$
$$u(L, t) = u(L - \Delta x, t). \tag{14}$$

Using our subscripted notation, Eqs. (13) and (14) translate to

$$u_{0,j} = u_{1,j}, \tag{15}$$
$$u_{n,j} = u_{n-1,j}, \tag{16}$$

for *every* time row j. Conditions (15) and (16) mean that the left column of the numerical solution table is the same as the next adjacent column to its right, and the right column is the same as the column immediately to its left. Therefore, in calculating a new time stage row, we first calculate the interior numerical approximation of the row according to Step 4 in our procedure. Then we employ conditions (15) and (16) to set the first and last entries of the row. With that row completed, we can then advance to the next row for the next time step, calculate the interior entries, enter the values of the first and last columns, and so forth. Here's an example.

EXAMPLE 2 Solve numerically the following heat-conduction model with insulated boundary conditions until $t = 0.10$:

PDE: $u_{xx} = u_t, \quad 0 < x < 1;$

BC: $u_x(0, t) = u_x(1, t) = 0, \quad t > 0;$

IC: $u(x, 0) = \begin{cases} 1, & 0 < x < 0.5, \\ 0, & 0.5 < x < 1. \end{cases}$

Solution. Steps 1 and 2 are the same as in Example 1. Thus

Time Interval t	Position Interval x					
	0.0	0.2	0.4	0.6	0.8	1.0
0	1	1	1	0	0	0

We skip Step 3 because we do not know the boundary values yet. Step 4 produces the interior entries of the next row as before:

$$u_{i,j+1} = \frac{u_{i-1,j} + 2u_{i,j} + u_{i+1,j}}{4}.$$

Time Interval t	Position Interval x					
	0.0	0.2	0.4	0.6	0.8	1.0
0	1	1	1	0	0	0
0.01		1	0.75	0.25	0	

Using the insulated boundary conditions, we finally employ Eqs. (15) and (16) to complete the first and last entries of the row:

Time Interval t	Position Interval x					
	0.0	0.2	0.4	0.6	0.8	1.0
0	1	1	1	0	0	0
0.01	1	1	0.75	0.25	0	0

Repeating this procedure row by row yields the following numerical solution table with entries rounded to two decimal places:

Time Interval t	Position Interval x					
	0.0	0.2	0.4	0.6	0.8	1.0
0	1	1	1	0	0	0
0.01	1	1	0.75	0.25	0	0
0.02	0.94	0.94	0.69	0.31	0.06	0.06
0.03	0.88	0.88	0.66	0.34	0.13	0.13
0.04	0.83	0.83	0.64	0.37	0.18	0.18
0.05	0.78	0.78	0.62	0.39	0.23	0.23
0.06	0.74	0.74	0.60	0.41	0.27	0.27
0.07	0.71	0.71	0.59	0.42	0.31	0.31
0.08	0.68	0.68	0.58	0.44	0.34	0.34
0.09	0.66	0.66	0.57	0.45	0.37	0.37
0.10	0.64	0.64	0.56	0.46	0.39	0.39

You can observe from the table how the left side of the insulated rod is gradually cooling while the right side of the rod is warming, both sides converging to the steady-state temperature of $u = 0.50$.

Given the crudeness of our approximation procedure and the roundoff error in our computations, the results of the $t = 0.10$ row compare reasonably well with the values in the graph of Fig. 10.23.

Other numerical methods exist for solving the heat-conduction model. Our discussion was only intended as a brief introduction.*

EXERCISES 10.4

In problems 1 – 10 solve numerically the given heat-conduction models until $t = 0.10$. Express your results in tabular form as illustrated in the text examples. Check that $0 < r \leq 0.5$ to ensure convergence.

1. $u_{xx} = 0.1u_t$, $u(0, t) = u(100, t) = 0$,

$$u(x, 0) = \begin{cases} 0, & 0 < x < 25 \\ 50, & 25 \leq x \leq 75 \\ 0, & 75 < x < 100 \end{cases}$$

2. $u_{xx} = 2u_t$, $u(0, t) = u(1, t) = 0$, $u(x, 0) = x$, $0 < x < 1$

3. $3u_{xx} = u_t$, $u(0, t) = u(\pi, t) = 0$, $u(x, 0) = \sin x$, $0 < x < \pi$

4. $u_{xx} = 0.25u_t$, $u(0, t) = u(2\pi, t) = 0$, $u(x, 0) = e^{2x}$, $0 < x < 2\pi$

5. $u_{xx} = 0.5u_t$, $u(0, t) = 100$, $u(1, t) = 50$, $u(x, 0) = 100$, $0 < x < 1$

6. $u_{xx} = u_t$, $u(0, t) = u(\pi, t) = 100$, $u(x, 0) = \sin 2x$, $0 < x < \pi$

7. $u_{xx} = u_t$, $u_x(0, t) = u_x(1, t) = 0$, $u(x, 0) = x$, $0 < x < 1$.

8. $u_{xx} = 2u_t$, $u_x(0, t) = u_x(\pi, t) = 0$, $u(x, 0) = \cos x$, $0 < x < \pi$

9. $u_{xx} = 3u_t$, $u_x(0, t) = u_x(2, t) = 0$, $u(x, 0) = 2 + \cos \pi x$, $0 < x < 2$

10. $u_{xx} = u_t$, $u_x(0, t) = u_x(2, t) = 0$,

$$u(x, 0) = \begin{cases} x, & 0 < x < 1 \\ 2 - x, & 1 < x < 2 \end{cases}$$

11. Derive the formula for the numerical calculation of the second derivative

$$u_{xx}(x_i, t_j) \approx \frac{u_{i-1,j} - 2u_{i,j} + u_{i+1,j}}{(\Delta x)^2}.$$

12. Consider a one-dimensional rod 50 cm long whose lateral surface is not insulated. Assume the rod is made with a material that absorbs heat by chemical reaction. (This same chemical reaction is employed in coldpacks used for treating injuries such as sprains or internal bleeding.) The chemical reaction absorbs heat equal to the temperature of the reaction at each point in the rod. Assume the *ends* of the rod are insulated and that initially the rod is at 100°C.

a) Construct a mathematical model that can be used to predict the temperature $u(x, t)$ at each point x of the rod at any time t.
b) Derive a finite difference formula for the partial differential equation given in your model.
c) Solve the model numerically until $t = 0.10$.

13. Write a computer program in a language of your choice to solve numerically the heat-conduction model with fixed boundary conditions.

* An excellent presentation of numerical methods for solving partial differential equations can be found in *Numerical Analysis* by Richard L. Burden and J. Douglas Faires (Boston: Prindle, Weber & Schmidt Publishers, 1985), chap. 11.

14. Write a computer program in a language of your choice to solve numerically the heat-conduction model with insulated boundary conditions.

15. Derive a numerical method for solving the heat-conduction problem in a ring. Test your method on the following model.

$$\begin{aligned} \text{PDE:}\quad & u_{xx} = u_t, \quad -\pi < x < \pi, \quad t > 0; \\ \text{BC:}\quad & u(-\pi, t) = u(\pi, t), \\ & u_x(-\pi, t) = u_x(\pi, t), \quad t > 0; \\ \text{IC:}\quad & u(x, 0) = \begin{cases} 0, & -\pi < x \leq 0 \\ 1, & 0 < x < \pi. \end{cases} \end{aligned}$$

10.5

ONE-DIMENSIONAL WAVE EQUATION

Consider a uniform, perfectly flexible string of constant density that is tightly stretched between two fixed points, 0 and L. We assume the equilibrium position of the string is horizontal, with the string aligned along the x-axis as displayed in Fig. 10.31. You can think of it as being one of the strings on a musical instrument, like a guitar string. Suppose the string is plucked at time $t = 0$ causing the string to vibrate. We are interested in studying the vertical motion of this vibrating string. More precisely, our problem is to determine the displacement y of any point on the string away from its equilibrium position, as shown in Fig. 10.32. Since the string moves in time, the displacement of the point on the string depends both on the equilibrium location x of the point on the string as well as time; that is, $y = y(x, t)$. Actually the point on the vibrating string has both vertical and horizontal displacements as it moves. We are going to assume that the horizontal displacement is so small relative to the vertical displacement as to be negligible. In other words, we assume that all the motion of the vibrating string is in the vertical y direction. We also assume that the maximum displacement of each point on the string is small in comparison with the length L of the string. Thus the slope of the curve defined by the vibrating string at a fixed point in time is the partial derivative $\partial y/\partial x$, as displayed in Fig. 10.33.

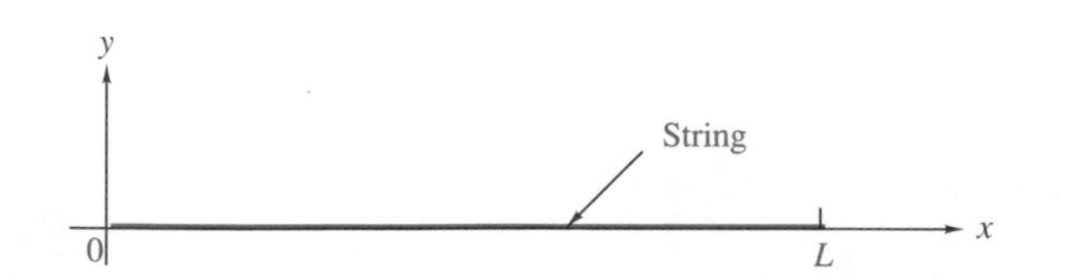

FIGURE 10.31

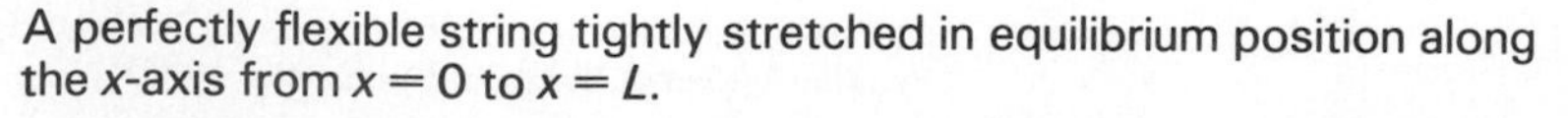
A perfectly flexible string tightly stretched in equilibrium position along the x-axis from $x = 0$ to $x = L$.

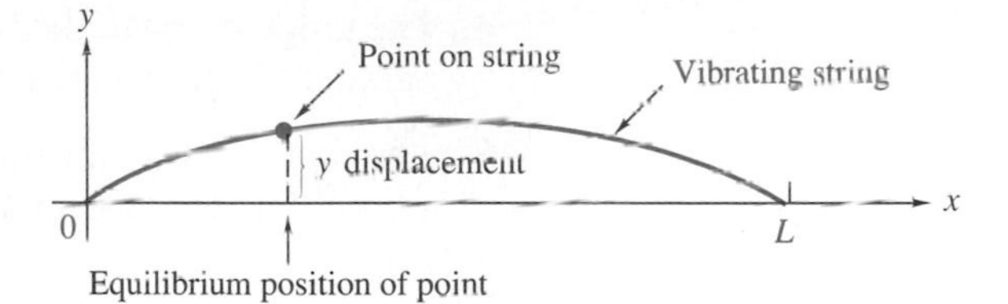

FIGURE 10.32
A point on the vibrating string displaced from its equilibrium position along the x-axis.

We next consider the forces acting on the vibrating string at the point (x, y). If we assume that the string offers no resistance to bending, then the force exerted by the rest of the string on the point is the **tension** in the string. The tension "tries to stretch" the string, pulling the point both to the left and right because the string is attached at its two ends. The tension acts in the direction of the tangent at the point. Since we assume small displacements in the string relative to the string's length, the slope of the string is nearly always zero. The tension also depends on both the location of the point and on time, but under our assumptions it is reasonable to assume that the tension is nearly constant: it is equal to the constant tension in the string in its horizontal equilibrium position. We also assume that the gravitational force acting on the string is small relative to the tensile force, so we can neglect gravitational forces.

Assumptions for the Vibrating String

Let us summarize the assumptions we have made so far in this section:

1. The string is uniform with constant linear density ρ (mass per unit length) and perfect flexibility.
2. The motion of the vibrating string is confined to the xy-plane and occurs in the y direction only.
3. The maximum displacement of each point on the string is small compared with the length of the string.

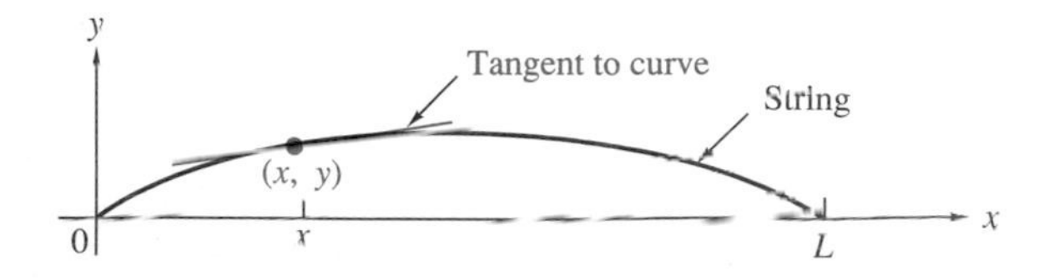

FIGURE 10.33
The slope of the tangent line to the curve defined by the vibrating string at a fixed instant of time is $\partial y/\partial x$.

4. The tangent line to the string at each point at any time is close to being parallel to the x-axis.
5. The only force acting on the string is the tension in the string, so there are no damping forces and gravity is negligible.
6. The tension T on the string is constant over the entire length of the string.

Constructing the Model for a Vibrating String

Consider an enlarged view of an infinitesimal segment Δs of the string at a fixed instant during its vibration, as displayed in Fig. 10.34. The angle θ between the tensile force and the x-axis depends both on location and time. This angle is nearly zero from our assumption that the tangent to the string is close to being parallel to the x-axis. We will use Newton's second law, $F = ma$, to calculate the total force acting on this infinitesimal string segment. Here m is the mass of the string segment, so

$$m = \rho \, \Delta s \tag{1}$$

from our assumption of constant linear density.

Since the motion at each point on the string is assumed to be vertical, the acceleration of the string segment is $\partial^2 y/\partial t^2$ (the second derivative of displacement with respect to time). Also, there are no resultant forces acting in the horizontal direction because the string has no horizontal motion; that is, the horizontal forces at each end of the segment are equal in magnitude but opposite in direction. Thus we need calculate only the vertical forces to obtain

$$\begin{aligned} F &= \text{sum of } y \text{ components of the tension applied to } \Delta s \\ &= T \sin[\theta(x + \Delta x, t)] - T \sin[\theta(x, t)]. \end{aligned} \tag{2}$$

Therefore, from Eqs. (1) and (2), $F = ma$ implies that

$$T \sin[\theta(x + \Delta x, t)] - T \sin[\theta(x, t)] = \rho \, \Delta s \frac{\partial^2 y}{\partial t^2}. \tag{3}$$

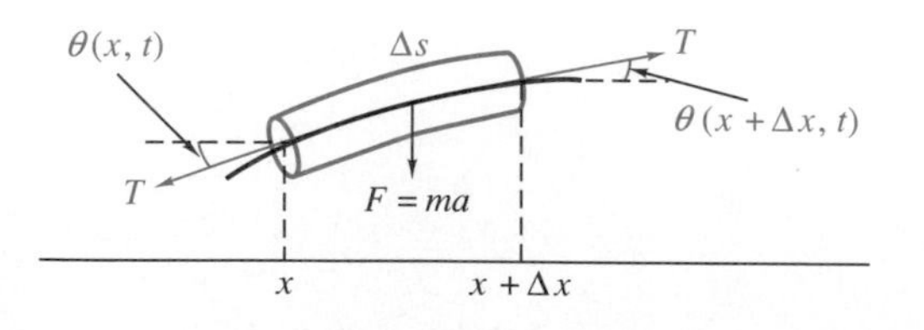

FIGURE 10.34
Enlarged infinitesimal segment Δs of the string showing the constant tension T at both ends of the segment and the angle θ between the tensile force and the horizontal equilibrium position.

Dividing both sides of Eq. (3) by Δx and taking the limit as $\Delta x \to 0$ gives us

$$\lim_{\Delta x \to 0} T \left\{ \frac{\sin[\theta(x + \Delta x, t)] - \sin[\theta(x, t)]}{\Delta x} \right\} = \lim_{\Delta x \to 0} \rho \frac{\Delta s}{\Delta x} y_{tt}$$

or

$$T \frac{\partial(\sin \theta)}{\partial x} = \rho \frac{\partial s}{\partial x} y_{tt}. \tag{4}$$

In Eq. (4) y_{tt} is the second derivative $\partial^2 y/\partial t^2$. From the elementary calculus we know that

$$\frac{\partial s}{\partial x} = \sqrt{1 + \left(\frac{\partial y}{\partial x}\right)^2},$$

so, from our assumption that the tangent line is nearly horizontal (or $\partial y/\partial x = 0$), we have

$$\frac{\partial s}{\partial x} \approx 1. \tag{5}$$

The same assumption tells us that

$$\begin{aligned} \sin \theta &= \frac{\tan \theta}{\sqrt{1 + \tan^2 \theta}} \\ &= \frac{y_x}{\sqrt{1 + (y_x)^2}} \\ &\approx y_x. \end{aligned} \tag{6}$$

Substituting Eqs. (5) and (6) into Eq. (4) results in

$$T \frac{\partial(y_x)}{\partial x} = \rho \cdot 1 \cdot y_{tt}$$

or

$$T y_{xx} = \rho y_{tt}.$$

After setting $a^2 = T/\rho$ we obtain

$$a^2 y_{xx} = y_{tt}. \tag{7}$$

Equation (7) is called the **one-dimensional wave equation** and provides a model for the vibrating string under our assumptions 1–6. It is another example of a homogeneous linear second-order partial differential equation. The constant a in Eq. (7) represents the velocity for which some disturbance is moving along the vibrating string in the positive direction (see problem 11, Exercises 10.5).

Initial Conditions

In addition to partial differential Eq. (7), what do we need to know in order to describe fully the displacement $y(x, t)$? Since the partial differential equation contains the second time derivative y_{tt}, our knowledge of ordinary differential equations suggests that we will need *two initial conditions* for the vibrating string: an initial position and an initial velocity for each segment of the string. Thus, at time $t = 0$ we need to know

$$y(x, 0) = f(x)$$

and

$$y_t(x, 0) = g(x).$$

Just about any functions can serve as $f(x)$ and $g(x)$ as long as they and their derivatives are piecewise continuous over the length of the string $0 < x < L$. Illustrative graphs showing one possible initial position and velocity are given in Fig. 10.35.

Boundary Conditions

The partial differential equation modeling the motion of the vibrating string also contains a second derivative with respect to position, y_{xx}. As was the case with the heat-conduction problem, we will need to know what happens at the two ends of the string, $x = 0$ and $x = L$, for all time $t > 0$, in order to predict the displacement function $y(x, t)$. These conditions are called *boundary conditions* and can be *fixed* or *elastic.*

Fixed Boundary Conditions In a fixed boundary condition the displacement of the end of the string in question is prescribed for all time. For instance,

$$y(L, t) = 0$$

fixes the displacement of the right end at the constant value of zero for all time; thus there is no displacement in the right boundary. The condition

$$y(L, t) = h(t)$$

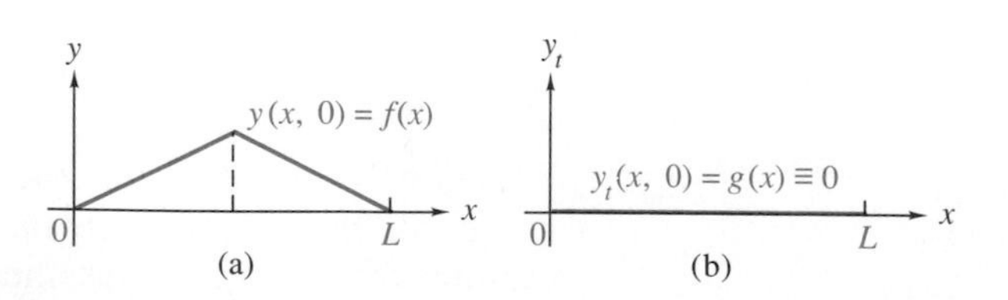

FIGURE 10.35
(a) Initial position and (b) initial velocity for a vibrating string satisfying the one-dimensional wave equation.

varies the displacement at the right end as determined by the values of some *known* function h. If $h(t) \equiv 0$, the fixed boundary condition is said to be *homogeneous;* otherwise it is *nonhomogeneous.*

Let us summarize our model for fixed boundary conditions.

VIBRATING STRING WITH FIXED HOMOGENEOUS ENDS

PDE:	$a^2 y_{xx} = y_{tt}$,	$0 < x < L, \quad t > 0;$	(8)
BC:	$y(0, t) = 0,$		(9)
	$y(L, t) = 0,$	$t > 0;$	
IC:	$y(x, 0) = f(x),$		(10)
	$y_t(x, 0) = g(x),$	$0 < x < L.$	

Solution to the Vibrating String Model with Fixed Homogeneous Ends

Observe first that the *trivial solution* $y(x, t) \equiv 0$ solves the partial differential equation and the fixed boundary conditions but *not* the initial conditions. Hence we will not consider this function any further since we are interested only in nontrivial solutions.

Since the partial differential equation and the boundary conditions are linear and homogeneous, we try to separate the variables just as we did with the heat equation. That is, we try the factorization

$$y(x, t) = X(x)T(t) \tag{11}$$

where X is a function only of position x and T a function only of time t. Then

$$y_{xx} = X''T$$

and

$$y_{tt} = XT''.$$

Substitution into wave Eq. (8) gives us

$$a^2 X''T = XT''$$

or

$$\underbrace{\frac{X''}{X}}_{\substack{\text{function}\\ \text{of } x\\ \text{alone}}} = \underbrace{\frac{1}{a^2}\frac{T''}{T}}_{\substack{\text{function}\\ \text{of } t\\ \text{alone}}} \tag{12}$$

As with the heat equation, Eq. (12) reduces to two *ordinary* differential equations, but now both are of second-order:

$$\frac{X''}{X} = \text{constant}, \tag{13}$$

$$\frac{T''}{a^2T} = \text{same constant}. \tag{14}$$

Again there are three cases to consider: whether the constant in Eqs. (13) and (14) is positive, zero, or negative. We investigate each case.

Case 1: Positive Constant $= \lambda^2$

In this case Eq. (13) becomes

$$X'' - \lambda^2 X = 0,$$

so

$$X = c_1 e^{\lambda x} + c_2 e^{-\lambda x}. \tag{15}$$

We apply the homogeneous boundary conditions to evaluate the constants c_1 and c_2:

$$y(0, t) = y(L, t) = 0$$

translates to

$$X(0) = X(L) = 0. \tag{16}$$

Substituting Eq. (16) into Eq. (15) implies that $c_1 = c_2 = 0$. We conclude that there is no nontrivial solution in this case.

Case 2: Zero Constant

Here Eq. (13) becomes

$$X'' = 0,$$

yielding

$$X = c_3 x + c_4.$$

Applying boundary conditions (16) again implies that $c_3 = c_4 = 0$, so no nontrivial solution exists in this case either.

Case 3: Negative Constant $= -\lambda^2$

In this case Eq. (13) becomes

$$X'' + \lambda^2 X = 0,$$

giving rise to the solution

$$X = c_5 \cos \lambda x + c_6 \sin \lambda x.$$

Applying the boundary conditions yields the following eigenvalues and eigenfunctions, as was the case with the heat equation:

$$\lambda = \frac{n\pi}{L}$$

and

$$X(x) = c_6 \sin \frac{n\pi x}{L}, \qquad n = 1, 2, 3, \ldots . \tag{17}$$

Now we determine the factor $T(t)$ associated with each eigenfunction $X(x)$. From Eq. (14)

$$T'' + a^2\lambda^2 T = 0$$

gives us the general solution

$$T = c_7 \cos a\lambda t + c_8 \sin a\lambda t. \tag{18}$$

From Eqs. (17) and (18) each value of the positive integer n yields a product solution of the form

$$y_n(x, t) = A_n \sin \frac{n\pi x}{L} \cos \frac{n\pi at}{L} + B_n \sin \frac{n\pi x}{L} \sin \frac{n\pi at}{L}.$$

From the superposition principle the infinite series

$$y(x, t) = \sum_{n=1}^{\infty} \left(A_n \cos \frac{n\pi at}{L} + B_n \sin \frac{n\pi at}{L} \right) \sin \frac{n\pi x}{L} \tag{19}$$

solves Eq. (8) subject to homogeneous fixed boundary conditions (9). There are two initial conditions and two families of coefficients, A_n and B_n, to determine in Eq. (19). The initial conditions are satisfied if

$$f(x) = \sum_{n=1}^{\infty} A_n \sin \frac{n\pi x}{L} \tag{20}$$

and

$$g(x) = \sum_{n=1}^{\infty} \frac{n\pi a}{L} B_n \sin \frac{n\pi x}{L}. \tag{21}$$

Therefore A_n represents the coefficients in the Fourier sine series for $f(x)$, and $(n\pi a/L)B_n$ represents the coefficients in the Fourier sine series for $g(x)$:

$$A_n = \frac{2}{L} \int_0^L f(x) \sin \frac{n\pi x}{L}\, dx, \tag{22}$$

$$\frac{n\pi a}{L} B_n = \frac{2}{L} \int_0^L g(x) \sin \frac{n\pi x}{L}\, dx. \tag{23}$$

Let us consider a specific example.

EXAMPLE 1 Find the solution to wave Eq. (8) subject to fixed homogeneous boundary conditions (9) when the initial conditions are given by

$$y(x, 0) = \begin{cases} bx, & 0 < x < \dfrac{L}{2}, \\ b(L - x), & \dfrac{L}{2} \leqslant x < L, \end{cases}$$

and

$$y_t(x, 0) = 0, \qquad 0 < x < L.$$

Solution. From Eq. (19) the solution has the form of the infinite series

$$y(x, t) = \sum_{n=1}^{\infty} \left(A_n \cos \frac{n\pi at}{L} + B_n \sin \frac{n\pi at}{L} \right) \sin \frac{n\pi x}{L}.$$

The A_n coefficients are given by Eq. (22):

$$\begin{aligned} A_n &= \frac{2}{L} \int_0^{L/2} bx \sin \frac{n\pi x}{L}\, dx + \frac{2}{L} \int_{L/2}^{L} b(L - x) \sin \frac{n\pi x}{L}\, dx \\ &= \frac{2b}{L} \left[-\frac{Lx}{n\pi} \cos \frac{n\pi x}{L} + \left(\frac{L}{n\pi} \right)^2 \sin \frac{n\pi x}{L} \right]_0^{L/2} \\ &\quad + \frac{2b}{L} \left[-(L - x) \frac{L}{n\pi} \cos \frac{n\pi x}{L} - \left(\frac{L}{n\pi} \right)^2 \sin \frac{n\pi x}{L} \right]_{L/2}^{L} \\ &= \left(-\frac{bL}{n\pi} \cos \frac{n\pi}{2} + \frac{2bL}{n^2\pi^2} \sin \frac{n\pi}{2} \right) \\ &\quad + \left[\frac{bL}{n\pi} \cos \frac{n\pi}{2} - \frac{2bL}{n^2\pi^2} \sin(n\pi) + \frac{2bL}{n^2\pi^2} \sin \left(\frac{n\pi}{2} \right) \right] \\ &= \frac{4bL}{n^2\pi^2} \sin \frac{n\pi}{2}. \end{aligned}$$

Since $g(x) = y_t(x, 0) \equiv 0$, we have $B_n = 0$ for all n. Therefore

$$y(x, t) = \sum_{n=1}^{\infty} \frac{4bL}{n^2\pi^2} \sin \frac{n\pi}{2} \cos \frac{n\pi at}{L} \sin \frac{n\pi x}{L}.$$

Elastic Boundary Conditions

There are other boundary conditions for the wave equation. In an elastic boundary condition the end or boundary of the string in question is attached to some kind of dynamical system that moves the end vertically up and down. This system could be a spring–mass system, for instance, that is moving inside a vertical track. Or the system may simply be a vertical track that is free to move up and down yet secures the string and prevents it from moving horizontally. As was the case for the heat equation, we can also have *mixed boundary conditions*, where one end of the string is fixed and the other is elastic. To study the wave equation subject to such boundary conditions, refer to any standard text on the subject of partial differential equations.

EXERCISES 10.5

1. Show that the following functions satisfy wave Eq. (7).

a) $y = \sin(x + at)$

b) $y = e^{x-at}$

c) $y = \ln(2x + 2at)$

In problems 2–9 solve the wave equation with fixed homogeneous boundary conditions given by Eqs. (8–10) with the specified data given in each case.

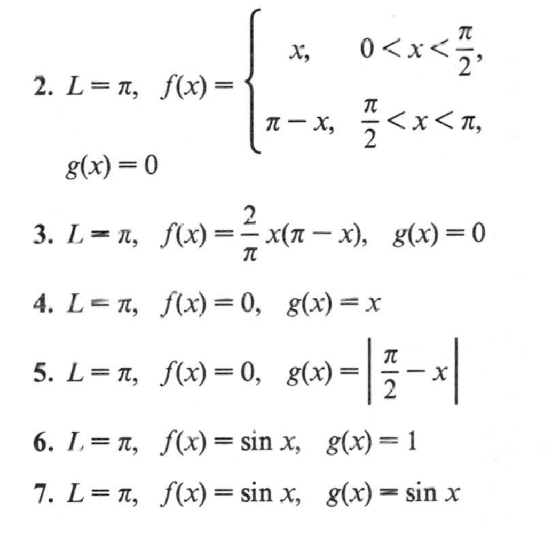

2. $L = \pi, \quad f(x) = \begin{cases} x, & 0 < x < \dfrac{\pi}{2}, \\ \pi - x, & \dfrac{\pi}{2} < x < \pi, \end{cases}$

$g(x) = 0$

3. $L = \pi, \quad f(x) = \dfrac{2}{\pi} x(\pi - x), \quad g(x) = 0$

4. $L = \pi, \quad f(x) = 0, \quad g(x) = x$

5. $L = \pi, \quad f(x) = 0, \quad g(x) = \left| \dfrac{\pi}{2} - x \right|$

6. $L = \pi, \quad f(x) = \sin x, \quad g(x) = 1$

7. $L = \pi, \quad f(x) = \sin x, \quad g(x) = \sin x$

8. $L = 1, \quad f(x) = 0, \quad g(x) = x(1 - x)$

9. $L = 1, \quad f(x) = x, \quad g(x) = x^2$

10. Assume the function $F(x)$ is twice-differentiable for all values of x. Verify that the two functions $y(x, t) = F(x + at)$ and $y(x, t) = F(x - at)$ both satisfy wave Eq. (7).

11. a) Using the substitutions $u = x + at$ and $v = x - at$, transform the equation $a^2 y_{xx} = y_{tt}$ into the partial differential equation $y_{uv} = 0$.

b) Solve $y_{uv} = 0$ from part (a). Conclude that

$$y(x, t) = F(x + at) + G(x - at)$$

is a solution to Eq. (7) for any two twice-differentiable functions F and G.

c) Plot the value of $G(x - at)$ for $t = 0$, $t = 1/a$, $t = 2/a$, and $t = 3/a$ if $G(u) = \cos u$. Note that for $t \neq 0$, the graph of $G(x - at)$ represents the graph of $G(x)$ displaced or translated a distance at in the positive x direction. Thus the constant a represents the velocity for which some disturbance is moving along the vibrating string in the positive direction.

10.6

LAPLACE'S EQUATION AND STEADY-STATE TEMPERATURE

Imagine heat being applied to the boundary of the *two-dimensional* rectangular region or plate R, as illustrated in Fig. 10.36. We assume the plate has no thickness and that the top and bottom faces of the plate are perfectly insulated. Thus heat can flow only in the x and y directions. As position (x, y) varies over the plate and as time t varies, the temperature u of the plate also varies. Thus $u = u(x, y, t)$. We are interested in knowing the temperature $u(x, y, t)$ at any time for any position along the plate. The one-dimensional version of this diffusion problem for a rod was modeled in Section 10.1.

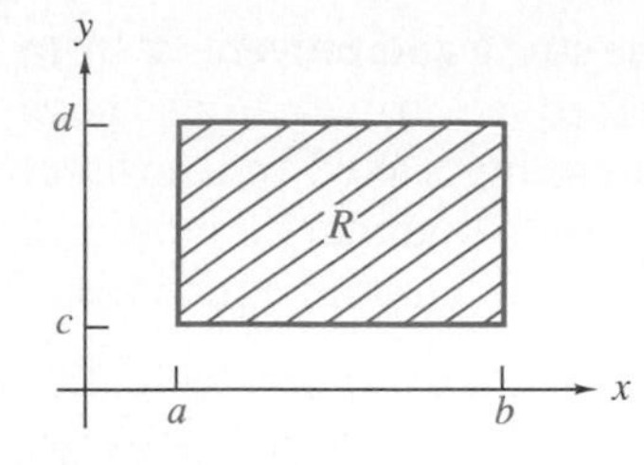

FIGURE 10.36
A two-dimensional insulated rectangular region R.

If k denotes the thermal diffusivity of the plate material (as discussed in Section 10.1), then it can be shown that the temperature function $u(x, y, t)$ satisfies the **two-dimensional heat equation**

$$\frac{\partial^2 u}{\partial x^2} + \frac{\partial^2 u}{\partial y^2} = \frac{1}{k}\frac{\partial u}{\partial t}. \tag{1}$$

The sum of the two second partial derivatives on the lefthand side of Eq. (1) is called the **Laplacian** of the function u and is denoted by

$$\nabla^2 u = \frac{\partial^2 u}{\partial x^2} + \frac{\partial^2 u}{\partial y^2}. \tag{2}$$

The symbol $\nabla^2 u$ is read "del squared of u."* Thus we can rewrite the heat equation as

$$\nabla^2 u = \frac{1}{k} u_t. \tag{3}$$

Comparing Eq. (3) with the one-dimensional heat equation $u_{xx} = (1/k)u_t$ we can easily see that in passing from one to two dimensions the second-order spatial derivative u_{xx} is replaced by the Laplacian $\nabla^2 u$. Similarly, if we think of the region R as a two-dimensional flexible membrane that is free to vibrate in the direction parallel to the z-axis (perpendicular to the plane of R), then the position function $z = z(x, y, t)$ satisfies the **two-dimensional wave equation**

$$a^2 \nabla^2 z = z_{tt}. \tag{4}$$

Observe that Eq. (4) has the same relation to one-dimensional wave Eq. (8) in Section 10.5 as two-dimensional heat Eq. (3) has to the one-dimensional heat equation in Section 10.1.

A question of particular interest in studying the two-dimensional heat-diffusion problem is that of *steady-state heat flow.* If a steady-state exists, then temperature u is a function of position (x, y) only and the time derivative u_t is zero. Therefore the two-dimensional steady-state temperature satisfies the equation

$$\nabla^2 u = 0 \tag{5}$$

or

$$u_{xx} + u_{yy} = 0. \tag{6}$$

Equation (5) is called **Laplace's equation.** In addition to the problem of steady-state heat flow, Laplace's equation occurs in physical and mathemati-

* The del operator $\nabla = \partial/\partial x\,\mathbf{i} + \partial/\partial y\,\mathbf{j}$ in vector analysis is applied to the scalar function u to produce the gradient vector $\nabla u = \partial u/\partial x\,\mathbf{i} + \partial u/\partial y\,\mathbf{j}$. James Clerk Maxwell (1831–1879) pointed out that $\nabla^2 = \nabla \cdot \nabla$ produces the scalar function in Eq. (2), and he called it the Laplacian, after the French mathematician Pierre Simon Laplace.

cal problems of hydrodynamics, gravitational attraction, elasticity, and certain motions of incompressible fluids.

Boundary Conditions

Because there is no time dependence in Laplace's Eq. (5), no initial conditions are required to be satisfied by the solutions $u(x, y)$. Nevertheless, certain boundary conditions on the boundary of the region R must be satisfied: *one condition must be specified at each point along the boundary.* A common way of specifying the boundary condition at a point on the boundary is to prescribe the value of the function u at that point. In the case of the steady-state heat equation this amounts to specifying the temperature $u(x, y)$ at each point (x, y) on the boundary. Sometimes the directional derivative of u normal to the boundary is specified instead. Some combination of these prescriptions, or mixed boundary conditions, might also be specified in a given situation: temperature values at some boundary points and derivatives at other points. The problem of finding the solution to Laplace's equation when the boundary conditions prescribe certain values for u is known as a **Dirichlet problem.*** The Dirichlet problem for a rectangle, as illustrated in Fig. 10.37, would be given as follows.

DIRICHLET PROBLEM FOR A RECTANGLE

PDE: $u_{xx} + u_{yy} = 0;$

BC: $u(x, 0) = f_1(x), \quad u(x, b) = f_2(x), \quad 0 < x < a;$

$$u(0, y) = g_1(y), \quad u(a, y) = g_2(y), \quad 0 < y < b. \tag{7}$$

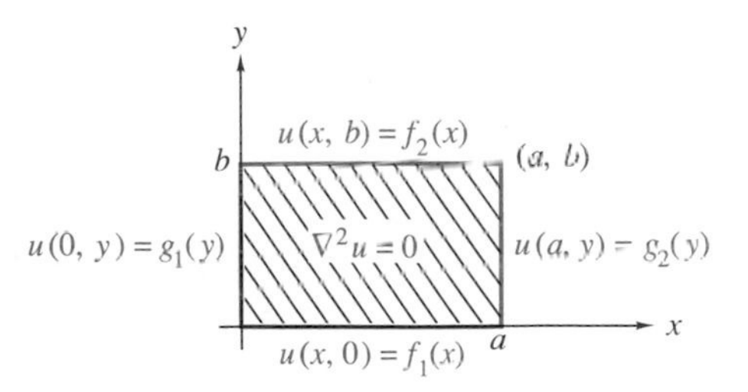

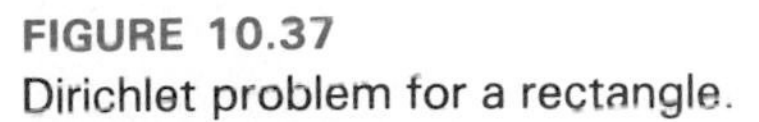

FIGURE 10.37
Dirichlet problem for a rectangle.

* P. G. L. Dirichlet (1805–1859) was a German mathematician. He made numerous important contributions to applied mathematics and is especially known for his work in differential equations.

Solving the Dirichlet Problem for a Rectangle

To solve the Dirichlet problem for a rectangle we divide the problem into four subproblems, solve each subproblem, and then apply the superposition principle by adding their solutions together to form $u(x, y)$. The four subproblems are defined by specifying the boundary condition on three sides of the rectangle to be zero while retaining the boundary condition on the fourth side as specified in Fig. 10.37. For instance,

$$u(x, 0) = u(0, y) = u(x, b) = 0 \qquad \text{and} \qquad u(a, y) = g_2(y). \tag{8}$$

Let us solve Laplace's equation subject to boundary conditions (8). Following the separation-of-variables technique, we assume the solution form

$$u(x, y) = X(x)Y(y). \tag{9}$$

Substitution into Laplace's equation (6) yields

$$\frac{X''}{X} = -\frac{Y''}{Y} = \lambda \tag{10}$$

where λ is the separation constant, as before. Hence we obtain the two second-order ordinary differential equations

$$X'' - \lambda X = 0, \tag{11}$$

$$Y'' + \lambda Y = 0. \tag{12}$$

Substitution of homogeneous boundary conditions (8) into Eq. (9) gives us

$$X(0) = 0 \tag{13}$$

and

$$Y(0) = Y(b) = 0. \tag{14}$$

The solution results are summarized as follows, leaving the details of the solution as an exercise:

1. The eigenvalues associated with Eq. (12) are

$$\lambda = \left(\frac{n\pi}{b}\right)^2, \qquad n = 1, 2, 3, \ldots$$

The corresponding eigenfunctions that satisfy boundary conditions (14) are

$$Y_n(y) = \sin \frac{n\pi y}{b}. \tag{15}$$

2. The solution of Eq. (11) for each eigenvalue λ is

$$X_n(x) = A_n \cosh \frac{n\pi x}{b} + B_n \sinh \frac{n\pi x}{b}.$$

Applying boundary condition (13) gives us $A_n = 0$ and

$$X_n(x) = B_n \sinh \frac{n\pi x}{b}. \tag{16}$$

3. The superposition principle gives us

$$u(x, y) = \sum_{n=1}^{\infty} B_n \sinh \frac{n\pi x}{b} \sin \frac{n\pi y}{b}. \tag{17}$$

4. The B_n coefficients are determined from the boundary condition

$$u(a, y) = \sum_{n=1}^{\infty} B_n \sinh \frac{n\pi a}{b} \sin \frac{n\pi y}{b} = g_2(y).$$

Therefore the quantity $B_n \sinh(n\pi a/b)$ is the Fourier sine coefficient of period $2b$ for the function $g_2(y)$:

$$B_n \sinh \frac{n\pi a}{b} = \frac{2}{b} \int_0^b g_2(y) \sin \frac{n\pi y}{b}\, dy. \tag{18}$$

In conclusion, the solution to Laplace's equation on a rectangle with boundary conditions (8) is the function $u(x, y)$ defined by Eq. (17) where the coefficients B_n satisfy Eq. (18).

EXERCISES 10.6

In problems 1–10 solve Laplace's equation for a rectangle $0 < x < a$, $0 < y < b$, subject to the specified boundary condition.

1. $u(x, 0) = u(0, y) = u(x, b) = 0, \quad u(a, y) = 10$

2. $u(x, 0) = u(0, y) = u(x, \pi) = 0, \quad u(\pi, y) = \cos y$, $a = b = \pi$

3. $u(x, 0) = u(0, y) = u(x, \pi) = 0, \quad u(\pi, y) = \sin y$, $a = b = \pi$

4. $u(x, 0) = u(0, y) = u(x, 1) = 0, \quad u(1, y) = -y$, $a = b = 1$

5. $u(x, 0) = u(0, y) = u(a, y) = 0, \quad u(x, b) = f_2(x)$

6. $u(x, 0) = u(0, y) = u(\pi, y) = 0, \quad u(x, \pi) = \cos x$, $a = b = \pi$

7. $u(x, 0) = u(0, y) = u(\pi, y) = 0, \quad u(x, \pi) = \sin x$, $a = b = \pi$

8. $u(x, 0) = u(0, y) = u(1, y) = 0, \quad u(x, 1) = x$, $a = b = 1$

9. $u(x, 0) = u(x, b) = u(a, y) = 0, \quad u(0, y) = g_1(y)$

10. $u(x, 0) = u(x, \pi) = u(\pi, y) = 0, \quad u(0, y) = \cos^2 y$, $a = b = \pi$

10.7

NUMERICAL SOLUTION TO THE ONE-DIMENSIONAL WAVE EQUATION (Optional)

In Section 10.4 we introduced the idea of numerical solutions to partial differential equations. Recall that a numerical solution of a partial differential equation involving two independent variables x and t consists of a two-dimensional table of values for the dependent variable y. If the independent variables are interpreted as position x and time t, then the first row of the table consists of the values of the initial-value function $y(x, 0)$ at selected position values x; the next row consists of the approximate values of the solution to the partial differential equation at the first time step, the next row consists of the approximations at the second time step, and so on. Moreover, the left column of the table contains the left-end boundary values $y(0, t)$ and the right column the right-end boundary values $y(L, t)$. The configuration for the numerical solution table is displayed below.

Numerical Solution Table for a Partial Differential Equation

	Position					
Time	**0**	x_1	x_2	x_3	. . .	L
0	$y(0, 0)$		. . . initial condition . . .			$y(L, 0)$
t_1	0	·	·	·	·	0
t_2	0	·	·	·	·	0
t_3	0	·	·	·	·	0
⋮	⋮	⋮	⋮	⋮	⋮	⋮
	0	·	·	·	·	0
	↑ BC $y(0, t)$					↑ BC $y(L, t)$

Next we investigate how to obtain numerical approximations for the interior values of the table in the case of the one-dimensional wave equation.

Numerical Formulation of the Wave Equation

Consider the partial differential equation modeling the vibrating string

$$a^2 y_{xx} = y_{tt}, \qquad 0 < x < L, \quad t > 0. \tag{1}$$

The model involves two second derivatives, each of which can be approximated by a finite divided difference.

As with the heat equation, we begin by subdividing the position interval $0 < x < L$ into n subdivisions of equal length $\Delta x = L/n$. We label the equally spaced subdivision points:

$$x_0 = 0, x_1 = x_0 + \Delta x, x_2 = x_1 + \Delta x, \ldots, x_n = x_{n-1} + \Delta x = L. \tag{2}$$

Next we choose the time increment Δt and label the time stages:

$$t_0 = 0, t_1 = t_0 + \Delta t, t_2 = t_1 + \Delta t, \ldots. \tag{3}$$

These choices for Δx and Δt are not independent if the numerical approximations are going to converge to the analytical solution of the vibrating string. We will present the condition for convergence further on. As before, we denote the grid-point solutions $y(x_i, t_j)$ by $y_{i,j}$ for simplicity. Using this notation the two second derivatives in Eq. (1) can be approximated by

$$y_{xx}(x_i, t_j) = \frac{y_{i-1,j} - 2y_{i,j} + y_{i+1,j}}{(\Delta x)^2} \tag{4}$$

and

$$y_{tt}(x_i, t_j) = \frac{y_{i,j-1} - 2y_{i,j} + y_{i,j+1}}{(\Delta t)^2}. \tag{5}$$

(See problem 11 in Exercises 10.4.) Substitution of these approximations into Eq. (1) yields

$$\left(\frac{a\Delta t}{\Delta x}\right)^2 [y_{i-1,j} - 2y_{i,j} + y_{i+1,j}] = y_{i,j-1} - 2y_{i,j} + y_{i,j+1}.$$

After setting $r = (a\Delta t/\Delta x)^2$ and solving the previous equation for $y_{i,j+1}$ we have the **PDE finite difference formula for the wave equation:**

$$y_{i,j+1} = [ry_{i-1,j} + 2(1-r)y_{i,j} + ry_{i+1,j}] - y_{i,j-1}. \tag{6}$$

Using the Finite Difference Formula for the Wave Equation

Let us interpret formula (6). Recall that the subscript i in Eq. (6) refers to the *position* or *column* number of the table and that j refers to the *time* or *row* number of the table. Thus the lefthand side of Eq. (6) gives the values of the entries in the $j + 1$st row of the table for all i columns (except the first and last columns, which we already know from the boundary conditions). The righthand side of Eq. (6) shows how to calculate the entry in the ith column of the new row. We must use the entries from previous known rows as illustrated in the configuration displayed in Fig. 10.38.

		Column i ↓	
Known row $j-1$ ⟶	·	$y_{i,j-1}$	·
Known row j ⟶	$y_{i-1,j}$	$y_{i,j}$	$y_{i+1,j}$
Unknown row $j+1$ ⟶	·	$y_{i,j+1}$	·

FIGURE 10.38
Grid entries involved in finite difference formula (6) for the wave equation.

To illustrate the calculation in Eq. (6), suppose $r = 0.25$ and we have the following configuration of entries:

Time row $j-1$ ⟶	·	0.200	·
Time row j ⟶	0.263	0.188	0
Time row $j+1$ ⟶	·	*	·

Then the value of the position marked * is computed as

$$0.25(0.263) + 2(1 - 0.25)(0.188) + 0.25(0) - 0.200 = 0.148.$$

More completely, suppose the entire *two* previously known rows are as follows:

	0	x_1	x_2	x_3	x_4	L
Time t_{j-1}	0	0.200	0.350	0.350	0.200	0
Time t_j	0	0.188	0.263	0.263	0.188	0

Then for $r = 0.25$ the next row in the table is

	0	x_1	x_2	x_3	x_4	L
Time t_{j+1}	0	0.148	0.157	0.157	0.148	0

Calculating the Second Row in the Numerical Solution Table

In using finite difference formula (6) for the wave equation we have observed that two previously known rows are required (see Fig. 10.38). This requirement creates a problem at the beginning: we only know the very first row from the initial condition $y(x, 0) = f(x)$, and to obtain the second row corresponding to $y(x_i, t_1)$, we cannot use formula (6). However, we have not used the *second* initial condition $y_t(x, 0) = g(x)$. From finite differences,

$$y_t(x_i, t_1) = \frac{y_{i,1} - y_{i,0}}{\Delta t}. \tag{7}$$

Therefore substitution of $y_t(x_i, t_1) = g(x_i)$ into Eq. (7) gives us

$$g(x_i) = \frac{y_{i,1} - y_{i,0}}{\Delta t}$$

or

$$y_{i,1} = y_{i,0} + g(x_i)\Delta t. \tag{8}$$

Formula (8) provides the method for calculating the second row of the solution table from the first row values $y_{i,0}$ and the known initial-condition function $g(x)$. For example, if the first row is given by

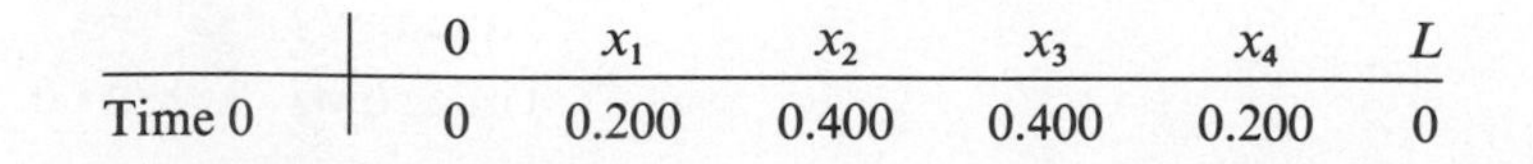

	0	x_1	x_2	x_3	x_4	L
Time 0	0	0.200	0.400	0.400	0.200	0

and if $y_t(x, 0) = g(x) = x$, then for $\Delta t = 0.1$ the second row is given by

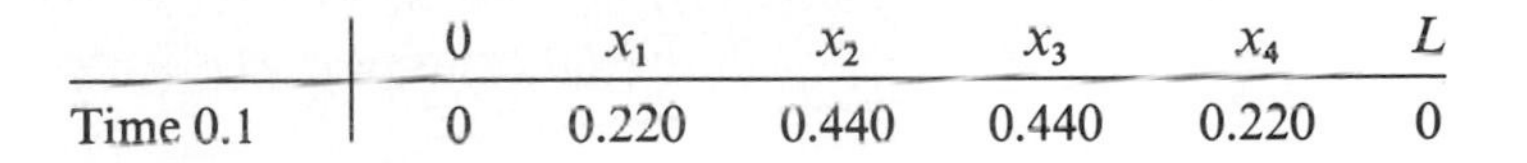

	0	x_1	x_2	x_3	x_4	L
Time 0.1	0	0.220	0.440	0.440	0.220	0

Once the first two rows of the numerical solution table have been entered, we can proceed to calculate the remaining interior values using finite difference formula (6) as discussed before.

Convergence of the Numerical Solutions

The numerical method we have just described for solving the wave equation does not necessarily produce numerical solutions that converge. Appropriate choices have to be made for the increments Δx and Δt that determine the value of the constant multiplier $r = (a\Delta t/\Delta x)^2$. We state the following result without proof.

THEOREM 10.3

If $0 < r \leq 1$ for the multiplier $r = (a\Delta t/\Delta x)^2$, then the $y_{i,j}$ approximations converge to the solution $y(x, t)$ of the wave equation.

Let us summarize the numerical method.

SUMMARY OF THE NUMERICAL SOLUTION METHOD TO SOLVE THE WAVE EQUATION WITH FIXED ENDS

Step 1. Subdivide the position interval $0 < x < L$ and choose the time increment Δt so that $r = (a\Delta t/\Delta x)^2$ satisfies $0 < r \leq 1$. Then $\Delta x = L/n$ and Δt yield the grid points

$$x_0 = 0,\ x_1 = x_0 + \Delta x,\ x_2 + \Delta x,\ \ldots,$$
$$x_n = x_{n-1} + \Delta x = L$$

$$t_0 = 0,\ \ t_1 = t_0 + \Delta t,\ t_2 = t_1 + \Delta t,\ \ldots,$$
$$t_j = t_{j-1} + \Delta t,\ \ldots.$$

Step 2. Evaluate the initial-condition function $y(x, 0) = f(x)$ at each subdivision point:

$$f(x_0), f(x_1), f(x_2), \ldots, f(x_{n-1}), f(x_n).$$

These values form the first row of the numerical solution table.

Step 3. Calculate the second row of the solution table using the second initial condition $y_t(x, 0) = g(x)$ according to the formula

$$y_{i,1} = y_{i,0} + g(x_i)\Delta t, \qquad i = 1, 2, \ldots, n-1.$$

Remember that $y_{i,0} = f(x_i)$ from Step 2.

Step 4. Set the first column and the last column of the solution table to the fixed homogeneous boundary-condition values:

$$y(0, t_j) = y(L, t_j) = 0$$

for every time row t_j.

Step 5. Calculate and record in the table the interior numerical solution approximations, row by complete row in succession, beginning with the third row, according to the formula

$$y_{i,j+1} = [ry_{i-1,j} + 2(1-r)y_{i,j} + ry_{i+1,j}] - y_{i,j-1}.$$

EXAMPLE 1 Solve numerically the following wave equation model with fixed homogeneous boundary conditions until time $t = 1.0$:

$$\begin{aligned} &\text{PDE:} \quad 4y_{xx} = y_{tt}, && 0 < x < 1, \quad t > 0; \\ &\text{BC:} \quad y(0, t) = y(1, t) = 0, && t > 0; \\ &\text{IC:} \quad y(x, 0) = \sin \pi x, && \\ &\qquad\; y_t(x, 0) = 0, && 0 \leq x \leq 1. \end{aligned}$$

Solution. The analytical solution to this model turns out to be

$$y(x, t) = \sin(\pi x) \cos(2\pi t).$$

Note that the solution repeats itself at all integer values of t. Let us calculate a numerical solution.

Step 1 We subdivide the interval $0 < x < 1$ into 10 subdivisions of equal length $\Delta x = 0.10$. Choose $\Delta t = 0.05$ for the time increment. Then $r = (a\Delta t/\Delta x)^2 = [2(0.05)/0.1]^2 = 1$ satisfies the convergence criterion.

Step 2 Evaluate the initial-value function $y(x, 0) = \sin(\pi x)$ to obtain the following first row of the solution table:

	Position Interval x										
Time Interval t	0	0.1	0.2	0.3	0.4	0.5	0.6	0.7	0.8	0.9	1
0	0	0.309	0.588	0.809	0.951	1.00	0.951	0.809	0.588	0.309	0

Step 3 To calculate the second row we observe that $y_t(x, 0) = g(x) = 0$ so that $y_{i,1} = y_{i,0}$. Thus the second row is simply a copy of the first.

Time Interval t	Position Interval x										
	0	0.1	0.2	0.3	0.4	0.5	0.6	0.7	0.8	0.9	1
0	0	0.309	0.588	0.809	0.951	1.00	0.951	0.809	0.588	0.309	0
0.05	0	0.309	0.588	0.809	0.951	1.00	0.951	0.809	0.588	0.309	0

Step 4 We enter the fixed homogeneous boundary conditions $y(0, t) = y(1, t) = 0$ throughout the first and last columns.

Time Interval t	Position Interval x										
	0	0.1	0.2	0.3	0.4	0.5	0.6	0.7	0.8	0.9	1
0	0	0.309	0.588	0.809	0.951	1.00	0.951	0.809	0.588	0.309	0
0.05	0	0.309	0.588	0.809	0.951	1.00	0.951	0.809	0.588	0.309	0
0.10	0										0
0.15	0										0
0.20	0										0
0.25	0										0
⋮	⋮										⋮

Step 5 For $r = 1$ finite difference formula (6) for the wave equation becomes

$$y_{i,j+1} = y_{i-1,j} + y_{i+1,j} - y_{i,j-1}.$$

Using this relation, we obtain the following numerical solution. Note that it is indeed periodic as expected from the analytical solution.

Time Interval t	Position Interval x										
	0	0.1	0.2	0.3	0.4	0.5	0.6	0.7	0.8	0.9	1
0	0	0.309	0.588	0.809	0.951	1.00	0.951	0.809	0.588	0.309	0
0.05	0	0.309	0.588	0.809	0.951	1.00	0.951	0.809	0.588	0.309	0
0.10	0	0.279	0.530	0.730	0.858	0.902	0.858	0.730	0.530	0.279	0
0.15	0	0.221	0.421	0.579	0.681	0.716	0.681	0.579	0.421	0.221	0
0.20	0	0.142	0.270	0.372	0.437	0.460	0.437	0.372	0.270	0.142	0
0.25	0	0.049	0.093	0.128	0.151	0.158	0.151	0.128	0.093	0.049	0
0.30	0	−0.049	−0.093	−0.128	−0.151	−0.158	−0.151	−0.128	−0.093	−0.049	0
0.35	0	−0.142	−0.270	−0.372	−0.437	−0.460	−0.437	−0.372	−0.270	−0.142	0
0.40	0	−0.221	−0.421	−0.579	−0.681	−0.716	−0.681	−0.579	−0.421	−0.221	0
0.45	0	−0.279	−0.530	−0.730	−0.858	−0.902	−0.858	−0.730	−0.530	−0.279	0
0.50	0	−0.309	−0.588	−0.809	−0.951	−1.00	−0.951	−0.809	−0.588	−0.309	0
0.55	0	−0.309	−0.588	−0.809	−0.951	−1.00	−0.951	−0.809	−0.588	−0.309	0
0.60	0	−0.279	−0.530	−0.730	−0.858	−0.902	−0.858	−0.730	−0.530	−0.279	0
0.65	0	−0.221	−0.421	−0.579	−0.681	−0.716	−0.681	−0.579	−0.421	−0.221	0
0.70	0	−0.142	−0.270	−0.372	−0.437	−0.460	−0.437	−0.372	−0.270	−0.142	0
0.75	0	−0.049	−0.093	−0.128	−0.151	−0.158	−0.151	−0.128	−0.093	−0.049	0
0.80	0	0.049	0.093	0.128	0.151	0.158	0.151	0.128	0.093	0.049	0
0.85	0	0.142	0.270	0.372	0.437	0.460	0.437	0.372	0.270	0.142	0
0.90	0	0.221	0.421	0.579	0.681	0.716	0.681	0.579	0.421	0.221	0
0.95	0	0.279	0.530	0.730	0.858	0.902	0.858	0.730	0.530	0.279	0
1.00	0	0.309	0.588	0.809	0.951	1.00	0.951	0.809	0.588	0.309	0

EXERCISES 10.7

In problems 1–5 solve numerically the wave equation models with the given fixed homogeneous boundary conditions until $t = 0.5$. Express your results in tabular form as illustrated in Example 1. Check that $0 < r \leq 1$ to ensure convergence.

1. $y_{xx} = y_{tt}$, $y(0, t) = y(1, t) = 0$, $y(x, 0) = x$, $y_t(x, 0) = x^2$, $0 < x < 1$

2. $4y_{xx} = y_{tt}$, $y(0, t) = y(1, t) = 0$, $y(x, 0) = 0$, $y_t(x, 0) = x$, $0 < x < 1$

3. $4y_{xx} = y_{tt}$, $y(0, t) = y(\pi, t) = 0$, $y(x, 0) = \sin x$, $y_t(x, 0) = 1$, $0 < x < \pi$

4. $y_{xx} = y_{tt}$, $y(0, t) = y(\pi, t) = 0$, $y(x, 0) = (2x/\pi)(\pi - x)$, $y_t(x, 0) = 0$, $0 < x < \pi$

5. $y_{xx} = y_{tt}$, $y(0, t) = y(\pi, t) = 0$, $y(x, 0) = \sin x$, $y_t(x, 0) = \sin x$, $0 < x < \pi$

CHAPTER 10 REVIEW EXERCISES

In problems 1–7, use the method of separation of variables to find all product solutions to the following partial differential equations. Solve the equations subject to the specified boundary conditions and initial conditions, if any.

1. $4\dfrac{\partial^2 y}{dx^2} = \dfrac{\partial^2 y}{dt^2}$

$y(0, t) = 0, \quad 0 \leq t < \infty$

$y(3, t) = 0, \quad 0 \leq t < \infty$

$$y(x, 0) = \begin{cases} x, & 0 \leq x < 1 \\ 1, & 1 \leq x < 2 \\ 3 - x, & 2 \leq x \leq 3 \end{cases}$$

$\dfrac{\partial y}{\partial t}(x, 0) = 0, \quad 0 \leq x \leq 3$

2. $u_t = 2u_{xx}, \quad 0 \leq t < \infty$

$u(0, t) = u(\pi, t) = 0$

$u(x, 0) = 2, \quad 0 \leq x < \pi$

3. $\dfrac{\partial^2 y}{\partial x^2} = \dfrac{\partial^2 y}{\partial t^2}$

$y(0, t) = 0, \quad 0 \leq t < \infty$

$y(\pi, t) = 0, \quad 0 \leq t < \infty$

$y(x, 0) = \sin 2x, \quad 0 \leq x \leq \pi$

$\dfrac{\partial y}{\partial t}(x, 0) = 0, \quad 0 \leq x \leq \pi$

4. $\dfrac{\partial^2 u}{\partial x^2} + \dfrac{\partial^2 u}{\partial y^2} = 0$

5. $u_t = 16u_{xx}, \quad 0 \leq x \leq \pi, \quad t > 0$

$u(0, t) = 0$

$u(\pi, t) = 0$

$$u(x, 0) = \begin{cases} 4, & 0 \leq x < \dfrac{\pi}{2} \\ 0, & \dfrac{\pi}{2} \leq x \leq \pi \end{cases}$$

6. $4\dfrac{\partial^2 u}{\partial x^2} = \dfrac{\partial u}{\partial t}$

$u(0, t) = 0, \quad 0 \leq t < \infty$

$u(10, t) = 0, \quad 0 \leq t < \infty$

$u(x, 0) = 50, \quad 0 \leq x \leq 10$

7. $w_{xx} = \dfrac{1}{2} w_t, \quad 0 < x < 3, \quad t > 0$

$w(0, t) = w(3, t) = 0$

$w(x, 0) = 30$

8. Suppose all the assumptions for the vibrating string presented in Section 10.5 are satisfied. Find the equation $y(x, t)$ that predicts the vertical displacement of the string by solving the wave equation

$$2500\, y_{xx} = y_{tt}, \quad 0 \leq x \leq 1, \quad t > 0$$

with conditions

$y(0, t) = y(1, t) = 0$

$$y(x, 0) = \begin{cases} \frac{x}{50}, & 0 \le x < \frac{1}{2} \\ -\frac{x}{50} + \frac{1}{50}, & \frac{1}{2} \le x \le 1 \end{cases}$$

$y_t(x, 0) = 0.$

9. Solve the wave equation in problem 8 with the new initial displacement function $y(x, 0)$ shown in Fig. 10.39.

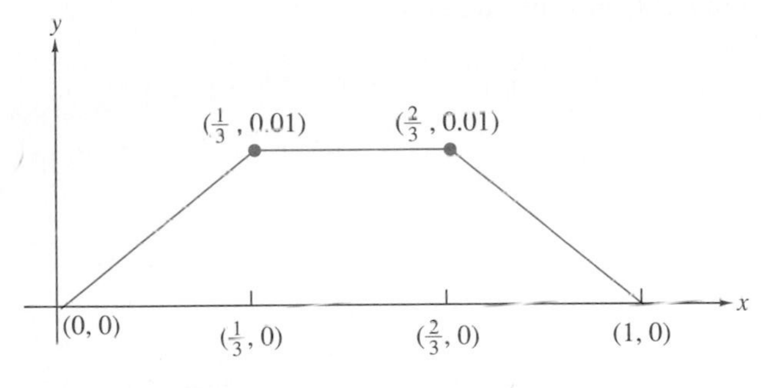

FIGURE 10.39

10. A one-dimensional uniform homogeneous bar of length 3 is oriented along the x-axis and is insulated on its lateral sides as well as on its ends. Let the temperature of the bar at any time t be $u(x, t)$. Assuming heat flows only in the x-direction and that the initial temperature is given by

$$u(x, 0) = \begin{cases} 2, & 0 \le x < 2 \\ 4, & 2 \le x \le 3 \end{cases}$$

find the temperature function $u(x, t)$ if the thermal diffusivity of the bar is $k = 4$.

11. A homogeneous iron rod 50 cm long is immersed in steam until its temperature reaches 100°C throughout. At time $t = 0$, its lateral surface is insulated and its two ends embedded in ice at 0°C. The thermal diffusivity of iron is 0.15 cm²/sec. Find the temperature function $u(x, t)$.

12. A bar whose surface is insulated has a length of 4 units and a diffusivity of 1 unit. Its ends are kept at temperature zero units. Its initial internal temperature is 30°C, uniform throughout the interior of the bar. Find the temperature function $u(x, t)$.

13. Suppose the vibrating string discussed in Section 10.5 is subjected to a damping force that is proportional at each instant in time to the velocity of the string at each point. Such a damping force might result from vibrations in a viscous fluid. The force resulting from the damping on the infinitesimal string segment Δs is derived as follows.

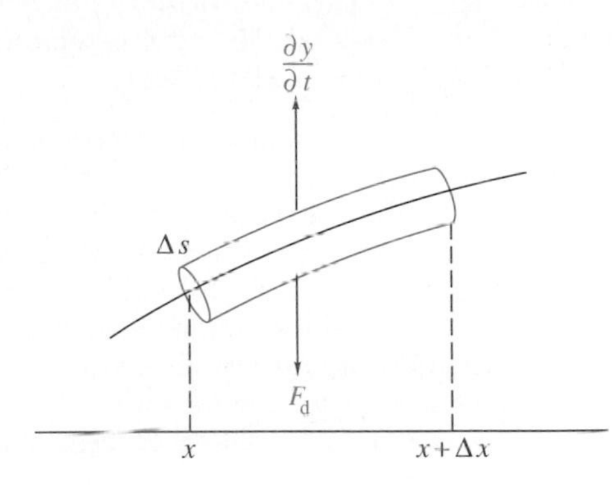

FIGURE 10.40

By making the assumption on the short segment Δs that the velocities of each point are nearly equal we find that the resultant damping force F_d on the segment Δs is proportional to both the velocity of Δs and length of Δs (the damping force on Δs increases as Δs gets longer and decreases as Δs gets shorter). Therefore F_d is proportional to

$$(\text{velocity}) \times (\text{length}) = \frac{\partial y}{\partial t} \Delta s$$

or

$$F_d = -K \frac{\partial y}{\partial t} \Delta s.$$

a) Assuming that the only forces acting on segment Δs are due to tension and damping, show that the partial differential equation that models this new situation is

$$\frac{\partial^2 y}{\partial t^2} + c \frac{\partial y}{\partial t} = a^2 \frac{\partial^2 y}{\partial x^2}$$

where $c = K/\rho$ and $a^2 = T/\rho$.

b) Suppose that we subject a vibrating string to a damping force that is proportional to the instantaneous velocity at each point. Find the equation of the displacement of each point, $y(x, t)$, by solving the following initial boundary value problem using the method of separation of variables:

$$\frac{\partial^2 y}{\partial t^2} + K \frac{\partial y}{\partial t} = a^2 \frac{\partial^2 y}{\partial x^2}, \quad 0 < x < L, \quad t > 0,$$

$$y(0, t) = 0, \quad y(L, t) = 0,$$

$$y(x, 0) = \begin{cases} bx, & 0 \le x < \frac{L}{2} \\ b(L - x), & \frac{L}{2} \le x < L, \end{cases}$$

$$\frac{\partial y}{\partial t}(x, 0) = 0, \quad 0 < x < L.$$

Assume that $K^2 - 4a^2\lambda^2 < 0$ and b is small compared with L. Solve for all coefficients.

14. A long one-dimensional rod of length L is of uniform cross section and homogeneous material. The rod is insulated along its entire length. The rod is made with a material that, by a chemical reaction, absorbs heat. This same chemical reaction is employed in coldpacks that are used for treating injuries such as sprains or internal bleeding. The chemical reaction absorbs heat in proportion to the temperature of the reaction at each point in the rod.

Using the principle of conservation of heat energy, which states

$$\text{Heat}_{\text{IN}} + \text{Heat}_{\text{GENERATED}} = \text{Heat}_{\text{OUT}} + \text{Heat}_{\text{STORED}},$$

we can construct a mathematical model that can be used to predict the temperature at each point in the rod at any time t. The following partial differential equation models the heat conduction problem described:

$$\frac{\partial^2 T}{\partial x^2} = \frac{\partial T}{\partial t} + aT$$

where a is the constant of proportionality for the heat of reaction of the chemical reaction that in this case is absorbing heat; T is temperature in degrees Centigrade. Find the equation that predicts the temperature at each point in the rod by solving the following initial boundary value problem using separation of variables.

$$\frac{\partial^2 T}{\partial x^2} = \frac{\partial T}{\partial t} + aT, \quad 0 < x < L, \quad t > 0,$$

$$T(0, t) = \frac{\partial T}{\partial x}(L, t) = 0,$$

$$T(x, 0) = 100°\text{C}.$$

Solve for all coefficients.

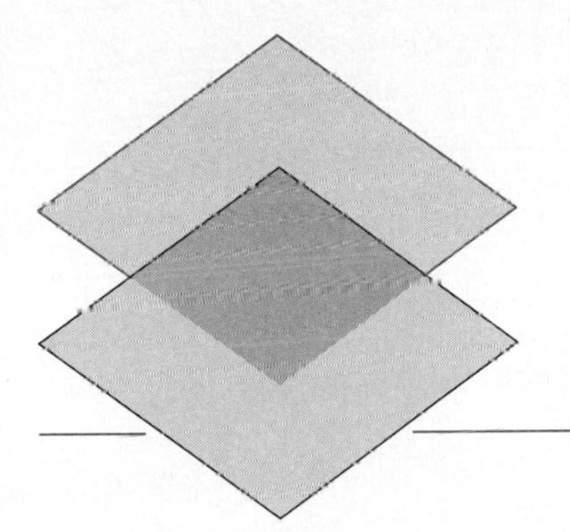

Appendixes

BRIEF REVIEW OF INTEGRATION TECHNIQUES

u-Substitution

The basic idea underlying u-substitution is to perform a simple substitution that converts the integral into a recognizable form ready for immediate integration. For example, given

$$\int \frac{\cos x}{1 + \sin x}\, dx,$$

let $u = 1 + \sin x$ and differentiate to find $du = \cos x\, dx$. Substitution then yields

$$\int \frac{\cos x}{1 + \sin x}\, dx = \int \frac{du}{u} = \ln|u| + C.$$

Substituting for u again in this last expression gives you

$$\int \frac{\cos x}{1 + \sin x}\, dx = \ln|1 + \sin x| + C.$$

Integration by Parts

Recall from calculus that

$$\int u\, dv = uv - \int v\, du.$$

In some cases it is necessary to apply the procedure several times before a form is obtained that can easily be integrated. In these and other situations it is useful to use the tabular method as follows:

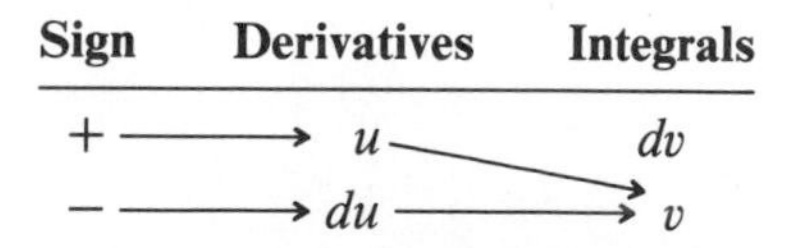

Diagonal arrows in the table indicate terms to be multiplied (uv in this case). The bottom row in the table has horizontal arrows to indicate the final integral to be evaluated ($\int v\, du$ in the above case). Finally, the sign column is associated with the differentiated term at each stage, beginning with a plus sign and alternating with the minus sign, as suggested by the table format.

Thus the table above would be read as follows:

$$\underbrace{\int u\,dv}_{\text{top row}} = \overbrace{\underbrace{+uv}_{\substack{\text{diagonal}\\ \text{arrow}}} - }^{\text{signs}} \underbrace{\int v\,du}_{\substack{\text{horizontal}\\ \text{arrow}}}$$

To apply integration by parts successively, build the table by repeatedly differentiating the derivatives (middle) column and integrating the integrals (right) column, while the sign (left) column alternates. Terminate the table with a horizontal arrow between the middle and right columns when you can readily integrate the product of the functions in the last row or when the last row simply repeats the terms in the first row (up to a multiplicative constant). Let us consider several examples.

EXAMPLE 1 Find the integral $\int xe^x\,dx$ by the tabular method.

Solution. We set up the table as follows:

Sign	Derivatives	Integrals
$+$	x	e^x
$-$	1	e^x
$+$	0	e^x

Interpreting the table, we get

$$\int xe^x\,dx = +xe^x - 1\cdot e^x + \int 0\cdot e^x\,dx + C$$
$$= (x-1)e^x + C.$$

EXAMPLE 2 Integrate $\int x^2e^{2x}\,dx$ by the tabular method.

Solution. We set up the table as before:

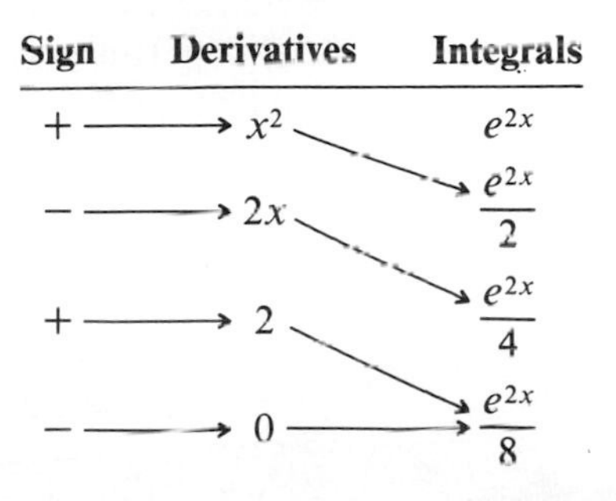

Sign	Derivatives	Integrals
$+$	x^2	e^{2x}
$-$	$2x$	$\frac{e^{2x}}{2}$
$+$	2	$\frac{e^{2x}}{4}$
$-$	0	$\frac{e^{2x}}{8}$

Thus

$$\int x^2e^{2x}\,dx = +\frac{x^2e^{2x}}{2} - \frac{2xe^{2x}}{4} + \frac{2e^{2x}}{8} - \int \frac{0\cdot e^{2x}}{8}\,dx + C$$

$$= \frac{e^{2x}}{4}(2x^2 - 2x + 1) + C.$$

EXAMPLE 3 Integrate $\int e^x \sin x\,dx$.

Solution. After filling in the table, we get

Sign	Derivatives	Integrals
+ ⟶	$\sin x$ ↘	e^x
− ⟶	$\cos x$ ↘	e^x
+ ⟶	$-\sin x$ ⟶	e^x

Thus

$$\int e^x \sin x\,dx = e^x \sin x - e^x \cos x + \int (-\sin x)e^x\,dx + C$$

or

$$\int e^x \sin x\,dx = \frac{e^x(\sin x - \cos x)}{2} + C_1.$$

Examples 1–3 illustrate the two basic strategies of integration by parts: (1) Choose a term to differentiate whose successive derivatives eventually become zero or repeat, and (2) continue to differentiate by parts until the integrand (up to a multiplicative constant) is repeated in the bottom row, as in Example 3. In choosing the term dv to integrate you may find the following mnemonic "detail ladder" useful:

dv
exponential
trigonometric
algebraic
inverse trigonometric
logarithmic

To use the ladder, choose the term dv to integrate in order of priority from the top to the bottom. Conversely, the term u to differentiate is chosen from bottom to top. For example, when integrating

$$\int x^2 e^x \, dx,$$

which involves a polynomial and an exponential, integrate the exponential $dv = e^x \, dx$ and differentiate the polynomial $u = x^2$. The above mnemonic device is a rule of thumb only and may not work in some cases.

Rational Functions

Given an algebraic fraction with a polynomial in both the numerator and denominator (that is, a **rational function**), division may lead to a simpler form. If the highest power in the numerator is equal to or greater than the highest power in the denominator, first perform polynomial division and then integrate the result. For example,

$$\frac{y+1}{y-1} = 1 + \frac{2}{y-1}$$

so

$$\int \frac{y+1}{y-1} \, dy = \int \left(1 + \frac{2}{y-1}\right) dy = y + 2 \ln|y-1| + C.$$

Partial Fractions

In algebra you learned to sum fractional expressions by finding a common denominator. For example,

$$\frac{2}{x-1} + \frac{4}{x+3} = \frac{2(x+3) + 4(x-1)}{(x-1)(x+3)} = \frac{6x+2}{x^2+2x-3}.$$

For purposes of integration we need to reverse this procedure. That is, given the integral

$$\int \frac{6x+2}{x^2+2x-3} \, dx$$

we use partial fraction decomposition to obtain a new expression that is readily integrable:

$$\int \left(\frac{2}{x-1} + \frac{4}{x+3}\right) dx = 2 \ln|x-1| + 4 \ln|x+3| + C.$$

This process of splitting a fraction $f(x)/g(x)$ into a sum of fractions with linear or quadratic denominators is called **partial fraction decomposition.** For the method to work the degree of the numerator $f(x)$ must always be less than the degree of the denominator $g(x)$; otherwise you must first perform polynomial long division. To use the method the denominator must be factored into linear and quadratic factors. In Examples 4–6 we review three cases that may exist for the factored denominator:

1. Distinct linear factors,
2. Repeated linear factors,
3. Quadratic factors.

EXAMPLE 4 DISTINCT LINEAR FACTORS

Find the integral $\displaystyle\int \frac{2x^2 - x + 1}{(x+1)(x-3)(x+2)}\,dx.$

Solution. We must find constants A, B, and C such that

$$\frac{2x^2 - x + 1}{(x+1)(x-3)(x+2)} = \frac{A}{x+1} + \frac{B}{x-3} + \frac{C}{x+2}. \tag{1}$$

Algebraic Method In this method you multiply through by the factored denominator to obtain

$$2x^2 - x + 1 = A(x-3)(x+2) + B(x+1)(x+2) + C(x+1)(x-3).$$

Then expand the righthand side and combine like powers of x:

$$2x^2 - x + 1 = (A + B + C)x^2 + (-A + 3B - 2C)x + (-6A + 2B - 3C).$$

Next equate the coefficients of like powers of x on both sides of this last equation. This procedure results in a system of linear algebraic equations involving our three unknowns:

$$\begin{aligned} A + B + C &= 2, \\ -A + 3B - 2C &= -1, \\ -6A + 2B - 3C &= 1. \end{aligned}$$

Solution of this system by elimination or by the method of determinants yields

$$A = -1, \qquad B = \frac{4}{5}, \qquad \text{and} \qquad C = \frac{11}{5}.$$

Thus

$$\int \frac{2x^2 - x + 1}{(x+1)(x-3)(x+2)}\,dx = -\int \frac{dx}{x+1} + \frac{4}{5}\int \frac{dx}{x-3} + \frac{11}{5}\int \frac{dx}{x+2}$$

$$= -\ln|x+1| + \frac{4}{5}\ln|x-3| + \frac{11}{5}\ln|x+2| + C.$$

Heaviside Method There is a shortcut method for finding the constants in the partial fraction decomposition of $f(x)/g(x)$. First, write the rational function with $g(x)$ completely factored into its linear terms:

$$\frac{f(x)}{g(x)} = \frac{f(x)}{(x-r_1)(x-r_2)\cdots(x-r_n)}. \tag{2}$$

To find the constant A_i associated with the term

$$\frac{A_i}{x-r_i}$$

in the partial fraction decomposition, cover the factor $x - r_i$ in the denominator of the righthand side of Eq. (2) and replace all the uncovered x's with the number r_i. For instance, to find the constant A in Eq. (1), cover the factor $x+1$ in the denominator and replace all the uncovered x's with $x = -1$:

$$A = \frac{2-(-1)+1}{\underbrace{(x+1)}_{\text{covered}}(-1-3)(-1+2)} = \frac{4}{(-4)(1)} = -1.$$

Likewise, we find B by covering the factor $x - 3$ and replacing all the uncovered x's with $x = 3$:

$$B = \frac{2(9)-3+1}{(3+1)\underbrace{(x-3)}_{\text{covered}}(3+2)} = \frac{16}{4(5)} = \frac{4}{5}.$$

Finally, C is determined when $x = -2$:

$$C = \frac{2(4)-(-2)+1}{(-2+1)(-2-3)\underbrace{(x+2)}_{\text{covered}}} = \frac{11}{(-1)(-5)} = \frac{11}{5}.$$

The integration is the same as before. We emphasize that *the Heaviside method can only be used with distinct linear factors.* In the next example we present another method for finding the constants when the linear factors are repeated. Of course, you can always resort to the more tedious algebraic method.

EXAMPLE 5 A REPEATED LINEAR FACTOR

Find the integral $\int \frac{3P}{(P+4)^2(P+1)}\,dP$.

Solution. We need to find constants A, B, and C such that

$$\frac{3P}{(P+4)^2(P+1)} = \frac{A}{P+4} + \frac{B}{(P+4)^2} + \frac{C}{P+1}$$

or

$$3P = A(P+4)(P+1) + B(P+1) + C(P+4)^2. \tag{3}$$

Substitution Method Since Eq. (3) is an identity, it holds for every value of P. Thus, to obtain three equations for finding the unknowns A, B, and C, we simply substitute convenient values for P:

$$\begin{aligned} P=-4: &\quad -12 = -3B, \\ P=-1: &\quad -3 = 9C, \\ P=\ \ 0: &\quad 0 = 4A + B + 16C, \end{aligned}$$

to give the solutions $A = 1/3$, $B = 4$, $C = -1/3$. Thus

$$\begin{aligned} \int \frac{3P}{(P+4)^2(P+1)}\,dP &= \int \left[\frac{1}{3(P+4)} + \frac{4}{(P+4)^2} - \frac{1}{3(P+1)}\right] dP \\ &= \frac{1}{3}\ln|P+4| - \frac{4}{P+4} - \frac{1}{3}\ln|P+1| + C. \end{aligned}$$

EXAMPLE 6 A QUADRATIC FACTOR

Find the integral $\int \frac{dP}{(P+1)(P^2+1)}$.

Solution. We must find constants A, B, and C such that

$$\frac{1}{(P+1)(P^2+1)} = \frac{A}{P+1} + \frac{BP+C}{P^2+1}.$$

Thus

$$1 = A(P^2+1) + (BP+C)(P+1).$$

Since this expression is to hold for all P, the coefficients of like powers of P on both sides of the equation must be equal. After collecting like powers of P on the righthand side, we get

$$0P^2 + 0P^1 + 1P^0 = (A + B)P^2 + (B + C)P + (A + C),$$

which yields the linear system

$$\begin{aligned} 0 &= A + B, \\ 0 &= \quad\; B + C, \\ 1 &= A + \quad\; C. \end{aligned}$$

The solution is $A = \frac{1}{2}$, $B = -\frac{1}{2}$, and $C = \frac{1}{2}$. Thus

$$\int \frac{dP}{(P+1)(P^2+1)} = \int \left[\frac{1}{2(P+1)} + \frac{-\frac{P}{2} + \frac{1}{2}}{P^2+1} \right] dP$$

$$= \frac{1}{2} \ln|P+1| - \frac{1}{4} \ln|P^2+1| + \frac{1}{2} \tan^{-1} P + C.$$

B

EXPONENTIAL FUNCTIONS

Let b denote a positive constant. Consider the class of functions of the form $y = b^{kx}$. Here k is an arbitrary constant and x is the independent variable. Let us find the derivative of $y = b^{kx}$. Applying the natural logarithm to both sides yields

$$\ln y = \ln b^{kx} = kx \ln b.$$

Implicit differentiation then gives us

$$\frac{1}{y}\frac{dy}{dx} = k \ln b$$

or

$$\frac{dy}{dx} = (k \ln b)y = k_1 b^{kx}.$$

If $b > 1$ and $k > 0$, then $k_1 > 0$ and the following proportionality holds:

$$\frac{d}{dx}(b^{kx}) \propto b^{kx}, \qquad b > 1. \tag{1}$$

Proportionality (1) can be expressed in words as follows:

> Each function of the form $y = b^{kx}$ for $b > 1$ has the property that its derivative is proportional to the function itself.

Note also that

$$b^{kx} = e^{(k \ln b)x} = e^{k_1 x}. \tag{2}$$

Thus

> Every function $y = b^{kx}$ is also expressible as an exponential function $y = e^{k_1 x}$ involving the number e for a suitable constant k_1.

On the other hand, suppose $y = f(x)$ is an everywhere-positive function that satisfies the relation

$$\frac{dy}{dx} = k_1 y. \tag{3}$$

Then

$$\frac{dy}{y} = k_1\, dx, \tag{4}$$

and integration of both sides of Eq. (4) gives us

$$\ln|y| = k_1 x + C.$$

Applying the exponential function to both sides of this last equation results in

$$|y| = e^{k_1 x + C} = e^C \cdot e^{k_1 x}.$$

Since $y > 0$, we then have

$$y = C_1 e^{k_1 x} \tag{5}$$

where $C_1 = e^C$. In words Eq. (5) translates to the following:

> Whenever a positive function has a derivative equal to a constant multiple of the function itself, then that function is an exponential $y = e^{k_1 x}$ (possibly multiplied by a constant).

C

SYNTHETIC DIVISION: HORNER'S ALGORITHM FOR NEW COEFFICIENTS

Let us consider the following problem for polynomials. Given

$$g(t) = a_n t^n + a_{n-1} t^{n-1} + \cdots + a_2 t^2 + a_1 t + a_0, \tag{1}$$

find the coefficients $b_n, b_{n-1}, \ldots, b_0$ so that

$$g(t) = b_n(t-a)^n + b_{n-1}(t-a)^{n-1} + \cdots + b_1(t-a) + b_0. \tag{2}$$

The method for finding the b_i coefficients, known as **Horner's algorithm,** is to divide the original polynomial (1) repeatedly by $t - a$. Each time a division is performed, a quotient and a remainder are obtained. The remainder is one of the coefficients we seek, and the quotient provides the new dividend for the next successive division by $t - a$. In this way the coefficients are found, one by one, in the order $b_0, b_1, \ldots, b_{n-1}, b_n$. The division process itself is performed using the method of synthetic division studied in elementary algebra.

EXAMPLE 1 Express $t^2 - t + 2$ in powers of $t - 2$ using Horner's algorithm.

Solution. Note that $a = 2$. First we write the coefficients of $t^2 - t + 2$, and then we perform the synthetic division process:

$$\begin{array}{ccc|l} 1 & -1 & 2 & \underline{2} \leftarrow \text{divisor is } a \\ & 2 & 2 & \\ \hline 1 & 1 & \textcircled{4} & \leftarrow \text{remainder} = b_0 \end{array}$$

(1 1: new dividend)

To execute the synthetic division indicated above, we start by bringing down the leading coefficient 1 to the third line. The 1 is multiplied by the divisor to obtain 2, and that result is written on the second line beneath the next coefficient -1. Then we add these results vertically $(-1 + 2 = 1)$, write the answer in the same column on the third line, and multiply the result by the divisor 2. Then the resulting product 2 is written on the second line beneath the next coefficient and summed to obtain the 4 on the third line. The result in the bottom right corner, in this case 4, is the final remainder in the division process and equals the first coefficient b_0 we seek. The remaining results on the third line, namely 1 and 1, are the coefficients of the new dividend polynomial. (If you perform polynomial long division to divide $t^2 - t + 2$ by $t - 2$, as studied in high school, you will readily see that the synthetic division procedure is just a shortcut to streamline the process.)

Now we repeat the division process by dividing the new dividend, corresponding to $t + 1$, by the divisor $t - 2$ in order to obtain the next coefficient b_1:

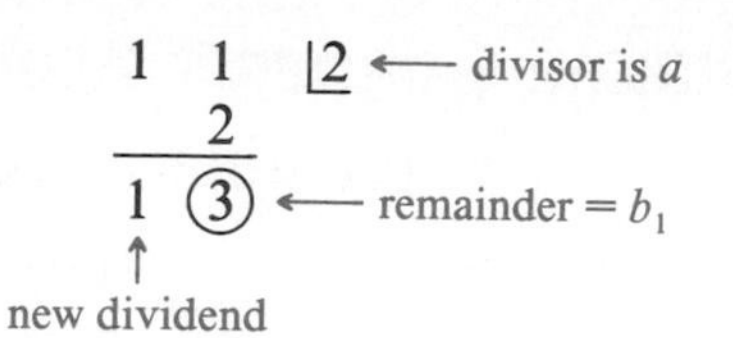

Since the new dividend is the constant term 1, the third division simply produces that constant for the remainder. (Division of a constant by $t - 2$ results in a quotient of zero and a remainder equal to the constant.) Thus $b_2 = 1$. Therefore we have

$$t^2 - t + 2 = (t - 2)^2 + 3(t - 2) + 4,$$

which is in agreement with Example 7 in Section 5.5.

Next we present an example using a higher-order polynomial.

EXAMPLE 2 Express $t^5 - 3t^3 + t^2 - 1$ in powers of $t + 2$.

Solution. First we note that $t + 2 = t - (-2)$, so $a = -2$. Next we perform the synthetic division (remembering to use zero for the coefficient of any missing power of t):

	1	0	−3	1	0	−1	⌊−2 ← always the divisor
		−2	4	−2	2	−4	
divide again by −2	1	−2	1	−1	2	(−5) $= b_0$	
		−2	8	−18	38		
	1	−4	9	−19	(40) $= b_1$		
divide again		−2	12	−42			
	1	−6	21	(−61) $= b_2$			
divide again		−2	16				
	1	−8	(37) $= b_3$				
divide again		−2					
divide again	1	(−10) $= b_4$					
	(1) $= b_5$						

Therefore

$$t^5 - 3t^3 + t^2 - 1 = (t + 2)^5 - 10(t + 2)^4 + 37(t + 2)^3 - 61(t + 2)^2 + 40(t + 2) - 5.$$

D

REVIEW OF MATRICES

A **matrix** is a rectangular array of numbers or functions. For example,

$$A = \begin{pmatrix} 1 & 0 & -1 \\ 3 & 2 & 4 \end{pmatrix} \quad \text{and} \quad \mathbf{X} = \begin{pmatrix} 2t \\ 1 - e^t \\ \sin t \end{pmatrix} \tag{1}$$

are matrices. Matrix A has two rows and three columns, so A is called a "2×3 matrix." Matrix $\mathbf{X}$ has three rows and only a single column, so $\mathbf{X}$ is called a **column matrix** or **vector.** (Note that vectors are boldfaced.)

The entry in the ith row and jth column of matrix A is denoted by a_{ij}, pronounced "aye eye jay." For matrix A given in Eq. (1) the entries are

$$\begin{aligned} a_{11} &= 1, & a_{12} &= 0, & a_{13} &= -1, \\ a_{21} &= 3, & a_{22} &= 2, & a_{23} &= 4. \end{aligned}$$

For the column matrix $\mathbf{X}$ the entries are

$$\begin{aligned} x_{11} &= 2t, \\ x_{21} &= 1 - e^t, \\ x_{31} &= \sin t. \end{aligned}$$

For column matrices we usually denote the entry in the ith row simply as x_i rather than x_{i1} (since the matrix consists only of a single column).

Equality of Matrices

Two matrices A and B are said to be **equal** if and only if they are of the exact same size and the corresponding entries in each position are equal. For example, if

$$B = \begin{pmatrix} b_{11} & b_{12} & b_{13} \\ b_{21} & b_{22} & b_{23} \end{pmatrix}$$

and $B = A$ for the matrix A in Eq. (1), then

$$\begin{aligned} b_{11} &= 1, & b_{12} &= 0, & b_{13} &= -1, \\ b_{21} &= 3, & b_{22} &= 2, & b_{23} &= 4. \end{aligned}$$

Multiplication of a Matrix by a Scalar

If A is a matrix and k is a real or complex scalar constant, then kA is the matrix obtained by multiplying each entry of A by the constant k. For example, if $k = -2$ and A is the matrix given by Eq. (1), then

$$-2A = -2\begin{pmatrix} 1 & 0 & -1 \\ 3 & 2 & 4 \end{pmatrix} = \begin{pmatrix} -2 & 0 & 2 \\ -6 & -4 & -8 \end{pmatrix}. \tag{2}$$

Addition of Matrices

Two matrices A and B can be added if and only if they have the same size. The **sum** $A + B$ is the matrix whose entry in the ith row and jth column is $a_{ij} + b_{ij}$. That is, $A + B$ is obtained by adding together the entries in corresponding positions from A and B. For example, if

$$A = \begin{pmatrix} 1 & 0 & -1 \\ 3 & 2 & 4 \end{pmatrix} \quad \text{and} \quad B = \begin{pmatrix} -1 & 5 & 3 \\ 0 & 1 & -2 \end{pmatrix},$$

then

$$A + B = \begin{pmatrix} 0 & 5 & 2 \\ 3 & 3 & 2 \end{pmatrix}. \tag{3}$$

Multiplication of Matrices

Two matrices A and B can be multiplied if and only if their sizes conform in a certain way. Specifically, if matrix A has m rows and n columns, it can premultiply any matrix B with n rows and p columns. The product $C = AB$ is the matrix whose entry in the ith row and jth column is given by

$$c_{ij} = a_{i1}b_{1j} + a_{i2}b_{2j} + \cdots + a_{in}b_{nj} = \sum_{k=1}^{n} a_{ik}b_{kj} \tag{4}$$

where $i = 1, 2, \ldots, m$ and $j = 1, 2, \ldots, p$. We also say that matrix B is **multiplied by** matrix A in the product AB.

For example, if

$$A = \begin{pmatrix} 1 & 0 & -1 \\ 3 & 2 & 4 \end{pmatrix}_{2\times 3} \quad \text{and} \quad B = \begin{pmatrix} a & b & c \\ d & e & f \\ u & v & w \end{pmatrix}_{3\times 3},$$

then

$$C = AB = \begin{pmatrix} a + 0d - u & b + 0e - v & c + 0f - w \\ 3a + 2d + 4u & 3b + 2e + 4v & 3c + 2f + 4w \end{pmatrix}_{2\times 3}.$$

We have subscripted each matrix with its size in the above example for emphasis. Note that the product of an $m \times n$ matrix with a $n \times p$ matrix (in that order) is an $m \times p$ matrix. For the matrices given above, the product BA is *not defined* because a 3×3 matrix can only premultiply to a $3 \times p$ matrix, whereas A is a 2×3 matrix.

From the definition of matrix multiplication it is easy to multiply a column matrix or vector by a matrix. Thus

$$\begin{pmatrix} 1 & 0 & -1 \\ 3 & 2 & 4 \end{pmatrix}\begin{pmatrix} x_1 \\ x_2 \\ x_3 \end{pmatrix} = \begin{pmatrix} x_1 - x_3 \\ 3x_1 + 2x_2 + 4x_3 \end{pmatrix}.$$

Therefore the system of simultaneous equations

$$\begin{aligned} a_{11}x_1 + a_{12}x_2 + a_{13}x_3 &= b_1, \\ a_{21}x_1 + a_{22}x_2 + a_{23}x_3 &= b_2 \end{aligned} \tag{5}$$

can be written in matrix form as

$$\begin{pmatrix} a_{11} & a_{12} & a_{13} \\ a_{21} & a_{22} & a_{23} \end{pmatrix}\begin{pmatrix} x_1 \\ x_2 \\ x_3 \end{pmatrix} = \begin{pmatrix} b_1 \\ b_2 \end{pmatrix}, \tag{6}$$

or more compactly, as

$$A\mathbf{X} = \mathbf{B}$$

where

$$A = \begin{pmatrix} a_{11} & a_{12} & a_{13} \\ a_{12} & a_{22} & a_{23} \end{pmatrix}, \qquad \mathbf{X} = \begin{pmatrix} x_1 \\ x_2 \\ x_3 \end{pmatrix}, \qquad \mathbf{B} = \begin{pmatrix} b_1 \\ b_2 \end{pmatrix}. \tag{7}$$

$n \times n$ Identity Matrix

For any positive integer n the $\boldsymbol{n \times n}$ **identity matrix** is the square matrix whose entries are 1 along the main diagonal (top left to bottom right) and zeros elsewhere:

$$I = \begin{pmatrix} 1 & 0 & 0 & \cdots & 0 \\ 0 & 1 & 0 & \cdots & 0 \\ 0 & 0 & 1 & \cdots & 0 \\ \vdots & \vdots & \vdots & & \vdots \\ 0 & 0 & 0 & \cdots & 1 \end{pmatrix}_{n\times n}. \tag{8}$$

If A is any *square* $n \times n$ matrix, then

$$AI = IA = A. \tag{9}$$

Properties of Matrix Operations

Whenever the matrices under consideration conform in size so the indicated operations are defined, the following properties are satisfied. If A, B, and C are matrices, then

1. Addition is **associative:** $A + (B + C) = (A + B) + C$.
2. Addition is **commutative:** $A + B = B + A$.
3. There is an **additive identity** or **zero matrix 0** consisting of all zero entries and satisfying $A + \mathbf{0} = A$.
4. For each matrix A there is an **additive inverse** $-A$ whose entries are the negatives of the corresponding entries in A. Moreover, $A + (-A) = \mathbf{0}$.
5. Multiplication is **associative:** $A(BC) = (AB)C$.
6. There is a **multiplicative identity** I satisfying the relation $AI = A$.
7. Multiplication is both **left distributive** and **right distributive** over matrix addition:

$$A(B + C) = AB + AC,$$
$$(A + B)C = AC + BC.$$

If k, k_1, and k_2 are scalars, then

8. Multiplication by scalars is **associative:**

$$(k_1 k_2)A = k_1(k_2 A),$$
$$k(AB) = (kA)B = A(kB).$$

9. Multiplication by scalars is **distributive** over matrix addition:

$$k(A + B) = kA + kB.$$

10. Multiplication by scalars is **distributive** over scalar addition:

$$(k_1 + k_2)A = k_1 A + k_2 A.$$

In general, matrix multiplication is ***not commutative:***

$$AB \neq BA. \tag{10}$$

Transpose of a Matrix

Given any matrix A, the **transpose** of A is the matrix A^T obtained from A by interchanging the rows and columns. Thus the element in the ith row and jth column of A^T is the entry a_{ji} from the jth row and ith column of A. For instance, if

$$A = \begin{pmatrix} 1 & 0 & -1 \\ 3 & 2 & 4 \end{pmatrix},$$

then

$$A^T = \begin{pmatrix} 1 & 3 \\ 0 & 2 \\ -1 & 4 \end{pmatrix}.$$

Determinant of a Square Matrix

If $A = (a_{ij})$ is a square $n \times n$ matrix, then the **determinant** of A, denoted "det A," is defined as follows:

1. If $n = 1$, then $\det A = a_{11}$.
2. If $n = 2$, then $\det A = a_{11}a_{22} - a_{21}a_{12}$.
3. If $n > 2$, then

$$\det A = \sum_{j=1}^{n} (-1)^{i+j} a_{ij} \det A_{ij}, \tag{11}$$

where i is any one of the fixed integers $1, 2, \ldots, n$, and A_{ij} is the $(n-1) \times (n-1)$ matrix obtained from A by deleting the ith row and the jth column.

The determinant can also be denoted by placing vertical rules around the elements in the matrix. For example, if

$$A = \begin{pmatrix} 1 & -1 & 0 \\ 2 & 4 & 5 \\ 3 & 7 & 2 \end{pmatrix}, \qquad \text{then } \det A = \begin{vmatrix} 1 & -1 & 0 \\ 2 & 4 & 5 \\ 3 & 7 & 2 \end{vmatrix}.$$

If we select the first row in A for the expansion, then part 3 of the definition tells us that

$$\begin{aligned} \det A &= \sum_{j=1}^{3} (-1)^{1+j} a_{1j} \det A_{1j} \\ &= a_{11}\begin{vmatrix} 4 & 5 \\ 7 & 2 \end{vmatrix} - a_{12}\begin{vmatrix} 2 & 5 \\ 3 & 2 \end{vmatrix} + a_{13}\begin{vmatrix} 2 & 4 \\ 3 & 7 \end{vmatrix} \\ &= 1(4\cdot 2 - 7\cdot 5) - (-1)(2\cdot 2 - 3\cdot 5) + 0(2\cdot 7 - 3\cdot 4) \\ &= -38. \end{aligned}$$

It is also possible to evaluate any determinant by selecting any column $j = 1, 2, \ldots, n$ for expansion:

$$\det A = \sum_{i=1}^{n} (-1)^{i+j} a_{ij} \det A_{ij}. \tag{12}$$

Properties of Determinants

The following results give the most important properties of determinants used in this text.

1. The determinant of any square matrix is independent of the row or column selected in Eqs. (11) or (12).
2. If two columns or two rows of matrix A are identical, then $\det A = 0$.

3. If matrix B is obtained from matrix A by interchanging two columns or two rows of A, then $\det B = -\det A$.

4. If matrix B is obtained from matrix A by multiplying a row or a column of A by a constant k, then $\det B = k \det A$.

5. If matrix B is obtained from matrix A by adding a multiple of one row to another row, or a multiple of one column to another column, then $\det B = \det A$.

6. The determinant of a product of two matrices is the product of the determinants: $\det AB = \det A \cdot \det B$.

7. For any matrix A, $\det A^T = \det A$.

8. **Cramer's rule:** If A is an $n \times n$ matrix and $\det A \neq 0$, then the unique solution to the system $A\mathbf{X} = \mathbf{B}$, where $\mathbf{B}$ is an $n \times 1$ matrix of constants and $\mathbf{X} = (x_j)$ is the $n \times 1$ matrix of unknowns, is given by

$$x_p = \frac{\det A_p}{\det A}, \qquad p = 1, 2, \ldots, n,$$

where A_p is the matrix obtained from A by replacing the pth column of A with $\mathbf{B}$.

Inverse of a Matrix

Suppose A is an $n \times n$ matrix. If B is a matrix with the property that

$$AB = I,$$

then B is called the **inverse** of A. It is conventional to denote the inverse by A^{-1}, read "A inverse."

As an example, it is straightforward to verify that

$$\begin{pmatrix} 1 & 3 \\ 2 & 5 \end{pmatrix} \begin{pmatrix} -5 & 3 \\ 2 & -1 \end{pmatrix} = \begin{pmatrix} 1 & 0 \\ 0 & 1 \end{pmatrix}.$$

Thus

$$\begin{pmatrix} 1 & 3 \\ 2 & 5 \end{pmatrix}^{-1} = \begin{pmatrix} -5 & 3 \\ 2 & -1 \end{pmatrix}.$$

There are several methods that can be used to calculate the inverse of a matrix whenever it exists. Moreover, it can be proven that a square matrix has an inverse if and only if $\det A \neq 0$. Results of this kind properly belong to a course in linear or matrix algebra. However, the following rule can be used to calculate the inverse of any 2×2 matrix with a nonzero determinant.

THEOREM D.1

If

$$A = \begin{pmatrix} a & b \\ c & d \end{pmatrix}$$

and $\det A = ad - bc \neq 0$, then

$$A^{-1} = \frac{1}{ad - bc} \begin{pmatrix} d & -b \\ -c & a \end{pmatrix}.$$

More generally, let $A = (a_{ij})$ be any $n \times n$ matrix. Define

$$\alpha_{ij} = (-1)^{i+j} \det A_{ij} \tag{13}$$

where A_{ij} is the $(n-1) \times (n-1)$ matrix obtained from A by deleting the ith row and jth column. Then the **adjoint** of A is the matrix

$$\text{Adj } A = (\alpha_{ij})^T. \tag{14}$$

EXAMPLE 1 Suppose that

$$A = \begin{pmatrix} 1 & 0 & 1 \\ -1 & 2 & 1 \\ 2 & 0 & 3 \end{pmatrix}.$$

Then we have the following determinants:

$$\det A_{11} = \begin{vmatrix} 2 & 1 \\ 0 & 3 \end{vmatrix} = 6, \quad \det A_{12} = \begin{vmatrix} -1 & 1 \\ 2 & 3 \end{vmatrix} = -5, \quad \det A_{13} = \begin{vmatrix} -1 & 2 \\ 2 & 0 \end{vmatrix} = -4,$$

$$\det A_{21} = \begin{vmatrix} 0 & 1 \\ 0 & 3 \end{vmatrix} = 0, \quad \det A_{22} = \begin{vmatrix} 1 & 1 \\ 2 & 3 \end{vmatrix} = 1, \quad \det A_{23} = \begin{vmatrix} 1 & 0 \\ 2 & 0 \end{vmatrix} = 0,$$

$$\det A_{31} = \begin{vmatrix} 0 & 1 \\ 2 & 1 \end{vmatrix} = -2, \quad \det A_{32} = \begin{vmatrix} 1 & 1 \\ -1 & 1 \end{vmatrix} = 2, \quad \det A_{33} = \begin{vmatrix} 1 & 0 \\ -1 & 2 \end{vmatrix} = 2.$$

Hence

$$\text{Adj } A = \begin{pmatrix} 6 & 5 & -4 \\ 0 & 1 & 0 \\ -2 & -2 & 2 \end{pmatrix}^T = \begin{pmatrix} 6 & 0 & -2 \\ 5 & 1 & -2 \\ -4 & 0 & 2 \end{pmatrix}.$$

THEOREM D.2

Let A be an $n \times n$ matrix. Then A is invertible if and only if $\det A \neq 0$. If A is invertible, then the inverse of A is given by

$$A^{-1} = \frac{1}{\det A} \operatorname{Adj} A.$$

From Example 1 we have

$$\det A = \begin{vmatrix} 1 & 0 & 1 \\ -1 & 2 & 1 \\ 2 & 0 & 3 \end{vmatrix} = 2.$$

Therefore

$$A^{-1} = \frac{1}{2}\begin{pmatrix} 6 & 0 & -2 \\ 5 & 1 & -2 \\ -4 & 0 & 2 \end{pmatrix} = \begin{pmatrix} 3 & 0 & -1 \\ \frac{5}{2} & \frac{1}{2} & -1 \\ -2 & 0 & 1 \end{pmatrix}.$$

Answers to Odd-numbered Exercises

CHAPTER 1

Section 1.1

5. Ordinary first-order

7. Ordinary second-order

9. Partial second-order

11. Ordinary first-order

13. $y' = (Ce^x - e^{2x}) - e^{2x}$ or $y' = y - e^{2x}$

15. $y' = -Ae^{-x} + Be^{-x} - Bxe^{-x}$
$y'' = (A - B)e^{-x} - Be^{-x} + Bxe^{-x}$

17. $2x - 8yy' = 0$ or $y' = \dfrac{x}{4y}$

19. $2xy + (x^2 + C)y' = 0$ and $x^2 + C = -\dfrac{2}{y}$ yields $y' = xy^2$.

21. $2e^{2x} = 2yy'(x^2 + 1) + 2xy^2$

23. $y'' = \frac{1}{2}\sec x \tan^2 x + \frac{1}{2}\sec^3 x$

25. $y' = (A + \cos e^{-x})e^x + 2(B - \sin e^{-x})e^{2x}$
$y'' = (A + 3\cos e^{-x})e^x + 4(B - \sin e^{-x})e^{2x} + \sin e^{-x}$

Section 1.2

1. For a particular individual skier, on a given day, particular mountain slope, and specific course, the predicted time to ski the course is given by

$$v = f(\text{propulsion, resistance}).$$

3. Total stopping distance = reaction distance + braking distance, reaction distance = f(response time, speed), $d_r = t_r v$, and $d_b \propto v^2$.

5. For a given species with a known current population, the predicted population P at some future time is given by

$$\frac{dP}{dt} = kP,$$

where k is a proportionality factor.

7. See discussion on radioactive decay in Section 1.3 of the text.

Section 1.3

1. Approximately 84% remains.

3. Approximately 79% remains.

5. a) $Q_k = \left(1 + \frac{r}{k}\right)^k Q_0$

b) $Q \approx \$2909.98$

7. 11,892, May 1986

9. 3 hr

11. $\frac{dv}{dt} = g(\sin\theta - \mu\cos\theta), \quad v(0) = v_0$

13. $\frac{125}{32} v' + kv = 125, \quad v(0) = 100$

15. $\frac{600}{g}\frac{dv}{dt} = F_p - 600\mu, \quad v(0) = 0$

17. $\frac{dS}{dt} + \frac{Sr_2}{V_0 + (r_1 - r_2)t} = cr_1$

19. $\frac{dy}{dt} = 12 - gt, \quad y(0) = 80$

21. $\frac{dI}{I} = -k\,dx, \quad k > 0$

Section 1.4

1. $y' = y$

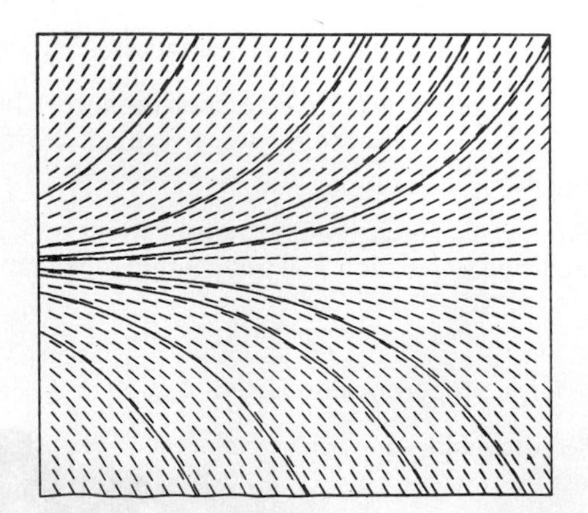

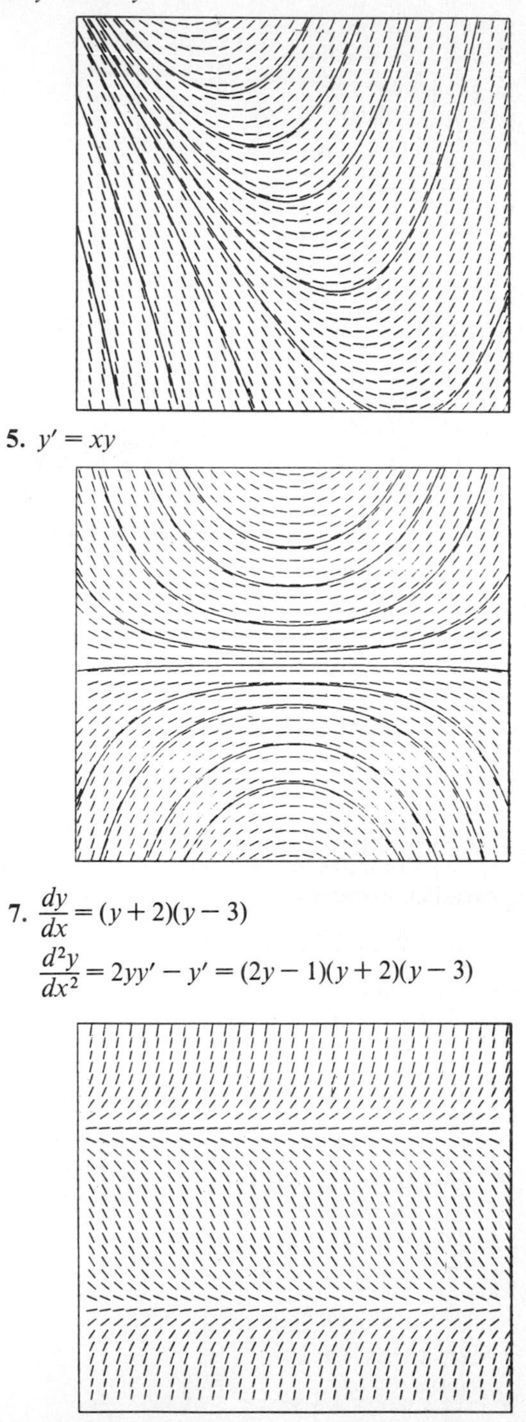

3. $y' = x + y$

5. $y' = xy$

7. $\frac{dy}{dx} = (y+2)(y-3)$

$\frac{d^2y}{dx^2} = 2yy' - y' = (2y-1)(y+2)(y-3)$

9. $y' = y(y+1)(y-1)$

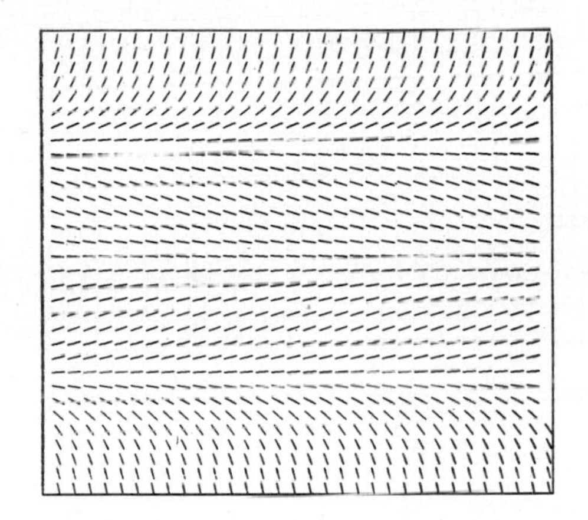

11. Damped exponential $y = Ce^{rt}$, $r < 0$. As $t \to \infty$, $y \to 0$.

13. $\frac{dP}{dt} = P(a - bP), \quad a,b > 0$

The rest points are at $P = 0$ and $P = a/b$. $P = 0$ is unstable, $P = a/b$ is stable (asymptotic), and $P = a/b$ for large t.

15. $\frac{dP}{dt} = kP(M - P)(P - m), \quad k,M,m > 0$

The rest points at $P = 0$ and $P = M$ are asymptotically stable while $P = m$ is unstable. For initial populations greater than m, the model predicts that P approaches M for large values of t. For initial populations of less than $P = m$, the model predicts extinction. Points of inflection occur at

$$P = \frac{M + m \pm \sqrt{M^2 - mM + m^2}}{3}.$$

17. a) The model is reasonable in the sense that if $P < m$, then $P \to 0$ as $t \to \infty$; if $m < P < M$, then $P \to M$ as $t \to \infty$; if $P > M$, then $P \to M$ as $t \to \infty$.
b) If the population falls below m, then $P \to 0$ as $t \to \infty$ (extinction).
c) For $P > M$,

$$\frac{dP}{dt} = kP(M - P)(P - m)$$

is negative.

19. $v = \sqrt{\frac{mg}{k}}$, terminal velocity

21. $v = \left(\frac{mg}{k}\right)^{1/r}$, terminal velocity

Chapter 1 Review Exercises

1. Nonlinear second-order
3. Nonlinear fourth-order
5. Nonlinear third-order
7. Nonlinear second-order
9. Linear second-order
11. Nonlinear first-order
13. Nonlinear second-order
15. Nonlinear first-order
17. Nonlinear second-order
19. Linear first-order
21. Nonlinear first-order
23. ≈ 0.00853
25. $d = 1.1v + 0.054v^2 \approx 650$ ft

CHAPTER 2

Section 2.1

1. $y = 1 - (x + C)^{-1}$
3. $y - \ln|y| = (x+1)^3 + C$
5. $y^2 = C - 2 \sec x$
7. $y = -\ln[C - \frac{2}{5}(x-2)^{5/2} - \frac{4}{3}(x-2)^{3/2}]$
9. $\tan y = -(x \sin x + \cos x) + C$
11. $-ye^{-y} - e^{-y} = \ln|x| + C$
13. $y - \ln(x - x \ln x + C)$
15. $\ln|y| = x \sin^{-1}x + \sqrt{1 - x^2} + C$
17. $|y| = C\left|\frac{x-1}{x+2}\right|$
19. $|y| = C\left|\frac{x-1}{x}\right|$
21. $y = C\left|\frac{x+2}{x-1}\right| e^{3/(1-x)}$

23. $\tan^{-1}(x-1)+\sin^{-1}y=C$

25. $2y\sqrt{2y}=3\sqrt{x}+C$

27. $y=\dfrac{C}{x}e^{-1/x}$

29. $\dfrac{y^3}{9}(3\ln y-1)=\frac{1}{2}x^2+2x+\ln|x|+C$

31. $y=\ln(C+x-\frac{1}{2}\sin x^2)$

33. $y^3-1=6\sinh x$

35. $y=\frac{1}{2}e^{2(x-2)}-\frac{1}{2}$

37. $\ln|P|=(1-t)(1-e^t)$

39. $(\sqrt{y}+1)^2=4|x|$

Section 2.2

1. $y=\frac{1}{2}+Ce^{-x^2}, \quad -\infty<x<\infty$

3. $y=\frac{1}{4}x^2e^{x/2}+Ce^{x/2}, \quad -\infty<x<\infty$

5. $y=\frac{1}{2}-x^{-1}+Cx^{-2}, \quad x\neq 0$

7. $y=Ce^x-e^{2x}, \quad -\infty<x<\infty$

9. $y=\dfrac{1}{|x|}\ln x^2+\dfrac{C}{|x|}, \quad x\neq 0$

11. $y=\dfrac{1}{2}x-2+\dfrac{1}{x}\ln|x|+\dfrac{C}{x}$

13. Linear in x and $x=(2y-1)e^y+Ce^{-y}$

15. Linear in x and $x=\dfrac{\sin y+C}{y^3}$

17. $e^{x^3}y=\frac{1}{3}(e^{x^3}-4)$

19. $y=3x^2(x-1)e^{-x}$

21. Approximately 99.5% oxygen

Section 2.3

1. $y'=-\dfrac{x}{4y}$

3. $y'=2(y-4)$

5. $y'=\dfrac{y(y^2-1)}{x(y^2+1)}$

7. $(y')^2-4y\cos^2 x=0$

9. $xy'=y-x\sqrt{x^2-y^2}$

11. Exact

13. Exact

15. Not exact

17. Exact

19. Exact

21. $\frac{1}{2}x^2-2xy+\frac{1}{3}y^3=C$

23. $x\sin y-x^2y+\frac{1}{3}x^3=C$

25. $-e^y\cos x-ye^{-x}-e^y=C$

27. $xy\ln y+e^x-e^{-y}=C$

29. $x^3y+4x^2y^2+x^2+3y^3=C$

31. $x^2y-2xy^2+x^2+y^2=1$

33. $y^2-e^{-x}\cos y+e^x+\dfrac{1}{2}x^2=1+\dfrac{\pi^2}{4}$

35. $\cos x\sin y+\dfrac{1}{2}y^2-\ln|\cos x|=1+\dfrac{\pi^2}{8}$

37. $M(x,y)=-\cos y-\frac{1}{2}y^2e^x+g(x)$, where $g(x)$ is *any* function of x.

39. $\frac{1}{2}x^2-2xy+\frac{1}{3}y^3=C$

Section 2.4

1. Bernoulli (in the variable y)

3. Linear in y

5. Homogeneous-type

7. Ricatti

9. Bernoulli in x

11. $|x|(x^2+3y^2)=C$

13. $|x|e^{y/x}=C(x+y)^2$

15. $|x|e^{y/x} = C$

17. $y = xe^{1+Cx}$

19. $xe^{2\sqrt{1-y/x}} = C$

21. $y = \dfrac{e^x}{e^x + C}$

23. $\dfrac{1}{y^2} = \dfrac{2}{5}x^{-1} + Cx^4$

25. $y = \dfrac{3e^{2x}}{e^{3x} + C}$

27. $x^2y^2 = 2y^3 + C$

29. $xy = \dfrac{1}{e^x + C}$

31. $\dfrac{1}{y} = 2 - x^2 + Ce^{-x^2/2}$

33. $y = \dfrac{-2C}{e^{2x} + C}$

35. $y = x + \dfrac{5x}{C - x^5}$

37. $y = x + \dfrac{2x}{Cx^2 - 1}$

39. **a)** $\tan^{-1}(x + y) = x + C$
b) $x + \tan\left(\dfrac{\pi}{4} - \dfrac{x+y}{2}\right) = C$
c) $2y - x - 3\ln|x + y + 1| = C$

Section 2.5

1. $\dfrac{dQ}{dt} = 0.10Q, \quad Q(t = 0) = 100, \quad 0 \le t \le 1$

3. **a)** $Q(t) = 100e^{0.10t}$
b) $Q(1) = 110.5170918$
c) $Q(1) = 110.3812891$ for $\Delta t = 0.25$
d) $0.1025 = 10.25\%$, semiannually;
$0.105155781 = 10.5155781\%$, daily.
e) For $n = 1000$: 1.105165393
$n = 100{,}000$: 1.105170863
f) $e^{0.10} = 1.105170918$

5.

x	Euler y	Actual $y = -\ln(2 - e^x)$
0	0.000	0.000
0.02	0.020	0.020
0.04	0.041	0.042
0.06	0.062	0.064
0.08	0.085	0.087
0.10	0.109	0.111

Section 2.6

5.

x	Runge–Kutta y
0	2.0000
0.1	2.1105
0.2	2.2443
0.3	2.4050
0.4	2.5967

7.

x	Runge–Kutta y
1.0	1.000
1.2	0.981
1.4	0.929
1.6	0.848
1.8	0.742
2.0	0.614

Section 2.7

1. $y_2 = 1 + x + \dfrac{1}{2}x^2$

$y_4 = 1 + x + \dfrac{1}{2}x^2 + \dfrac{1}{6}x^3 + \dfrac{1}{24}x^4$

3. $y_2 = \dfrac{1}{2}x^2 - \dfrac{1}{3!}x^3$

$y_4 = \dfrac{1}{2}x^2 - \dfrac{1}{3!}x^3 + \dfrac{1}{4!}x^4 - \dfrac{1}{5!}x^5$

5. $y_2 = \dfrac{1}{3}x^3 + \dfrac{1}{3\cdot 4}x^4$

$y_4 = \dfrac{1}{3}x^3 + \dfrac{1}{3\cdot 4}x^4 + \dfrac{1}{3\cdot 4\cdot 5}x^5 + \dfrac{1}{3\cdot 4\cdot 5\cdot 6}x^6$

7. A unique solution does exist.

9. Not guaranteed of a unique solution

11. Unique solution

Section 2.8

1. $\frac{F_E}{F_M} = 1$

3. $r = cx + r_0$

Section 2.9

1. a)

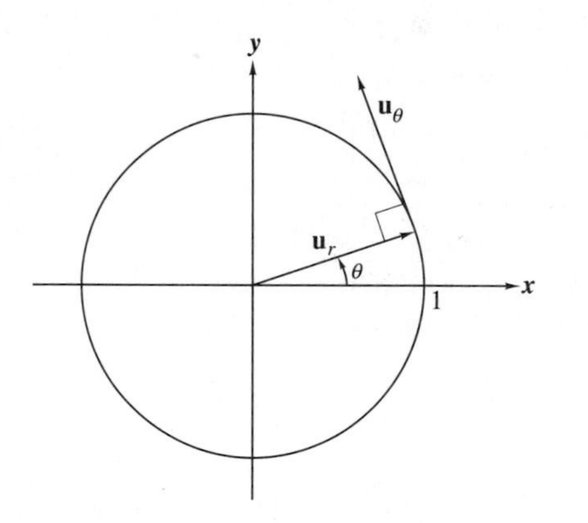

b) $\mathbf{u}_r \cdot \mathbf{u}_\theta = 0$ implies $\mathbf{u}_r$ and $\mathbf{u}_\theta$ are perpendicular, where $\mathbf{r} = r\mathbf{u}_r$.

c) $\mathbf{v} = r\frac{d\theta}{dt}\mathbf{u}_\theta + \frac{dr}{dt}\mathbf{u}_r$

d) $\mathbf{a} = \left[\frac{d^2r}{dt^2} - r\left(\frac{d\theta}{dt}\right)^2\right]\mathbf{u}_r + \left(r\frac{d^2\theta}{dt^2} + 2\frac{dr}{dt}\frac{d\theta}{dt}\right)\mathbf{u}_\theta$

3. a) $A = \frac{1}{2}Tr_0v_0$
b) $A = \frac{1}{2}Tr_0v_0$,
$T = \frac{2\pi a^2}{r_0v_0}\sqrt{1 - e^2}$

Chapter 2 Review Exercises

1. $\frac{x^3}{3} + 3x^2y + y^2x + \frac{y^2}{2} = C$

3. $y = \frac{x^2}{2} + 2x + \ln(x-2)^{10} + C$

5. $x^2y + \frac{3x^2}{2} + y^2x = \frac{15}{2}$

7. $y \sin x + x^2e^y + 2y = \pi^2$

9. $\frac{x^4}{2} - \frac{x^2y^2}{2} - 2xy + 3x = 8$

11. $x^3y^2 - xy^3 - x^2 + y = -3$

13. $x^2y + \frac{x^3}{3} + \frac{y^2}{2} = 9$

15. $y = x^3 + Cx^{-3}$

17. $y = -\frac{1}{2}x + Cx^3$

19. $x^4y^3 - x^2y + 3y = C$

21. $y = 2 + Ce^{-x^2}$

23. $y = \frac{e^{x^2}}{2x} + \frac{C}{x}$

25. $y = \frac{-\cos x}{x} + \frac{C}{x}$

27. $y - 2\ln|y+2| = \ln|x+4|^{-1} + C$

29. $y = -\frac{1}{2}x + Cx^3$

31. $y = \frac{2x^2 + 4x + C}{(x+1)^3}$

33. $r^2 \sin\theta + r\cos\theta = C$

35. a) 108.3042 hr
b) 251.4747 hr

37. 117 yr

39. 8:30 PM

41. 1385.75 lb

43. $y(0.5) = 0.5640$

45. $y(1.2) = 3.1119$, $y(1.4) = 3.2254$, $y(1.6) = 3.3404$

47. $y(3.4) = 43.2511$, $y(3.8) = 1184.4000$

49. a) $\frac{dv}{dt} = -9.8 - 0.02v^2$, $v(0) = 8$
b) $v(0.05) = 7.4498$ m/sec, $v(0.10) = 6.9078$ m/sec, $v(0.15) = 6.3789$ m/sec

51. $x(1) = 3.2089$ m, $x(2) = 13.2892$ m

CHAPTER 3

Section 3.1

1. $y'' + \frac{6}{5}y' = 32, \quad y(0) = 0, \quad y'(0) = 0$

3. 9-lb force, $y \approx 0.0198$ ft
   22-lb force, $y \approx 0.0485$ ft

5. $my'' + y' + y = 0, \quad y(0) = 2, \quad y'(0) = 2$

7. $\frac{25}{32}y'' + 40y = 0, \quad y(0) = \frac{7}{24}, \quad y'(0) = \frac{v_0}{12}$

9. $\frac{8}{32}y'' + 2y' + 32y = 0, \quad y(0) = 0, \quad y'(0) = \frac{4}{12}$

11. $2q'' + 4q' + 10q = 20\cos t, \quad q(0) = 2, \quad q'(0) = 3$

13. $5q'' + 10q' + 30q = 100\cos 8t$

Section 3.2

1. $y_3'' - 7y_3' + 12y_3$
   $= 9c_1e^{3x} + 16c_2e^{4x} - 21c_1e^{3x} - 28c_2e^{4x} + 12c_1e^{3x} + 12c_2e^{4x} = 0$

3. $y_2'' + y_2' - 6y_2$
   $= 4e^{2x} - 9e^{-3x} + 2e^{2x} + 3e^{-3x} - 6e^{2x} + 6e^{-3x} = 0$

5. $y_3'' + y_3$
   $= (-c_1\sin x - c_2\cos x) + (c_1\sin x + c_2\cos x) = 0$

7. $y_2'' + 2y_2' + y_2$
   $= e^{-x}(x-2) + 2e^{-x}(1-x) + xe^{-x} = 0$

11. Nonhomogeneous linear

13. Nonlinear because of $\sqrt{y}$ term

15. Nonhomogeneous linear

19–35. Linearly independent

37. $y(x) = \frac{3}{5}e^{2x} + \frac{2}{5}e^{-3x}$

39. $y(x) = e^{-x}\sin 2x + e^{-x}\cos 2x$

41. $y(x) = (1 + 2x)e^{-2x}$

43. $y(x) = x(1 - 2\ln x)$

47. $y_1(x)y_2'(x) - y_1'(x)y_2(x)$
    $= -4\sin x\cos^3 x$
    $+ 2\cos x\sin x + 2\cos x\sin x\cos 2x = 0$

Section 3.3

1. $y = c_1e^{-3x} + c_2e^{4x}$

3. $y = c_1e^{-4x} + c_2e^{x}$

5. $y = c_1e^{-2x} + c_2e^{2x}$

7. $y = c_1e^{-x} + c_2e^{3x/2}$

9. $y = c_1e^{-x/4} + c_2e^{3x/2}$

11. $y = c_1\cos 3x + c_2\sin 3x$

13. $y = c_1\cos 5x + c_2\sin 5x$

15. $y = e^{x}(c_1\cos 2x + c_2\sin 2x)$

17. $y = e^{-x}(c_1\cos\sqrt{3}x + c_2\sin\sqrt{3}x)$

19. $y = e^{-2x}(c_1\cos\sqrt{5}x + c_2\sin\sqrt{5}x)$

21. $y = c_1 + c_2x$

23. $y = c_1e^{-2x} + c_2xe^{-2x}$

25. $y = c_1e^{-3x} + c_2xe^{-3x}$

27. $y = c_1e^{-x/2} + c_2xe^{-x/2}$

29. $y = c_1e^{-x/3} + c_2xe^{-x/3}$

31. $y = -\frac{3}{4}e^{-5x} + \frac{3}{4}e^{-x}$

33. $y = \frac{1}{2\sqrt{3}}\sin 2\sqrt{3}x$

35. $y = -\cos 2\sqrt{2}x + \frac{1}{\sqrt{2}}\sin 2\sqrt{2}x$

37. $y = (1 - 2x)e^{2x}$

39. $y = 2(1 + 2x)e^{-3x/2}$

41. $y = c_1e^{-x} + c_2e^{3x}$

43. $y = c_1e^{-x/2} + c_2xe^{-x/2}$

45. $y = c_1\cos\sqrt{5}x + c_2\sin\sqrt{5}x$

47. $y = c_1e^{-x/5} + c_2xe^{-x/5}$

49. $y = e^{-x/2}(c_1\cos x + c_2\sin x)$

51. $y = c_1e^{3x/4} + c_2xe^{3x/4}$

53. $y = c_1e^{-4x/3} + c_2xe^{-4x/3}$

55. $y = c_1e^{-x/2} + c_2e^{4x/3}$

57. $y = (1 + 2x)e^{-x}$

59. $y = \frac{15}{13}e^{-7x/3} + \frac{11}{13}e^{2x}$

Section 3.4

1. 0.0864 ft (above equilibrium)

3. $y(t) = 0.2917 \cos(7.1552t) + \dfrac{v_0}{85.8623}\sin(7.1552t)$ (in feet), or $y = 3.5 \cos(7.1552t) + \dfrac{v_0}{7.1552}\sin(7.1552t)$ (in inches).

5. 0.316 sec

7. 8.8334 lb

9. 24.4949 ft/sec

11. -1.56 ft/sec^2 (acceleration upward)

13. $q(t) = -8e^{-3t} + 10e^{-2t}$, $\lim_{t\to\infty} q(t) = 0$

15. $q'' + 6q' + 8q = 0$, $q(0) = 10$, $q'(0) = 0$
Solution: $q(t) = 20e^{-2t} - 10e^{-4t}$.

17. $\dfrac{d\theta}{dt} \approx -0.1164\pi$ rad/sec or -20.9520 deg/sec

19. $\theta(t) = \frac{4}{9}\pi e^{-t} - \frac{1}{9}\pi e^{-4t}$; $\lim_{t\to\infty} \theta(t) = 0$
The model predicts door will approach the closed position as t gets large.

Section 3.5

1. $r^3 - r^2 - r + 1 = 0$

3. $r^4 - 5r^2 + 4 = 0$

5. $r^4 + 1 = 0$

7. $r = -1, \frac{1}{2}(3 \pm \sqrt{5})$

9. $r = 1, -1, \pm i$

11. $r = 1, 1, -1, -1$

13. $r = \frac{1}{2}, -\frac{1}{2}, \pm i$

15. $r = 0, \pm 2i, -1, 1$

17. $y = (c_1 + c_2x + c_3x^2)e^{-2x}$

19. $y = c_1 + c_2x + c_3e^{3x}$

21. $y = c_1 + c_2e^{3x} + c_3xe^{3x} + c_4e^{-x}$

23. $y = c_1\cos x + c_2\sin x + c_3x\cos x + c_4x\sin x$

25. $y = c_1 + c_2x + c_3\cos x + c_4\sin x + c_5x\cos x + c_6x\sin x + c_7x^2\cos x + c_8x^2\sin x$

27. $y = c_1e^{-x} + c_2xe^{-x} + c_3x^2e^{-x}$

29. $y = c_1e^{-x} + c_2e^{x} + c_3\cos x + c_4\sin x$

31. $y = c_1e^{-2x} + c_2e^{x} + c_3e^{2x}$

33. $y = c_1e^{-x} + e^{-x}(c_2\cos x + c_3\sin x)$

35. $y = c_1e^{-x} + e^{-x/2}(c_2\cos\sqrt{2}x + c_3\sin\sqrt{2}x)$

37. $y = c_1e^{-\sqrt{6}x} + c_2e^{\sqrt{6}x} + c_3\cos x + c_4\sin x$

39. $y = c_1e^{-x} + c_2e^{x} + e^{2x}(c_3\cos x + c_4\sin x)$

41. $y = -2 + 2x + 3e^{-x}$

43. $y = \frac{17}{8}e^{-x} + \frac{1}{4}e^{x} - \frac{3}{8}e^{3x}$

45. $y = x + \sin x$

Chapter 3 Review Exercises

1. $W = 3 \neq 0$

3. $W = x^{-2} \neq 0$

5. $y = c_1e^{-x} + c_2xe^{-x} + c_3e^{2x}$

7. $y = c_1e^{x} + c_2e^{-2x} + c_3xe^{-2x}$

9. $y = 2\cos 12x + 3\sin 12x$

11. $y = e^{2x}\cos 3x$

13. $y = e^{-2x} - \dfrac{1}{\pi}xe^{-2x}$

15. $y = c_1 + c_2e^{-4x} + c_3e^{4x}$

17. **a)** -4.3584×10^{-7} (above)
b) Underdamped

19. $350y'' + 3500y' + 140{,}000y = 0$, $y(0) = 0$, $y'(0) = 75$, $y(0.0681) = 2.67$ cm, underdamped

21. 0.5403 rad

23. **a)** $q = 60e^{-2t} - 40e^{-3t}$
b) 6.13 C
c) -10.27 A

CHAPTER 4

Section 4.1

1. $y_p = -2 - x$

3. $y_p = \frac{1}{2} - x + \frac{1}{2}x^2$

5. $y_p = 5 + 6x + 3x^2$

7. 0, 0, 0

9. 0, -1

11. 1, -1, -1

13. $1 \pm i$, 0, 0

15. 3, $-2 \pm 2i$, 0, 0

17. $y_h = c_1 e^{-x} + c_2 x e^{-x} + c_3 e^{x}$

19. $y_h = c_1 e^{-2x} + c_2 x e^{-2x} + c_3 e^{x} + c_4 x e^{x}$

21. $y_h = c_1 e^x + c_2 x e^x + c_3 x^2 e^x + c_4 \cos x + c_5 \sin x$

23. $y_h = c_1 + c_2 x + c_3 x^2 + c_4 x^3 + c_5 e^x$

25. $y_h = c_1 + c_2 e^x + c_3 x e^x + c_4 \cos x + c_5 \sin x$

27. $y_h = c_1 + c_2 \cos\sqrt{3}x + c_3 \sin\sqrt{3}x + c_4 x \cos\sqrt{3}x + c_5 x \sin\sqrt{3}x$

29. $y_h = e^x(c_1 \cos x + c_2 \sin x + c_3 x \cos x + c_4 x \sin x)$

31. $y_h = c_1 \cos x + c_2 \sin x + c_3 x \cos x + c_4 x \sin x + c_5 x^2 \cos x + c_6 x^2 \sin x$

33. $y_h = e^{-x}(c_1 \cos x + c_2 \sin x) + c_3 + c_4 x + c_5 x^2 + c_6 \cos x + c_7 \sin x$

35. $y_h = e^{-x}(c_1 \cos 2x + c_2 \sin 2x) + c_3 + c_4 x + c_5 x^2 + c_6 \cos x + c_7 \sin x + c_8 x \cos x + c_9 x \sin x$

37. $y = -e^{-x} + 1 + \dfrac{1}{2}x^2 - x$

39. $y = 2(e^x - e^{-x})\cos x - 3e^{-x} \sin x$

41. $y = (1 - x + x^2)e^x$

Section 4.2

1. $y = c_1 e^{-3x} + c_2 e^{2x} + x e^{2x}$

3. $y = c_1 e^{-3x} + \left(c_2 + \dfrac{x}{3}\right)e^{3x} - \dfrac{1}{9}x$

5. $y = c_1 e^{2x} + c_2 x e^{2x} + \frac{1}{2} - \frac{1}{2}e^x \cos x$

7. $y = c_1 e^{-x} + c_2 e^{x/2} + \frac{1}{5}e^x \sin x - 1$

9. $y = c_1 e^{-x} \cos x + c_2 e^{-x} \sin x + \frac{1}{5}e^x - \frac{2}{5}\cos x + \frac{1}{5}\sin x$

11. $y = c_1 \cos 2x + c_2 \sin 2x + \frac{1}{4} + \frac{2}{25}e^x - \frac{1}{5}x e^x$

13. $y = c_1 \cos 3x + c_2 \sin 3x + \frac{1}{6}x \sin 3x + \frac{1}{9}x$

15. $y = c_1 + c_2 e^{-x} + c_3 e^x + \frac{1}{2}x e^x - 2x$

17. $y = c_1 + c_2 e^{2x} + c_3 x e^{2x} + \frac{3}{8}x + \frac{1}{4}x^2 + \frac{1}{12}x^3 + \frac{3}{25}\cos x - \frac{4}{25}\sin x$

19. $y = c_1 e^{-x} + c_2 x e^{-x} + c_3 e^x + c_4 x e^x - \frac{1}{4}x^2 e^{-x} + x$

21. $y = -e^{-x} + 2 + \frac{1}{2}x^2 - x$

23. $y = -e^{x} + \frac{1}{2}e^{2x} + x e^x + \frac{1}{2}$

25. $y = -\frac{1}{16}e^{-2x} + \frac{17}{16} + \frac{7}{8}x + \frac{1}{8}x^2 - \frac{1}{12}x^3 + \frac{1}{24}x^4$

27. $y = \left(3 - \dfrac{\pi^2}{4}\right)\cos x + \pi \sin x + x^2 - 2$

29. $y = -\frac{1}{4}\cosh x + \frac{1}{4}\cos x + \frac{3}{4}\sin x + \frac{1}{4}x e^x - x$

Section 4.3

1. $y = Ce^{-x^2} + \frac{1}{2}(x^2 - 1)$

3. $y = Ce^{-2x} + e^{-x}(\sin x - \cos x)$

5. $y = Cx^{-4} + \frac{1}{5}x - \frac{1}{7}x^3$

7. $y = Ce^x - e^{2x}$

9. $y = \dfrac{C}{x} + \dfrac{2}{x}\ln x, \quad x > 0$

11. $y = \dfrac{C}{x} + \dfrac{1}{2}x - 2 + \dfrac{\ln x}{x}, \quad x > 0$

13. $x = Ce^{-y} + (2y - 1)e^y$

15. $xy^3 = \sin y + C, \quad y > 0$

17. $y = \frac{1}{3}(x^3 - 1) - \frac{2}{3}e^{-x^3}$

19. $y = 3x^2(x-1)e^{-x}, \quad x > 0$

Section 4.4

1. $y = c_1 \cos x + c_2 \sin x + \cos x \ln(\cos x) + x \sin x$

3. $y = c_3 e^x + c_2 e^{2x} + (e^x + e^{2x})\ln(1 + e^{-x})$

5. $y = c_1 e^{-x} + c_2 x e^{-x} + x^2 e^{-x}(2 \ln x - 3)$

7. $y = c_1 e^x + c_2 e^{2x} - e^{2x} \sin e^{-x}$

9. $y = c_1 e^{-4x} + c_2 x e^{-4x} - e^{-4x}(\ln x + 1)$

11. $y = c_1 e^{-x} + c_2 e^x - \frac{2}{5} - \frac{1}{5}\sin^2 x$

13. $y = c_1 e^{-x} \cos x + c_2 e^{-x} \sin x + e^{-x}(2 - \sin^2 x)$

15. $y = c_1 e^{-x} + c_2 e^x - \sqrt{1 - e^{-2x}} - e^x \sin^{-1} e^{-x}$

17. $y = c_1 e^{-x} \cos x + c_2 e^{-x} \sin x +$
$2e^{-x} \sin x \ln|\sec x + \tan x| - 4e^{-x}$

19. $y = c_1 \cos x + c_3 \sin x + \sin x \ln(\cot x)$

21. $y_p = -\frac{3}{4} \ln|\csc 2x + \cot 2x| - \frac{3}{4} \cos 2x \ln|\sin 2x| -$
$\frac{3}{4} x \sin 2x$

23. $y_p = \frac{1}{2} e^x (x - 1)$

25. $y_p = -\cos x \ln|\sec x + \tan x|$

27. $y = \cos 3x + (\sin 3x) \ln|\sec 3x + \tan 3x| - 1$

29. $y = \sin x - \cos x + \cos^4 x + \sin^2 x - \frac{1}{3}\sin^4 x$

31. $y = \cos x - \frac{1}{2} \sec x \cos 2x$

33. $y_p = \frac{1}{4} x^2$

Section 4.5

1. 125.5 sec

3. $y(\pi) = -2$ m (above equilibrium)

5. $y(t) = 1 + 2e^{-t} - \frac{1}{3}e^{-2t} - \frac{2}{3}e^{-8t}$

7. $y(t) = (-\frac{9}{20} \sin 2t - \frac{13}{5} \cos 2t)e^{-3t/2} + \frac{8}{5}$

9. $y_p = \frac{60}{901} \cos 2t + \frac{2}{901} \sin 2t$

11. b) $A = \dfrac{F_0}{\omega\sqrt{\omega^2 m^2 + \delta^2}};$

$\phi = \tan^{-1}\left(\dfrac{c_2}{c_1}\right)$, where

$c_1 = \dfrac{-mF_0}{\omega^2 m^2 + \delta^2}, \; c_2 = \dfrac{\delta F_0}{\omega(\omega^2 m^2 + \delta^2)}$

13. $q(t) = \dfrac{1}{5} + \left(\dfrac{49\sqrt{199}}{995} \sin \dfrac{\sqrt{199}}{2} t + \dfrac{49}{5} \cos \dfrac{\sqrt{199}}{2} t\right)e^{-t/2}$

15. $\lim_{x \to \infty} q(t) = \frac{5}{6}$

17. $x'' + 4x' + 20x = 10 \cos 2t, \; x(0) = 5, \; x'(0) = 10;$

Solution: $x(t) = e^{-2t}\left(\dfrac{37}{8} \sin 4t + \dfrac{9}{2} \cos 4t\right)$
$+ \dfrac{1}{2} \cos 2t + \dfrac{1}{4} \sin 2t.$

19. Resonance

Section 4.6

1. $x = 0.3$, Euler $y = 2.06000$, Runge–Kutta $y = 2.09068$, and actual $= 2.09068$, where $y = 2 \cosh x$.

3. $x = 0.3$, Euler $y = 1.29800$, Runge–Kutta $y = 1.28957$, and actual $= 1.28957$, where $y = e^x \cos x$.

5. $x = 0.3$, Euler $y = 1.03405$, Runge–Kutta $y = 1.06832$, and actual $= 1.06834$, where $y = \frac{4}{3} + \frac{1}{6}e^{3x} - \frac{1}{2}e^x$.

7. $x = 0.6$, Euler $y = 0.59360$, Runge–Kutta $y = 0.57531$, and actual $= 0.57531$, where $y = 2 \cos x + \sin x + x^2 - 2$.

9. $x = 0.3$, Euler $y = 1.00005$, Runge–Kutta $y = 1.00034$, and actual $= 1.00034$, where $y = (\cos x)[1 + \ln(\cos x)] + x \sin x$.

11.

x	Runge–Kutta y
1.0	1.00000
1.1	1.10517
1.2	1.22140
1.3	1.34986
1.4	1.49182
1.5	1.64872

Chapter 4 Review Exercises

1. $y = \frac{1}{2}e^{-3x} - \frac{1}{2}e^{x} + 3xe^{x}$

3. $y = xe^{-2x} + \frac{1}{6}x^3e^{-2x}$

5. $y = c_1e^{x} + c_2xe^{x} + xe^{-x} + e^{-x}$

7. $y = c_1e^{-x} + c_2xe^{-x} - \frac{2}{5}e^{4x} + xe^{4x}$

9. $y = c_1e^{4x} + c_2e^{-x} + \frac{3}{17}\cos x - \frac{5}{17}\sin x$

11. $\theta = \frac{129}{128}\cos 4t + \frac{5}{4}\sin 4t - \frac{1}{128} + \frac{1}{16}t^2$

13. $x = c_1\cos 2t + c_2\sin 2t - 0.75t\cos 2t$

15. $y = c_1\cos 4x +$
$c_2\sin 4x - \frac{1}{16}\ln|\cos 4x|\cos 4x - \frac{1}{4}x\sin 4x$

17. $y = c_1\cos 2x + c_2\sin 2x +$
$\frac{1}{4}\ln|\cos 2x|\cos 2x + \frac{x}{2}\sin 2x$

19. $x = c_1\cos 8t + c_2\sin 8t + 0.0154e^{-t} + 0.0159\cos t$

21. $y = e^{-5t}(0.0006\cos 19.3649t -$
$0.0019\sin 19.3649t) + e^{-2t}(-0.0006\cos 3t +$
$0.0127\sin 3t)$, 0.0056 cm, underdamped

23. 0.7459 rad, critically damped

25. **a)** $q(t) = 58e^{-3t} - 51e^{-4t} - 7\cos 4t - \sin 4t$
b) $q(1) = 7.2859$ C
c) $i(1) = -23.5025$ A

27. **a)** $q(t) = \frac{12}{5}e^{-3t} - \frac{13}{5}e^{-2t} + \frac{1}{5}\cos 2t + \sin 2t$
b) 0.5937 C
c) -0.8507 A
d) No

29. $\theta = e^{-t}(0.3929\cos 2t + 7.8490\sin 2t) +$
$e^{-4t}(1.6071\cos 2t - 1.2535\sin 2t) +$
$te^{-4t}(1.4793\cos 2t - 3.5503\sin 2t)$,
2.4619 (same side), underdamped

CHAPTER 5

Section 5.1

1. $\frac{5}{s}, \quad s > 0$

3. $\frac{2}{s^2} - \frac{1}{s}, \quad s > 0$

5. $\frac{e}{s-1}, \quad s > 1$

7. $\frac{1}{s^2} + \frac{2}{s-1}, \quad s > 1$

9. $\frac{1}{s+2} + \frac{2}{s^3}, \quad s > 0$

11. $\frac{s+2}{s^2+1}, \quad s > 0$

13. $\frac{2 + 2s + s^2}{s^3}, \quad s > 0$

15. $\frac{3}{s^2+36}, \quad s > 0$

17. $y(t) = e^{-2t}$

19. $y(t) = -\frac{1}{4} - \frac{1}{2}t + \frac{5}{4}e^{2t}$

21. $y(t) = -\frac{1}{3}e^{-t} + \frac{1}{3}e^{2t} - 1 + t$

23. $y(t) = \frac{9}{5}e^{-4t} + \frac{1}{5}\cos 2t + \frac{1}{10}\sin 2t$

25. $y(t) = -\frac{2}{15}e^{-t/2} + \frac{1}{3}e^{t} - \frac{1}{5}\cos t + \frac{2}{5}\sin t$

27. $i(t) = 0.05(e^{-0.02t} - \cos 2t) + 0.0005\sin 2t$

29. $q(t) = 50 - 50e^{-t/125}, \quad q(60) \approx 19$ min

31. $q(t) = 30912 - 24{,}000e^{-t/10}, \quad 0.179\%$ CO_2

Section 5.2

1. $\frac{1}{s+4}, \quad s > -4$

3. $\frac{3}{s^2-9}, \quad s > 3$

5. $\frac{4!}{s^5}, \quad s > 0$

7. $\dfrac{2}{(s+1)^3}, \quad s>-1$

9. $\dfrac{s+1}{(s+1)^2+1}, \quad s>-1$

11. $\dfrac{1}{s}(2e^{-2s}-1), \quad s>0$

13. $\dfrac{s}{s^2+1}(1+e^{-\pi s})+e^{-\pi s}\left(\dfrac{\pi}{s}+\dfrac{1}{s^2}\right), \quad s>0$

15. $\dfrac{2s^2+3s-2}{s^3}$

17. $\dfrac{3s^2+2s-2}{s^3}$

19. $\dfrac{2}{s^2-1}$

21. $\dfrac{s^4+10s^2+40}{s^3(s^2+4)}$

23. $\dfrac{s^2+1}{s(1-s^2)}$

25. $\dfrac{3s^4-5s^3+12s+24}{s^4(s+2)}$

27. $\dfrac{s^2+4es+(1-20e)}{e(s-5)(s^2+1)}$

29. $\dfrac{4}{s(s^2+4)}$

31. $M=1, \quad c=n, \quad T>0$

35. $\dfrac{s-3}{s^2-6s+10}$

37. $\dfrac{s+2}{s^2+4s+3}$

39. $\dfrac{5!}{(s+1)^6}$

41. $\dfrac{s^2+4s+5}{(s+1)^3}$

43. $\dfrac{4}{(s+1)(s^2+2s+5)}$

Section 5.3

1. $2e^{-4t}$

3. $3\cos\sqrt{2}t$

5. $\frac{1}{2}\sin 2t-\frac{1}{3}t$

7. $2\sinh t-3\cos t$

9. $\frac{1}{5}e^{-2t/5}-\frac{1}{4}t^2$

11. $\dfrac{1}{\sqrt{2}}\sin\sqrt{2}t-t$

13. $-\frac{1}{18}e^{-2t}+\frac{1}{54}e^{4t}-\frac{27}{8}e^{-t/2}$

15. $2-3\cos\sqrt{2}t$

17. $2\cos\sqrt{2}t-e^{4t}$

19. $3e^{t}-\frac{1}{2}e^{-t/2}$

21. $\frac{2}{3}e^{t/3}-\frac{1}{2}e^{-t/2}$

23. $2\sin\dfrac{t}{2}-\dfrac{1}{3}e^{t/3}$

25. $t-2e^{-t}+\dfrac{1}{2}\cos\dfrac{1}{\sqrt{2}}t$

27. $3te^{-2t}$

29. $e^{-t}(\cos t-2\sin t)$

31. $1-te^{-2t}$

33. $\frac{5}{8}e^{5t/2}+\frac{3}{8}e^{-3t/2}$

35. $e^{t}(2-t)+t$

37. $\dfrac{e^{-at}}{b}(b\cos bt-a\sin bt)$

Section 5.4

1. $y(t)=\frac{1}{4}-\frac{1}{4}\cos 2t$

3. $y(t)=2-t+\frac{1}{2}t^2-e^{-t}$

5. $y(t)=\frac{1}{5}e^{-3t}+\frac{1}{5}(5t-1)e^{2t}$

7. $y(t)=-\frac{9}{25}e^{-2t}-\frac{16}{25}e^{3t}+\frac{1}{5}te^{3t}$

9. $y(t)=\frac{3}{2}e^{-t}-\frac{1}{2}(\cos t-\sin t)-1$

11. $y(t) = \frac{1}{2}e^t - \frac{1}{2}te^t + \frac{1}{2}\cos t$

13. $y(t) = \frac{3}{65}e^{3t} + \frac{1}{20}e^{-2t} - \frac{1}{52}\sin 2t - \frac{5}{52}\cos 2t$

15. $y(t) = -6e^{2t} + 2te^{2t} + 6e^t + 4te^t + t^2e^t$

17. $y(t) = \frac{5}{36} - \frac{1}{6}t - \frac{1}{12}e^{3t} + \frac{121}{252}e^{6t} + \frac{369}{252}e^{-t}$

19. $y(t) = \frac{22}{25}e^{2t} - \frac{13}{5}te^{2t} + t^2e^{2t} + \frac{3}{25}\cos t - \frac{4}{25}\sin t$

21. $y(t) = -\frac{1}{6}e^t + \frac{1}{12}e^{2t} - \frac{1}{12}e^{-2t} + \frac{1}{6}e^{-t}$

23. $y(t) = -\frac{5}{12}e^t + \frac{5}{18}e^{2t} + \frac{1}{6}e^{-2t} + \frac{1}{6}te^{-t} - \frac{1}{36}e^{-t}$

25. $y(t) = -2t - t^2 + \frac{3}{2}e^t - \frac{3}{2}\cos t - \frac{1}{2}\sin t$

27. $\mathscr{L}\{t \sin kt\} = \dfrac{2ks}{(s^2 + k^2)^2}$

29. $\mathscr{L}\{\cosh t + t \sinh t\} = \dfrac{s(s^2 + 1)}{(s^2 - 1)^2}$

31. $\mathscr{L}\{te^t \sin t\} = \dfrac{2(s - 1)}{(s^2 - 2s + 2)^2}$

37. $y(t) = \frac{99}{500}e^{-7t} - \frac{9}{50}e^{-2t} + \frac{13}{500}\sin t - \frac{9}{500}\cos t$

39. $q(t) = \frac{75}{4} - \frac{35}{2}e^{-2t} + \frac{35}{4}e^{-4t}$

41. $y(t) = \frac{1}{8}te^{-t}\sin 4t$

Section 5.5

1. $f(t) = U(t - 2)$

3. $f(t) = tU(t - \frac{1}{2})$

5. $f(t) = t + U(t - 1)(2t^2 - t)$

7. $f(t) = 3e^{-t}[1 - U(t - 1)] - U(t - 1)$

9. $f(t) = \cos t + U\left(t - \dfrac{\pi}{2}\right)(1 - \cos t) +$
$U(t - \pi)(-\sin t - 1)$

11. $\dfrac{2}{s} - \dfrac{3e^{-s}}{s}$

13. $\dfrac{3}{s} + \dfrac{e^{-2s}}{s^2} - \dfrac{e^{-2s}}{s}$

15. $\dfrac{1}{s}\left(1 - \dfrac{2}{s^2}\right) + e^{-s}\left(\dfrac{2}{s^3} + \dfrac{2}{s^2} + \dfrac{1}{s}\right)$

17. $\dfrac{s + e^{-2\pi s}(s^2 - s + 1)}{s(s^2 + 1)}$

19. $\dfrac{e^{-s}}{s - 1}$

21. $y(t) = \frac{1}{2} - \frac{1}{2}e^{-2t} - U(t - 3)[\frac{1}{2} - \frac{1}{2}e^{-2t+6}]$

23. $y(t) = 1 - e^{-t} + U(t - 2)(t - 2) +$
$U(t - 4)[t - 1 - 3e^{4-t}]$

25. $y(t) = -\frac{1}{5} + \frac{6}{5}e^{5t} -$
$\frac{1}{625}U(t - 1)[39 + 430(t - 1) + 450(t - 1)^2 +$
$250(t - 1)^3 - 39e^{5t-5}]$

27. $y(t) = \frac{1}{2} + e^t - \frac{1}{2}e^{2t} - U(t - 2)[1 - 2e^{t-2} + e^{2t-4}]$

29. $y(t) = -\frac{1}{3} + \frac{1}{3}e^{-3t} - \frac{1}{12}U(t - 1)[-4 + e^{-3t+3} + 3e^{t-1}]$

31. $y(t) = \dfrac{1}{4}\sinh t - \dfrac{1}{2}te^{-t} -$
$\dfrac{e}{2}U(t - 1)[\sinh(t - 1) - (t - 1)e^{1-t}]$

33. $y(t) = t - \sinh t +$
$U(t - 1)[\sinh(t - 1) + \cosh(t - 1) - t]$

35. $y(t) = -3 + e^t + \cosh t + e^3U(t - 3)[t - 2 - e^{t-3}]$

37. $y(t) = \frac{5}{8}\sin 2t - \frac{1}{4}t\cos 2t -$
$\frac{1}{8}U(t - \pi)[\sin 2t - 2(t - \pi)\cos 2t]$

39. $t^3 - 2t^2 + t - 7$
$= (t + 1)^3 - 5(t + 1)^2 + 8(t + 1) - 11$

41. $t^4 + t^3 + t^2 + t + 1$
$= (t + 2)^4 - 7(t + 2)^3 + 19(t + 2)^2 - 23(t + 2) + 11$

43. $t^5 = (t + 1)^5 - 5(t + 1)^4 + 10(t + 1)^3 -$
$10(t + 1)^2 + 5(t + 1) - 1$

45. $\mathscr{L}\{U(a - t)\} = \dfrac{1 - e^{-as}}{s}$

Chapter 5 Review Exercises

1. $\dfrac{2}{s} + \dfrac{4e^{-7s}}{s}$

3. $\dfrac{1}{s^2} + \dfrac{1}{s} - \dfrac{2e^{-s}}{s} - \dfrac{e^{-s}}{s^2}$

5. $\dfrac{e}{s - \frac{1}{3}}e^{-3s}$

7. $\frac{1}{s} - \frac{2}{s}e^{-s} + \frac{3}{s}e^{-2s} + \frac{1}{s^2}e^{-2s}$

11. $f(t) = t + (1-t)U(t-1)$, $\mathscr{L}\{f(t)\} = \frac{1}{s^2} - \frac{e^{-s}}{s^2}$

13. $f(t) = e^t - e^tU(t-2)$, $\mathscr{L}\{f(t)\} = \frac{1}{s-1} - \frac{e^{2-2s}}{s-1}$

15. $f(t) = 2 - 4U(t-1)$, $\mathscr{L}\{f(t)\} = \frac{2}{s} - \frac{4}{s}e^{-s}$

17. $\frac{14(s+4)}{[(s+4)^2+49]^2}$

19. $\frac{(s+1)^2-9}{[(s+1)^2+9]^2}$

21. $2 + \cos 2t$

23. $t + e^{-3t}$

25. $[\cos(3t-15)]U(t-5)$

27. $\frac{t}{2}e^{-t}\sin t$

29. $2e^{-6t}\cos 2t - e^{-6t}\sin 2t$

31. $\frac{1}{5}e^t - \frac{1}{5}e^{-4t}$

33. $y = -\frac{2}{15}e^{3t} + \frac{3}{10}e^{-2t} - \frac{1}{6} - 3U(t-5)\left[\frac{1}{6} - \frac{1}{15}e^{3(t-5)} - \frac{1}{10}e^{-2(t-5)}\right]$

35. $y = -\frac{1}{16} + \frac{1}{8}t + \frac{1}{16}e^{-2t}\cos 2t + \frac{1}{2}e^{-2t}\sin 2t + U(t-2)\left[\frac{1}{16} - \frac{1}{8}t + \frac{3}{16}e^{4-2t}\cos(2t-4) + \frac{1}{4}e^{4-2t}\sin(2t-4)\right]$

37. $x(t) = -\frac{1}{8}e^{-t} + \frac{5}{104}e^{-5t} + \frac{1}{13}\cos t + \frac{3}{26}\sin t$

39. a) $x = \frac{500}{14}(\sin 4t - \sin\sqrt{30}t) + \frac{125}{4}\left[\cos\left(4t - \frac{3\pi}{2}\right) - \cos\sqrt{30}\left(t - \frac{3\pi}{8}\right)\right]U\left(t - \frac{3\pi}{8}\right)$

b) $x = -0.9954\cos 4t + 0.1124\sin 4t - 0.01882e^{-30t} + 1.0142e^{-t} + \left[0.1124\cos\left(4t - \frac{3\pi}{2}\right) + 0.9954\sin\left(4t - \frac{3\pi}{2}\right) + 0.1412e^{-30(t-3\pi/8)} - 0.2536e^{-(t-3\pi/8)}\right]U\left(t - \frac{3\pi}{8}\right)$

41. $x = \frac{27}{13} - \frac{27}{13}e^{-3t}\cos 2t - \frac{81}{26}e^{-3t}\sin 2t - \left[\frac{27}{13} - \frac{27}{13}e^{-3(t-0.2)}\cos 2(t-0.2) - \frac{81}{26}e^{-3(t-0.2)}\sin 2(t-0.2)\right]U(t-0.2)$

43. $q(t) = -\frac{1}{5}e^{-2t} + \frac{2}{17}e^{-4t} + \frac{7}{85}\cos t + \frac{6}{85}\sin t$

45. a) $E(t) = 25t + (75 - 25t)U(t-3)$
b) $q = t - \frac{1}{5}\sin 5t + [t - 3 - \frac{1}{5}\sin 5(t-3)]U(t-3)$

47. a) $E(t) = 24 + (104\cos 2t - 24)U(t - 2\pi)$
b) $q = 0.4 - 1.2e^{-2t} + 0.8e^{-3t} + [-2.6e^{-2(t-2\pi)} + 2.4e^{-3(t-2\pi)} + 0.2\cos 2(t-2\pi) + \sin 2(t-2\pi) - 0.4 - 1.2e^{-3(t-2\pi)} + 0.8e^{-3(t-2\pi)}]U(t-2\pi)$
c) $q(8) = -0.5470$ C

49. a) $F(t) = A\cos t + (Be^{-t}\cos t)U\left(t - \frac{\pi}{2}\right)$
b) $x = \frac{A}{10}(3e^{-3t} - 4e^{-2t} + \cos t + \sin t) + \frac{B}{10}e^{-\pi/2}\left[2e^{-3(t-\pi/2)} - 5e^{-2(t-\pi/2)} + 3e^{-(t-\pi/2)}\cos\left(t - \frac{\pi}{2}\right) - e^{-(t-\pi/2)}\sin\left(t - \frac{\pi}{2}\right)\right]U\left(t - \frac{\pi}{2}\right)$
c) $x\left(\frac{\pi}{4}\right) = 0.08670A$
$x(\pi) = -0.1007A - 0.008440B$

CHAPTER 6

Section 6.1

1. $\frac{1}{s(1+e^{-s})}$

3. $\frac{1}{s^2(1+e^{-s})} - \frac{e^{-s}}{s(1-e^{-2s})}$

5. $\frac{1-e^{-s}}{s^2(1+e^{-s})}$

7. $\frac{1+e^{-\pi s}}{1+s^2}$

9. $\frac{s}{s^2+1}$

11. $i(t) = \dfrac{1}{10}(1 - e^{-10t}) + \dfrac{1}{10}\sum_{k=1}^{\infty}(-1)^k U(t-k)[1 - e^{-10(t-k)}]$

13. $y(t) = \dfrac{1}{101}[e^{-10t} - \cos t + 10\sin t + U(t-\pi)(e^{-10(t-\pi)} + \cos t - 10\sin t)]$

15. $y(t) = -2 + t + 2e^{-t} + te^{-t} + \sum_{k=1}^{\infty}(-1)^k(t-k)U(t-k) + \sum_{k=1}^{\infty}[-2 + 2e^{-(t-2k)} + (t-2k)e^{-(t-2k)}]U(t-2k) + \sum_{k=1}^{\infty}[1 + e^{-(t-2k+1)}]U(t-2k+1)$

17. $\mathscr{L}\{\sin kt\} = \dfrac{k}{s^2 + k^2}$

Section 6.2

1. $y(t) = e^{2(t-3)}U(t-3)$

3. $y(t) = -e^{3t} + e^{3(t-2)}U(t-2)$

5. $y(t) = 2e^{-t+2}U(t-2)$

7. $y(t) = \frac{2}{5}(e^t - e^{-4t}) + \frac{1}{5}(e^{t-2} - e^{-4t+8})U(t-2)$

9. $y(t) = -\frac{1}{6}(e^{5t} - 5e^{-t}) + \frac{1}{3}(e^{5t-10} - e^{-t+2})U(t-2)$

11. $y(t) = -te^{-2t} + (t - \frac{1}{2})e^{-2t+1}U(t - \frac{1}{2})$

13. $y(t) = \frac{3}{4}(e^{2t} - 2t - 1) + \frac{1}{2}(e^{2t-2} - 1)U(t-1)$

15. $y(t) = \frac{1}{2}\sinh 2t + U(t-2)\sinh 2(t-2)$

Section 6.3

1. $\dfrac{1}{(s-1)(s^2+1)}$

3. $\dfrac{1}{s^2(s-2)}$

5. $\dfrac{2}{s^3(s^2+1)}$

7. $\dfrac{2}{s^4}$

9. $\dfrac{6}{s^4(s+1)}$

11. $\dfrac{s}{(s-2)(s^2+1)}$

13. $\frac{1}{3}(e^{2t} - e^{-t})$

15. $\frac{1}{2}(e^t - \sin t - \cos t)$

17. $\frac{1}{6}(2\cos\sqrt{2}t + \sqrt{2}\sin\sqrt{2}t - 2e^{-2t})$

19. $\frac{1}{5}e^t(2\sin t - \cos t) - \frac{1}{5}e^{-t}$

21. $\frac{1}{13}e^{2t}(2\cos 3t + 3\sin 3t) - \frac{2}{13}$

23. $\dfrac{1}{s(s-2)^2}$

25. $\dfrac{s-1}{s(s^2-2s+5)}$

27. $\left(\dfrac{1}{se}\right)e^{-s}$

29. $\dfrac{2s}{(s^2+1)^2}$

31. $y(t) = 2$

33. $y(t) = 1 - t$

35. $y(t) = t + \frac{3}{2}\sinh 2t$

37. $y(t) = -\frac{1}{8}e^{-t} + \frac{3}{8}e^t + \frac{1}{2}te^t + \frac{1}{4}t^2e^t$

39. $y(t) = 1 - t^2 - \cos 2t$

41. $y(t) = 1 - e^{-t}$

43. $y(t) = 1 - t$

45. $y(t) = \sinh t$

47. $e^{2t}(\sin 1 + \cos 1 - \cos e^{-t}) - e^t \sin 1$

49. $\frac{1}{2}t\sin t$

Section 6.4

1. $H(s) = \dfrac{1}{s+2} + \dfrac{-1}{s+3},\quad y_p(t) = \dfrac{1}{6}e^t - \dfrac{2}{3}e^{-2t} + \dfrac{1}{2}e^{-3t}$

3. $H(s) = \frac{\frac{1}{2}}{s+1} + \frac{-\frac{1}{2}}{s+3}$,
$y_p(t) = \frac{1}{8}(2t^2 + 2t + 1)e^{-t} - \frac{1}{8}e^{-3t}$

5. $H(s) = \frac{\frac{2}{5}}{2s+1} + \frac{-\frac{1}{5}}{s+3}$,
$y_p(t) = \frac{7}{25}\sin t + \frac{1}{25}\cos t - \frac{4}{25}e^{-t/2} + \frac{3}{25}e^{-3t}$

7. $H(s) = \frac{1}{s^2+4}$, $y_p(t) = \frac{5}{6}(2\sin t - \sin 2t)$

9. $H(s) = \frac{1}{s^2+16}$, $y_p(t) = -\frac{4}{7}\cos 4t + \frac{4}{7}\cos 3t$

11. $H(s) = \frac{1}{s^2+\omega^2}$, $y_p(t) = \frac{F_0}{2m\omega} t \sin \omega t$

Chapter 6 Review Exercises

1. $y = -\frac{1}{2}e^{-3t} + \frac{1}{2}e^{-t} - (e^{-3(t-2)} + e^{-(t-2)})U(t-2) + (-\frac{1}{2}e^{-3(t-4)} + \frac{1}{2}e^{-(t-4)})U(t-4)$

3. **a)** $f(t) = 5250\delta(t)$
b) $x = 0.7746e^{-5t}\sin\sqrt{375}t$
d) 0.5336 cm

5. **a)** $x(t) = 5e^{-3t} - 10e^{-2t} + 5e^{-t} + \sum_{k=1}^{\infty}(4.3233e^{-3(t-2k)} - 8.6466e^{-2(t-2k)} + 4.3233e^{-(t-2k)})U(t-2k)$
b) 0.7350, 0.8604

CHAPTER 7

Section 7.1

1. $\frac{dx}{dt} = 1.03x$, $\frac{dy}{dt} = kx$ (k a constant)

3. $$\frac{dx}{dt} = -0.05x \qquad + 10$$
$$\frac{dy}{dt} = 0.02x - 0.12y + 0.04z$$
$$\frac{dz}{dt} = \qquad 0.02y - 0.04z$$

5. $(R_1 + R_2)i_1 + L_1 i_2' = E(t)$, $L_2(i_1 - i_2)' - L_1 i_2' = 0$, where $i_3 = i_1 - i_2$. Alternative formulations with the same solutions exist, depending on the loops selected.

7. $Li_1'' + \frac{1}{C}i_2 = E'(t)$, $R(i_1 - i_2)' - \frac{1}{C}i_2 = 0$,
where $i_3 = i_1 - i_2$. Equivalent formulations are possible.

9. $R_1 i_1 + L_2 i_1' + R_2 i_2 + L_1 i_2' = E(t)$,
$R_3(i_1 - i_2) + L_3(i_1 - i_2)' - R_2 i_2 - L_1 i_2' = 0$,
where $i_3 = i_1 - i_2$. Equivalent formulations are possible.

11. $$\frac{dx}{dt} = ax - bxy \quad (a < 0, b > 0)$$
$$\frac{dy}{dt} = m\left(1 - \frac{y}{M}\right)y - nxy \quad (m, n > 0)$$

13. For a fixed price, as Q increases, dP/dt decreases and possibly also becomes negative. This observation implies that as the quantity supplied increases, the price will not rise as fast. If Q becomes high enough, then the price will decrease.

Next, consider dQ/dt. For a fixed quantity, as P increases, dQ/dt gets larger. Thus, as the market price increases, the quantity supplied will increase at a faster rate. If P is too small, dQ/dt will be negative and the quantity supplied will decrease. This observation is the traditional explanation of the effect of market price levels on the quantity supplied.

Section 7.2

1. $\frac{dx}{dt} = -e^t = -y$, $\frac{dy}{dt} = e^t = -x$

3. $\frac{dx}{dt} = e^{2t}$, $\frac{dy}{dt} = \frac{3}{4}e^{2t} - \frac{3}{4}e^{-2t}$

5. $\frac{dx}{dt} = \sinh t + \cos t$, $\frac{dy}{dt} = \cosh t$

7. $\frac{dx}{dt} = (22 + 104t)e^{4t}$, $\frac{dy}{dt} = (37 + 52t)e^{4t}$

9. $\frac{dx}{dt} = 2e^{2t}$, $\frac{dy}{dt} = e^t$

11. The rest point (0, 0) is stable.

13. All the points along the horizontal line $y = 1$ are rest points, and (1, 0) is a rest point. The point (1, 0) is stable; the rest points $(x, 1)$ are all unstable.

15. $\dfrac{dx_1}{dt} = x_2, \quad \dfrac{dx_2}{dt} = -Q(t)x_1 - P(t)x_2$

17. a) $\dfrac{dx_1}{dt} = x_2, \quad \dfrac{dx_2}{dt} = -\dfrac{g}{\ell} \sin x_1$

b) The rest points are $(0, 0)$, $(\pm\pi, 0)$, $(\pm 2\pi, 0)$, $(\pm 3\pi, 0)$, . . .

Section 7.3

1. $\mathbf{X}' = \begin{pmatrix} 3 & -1 \\ 1 & 3 \end{pmatrix}\mathbf{X}$

3. $\mathbf{X}' = \begin{pmatrix} 3 & -\frac{1}{2} \\ 8 & -1 \end{pmatrix}\mathbf{X}$

5. $\mathbf{X}' = \begin{pmatrix} 2 & -1 \\ 4 & -3 \end{pmatrix}\mathbf{X}$

7. $\dfrac{dx}{dt} = 5x - y, \quad \dfrac{dy}{dt} = 4x - y$

9. $\dfrac{dx}{dt} = 2x + 3y, \quad \dfrac{dy}{dt} = 4y$

11. $\dfrac{d\mathbf{X}}{dt} = \begin{pmatrix} 2e^{2t} \\ -2e^{-2t} \end{pmatrix}$

13. $\dfrac{d\mathbf{X}}{dt} = \begin{pmatrix} e^t \cos t - e^t \sin t \\ -e^t \sin t - e^t \cos t \end{pmatrix}$

15. $\dfrac{d\mathbf{X}}{dt} = \begin{pmatrix} -3e^{3t} \\ e^{3t} + 3te^{3t} \end{pmatrix}$

17. $\dfrac{d}{dt}(c\mathbf{X}) = \begin{pmatrix} c - 2cte^{-t} \\ -cte^{-t} \end{pmatrix}$

19. $\dfrac{d}{dt}(c_1\mathbf{X}_1 + c_2\mathbf{X}_2)$

$$= \begin{pmatrix} (2c_1 + 3c_2)e^{2t} \cos 3t + (-3c_1 + 2c_2)e^{2t} \sin 3t \\ (-3c_1 + 2c_2)e^{2t} \cos 3t + (-2c_1 - 3c_2)e^{2t} \sin 3t \end{pmatrix}$$

21. Linearly independent

23. Linearly independent

25. Linearly independent

27. $\lambda^2 - 4 = 0$, characteristic equation; eigenvalues are $\lambda = -2, 2$.

29. $\lambda^2 - 2\lambda + 2 = 0$, characteristic equation; eigenvalues are $\lambda = 1 \pm i$.

31. $\lambda^2 = 0$, characteristic equation; eigenvalues are $\lambda = 0, 0$.

33. $\lambda^2 - 9\lambda + 21 = 0$, characteristic equation; eigenvalues are $\lambda = \frac{9}{2} \pm \frac{1}{2}\sqrt{3}i$.

Section 7.4

1. $\mathbf{X} = c_1\begin{pmatrix} 0 \\ 1 \end{pmatrix}e^{-t} + c_2\begin{pmatrix} 2 \\ 1 \end{pmatrix}e^{3t}$

3. $\mathbf{X} = c_1\begin{pmatrix} 1 \\ 2 \end{pmatrix}e^{-t} + c_2\begin{pmatrix} 2 \\ 1 \end{pmatrix}e^{2t}$

5. $\mathbf{X} = c_1\begin{pmatrix} 1 \\ -9 \end{pmatrix}e^{-2t} + c_2\begin{pmatrix} 1 \\ -1 \end{pmatrix}e^{6t}$

7. $\mathbf{X} = c_1\begin{pmatrix} 1 \\ -2 \end{pmatrix}e^{-3t} + c_2\begin{pmatrix} -2 \\ 1 \end{pmatrix}e^{3t}$

9. $\mathbf{X} = c_1\begin{pmatrix} 1 \\ -1 \end{pmatrix}e^{2t} + c_2\begin{pmatrix} 1 \\ -2 \end{pmatrix}e^{3t}$

11. $\mathbf{X} = c_1\begin{pmatrix} 6 \\ -5 \end{pmatrix}e^{-3t} + c_2\begin{pmatrix} 1 \\ 2 \end{pmatrix}e^{11t/2}$

13. $\mathbf{X} = c_1\begin{pmatrix} 1 \\ -2 \end{pmatrix} + c_2\begin{pmatrix} 3 \\ 1 \end{pmatrix}e^{7t}$

15. $\mathbf{X} = c_1\begin{pmatrix} -2 \\ 1 \end{pmatrix}e^{-t} + c_2\begin{pmatrix} 2 \\ 1 \end{pmatrix}e^{3t}$

17. $\mathbf{X} = c_1 e^{-t}\begin{pmatrix} -\sin t \\ \cos t \end{pmatrix} + c_2 e^{-t}\begin{pmatrix} -\cos t \\ -\sin t \end{pmatrix}$

19. $\mathbf{X} = c_1 e^t\begin{pmatrix} -2 \cos \sqrt{3}t - \sqrt{3} \sin \sqrt{3}t \\ 4 \cos \sqrt{3}t \end{pmatrix}$

$$+ c_2 e^t\begin{pmatrix} -\sqrt{3} \cos \sqrt{3}t + 2 \sin \sqrt{3}t \\ -4 \sin \sqrt{3}t \end{pmatrix}$$

21. $\mathbf{X} = c_1\begin{pmatrix} 2 \cos \sqrt{5}t + \sqrt{5} \sin \sqrt{5}t \\ \cos \sqrt{5}t \end{pmatrix}$

$$+ c_2\begin{pmatrix} \sqrt{5} \cos \sqrt{5}t - 2 \sin \sqrt{5}t \\ -\sin \sqrt{5}t \end{pmatrix}$$

23. $\mathbf{X} = c_1 e^{2t}\begin{pmatrix} 2\cos\sqrt{3}t \\ -\cos\sqrt{3}t - \sqrt{3}\sin\sqrt{3}t \end{pmatrix}$

$+ c_2 e^{2t}\begin{pmatrix} -2\sin\sqrt{3}t \\ -\sqrt{3}\cos\sqrt{3}t + \sin\sqrt{3}t \end{pmatrix}$

25. $\mathbf{X} = c_1 e^{4t}\begin{pmatrix} \cos t \\ -\sin t \end{pmatrix} + c_2 e^{4t}\begin{pmatrix} -\sin t \\ -\cos t \end{pmatrix}$

27. $\mathbf{X} = c_1 e^{2t}\begin{pmatrix} 3\cos\sqrt{7}t - \sqrt{7}\sin\sqrt{7}\,t \\ 2\cos\sqrt{7}t \end{pmatrix}$

$+ c_2 e^{2t}\begin{pmatrix} -\sqrt{7}\cos\sqrt{7}t - 3\sin\sqrt{7}t \\ -2\sin\sqrt{7}t \end{pmatrix}$

29. $\mathbf{X} = c_1 e^{3t/2}\begin{pmatrix} \cos\frac{t}{2} + \sin\frac{t}{2} \\ 6\cos\frac{t}{2} \end{pmatrix}$

$+ c_2 e^{3t/2}\begin{pmatrix} \cos\frac{t}{2} - \sin\frac{t}{2} \\ -6\sin\frac{t}{2} \end{pmatrix}$

31. $\mathbf{X} = \begin{pmatrix} 1 \\ 2 \end{pmatrix}e^{t} - \begin{pmatrix} 1 \\ 1 \end{pmatrix}e^{3t} = \begin{pmatrix} e^{t} - e^{3t} \\ 2e^{t} - e^{3t} \end{pmatrix}$

33. $\mathbf{X} = e^{4t}\begin{pmatrix} \cos t - \sin t \\ -\sin t \end{pmatrix}$

35. $\mathbf{X} = e^{5t/2}\begin{pmatrix} \frac{2\sqrt{3}}{3}\sin\frac{\sqrt{3}t}{2} \\ -\cos\frac{\sqrt{3}t}{2} - \frac{1}{\sqrt{3}}\sin\frac{\sqrt{3}t}{2} \end{pmatrix}$

Section 7.5

1. $\mathbf{X} = c_1\begin{pmatrix} 1 \\ 1 \end{pmatrix}e^{3t} + c_2\begin{pmatrix} t+1 \\ 1 \end{pmatrix}e^{3t}$

3. $\mathbf{X} = c_1\begin{pmatrix} 1 \\ 2 \end{pmatrix} + c_2\begin{pmatrix} t \\ 2t-1 \end{pmatrix}$

5. $\mathbf{X} = c_1\begin{pmatrix} 1 \\ 1 \end{pmatrix}e^{t} + c_2\begin{pmatrix} t+\frac{1}{3} \\ t \end{pmatrix}e^{t}$

7. $\mathbf{X} = c_1\begin{pmatrix} 1 \\ 1 \end{pmatrix}e^{2t} + c_2\begin{pmatrix} t-\frac{1}{3} \\ t \end{pmatrix}e^{2t}$

9. $\mathbf{X} = c_1\begin{pmatrix} 1 \\ -1 \end{pmatrix}e^{-3t} + c_2\begin{pmatrix} t \\ -t-1 \end{pmatrix}e^{-3t}$

11. $\mathbf{X} = c_1\begin{pmatrix} -3 \\ 1 \end{pmatrix}e^{-2t} + c_2\begin{pmatrix} -3t-1 \\ t \end{pmatrix}e^{-2t}$

13. $\mathbf{X} = c_1\begin{pmatrix} 1 \\ 4 \end{pmatrix}e^{t} + c_2\begin{pmatrix} t \\ 4t-2 \end{pmatrix}e^{t}$

15. $\mathbf{X} = c_1\begin{pmatrix} 1 \\ 1 \end{pmatrix}e^{5t} + c_2\begin{pmatrix} t \\ t-\frac{1}{2} \end{pmatrix}e^{5t}$

17. $\mathbf{X} = c_1\begin{pmatrix} 1 \\ -1 \end{pmatrix}e^{4t} + c_2\begin{pmatrix} t \\ -t-\frac{1}{3} \end{pmatrix}e^{4t}$

19. $\mathbf{X} = c_1\begin{pmatrix} -4 \\ 1 \end{pmatrix}e^{t} + c_2\begin{pmatrix} -4t+2 \\ t \end{pmatrix}e^{t}$

21. $\mathbf{X} = \begin{pmatrix} 3t+1 \\ 3t-2 \end{pmatrix}e^{-2t}$

23. $\mathbf{X}\begin{pmatrix} 8t \\ -4t-1 \end{pmatrix}e^{2t}$

25. $\mathbf{X} = \begin{pmatrix} -2-10t \\ 3-10t \end{pmatrix}e^{3t}$

Section 7.6

1. Seasonal variations, nonconformity of the environment, effects of other interactions, unexpected disasters, etc.

3. This model assumes the number of interactions is proportional to the product of x and y:

$$\frac{dx}{dt} = (a - by)x, \quad a < 0,$$

$$\frac{dy}{dt} = m\left(1 - \frac{y}{M}\right)y - nxy = y\left(m - \frac{m}{M}y - nx\right).$$

Rest points are (0, 0), unstable, and (0, M), stable.

5. **a)** Logistic growth occurs in the absence of the competitor, and involves a simple interaction between the species: growth dominates the competition when either population is small, so it is difficult to drive either species to extinction.

b) a: per capita growth rate for trout
m: per capita growth rate for bass
b: intensity of competition to the trout
n: intensity of competition to the bass
k_1: environmental carrying capacity for the trout
k_2: environmental carrying capacity for the bass
$\frac{a}{b}$: growth versus competition or net growth of trout
$\frac{m}{n}$: relative survival of bass

c) $\frac{dx}{dt}=0$ when $x=0$ or $y=\frac{a}{b}-\frac{a}{bk_1}x$,

$\frac{dy}{dt}=0$ when $y=0$ or $y=k_2-\frac{k_2 n}{m}x$.

By picking $a/b > k_2$ and $m/n > k_1$ we insure that an equilibrium point exists inside the first quadrant.

Section 7.7

1. $f'(y) = ay^{a-1}e^{-by} - by^a e^{-by} = 0$ yields $y = a/b$.

3. $\frac{dx}{dt} = (a - by)x - kx$, $\frac{dy}{dt} = (-m + nx)y - ky$, where k is a common rate of insecticide application. If $a > k$, the new equilibrium levels are

$$\bar{x} = \frac{m+k}{n} \quad \text{and} \quad \bar{y} = \frac{a-k}{b}.$$

5. a) In the absence of cooperation,

$$\frac{dx}{dt} = -ax \quad \text{and} \quad \frac{dy}{dt} = -my,$$

and both population levels decline exponentially.

b) a: self-regulation of species x or its constant rate of decline
b: intensity of cooperation of species x with species y
m: self-regulation of species y or its constant rate of decline
n: intensity of cooperation of species y with species x

c) Equilibrium levels: $(0, 0)$ is stable and $(m/n, a/b)$ is unstable.

e) If (x_0, y_0) lies on a trajectory, then

$$b(y - y_0) - a\ln\left(\frac{y}{y_0}\right) = n(x - x_0) - m\ln\left(\frac{x}{x_0}\right).$$

f) The model predicts that if $x > m/n$ and $y > a/b$, then both populations will grow and flourish.

Section 7.8

1.

	h = 0.250		h = 0.125		Analytical Solution	
t	$x(t)$	$y(t)$	$x(t)$	$y(t)$	$x(t)$	$y(t)$
0.0	1.0	0.0	1.0	0.0	1.0	0.0
0.250	2.119	1.340	2.133	1.354	2.135	1.356
0.750	20.920	20.448	21.436	20.964	21.497	21.024

3.

	h = 0.250		h = 0.125		Analytical Solution	
t	$x(t)$	$y(t)$	$x(t)$	$y(t)$	$x(t)$	$y(t)$
0.0	1.0	1.0	1.0	1.0	1.0	1.0
0.250	1.237	1.648	1.237	1.649	1.237	1.649
0.750	1.125	4.479	1.121	4.482	1.120	4.482

5.

t	h = 0.250 x(t)	h = 0.250 y(t)	h = 0.125 x(t)	h = 0.125 y(t)	Analytical Solution x(t)	Analytical Solution y(t)
0.0	0.0	2.0	0.0	2.0	0.0	2.0
0.250	1.771	2.255	1.772	2.255	1.772	2.255
0.750	8.981	4.703	8.987	4.705	8.988	4.705

7.

t	h = 0.250 x(t)	h = 0.250 y(t)	h = 0.125 x(t)	h = 0.125 y(t)	Analytical Solution x(t)	Analytical Solution y(t)
0.0	2.0	−1.0	2.0	−1.0	2.0	−1.0
0.250	4.697	−3.169	4.702	−3.180	4.702	−3.180
0.750	23.901	−23.994	23.975	−24.167	23.982	−24.184

9.

t	h = 0.250 x(t)	h = 0.250 y(t)	h = 0.125 x(t)	h = 0.125 y(t)
0.0	2.0	1.0	2.0	1.0
0.250	1.542	1.060	1.542	1.060
0.750	0.703	1.648	0.702	1.648

11.

t	h = 0.250 x(t)	h = 0.250 y(t)	h = 0.125 x(t)	h = 0.125 y(t)
0.0	2.0	1.0	2.0	1.0
0.250	1.409	0.581	1.408	0.584
0.750	1.034	0.343	1.033	0.344

Section 7.9

1. $x(t) = -2e^t + \frac{3}{2}e^{2t} + \frac{1}{2}e^{-2t}, \quad y(t) = \frac{1}{2}e^{2t} - \frac{1}{2}e^{-2t}$

3. $x(t) = e^t, \quad y(t) = -e^{-t}$

5. $x(t) = -\frac{1}{5}e^{2t} - \frac{4}{5}e^{-3t}, \quad y(t) = -\frac{3}{5}e^{2t} + \frac{8}{5}e^{-3t}$

7. $x(t) = \frac{5}{8}e^{3t} + \frac{1}{8}e^{-3t} - \frac{1}{8}e^t - \frac{5}{8}e^{-t},$
$y(t) = \frac{1}{8}e^{3t} + \frac{3}{8}e^{-3t} + \frac{1}{8}e^t - \frac{5}{8}e^{-t}$

9. $x(t) = \frac{7}{36} + \frac{1}{6}t + \frac{1}{4}e^{-2t} - \frac{4}{9}e^{-3t},$
$y(t) = \frac{17}{18} - \frac{1}{3}t + \frac{1}{2}e^{-2t} - \frac{4}{9}e^{-3t}$

11. $x(t) = \cosh t - \cos t, \quad y(t) = \cosh t - \sin t$

13. $x(t) = -6t - \frac{17}{2}e^{-t/3} + \frac{19}{2}e^{t/3},$
$y(t) = 9 + 9t + \frac{1}{2}e^t + \frac{17}{2}e^{-t/3} - 19e^{t/3}$

15. $x(t) = e^{-2t/3}, \quad y(t) = -\frac{1}{5}e^{-2t/3} + \frac{6}{5}e^t$

17. $x(t) = t + 3e^{-t} - 2e^{-3t}, \quad y(t) = 1 - t + 2e^{-3t}$

Chapter 7 Review Exercises

1. b) Linearly independent

c) $\mathbf{X} = c_1\begin{pmatrix}3\\1\end{pmatrix}e^t + c_2\begin{pmatrix}1\\1\end{pmatrix}e^{-t}$

3. $\mathbf{X} = c_1\begin{pmatrix}1\\1\end{pmatrix}e^{-3t} + c_2\left[\begin{pmatrix}1\\1\end{pmatrix}t + \begin{pmatrix}1\\\frac{6}{5}\end{pmatrix}\right]e^{-3t}$

5. $\mathbf{X} = c_1\left[\begin{pmatrix}2\\-1\end{pmatrix}\cos t + \begin{pmatrix}0\\-1\end{pmatrix}\sin t\right]e^{2t} + c_2\left[\begin{pmatrix}0\\-1\end{pmatrix}\cos t - \begin{pmatrix}2\\-1\end{pmatrix}\sin t\right]e^{2t}$

7. $\mathbf{X} = c_1\left[\begin{pmatrix}1\\0\end{pmatrix}\cos 2t + \begin{pmatrix}0\\-1\end{pmatrix}\sin 2t\right]e^{2t} + c_2\left[\begin{pmatrix}0\\-1\end{pmatrix}\cos 2t - \begin{pmatrix}1\\0\end{pmatrix}\sin 2t\right]e^{2t}$

9. $\mathbf{X} = c_1\begin{pmatrix}1\\-1\end{pmatrix}e^{-5t} + c_2\begin{pmatrix}1\\1\end{pmatrix}e^t$

11. $\mathbf{X} = c_1\left[\begin{pmatrix}1\\0\end{pmatrix}\cos 3t + \begin{pmatrix}0\\3\end{pmatrix}\sin 3t\right]e^{2t} + c_2\left[\begin{pmatrix}0\\3\end{pmatrix}\cos 3t - \begin{pmatrix}1\\0\end{pmatrix}\sin 3t\right]e^{2t}$

13. $\mathbf{X} = c_1\begin{pmatrix}\frac{5}{2}\\ \frac{3}{2} - \frac{1}{2}\sqrt{34}\end{pmatrix}e^{\sqrt{34}t/2} + c_2\begin{pmatrix}\frac{5}{2}\\ \frac{3}{2} + \frac{1}{2}\sqrt{34}\end{pmatrix}e^{-\sqrt{34}t/2}$

15. $\mathbf{X} = c_1\begin{pmatrix}1\\1\end{pmatrix}e^{-3t} + c_2\left[\begin{pmatrix}1\\1\end{pmatrix}t + \begin{pmatrix}\frac{1}{4}\\0\end{pmatrix}\right]e^{-3t}$

17. $\mathbf{X} = c_1\begin{pmatrix}1\\2\end{pmatrix} + c_2\left[\begin{pmatrix}1\\2\end{pmatrix}t + \begin{pmatrix}\frac{1}{4}\\0\end{pmatrix}\right]$

19. $\mathbf{X} = c_1\left[\begin{pmatrix}1\\0\end{pmatrix}\cos t + \begin{pmatrix}0\\1\end{pmatrix}\sin t\right]e^{2t} + c_2\left[\begin{pmatrix}0\\1\end{pmatrix}\cos t - \begin{pmatrix}1\\0\end{pmatrix}\sin t\right]e^{2t}$

21. $\mathbf{X} = c_1\left[\begin{pmatrix}2\\0\end{pmatrix}\cos t + \begin{pmatrix}0\\1\end{pmatrix}\sin t\right]e^{-t} + c_2\left[\begin{pmatrix}0\\1\end{pmatrix}\cos t - \begin{pmatrix}2\\0\end{pmatrix}\sin t\right]e^{-t}$

23. $x(t) = -e^{7t} + e^{8t}$
$y(t) = -e^{7t} + e^{8t}$

25. $\mathbf{X} = \left[\begin{pmatrix}13\\-3\end{pmatrix}\cos 2t + \begin{pmatrix}0\\-2\end{pmatrix}\sin 2t\right]e^{t} + \left[\begin{pmatrix}0\\12\end{pmatrix}\cos 2t + \begin{pmatrix}78\\-18\end{pmatrix}\sin 2t\right]e^{t}$

27. $\mathbf{X} = \left[\begin{pmatrix}17\\5\end{pmatrix}\cos 3t + \begin{pmatrix}0\\3\end{pmatrix}\sin 3t\right]e^{-t} + \left[\begin{pmatrix}0\\15\end{pmatrix}\cos 3t - \begin{pmatrix}85\\25\end{pmatrix}\sin 3t\right]e^{-t}$

29. $\mathbf{X} = \begin{pmatrix}2\\1\end{pmatrix}e^{10t} + 2\begin{pmatrix}3\\2\end{pmatrix}e^{3t}$

CHAPTER 8

Section 8.1

1. $\dfrac{d\mathbf{X}}{dt} = \begin{pmatrix}3 & -4\\2 & -3\end{pmatrix}\mathbf{X} + \begin{pmatrix}1\\t\end{pmatrix}$

3. $\dfrac{d\mathbf{X}}{dt} = \begin{pmatrix}0 & -1\\-1 & 0\end{pmatrix}\mathbf{X} + \begin{pmatrix}2e^{3t}\\e^{3t} - e^{-3t}\end{pmatrix}$

5. $\dfrac{d\mathbf{X}}{dt} = \begin{pmatrix}-1 & 1\\0 & -1\end{pmatrix}\mathbf{X} + \begin{pmatrix}e^{-t}\cos t\\te^{-t}\end{pmatrix}$

7. $\dfrac{dx}{dt} = 2x + 3y + te^{t}, \quad \dfrac{dy}{dt} = -3x + 2y + e^{t}$

9. $\dfrac{dx}{dt} = -x, \quad \dfrac{dy}{dt} = -y - 4z, \quad \dfrac{dz}{dt} = y - z$

11. $\dfrac{d\mathbf{X}}{dt} = \begin{pmatrix}4e^{t} + 4\\2e^{t} + 3\end{pmatrix}$

13. $\dfrac{d\mathbf{X}}{dt} = \begin{pmatrix}7e^{t} + 4te^{t}\\5e^{t} + 2te^{t}\end{pmatrix}$

15. $\dfrac{d\mathbf{X}}{dt} = \begin{pmatrix}\frac{1}{2}\cos t\\-\frac{1}{2}\sin t\end{pmatrix}$

17. $\dfrac{dx_1}{dt} = x_2, \quad \dfrac{dx_2}{dt} = 3x_2 + 5e^{-2t} + \sin t$

19. $\dfrac{d\mathbf{X}}{dt} = \begin{pmatrix}0 & 1\\1 & 0\end{pmatrix}\mathbf{X} + \begin{pmatrix}0\\\sec t\end{pmatrix}$

21. $W[\mathbf{X}_1, \mathbf{X}_2, \mathbf{X}_3] = -e^{6t} \neq 0$; fundamental set

23. $W[\mathbf{X}_1, \mathbf{X}_2, \mathbf{X}_3] = 6e^{2t} \neq 0$; fundamental set

25. $W[\mathbf{X}_1, \mathbf{X}_2, \mathbf{X}_3] = e^{-6t} \neq 0$; fundamental set

27. $x(t) = e^{t}, \quad y(t) = -e^{t}$

29. $x(t) = (1 + 6t)e^{2t}, \quad y(t) = 3(1 + 2t)e^{2t}$

Section 8.2

1. $\mathbf{X} = c_1\begin{pmatrix}-1\\1\end{pmatrix}e^{-2t} + c_2\begin{pmatrix}3\\1\end{pmatrix}e^{2t} + \begin{pmatrix}-2e^{t}\\0\end{pmatrix}$

3. $\mathbf{X} = c_1\begin{pmatrix}1\\1\end{pmatrix}e^{-t} + c_2\begin{pmatrix}-1\\1\end{pmatrix}e^{t} + \begin{pmatrix}\frac{1}{4} - \frac{1}{2}t\\ \frac{1}{4} + \frac{1}{2}t\end{pmatrix}e^{t} + \begin{pmatrix}0\\1\end{pmatrix}$

5. $\mathbf{X} = c_1\begin{pmatrix}1\\-2\end{pmatrix}e^{-3t} + c_2\begin{pmatrix}1\\1\end{pmatrix}e^{3t} - \frac{1}{5}\begin{pmatrix}1\\3\end{pmatrix}e^{2t}$

7. $\mathbf{X} = c_1\begin{pmatrix}1\\1\end{pmatrix}e^{-t} + c_2\begin{pmatrix}-1\\1\end{pmatrix}e^{t} + \begin{pmatrix}\frac{5}{8}\\ \frac{1}{8}\end{pmatrix}e^{3t} + \begin{pmatrix}\frac{1}{8}\\ \frac{3}{8}\end{pmatrix}e^{-3t}$

9. $\mathbf{X}=c_1\begin{pmatrix}\cos t\\ \sin t\end{pmatrix}+c_2\begin{pmatrix}-\sin t\\ \cos t\end{pmatrix}+\frac{1}{2}\begin{pmatrix}\sin t\\ \cos t\end{pmatrix}$

11. $\mathbf{X}=c_1e^{2t}\begin{pmatrix}\cos 3t\\ -\sin 3t\end{pmatrix}+c_2e^{2t}\begin{pmatrix}\sin 3t\\ \cos 3t\end{pmatrix}+\frac{1}{9}e^{2t}\begin{pmatrix}3t\\ 4\end{pmatrix}$

13. $\mathbf{X}=\left[c_1\begin{pmatrix}\cos t\\ -\sin t\end{pmatrix}+c_2\begin{pmatrix}-\sin t\\ -\cos t\end{pmatrix}+\begin{pmatrix}1\\ -t\end{pmatrix}+\begin{pmatrix}\sin t\\ \cos t\end{pmatrix}\frac{t}{2}+\frac{1}{4}\begin{pmatrix}\cos t\\ \sin t\end{pmatrix}\right]e^t$

15. $\mathbf{X}=c_1\begin{pmatrix}1\\ 1\end{pmatrix}e^{2t}+c_2\left[\begin{pmatrix}1\\ 1\end{pmatrix}t+\begin{pmatrix}0\\ \frac{1}{3}\end{pmatrix}\right]e^{2t}-\begin{pmatrix}3t+1\\ 3t+1\end{pmatrix}e^t$

17. $\mathbf{X}=c_1\begin{pmatrix}1\\ 0\end{pmatrix}e^{-t}+c_2\left[\begin{pmatrix}1\\ 0\end{pmatrix}t+\begin{pmatrix}0\\ 1\end{pmatrix}\right]e^{-t}+\begin{pmatrix}\sin t\\ 0\end{pmatrix}e^{-t}+\begin{pmatrix}\frac{1}{6}t^3\\ \frac{1}{2}t^2\end{pmatrix}e^{-t}$

19. $\mathbf{X}=c_1\begin{pmatrix}1\\ -1\end{pmatrix}e^{2t}+c_2\left[\begin{pmatrix}1\\ -1\end{pmatrix}t+\begin{pmatrix}1\\ 0\end{pmatrix}\right]e^{2t}+\frac{1}{2}\begin{pmatrix}t^3+t^2\\ -t^3+2t^2-4t\end{pmatrix}e^{2t}$

21. $x(t)=-\frac{75}{4}e^{-2t/5}-\frac{75}{2}e^{-t/5}+\frac{365}{4}$, $t\approx 3.5$ hr

Section 8.3

1. $\mathbf{X}=c_1\begin{pmatrix}-1\\ 4\\ 1\end{pmatrix}e^t+c_2\begin{pmatrix}1\\ 2\\ 1\end{pmatrix}e^{3t}+c_3\begin{pmatrix}-1\\ 1\\ 1\end{pmatrix}e^{-2t}$

3. $\mathbf{X}=c_1\begin{pmatrix}1\\ 0\\ 0\end{pmatrix}e^{-t}+c_2\begin{pmatrix}0\\ \sqrt{6}\\ 3\end{pmatrix}e^{(1+\sqrt{6})t}+c_3\begin{pmatrix}0\\ -\sqrt{6}\\ 3\end{pmatrix}e^{(1-\sqrt{6})t}$

5. $\mathbf{X}=c_1\begin{pmatrix}7\\ -2\\ 13\end{pmatrix}e^{-t}+c_2\begin{pmatrix}1\\ 0\\ 1\end{pmatrix}e^t+c_3\begin{pmatrix}1\\ 1\\ 1\end{pmatrix}e^{2t}$

7. $\mathbf{X}=c_1\begin{pmatrix}1\\ 0\\ 0\end{pmatrix}e^{-t}+c_2\begin{pmatrix}0\\ -2\sin 2t\\ \cos 2t\end{pmatrix}e^{-t}+c_3\begin{pmatrix}0\\ 2\cos 2t\\ \sin 2t\end{pmatrix}e^{-t}$

9. $\mathbf{X}=c_1\begin{pmatrix}7\\ 2\\ 3\end{pmatrix}e^t+c_2\begin{pmatrix}\cos 2t+\sin 2t\\ \cos 2t+\sin 2t\\ 2\cos 2t\end{pmatrix}+c_3\begin{pmatrix}\cos 2t-\sin 2t\\ \cos 2t-\sin 2t\\ -2\sin 2t\end{pmatrix}$

11. $\mathbf{X}=c_1\begin{pmatrix}0\\ 2\\ 1\end{pmatrix}e^t+c_2\begin{pmatrix}\sin t\\ \cos t\\ \cos t\end{pmatrix}e^t+c_3\begin{pmatrix}\cos t\\ -\sin t\\ -\sin t\end{pmatrix}e^t$

13. $\mathbf{X}=c_1\begin{pmatrix}0\\ 1\\ 1\end{pmatrix}e^{2t}+c_2\begin{pmatrix}0\\ t\\ t+\frac{1}{3}\end{pmatrix}e^{2t}+c_3\begin{pmatrix}1\\ 0\\ 0\end{pmatrix}e^{3t}$

15. $\mathbf{X}=c_1\begin{pmatrix}0\\ -1\\ 1\end{pmatrix}e^{2t}+c_2\begin{pmatrix}1\\ t\\ 1-t\end{pmatrix}e^{2t}+c_3\begin{pmatrix}2+2t\\ t^2\\ 4+2t-t^2\end{pmatrix}e^{2t}$

17. $\mathbf{X}=c_1\begin{pmatrix}1\\ 0\\ 0\end{pmatrix}e^{2t}+c_2\begin{pmatrix}t\\ 1\\ 0\end{pmatrix}e^{2t}+c_3\begin{pmatrix}0\\ 0\\ 1\end{pmatrix}e^t$

19. $\mathbf{X}=c_1\begin{pmatrix}1\\ 1\\ -1\end{pmatrix}e^t+c_2\begin{pmatrix}0\\ 1\\ 0\end{pmatrix}e^{-2t}+c_3\begin{pmatrix}3\\ 8t\\ 1\end{pmatrix}e^{-2t}$

21. $\mathbf{X}=c_1\begin{pmatrix}1\\ 1\\ 0\end{pmatrix}e^{-t}+c_2\begin{pmatrix}-2\\ 1\\ 0\end{pmatrix}e^{2t}+c_3\begin{pmatrix}0\\ 0\\ 1\end{pmatrix}e^{2t}+\begin{pmatrix}3t+\frac{3}{2}\\ -\frac{3}{2}\\ -1\end{pmatrix}e^t$

23. $\mathbf{X} = c_1 \begin{pmatrix} 1 \\ 0 \\ -1 \end{pmatrix} e^t + c_2 \begin{pmatrix} t-1 \\ -1 \\ -t \end{pmatrix} e^t + c_3 \begin{pmatrix} 1 \\ -1 \\ -1 \end{pmatrix} e^{2t}$
$+ \begin{pmatrix} te^t + e^t - \frac{3}{2} \\ -e^t - \frac{1}{2} \\ -te^t - e^t + \frac{1}{2} \end{pmatrix}$

25. $\mathbf{X} = c_1 \begin{pmatrix} 1 \\ -1 \\ 0 \end{pmatrix} e^{-t} + c_2 \begin{pmatrix} 0 \\ 0 \\ 1 \end{pmatrix} e^{-t} + c_3 \begin{pmatrix} 1 \\ 1 \\ 0 \end{pmatrix} e^t$
$- \begin{pmatrix} t \\ 2 \\ te^{-t} \end{pmatrix}$

27. $\mathbf{X} = c_1 e^t \begin{pmatrix} \cos t \\ \sin t \\ 0 \end{pmatrix} + c_2 e^t \begin{pmatrix} -\sin t \\ \cos t \\ 0 \end{pmatrix} + c_3 e^{-t} \begin{pmatrix} 0 \\ 0 \\ 1 \end{pmatrix}$
$+ \begin{pmatrix} -2e^t \\ -e^t \\ t^3 e^{-t} \end{pmatrix}$

29. $\mathbf{X} = \begin{pmatrix} 1 + 3t + \dfrac{t^2}{2} \\ t - 3 \\ 4 \end{pmatrix} e^{-2t}$

31. $\mathbf{X} = \begin{pmatrix} \dfrac{1}{2} + \dfrac{3e^{2t}}{8} + \dfrac{t^2}{4} + \dfrac{t}{4} + \dfrac{1}{8} \\ \dfrac{1}{2} - \dfrac{3e^{2t}}{8} + \dfrac{t^2}{4} - \dfrac{t}{4} - \dfrac{1}{8} \\ e^{4t} - e^{3t} \end{pmatrix}$

33. $x_1' = \frac{102}{204} x_3 + 5e^{0.07t} - \frac{34}{340} x_1 - \frac{68}{340} x_1$,
$x_2' = \frac{68}{340} x_1 - \frac{68}{680} x_2$, $x_3' = \frac{34}{340} x_1 + \frac{68}{680} x_2 - \frac{102}{204} x_3$
$\lambda = 0, -\frac{3}{10}, -\frac{9}{10}$

35. $x_1 = \dfrac{a}{2}(\cos \omega_0 t + \cos \sqrt{3}\omega_0 t)$,
$x_2 = \dfrac{a}{2}(\cos \omega_0 t - \cos \sqrt{3}\omega_0 t)$,
where $\omega_0 = \sqrt{k/m}$.

Chapter 8 Review Exercises

1. $\mathbf{X}' = \begin{pmatrix} 0 & 1 & 0 \\ 0 & 0 & 1 \\ -1 & -4 & 2 \end{pmatrix} \mathbf{X} + \begin{pmatrix} 0 \\ 0 \\ 4 \sin 2t + 2e^{-t} \end{pmatrix}$

3. $\dfrac{d\mathbf{X}}{dt} = \begin{pmatrix} 0 & 1 & 0 \\ 0 & 0 & 1 \\ -1 & 6 & -\frac{1}{2} \end{pmatrix} \mathbf{X} + \begin{pmatrix} 0 \\ 0 \\ \cos t + \frac{1}{2} \sin t \end{pmatrix}$

5. $\mathbf{X}_p' = \begin{pmatrix} 2t - 2 \\ 4 \end{pmatrix}$

7. $\mathbf{X}_p' = \begin{pmatrix} \dfrac{8}{5} + \dfrac{1}{t} \\ \dfrac{16}{5} + \dfrac{2}{t} \end{pmatrix}$

9. $\mathbf{X} = c_1 \begin{pmatrix} -\cos 3t + \sin 3t \\ 2 \cos 3t \end{pmatrix} + c_2 \begin{pmatrix} 3 \cos 3t + \sin 3t \\ -2 \sin 3t \end{pmatrix} - \frac{1}{3} \begin{pmatrix} 1 \\ 1 \end{pmatrix}$

11. $\mathbf{X} = c_1 \begin{pmatrix} 1 \\ -1 \end{pmatrix} e^t + c_2 \begin{pmatrix} 3 \\ 1 \end{pmatrix} e^{5t} + \begin{pmatrix} -\frac{8}{3} e^{4t} + \frac{24}{5} \\ -\frac{4}{3} e^{4t} - \frac{32}{5} \end{pmatrix}$

13. $\mathbf{X}(t) = \begin{pmatrix} t \\ 3t + 1 \end{pmatrix} e^{3t}$

15. $\mathbf{X} = \left[-\frac{26}{9} \begin{pmatrix} 3 \sin 3t \\ \cos 3t \end{pmatrix} + \frac{2}{3} \begin{pmatrix} 3 \cos 3t \\ -\sin 3t \end{pmatrix} + \begin{pmatrix} 0 \\ -\frac{1}{9} \end{pmatrix} \right] e^t$

17. $\mathbf{X} = 11 \begin{pmatrix} 2 \\ 1 \end{pmatrix} e^t - 6 \left[\begin{pmatrix} 2 \\ 1 \end{pmatrix} t + \begin{pmatrix} 3 \\ 1 \end{pmatrix} \right] e^t$
$- \begin{pmatrix} 3 \\ \frac{3}{2} \end{pmatrix} t^2 e^t + \begin{pmatrix} 1 \\ 2 \end{pmatrix} te^t$

19. $\mathbf{X} = c_1 \begin{pmatrix} 2 \\ -2 \\ 3 \end{pmatrix} e^t + \left[c_2 \begin{pmatrix} 0 \\ -\sin 2t \\ \cos 2t \end{pmatrix} + c_3 \begin{pmatrix} 0 \\ \cos 2t \\ -\sin 2t \end{pmatrix} \right] e^t$

21. $\mathbf{X} = c_1 \begin{pmatrix} 0 \\ 2 \\ 1 \end{pmatrix} e^t + \left[c_2 \begin{pmatrix} \sin t \\ \cos t \\ \cos t \end{pmatrix} + c_3 \begin{pmatrix} \cos t \\ -\sin t \\ -\sin t \end{pmatrix} \right] e^t$

23. $\mathbf{X}_1 = \begin{pmatrix} 1 \\ 1 \\ 1 \end{pmatrix} e^t$

$\mathbf{X}_2 = \begin{pmatrix} 1 \\ 1 \\ 1 \end{pmatrix} te^t + \begin{pmatrix} 1 \\ 2 \\ 3 \end{pmatrix} e^t$

$\mathbf{X}_3 = \begin{pmatrix} 1 \\ 1 \\ 1 \end{pmatrix} \frac{t^2}{2} e^t + \begin{pmatrix} 1 \\ 2 \\ 3 \end{pmatrix} te^t + \begin{pmatrix} 1 \\ 2 \\ 4 \end{pmatrix} e^t$

25. $\mathbf{X} = c_1 \begin{pmatrix} 0 \\ 1 \\ -1 \end{pmatrix} e^{-2t}$

$+ c_2 \left[\begin{pmatrix} -2 \\ 1 \\ 1 \end{pmatrix} \cos 2t + \begin{pmatrix} 2 \\ 0 \\ 0 \end{pmatrix} \sin 2t \right]$

$+ c_3 \left[\begin{pmatrix} 2 \\ 0 \\ 0 \end{pmatrix} \cos 2t - \begin{pmatrix} -2 \\ 1 \\ 1 \end{pmatrix} \sin 2t \right]$

27. $\mathbf{X} = \frac{23}{4} \begin{pmatrix} 2 \\ 1 \end{pmatrix} e^{3t} - \begin{pmatrix} 2 \\ 5 \end{pmatrix} e^{-t} - \begin{pmatrix} 10 \\ 5 \end{pmatrix} te^{3t} + \begin{pmatrix} \frac{1}{2} \\ \frac{5}{4} \end{pmatrix} e^{3t}$

29. **a)** $\mathbf{I}' = \begin{pmatrix} 0 & -100 \\ 200 & -200 \end{pmatrix} \mathbf{I} + \begin{pmatrix} 120 \\ 0 \end{pmatrix}$ where $\mathbf{I} = \begin{pmatrix} i_1 \\ i_2 \end{pmatrix}$.

b) $\mathbf{I}_c = c_1 \begin{pmatrix} \cos 100t + \sin 100t \\ 2 \sin 100t \end{pmatrix} e^{-100t}$

$+ c_2 \begin{pmatrix} -\sin 100t \\ \cos 100t - \sin 100t \end{pmatrix} e^{-100t}$

c) $i_1 \to \frac{6}{5}, \quad i_2 \to \frac{6}{5}, \quad i_3 \to 0$

31. **a)** $\theta_1 = -\frac{1}{2} \cos 2t - \frac{1}{2} \sin 2t$,

$\theta_2 = \cos 2t + \sin 2t$

c) Lower pallet will be lost if $\theta_2(t) \geq 60° = \pi/3 \approx 1.04$.

CHAPTER 9

Section 9.1

1. $y = c_1 e^{-x} + c_2 e^{3x}$

3. $y = c_1 e^{2x} + c_2 x e^{2x}$

5. $y = c_1 \sin 2x + c_2 \cos 2x$

7. $y = c_1 e^{-x} + c_2 x$

9. $y = c_1 x^3 + c_2 x^{-2}$

11. $y = c_1 + c_2 x^2$

13. $y = c_1 + c_2(x + \frac{1}{3}x^3)$

15. $y = c_1 x + c_2 x \ln x$

17. $y = c_1 \sqrt{x} + c_2 x^{-1}$

19. $y = c_1 x^3 + c_2 x^3 \ln x$

21. $y = c_1 \cos(\ln x) + c_2 \sin(\ln x)$

Section 9.2

1. $y = \frac{c_1}{|x|^2} + c_2 |x|$

3. $y = \frac{c_1}{|x|^2} + c_2 |x|^3$

5. $y = c_1 |x|^2 + c_2 |x|^4$

7. $y = c_1 |x|^{-1/3} + c_2$

9. $y = |x|(c_1 + c_2 \ln|x|)$

11. $y = |x|[c_1 \cos(2 \ln|x|) + c_2 \sin(2 \ln|x|)]$

13. $y = \frac{1}{|x|}[c_1 \cos(3 \ln|x|) + c_2 \sin(3 \ln|x|)]$

15. $y = \frac{1}{\sqrt{|x|}}[c_1 \cos(\ln|x|) + c_2 \sin(\ln|x|)]$

17. $y = \frac{1}{|x|}(c_1 + c_2 \ln|x|)$

19. $y = c_1 + c_2 \ln|x|$

21. $y = \frac{1}{\sqrt[3]{|x|}}(c_1 + c_2 \ln|x|)$

23. $y = |x|^{-5/4}(c_1 + c_2 \ln|x|)$

25. $y = \frac{1}{2|x|^3} + \frac{|x|}{2}$

27. $y = |x|$

29. $y = |x|[-\cos(\ln|x|) + 2 \sin(\ln|x|)]$

31. $y = c_1 x + \frac{c_2}{x} + \frac{x \ln x}{2} - \frac{x}{4}$

33. $y = c_1 x^2 + c_2 x^2 \ln x + x^3$

35. $y = c_1 \cos(\ln x) + c_2 \sin(\ln x) + \dfrac{1}{2x}$

Section 9.3

1. $y = c_0 \sum_{n=0}^{\infty} \dfrac{2^n x^n}{n!}, \quad R \to \infty$

3. $y = c_0 \sum_{n=0}^{\infty} \dfrac{x^{2n}}{2^n n!}, \quad R \to \infty$

5. $y = c_0 \sum_{n=0}^{\infty} (-1)^n x^{2n}, \quad R = 1$

7. $y = c_0 + c_1(x - x^2 + \frac{2}{3}x^3 - \cdots)$
$= c_0 - \dfrac{c_1}{2} e^{-2x}$

9. $y = c_0(1 - 2x^2 + \cdots) + c_1(x - \frac{2}{3}x^3 + \cdots)$
$= c_0 \cos 2x + c_1 \sin 2x$

11. $y = c_1 x + c_2 x^2$

13. $y = c_0(1 + \frac{1}{2}x^2 - \frac{1}{6}x^3 + \cdots) + c_1(x + \frac{1}{6}x^3 + \cdots)$

15. $y = c_0(1 - x^2 + \frac{5}{12}x^4 - \cdots) + c_1 x$
$= a\left[\dfrac{x}{2} \ln\left(\dfrac{1+x}{1-x}\right) - 1\right] + bx$

17. $y = c_0 + c_1(x - \frac{1}{6}x^3 + \frac{1}{40}x^5 - \cdots)$

19. $y = c_0(1 - 3x^2 + \cdots) + c_1(x - x^3)$

21. $y = c_0(1 + x^2 + \frac{2}{3}x^4 + \cdots)$
$+ c_1(x + x^3 + \frac{3}{5}x^5 + \cdots)$

23. $y = c_0(1 - \frac{3}{2}x^2 + \cdots) + c_1(x - \frac{1}{2}x^3 + \cdots)$

25. $y = c_0(1 - \frac{3}{2}x^2 + \frac{1}{6}x^4 + \cdots) + c_1(x - \frac{1}{3}x^3)$

27. $y = c_0(1 - \frac{1}{2}x^2 + \frac{1}{24}x^4 - \cdots)$
$+ c_1(x - \frac{1}{6}x^3 + \frac{1}{120}x^5 - \cdots)$
$+ (\frac{1}{6}x^3 - \frac{1}{720}x^5 + \cdots)$
$= a \cos x + b \sin x + \dfrac{1}{2} \sin x - \dfrac{x}{2} \cos x$

29. $y = c_1(x + \frac{1}{2}x^3 + \cdots) + c_2(x^2 + \frac{1}{6}x^4 + \cdots)$
$= c_1 x \cosh x + c_2 x \sinh x$

31. $y = 1 - x + \frac{1}{6}x^3 + \frac{1}{12}x^4 + \cdots$

33. $y = x - 2x^3 + 3x^5 - 4x^7 + \cdots$

35. **a)** $y = c_0(1 - \alpha x^2 + \cdots)$
$+ c_1\left[x + \dfrac{2(1-\alpha)}{(n+2)(n+1)} x^3 + \cdots\right]$

b) $\alpha = 0$: $y_1 = c_0(1)$
$\alpha = 1$: $y_2 = c_1 x$
$\alpha = 2$: $y_1 = c_0(1 - 2x^2)$
$\alpha = 3$: $y_2 = c_1\left(x - \dfrac{2}{3}x^3\right)$

Section 9.4

1. $x = 0$, irregular singular point

3. $x = 1$, regular singular point

5. $x = \pm i$, both regular singular points

7. $x = 0$, irregular singular point; $x = 1$, regular singular point

9. $x = -1$ and $x = 2$, both regular singular points

11. $y = c_0 x^{-2} + c_1 x$

13. $y = c_0 x^{-4} + c_1 x^{-1}$

15. $y = b_0 + c_0 x^{-1/3}$

17. $y = c_0 x + c_1 x \ln x$

19. $y = c_0 x^{-1} + c_1 x^{-1} \ln x$

21. $y_1 = c_0 \sum_{n=0}^{\infty} \dfrac{(-1)^n}{(2n)!} (x^{1/2})^{2n} = c_0 \cos \sqrt{x}$,
$y_2 = b_0 \sum_{n=0}^{\infty} \dfrac{(-1)^n}{(2n+1)!} (x^{1/2})^{2n+1} = b_0 \sin \sqrt{x}$

23. $y = c_0 x^{-2} + c_2 \sum_{n=2}^{\infty} \dfrac{(-1)^n 2^{n-1}}{n(n-2)!} x^{n-2}$

25. $y = x^{-1}\left[c_0 \sum_{n=0}^{\infty} \dfrac{1}{(2n)!} x^{2n} + c_1 \sum_{n=0}^{\infty} \dfrac{1}{(2n+1)!} x^{2n+1}\right]$
$= x^{-1}(c_0 \cosh x + c_1 \sinh x)$

27. $y = C_1\left[x^{-1} - \sum_{n=1}^{\infty} \dfrac{x^{n-1}}{n!\, 3 \cdot 5 \cdots (2n-3)}\right]$
$+ C_2\left[x^{1/2} + \sum_{n=1}^{\infty} \dfrac{x^{n+1/2}}{n!\, 5 \cdot 7 \cdots (2n+3)}\right]$

29. $y = C_0(x^{-1} + 1) + C_1 \sqrt{x}$

31. $y = x\left[c_0 \sum_{n=0}^{\infty} \frac{(-1)^n}{(2n)!} x^{2n} + c_1 \sum_{n=0}^{\infty} \frac{(-1)^n}{(2n+1)!} x^{2n+1}\right]$
$= c_0 x \cos x + c_1 x \sin x$

33. $y = \sqrt{x}\left(C_1 + C_2 \sum_{n=1}^{\infty} \frac{x^n}{2^n n!}\right)$

35. $y = C_1\sqrt{x} +$
$C_2\sqrt{x}\left(\ln x + x + \frac{x^2}{2\cdot 2!} + \frac{x^3}{3\cdot 3!} + \frac{x^4}{4\cdot 4!} + \cdots\right)$

Section 9.5

1. a) $J_0(\sqrt{x}) = \sum_{n=0}^{\infty} \frac{(-1)^n x^n}{2^{2n}(n!)^2}$

b) $\sqrt{x}\, J_1(\sqrt{x}) = \sum_{n=0}^{\infty} \frac{(-1)^n x^{2n+1}}{2^{2n+1} n!(n+1)!}$

3. $y' = xJ_0, \quad y'' = J_0 - xJ_1$

5. $y' = -\frac{2}{x^2} J_1 + \frac{1}{x} J_0, \quad y'' = \left(\frac{6}{x^3} - \frac{1}{x}\right) J_1 - \frac{3}{x^2} J_0$

9. $\frac{dy}{dx} = [J_0(u) - uJ_1(u)]\frac{du}{dx}, \quad u = \sqrt{x}$

$\frac{d^2y}{dx^2} = [J_0(u) - uJ_1(u)]\frac{d^2u}{dx^2}$
$- J_1(u)]\left(\frac{du}{dx}\right)^2 - uJ_0(u)\left(\frac{du}{dx}\right)^2$

$y = 2.23323\sqrt{x}J_0(\sqrt{x})$

11. $u^2\frac{d^2y}{du^2} + u\frac{dy}{du} + (u^2 - 1)y = 0, \quad u = \sqrt{x}$
$y \approx 1.734 J_1(\sqrt{x})$

21. $\frac{dy}{dx} = px^{p-1}J_p(x) + x^p J_p'(x)$

$\frac{d^2y}{dx^2} = p(p-1)x^{p-2}J_p(x) + 2px^{p-1}J_p'(x) + x^p J_p''(x)$

23. $\frac{d}{dx}[xJ_1(x)] = xJ_0(x)$, and integrate from 0 to x.

Section 9.6

1. a) $P_2(x) = \frac{1}{2}(3x^2 - 1)$
b) $P_4(x) = \frac{1}{8}(35x^4 - 30x^2 + 3)$
c) $P_5(x) = \frac{1}{8}(63x^5 - 70x^3 + 15x)$
d) $P_6(x) = \frac{1}{32}(462x^6 - 630x^4 + 35x^2 - 10)$
e) $P_7(x) = \frac{1}{16}(858x^7 - 99x^5 + 315x^3 - 35x)$
f) $P_8(x) = \frac{1}{128}(6{,}435x^8 - 12{,}012x^6 + 6{,}930x^4 - 1{,}260x^2 + 35)$

5. $f(x) = P_0(x) + \frac{3}{2}P_1(x) - \frac{7}{8}P_3(x) + \frac{11}{16}P_5(x) - \cdots$

7. $f(x) = \frac{1}{4}P_0(x) + \frac{1}{2}P_1(x) + \frac{5}{16}P_2(x) - \frac{3}{32}P_4(x) + \cdots$

9. $f(x) = P_0(x) + \frac{5}{8}P_2(x) - \frac{3}{16}P_4(x) - \frac{1131}{64}P_6(x) + \cdots$

11. $f(x) = \frac{1}{5}P_0(x) + \frac{4}{7}P_2(x) + \frac{8}{35}P_4(x) - \frac{325}{32}P_6(x) + \cdots$

13. 1.3333333333

15. 0

17. 1.933421497

Chapter 9 Review Exercises

1. Ordinary point, $\sum_{n=0}^{\infty} c_n(x-2)^n, \; 2-\sqrt{2} < x < 2+\sqrt{2}$

3. Regular singular point, $\sum_{n=0}^{\infty} c_n(x+1)^{n+r}$,
$-2 < x < 0$

5. Ordinary point, $\sum_{n=0}^{\infty} c_n x^n, \; -6 < x < 6$

7. $y = c_1 x^{1/2+\sqrt{5}/2} + c_2 x^{1/2-\sqrt{5}/2}$

9. $y = c_1 x^{-2} + c_2 x^{-2}\ln x$

11. $y = c_0 x\left(1 + x + \frac{x^2}{4} + \frac{x^3}{36} + \cdots\right)$

13. $y = c_0(1 - \frac{3}{2}x^2 + \frac{21}{24}x^4 + \cdots)$
$+ c_1(x - \frac{2}{3}x^3 + \frac{24}{60}x^5 + \cdots)$

15. $y = c_0 x^{-1}\left(1 - x + \frac{x^2}{4} - \frac{x^3}{36} + \cdots\right)$

17. $y_1(x) = c_0 x^{3/2}(1 - \frac{1}{2}x + \frac{5}{32}x^2 - \frac{7}{192}x^3 + \cdots)$

19. $y = c_0\left(1 + x^2 + \frac{x^4}{2} + \cdots\right)$
$+ c_1\left(x + \frac{x^3}{2} + \frac{11}{40}x^5 + \cdots\right)$

21. $y = c_0\left(1 - \frac{x^2}{2} - \frac{x^4}{8} - \cdots\right) + c_1\left(x + \frac{x^3}{3} + \frac{x^5}{30} + \cdots\right)$

23. a) $T(t) = Ce^{-\lambda^2 kt}$

c) $R_2(\lambda\rho) = J_0(\lambda\rho)\ln\rho + \sum_{n=1}^{\infty} b_n(\lambda\rho)^n$

d) $u(\rho, t) =$

$$e^{-\lambda^2 kt}\left\{C_1 J_0(\lambda\rho) + C_2\left[J_0(\lambda\rho)\ln\rho + \sum_{n=1}^{\infty} b_n(\lambda\rho)^n\right]\right\}$$

g) $R_2(\lambda\rho) = J_0(\lambda\rho)\ln\rho + \frac{(\lambda\rho)^2}{4} - \frac{3(\lambda\rho)^4}{128} + \frac{11(\lambda\rho)^6}{13{,}824} - \cdots$

CHAPTER 10

Section 10.1

1. Let x be expressed in centimeters, t in seconds, u in degrees Celsius. Then

PDE: $\frac{\partial^2 u}{\partial x^2} = \frac{20}{3}\frac{\partial u}{\partial t}$

BC: $u(0, t) = u(50, t) = 0 \quad (t > 0)$

IC: $u(x, 0) = 100 \quad (0 < x < 50)$

3. $X'' + \lambda^2 X = 0, \quad T'' + \lambda^2 T = 0$

5. $xX'' + \lambda^2 X = 0, \quad T' - \lambda^2 T = 0$

7. $X'' - X' + \lambda^2 X = 0, \quad T' + \lambda^2 T = 0$

9. Not separable

Section 10.2

1. $f(x) = 1$

3. $f(x) = \sum_{n=1}^{\infty} \frac{2(-1)^{n+1}}{n} \sin nx$

5. $f(x) = \frac{\pi^2}{12} + \sum_{n=1}^{\infty} \frac{(-1)^n}{n^2} \cos nx$

7. $f(x) = \frac{2\sinh\pi}{\pi}\left[\frac{1}{2} + \sum_{n=1}^{\infty} \frac{(-1)^n}{n^2+1}(\cos nx - n\sin nx)\right]$

9. $f(x) = \frac{1}{2}\cos x + \frac{1}{\pi}\sum_{n=2}^{\infty} \frac{n(-1)^n}{n^2-1}\sin nx$

11. $f(x) = \frac{1}{2} + \frac{2}{\pi}\sum_{k=0}^{\infty} \frac{(-1)^k}{2k+1}\cos(2k+1)x$

13. $f(x) = \frac{5}{4} + \frac{4}{\pi^2}\sum_{n=1}^{\infty} \frac{1}{n^2}\left[(-1)^n - \cos\frac{n\pi}{2}\right]\cos(n\pi x) + \frac{2}{\pi}\sum_{n=1}^{\infty} \frac{1}{n}\left[(-1)^n - \frac{2}{n\pi}\sin\frac{n\pi}{2}\right]\sin(n\pi x)$

15. $f(x) = \frac{\pi}{2} + \frac{2}{\pi}\sum_{n=1}^{\infty} \frac{[(-1)^n - 1]}{n^2}\cos nx$

17. $f(x) = e - 1 + 2\sum_{n=1}^{\infty} \frac{[e(-1)^n - 1]}{1 + n^2\pi^2}\cos n\pi x$

19. $f(x) = -\frac{1}{4} + \frac{4}{\pi}\sum_{n=1}^{\infty}\left[\frac{1}{n}\sin\frac{n\pi}{2} + \frac{1}{\pi n^2}\left((-1)^{n+1} + \cos\frac{n\pi}{2}\right)\right]\cos\frac{n\pi x}{2}$

21. $f(x) = \frac{1}{2} + \sum_{n=1}^{\infty} \frac{4}{n^2\pi^2}\left[(-1)^n - \cos\frac{n\pi}{2}\right]\cos n\pi x$

23. $f(x) = 2\sum_{n=1}^{\infty} \frac{(-1)^n}{n\pi}\sin n\pi x$

25. $f(x) = \frac{8}{\pi}\sum_{k=1}^{\infty} \frac{k}{4k^2-1}\sin 2kx$

27. $f(x) = \sin x$

29. $f(x) = \frac{2}{\pi}\sum_{n=1}^{\infty}\left[\frac{1}{n} + \frac{4}{n^2\pi}\sin\frac{n\pi}{2}\right]\sin\frac{n\pi x}{2}$

31. $\int_{-L}^{L} \cos\frac{m\pi x}{L}\,dx = \frac{L}{m\pi}[\sin m\pi - \sin(-m\pi)]$

33. If $m \neq n$, then

$$\int_{-L}^{L} \cos\frac{n\pi x}{L}\cos\frac{m\pi x}{L}\,dx = \frac{1}{2}\int_{-L}^{L}\left[\cos\frac{(m+n)\pi x}{L} + \cos\frac{(n-m)\pi x}{L}\right]dx;$$

if $m = n$,

$$\int_{-L}^{L} \cos\frac{n\pi x}{L}\cos\frac{m\pi x}{L}\,dx = \frac{1}{2}\int_{-L}^{L}\left(\cos\frac{2m\pi x}{L} + 1\right)dx.$$

35. If $m \neq n$, then

$$\int_{-L}^{L} \sin\frac{n\pi x}{L}\cos\frac{m\pi x}{L}\,dx = \frac{1}{2}\int_{-L}^{L}\left[\sin\frac{(n+m)\pi x}{L} + \sin\frac{(n-m)\pi x}{L}\right]dx;$$

if $m = n$,

$$\int_{-L}^{L} \sin\frac{m\pi x}{L}\cos\frac{m\pi x}{L}\,dx = \frac{1}{2}\int_{-L}^{L}\left(\sin\frac{2\pi x}{L} + 0\right)dx.$$

Section 10.3

1. $u(x, t) = e^{-\pi^2 t/2}\sin \pi x$

$\lim_{t\to\infty} u(\frac{3}{2}, t) = 0$, steady-state temperature

3. $u(x, t) = e^{-t}\sin x$

$\lim_{t\to\infty} u\left(\frac{\pi}{2}, t\right) = 0$, steady-state temperature

5. $u(x, t) = e^{-t}\cos x$

$\lim_{t\to\infty} u\left(\frac{\pi}{2}, t\right) = 0$, steady-state temperature

7. $u(x, t) = 2 + e^{-\pi^2 t/3}\cos \pi x$

$\lim_{t\to\infty} u(1, t) = 2$, steady-state temperature

9. $u(x, t) = \frac{100\sqrt{2}}{\pi}\sum_{k=0}^{\infty} e^{-(2k+1)^2\pi^2 t/10^3}\sin\frac{(2k+1)\pi x}{100}$

$\lim_{t\to\infty} u(50, t) = 0$, steady-state temperature

11. $u(x, t)$

$$= \frac{1}{2} - \frac{16}{\pi^2}\sum_{k=0}^{\infty}\frac{1}{(4k+2)^2}e^{-(4k+2)^2\pi^2 t/4}\cos(2k+1)\pi x$$

$\lim_{t\to\infty} u(1, t) = \frac{1}{2}$, steady-state temperature

13. $u(x, t)$

$$= 100 + \frac{1}{\pi}\sum_{n=1}^{\infty}\frac{1}{n}\left[198(-1)^n - 200 + 2\cos\frac{n\pi}{2}\right]e^{-n^2 t/2}\sin nx$$

$\lim_{t\to\infty} u\left(\frac{\pi}{2}, t\right) = 100$, steady-state solution

15. $u(x, t) = 10x + \frac{60}{\pi}\sum_{n=1}^{\infty}\frac{1}{n}e^{-n^2\pi^2 t/18}\sin\frac{n\pi x}{3}$

$\lim_{t\to\infty} u(\frac{3}{2}, t) = 10(\frac{3}{2}) = 15$, steady-state temperature

17. $u(x, t) = 100 - 50x - \frac{100}{\pi}\sum_{n=1}^{\infty}\frac{1}{n}e^{-2n^2\pi^2 t}\sin n\pi x$

$\lim_{t\to\infty} u(\frac{1}{2}, t) = 75$, steady-state temperature at center

19. $u(x, t) = \frac{1}{2} + \sum_{n=1}^{\infty}\frac{2}{n^2\pi^2}[(-1)^n - 1]e^{-2n^2\pi^2 t}\cos n\pi x$

$\lim_{t\to\infty} u(0, t) = \frac{1}{2}$, steady-state temperature

21. $u(x, t) = \frac{3\pi^2}{2} + \sum_{n=1}^{\infty}\left\{\frac{2}{n^2}[(-1)^n - 1]\cos nx + \frac{2\pi}{n}[1 - 2(-1)^n]\sin nx\right\}e^{-3n^2 t}$

$\lim_{t\to\infty} u(0, t) = \frac{3\pi^2}{2}$, steady-state temperature

23. $E(x, t) = \sum_{n=1}^{\infty}\frac{4800}{n\pi}[1 - (-1)^n]e^{-n^2\pi^2 t/L^2}\sin\frac{n\pi x}{L}$

25. $u(x, t) = \sum_{n=1}^{\infty}\frac{200}{n\pi}[1 - (-1)^n]e^{-1.14n^2\pi^2 t/10^4}\sin\frac{n\pi x}{100}$

a) $u(50, \frac{1}{2}) = 100°$
b) $t \approx 4.31$ hr

27. a) $25°$, steady-state temperature
b) $t \approx 2$ hr

29. $u(x, t) = \sum_{n=1}^{\infty}\frac{2}{n\pi}\left(1 - \cos\frac{n\pi}{2}\right)e^{-(1+n^2\pi^2)t}\sin n\pi x$

Section 10.4

1. $r = \frac{10(0.01)}{100} = 0.001$

	0	10	20	30	40	50	60	70	80	90	100
0.10	0	0	0.50	49.50	50	50	50	49.50	0.50	0	0

3. $r = \dfrac{3(0.01)}{(0.52)^2} = 0.109$

	0	0.52	1.05	1.57	2.09	2.62	3.14
0.10	0	0.39	0.65	0.76	0.65	0.39	0

5. $r = \dfrac{2(0.01)}{0.04} = 0.5$

	0	0.2	0.4	0.6	0.8	1.0
0.10	100	91.95	83.89	73.15	62.40	50

7. $r = \dfrac{0.01}{0.04} = 0.25$

	0.0	0.2	0.4	0.6	0.8	1.0
0.10	0.45	0.45	0.49	0.54	0.57	0.57

9. $r = \dfrac{0.02}{3(0.16)} = 0.042$

	0	0.4	0.8	1.2	1.6	2.0
0.10	2.11	2.11	1.39	1.39	2.11	2.11

15. Let $\Delta x = 2L/n$ such that $-L < x < L$:
a) $u_{0,j} = u_{n,j}$
b) $u_{n,j} = \frac{1}{2}(u_{1,j} + u_{n-1,j})$

	−3.14	−2.62	−2.09	−1.57	−1.05	−0.52	0	0.52	1.05	1.57	2.09	2.62	3.14
0	0	0	0	0	0	0	0	1.00	1.00	1.00	1.00	1.00	1.00
0.01	0.50	0	0	0	0	0	0.04	0.96	1.00	1.00	1.00	1.00	0.50
0.10	0.50	0.12	0	0	0	0.05	0.24	0.78	0.96	1.00	1.00	0.88	0.50
0.20	0.50	0.21	0.04	0	0	0.10	0.33	0.68	0.91	1.00	0.96	0.79	0.50

Average these two values to fill end columns; then go to next row.

Section 10.5

3. $y(x, t) = \dfrac{16}{\pi^2} \displaystyle\sum_{n=1}^{\infty} \frac{1}{(2n-1)^3} \sin(2n-1)x \cos(2n-1)at$

5. $y(x, t) =$
$$\frac{1}{a} \sum_{n=1}^{\infty} \left[\frac{1-(-1)^n}{n^2} - \frac{2}{n^3\pi}\left(1 + \sin\frac{n\pi}{2}\right)\right] \sin nx \sin nat$$

7. $y(x, t) = (\sin x)(\cos at + \dfrac{1}{a} \sin at)$

9. $y(x, t) = 2 \displaystyle\sum_{n=1}^{\infty} \left(\frac{(-1)^{n+1}}{n\pi} \cos n\pi at + \left\{ \frac{2[(-1)^n - 1]}{n^4\pi^4 a} - \frac{(-1)^n}{n^2\pi^2 a} \right\} \sin n\pi at \right) \sin n\pi x$

11. c)

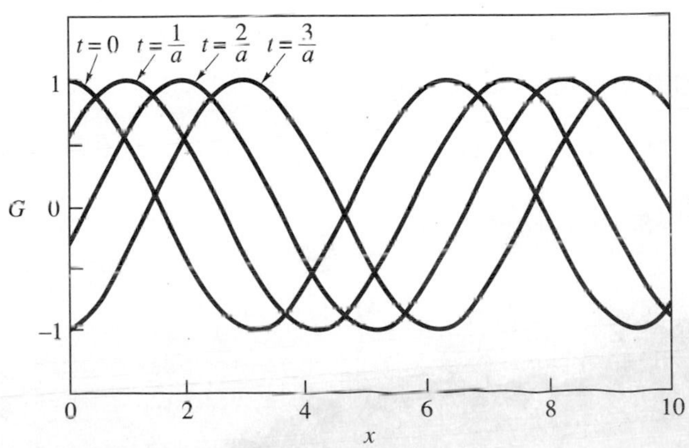

Section 10.6

1. $$u(x, y) = \frac{40}{\pi} \sum_{n=1}^{\infty} \frac{1}{(2n-1)\sinh \frac{(2n-1)\pi a}{b}} \sinh \frac{(2n-1)\pi x}{n} \sin \frac{(2n-1)\pi y}{b}$$

3. $$u(x, y) = \frac{\sinh x \sin y}{\sinh \pi}$$

5. $$u(x, y) = \frac{2}{a} \sum_{n=1}^{\infty} \frac{\int_0^a f_2(x) \sin \frac{n\pi x}{a}\, dx}{\sinh \frac{n\pi b}{a}} \sin \frac{n\pi x}{a} \sinh \frac{n\pi y}{a}$$

7. $$u(x, y) = \frac{\sin x \sinh y}{\sinh \pi}$$

9. $$u(x, y) = \frac{2}{b} \sum_{n=1}^{\infty} \frac{\int_0^b g_1(y) \sin \frac{n\pi y}{b}\, dy}{\sinh \frac{n\pi a}{b}} \sinh \frac{n\pi(a-x)}{b} \sin \frac{n\pi y}{b}$$

Section 10.7

1. $r = \left(\frac{0.05}{0.1}\right)^2 = 0.25$

	0	0.1	0.2	0.3	0.4	0.5	0.6	0.7	0.8	0.9	1.0
0	0	0.10	0.20	0.30	0.40	0.50	0.60	0.70	0.80	0.90	0
0.05	0	0.10	0.20	0.30	0.41	0.51	0.62	0.72	0.83	0.94	0
0.50	0	0.14	0.26	0.40	0.50	0.43	−0.01	−0.37	−0.11	−0.04	0

3. $r = 4\left(\frac{0.10}{0.52}\right)^2 = 0.15$

	0	0.52	1.05	1.57	2.09	2.62	3.14
0	0	0.50	0.87	1.00	0.87	0.50	0
0.1	0	0.65	1.02	1.15	1.02	0.65	0
0.5	0	0.74	1.22	1.35	1.22	0.74	0

5. $r = \left(\frac{0.25}{\pi/10}\right)^2 \approx 0.633$

	0.00	0.31	0.63	0.94	1.26	1.57	1.89	2.20	2.51	2.83	3.14
0	0.00	0.31	0.59	0.81	0.95	1.00	0.95	0.81	0.59	0.31	0
0.25	0.00	0.39	0.74	1.01	1.19	1.25	1.19	1.01	0.74	0.39	0
0.50	0.00	0.44	0.84	1.15	1.35	1.42	1.35	1.15	0.84	0.44	0

Chapter 10 Review Exercises

1. $y(x, t) = \sum_{n=1}^{\infty} \frac{6}{n^2\pi^2}\left(\sin\frac{n\pi}{3} + \sin\frac{2n\pi}{3}\right)\sin\frac{n\pi x}{3}\cos\frac{2n\pi t}{3}$

3. $y(x, t) = \sin 2x \cos 2t$

5. $u(x, t) = \sum_{n=1}^{\infty} \frac{-8}{n\pi}\left(\cos\frac{n\pi}{2} - 1\right)e^{-16n^2 t}\sin nx$

7. $w(x, t) = \frac{120}{\pi}\left[\left(\sin\frac{\pi x}{3}\right)e^{-2(\pi/3)^2 t} + \frac{1}{3}(\sin \pi x)e^{-2\pi^2 t} + \frac{1}{5}\left(\sin\frac{5\pi x}{3}\right)e^{-2(5\pi/3)^2 t} \ldots\right]$

9. $y(x, t) = \frac{0.03}{25\pi^2}\sum_{n=1}^{\infty}\frac{1}{n^2}\left(\sin\frac{n\pi}{3} + \sin\frac{2n\pi}{3}\right)\sin n\pi x \cos 50n\pi t$

11. $u(x, t) = \sum_{n=1}^{\infty} A_n e^{-0.15(n^2\pi^2/2500)t}\sin\frac{n\pi x}{50}$, $A_n = \frac{-200}{n\pi}(\cos n\pi - 1)$

13. b) $y(x, t) = \sum_{n=1}^{\infty} e^{\alpha t}\left(\frac{4bL}{n^2\pi^2}\sin\frac{n\pi}{2}\right)\sin\frac{n\pi x}{L}\left(\cos\beta t - \frac{\alpha}{\beta}\sin\beta t\right)$

where $\alpha = -\frac{K}{2}$, $\beta = \frac{\sqrt{4a^2\lambda^2 - K^2}}{2}$, $\lambda = \frac{n\pi}{L}$ $(n = 1, 2, \ldots)$

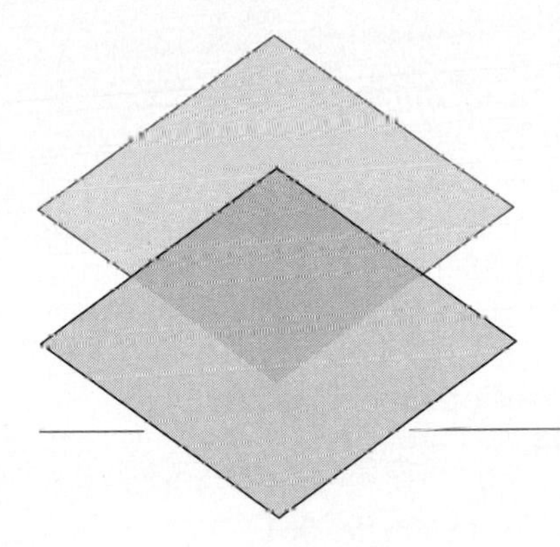

Index

M